Biotechnology

Second Edition

Volume 11b
Environmental Processes II

WILEY-VCH

Biotechnology

Second Edition

All volumes are also displayed on our Biotech Website:
http://www.wiley-vch.de/home/biotech

A Multi-Volume Comprehensive Treatise

Biotechnology

Second, Completely Revised Edition

Edited by
H.-J. Rehm and G. Reed
in cooperation with
A. Pühler and P. Stadler

Volume 11b

Environmental Processes II
Soil Decontamination

Edited by
J. Klein

 **WILEY-VCH**

Weinheim · New York · Chichester · Brisbane · Singapore · Toronto

Series Editors:
Prof. Dr. H.-J. Rehm
Institut für Mikrobiologie
Universität Münster
Corrensstraße 3
D-48149 Münster
FRG

Prof. Dr. A. Pühler
Biologie VI (Genetik)
Universität Bielefeld
P.O. Box 100131
D-33501 Bielefeld
FRG

Dr. G. Reed
1029 N. Jackson St. #501-A
Milwaukee, WI 53202-3226
USA

Prof. Dr. P. I W. Stadler
Artemis Pharmaceuticals
Geschäftsführung
Pharmazentrum Köln
Neurather Ring
D-51063 Köln
FRG

Volume Editor:
Prof. Dr. J. Klein
DMT-Gesellschaft für
Forschung und Prüfung GmbH
Franz-Fischer-Weg 61
Postfach 6980
D-45307 Essen
FRG

Library of Congress Card No.: applied for

British Library Cataloguing-in-Publication Data:
A catalogue record for this book is available from the British Library

Die Deutsche Bibliothek – CIP-Einheitsaufnahme

A catalogue record for this book
is available from Der Deutschen Bibliothek
ISBN 3-527-28323-4

Preface

In recognition of the enormous advances in biotechnology in recent years, we are pleased to present this Second Edition of "Biotechnology" relatively soon after the introduction of the First Edition of this multi-volume comprehensive treatise. Since this series was extremely well accepted by the scientific community, we have maintained the overall goal of creating a number of volumes, each devoted to a certain topic, which provide scientists in academia, industry, and public institutions with a well-balanced and comprehensive overview of this growing field. We have fully revised the Second Edition and expanded it from ten to twelve volumes in order to take all recent developments into account.

These twelve volumes are organized into three sections. The first four volumes consider the fundamentals of biotechnology from biological, biochemical, molecular biological, and chemical engineering perspectives. The next four volumes are devoted to products of industrial relevance. Special attention is given here to products derived from genetically engineered microorganisms and mammalian cells. The last four volumes are dedicated to the description of special topics.

The new "Biotechnology" is a reference work, a comprehensive description of the state-of-the-art, and a guide to the original literature. It is specifically directed to microbiologists, biochemists, molecular biologists, bioengineers, chemical engineers, and food and pharmaceutical chemists working in industry, at universities or at public institutions.

A carefully selected and distinguished Scientific Advisory Board stands behind the series. Its members come from key institutions representing scientific input from about twenty countries.

The volume editors and the authors of the individual chapters have been chosen for their recognized expertise and their contributions to the various fields of biotechnology. Their willingness to impart this knowledge to their colleagues forms the basis of "Biotechnology" and is gratefully acknowledged. Moreover, this work could not have been brought to fruition without the foresight and the constant and diligent support of the publisher. We are grateful to VCH for publishing "Biotechnology" with their customary excellence. Special thanks are due to Dr. Hans-Joachim Kraus and Karin Dembowsky, without whose constant efforts the series could not be published. Finally, the editors wish to thank the members of the Scientific Advisory Board for their encouragement, their helpful suggestions, and their constructive criticism.

H.-J. Rehm
G. Reed
A. Pühler
P. Stadler

Scientific Advisory Board

Contributors

Dr. Alan J. M. Baker
Deptartment of Animal and Plant Sciences
University of Sheffield
Western Bank
Sheffield, S10 2TN
UK
Chapter 17

Prof. Dr. Karl-Heinz Blotevogel
Fachbereich Biologie/Mikrobiologie
Universität Oldenburg
Postfach 2503
D-26111 Oldenburg
Germany
Chapter 11

Prof. Dr. Wilhelm G. Coldewey
Universität Münster
Corrensstrafle 3
D-48149 Münster
Germany
Chapter 2

Dr. Mary F. DeFlaun
Envirogen, Inc.
41000 Quakerbridge Road
Lawrenceville, NJ 08648
USA
Chapter 18

Dr. rer. nat. Helmut Dörr
Arcadis Trischler & Partner GmbH
Berliner Allee 6
D-64295 Darmstadt
Germany
Chapter 15

Dr. James P. Easter
Center for Environmental Biotechnology
University of Tennessee
676 Dabney Hall
Knoxville, TN 37996
USA
Chapters 20, 21

Dr. Adolf Eisenträger
Institut für Hygiene und Umweltmedizin
RWTH Aachen
Pauwelstr. 30
D-52057 Aachen
Germany
Chapter 5

Prof. Dr. Wolfgang Fritsche
Institut für Mikrobiologie
Universität Jena
Philosophenweg 12
D-07743 Jena
Germany
Chapter 6

Dr. Thomas Gorontzy
Fachbereich Biologie/Mikrobiologie
Universität Oldenburg
Postfach 2503
D-26111 Oldenburg
Germany
Chapter 11

Dr. Thomas Held
Arcadis Trischler & Partner GmbH
Berliner Allee 6
D-64295 Darmstadt
Germany
Chapter 15

Dr. Martin Hofrichter
Institut für Mikrobiologie
Universität Jena
Philosophenweg 12
D-07743 Jena
Germany
Chapter 6

Dr. Kerstin Hund
FH-Institut für Umweltchemie
D-57392 Schmallenberg-Grafschaft
Germany
Chapter 5

Dr.-Ing. Karsten Hupe
Arbeitsbereich Abfallwirtschaft
TU Hamburg-Harburg
Harburger Schloßstr. 37
D-21079 Hamburg
Germany
Chapter 12

Dr. Johan E.T. van Hylckama Vlieg
University of Groningen
Nijenborgh 4
NL-9747 AG Groningen
The Netherlands
Chapter 8

Prof. Dr. Dick B. Janssen
University of Groningen
Nijenborgh 4
NL-9747 AG Groningen
The Netherlands
Chapter 8

Dr. Matthias Kästner
UFZ – Umweltforschungszentrum Halle-
Leipzig
Sektion Sanierungsforschung
Permoserstr. 15
D-04318 Leipzig
Germany
Chapters 4, 9

Prof. Dr. Jürgen Klein
DMT-Gesellschaft für
Forschung und Prüfung GmbH
Franz-Fischer-Weg 61
D-45307 Essen
Germany
Chapter 22

Dr. René H. Kleijntjens
Partners in Milieutechnik b.v.
Mercuriusweg 4
NL-2516 AW's-Gravenhage
The Netherlands
Chapter 14

Dr. Christoph Klinger
DMT-Gesellschaft für Forschung und
Geschäftsbereich GUC
Am Technologiepark 1
D-45307 Essen
Germany
Chapter 2

Dr.-Ing. Michael Koning
Arbeitsbereich Abfallwirtschaft
TU Hamburg-Harburg
Harburger Schloßstr. 37
D-21079 Hamburg
Germany
Chapter 12

Dr. Fu-Min Menn
Center for Environmental Biotechnology
University of Tennessee
676 Dabney Hall
Knoxville, TN 37996
USA
Chapters 20, 21

Prof. Dr. Karel Ch. A. M. Luyben
Technical University of Delft
Julianalaan 67a
NL-2628 Delft 8
The Netherlands
Chapter 14

Prof. Dr. Bernd Mahro
Fachbereich 3
Hochschule Bremen
Neustadtswall 30
D-28199 Bremen
Germany
Chapter 3

Prof. Dr. Rudolf Müller
Arbeitsbereich 2-100
TU Hamburg-Harburg
Denickestr. 15
D-21073 Hamburg
Germany
Chapter 10

Prof. Dr. Hans-Jürgen Rehm
Institut für Mikrobiologie
Universität Münster
Corrensstraße 3
D-48149 Münster
Germany
Chapter 19

Dr. Michael Roemer
Melatener Str. 79
D-52074 Aachen
Germany
Chapter 23

Dr. David E. Salt
Department of Chemistry
Northern Arizona University
Flagstaff, AZ 86011-5698
USA
Chapter 17

Dr. Gary S. Sayler
Center for Environmental Biotechnology
University of Tennessee
676 Dabney Hall
Knoxville, TN 37996
USA
Chapters 20, 21

Prof. Dr. Bernhard Schink
Institut für Mikrobielle Ökologie
Universität Konstanz
Universitätsstr. 10
D-78457 Konstanz
Germany
Chapter 7

Prof. Dr. Jerald L. Schnoor
116 Engineering Research Facility
The University of Iowa
330 S. Madison St.
Iowa City, Iowa 52242-1000
USA
Chapter 16

Dr. Volker Schulz-Berendt
Umweltschutz Nord GmbH & Co.
Industriepark 6
D-27767 Ganderkesee
Germany
Chapter 13

Dr. Robert J. Steffan
Envirogen, Inc.
41000 Quakerbridge Road
Lawrenceville, NJ 08648
USA
Chapter 18

Prof. Dr. Rainer Stegmann
Arbeitsbereich Abfallwirtschaft
TU Hamburg-Harburg
Harburger Schloßstr. 37
D-21079 Hamburg
Germany
Chapter 12

Dr. Wolfgang Ulrici
Blücherstr. 17
D-53115 Bonn
Germany
Chapter 1

Dr. Ronald Unterman
Envirogen, Inc.
4100 Quakerbridge Road
Lawrenceville, NJ 08648
USA
Chapter 18

Dr. Catrin Wischnak
Arbeitsbereich 2-100
TU Hamburg-Harburg
Denickestr. 15
D-21073 Hamburg
Germany
Chapter 10

Contents

Introduction

JÜRGEN KLEIN

Essen, Germany

The use of biological processes for treatment of liquid wastes from human activities is an established technology dating back at least 4000 years. For more than a century natural biochemical processes (nature's self-cleaning abilities) have been utilized to treat effluent, and reactors and plant systems had been adapted with increasing effect to cope with the difficult conditions. The knowledge of the biodegradation mechanisms of organic pollutants and especially of synthetic compounds, however, is more recent and has only been developed in the second part of this century.

Over the past few decades environmental protection has primarily meant protection of air and water. Only with the increasing use of land in industrialized societies and the highlighting of possible hazards from contaminated soil did the public become aware of soil protection in the early 1980s. This has also prompted industry to take up this market segment. Engineers and scientists have thus been spurred on to look for technically optimized, ecologically sound, and economically appropriate solutions.

The experience accumulated in biological soil clean-up in the first few years was characterized by both, success and failure, including those of unprofessional suppliers. Initially, therefore, acceptance of biological soil clean-up was only limited. But now, due to intensive and interdisciplinary work, impressive success is in evidence. Consequently, biological soil clean-up methods now enjoy a high technical level and a broad acceptance.

The aim is to maintain and apply this level, even if at present the prospects for soil clean-up are seen from a more modest point of view. The enthusiasm that has accompanied soil decontamination in the last few years has now given way to a realistic attitude of more appropriate proportions. The discussion on the equivalence of securing and decontamination techniques must not lead to clean-up on a low level, but rather to ecologically and economically appropriate solutions.

The original objective of multifunctional use through the restoration of a "natural" soil is not feasible in most cases for technical and financial reasons. In view of this, biological clean-up techniques still have to be improved and optimized in order to provide cost-effective, technically simple, and near-natural processes.

In the First Edition of *Biotechnology* there was no contribution regarding biological processes for the remediation of contaminated soil. But the increasing need of remediation techniques and the increasing scientific and industrial interest in biological methods for soil remediation generated the necessity of a special volume on the topic: *Environmental Processes II – Soil Decontamination*.

The field of biodegradation and thereby bioremediation has experienced a dynamic evolution and remarkable developments over the past few years. It seems to have entered its most interesting and intense phase yet. The isolation and characterization of new microorganisms with novel catabolic activities continues unabated, and the use of plants and plant–microbe associations in bioremediation is expanding strongly. The continuously growing knowledge on catabolic pathways and critical enzymes provides the basis for the rational genetic design of new and improved enzymes and pathways for the development of more performant processes.

This volume *Environmental Processes II* summarizes the state of the art of scientific research in the field of biodegradation of xenobiotics, of transferring the knowlegde obtained into commercial application, and the future developments necessary to cope with the requirements of sustainable strategies for ecological, but ecomically appropiate solutions.

The first part is dedicated to more general aspects of bioremediation: firstly, a survey on the different international approaches of dealing with soil contamination problems, then elucidating inherent aspects of soil, i.e., geochemical and hydrogeological features as well as bioavailibility and humification processes influencing the interaction between microorganisms and contaminants in soil and methods to assess the ecotoxic potential of contaminated and the treated soil.

The second part of this volume summarizes the results of scientific research on the biodegradation of those substance groups which count as relevant in terms of contaminated sites, e.g., aliphatic and aromatic hydrocarbons and their halogenated derivatives, pesticides, and nitro compounds.

In the third part the different commercially methods of *in situ* and *ex situ* bioremediation are described, including advanced strategies proposing to use the abilities of plant–microbe associations called phytoremediation, and the enhancement of natural attenuation processes, especially biostimulation and bioaugmentation, and, last but not least, the potential of genetic modifications of the microorganisms applied. In the final chapter methods for reliable and representative sampling and chemical analysis of the processed samples are explained.

It is beyond the scope of any book to cover all aspects of soil bioremediation and to have considered all relevant results which are obtained in the huge community of researchers in science and industry. But the Volume Editor and the Series Editors hope that this book will give sufficient, elucidating and stimulating insight into the biology and technology of soil decontamination processes.

Essen, November 1999   Jürgen Klein

I General Aspects

1 Contaminated Soil Areas, Different Countries and Contaminants, Monitoring of Contaminants

WOLFGANG ULRICI

Bonn, Germany

List of Abbreviations

ADR	alternative dispute resolution
ALSAG	Altlastensanierungsgesetz, abandoned contaminated site remediation act (Austria)
BSB	Bodemsanering van in gebruik zijnde bedrijfsterreinen, clean-up of industrial sites in use (The Netherlands)
CALTOX	Californian model for assessing toxic effects from soil contamination
CCME	Canadian Council of Ministers of the Environment
CERCLA	Comprehensive Environmental Response, Compensation, and Liability Act (Superfund Act, USA)
CERCLIS	CERCLA Information System
CLEA	contaminated land exposure assessment (English scheme)
CP	physico-chemical (treatment)
CSOIL	Dutch model for assessing the risks to human health caused by hazardous substances on and in land, via various exposure pathways
DEC	New York State Department of Environmental Conservation
DoD	Department of Defence (Commonwealth of Australia)
DRIRE	Direction Régionale de l'Industrie, de la Recherche, et de l'Environnement, regional plant inspectorate (France)
DTSC	Department of Toxic Substances Control (California)
ECL	Environmental Conservation Law (New York)
ECRA	Environmental Clean-Up and Responsibility Act (New Jersey)
EPA	Environment Protection Authority (Australia/Victoria)
HESP	human exposure to soil pollutants, CSOIL-based computer program (program management by Shell International)
HRS	hazard ranking system
IBC	isolatie, beheer, controle; isolation, control, monitoring
ICRCL	Interdepartmental Committee on the Redevelopment of Contaminated Land (UK)
LAGA	Länderarbeitsgemeinschaft Abfall, joint Länder working group on waste (FRG)
MERLA	Minnesota Environmental Response and Liability Act
MN	Minnesota
MOA	memorandum of agreement (in the context of the NJ voluntary clean-up program)
NCS	national classification system (Canada)
NCSRP	national contaminated sites remediation program (Canada)
NJDEP	New Jersey Department of Environmental Protection
NPL	national priorities list (USA)
OECD	Organization for Economic Cooperation and Development
OVAM	Openbare Vlaamse Afvalstoffen Maatschappij, Flemish regional waste authority
PEA	preliminary endangerment assessment (California)
PR	public relations
PRESCORE	simplified program derived from HRS, using largely default input parameters, in order to rapidly assess eligibility of site for NPL
R&D	research and development
RAGS-B	risk assessment guidance Superfund B
RCRA	Resource Conservation and Recovery Act (USA)
SARA	Superfund Amendments and Reauthorization Act (USA)
SCG	Service Centrum Grondreiniging, Dutch soil clean-up service center
SEG	soil exposure guideline (Minnesota)

SITE	superfund innovative technology evaluation
SSL	soil screening levels (USA)
SUBAT	Stichting Uitfoering Bedemsanering Amovering Tankstations, voluntary fund created by Dutch petroleum industry for the clean-up of out-of-service gas stations
USEPA	US Environmental Protection Agency
VIC	voluntary investigation and clean-up program (Minnesota)

1 Introduction

Soil contamination is a typical unwanted side effect of industrial activity. The problem is especially grave in highly industrialized countries, where the opportunities offered by natural resources are exploited in a particularly sophisticated manner and on a very intensive scale. This applies in particular to the nations of the OECD. However, densely populated areas in other countries do suffer from increasing soil quality deterioration as well.

The problems posed by soil contamination are not identical in different countries, and a wide range of different approaches and practices for cleaning up contaminated land have been developed. To a certain extent, these differences may be attributed to the wide variety of political and administrative cultures. External circumstances also differ; e.g., the problems facing countries dominated by vast, open spaces are different from those of densely populated and highly industrialized regions such as the central European conurbations.

Against the background of the variety of approaches to the registration, investigation, assessment and remediation of contaminated land, the regulations and experiences of the following countries were studied in 1994/95 on behalf of the German Ministry for Education, Science, Research and Technology (ULRICI, 1995):

- Australia (national, Queensland, Victoria),
- Austria,
- Belgium/Flanders,
- Canada (national, Alberta, British Columbia, Ontario),
- Denmark,
- France,
- Federal Republic of Germany,
- Japan,
- The Netherlands,
- Norway,
- United Kingdom,
- USA (federal, California, Minnesota, New Jersey, New York).

An updated summary of this information taking into account of the legislative environment, in particular, is given in this chapter. The information provided is largely based on personal interviews with competent authorities.

An attempt was made to take into account and assess all important influencing factors and conditions. Thus, the investigations focused on those aspects which largely determine the approach to contaminated site management, such as:

- specific problems and political will to act on solving these problems,
- legal regulations and administrative organization to implement the rules,
- financing and liability questions,
- identification and registration, risk assessment, decision making towards intervening and priority setting, clean-up goals and clean-up work standards,
- approaches to conflict management.

It turned out that for an efficient management of contaminated sites, the following aspects are of major importance and should thus be included in any system regulating contaminated site management:

(1) careful balancing of liability and financing duties, as those ordered to clean up a site will regard the related allocation of responsibility as personal punishment, because of the financial consequences, regardless of jurisprudential attempts towards other, less discriminating interpretations, which are virtually not communicable;

(2) stipulation of strict and/or retrospective liability of the polluter;

(3) allocation of liabilities following ground transactions;

(4) a consistent technical decision making scheme;

(5) the State's share of clean-up responsibilities.

A synopsis of the different systems is given in Tabs. 1–11. Finally, system elements which proved efficient within specific systems are shown in Tab. 12. The observations may be summarized as follows:

Tab. 1. Specific Problems and Status of Technical and Legal Development of the Investigated Systems for Contaminated Site Management

State	Specific Problems	Preferred Treatment Techniques	Legislative Status	Remarks
Australia (national)	obstacles to city development, pollution of ground and surface	soil exchange, soil vapor extraction, soil ventilation, bioreme-	national harmonization directives	intergovernmental agreement among central, state and
Queensland	water	diation, land farming	Contaminated Land Act 1991, Local Government (Planning and Environment) Act 1990	local governments
Victoria			regulations in Environment Protection Act 1970 and amendments	EPA independent of state government
Austria	groundwater contamination	containment, bioremediation, CP treatment, solidification	Abandoned Contaminated Site Remediation Act 1989 and amendments Water Act 1959 and Amendments	

Tab. 1. Continued

State	Specific Problems	Preferred Treatment Techniques	Legislative Status	Remarks
Belgium/ Flanders	uncontrolled dumps, groundwater contamination	containment, soil exchange	Flemish Decree of February 22, 1995 on soil decontamination	
Canada (national)	groundwater contamination	soil exchange, bioremediation, land farming,	national harmonization directives	cooperation of provincial and central governments within Canadian Council of Ministers of the Environment (CCME)
Alberta		thermal desorption, pump-and-treat, soil vapor extraction, containment	regulations in Environmental Protection and Enhancement Act 1993	
British Columbia			regulations in Environment Management Act 1982	
Ontario			regulations in Environmental Protection Act 1970 and amendments	
Denmark	groundwater contamination	pump-and-treat, soil exchange	regulations in Waste Disposal Act 1983/ 1990, Loss-of-Value Act 1993	
France	abandoned waste dumps, groundwater contamination	containment, thermal treatment soil vapor extraction, bioremediation	no specific legislation; circulars, amendments to 1976 Act on Classified Plants	
Federal Republic of Germany	housing on contaminated sites, pollution of ground and surface water, obstacles to city development	soil exchange, safeguarding by containment, surface sealing, immobilization, hydraulic measures, thermal treatment, CP treatment, soil washing, soil vapor extraction, bioremediation	Federal Soil Protection Act 1998	Cooperation of federal and Länder governments within LAGA working groups
Japan	polluted agricultural land	soil cover, containment	Agricultural Land Law 1970	

Tab. 1. Continued

State	Specific Problems	Preferred Treatment Techniques	Legislative Status	Remarks
The Netherlands	housing on contaminated sites, groundwater contamination, sediment pollution	soil excavation and thermal or CP treatment in central plants, rarely *in situ* and on site pump-and-treat	Soil Protection Act 1987, amended 1994	
Norway	fjord and river pollution, obstacles to city development	soil excavation and treatment, containment, bioremediation	no specific legislation	
United Kingdom	obstacles to development of industrial areas	containment, bioremediation, solidification, thermal treatment	Environment Act 1995	
USA (national)	hazardous waste dumps, groundwater contamination, obstacles to city development	oil exchange, safeguarding by containment, surface sealing, immobilization, hydraulic measures, thermal treatment, CP treatment, soil washing, soil vapor extraction, bioremediation	RCRA 84, CERCLA 80, SARA 86	USEPA; cooperation between USEPA and federal state governments under bilateral agreements
California			regulations in Hazardous Waste/Substance Acts 1972, 1991	
Minnesota			MERLA 1983, Comprehensive Ground Water Protection Act 1989	
New Jersey			ECRA 1983, Superfund Act 1977, ISRA 1993	
New York			ECL Acts 1979, 1982, 1986	

Tab. 2. Definitions and Principles of Contaminated Site Management

State	Definitions	Principles
Australia	land is contaminated, if hazardous substances occur at concentrations above background levels and where assessment indicates it poses, or is likely to pose, an immediate or long-term hazard to human health or the environment	ensure that the health, diversity, and productivity of the environment is maintained or enhanced for the benefit of future generations; integrate economic and environmental considerations in decision making processes in order to improve community well-being; establish incentive structures and market mechanisms; remediation goals depending on use and objects to be protected
Austria	contaminated sites: waste sites and industrial sites including the potentially polluted soils and aquifers, which, according to the results of a risk assessment, pose a considerable threat to human health and the environment	preference for permanent remediation; hazard prevention
Belgium/ Flanders	contamination: man-made presence of substances at concentrations directly or indirectly impairing, or capable of impairing, soil quality	aim to achieve level of soil quality where no adverse effects are observed
Canada	contaminated site: area of land in which the soil or any groundwater lying beneath it, or the water or the underlying sediment, contains a special waste or another prescribed substance in quantities or concentrations exceeding prescribed criteria, standards or conditions	remediation goals dependent on use and objects of protection; remediation to the lowest level practicable; make the polluter pay, if he is traceable and solvent; contribution from those who have benefitted or still benefit from the contamination or from remediation; specific rules for orphan sites
Denmark	contamination: hazardous substances appearing in concentrations which are of concern for sensitive uses	everybody should be able to move and live anywhere without having to fear risks from pollutants; safeguard nature and environment, thus ensuring a sustainable social development in respect of human conditions of life, protecting flora and fauna
France	diffentiation between sources of pollution and sites in the vicinity with secondary contamination	soil contamination is not critical as long as it can be kept under control and no detrimental effects occur; hazard prevention
Federal Republic of Germany	contaminated sites: waste disposal or industrial sites which cause harmful soil modifications impairing the soil functions in the long term or which do not comply with the character of the area or with planning, or pose other risks to individuals and/or the public	remediation goals dependent on use and objects of protection

Tab. 2. Continued

State	Definitions	Principles
Japan	contaminated site definition restricted to agricultural land	reduction of pollutant concentrations in soil used for agricultural purpose
The Netherlands	seriously contaminated site: site where the soil is or threatens to be contaminated so that the functional properties which the soil has for man, flora, and fauna have been, or are in danger of being, seriously reduced	preserve or restore soil multifunctionality; if this is not possible for reasons of environmental hygiene, or for technical or financial reasons, reduce negative effects by observing the IBC criteria for the harmful substances remaining in the soil
Norway	pollution: discharge of solid matter, fluid or gas into air, water or ground, which cause or may cause damage or disamenity to the environment	achieve satisfactory environmental quality on the basis of a total appraisal of health, welfare, the natural environment, costs related to control measures and economic considerations
United Kingdom	contaminated land: any land that appears to the local authority in whose area it is situated to be in such a condition, by reason of substances in, on or under land, that significant harm, or pollution of controlled waters, is being or is likely to be, caused	remedial action required only where the contamination poses unacceptable actual or potential risks to health or the environment, and there are appropriate and cost-effective means available to do so, taking into account the actual or intended use of the site; establish market to put wasteland in inner cities and brownfields to productive use wherever possible
USA	contaminated site: pollutant concentrations are above applicable remediation standards	protect human health and the environment on a permanent basis; treat the most dangerous contaminated sites under national management using decision making procedures that are highly formalized in view of their legal strength; priority for financial contribution by the polluter; detailed procedures elaborated to inform and involve the public

Tab. 3. Liability Rules

State	Type of Liability			Potentially Responsible Parties					Remarks
	Joint & Several	Strict	Retro-active	Pol-luter	Holder of Land	Lender	Other	Innocent Owner	
Australia (national)	x	x	x	x	x			(x)	
Queensland	x	x	x	x	(x)	x	x		innocent private owner/occupier not liable; authorities liable for fault
Victoria	x	x	x	x	x	(x)	x	(x)	innocent private owner not liable
Austria				(x)	(x)				liability based on fault, albeit for ancient contamination restricted by admitting state's lack of apprehension of soil contamination
Belgium/ Flanders	x	x		x	x		x		
Canada (national)	x	(x)		x		(x)	(x)		CCME recommendations
Alberta	x	x	x	x	x	x	x	x	whoever contributed, in Environment Alberta's opinion, may be held liable
British Columbia	x	x	x	x	x				whoever dealt with pollutants may be held liable
Ontario	x	x	x	x	x			x	affected municipalities liable as well
Denmark				(x)	(x)				liability based on fault after legislation came into force; statutory period of limitation 20 years
France		x	x	x	x		x		owner liable only if he drew benefit from holding pollutants

Tab. 3. Continued

State	Type of Liability			Potentially Responsible Parties					Remarks
	Joint & Several	Strict	Retro-active	Pol-luter	Holder of Land	Lender	Other	Innocent Owner	
Federal Republic of Germany	x	x	x	x	x		x	x	
Japan		x	x	x					
The Netherlands	(x)	(x)	(x)	x	x		x		
Norway	x	x		x	x			(x)	financial contribution of innocent owner limited to value of ground
United Kingdom		x		(x)	x		x		liability essentially fault-based
USA (national)	x	x	x	x	x	(x)	x	x	lenders *de facto* exempted
California	x	x	x	x	x		x	x	allocation of liability share by DTSC
Minnesota	x	x	x	x	x		x		various exceptions to liability
New Jersey	x	x	x	x	x		x	x	allocation of liability share by NJDEP
New York	x	x	x	x	x		x	x	allocation of liability share by DEC

x: item fully applies to every case
(x): item is restricted to particular cases such as:
   – recommended, but not generally practiced
   – applied only to particular affected groups, exempting others
   – applied only under certain specific circumstances

Tab. 4. Financing Schemes if Responsible Party is Not Available or Not Solvent

State	Financing Schemes				Remarks
	Taxes/ Community	Levies/Group Burden	Mixed Funding	Other	
Australia (national)	x				defence monies for DoD sites
Queensland	(x)				state assistance in exceptional cases
Victoria	x				civil protection monies
Austria	x	x			waste levy
Belgium/ Flanders	x		(x)		OVAM budget; financial assistance for oil tank owners funded by mixed state and private industry contributions (pilot study)
Canada (national)	(x)				NCSRP monies: equal shares taken by central and affected provincial governments; program was limited in time and has not been renewed
Alberta	x				
British Columbia	x				civil protection monies
Ontario	x				special programs for large sites, gas works sites
Denmark	x	x		x	state, regional, communal monies; oil industry's environment fund; site owner's contribution under Value Loss Act
France	x	x			state monies; special waste tax
Federal Republic of Germany	x	(x)			assistance to remediation in new Länder; in some Länder, levies on waste or landfilling were raised which now are jeopardized on principle grounds; case-by-case assistance, if there is government interest in developing specific new technologies
Japan	x				
The Netherlands	x			x	state, province, communal monies; industry remediation under BSB and SUBAT covenants

Tab. 4. Continued

State	Financing Schemes				Remarks
	Taxes/ Community	Levies/Group Burden	Mixed Funding	Other	
Norway					
United Kingdom	x				government credit approvals to municipalities, remediation assistance by English partnerships
USA (national)			x		superfund
California	x	x		x	civil protection monies; Hazardous Substance Account, Hazardous Substance Cleanup Fund, Expedited Site Remediation Trust Fund, Hazardous Substance Victim's Compensation Fund
Minnesota	x	x			MN superfund; landfill clean-up program
New Jersey		x		x	Hazardous Discharge Site Clean up Fund (bonds), Spill Compensation Fund
New York	x				Environmental Quality Bond

x: item fully applies to every case
(x): item is restricted to particular cases such as:
 – recommended, but not generally practiced
 – applied only to particular affected groups, exempting others
 – applied only under certain specific circumstances

Tab. 5. Identification and Registration of Contaminated Sites

State	Active Search	Regular Check by Authority	Data Supplied on Request	Publication of Data	Data in Land Register	Delisting upon Clean-Up	Remarks
Australia (national)	(x)	(x)					(contaminated military sites)
Queensland	x		x		x		
Victoria		x	x	x		x	publication of risk sites whose owners do not cooperate; delisting upon audit certificate
Austria	x	x	x	x			"contaminated site atlas"
Belgium/ Flanders		x	x	x	x		
Canada (national)							
Alberta	(x)			x			identification of abandoned sites
British Columbia			x	(x)	x		publication of sites which are not properly remediated
Ontario	(x)	(x)					identification of abandoned sites
Denmark	x	x	x			x	
France		x	x				
Federal Republic of Germany	x	x	x				
Japan	(x)	(x)					(contaminated agricultural land)
The Netherlands	x	x	x		x	x	delisting when multifunctionality has been achieved
Norway	x		x				national database
United Kingdom		x	x			x	public registers of regulatory action

Tab. 5. Continued

State	Active Search	Regular Check by Authority	Data Supplied on Request	Publication of Data	Data in Land Register	Delisting upon Clean-Up	Remarks
USA (national)		x	x			x	CERCLIS database; delisting of cleaned up NPL sites
California	x		x		x		
Minnesota	(x)		x		x	x	identification of sites belonging to certain categories
New Jersey	x	x	x	x	x		
New York			x		x	x	

x: item fully applies to every case
(x): item is restricted to particular cases such as:
    – recommended, but not generally practiced
    – applied only to particular affected groups, exempting others
    – applied only under certain specific circumstances

Tab. 6. Risk Assessment

State	Assessment by		Assessment based on			Procedure Fixed	Remarks
	Authority	Other	Prescribed Values	Risk Assessment	Computer Models		
Australia (national)		(x)	x	x		x	national recommended trigger values
Queensland		x	x	x	x	x	HESP program in use
Victoria		x	x	x		x	
Austria	x	x	(x)	x		x	multi-level decision making scheme
Belgium/ Flanders	x		(x)	x	x		HESP program in use
Canada (national)	x	x	x	x		x	NCS, use-dependent environmental quality criteria
Alberta		x	x	x		x	
British Columbia	x	x	x	x		x	own set of quality criteria
Ontario	x	x	x	x		x	own set of quality criteria
Denmark	x			x		x	
France	x	x	(x)	x			
Federal Republic of Germany	(x)	x	x	x			
Japan	x		x				
The Netherlands	x	x	x	x	x	x	detailed decision making scheme; use-dependent to a certain extent; HESP program based on CSOIL model
Norway		x	(x)	x			
United Kingdom	x	x	x	x	x	x	use-dependent quality criteria; CLEA model

Tab. 6. Continued

State	Assessment by		Assessment based on			Proce-dure Fixed	Remarks
	Author-ity	Other	Prescribed Values	Risk Assess-ment	Computer Models		
USA (national)	x	x	x	x		x	SSL, RAGS-B
California	x	x		x	x	x	PEA, CALTOX com-puter models
Minnesota		x		x		x	SEG guideline
New Jersey		x	x	x		x	use-dependent soil quality targets
New York		x		x	x	x	

x: item fully applies to every case
(x): item is restricted to particular cases such as:
   – recommended, but not generally practiced
   – applied only to particular affected groups, exempting others
   – applied only under certain specific circumstances

Tab. 7. Action Thresholds

State	Action Triggered by				Triggered Action		Remarks
	Exceeding of Threshold Value	Classification	Change of Use	Ground Transaction	Detailed Investigation	Remediation	
Australia (national)	x				x		set of trigger values; for pollutants for which no values are given, use of Dutch list values
Queensland	x		x		x		
Victoria	x		x		x		
Austria	x	x			x		multi-level decision making scheme
Belgium/ Flanders	x			x	x	x	
Canada	x	x			x	x	NCS; environmental quality criteria; class 1 sites scoring more than 70 need action urgently
Denmark	(x)			x	x		non-binding quality criteria
France	(x)				x		set of threshold values planned
Federal Republic of Germany	x				x		set of use-dependent threshold values
Japan	x				x	x	soil quality standards; government owned land must be cleaned up
The Netherlands	x		x	x	x	x	decision points at I, S, $(I+S)/2$ values
Norway	(x)						
United Kingdom	x		x		x		non-binding ICRCL criteria

Tab. 7. Continued

State	Action Triggered by				Triggered action		Remarks
	Exceeding of Threshold Value	Classification	Change of Use	Ground Transaction	Detailed Investigation	Remediation	
USA (national)	(x)	x			x	x	internal SSL; HRS; sites scoring more than 28.5 are put on NPL and must be cleaned up
California			x		x		
Minnesota	x				x	x	own set of intervention values
New Jersey	x		x	x	x	x	use-dependent intervention values
New York	x		x	x	x		

x: item fully applies to every case
(x): item is restricted to particular cases such as:
    – recommended, but not generally practiced
    – applied only to particular affected groups, exempting others
    – applied only under certain specific circumstances

Tab. 8. Priority Setting

State	Decision Making					Remarks
	Technical	Political	Techno-Political	Market	Computer-Assisted	
Australia			x	x		
Austria	x		x			multi-level decision making scheme
Belgium/Flanders			x	x		CSOIL model in use
Canada	x		x		(x)	NCS
Denmark	x					change of priorities possible according to Loss-of-Value Act
France		x	x			
Federal Republic of Germany		x	x			
Japan	x					priority for agricultural land
The Netherlands	x		x		x	CSOIL model; HESP program based on CSOIL
Norway	x					
United Kingdom			x	x		market driven by needs of developing industry and commerce
USA (national)	x				x	HRS; PRESCORE
California	x		x	x	x	PEA; CALTOX; walk-in program
Minnesota	x			x		VIC program
New Jersey	x			x		market established through ECRA; MOA voluntary clean-up program
New York	x			x		voluntary clean-up program

x: item fully applies to every case
(x): item is restricted to particular cases such as:
- recommended, but not generally practiced
- applied only to particular affected groups, exempting others
- applied only under certain specific circumstances

Tab. 9. Clean-Up Goals, Clean-Up Work and Verification of Clean-Up Success, Aftercare

State	Remediation Target				Containment Accepted as Clean-Up	Certificate	Aftercare Regulations	Remarks
	Below Trigger Value	Multi-functionality	Set of Values	Separate Evaluation				
Australia (national)	x		x	x				in the regular case, contamination should be kept below given trigger values; in special cases, risk assessment to fix clean-up target; Victorian certificate will be guaranteed by authority
Queensland	x		x	x				
Victoria	x		x	x	(x)	x		
Austria		(x)	x	x				
Belgium/ Flanders	x		x	x	x	(x)		
Canada (national)	x		x	x				in the regular case, contamination should be kept below given trigger values; in special cases, risk assessment to fix clean-up targets; certificate will not establish authority responsibility
Alberta	x		x	x	x		x	
British Columbia	x		x	x		x		
Ontario	x	x	x	x				
Denmark		x	x	x	x	x		site-specific departures from multifunctionality allowed
France		(x)	x	x	x			aftercare by DRIRE
Federal Republic of Germany	x		x	x	x			use-dependent targets
Japan	x		x		x			
The Netherlands		x	x	(x)	x	x	x	securing a site accepted as equivalent if IBC criteria are kept; soil quality certificate from SCG
Norway	(x)		(x)	x	x			
United Kingdom	x		x	x	x		x	

Tab. 9. Continued

State	Remediation Target				Contain-ment Ac-cepted as Clean-Up	Certi-ficate	After-care Regula-tions	Remarks
	Below Trigger Value	Multi-func-tionality	Set of Values	Separate Evalua-tion				
USA (national)	x		(x)	x			x	SSL procedure to determine clean-up targets; containment functioning to be checked every 5 years
California	x			x		x	x	certificates will not establish responsibility of authorities
Minnesota		x		x		x	x	
New Jersey	x	x	x	x		x	x	decontamination is preferred option
New York	x			x	x	x	x	

x: item fully applies to every case
(x): item is restricted to particular cases such as:
  – recommended, but not generally practiced
  – applied only to particular affected groups, exempting others
  – applied only under certain specific circumstances

Tab. 10. Organization and Enforcement

State	Responsibility		Allocation of Liability by			Technical Decisions by		Enforcement			Remarks
	Centralized	Decentralized	Authority	Courts	Arbitration	Authority	Responsible Party	Criminal Justice	Market Forces	Incentive System	
Australia (national)	(x)	x				x	x				national government responsible only for military sites (DoD)
Queensland	x		x	x		x		(x)	x		
Victoria	x		x	x		x		x	x		EPA independent of government; system of auditors
Austria	x	x	x	x		x				x	financing regulated by ALSAG
Belgium/ Flanders	x		x			x		(x)			
Canada (national)	x	x				x					national and provincial governments shared responsibility for action within former NCSRP
Alberta	x		x	x	x	x	x	x			
British Columbia	x		x	x	x	x	x	x			
Ontario	x		x	x	x	x	x	x			
Denmark	x	x	x	x		x				x	
France	x		x			x		x	x		
Federal Republic of Germany	(x)	x	x			x	x				federal government responsibility only for abandoned military and new Länder sites
Japan	x	x	x			(x)					

Tab. 10. Continued

State	Responsibility		Allocation of Liability by			Technical Decisions by		Enforcement			Remarks
	Centralized	Decentralized	Authority	Courts	Arbitration	Authority	Responsible Party	Criminal Justice	Market Forces	Incentive System	
The Netherlands	x	x	x	x		x	x	x	x		
Norway	x		x			x		x			
United Kingdom		x	x	x		x	x		x	x	financial incentives by English partnerships
USA (national)	x	x	x	x		x		x			to-date, litigation-oriented enforcement strategy
California	x		x	x	x	x	x	x	x	x	walk-in program
Minnesota	x		x	x	x	x	x	x	x	x	VIC program
New Jersey	x		x	x	x	x	x	x	x	x	voluntary MOA process
New York	x		x	x	x	x		x	x	x	voluntary clean-up program

x: item fully applies to every case
(x): item is restricted to particular cases such as:
     – recommended, but not generally practiced
     – applied only to particular affected groups, exempting others
     – applied only under certain specific circumstances

Tab. 11. Conflict Management

State	Codification				Remarks
	Financing	Compensation	Information	Involvement	
Australia (national)			x		focus on communicating with affected people
Queensland	(x)		x		government may take over contaminated land from
Victoria	x		x		innocent private owners
Austria	x	x	x		compensation for restrictions on using contaminated land during remediation
Belgium/ Flanders					
Canada	x		x	x	CCME guidance for PR work
Denmark	x	x	x	x	possibility of shifting priorities under Loss-of-Value Act
France	x		x		principle not to shake the trust of innocent parties in the legality of state's actions
Federal Republic of Germany			x		information of affected parties on remedial planning
Japan	x				
The Netherlands	x		x	x	decision making involving citizens' votes
Norway	x		x		administrative files are open to public inspection
United Kingdom					

Tab. 11. Continued

State	Codification				Remarks
	Financing	Compensation	Information	Involvement	
USA (national)	x		x	x	structured procedure involving:
					– community interviews
California	x	x	x	x	– community relations plan
					– community relations co-
Minnesota	x	x	x	x	ordinator
					– external consultants, con-
New Jersey	x	x	x	x	flict resolution professio-
					nals and mediators
New York	x		x	x	– community working
					groups
					– citizens' information
					access offices
					– record of decision
					including citizens' votes

x: item fully applies to every case
(x): item is restricted to particular cases such as:
  – recommended, but not generally practiced
  – applied only to particular affected groups, exempting others
  – applied only under certain specific circumstances

Tab. 12. System Elements which Proved Efficient within Specific Systems

System Elements	Remarks
(1) Non-Polluting Sites	
Open declaration of the caveat emptor principle for ground transactions	open declaration of practiced liability and financing rule
Obligation of site auditing when planning ground transactions	corrective element in the interest of protecting the buyer of ground; the audit may be waived if a certificate of multifunctionality can be produced; possibly, limitation of certificate validity for industrial ground
Establishment of site auditors	relief of authorities, establishing, at the same time, market for auditors; if certification procedures for site auditors are sufficiently demanding, authorities may even take over responsibility for expertise which might not stand the demands for correctness
Tying systematical administrative identification and registration activities to decisions that are preset in a binding manner, depending only on the data obtained	relief of authorities from data raising activities for unclear or unnecessary purposes; this appears to be particularly justified by the fact that completeness of data will never be guaranteed, and implementation deficits will not allow to control observance of soil protection requirements to any extent of completeness; thus, identification activities make sense only if the data are definitely needed for administrative decision making; the demands for security from soil pollution hazards may be satisfied by a site audit system
Reactive registration of non-polluting sites	relief of authorities from book keeping
Release of non-polluting sites from administrative control	relief of authorities from control activities, a site audit system showing the risks of incomplete remediation for the buyer as well as for the vendor, and a restrictive delisting practice refusing certificates if soil cannot be put to multifunctional use
Administrative action triggered by change of site use	no need for action if there is no nuisance or damage to third parties; incentive for demanding remediation if disadvantages to ground vendors were too high
Fixing of minimum remediation goals on the basis of assessment of risks, taking the intended use and potentially affected objects of protection into account	minimum requirement of protecting human health in the interest of implementing the system; may be achieved by decontamination or containment for which equivalence is stated using the Dutch IBC criteria
Delisting of sites restored to multifunctionality	incentive for restoring multifunctionality; reduces the apparent economic advantage of containment measures
Listing of site use restrictions within the competent authority's register and/or the land register	in the interest of buyers; incentive for restoring multifunctionality
Use of partially decontaminated soil for non-sensitive purposes	saving clean soil resources if activities, such as road construction, tend to contaminate soil anyway; need for development of minimum quality standards by construction industry

Tab. 12. Continued

System Elements	Remarks
(2) Polluting Sites	
• Liability and Financing	
Explicit definition of priorities for drawing on potentially responsible parties for financing clean-up; polluter to be held responsible in the first place	enforcement of polluter pays principle in accordance with common understanding; element of fault in liability
Burden of proof on alleged polluter	improvement of implementability of polluter pays principle
Right of called-on responsible party to seek contribution from other potentially responsible parties	on grounds of fairness towards non-polluters; assertion of right of seeking contribution may be denied to polluters
alternatively Allocation of liability shares by authority, e.g., following the Canadian deliberation pattern	reasonable in order to prevent litigation; may be flanked by restrictions on right of called-on responsible parties to object
State to take over orphan shares	relief for remaining responsible parties; may be communicated as corrective measure balancing authority's allocation of liability shares
De minimis and *de micronis* schemes to relieve small-share responsible parties from further involvement	saves authorities administrative and political efforts
If owners are to be held liable: extension of responsible parties from present owners to past owners of a contaminated site, thus relieving the present owner	works in the sense of the polluter pays principle and is easier to communicate than allocation of liability to and calling on present owners alone; prevention against accusation of administrative support to concealing site defects during ground transactions
Regulation of liability of heirs, donees, creditors holding security over a contaminated site, following, e.g., the Australian scheme of options	sensible in order to prevent troubles for financial markets; works in the same sense as take over of orphan shares by the state
Limitation of owner's financial liability using technical rather than financial criteria, such as prevention of existing and foreseeable hazards from third parties	technical criteria are more easily assessed than, e.g., ground value; may be communicated as expressing owner's social responsibility
Explicit definition of cause and extent of state's potential liability and share	fixing the rules of the game; definition must be simple and easy to understand by everyone in order to prevent misleading interpretation and abuse
Reimbursement and compensation for owner of suspect contaminated site if investigations do not confirm suspicion	on grounds of fairness
Exemption of innocent/bona fide potentially responsible parties	element of fault in determining liability; criteria for claims of innocence/action in good faith may be exacting in order to prevent abuse

Tab. 12. Continued

System Elements	Remarks
Statutory period of limitation for responsibility of potentially responsible parties	fixing the gravity of polluting soil in comparison with other civil, criminal, and economical offence
Take over of responsibility for remediation of contaminated sites which occurred before relevant legislation came into force, by the state	prevention of retrospective liability if the state of knowledge had not been sufficient to apprehend the problem of soil contamination in general
Take over of responsibility for remediation of contaminated sites which occurred with authorities' consent, by the state	preservation of confidence in the responsiveness of authorities' actions
Financing of orphan contaminated sites from tax monies, priorities being annually updated	tax monies for state business
Take over of responsibility for hazard prevention by the state if all potentially responsible parties are innocent or acted in good faith	hazard prevention, alike civil protection, as minimum state business; calling on innocent/bona fide potentially responsible parties will lead to politically colored conflicts
Compensation for defective administrative planning, even retrospectively	preservation of confidence in the responsiveness of authorities' actions
Registration of a charge in the deed register, if the owner of a contaminated site benefits from remediation activities on his site; this charge should have to be redeemed before selling the ground	long-term security for state's claims for reimbursement; however, present owner will not be obliged to immediately give up elementary rights of using his property

- Identification and Registration

System Elements	Remarks
Definition of soil contamination on direct proof of soil toxicity	direct assessment of toxicity is more adequate than presently practiced indirect methods, reflecting the objective of removing and preventing health hazards; soil contamination, as defined by concentration of hazardous substances, will describe health hazards only in an indirect and provisional manner
Publication of information in the register of confirmed contaminated sites	public access to environmental information

- Risk Assessment

System Elements	Remarks
Human/public health as only object of protection	due to the lack of commonly accepted criteria for assessing ecological damage
Use of bioindicators for assessing risk	direct assessment preferable, reflecting the soil protection objective of removing and preventing health hazards

Tab. 12. Continued

System Elements	Remarks
• Risk Assessment	
Fixed action thresholds	to be used in the regular case; easy to handle
Risk assessment only if individual case justifies the effort	to keep flexibility and avoid undue hardship
Accounting for insufficient information through error margins following, e.g., the Canadian NCS model scheme	giving a clear idea of the amount of information deficits; works best with clear conception of priority setting
Structuring the political procedure for setting priorities	prestructuring decision making procedures in order to improve transparency
• Targets and Execution of Remediation	
Equality of remediation goal to action threshold, as a rule, remediation goal to be modified only by risk assessment accounting for intended use of the site	reflects the traditional approach of combatting the problem causing risks as far as to keep remaining risks acceptable
Equivalence of decontamination and containment if Dutch IBC criteria are observed	definition of demanding quality criteria for containment, which may provide an incentive for permanent solution
Check of containment efficiency every 5 years	minimum requirement
Registration of use restrictions in land register	incentive for full remediation
Conception of dependence of remediation goals on layers to be cleaned up	decontamination of upper layer with a view to the intended use; clean-up of deeper layers to levels complying with industrial use; groundwater should be protected in any case
Principle of assigning first priority to preventing hazards to residents, second priority to preventing hazards to workers	in pursuit of the objective of remediation, namely to protect human health
In emergencies, facilitation of authorization procedure for time-limited interim storage for contaminated soil	facilitation of administrative procedures for time limited emergency measure
Certificating and delisting of the site if multifunctionality of contaminated soil has been restored	relief of authorities from book keeping on the one hand, proof of success on the other hand; if maximum remediation goals are achieved, certificate will put the authorities at risk on equal terms with private economy and citizens

Tab. 12. Continued

System Elements	Remarks
Assessment of remediation success using concentration percentiles	less demanding than separate keeping of goals within each sample taken for naturally occurring contamination – which should be assumed after clean-up – the percentile method has proven effective in geology; if deemed necessary, upper limits to excessive concentrations

● Conflict Management

Detailed and continuous information of the public	confidence creating measure
Most detailed information of affected persons	confidence creating measure
Financial assistance to citizens' information gathering activities	confidence creating mesure; basis for reducing misunderstandings and conflicts during negotiations
Financial assistance to citizens' self-organizing activities	confidence creating measure making sure parties negotiate on equal terms
Structuring PR contents and procedures	professionalization of authorities' PR work
Definition of data protection scheme to be applied	to remove unclear and ambiguous ideas regarding information duties and rights
Giving affected persons a share in preparing decisions	confidence creating measure; authorities' last instance responsibility should be observed
Calling-in of conflict negotiators and mediators	useful, as in the regular case, contaminated site management is outside the day-to-day routine
Clean-up of contaminated sites in residential use at state expense	confidence creating measure; calling on innocent or bona fide owners will regularly result in political crises undermining authorities' credits and image losses, eventually driving authorities to pay for clean-up anyway
Refusal to compensate for intangibles, such as losses on use due to use restrictions	prevention of jealousy and conflicts with interests that are not related to remediation

● Enforcement and Organization

Environmental protection body with judicial powers and government independence	if political will supports the establishment of an independent environmental protection body, this body may prove highly efficient in implementing and enforcing environment protection, as it makes sure administrative decisions affecting environment will not depend on government interests. In some states, to create such a body may require modification of the basic law
Management of the most serious contamination cases by national government	in the interest of setting standards and of harmonizing treatment of the issue by lower-level authorities

Tab. 12. Continued

System Elements	Remarks
Grants of assistance for remedial measures only if potential beneficiary complies with certain requirements, such as obligation to immediately inform authorities, securities for aftercare, etc.	incentive to cooperate with authorities
Registration and publication of contaminated site data as informal punishment if remediation plans are not elaborated in due time	seizing the responsible party by its most sensitive point
Out-of-court settlements with authorities only if responsible party recognizes its liability on principle	enforcement of observance of terms of settlement by securing submission of responsible party

2 Specific Problems

Generally, the main issue of soil contamination is ground and surface water pollution which may adversely affect the use of water for drinking and technical purposes and impair fishery. The impairment of municipal and industrial development appears to be of secondary importance; it will play a relatively major role in those countries, in particular where water resources are abundant.

The problem of residential buildings on contaminated ground appears to be well-known to all of the studied countries. It strongly contributes to building up political sensitiveness for the soil contamination issue. However, in terms of quantity, the problem generally appears to be of rather minor importance.

3 Preferred Treatment Techniques

Relatively simple and cost-effective techniques, such as the safeguarding of contaminated sites by inertization and encapsulation as well as soil exchange and landfilling of the contaminated soil, are widely applied. Only some countries give preference to treatment which permanently removes the risk potential of hazardous soil contaminants; as an example, The Netherlands give priority to thermal and physico-chemical soil treatment; other countries regularly apply pump-and-treat techniques.

In situ and on site treatment technologies are developed in several countries. Development focuses on bioremediation techniques, in particular within reactors and through landfarming. Denmark has issued specific short-term bioremediation goals which are less stringent than goals to be achieved by using other non-biological techniques, taking the characteristics of microbial action into account which will continue in the long term. On the contrary, however, the Dutch do not favor bioremediation, as its efficiency is considered insufficient in the short term.

One of the problems new technologies have to overcome in particular is the reluctance of investors and authorities to try them out; in general, they are interested in the safe success of the remediation measures rather than in starting costly and uncertain developments, taking into account the potential consequences of poor performance for liability and economy. A large-scale explicit technology development program to overcome this bottleneck is the US SITE (Superfund Innovative Technology Evaluation) program which was created within the frame of the Superfund Amendments and Reauthorization Act. Other countries generally run rather dispersed government programs to support R&D for the treatment of contaminated sites.

4   Legislative Status

The legislative status varies considerably from one state to the next. It largely reflects the political sensitiveness for the soil contamination issue. In this context, one should note the situation in the states of Australia and Canada which are organized along federal lines. Here, the federal states/provinces agreed on common principles for dealing with contaminated sites, at the level of the state environment ministers that are relatively loosely coordinated by central environmental agencies. In the USA, on the other hand, the most dangerous contaminated sites are tackled by the federal government's USEPA (US Environmental Protection Agency), which is given important financial and administrative power.

5   Definitions

The definitions of soil contamination show the common technical understanding of the selected countries regarding the risk potential of soil contamination. However, the definitions vary according to the different ways of considering risks as significant – through exceeding of generally fixed soil quality standards, of certain intervention values, or of standards set at the discretion of the authorities, on the one hand, through individual risk assessments, on the other – and to the type of sites not falling under the definition: e.g., military sites are frequently excluded from the general definition and from contaminated site legislation. In Japan, only contaminated agricultural sites are covered by present legislation.

6   Principles

Beyond danger prevention, the goals of treating contaminated sites vary considerably. Extremes are found in The Netherlands, on the one hand, where first priority is given to restoring and conserving soil multifunctionality – waivers are possible only if this goal cannot be reached on economical and technical grounds, in which case remediation goals are to be defined so as to protect sensitive targets, taking into account the intended use of the site – and in the United Kingdom, on the other hand, where remediation is required only if appropriate and cost-efficient possibilities exist, taking into account the intended use. The first type of systems favors permanent solutions against use-specific solutions, the second type of systems gives first priority to use-specific solutions. Some countries have established mechanisms that aim at obtaining permanent solutions in the long term if this should not be possible from the outset. These mechanisms usually work by taking advantage of market forces related to ground transactions.

7   Liability

The majority of the selected countries hold the responsible parties strictly, jointly, and severally liable, with the exception of Austria, Denmark, and the United Kingdom. France applies strict, but not joint and several liability. Liability is retrospective in most of the states studied. Frequently, certain groups of responsible persons, such as innocent or *bona fide* private houseowners are exempted from the duty of paying for remediation of their ground. In some cases, authorities are explicitly held responsible for faulty decisions concerning contaminated sites.

In virtually all systems, polluters will be held responsible; exceptions are the Austrian and English systems where polluters have the defence of insufficient state of knowledge in the past, and the Danish system in which the polluter will be held liable only if the pollution occurred after the Environmental Protection Act came into force. This is barred by the statute of limitations after 20 years.

In most systems, another potentially responsible party is the holder of power over the site, in some cases even if he did not have or ought not to have had any knowledge of the contamination at the time he bought the ground. Occasionally, lenders will be held liable as well, in particular if their role within the enterprise that caused the contamination went beyond purely financial interests.

Liability shares, in the case of several responsible parties, will be occasionally allocated at the discretion of the authorities, or else by the courts. Typically, parties held financially liable under the regime of joint and several liability will be entitled to seek contributions from their fellow responsible parties.

8 Financing Models and Assistance Schemes

For those cases in which responsible parties cannot be found or are not sufficiently solvent, states have established financing models of varying type and complexity. They range from government funds that are financed from tax revenues – occasionally, civil protection means are used for the clean-up of contaminated sites – to funds financed by levies charged to specific groups, such as levies on hazardous wastes or on potentially hazardous raw chemicals (US Superfund). Frequently, mixed financing models are in use. A special case is the owner's contribution to remediation costs according to the Danish loss-of-value regulation, which may be used by the owner to accelerate the treatment of his contaminated site, which is the responsibility of the Government.

Most of the states which hold polluters and owners liable for clean-up have developed more or less comprehensive and elaborated assistance programs in order to prevent hardship and to provide an incentive to the responsible parties to take the initiative for cleaning up their contaminated sites.

The traditional reason for granting assistance being political pressure from politically influential responsible parties (house building and renting companies, citizen's opposition groups), more advanced schemes try to mobilize the initiative of responsible parties by technical help schemes combined with some financial assistance if necessary, e.g., in the respective programs in some states of the USA.

9 Identification and Registration

Not all countries require their authorities to actively search for contaminated sites. On the contrary, in some countries, such as Victoria/Australia, and the United Kingdom, in particular the strategy is to care for identification and treatment only if danger is imminent or may be a concern if the use of the site changes. It is left to the market to decide if a use change is intended and action may thus be required.

If a site is suspect of being contaminated, the duty to investigate is with the authorities, at the one extreme, or a potentially responsible party may be ordered to carry out the investigation, at the other. In Germany, e.g., the practice is to order the potentially responsible party to investigate his ground and reimburse the costs, if the suspicion is not verified.

The passing-on of information on site contamination is subject to varying rules according to the respective data protection schemes. The range of rules starts with public contaminated site registers providing the full extent of available information and ends with the right of inspecting the files containing only administrative action on the respective site. Between these two extremes, extracts from the complete contaminated site register may be provided. Some of these contain those sites in particular, where responsible parties have refused to take the necessary remediary action.

10 Listing and Delisting

After listing a contaminated site or a site suspected of contamination, procedures to publish the listed data vary according to data protection schemes and to legal links between responsibility for remediation and ground transactions. In some states, the land register will contain a copy of the listed information or, at least, a reference to the list of contaminated sites. Other states will separately publish the full or partial contents of the list; New Jersey uses both ways, presumably due to the necessity to remediate as a precondition for ground transactions and to the will to make every effort to inform potential buyers of contaminations.

Delisting after successful remediation is handled much more restrictively. Most states tend to keep a site once registered on the list or in the land register for an indefinite time, albeit with a note referring to the remediation. Only few states will remove remediated sites from the registers; Victoria/Australia, will require an auditor's certificate stating that the site has been thoroughly cleaned up and is ready for multifunctional use.

11 Risk Assessment

In some states, risk assessments will be carried out by the administration in principle, other states will have them carried out by external consultants only. Victoria/Australia has established a specific auditors' system for this purpose. Most states, however, use a mixed approach, external consultants regularly assessing the risks and authorities performing an assessment of their own only from time to time.

Lists of intervention or soil quality values, such as the Dutch list, are widely used, at least for decisions based on initial investigations. Some countries explicitly state that their approach is to regularly use such lists and take recourse to detailed risk assessment only if this appears to be justified by specific circumstances, such as the size and complexity of the contamination case.

Other countries, however, rely only on detailed risk assessment for each individual case.

In those cases where they are defined, the decision models are of varying complexity. The extremes are USEPA's highly complex HRS

(Hazard Ranking System), on the one hand, and the Canadian NCS (National Classification System), lending itself to manual evaluation, on the other.

Some of the decision models are computerized; whereas in the Canadian NCS, computerization serves only as a tool to simplify the already simple bookkeeping and report preparation, other models need computerization for the calculation of risk potentials along the various exposure pathways. Again, complex extremes are the HRS and the British CLEA (Contaminated Land Exposure Assessment) model which applies Monte Carlo methods to evaluate risk through the distribution of the different exposure pathways. Beyond national borders, the Dutch CSOIL-based HESP (Human Exposure to Soil Pollutants) model is in use, which is established and updated by Shell International.

12 Action Thresholds

Measures to deal with contaminated sites are triggered by:

(1) exceeding of certain threshold values; this will cause further investigative action, at least;
(2) the classification of certain sites as potentially hazardous; in Flanders/Belgium, a survey of an area of land is compulsory if a potentially problematic industrial use is discontinued or if pollution is suspected, and it is to be sold;
(3) the intention to use a site for more sensitive purposes; this method is used by those countries in particular which do not require authorities to actively search for sites suspect of contamination;
(4) ground transactions (New Jersey, USA, Flanders/Belgium).

Some countries use more than one of these thresholds.

In most countries, the triggered action will consist of a deeper investigation of the case. Only few systems stringently require remediation, if action thresholds are exceeded. In most systems, however, it is left to the discretion of the authorities to decide upon the need for remediation.

13 Priority Setting

Priorities are mostly set on technical considerations. However, decisions based on purely technical reasoning, such as within USEPA's computer-assisted HRS, are rather rare, political considerations playing a significant role as well. Some countries such as Australia in particular rely, partially at least, on market forces within their systems in order to set priorities. In some cases, in the USA in particular, there are additional systems established to provide incentives for voluntary remediation.

14 Remediation Goals

The most ambitious goal of restoring soil multifunctionality is pursued only in few countries, above all in The Netherlands where this is explicitly stated as policy. Most countries regard as sufficient compliance with action thresholds and in particular reaching soil quality as defined by the intended use and the protection of sensitive targets. Remediation goals are usually based on individual assessments which are frequently supported by lists of soil quality values.

However, in some cases a certain preference for thorough clean-up is established by issuing remediation certificates and, more important, by delisting remediated sites only on condition that multifunctionality has been restored.

Containment and decontamination measures are treated as equivalent by many countries, containment being economically preferable to decontamination in most cases. The Dutch model explicitly defines how to balance the decision. Technically, the containment is required to comply with the IBC (isolation, control, monitoring) criteria, and economically, not only the investment costs but also the costs

of monitoring and maintenance of the containment have to be taken into account. In some cases, containment will imply the need to monitor the site every 5 years. In the USA, the costs of aftercare for the sites on the National Priority List that have been merely contained will have to be borne by the affected federal states, whereas for thorough clean-up, they would have to bear only part of the total cost, thus providing a powerful incentive for thorough decontamination.

15 Monitoring

Monitoring comprises sampling and analysis. The statistical criteria to be applied govern sampling, the definition of toxicity analytical methods.

The statistical criteria in use for acceptable soil contamination range among

- keeping the soil quality criteria for each individual sample point by point; this criterion is the most stringent and adequate only for homogeneous soil which may be obtained by certain treatment, such as thermal treatment, or in cases of homogeneous sludges; random grainy impurities may yield strongly misleading results;
- keeping the soil quality criteria in mixed samples or by averaging over a certain number of samples; this criterion overcompensates for grainy impurities and may not keep track of clustered contaminants;
- keeping the soil quality criteria in a certain percentile – usually 90–95% – of the samples taken; the origin of this criterion is in geology where systematically elevated concentrations of certain minerals are looked for; this method appears to be most appropriate for the statistics of remediated soil which is considered to be in a state close to natural.

In virtually all states, soil quality is defined in terms of concentrations of individual chemicals or groups of chemicals for which the toxicity has been determined by animal tests. Chemically monitoring only these may lead to overlooking hazards not covered by the selected chemicals. The direct determination of soil toxicity – which basically describes soil quality – is generally not pursued. However, reliable and fast test batteries whose results may be compared to animal tests appear to exist, e.g., the dormant larvae assay developed by PERSOONE at Ghent University (PERSOONE et al., 1993). The philosophy behind them would be to study toxicity first and then analyze the soil for the agent causing it in order to systematically remove the hazard.

consensus of the parties appears to meet with problems despite relatively generous financial assistance. A special case is provided by the system of Victoria/Australia where an Environment Protection Authority was established which is independent of Government and which may and does take government institutions to court for environmental offence.

In addition, some countries purposely mobilize market forces, e.g., by authorizing ground transactions only on condition that soil quality will not pose risks to the intended use, or by establishing incentive systems for voluntary remediation.

16 Organization

The responsibility for managing contaminated sites is with the central authorities in most countries. In states organized along federal lines, usually relatively efficient institutions harmonize the guidelines for tackling the problem, at least. In the USA and Canada they actively participate(d) in programs for the remediation of the most severe contamination cases.

The responsible party will be defined by the authorities, in most systems. Only complex cases with several potentially responsible parties will be left to the courts. Appeals against the decision of the authorities may be brought into court. In the USA and in Canada, additional arbitration bodies are established, most of them at high levels, to reach consensus and settle disputes out of court.

In most cases, technical decisions are made, or confirmed at least, by the authorities. Beyond financial responsibility, the responsible party will participate in the technical decisions only in few cases; this applies to those cases, in particular, where market forces are mobilized for remediation.

17 Enforcement

Most countries enforce compliance by penalties; the Austrian approach appealing to the

18 Conflict Management

Most countries have codified the authorities' duty to inform those affected by the contaminated site. In addition, most countries have established funds to avoid hardship and financially assist remediation. In some cases, even third parties affected by remediation activities may claim compensation for loss and damage.

However, mutual communication with the affected parties and their active participation in the making of those decisions which directly affect them is practised only in few countries, above all in the USA where efficient basic democratic opposition caused the authorities at an early stage to develop and apply alternative dispute resolution (ADR) methods.

19 Efficient System Elements

Analyzing the systems, a number of system elements were identified which proved particularly efficient within their specific systems. These elements are listed in Tab. 12.

The analysis suggests the conclusion that due to the dimension of the problem of contaminated sites and because of the implementation deficiencies inherent in an inflexible administrative system along police–law lines, a promising approach to tackle the issue and solve it in the long term may consist in

- treating non-polluting sites that do not pose any risks to third parties separately from polluting hazardous sites;
- taking emergency measures to clean up polluting hazardous sites;
- mobilizing market forces involving ground transactions in particular, to clean up non-polluting sites in the long term.

20 References

PERSOONE, G., GOYVAERTS, M., JANSSEN, C., DE COEN, W., VANGHELUWE, M. (1993), Cost-effective acute hazard monitoring of polluted waters and waste dumps with the aid of Toxkits, *Final Report for the Project No. ACE 89/BE 2/D3* supported by the Commission of the European Communities, DG XI, Ghent.

ULRICI, W. (1995), International experience in remediation of contaminated sites, *Final Report for the Project No. 1470902* (English version edited by the German Federal Environmental Agency on behalf of the federal Ministry of Education, Science, Research and Technology, Berlin).

2 Characterization of the Geological and Hydrogeological Situation, Effects on Natural Geochemical Barriers and Remediation

WILHELM G. COLDEWEY

Essen, Germany

CHRISTOPH KLINGER

Münster, Germany

1 Introduction

For man groundwater is one of the most valuable natural resources and very important for life and economy. So it is essential that everyone involved in groundwater issues should have some basic knowledge of the occurrence, movement, use, and protection of groundwater.

As the groundwater environment is hidden from view, it is difficult for a non-hydrogeologist to understand the complex sciences involved. So it is intented to present in this chapter basic information on the occurrence, movement, and interaction of groundwater with the soil, through which it flows.

The transport of pollutants in the groundwater is influenced by numerous geochemical processes. The efficiency of such processes can be observed in natural material cycles. Different geochemical barrier functions in the flow path of subsurface water from a contaminant source to the biosphere are often closely interlinked. The environment, therefore, provides numerous mechanisms for retardation and natural attenuation of anthropogenically introduced contaminants.

2 Geology and Hydrogeology

2.1 Hydrologic Cycle

The Earth's water follows an unending sequence called the hydrologic cycle. Because groundwater is one of the components of this cycle, some preliminary information is necessary.

A schematic presentation of the movement of water is shown in Fig. 1. In the atmosphere vapor is condensed into ice crystals or water droplets, which precipitate as snow, hail, or rain. A part of this precipitation evaporates and returns to the atmosphere. Another part flows across the ground surface to the next river and to the ocean at last. The rest infiltrates the ground and creates the groundwater body. This groundwater can return to ground surface through springs or it can flow to the next stream and with this to the ocean. During this cycle the form of water may change, but the total amount of water in or on the Earth and in the atmosphere remains the same.

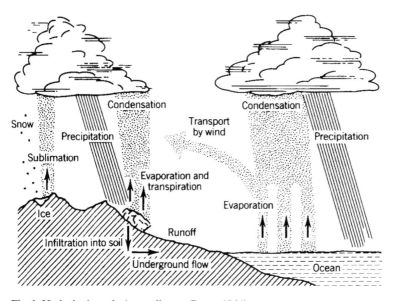

Fig. 1. Hydrologic cycle (according to Chow, 1964).

2.2 Subsurface Water

Subsurface water or undergroundwater is referred as all water beneath the surface. After MEINZER (1923) groundwater is defined as "water in the ground that is in the saturated zone from which wells, springs, and groundwater run off are supplied".

In Fig. 2 the two different groundwater zones are shown. The aerated zone is found directly below the ground surface. This zone contains both air and water. The aerated zone is underlain by the saturated zone in which all openings are filled with water. The zone of saturation is most important for water supply.

Recharge of the saturated zone results from the percolation of water through the aerated zone. This is why the aerated zone is of importance to groundwater hydrology.

At the interface between the aerated zone and the saturated zone the water rises to a certain height. This part is called capillary fringe.

Groundwater flows in response to a hydraulic gradient in the same way as water flows in an open channel. This movement is only possible when the soil or rock has pores or interconnecting openings. Near the Earth's surface most rocks are composed of both solids and voids.

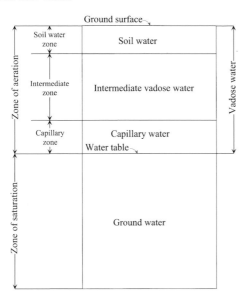

Fig. 2. Divisions of subsurface water (according to Ground Water Manual, 1981).

The water bearing and permeable rocks or aquifers can be divided into unconsolidated deposits or consolidated rock. Unconsolidated deposits can be composed of particles of minerals or rock, which can vary greatly in size from µm (in clay) to m (in boulder clay). These deposits are very important for the water supply. Consolidated or hard rocks can be classified as sedimentary, ingenious, or metamorphic. Of the sedimentary rocks, limestone, sandstone, and conglomerates can be particularly important for the supply of water. Layers of lower permeability are called aquitards and retard the movement of groundwater. These can contain water, but do not yield water to pump wells.

2.3 Unconfined and Confined Aquifers

All rocks can be classified under the consideration of groundwater occurrence as unconfined or confined aquifers (Fig. 3).

In the unconfined aquifer water fills only a part of the ground mass and the groundwater table is free to rise and decline. The water level in a well will indicate the approximate position of the water table in the surroundings. A confined aquifer is completely filled with water and is overlain by confining layers. The water level in a well in a confined aquifer will rise above the bottom of the confining layer to a height at which it is in balance with atmospheric pressure (piezometric surface). This well is called a non-flowing well. In certain cases the water can stand above the local ground surface, in which case the well is called a flowing artesian well. When layers of material with low permeability, e.g., clay, are located above the water table, percolating water can form a saturated zone. This zone is called a perched aquifer with a perched water table.

2.4 Porosity

In order to store and transmit water there must be open spaces, voids, pores, and interstices in the soil or rock. Transmissibility depends on the type, size, and arrangement of these voids, pores, etc.

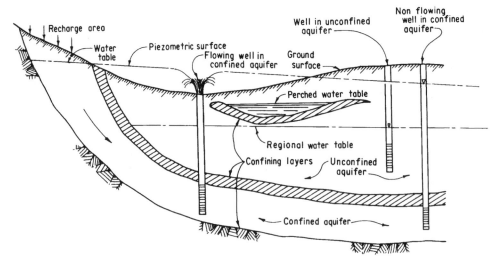

Fig. 3. Types of aquifers (United State Department of Interior, 1981).

In Fig. 4 different types of voids and interstices are shown, which can be classified in three groups:

- intergranular pore spaces, which are characteristic of loose rocks (unconsolidated) and consolidated sedimentary rocks,
- fractures, joints, and bedding planes are typical of hard rocks (Fig. 5),
- solution channels can occur in carbonates and evaporates.

The total porosity of intergranular pores and interstices depends on the grain size, different arrangements, and packing of the material. In general, total porosity (expressed as a percentage) is high when the particle size is small. Thus, total porosity of clay is higher than the porosity of sand (Tab. 1).

On the other hand, fine material (e.g., clay) is able to retain water by molecular and surface tension forces so that the effective porosity is very low (Tab. 1). More details are described in Sect. 3.

2.5 Permeability

As a result of natural and man-made influences most groundwater is in motion. The factors controlling groundwater movement were expressed by HENRY DARCY in 1856 in the form of Eq. (1):

$$\dot{V} = k_f \cdot A \frac{dh}{dL} = k_f \cdot A \cdot i \qquad (1)$$

$\dot{V}$: rate of flow, k_f: coefficient of permeability, A: area normal to the direction of flow, i: hydraulic gradient, h: headloss at distance L, L: length of the flowpath.

The coefficient of permeability can be obtained from formulae based on physical properties of the porous media, from laboratory and field tests. Representative values are given in Tab. 1 for different types of soils and rocks.

2.6 Unsaturated Flow

In most cases recharge of the groundwater system is caused by percolation through the aerated zone.

This movement of water is influenced by gravitational and capillary forces. Cohesion (mutual attraction between water molecules) and adhesion (attraction between water and soil particles) make up the capillary effects. So in most pores of capillary size (e.g., in granular material) water is pulled upward into the capillary zone. The capillary gradient can be measured by field tests with a tensiometer.

Porous aquifer

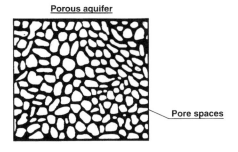

Pore spaces

Joint aquifer

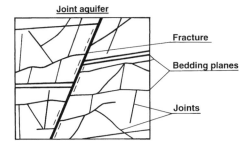

Fracture

Bedding planes

Joints

Karst aquifer

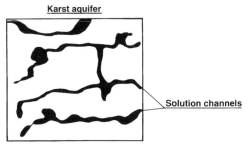

Solution channels

Fig. 4. Types of interstices (COLDEWEY and KRAHN, 1991).

Most sediments are deposited in layers. Each layer has its own grain size and grading of material. When there is a difference in the grain size and grading of adjacent layers, there is also a difference in the permeability. These differences affect the percolation of water through the aerated zone and the movement of groundwater. In most cases the aerated zone is built up of horizontal layers and the movement of water is in a vertical direction. Thus, the movement of liquids through the aerated zone is of great importance in many groundwater problems, especially with respect to mobile pollutants.

In these cases it is important to have a good understanding of the stratification in order to get a concept of potential pollution migration and the possible need for clean-up remediation.

2.7 Saturated Flow

In the saturated zone all interstices and pores are filled with water. The groundwater flows down the gradient through these openings. In relation to the hydraulic gradient and the magnitude of the openings, the movement may be either laminar or turbulent.

In laminar flow the water moves in a definite path and there is no transverse mixing. Flows are laminar in most granular deposits. In turbulent flow the water moves in a disordered manner. Turbulent flows are typical for large openings such as in limestone cavities or gravels.

In a granular deposit under laminar flow conditions, the different streamlines converge in narrow necks between the soil particles and diverge in larger pores. This effect causes transverse dispersion. The friction between the water and the soil particles causes differences in velocity. The fastest movement occurs in the center of pores, the slowest near the soil particles. This causes a dispersion in the direction of the flow (so-called longitudinal dispersion).

In cases of contamination of groundwater knowledge of this dispersion potential is very important. Unfortunately, dispersion is difficult to measure in the field, because the flow direction of pollutants is also affected by other conditions and processes like stratification and ion exchange (see Sect. 3).

2.8 Groundwater Movement

Groundwater movement is controlled by vertical and horizontal permeability and the thickness of the aquifer and the confining layers.

The dominant driving force for groundwater movement is gravity. Under natural conditions groundwater moves from the recharge area downwards until it reaches ground surface at a spring or it infiltrates a river or the sea. The dif-

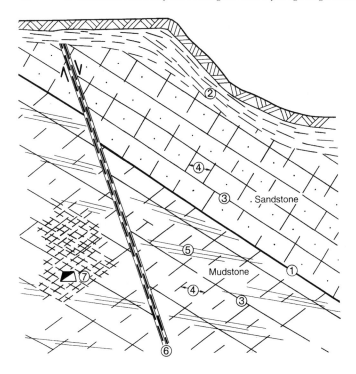

Fig. 5. Occurrence of permeability zones in hard rocks (according to HÄHNE and FRANKE; from COLDEWEY and KRAHN, 1991), 1: Boundary surface between different rocks, 2: release joints, 3: bedding planes, 4: joints, 5: foliation joints, 6: fracture, 7: cavity caused by mining.

Tab. 1. Range of Porosity, Effective Porosity, and Permeability (according to CHOW, 1964; DOMENICO and SCHWARTZ, 1990; HÖLTING, 1995)

	Total Porosity [%]	Effective Porosity [%]	Permeability [m s^{-1}]
	Sedimentary		
Gravel, coarse	24–36	14–24	10^{-1}–10^{-2}
Gravel, fine	25–38	15–25	10^{-2}–10^{-4}
Sand, coarse	31–46	15–30	10^{-3}–10^{-4}
Sand, fine	26–53	10–20	10^{-4}–10^{-5}
Silt	34–61		10^{-6}–10^{-9}
Clay	34–60	<5	$<10^{-9}$
	Sedimentary Rocks		
Sandstone	5–30		10^{-6}–10^{-9}
Siltstone	21–41		10^{-8}–10^{-11}
Shale	0–10		10^{-9}–10^{-13}
Limestone	0–20		10^{-6}–10^{-9}
	Crystalline Rocks		
Basalt	3–35		10^{-6}–10^{-8}
Granite	0,1		10^{-5}–10^{-6}

ference in level between two parts of a water body system is called hydraulic head.

The greatest difference in head exists between the recharge area and the discharge area.

The greatest head losses occur while water flows across confining layers. With greater depth groundwater movement becomes slower. The most dynamic groundwater movement occurs through shallow aquifers.

2.9 Groundwater Velocity

The velocity of groundwater is important in all problems related to pollution of groundwater and soil. The velocity of groundwater is influenced by ground permeability and the hydraulic gradient. The equation for groundwater velocity can be derived from the combination of Darcy's law (see Sect. 2.5) and the velocity equation for hydraulics

$$\dot{V} = k_f \cdot A \cdot i \qquad (2)$$

$$\dot{V} = A \cdot v \qquad (3)$$

v: velocity.

Combination of Eqs. (2) and (3) yields Eq. (4).

$$A \cdot v = k_f \cdot A \cdot i \qquad (4)$$

Changing and canceling the area terms result in Eq. (5).

$$v = k_f \cdot i \qquad (5)$$

Because water flows through the openings of a soil or rock, which are defined by porosity n, the complete equation is:

$$v = \frac{k_f \cdot i}{n} \qquad (6)$$

Normally, groundwater velocity is very low. In sand, e.g., the velocity is of the order of cm d^{-1}, while the velocity in clay is cm a^{-1}. In limestone cavities groundwater velocity can be similar to that in rivers. Groundwater velocity can be measured in the field by tracer tests.

3 Subsurface as Reaction Zone for Geochemical Processes

The ground provides, apart from purely hydraulic aspects, a wide spectrum of interactions which can influence contaminant movement. This applies to organic as well as inorganic components. A close connection can exist between the characteristics of the mineral components of the rocks, the chemical-physical conditions of the environment, and biological processes.

Transport of contaminants in the subsurface takes place mainly by groundwater, but volatile organic compounds can spread in the aerated zone via the vapor phase. The mobility of pollutants in groundwater is influenced by numerous processes which can cause immobilization of these dissolved substances. The efficiency of such processes can be observed in natural material cycles, e.g., in an area of deposit formation. Rocks, therefore, provide a natural barrier function, which also can effectively prevent the spreading of anthropogenically introduced contaminants.

The different reactions and their contribution to the processes during substance transport will be described in Sects. 3.1–4.2.

3.1 Structure of the Transport Matrix

Different processes contribute in variable amounts to the transport of contaminants, depending on the properties of the aquifer. In principle, transport by general movement of the groundwater (advection, dispersion), and substance transport due to a concentration gradient (diffusion) are differentiated (see Sect. 2.7).

Diffusive processes are of interest, particularly in very slightly water-permeable rocks and residual materials, or in stagnant water conditions. Here they can exceed flow, depending on transport rates. The spatial-physical conditions of the transport zone consequently influence not only the hydraulic processes, but also the barrier function of the subsurface.

In joint aquifers reactive surfaces are primarily limited to the joint surfaces. Yet a diffusive substance transport takes place from the joint surface to the pore spaces, which is called matrix diffusion. Matrix diffusion results in storage, or intermediate storage of the dissolved components and thus reduces the quantity of contaminants transported as well as their concentration. Furthermore, the pore space provides a reactive grain surface for sorptive fixation (see Sect. 4.1) which consequently reduces the concentrations in the pore water, facilitating further matrix diffusion.

In loose rocks or porous media, almost the entire grain surface is available for interaction with contaminants. Particularly fine-grained materials such as silts and clays provide extensive bonding places and geochemically reactive potential.

3.2 Lithology

Mineral rock components show substantial differences in their bonding capacity for contaminants. The bonding capacity, especially for inorganic pollutants like heavy metals, is characterized by the cation exchange capacity, which represents a measure of adsorption places available. Clay minerals constitute a particularly effective sorbent due to their crystalline structure and their large grain surface. Sorption of organic contaminants is closely correlated to the content of organic carbon.

There are numerous processes based on specific bondings of dissolved contaminants with minerals. Carbonates can integrate heavy metals into their crystal lattices and iron hydroxides in particular have specific bonding places for the adsorption of heavy metals. During formation of iron hydroxides coprecipitation can occur as well.

Organic substances, which are contained in numerous rocks in contents of up to several percent, bond preferably with organic contaminants and mercury in spite of sometimes a high degree of carbonization. Thus, bituminous coal constitutes a geochemical barrier, although a very selective one.

3.3 Influence of Soil Properties

Processes in the upper soil layers influence the composition of the percolating precipitation water, among them hydrolysis reactions of the silicate particles. Also, there can be an enrichment with biological decomposition products like carbon dioxide gas (CO_2) and humic acids. These processes influence the hydrochemical conditions in the percolating water and, consequently, the mobilization properties for contaminants, particularly in the unsaturated zone.

In this context the influence of compounds from the atmosphere, e.g., acid rain, is also of importance. The solution of buffering compounds such as carbonates leads to decreased pH values in the pore and seepage water and thus to an increased mobilization potential for most inorganic contaminants.

3.4 Chemical-Physical Influences

The general chemical-physical conditions in the transport medium determine type and kinetic properties of mobilization and transport processes of contaminants as well as of biological degradation. The pH value serves as a master variable determining the solubility and the sorption properties of almost all heavy metals (CREMER, 1991). Most elements display distinct mobilities in an acidic environment with substance or system-specific pH limits (CALMANO, 1989; HERMS and BRÜMMER, 1978).

Moreover, many heavy metals possess amphoteric qualities, i.e., they become mobile also in intensely alkaline pH conditions, a characteristic they share with numerous organic compounds.

A distinct pH-dependent mobility can also be noted for cyanides, which, under alkaline conditions, are transformed from their complex bonding into a mobile form. If material containing cyanide is covered with building rubble, e.g., a quick release of cyanide can be caused because of the changed environmental conditions (unpublished study).

The redox (i.e., reduction–oxidation) potential (*Eh*) constitutes the second important factor for physical-chemical influences. The redox environment has an effect on the precipitation of many heavy metals, provided that the conditions are below the stability limits from sulfate to sulfide. This depends on the availability of oxygen on the one hand, and on biological decomposition processes on the other hand. Almost all heavy metals form limited soluble sulfides in environments with highly reducing conditions. Such anaerobic conditions hence rank as the ideal heavy metal trap. Likewise, most biological/bacterial decomposition processes are highly sensitive to the hydrochemical redox environment.

Iron hydroxides play an important role as sorbents for contaminants because of their large surface and their surface structure. They form under oxidizing conditions, particularly in the aerated zone. The stability of iron hydroxides depends on the pH as well as on the *Eh* value. The iron compounds consequently constitute a valuable indicator for the geochemical environment. Soils with perched groundwater very often show this dependency clearly by an (oxidizing) yellow-brown colored area in the unsaturated zone and (reducing) grey-green colors in the saturated zone.

In the context of laboratory tests the environmental conditions relating to the examined sediments must be observed meticulously. Access of oxygen to samples actually stable under reducing conditions leads to quick oxidation of iron compounds (e.g., fine grained sulfides) and formation of iron hydroxides, in turn effecting distinctly greater sorption rates than in the original material (see Sect. 4.3).

Adsorption processes are classified as exothermal processes (WAGNER, 1992). Therefore, decreasing temperatures should lead to higher levels of adsorption. However, experiments frequently show effects to the contrary. This can be traced back to an overlap of plain sorption processes with kinetic effects, e.g., diffusion. Higher temperatures are also conductive to the solubility of most substances. Furthermore, the viscosity of water decreases with rising temperatures, so that the diffusive substance transport increases as well.

3.5 Composition of the Subsurface Water

The concentration of dissolved salts influences the mobility and sorption properties of contaminants through anions as well as cations. Dissolved cations (particularly alkali and earth alkali elements) act predominantly as competitors for the available sorption places on the exchangers. Anions (e.g., chloride, sulfate, carbonate), which are frequent in groundwaters, form complexes with heavy metals, influencing substance solubilities, sorption properties, and transport characteristics to a high degree.

In alkaline conditions hydroxo–metal complexes are important for the specific bonding of heavy metals on the surface of sorbents. Also complexes are formed with the anionic ligands mentioned above. Every species has different chemical and physical characteristics, which are caused by different ion sizes and charges in particular. The type of complex species formed depends on specific affinities between the bonding partners, but may also be superimposed by high concentrations. The speciation of dissolved metals can be calculated, if the composition of the solution and the stability constant of the respective species are known. For this purpose thermodynamic equilibrium-type models are available (see Sect. 4.4).

Increasing chloride content in the water causes a shift from a species composition at first dominated by free cations, then monochlorine complexes with a positive charge to chlorine complexes with a negative charge, so that the sorption characteristics are also changed: sorption is dominated by the properties of the ligands and not any more by the central metal cation, which in this case leads to notably reduced sorption and increased mobility.

The formation of colloids also belongs to this context. Colloids, because of their intermediate position between dissolved ions and solid particles, can cause a substance transport beyond the theoretical solubilities (see Sect. 4.2).

4 Interaction between Contaminants and Subsurface Matrix

In most cases of contamination the dominant migration of pollutants occurs by water, as already mentioned. The type and velocity of groundwater transport have significant influence on the available reaction time between contaminants and sorbents. Beyond that, there is sorption of volatile organic compounds onto soil minerals from the gas phase as well.

The conditions at the source of contamination are determined by the deposited material itself, the surrounding rock, and the hydrochemical and hydraulic conditions of the site. Contamination input occurs very often by percolating water in the aerated zone (vertical water movement, see Sect. 2.6). These conditions influence the time-dependent mobilization of contaminants.

An essential precondition for the recording and evaluation of the mobilization behavior of contaminants in the medium to long term are the simulation of environmental conditions as close as possible to reality (under consideration of the above mentioned factors), as well as the application of suitable experimental methods.

4.1 Geochemical Barriers

Apart from the theoretical solubilities of different components, interactions with the rock matrix can codetermine or dominate the mobility of contaminants, resulting in retardation and attenuation. We can distinguish different effects which, however, can superimpose or complement each other:

- hydrogeological barrier,
- geochemical barrier,
- hydrochemical barrier,
- hydraulic barrier.

Hydraulic, hydrochemical, and geochemical barrier functions are often closely interlinked. The listed barrier types need to be understood more as function principles rather than spatial units. Some of the connections and influences are illustrated in Fig. 6.

Apart from interactions between dissolved substances and rock, exceeding the solubility of minerals in mixing waters and changing the chemical environment of the subsurface water, can contribute to a barrier function. This hydrochemical barrier is consequently based essentially on reversible precipitation reactions.

All such barriers, and the conditions influencing them, change over the course of time. Almost all fixations by geochemical barriers are easily reversible (except coprecipitation into newly formed minerals). For this reason the influence of natural geological processes, and particularly the anthropogenically caused changes in the natural systems on the retardation of contaminants, has to be taken into account for long-term safety studies. This aspect is particularly relevant for heavy metals which, in contrast to organic pollutants, are generally subject to neither decomposition nor decay.

The efficacy of barriers is partly substance-specific. Highly soluble substances, such as chloride in water, can hardly be removed from the water by geochemical barriers. In such cases only hydraulic effects have to be considered.

4.2 Sorption Processes

The concurrent adsorption of cations on sorbent surfaces with a negative charge and desorption of equivalent quantities of other cations is called cation exchange. Cation exchange processes show the following characteristics (WILD, 1995):

- The reactions are quick.
- The reactions are reversible.
- The exchange is selective.
- The distribution of two kinds of cations between sorbents and solution depends on their reactive concentrations in the solution.

The general term sorption comprises adsorption, absorption, and surface precipitation processes. Fixation can, therefore, depend on physical interactions or on chemical bondings. WAGNER (1992), among others, provides a comprehensive survey of the different sorption processes. All solid substances are princi-

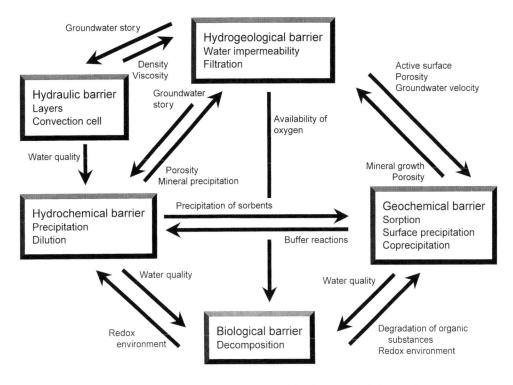

Fig. 6. Compilation of some active mechanisms of natural barriers and their interactions.

pally capable of sorption processes. Essential for the sorption capacity of a particular substance is the size of its specific surface as well as the type and value of its charge.

Surfaces with a negative charge preferably adsorb cations. The negative surface charge can be created by either an isomorph replacement in the crystal lattice or by dissociation of H^+ ions in functional surface complexes. This first kind of charge is called permanent charge, because the isomorph replacement is largely independent of the pH value. It prevails in clay minerals. In contrast to this, the charge of functional surface complexes depends on the pH value of the solution. It is called variable charge and occurs in Fe and Mn oxides in particular, as well as in Fe and Mn hydroxides. These bonding places can have a positive, neutral, or negative charge. In addition to the minerals listed, organic substances (such as humates, peat, and carbonized substances) also present important potential sorbents.

Sorption reactions are described with either the help of empirical functions (Kd concept, Freundlich isotherm), or by mechanistic models (e.g., surface complexation model). Sorption isotherms provide basic data for mass transport modeling in connection with models describing the fluid flow.

Sorption isotherms describe the empirically established functional correlation between dissolved and sorbed substance for the state of equilibrum at constant temperature conditions and defined rock–water systems. They cannot provide direct information concerning the type and mechanism of the process.

In contrast to the Henry isotherm, which shows a linear correlation between the sorbed quantity and the equilibrium concentration in the solution and only applies to very low concentration levels, the non-linear, empirically established Freundlich isotherm describes an exponentially decreasing level of sorption with increasing concentrations in the equilibrium solution (Fig. 7). This means that, with a progressive decrease of available sorption places, further sorption becomes more and more difficult, without achieving a sorption maximum. Frequently, good adjustment of the sorptive behavior by only one isotherm is hardly possible. A markedly better description of the correlation is achieved by consideration of two partial isotherms for different concentration ranges (GERTH and BRÜMMER, 1977; BENJAMIN and LECKIE, 1981).

The Langmuir isotherm was developed according to a concept allowing only for a limited number of sorption places. If these are occupied, no more additional substances can be adsorbed, even at increasing supply.

Ultimately we can assume a model in which the single bonding places behave according to the Langmuir model. The entire isotherm, however, can be approximated by the Freundlich isotherm.

A basic disadvantage of sorption isotherms ascertained in laboratory tests lies in their exclusively empirical deduction. The effects of all processes, encompassed by the term sorption, are integrated by derivation of characteristic sorption values. This implies that the isotherms reflect the sorption behavior only of the examined physical-chemical system. A spatial or temporal interpolation for other geochemical conditions is not permissible because the influence of single parameters of solution and solid matter on the adsorption cannot be established. Therefore, it follows that numerous preconditions apply if sorption isotherms are employed for substance transport models (see Sect. 4.4).

In a different approach, sorption processes are described mechanistically as surface complexations. This assumes the formation of a complex during the reaction of the pH-dependent, variable charges of sorbent surfaces with substances in the solution. Such a fixation of a metal ion is defined as a surface complex of the metal. Thus it follows that in the surface complexation model sorption reactions are regarded and calculated in analogy to complexation in solutions.

A precondition for this description of sorption processes close to reality are comprehensive basic data, which only in the rarest cases are available to the necessary extent. Furthermore, most thermodynamic data are not available as yet. Limited applicability and validity of the models are the consequences of indispensable practice oriented simplifications.

Difficulties with any model calculation arise when colloids act as a medium for the transport of contaminants. Different types are

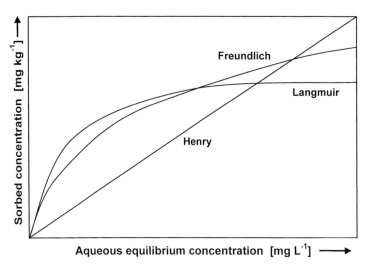

Fig. 7. Schematic course of the different sorption isotherms.

- silicate colloids, i.e., clay minerals,
- oxides and hydroxides, mainly iron compounds,
- organic colloids, mainly humic acid,
- agglomerations of metal complexes, which can reach colloidal size.

Colloids are found in almost all subsurface waters in varying levels. They are defined by their diameter between 1 nm and 1 μm. Because of their generally very large specific surfaces they can fix and transport proportionately high amounts of contaminants. As long as there is no interaction between sediment and colloid, the colloid-fixed contaminants can be transported with the advective velocity of the groundwater. This constitutes the importance of the colloids for migration of pollutants.

4.3 Experimental Methods

Sorption capacities measured in the laboratory considerably depend on methodical differences in the set-up of the experiments (GERTH, 1985). There are no standardized procedures as yet. Different approaches and sequential extractions allow at least in part the description of active bonding mechanisms and the derivation of sorption capacities fairly close to the complex situation in reality.

It is essential to use site-specific solutions in the laboratory test to ascertain transferable sorption data. Information about the matrix composition of the water and of the dissolved contaminants and their concentrations has to be available.

In reality, the amount of water in the pore space is generally much lower than in laboratory tests, and the retention and reaction times are much longer. Solution and precipitation reactions can take place in the contact zone of pore space–solid matter. Long-term tests in large scale columns and field experiments have shown repeatedly that contaminant discharges are markedly lower and retardation correspondingly greater than in the laboratory tests.

Batch tests, however, provide a good method to acquire comparable data at standardized conditions because of the comparatively low expenditure and their good reproducibility. Yet there are numerous disadvantages. In batch test samples the natural rock is destroyed and the reactive surfaces are very large due to crushing. It should also be taken into account that completed equilibrium conditions, which are influenced by diffusion processes in the sediment components, e.g., are only achieved after very long contact times of several weeks or more. Generally, sorption isotherms (see Sect. 4.2) are determined in batch tests by adding solutions with different concentrations of a dissolved substance to a defined amount of the sorbent, and determining the remaining concentrations in the solutions after adjusting an "equilibrium".

WAGNER (1992) mentions several studies by other authors dealing with the transferability of percolation test data to batch tests. According to him, some authors describe a very good correlation, whereas others frequently found a markedly lower sorption capacity in percolation tests. Some reasons for poor correlation are non-linearity and reversibility of the sorption process, accessibility of the mineral surfaces, sorption kinetics, water–solid matter ratio, and pH.

The transfer of laboratory results to large-scale conditions of the research site presents several problems. It is unlikely that the sample material represents the subsurface composition. The duration of experiments is often too short in comparison with the natural time scale. Kinetic processes cannot be considered at all or only insufficiently.

The scaling up in time, e.g., the prognosis for future geochemical changes in a groundwater system, is particularly relevant for long-term safety analysis. Laboratory test results have to be extrapolated over several 1,000 years into the future. Such long periods contain too many possible anthropogenetic or naturally caused changes of geochemical conditions than could possibly be considered or calculated.

4.4 Application of Geochemical Models

There are now numerous computer codes available which, given the chemical-thermodynamic data, are able to calculate geochemical reactions even in solutions of complex composition.

The PHREEQE codes (PARKHURST, 1995), developed by the U.S. Geological Survey prevail as the standards for geochemical calculations. These programs include in their recent versions capabilities for 1 D transport calculations, modeling of kinetic effects, cation exchange, and surface complexation. Numerous other programs based on similar methods and data are available (BALL et al., 1987; WIWCHAR et al., 1988; SCHULZ and KÖLLING, 1992). However, only codes considering the realized chemical conditions should be employed. This becomes evident considering that, e.g., up to now there is a very limited thermodynamic data base for saline solutions (see Sect. 4.2). In particular, the application to organic substances is very limited in most model codes.

In addition, as mentioned above, retardation, in terms of a lasting elimination of contaminants from the solution, is influenced by many interlinked factors which should be considered in transport model calculations:

- active rock surface,
- sorption capacity,
- transport velocity,
- porosity, in conjunction with diffusive transport in joint aquifers,
- geochemical environment,
- buffer and redox reactions.

Fig. 8 shows geochemical processes in the flow path of subsurface water from a contaminant source to the biosphere. The environment, therefore, provides numerous mechanisms for natural attenuation of anthropogenically introduced contaminants.

Thus, for complex multicomponent systems such as contaminated sites, a precise quantification, mathematical description, and modeling of the physical-chemical processes is hardly possible. Such uncertainties in the quantification of these processes, however, considerably impair the validity of a prognosis for pollutant transport more than an inaccurate description of the hydraulic characteristics (MAIER and DÖRHÖFER, 1994).

Therefore, most of the available substance transport codes work according to the Kd isotherm based data, neglecting the non-constant geochemical conditions in reality.

There are various approaches to models coupling substance transport and geochemical or biological reactions. Mechanistic and empirical sorption models, which in simplified form are integrated in substance transport models, require a large computer effort even without consideration of geochemical reactions.

A precondition for the application of simplified geochemical codes in transport models is the comprehensive description of the relevant geochemical reactions, in such a way as to facilitate a conservative prediction of discharge and migration of pollutants. An inclusive consideration of the complex processes and their transformation into transport calculation codes, however, does not seem likely in the foreseeable future.

5 Geological-Hydrogeological Aspects for Biological Remediation

Effective biological degradation demands optimal chemical-physical conditions. In addition in situ bioremediation requires suitable geological-hydrogeological properties of the subsurface in particular.

So geological-hydrogeological properties of a site are decisive for the choice of the remediation technology for soil or subsurface. To realize economic microbiological reaction rates a sufficient supply with nutrient salts and electron acceptors or donors has to be guaranteed. For these processes the following properties of soil or subsurface are of importance:

- permeability of water (and, therefore, transported salts) and gas,
- grain size distribution.

For these properties the coefficient of permeability (k_f, see Sect. 2.5) can be used as guiding parameter. Practice shows that hydraulic permeabilities of $k_f > 10^{-4}$ ms^{-1} are required for in situ bioremediation. These values are realized in course grained sediments.

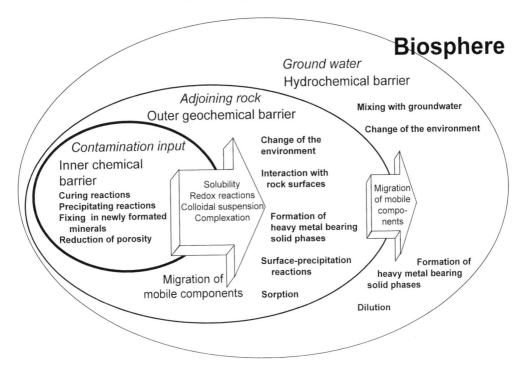

Fig. 8. Compilation of geochemical processes during groundwater flow.

In addition, the technology used has to be adjusted to the position of the contamination in relation to the water table and the hydraulic setting (saturated-unsaturated) (see Sects. 2.6, 2.7).

In case of iron or manganese containing groundwater *in situ* bioremediation under aerobic conditions can cause precipitation of oxides and hydroxides. Such precipitates change the permeability of the subsurface and the microbiological environment.

In addition, active *in situ* technologies require a homogeneity of the contaminated subsurface area. In case of inhomogeneities as well as low contaminant solubility and limited bioavailability the use of passive *in situ* technologies (reactive walls, etc.) may be practical.

Therefore, in most cases an *ex situ* treatment of soil in beds or reactors will be necessary. As well as *in situ* – in addition to the above listed characteristics – for these techniques soil physical properties as

- sorption of contaminants on soil particles,
- solubility, and
- diffusive mass transfer and capillary forces in the soil matrix

have substantial influence on degradation and degradation rates. These are the same parameters influencing contaminant migration and natural attenuation processes (see Sects. 3 and 4). The mentioned properties affect the availability of contaminants for microorganisms (bioavailability). Therefore, the nature of the composite soil, i.e., mineral composition and grain size distribution, plays a part not only for the hydraulic *in situ* situation. In case of insufficient bioavailability a microbiological treatment of soil is not possible in spite of degradability of the organic contaminants.

For bioremediation in solid matter reactors unfavorable physical properties of the soil can result from optimal conditions (e.g., water content) for the microorganisms. As material-dependent limiting factors for this method plasticity, flowability, aeration, and wall adhesion can be mentioned.

Therefore, spreading of contaminant plumes and microbiological remediation are both influenced by hydraulic general conditions and sorption and transport properties of the subsurface. Low hydraulic conductivity and high sorption rates prevent quick migration just as high efficiency of natural degradation and of remediation technologies.

6 References

BALL, J. W., NORDSTROM, D. K., ZACHMANN, D. W. (1987), WATEQ4F – A Personal Computer FORTRAN Translation of the Geochemical Model WATEQ2 with Revised Database, *U.S. Geological Survey Open File Report*.

BENJAMIN, M. M., LECKIE, J. O. (1981), Multiple-site adsorption of Cd, Cu, Zn and Pb on amorphous iron oxyhydroxide, *J. Colloid Interface Sci.* **79**, 209–221.

BUSCH, K.-F., LUCKNER, L., TIEMER, K. (1993), Geohydraulik, in: *Lehrbuch der Hydrogeologie* (MATTHESS, G., Ed.). Berlin: Gebrüder Borntraeger.

CHOW, V.-T. (1964), *Handbook of Applied Hydrology: A Compendium of Water-Resources Technology.* New York: McGraw-Hill.

CALMANO, W. (1989), *Schwermetalle in kontaminierten Feststoffen: chemische Reaktionen, Bewertung der Umweltverträglichkeit, Behandlungsmethoden am Beispiel von Baggerschlämmen.* Köln: Verlag TÜV Rheinland.

COLDEWEY, W. G., KRAHN, L. (1991), *Leitfaden zur Grundwasseruntersuchung in Festgesteinen bei Altablagerungen und Altstandorten* (Ministerium für Umwelt, Raumordnung und Landwirtschaft des Landes Nordrhein-Westfalen, Eds.), Düsseldorf.

CREMER, S. (1991), Mobilisierung von Schwermetallen in Porenwässern von Deponien und belasteten Böden – Ermittlung relevanter Parameter des hydro-geochemischen Milieus für die Entwicklung eines aussagekräftigen Elutionsverfahrens, *Thesis*, University of Bochum.

DARCY, H. (1856), *Les fontaines publiques de la ville de Dijon.* Paris: Victor Dalmont.

DOMENICO, P. A., SCHWARTZ, F. W. (1990), *Physical and Chemical Hydrogeology.* New York: John Wiley & Sons.

FETTER, C. W. (1994), *Applied Hydrogeology.* Englewood Cliffs, NY: Prentice Hall.

GERTH, J. (1985), Untersuchungen zur Adsorption von Nickel, Zink und Cadmium durch Bodentonfraktionen unterschiedlichen Stoffbestandes und verschiedene Bodenkomponenten, *Thesis*, University of Kiel.

GERTH, J., BRÜMMER, G. (1977), Quantitäts-Intensitäts-Beziehungen von Cadmium, Zink und Nickel in Böden unterschiedlichen Stoffbestandes, *Mitt. Dtsch. Bodenkundl. Ges.* **29**, 555–556 (Göttingen).

Ground Water Manual (1981), A guide for the investigation development and management of ground water resources, U.S. Department of the Interior, *A Water Resources Technical Publication.* Denver, CO: United States Government Printing Office.

HÄHNE, R., FRANKE, V. (1983), Bestimmung anisotroper Gebirgsdurchlässigkeiten *in situ* im grundwasserfreien Festgestein, *Z. Angew. Geol.* **29**, 219–226.

HERMS, U., BRÜMMER, G. (1978), Löslichkeit von Schwermetallen in Siedlungsabfällen und Böden in Abhängigkeit von pH-Wert, Redoxbedingungen und Stoffbestand. *Mitt. Dtsch. Bodenkundl. Ges.* **27**, 23–34 (Göttingen).

HÖLTING, B. (1995), *Hydrogeologie – Einführung in die Allgemeine und Angewandte Hydrogeologie.* Stuttgart: Ferdinand Enke Verlag.

LANGUTH, H., VOIGT, R. (1980), *Hydrogeologische Methoden.* Berlin: Springer-Verlag.

LUCKNER, L., SCHESTAKOW, W. M. (1991), *Migration Processes in the Soil and Groundwater Zone.* Chelsea, MI: Lewis Publishers.

MAIER, J., DÖRHÖFER, G. (1994), Schadstofftransport in klüftigen Tongesteinen. Feld- und Laboruntersuchungen unter besonderer Berücksichtigung organischer Schadstoffe, in: *Umweltgeologie heute 4* (DÖRHÖFER, G., THEIN, J., WIGGERING, H., Eds.), pp. 119–127. Berlin: Ernst & Sohn.

MEINZER, O. E. (1923), The occurrence of ground water in the United States with a discussion of principles, *U.S. Geological Survey Water Supply Paper 489.*

PALMER, C. M. (1996), *Principles of Contaminated Hydrogeology.* Boca Raton, FL: Lewis Publishers.

PARKHURST, D. L. (1995), User's guide to PHREEQC – a computer program for speciation, reaction-path, advective-transport and inverse geochemical calculations, U.S. Geological Survey, *Water-Recources Investigations Report 95-4227*, 143 p. Lakewood, CO.

SCHULZ, H. D., KÖLLING, M. (1992), Grundlagen und Anwendungsmöglichkeiten hydrogeochemischer Modellprogramme, *DVWK-Schriften 100*, pp. 1–96. Hamburg: Parey.

TODD, D. K. (1959), *Ground Water Hydrology.* New

York: John Wiley & Sons.

United State Geological Survey (1983), Basic ground-water hydrology. *Water Supply Paper 2220.*

United State Department of Interior (1981), *Ground Water Manual* – A Water Resources Technical Publication – A Guide for the Investigation, Development, and Management of Groundwater Resources. Denver, CO: United States Government Office.

WILD, A. (1995), *Umweltorientierte Bodenkunde.* Heidelberg: Spektrum Akademischer Verlag.

WAGNER, J.-F. (1992), Verlagerung und Festlegung von Schwermetallen in tonigen Deponieabdichtungen. Ein Vergleich Labor- und Geländestudien, *Schriftenreihe Angewandte Geologie Karlsruhe* **22** (Karlsruhe).

DE WIEST, R. J. (1965), *Geohydrology.* New York: John Wiley & Sons.

WIWCHAR, B. W., PERKINS, E. H., GUNTER, W. D. (1988), SOLMINEQ88 PC/Shell. *Alberta Research Council Report,* Edmonton, Alberta, Canada.

3 Bioavailability of Contaminants

BERND MAHRO

Bremen, Germany

List of Abbreviations

CMC   critical micellar concentration
DNAPL   dense non-aqueous phase liquids
HCH   hexachlorocyclohexane
HLB   hydrophily–lipophily balance
HPLC   high-pressure liquid chromatography
IUPAC   international union of pure and applied chemistry
NAPL   non-aqueous phase liquid
PAH   polycyclic aromatic hydrocarbons
TNT   trinitrotoluene

1 Introduction

The bioremediation of contaminated former industrial sites or waste deposits became a well established field of environmental biotechnology business during the recent decade. The growing importance of the field is well reflected, e.g., by a number of international conferences like those sponsored and organized by Battelle (ALLEMAN and LEESON, 1997), DECHEMA (1992), FZK/TNO (1998) or OECD (1996). However, although many sites were successfully cleaned up by bioremediation treatments a number of field reports indicated also that biological treaments may sometimes be very slow or incomplete. Especially sites contaminated with polycyclic aromatic hydrocarbons (PAH) frequently turned out to resist a fast and complete clean-up (AFFERDEN et al., 1992; BEWLEY and THEILE, 1988; ERICKSON et al., 1993; SCHAEFER et al., 1995; STEILEN et al., 1993), but similar experi-

ences were also reported for other types of hydrocarbon pollution (BOSSERT et al., 1984; MIETHE et al., 1996; MUELLER et al., 1991a). Fig. 1 shows a typical pattern of incomplete degradation with soil from long-term contaminated sites. The initial concentration of PAH dropped rapidly in the supplemented soil batch, indicating a strong microbial PAH degradation. However, the degradation slowed down significantly in the course of the treatment, leaving a residual PAH concentration that could not be reduced much further even if the soil was incubated for many more months.

. Although the reasons for the attenuation and incomplete pollutant degradation with soil from long-term contaminated sites and in the field can be manifold (see below) the lack of bioavailability has been identified as one of the most prominent factors that may inhibit a fast and complete pollutant degradation (LUTHY et al., 1994; MAHRO and SCHAEFER, 1998; MULDER et al., 1998).

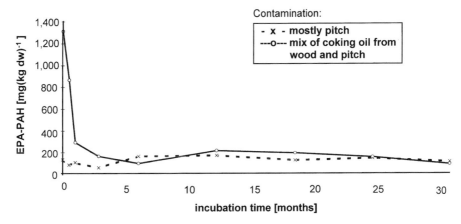

Fig. 1. Degradation of PAH in a bioreactor containing a tar oil contaminated soil that was supplemented with the white-rot fungus *Pleurotus ostreatus* on straw (SCHAEFER et al., 1995).

The above mentioned reasons for an attenuating and incomplete degradation of xenobiotics in soil can be as follows:

- Toxic metabolites accumulate during the degradation causing a self-inhibition of the process.
- Conditions for optimal degradation worsen during the process (e.g., depletion of particular nutrients, pH shift.
- Residual concentrations of xenobiotics comprise those compounds of a pollutant mixture that are not biodegradable for biochemical reasons.

The term "bioavailability" is used in very different fields of applied biology (Tab. 1), in general to describe the fact that a chemical compound must get into contact with the biological system in order to induce any effect. A more quantifiable definition of bioavailability that had been originally coined by KLEIN and KÖRDEL (1984) and that has recently been adopted from the *"Commission on Soil and Water Chemistry"* of the IUPAC describes bioavailability as *"the portion of a present contaminant which can be taken up by the organism from its environment and food and is subsequently transported, distributed and metabolized by the organism"* (KÖRDEL et al., 1997). The definition clearly indicates that only a portion of the compound present in any soil compartment is available to the organism, the exact portion depending on both, the physiology and habitat of the organism. This defini-

Tab. 1. Use of the Term Bioavailability in Different Fields of Applied Biological Science

Agriculture	accessibility of the soil nutrients for plants in order to grow
Pharmacy	possibility of a drug to reach the receptor at the target cells in order to cause positive effects
Toxicology	possibility of a drug to reach the receptor at the target cells and to cause harmful effects
Nutrition physiology	suitability of the food to become processed by the human digestive tract
Biotechnology	status of accessibility of the fed substrates for microbial transformation into useful products
Environmental biotechnology	status of accessibility of pollutants in their environmental compartment for microbial degradation

tion of bioavailability reflects especially the ecotoxicological requirement to carry out hazard and risk assessment of chemicals in soil with the "right and representative organism". The IUPAC definition of bioavailability is in its general meaning certainly also important for environmental biotechnology, but for the special purpose of the environmental bioremediation business it is also necessary to include the factor "time" more expressively. Especially in the context of soil bioremediation bioavailability can, therefore, be considered as the chemical and physical status and compartmental localization of an environmental pollution that determines the extent and velocity by which a degrading microbial community may make metabolic contact with the xenobiotic substrates in order to transform and detoxify them efficiently.

This chapter will summarize and discuss the major factors that may contribute to the bioavailability of pollutants in bioremediation and will also point to some strategies to overcome or reduce the problem of a lack of bioavailability.

2 Factors that Influence Bioavailability

2.1 Factors that Influence Bioavailability in the Water Phase

A general requirement for microbial growth is that the microorganisms get continuous access to the substrates they live on. Since the whole machinery for energy metabolism and biosynthesis is located within the cell it is also necessary that the particular nutrients and substrates can be transported to and into the cell. The necessity to import substrates into the cell requires that the substrates are provided in a form that makes it possible to transport them across the membrane without disturbing or destroying the integrity of the cell membrane. The molecules to be taken up must, therefore, be both: small enough to cross the outer cell envelope and in a dissolved state

(SIKKEMA et al., 1995; WODZINSKI and BERTOLINI, 1972; WODZINSKI and COYLE, 1974). Since water is the only relevant natural solvent for all microbial nutrients and substrates the *solubility of a substrate in water* is, therefore, the first important factor that governs its bioavailability. Due to their hydrocarbon origin many soil pollutants, especially those being relevant for bioremediation, however show a rather poor water solubility (Tab. 2). It is shown in Tab. 2 that the solubilities of most of the xenobiotic substrates vary between mg L^{-1} and µg L^{-1} while many natural substances show water solubilities in the range of g L^{-1}. The availability of a pollutant substrate for microbial growth is, nevertheless very often less a problem of the absolute amount of a substrate present in the environment of the cell rather than of the velocity by which a substrate becomes dissolved or depolymerized and transported into the cell (BOSMA et al., 1997; STUCKI and ALEXANDER, 1987). The low aqueous solubility of many hydrophobic pollutants leads to a *crystallization of the compound* once the maximum solubility is exceeded. Since a direct uptake of crystals into the cell has not been observed yet (WODZINSKI and BERTOLINI, 1972; WODZINSKI and COYLE, 1974; WODZINSKI and LAROCCA, 1977) the bioavailability of the crystallized compound is mainly determined by the rate the crystals become (re-)dissolved in the aqueous phase. Due to the dissolution surface the most important factor that governs the rate of dissolution and thereby the bioavailability is the size of the crystal (VOLKERING et al., 1992; KÖHLER et al., 1992; THOMAS et al., 1986; Fig. 2). The rate of mass transfer from the solid or crystal phase to the aqueous phase can, therefore, be described as

$$-dC_S (dt)^{-1} = k_{diss} \cdot A \cdot (C_E - C_{WS}) \qquad (1)$$

with k_{diss} as the mass transfer constant governing dissolution (mol time^{-1}), A as the dissolution contact surface between pollutant source and water (area), C_E and C_{WS} as the equilibrium and the actual concentration of dissolved pollutants in the water phase, respectively (both in mol volume^{-1}). The central role of water solubility as bioavailability factor is also relevant for those pollutants that *become*

Tab. 2. Aqueous Solubilities of Some Important Environmental Pollutants. The Aqueous Solubility of Some Common Natural Nutrients is Given for Comparison

Substrates	Maximum Solubility in Water [mg L^{-1}]
Benzene	1,780[a]
Toluene	515[a]
1,2-Dimethylbenzene (*o*-Xylene)	175
Cyclohexane	58
Naphthalene	30
Anthracene	0.07
Benzo(a)pyrene	0.004
1,2-Dichlorobenzene	92–145
2,3,4,6-Tetrachlorophenol	100
Trichloroethane	950[a]
Tetrachloroethene	150[b]
Trinitrotoluene	140[b]
Some natural nutrients for comparison	
Sucrose	2,039,000
Alanine	167,000
Glycine	251,000
Valeric acid (pentanoic acid)	32,000
NH$_4$Cl	372,000
Urea	519,000
FeSO$_4$	266,000

[a] Data given for 20 °C.
[b] Data given for 25 °C; other data without specification of temperature; data were compiled from different chemistry textbooks or publications.

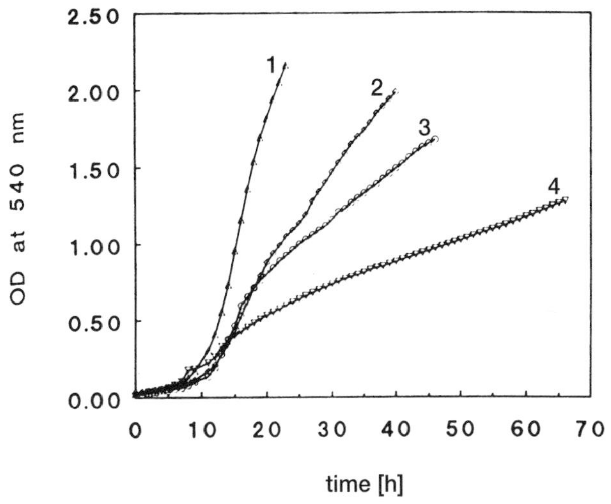

Fig. 2. Impact of crystal size of a pollutant on biological growth rate. Batch growth of a bacterial strain on sieved fractions of naphthalene with different diameters: 1: 600–1,000 µm; 2: 1,000–2,400 µm; 3: 2,400–3,350 µm; 4: >3,350 µm (VOLKERING et al., 1992) modified.

transformed co-metabolically, since this process also takes place within the cell. The role of water solubility is only less relevant for pollutants that are attacked primarily by enzymes that are excreted by the cell (e.g., extracellular enzymes of white-rot fungi).

A particular problem of bioavailability exists for pollutants that *constitute or are part of non-aqueous phase liquids (NAPL)* as, e.g., mineral oil, tar oil or ground water plumes of chlorinated aliphatic hydrocarbons (EFROYMSON and ALEXANDER, 1995; see ALEXANDER, 1994 or LUTHY et al., 1994 for review). The maximum amount of molecules soluble in the aqueous phase is under these circumstances governed by the specific equilibrium constant of the NAPL–water system (examples of some measured constants are given in BOYD and SUN, 1990; LEE et al., 1992a, b), while the velocity by which the equilibrium can be reached is – comparable to the solid crystal dissolution – determined by the size of the exchange surface of the non-aqueous phase droplets and the concentration difference between the NAPL and the water phase (GHOSHAL and LUTHY, 1996; MIHELCIC et al., 1993; SOUTHWORTH et al., 1983). Since – due to the lack of water and oxygen – the microorganisms cannot colonize the inner parts of the NAPL droplets, the overall biodegradation of the organic pollutant phase is, therefore, the slower the smaller the surface–volume ratio of the organic phase. For the same reason the toxicity of a pollutant may, however, also be reduced in the presence of a NAPL (ROBERTSON and ALEXANDER, 1996).

Since it is known that a number of microbial strains are able to adhere to and to grow at the surface of non-aqueous phase liquids like mineral oil (KÄPPELI et al., 1984; ROSENBERG and ROSENBERG, 1981; MARIN et al., 1996; Fig. 3), the question was also investigated whether the hydrophobic liquid compounds could eventually also be *taken up by the cell directly through the lipid membrane* without any "detour" through the aqueous phase.

The avoidance of the water phase by direct membrane–NAPL contact would theoretically help to circumvent both; the severe bottleneck caused by the low aqueous solubility of the pollutants and the time loss caused by the need of a long-distance substrate transport to the

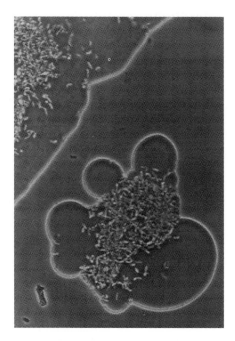

Fig. 3. Bacteria colonizing on diesel oil.

cell. This hypothesis is backed by few reports in the literature that describe a direct microbial uptake mechanism for hydrophobic pollutants. Thus it was found with heptamethyl nonane, being selected as an inert hydrophobic model NAPL, that naphthalene and hexadecane can be used directly from the NAPL rather than from the aqueous phase (EFROYMSON and ALEXANDER, 1991; WODZINSKI and LAROCCA, 1977). Direct uptake mechanisms may be supported by special cell surface features, cellular appendages like fimbriae (KÄPPELI et al., 1984; ROSENBERG et al., 1982) or by liposome-mediated transport (SCOTT and FINNERTY, 1976). However, it cannot be excluded unequivocally in these experiments that the cells did not need a minimum layer of water between the cell and the NAPL at least. The more widely accepted model of growth on hydrocarbon NAPL is, therefore, based on two non-direct uptake mechanisms (SIKKEMA et al., 1995):

(1) The bacterial cells are able to colonize the hydrophobic oil phases with the help of particular cell surface compounds and take up primarily those molecules that go into solution close to the NAPL source. This helps to avoid bioavailability problems due to diffusion distance or dilution.

(2) The bacteria can actively dissolve hydrocarbons due to the excretion of emulsifiers and surfactants (e.g., glycolipids, lipoproteins). The emulsifying agents can help to split the phase layer in small-sized droplets while the surfactants help to increase the maximum aqueous solubility of the hydrophobic substrates by their entrapment in surfactant micelles. An illustration of that model is summarized in Fig. 4.

2.2 Factors Influencing Bioavailability of Organic Substrates in Soil

It was shown in Sect. 2.1 that water can be considered as the major bottleneck all substrates have to pass before they can be metabolized by microorganisms. This is also true for the soil environment but the situation becomes more complex by the fact that the transfer of the solute to the cell can be hindered and slowed down by several *additional* factors like entrapment, sequestration, or sorption–desorption processes (ALEXANDER, 1994; BECK et al., 1993; LUTHY et al., 1997; SAWHNEY and BROWN, 1989). The presumed retardation of biodegradation in soil (compared to liquid cultures) was in fact shown by several different laboratories during the last years (GORDON and MILLERO, 1985; GUERIN and BOYD, 1992; MANILAL and ALEXANDER, 1991; OGRAM et al., 1985; WHITE et al., 1997).

The first possible explanation for this phenomenon is that the dissolved pollutant molecules are much smaller than the degrading bacteria or fungi and may, therefore, *diffuse into soil microregions to which the microorganisms have no access* (Fig. 5). The inaccessibility of pollutans is – due to the higher amount of small pores – more pronounced with clay soils than on sand (KNAEBEL et al., 1994; WHITE et al., 1997), but the degree of hydrophobicity of the soil material may also play an important role (NAM and ALEXANDER, 1998). Although all pollutant *molecules may diffuse back* from the inaccessible into the bioaccessible pore space according to the concentration equilibrium, this may take quite a long time depending on the diffusion distance, the pore tortuosity and the velocity by which the accessible molecules are eliminated in the pore space (GUERIN and BOYD, 1992; HARMS, 1996; HARMS and ZEHNDER, 1994, 1995). Residual pollutant concentrations – as they are frequently observed in the field (see Sect. 1) – represent,

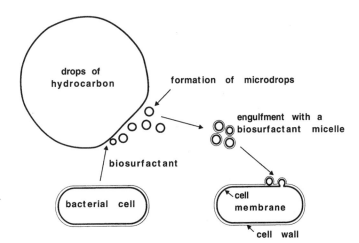

Fig. 4. Mechanisms of a biosurfactant catalyzed microbial hydrocarbon uptake from non-aqueous phase liquids (FRITSCHE, 1998, modified).

drops of hydrocarbon

formation of microdrops

engulfment with a biosurfactant micelle

biosurfactant

bacterial cell

cell membrane

cell wall

68   *3 Bioavailability of Contaminants*

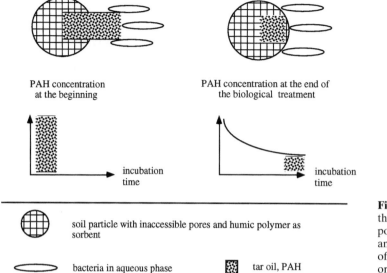

PAH concentration
at the beginning

PAH concentration at the end of
the biological treatment

incubation
time

incubation
time

soil particle with inaccessible pores and humic polymer as
sorbent

bacteria in aqueous phase   tar oil, PAH

Fig. 5. Correlation between the occurrence of residual pollutant concentrations and a lack of accessibility of soil pores for micro-organisms.

therefore, more often the lack of time for diffusion than a true biochemical failure of degradability (MAHRO et al., 1996).

The pollutant diffusion into the bioaccessible pore space may become further slowed down by continous *sorption–desorption processes along the diffusion pathway, a process also called sorption retarded diffusion* (LUTHY et al., 1997; PIGNATELLO and XING, 1996). It was found in many experimental investigations that the sorption affinity of an organic substrate in soil usually increases with its hydrophobicity, a fact that is also reflected by the high correlation of the respective K_{ow} and K_{oc} values of many organic compounds (MEANS et al., 1980; HASSETT and BANWART, 1989; Tab. 3). It is, therefore, widely accepted now that the hydrophobic humic polymer represents the major sorption site for organic chemicals in soil (CHIOU, 1989; KÖRDEL et al., 1997; SCHACHTSCHABEL et al., 1992) although inorganic soil components like clay minerals may contribute to the pollutant sorption to some extent as well (LUTHY et al., 1997; ZIECH-MANN and MÜLLER-WEGENER, 1990). It was assumed for a long time that the pollutant sorption in the organic soil material is primarily based on an equilibrium-governed partitioning process of the solute, taking place

between the aqueous and the organic humic phase of the soil (CHIOU, 1989; HASSETT and BANWART, 1989; LANE and LOEHR, 1992). However, the simple "partitioning model" has now also been questioned repeatedly, since some findings on pollutant transport in soil, e.g., the hysteresis phenomenon are not in accordance with the assumed linear sorption–desorption kinetics (BECK et al., 1993; LUTHY et al., 1997; PIGNATELLO and XING, 1996).

Another process of pollutant immobilization that cannot be explained by partitioning mechanisms is the fact that pesticides and other xenobiotics like PAH or TNT may become bound to the humic matrix in a way that makes them non-extractable by organic solvents thereafter (DFG, 1998; MAHRO and KÄSTNER, 1993; MAHRO et al., 1994; RIEGER and KNACK-MUSS, 1995). The process called pollutant humification is described in more detail in Chapter 4, this volume). The formation of non-extractable residues may severely reduce immediate bioavailability, but its particular impact on the long-term bioavailability is also hard to predict, since the underlying mechanisms are not well understood yet (DFG, 1998; Umweltbundesamt, 1999). Thus it is, e.g., not clear yet, whether non-extractable xenobiotic molecules

Tab. 3. Sorption-Related Parameters for Some Important Soil Contaminants

Substance	Logarithm of Partition Coefficient in n-Octanol–Water (log P_{ow})	Logarithm of Soil Adsorption Coefficient (log K_{oc})
Phenol	1.1[a]	1.9[a]
Benzene	1.8–2.1	0.8
Trinitrotoluene	2.0	1.6[a]
Trichloroethane	2.4	1.6
Toluene	2.4[a]–2.7	1.0
Tetrachloroethene	2.6–3.0	1.8[a]
1,2-Dimethylbenzene (o-Xylene)	2.8	1.2[a]
1,2-Dichlorobenzene	3.2–4.4	2.5
Cyclohexane	3.4	3.1[a]
Naphthalene	3.4	3.1
2,3,4,6-Tetrachlorophenol	4.1	3.8
Anthracene	4.5	4.3
Benzo(a)pyrene	6.0	5.3

[a] Given in literature as calculated data; data compiled from different soil chemistry textbooks or publications.

can be returned to the water phase in their original molecular structure, or whether the transformation may also be part of the natural humic turnover. The major parts of soil-bound PAH residues seem, however, to be very stable even under environmental stress conditions (ESCHENBACH et al., 1998).

The bioavailability of pollutants in soil can of course be also influenced by other environmental factors such as pH or water content. The *soil pH* is of particular importance for the bioavailability of heavy metals, since it determines both the relative amounts of sorbed and desorbed metal ions and the ability to form precipitates (BOURG, 1988; SCHACHTSCHABEL et al., 1992). The aqueous solubility of many of these precipitates is so low that they are practically insoluble, an ability which can also be used intentionally in geochemical engineering concepts of soil remediation (FÖRSTNER, 1997). The soil pH, however, is also of importance for the bioavailability of some organic compounds like those that can become ionized like, e.g., acidic or phenolic compounds (SCHACHTSCHABEL et al., 1992).

The *water content of the soil* may also have a somewhat different impact on the pollutant bioavailability in the soil. The first impact is that a drastic reduction of the soil water content would directly inhibit the water-bound mass transfer processes into the cell (see Sect. 2.2). An excess amount of water would, on the other hand, also inhibit the oxygen flow within the soil, eventually slowing down the microbially induced pollutant degradation. A decrease of the velocity by which the concentration gradient is held up as shown in Eq. (1) has, however, also an indirect influence on the overall velocity of the mass transfer in the soil pores. The soil water content is also important for the transport velocity of volatile compounds, since a lower gas pore volume in the soil would also reduce the transport of volatile compounds through the soil to the cells (HARMS, 1996). On the other hand, a higher soil water content can also suppress an undesired, quick evaporation of the pollutants from the soil.

2.3 Particular Problems of Bioavailability in Soil from Long-Term Contaminated Sites

It was shown in Sect. 2.2 that soil microorganisms may be confronted with severe problems of bioavailability already in their natural environment. Their situation may become worse, however, in soil environments of long-term contaminated sites.

The first problem in this regard is that both, the diffusion distance and the sorption intensity may increase in the course of time, a process that is also addressed in the literature as the *"aging" of a contamination* (CARMICHAEL et al., 1997; KELSEY and ALEXANDER, 1997; KELSEY et al., 1997). The dynamics of soil, especially the growth or rearrangement of the humic polymer in time may, e.g., stabilize the sequestration of the xenobiotics in the soil pores or within organo-mineral complexes and thereby slow down further the diffusion velocity of the pollutant. A particular type of aging occurs with hydrocarbon pollutants that are present in soil as a non-aqueous phase liquid. Such fluid NAPL may easily become translocated in small soil pores (LUTHY et al., 1997; SEITINGER et al., 1994) from where the diffusion of the translocated hydrocarbon molecules back into the accessible pore space may become severely retarded due to physical and chemical alterations of the oil–water interface (LUTHY et al., 1993). Such oil *aging or weathering processes* may start at the outer surface *of the NAPL* by evaporation of the more volatile pollutants from the complex hydrocarbon phase, ending in the formation of a rather viscous interface layer which may inhibit or prevent any pollutant diffusion from the NAPL into the aqueous phase (LUTHY et al., 1993; NELSON et al., 1996; Fig. 6).

The reduction of diffusion velocity represents also the major problem at soil sites that had been contaminated with *highly viscous tar oil or tar pitch* (LUTHY et al., 1994; MAHRO and SCHAEFER, 1998). Such high-viscosity types of NAPL are sometimes also called dense non-aqueous phase liquids (DNAPL) (LUTHY et al., 1993) in order to distinguish them from the more fluid NAPL mentioned above. In many cases, the presence of DNAPL may, especially at PAH contaminated sites, constitute the major reason for the persistence of residual pollutant concentrations at the end of bioremediation treatments (Fig. 1; MAHRO and SCHAEFER, 1998). The bioremediation of PAH-contaminated sites must, however, not necessarily end with residual pollutant concentrations. Experiments with a soil containing PAH formed during the fire of subsurface diesel oil containers

Fig. 6. Photograph of coal tar drop suspended in water from the tip of a syringe needle. (**a**) Freshly formed drop, (**b**) coal tar drop aged in water for 3 d (LUTHY et al., 1993).

showed, e.g., that even about 3 g of PAH kg^{-1} of soil could be completely eliminated, if the proper biological treatment was provided

(MAHRO et al., 1999). The chance to clean up soil sites that had been contaminated with PAH from *low viscosity carriers* as, e.g., mineral oil or coking oil from wood seems considerably higher than the chance to clean up sites that had been contaminated with DNAPL such as tar oil or pitch.

Another particular bioavailability problem that may occur at long-term contaminated sites is that the ground may also contain large amounts of *coal particles* (AFFERDEN et al., 1992; WEISSENFELS et al., 1992). This situation is very common, especially at sites where gas works or coking facilities had been run for a long time. Coal particles have an even higher capacity for the binding of hydrophobic pollutants like PAH than humic material, and the chance to reverse this adsorption in such a way that bacteria are provided with sufficient amounts of their "pollutant nutrients" in time is, therefore, very small. The problem of the strong binding capacity of a coal contaminated soil was shown in a very illustrative experiment by WEISSENFELS et al. (1992) (Fig. 7). WEISSENFELS et al. (1992) extracted PAH repeatedly from a coal dust containing soil by organic solvents and returned the extract subsequently to the soil. Parts of the respiked, but still bioavailable PAH were then degraded instantaneously by the bacteria present in the soil while other parts of the respiked PAH were readsorbed to the soil–coal particles. The biodegradation stopped as soon as the adsorption was complete and could only be started again by another extraction–respiking cycle.

3 Theoretical and Experimental Measures to Describe and Quantify the Relationship of Bioavailability and Biodegradation

3.1 Kinetic Description of Separate Steps of Biodegradation

The technical exploitation of biological processes for the clean-up of polluted soils demands development of *tools that are able to predict* or at least to estimate the time requirement and efficiency (transformation yield per time and volume) of a bioremediation process as precisely as possible. The individual processes that may influence the kinetics of the overall degradation process and which must, therefore, be taken into consideration in that regard are illustrated and summarized in Fig. 8. Each of the partial steps of the process summarized in Fig. 8 can be described by its own velocity equation (v_{diss}, v_{des}, v_{diff}, v_{up}, v_{trans}) according to the respective mechanism upon which one assumes it is based. While dissolution velocity (v_{diss}) could, e.g., be calculated according to the dissolution equation (Eq. 1), the desorption kinetics could be described by a variety of

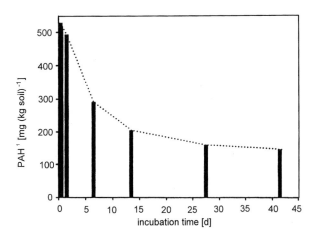

Fig. 7. Microbial PAH degradation in soil that contained a high amount of coal dust and coke particles (WEISSENFELS et al., 1992).

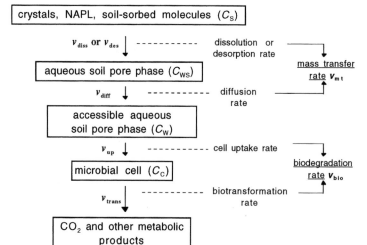

Fig. 8. Model of different kinetic steps in the overall process of pollutant biodegradation. The pollutant concentration C is different in the different pools, indicated by separate indices (C_S: source or soil pollutant concentration; C_{WS}: pollutant concentration in the aqueous phase close to the source; C_W: pollutant concentration in the accessible aqueous phase; C_C: pollutant concentration at or in the cell).

different equations, each reflecting a different model of sorption–desorption mechanism. Since the discussion on sorption–desorption modeling has now become a research field on its own, it will not be covered here. But appropriate reviews describing the field have been published, e.g., by BECK et al. (1993), LUTHY et al. (1997) or PIGNATELLO and XING (1996). Less controversial is it to describe the next step of the pollutant migration, namely the transfer of the pollutant from the inaccessible soil pores to the accessible aqueous phase and to the microbial cells (V_{diff}). This step may at least as a first approximation be calculated according to Fick's first law of diffusion as

$$v_{diff} = dp\,(dt)^{-1} = -D \cdot F \cdot dc\,(ds)^{-1} \qquad (2)$$

where $dp\,(dt)^{-1}$ represents the pollutant flow per time [mol time^{-1}] from C_{ws} to C_W, F the diffusion surface (area), dc the pollutant concentration gradient (mol volume^{-1}), ds the diffusion distance (length) and D the diffusion coefficient (cm^2 s^{-1}). To calculate the diffusion coefficients one can use data given in the literature for defined systems, but one can also estimate D if one has data on the diffusible pore space, phase viscosity, and size of·the diffusing molecule.

The third process that needs to be quantified in order to calculate the overall pollutant turnover rate in the soil is the intrinsic degra-

dation potential of the microorganisms (v_{bio}). The velocity of this process is mainly dependent on the intrinsic transformation activity of the cells, their affinity for the pollutants, and the actual concentration of the pollutants. A realistic way, e.g., to describe the *biokinetic potential for the degradation of pollutants with a very low aqueous solubility* and which, therefore, does not allow the cells to grow on these pollutants, is to use the Michaelis–Menten equation

$$v_{bio} = -dC_w\,(dt)^{-1} = v_{max} \cdot C_c\,K_P^{-1} + C_C \qquad (3)$$

where the terms v_{bio} and v_{max} represent the actual and the maximum intrinsic biodegradation rate, respectively (given as mol time^{-1}), C_c the actual concentration of the pollutant at the cell surface or within the cell (mol volume^{-1}) and K_P the cell surface pollutant concentration yielding 1/2 v_{max} (mol volume^{-1}). Since the pollutant concentration at the cell surface or within the cell is hard to monitor one can also replace C_c by C_w. It must be pointed out, however, that the given equation for v_{bio} reflects *just one possible type of kinetic of biodegradation* in soil. A number of other possibilities to describe the kinetics of pollutant degradation, which differ mainly with regard to the actual substrate concentration, the number of cells taking part in the transformation process and the presence or absence of growth have been described and summarized by ALEXANDER (1994), ALEXANDER and SCOW (1989) or KOCH (1982). Six typical pollutant disappearance curves, each representing a different type of kinetic, are given in Fig. 9.

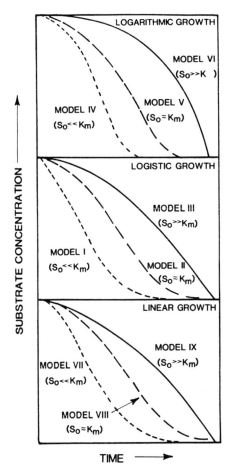

Fig. 9. Different types of pollutant degradation kinetics in relation to different types of growth (ALEXANDER and SCOW, 1989).

3.2 Experimental Strategies to Determine Kinetic Data on Mass Transfer and Biotransformation in Soil

The major problem to be solved, however, in order to use the given kinetic equation for calculation is to obtain data for each of the respective steps. Usually one has to elaborate v_{bio} and v_{diff} from actual experimental or field data right with the soil under investigation rather than from literature data, since these are either not available or were derived from other experimental situations. Each of the given processes must be measured separately, and it is also important to exclude all other processes than diffusion or biodegradation, that might contribute to the change of the concentration of the pollutant (e.g., evaporation, sorption at the glassware). The most basic parameter that can be determined experimentally is the dissolution kinetic. The dissolution rates for crystals of different sizes can thus be measured directly in liquid systems by determination of the time required until equilibrium is reached (analytical determination of the concentration, e.g., with HPLC or radiolabeled compounds). A possibility to determine the mass transfer rate coefficients with solid matrix systems is to measure the time that is necessary to compensate for any induced change of the equilibrium concentration in the system. If the system is run as elution or flow-through tests, where concentration changes in the eluate can directly be correlated with mass transport processes, one can also obtain data on the convective mass transfer potential. The time of diffusion needed by particular compounds in a given matrix can also be measured directly in defined diffusion tubes, as described, e.g., by VAN DER SLOOT and DE GROOT (1988) for the diffusion of radiolabeled compounds (Fig. 10). Another more indirect possibility to determine the influence of the matrix length or of the structure on the mass transfer is to measure the growth kinetics of bacterial cells that are separated from the source of the nutritive pollutants by the respective matrix (HARMS, 1996). The setup of such an experimental system is shown in Fig. 11.

The *determination of the second process, the biodegradation rate*, can be based either on pure liquid culture systems or on experiments with the complex soil system itself. The first option has the advantage that interfering mass transfer and other problems can be held to a minimum and that one can clearly recognize when the substrate becomes limiting. Thus it was shown, e.g., by VOLKERING et al. (1992) that the growth kinetics of PAH-degrading bacteria changed from an exponential to a linear first order kinetic as soon as the dissolution rate of the pollutant became limiting for the growth of cultures with increasing high cell densities (Fig. 12). A major drawback of the use of liquid culture measurements as indicator of the biokinetic potential in soil is, however, that this technique neglects the fact that the *presence of solid surfaces* may also have a direct impact on microbial activity (DAGASTINO et al., 1991; DAVIES et al., 1993; GUERIN and BOYD, 1992). Thus it was shown that bacterial cells may modify their metabolism once they get contact to surfaces, e.g., by triggering the formation of biofilm forming molecules (DAVIES et al. (1993) or of surface-active compounds (GUERIN and BOYD, 1992), both measures which can also change the mass transfer process in the immediate environment of the cell. The specific biological pollutant turnover constant k_{bio} of the organism or the microbial community should, therefore, preferentially be

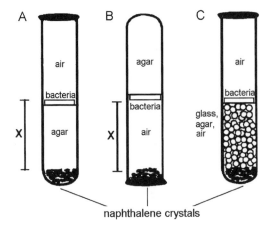

Fig. 11. Test tubes to measure bacterial growth on distant pollutant diffusing through aqueous agar phase, air or porous media (HARMS, 1996).

measured in complex soil systems, on condition that one can make sure that the overall degradation rates are corrected as well as possible for the particular contribution of mass transfer (this needs to be evaluated by separate measurements). It should also be made sure that the measured parameter represents true biodegradation (e.g., CO_2 release) rather than pollutant depletion, a process that may reflect the extraction efficiency more than the biodegradation (ESCHENBACH et al., 1994).

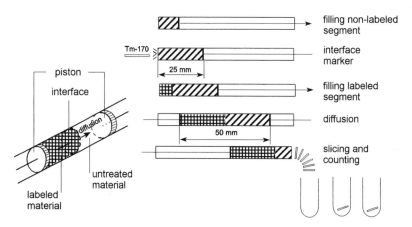

Fig. 10. Monitoring tubes for the measurement of pollutant diffusion in soil (VAN DER SLOOT and DE GROOT, 1988) (redrawn and modified by R. WIENBERG).

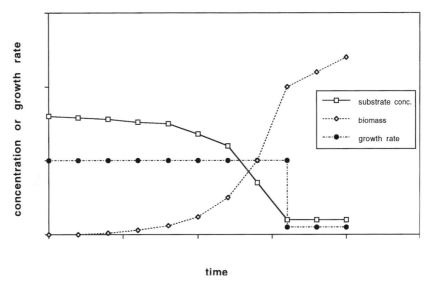

Fig. 12. Model of biphasic growth due to pollutant dissolution as growth limiting factor at higher cell numbers (modified and redrawn, based on VOLKERING et al., 1992).

GHOSHAL and LUTHY (1996) derived their k_{bio} values, e.g., by measuring the transformation of ^{14}C-labeled pollutants that had been spiked in two different tars (GHOSHAL and LUTHY, 1996). The biodegradation constants were calculated from the [14]CO_2 evolution that occurred during the initial part of the mineralization curve given in Fig. 13. The graphs of the two different transformation processes also give a good illustration of how the degradation kinetics can differ in soil batches where the overall transformation is either controlled by mass transfer or biokinetics. It can be seen in Fig. 13 that the degradation of naphthalene from tars that were spread on microporous silica particles (size of particle diameter 0.25 mm) were – due to the higher exchange surface of the tar – initially controlled biokinetically while the degradation of naphthalene from tar globuli with a size of about 11 mm was mainly determined by mass transfer. This conclusion was also in accordance with the calculated Damköhler numbers (ranging from 0.004–0.007 for the microporous media and from 9–13 for the tar globuli; for an explanation of the Damköhler number see Sect. 3.3).

3.3 Approaches to Describe and Model the Overall Biodegradation Process

The data obtained from the kinetic measurements of separate steps of the biodegradation process may also be combined to describe and model the overall kinetics of the mass transfer-related biodegradation as a common term. One common way to identify the relationship between the mass transfer-related processes and the biodegradation process is to express them in a simple quotient, a term also known as Damköhler number Da (DAMKÖHLER, 1937):

$$Da_{II} = \frac{\text{(number of pollutant molecules that are degraded by the cells)}}{\text{(number of pollutant molecules that diffuse to the cells)}^{-1}} \quad (4)$$

(without dimension)

The results obtained that way help to identify the process step that actually determines the overall rate of the degradation kinetics and which must, therefore, be optimized preferen-

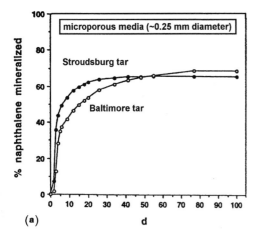

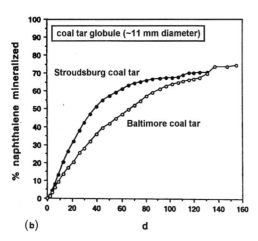

Fig. 13. Typical naphthalene mineralization profiles.
(a) System with coal tar imbibed in 0.25 mm diameter microporous silica beads (biokinetic limited; $Da<1$), (b) system with a 0.7 mL, 11 mm diameter coal tar globule (mass transfer limited; $Da>1$) (GHOSHAL and LUTHY, 1996).

tially. Thus, if the Damköhler number obtained from separate kinetic measurements is larger than unity ($Da>1$), one knows that mass transfer is limiting the process, while in contrast the biological transformation process needs to be optimized first if the Damköhler number is considerably smaller than unity ($Da<1$).

The kinetic equations of the different partial processes can also be combined in one common velocity equation. One possibility to construct such an overall velocity equation is to combine the first and the last two steps of the overall degradation process in a mass transfer and a biodegradation-related rate, respectively, which gives the overall velocity equation (Eq. 5):

$$dC_W (dt)^{-1} = [k_{mt} \cdot (C_{WS} - C_W)] \\ - [k_{bio} \cdot C_W] \tag{5}$$

with C_W actual concentration of the pollutant in the accessible aqueous bulk phase (given at particular time t, in mole volume^{-1}), C_{WS} equilibrium concentration of the pollutant in the pore water phase close to the source (given at particular time t in mole volume^{-1}), k_{mt} mass transfer constant for the diffusion of the pollutant to the cells (given as volume time^{-1}), k_{bio} specific biological turnover constant of the organism or the microbial community [given as mole (volume time)$^{-1}$].

Eq. (5) states that the total amount of pollutant molecules which are actually transformed per time unit by the microbial population is dependent on both, an intrinsic biodegradation rate and the velocity of a mass transfer process. The reaction constant k_{bio} in Eq. (5) is equal to the ratio of $v_{max} K_P^{-1}$ which was used in Eq. (3). It is a first-order reaction constant as long as the concentration of C_c is much lower than K_P (see discussion above). The mass transfer constant k_{mt} should comprise all different parameters that may have an impact on the respective mass transfer process itself like the dissolution or desorption rate, the diffusion surface area, the diffusion coefficient D for a given pollutant in a defined matrix and the diffusion length. Equations that include several independent kinetic parameters at once can only be solved by complex numerical approximation methods. The basic equations and theoretical framework for such approaches have been explained in more detail elsewhere (e.g., by BOSMA et al., 1997; KOCH, 1982, 1990; GHOSHAL and LUTHY, 1996; MULDER et al. 1998, or TABAK et al. 1997). An interesting illustration of the utility of a modeling approach has recently been described by BOSMA et al. (1997). They applied a formula that was originally described by Koch (1990) as the Best equation (Eq. 6) to calculate and simulate the combined action of mass transfer and microbial conversion on the turnover of two model pollutants in different soil systems (α-hexachloro cyclohexane, 3-chloro dibenzofuran),

$$q = q_{max} \frac{C_d + K_m + q_{max}\, k^{-1}}{2\, q_{max}\, k^{-1}}$$

$$\left\{ 1 - \left[1 - \frac{4\, C_d\, q_{max}\, k^{-1}}{(C_d + K_m + q_{max}\, k^{-1})^2} \right]^{1/2} \right\} \qquad (6)$$

with q quantity of pollutant which is transformed by the combined action of mass transfer and microbial conversion (mol time^{-1}), q_{max} maximum conversion flux that can be achieved by a cell (mol time^{-1}), C_d pollutant bulk concentration distant from cell surface (mol volume^{-1}), K_m concentration at cell surface yielding $1/2\, q_{max}$ (mol volume^{-1}), k exchange constant (volume t^{-1}).

The graph given in Fig. 14 shows that the Best equation matched very well with the experimental data obtained with α-HCH, proving its validity for the description of desorption limited biodegradation. The graph shows also that the overall activity of the microorganisms may be easily overestimated, if mass transfer limitation is neglected and the kinetic transformation model is based only on biokinetic terms. Based on a rearrangement of the

Best equation (not shown here, but explained in more detail by Bosma et al., 1997) the authors also proposed three terms that could be used to directly correlate the bioavailability status (expressed as bioavailability number Bn) with the bioavailable pollutant concentration C^* and the biological conversion rate Q^*. The terms Q^*, C^* and Bn were defined as follows:

$$Q^* = q\, q_{max}^{-1} \qquad (7)$$

dimensionless conversion rate, varying between 0 and 1,

$$C^* = C_d (C_d + K_m + q_{max} \cdot k^{-1})^{-1} \qquad (8)$$

bioavailable concentration, dimensionless; varying between 0 and 1,

$$Bn = k\, (q_{max}\, K_m^{-1})^{-1} \qquad (9)$$

bioavailability number.

Fig. 15 illustrates the relationship between Q^*, C^* and Bn for three different conditions of bioavailability. The graph indicates that when C^* equals 1, the conversion rate is at its maximum ($Q^* = 1$, i.e., the pollutant is fully available for biodegradation). Lower values of C^* indicate less bioavailability (Bosma et al., 1997). If one looks more closely at the newly introduced term, the bioavailability number Bn, it can be seen that its definition [(ratio of the mass exchange constant) · (the specific bio-

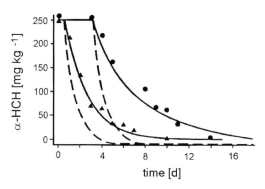

Fig. 14. Simulation of the biodegradation of α-HCH in stirred (triangle) and end-over-end mixed soil slurries (circles) using the Best equation (solid lines) and the Michaelis–Menten equation without mass transfer calculation (dashed lines) (Bosma et al., 1997).

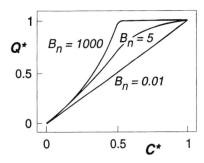

Fig. 15. Plot of the dimensionless conversion rate Q^* vs. the dimensionless concentration C^* at different Bn values. Q^* express the conversion rate normalized to q_{max} while C^* can be viewed as the bioavailable concentration (Bosma et al., 1997).

logical pollutant turnover constant $k_{bio})^{-1}$] is actually the inverse expression of the Damköhler number

$$Bn = (Da)^{-1} \qquad (10)$$

The use of the bioavailability number does not offer a principal advantage over the use of the Damköhler number but it is more easily recognized in its meaning, since the numbers are getting larger if the bioavailability becomes better (rather than approaching zero as with *Da*).

All attempts to predict and model the presumable degradation kinetics of pollutants in soil become considerably more complicated, if one turns from experiments with artificial soil to soil from long-term contaminated sites. A particular problem relevant for modeling approaches is, e.g., that the soil may contain pools of pollutants with different bioavailability status, one pool being accessible immediately, a second one limited by matrix diffusion and a third pool of "aged contaminations" which releases its pollutants only very slowly (see Sects. 2.2, 2.3). It should, therefore, be tried in the future to also implement these real-site problems in modeling scenarios as well as possible.

4 Technical Options to Improve Bioavailability

As shown in Sects. 2 and 3 the lack of bioavailability may cause a significant slowdown of the bioremediation process. Sect. 4 will, therefore, summarize some technical options that can be considered in order to overcome limits caused by a lack of bioavailability. The major parameters that may be influenced by technical means to increase the bioavailability are

(1) the *maximum amount of pollutants* that can become dissolved in the accessible pore space
(2) the *velocity* by which a pollutant particle may be transferred into the accessible pore space and to the cell.

4.1 Measures to Increase the Solubility of Pollutants in the Water Phase

The amount of dissolved pollutants that can be degraded by the microorganisms in the aqueous phase can be severely limited by the low aqueous solubility of many hydrophobic pollutants. The use of surfactants might, therefore, provide a suitable tool to increase the solubility of hydrophobic pollutants in the aqueous phase (ROUSE et al., 1994). If the amount of surfactant added to a solution exceeds a *critical micellar concentration* (CMC) the surfactant molecules form micelles in the aqueous phase in which hydrophobic pollutants may become concentrated in amounts that are larger than those that can be obtained in the aqueous environment (Fig. 16; GRIMBERG et al., 1996; LIU et al., 1995; VOLKERING et al., 1995).

However, the results of surfactant-based biodegradation experiments reported so far are, nevertheless, rather controversial. The *results range from* inhibition of degradation (DESCHENES et al., 1996; FOGHT et al., 1989; JIMENEZ and BARTHA, 1996; LAHA and LUTHY, 1991) over little or contradictory effects (PROVIDENTI et al., 1995; SOEDER et al., 1996) to findings that describe a clear increase of the degradation velocity by the use of surfactants (THIBAULT et al., 1996; NELSON et al., 1996; SCHWARTZ and BAR, 1995; TIEHM, 1994; ZHANG and MILLER, 1995).

Looking for an explanation for the contradictory results one must recognize first that molecules that are already dissolved in the aqueous phase itself are of course more bioavailable to microorganisms than molecules that have to pass across an interface (here micelle–water). The amount of pollutants directly accessible in the aqueous phase may, therefore, be even lower than in systems without surfactants (LAHA and LUTHY, 1991; VOLKERING et al., 1995) so that the K_m of the pollutant degradation may also be higher and V_{max} lower than in systems without surfactants (VOLKERING et al., 1995). It was, therefore, postulated that surfactants may only have a positive impact, if the pollutant concentrations are very high (GRIMBERG et al., 1996) or if the toxicity

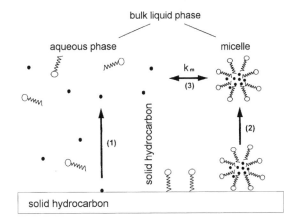

Fig. 16. Model of different phase transfer steps in surfactant mediated hydrocarbon uptake into an overlying liquid phase (GRIMBERG et al., 1995).

of the pollutant may become reduced due to its protective covering in the micelles (SCHWARTZ and BAR, 1995). The bioaccessibility of the micellar pollutants depends primarily on the velocity with which they are able to diffuse through the respective micellar interface into the aqueous phase, a relation that can be expressed and quantified by the respective equilibrium and exchange coefficients and by the concentration ratio of both, the pollutant and the surfactant molecules (GRIMBERG et al., 1995; WILLUMSEN and ARVIN, in press; Fig. 16). However, it was also reported by GUHA and JAFFÉ (1996a,b) that in some instances at least a fraction of the pollutants in the micellar surfactant phase could also be made bioavailable to the degrading microorganisms by direct micelle–membrane contact. Close contacts between the surfactants and the cell membrane may, however, also lead to toxic effects on the degrading cell itself and thereby reduce the overall degradation rate – as it was also reported repeatedly (LAHA and LUTHY, 1991; TIEHM, 1994; WILLUMSEN et al., 1998). The possible toxicity seems to be highly correlated with the hydrophily–lipophily balance of the surfactant (HLB) (TIEHM, 1994). If the surfactants are too lipophilic, i.e., have a low HLB, they have obviously a higher tendency to permeabilize the cell membrane and thereby to become toxic, than surfactants that are more water soluble (i.e., having a high HLB). Other reasons for a negative impact of surfactant additions may be that the attachment of the degrading bacteria might become dis-

turbed by the emulsifier (FOGHT et al., 1989) or that the surfactant is used as nutrient substrate itself, a problem that might be relevant especially with biosurfactants (DESCHENES et al., 1996; SOEDER et al., 1996). The latter problem might be overcome, however, by applying surfactant-producing microorganisms rather than a limited amount of surfactants to the soil, since such surfactant-producing microorganisms may constantly fill up the surfactant pool "*in situ*" (DEZIEL et al., 1996; PROVIDENTI et al., 1995).

4.2 Measures to Increase the Desorption and Mass Transfer Rate of the Soil Pollutants

The optimization of the water solubility does not solve the problem caused by slow desorption and mass transfer within the soil matrix yet, but it was shown that surfactants may also be used to accelerate the desorption process itself. Such a facilitated transfer of the pollutant from the solid phase to the aqueous phase can be caused by phenomena such as the lowering of the surface tension of the pore water in soil particles, interactions of the surfactant with solid interfaces and the interaction of the pollutant with single surfactant molecules (SUN et al., 1995; VOLKERING et al., 1995; YEOM et al., 1996). SUN et al. (1995) found that surfactant molecules may become effective as desorbing agents preferentially, if their concentration is considerably higher than the CMC and assumed that the surfactant micelles may then compete with the hydrophobic soil surface areas for sorbed pollutants, thereby "attracting" parts of the pollutant fraction into the aqueous phase. A facilitated passive transport and enhanced desorption of PAH may also be obtained if the soil is overlayed with vegetable oil, as can be seen in Fig. 17. Probably due to a better phase partitioning the use of vegetable oil as overlay phase induced a PAH transfer that was as high as one that could be obtained with acetone (Fig. 17).

The velocity of both, the desorption and the mass transfer can be *enhanced further by soil flushing and extraction (convective mass flow*

rather than diffusion). A number of different soil extraction and desorption fluids are currently under investigation comprising, e.g., numerous different synthetic surfactants (e.g., MARTEL et al., 1993; PETERS et al., 1992; STIEBER et al., 1994; SUN and BOYD, 1993; YEOM et al., 1996), biosurfactants (BRUSSEAU et al., 1994; HERMAN et al., 1997), humic acids (ABDUL et al., 1990; ENGEBRETSON et al., 1996; JOHNSON and AMY, 1995), vegetable oil-based micro emulsions (BONKHOFF et al., 1996) or lipid-containing waste fluids (MARBURG and MAHRO, unpublished results). The application of actively desorbing fluids is, however, in general very critical in *in situ* treatments in the field, since the increase of the pollutant load in

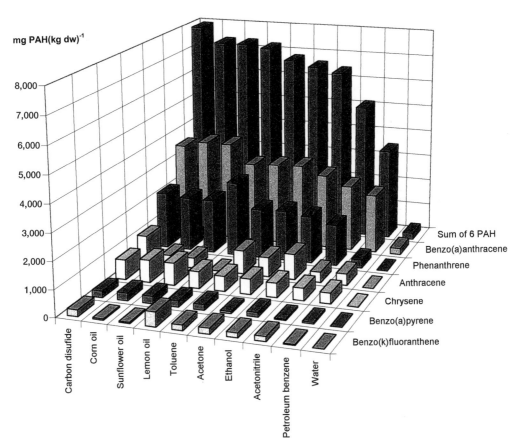

Fig. 17. Passive diffusion of PAH from tar–soil particles into a liquid overlay phase, consisting of either water, acetone, or vegetable oil (MARBURG and MAHRO, unpublished results).

the soil fluid phase may also lead to an increase of the pollutant load in the subsurface aquifer (OUYANG et al., 1995). It is, therefore, necessary to control and treat carefully the soil extracts that are carried out from the site under these conditions, e.g., by the use of soil slurry batch reactors (see Chapter 14, this volume).

A number of other measures can also be taken to increase the velocity of the mass transfer in the contaminated soil aggregates themselves (i.e., to improve the transfer from the contaminant source into the bioaccessible aqueous soil pores). According to Fick's diffusion law the velocity of mass transfer in a given matrix is dependent on the diffusion surface, the concentration gradient, the diffusion distance, and the specific diffusion coefficient [see Eq. (11) and Tab. 4]. The diffusion coefficient D is defined as

$$D = k \cdot T \cdot n_p (6 \pi \cdot \eta \cdot r \cdot \tau)^{-1} \qquad (11)$$

with k Boltzmann constant, T absolute temperature, n_p diffusible pore space, r size of the diffusing molecule, η viscosity of the diffusion phase, τ tortuosity pore geometry). Eqs. 12 and 13 clearly indicate the principal options one has to improve diffusion processes in soil. Tab. 4 summarizes these options.

The most effective option to optimize the pollutant diffusion from inaccessible soil pores into the bioaccessible pore space is to *reduce the length of the diffusion distance*. This can be done mechanically by breaking up soil clumps, e.g., by grinding (GEERDINK et al., 1993, 1996)

or by dispersing the soil clumps in a soil slurry (MUELLER et al., 1991b). The amount and length of soil pores a pollutant has to cross before it makes contact with the degrading microorganisms can also be reduced by the mechanical separation of the contaminated and the less contaminated soil fractions. Such a pretreatment was used, e.g., for the clean-up of a soil with large amounts of peat and clay (MANN et al., 1996) and may help to reduce the volume and costs for a subsequent soil reactor treatment. The reduction of the diffusion length can also be obtained with NAPL by the application of emulsifiers (OUYANG et al., 1995; ROUSE et al., 1994). The large *coherent NAPL are split up by the addition of emulsifying agents in little droplets* (Fig. 18), thereby reducing both, the diffusion length and the exchange surface–volume ratio.

Another diffusion-related parameter that can be optimized is the *viscosity of the carrier matrix*, especially if one deals with DNAPL. An effective measure to lower the viscosity in a given diffusion phase is to *increase the temperature in the DNAPL environment*, increasing the solubility of the hydrophobic pollutant at the same time. A prerequisite for the use of higher temperatures for soil bioremediation is, however, to have microorganisms able to tolerate these temperatures that can grow at higher temperatures at the expense of the pollutant. First reports on the isolation of strains able to grow on PAH at temperatures of up to 70 °C show that the approach is basically valid (FEITKENHAUER et al., 1996) although it will face severe limits in relation to cost effectiveness.

Tab. 4. Possibilities to Optimize the Diffusive Mass Transfer According to Fick's Diffusion Law

$dm \cdot dt^{-1} = D \cdot F \cdot dc \cdot ds^{-1}$ (12)	$D_B = k \cdot T \cdot n_p \cdot (6 \pi \cdot \eta \cdot r \cdot \tau)^{-1}$ (13)
increase of diffusion surface F by reduction of soil particle size	increase of temperature T
maintenance of the concentration gradient dc by continuous removal of c from the soil pore space by biological degradation	increase of the diffusible pore space n_p (e.g., by soil loosening)
reduction of the diffusion distance s by reduction of soil particle size	reduction of the size r of the diffusing molecule (not applicable in this case)
increase of diffusion velocity by optimization of parameters of the diffusion coefficient D	reduction of viscosity (η) in the diffusion phase (e.g., solubilization of tar pitch?)
	reduction of tortuosity τ (pore geometry), e.g., by soil floating (soil slurry)

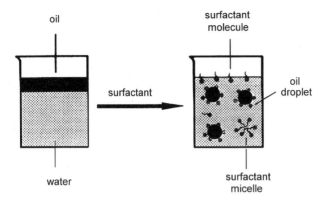

Fig. 18. Effect of emulsifier on the dispersion of a non-aqueous phase liquid (NAPL) (QOYANG et al., 1995).

5 Different Ways to Consider the Lack of Bioavailability

The lack of bioavailability has been considered so far primarily as a severe drawback for bioremediation. However, one should also be aware of the fact that pollutants that are not bioavailable in the aqueous phase of the soil pores can also not cause harmful effects in soil or ground water. The *degrading microorganisms themselves* may be the first *beneficiaries* of the lack of bioavailability since the inhibitory toxic effects of excess amounts of pollutants may also be reduced. Low pollutant feeding rates may, therefore, also help to maintain the biodegradation under certain conditions. But *also from an ecotoxicological point* of view the lack of bioavailability can be considered more an advantage than a problem. Supposing that the bioavailable fraction of the pollutant pool is eliminated first by active bioremediation treatment, one can assume that the residual non-available pollutant fraction is so stable or "slow" that the immediate danger of (eco-) toxicity is drastically reduced. Even if small amounts of pollutants are constantly released by the residual pollutant pool sources (e.g., sorption places, tar balls, crystals) one can assume that the "soil bioreactor" can handle such inputs by its permanent intrinsic bioremediation mechanisms.

The permanent presence of pollutants in soil that are unavailable for immediate biodegradation *requires, however, continuous tests of the fate* and mobility of this residual pollutant pool. An indirect approach to monitor the status of bioavailability is to use soil-related (eco-)toxicity tests as they have been decribed by a DECHEMA working group (DOTT and HUND, 1995). *Ecotoxicological monitoring* can be used to check for the presence of toxic pollutants or pollutant derivatives in the aqueous eluates, since only compounds that are available to the target organisms can be taken up and elicit toxic effects in the test organisms. Another, more direct possibility to test bioavailability is to *use reporter bacteria* that carry specific reporter gene operons. The rationale of that approach is, that the reporter gene operon is made sensitive to regulation by at least one of the xenobiotics to be degraded. Only those compounds that are bioavailable in the aqueous phase and that can, therefore, be taken up by the cell can then elicit the reporter gene signal by activating the corresponding gene operon. The reporter gene used in that way most frequently is the bioluminescence gene operon (HEITZER et al., 1994; KING et al., 1990; PEITZSCH et al., 1998; PRESTON et al., 1998; SELIFONOVA et al., 1993; STICHER et al., 1997; TAURIANEN et al., 1997). The advantage of the bioluminescence reporter gene is that its activity can be measured routinely by different techniques like luminometry or photon imaging (MAHRO et al., 1992; PETERMANN et al., 1993; PROSSER et al., 1996). The *lux* operon has meanwhile been coupled to a number of different promoters of degradation operons and has also been transformed in several different host bacteria, so that one can choose from a number of different reporter bacteria (PROSSER, 1994).

However, a necessary prerequisite to consider the lack of bioavailability more as a friend than a foe is that the authorities are willing to *accept small but stabilized residual pollutant levels as sufficiently safe,* although they are still extractable by strong organic solvents. The experience that has been made so far in the context of bioavailability research on soil pollutants suggests, that it is sufficient to assess the actual risk of non-available residual soil pollutants by the combined use of both, ecotoxicity measurements and the application of more natural elution methods. *Mild extraction methods* reflect the actual status of bioavailability and risk better (KELSEY and ALEXANDER, 1997; KELSEY et al., 1997; KÖRDEL and HUND, 1998; VOLKERING et al., 1998) than exhaustive organic soil extractions and can, therefore, also be used to define the acceptable residual pollutant concentration in soil. Such a more pragmatic assessment of the presence, fate, and actual risk of residual soil pollutants has meanwhile become the base for more passive soil remediation concepts called "intrinsic bioremediation" or "natural attenuation". These concepts are presented and discussed in more detail in Chapter 18, this volume.

6 References

ABDUL, S. A., GIBSON, T. L., RAI, D. N. (1990), Use of humic acid solution to remove organic contaminants from hydrogeologic systems, *Environ. Sci. Technol.* **24**, 328–333.

AFFERDEN, M. VAN, BEYER, M., KLEIN, J. (1992), Significance of bioavailability for the microbial remediation of PAH-contaminated soils, in: *Preprints of the International Symposium Soil Decontamination Using Biological Processes* (DECHEMA, Ed.), pp. 605–610. Frankfurt: DECHEMA.

ALEXANDER, M. (1994), *Biodegradation and Bioremediation.* San Diego, CA: Academic Press.

ALEXANDER, M., SCOW, K. M. (1989), Kinetics of biodegradation in soil, in: *Reactions and Movement of Organic Chemicals in Soils* (SAWHNEY, B. L., BROWN, K., Eds.), Soil Science Society of America (SSSA), Special Publication No. 22., pp. 243–269. Madison, WI: SSSA.

ALLEMAN, B. C., LEESON, A. (Eds.) (1997), *In situ* and on site bioremediation. 4th Int. *In Situ* and On Site Bioremediation Symp. Columbus, OH: Battelle Press.

BECK, A. J., JOHNSTON, A. E. J., JONES, K. C. (1993), Movement of non-ionic organic chemicals in agricultural soils, *Crit. Rev. Environ. Sci. Technol.* **23**, 219–248.

BEWLEY, R. J. F., THEILE, P. (1988), Dekontaminierung eines Gaswerksgeländes durch Einsatz von Mikroorganismen, in: *Altlastensanierung '88.* (WOLF, K., VAN DEN BRINK, W. J., COLON, F. J., Eds.), pp. 755–758. Dordrecht: Kluwer Academic Publisher.

BONKHOFF, K., HAEGEL, F. H., THIELE, P., SUBKLEW, G., GERHEIM, J. et al. (1996), Bodensanierung mit Mikroemulsionen – von der Extraktion bis zur biologischen Nachsorge, in: *Schriftenreihe Biologische Abwasserreinigung* Vol. 7 (TU Berlin, Ed.), pp. 123–140. Berlin: Technical University Berlin, Germany.

BOSMA, T. N. P., MIDDELDORP, P. J. M., ZEHNDER, A. J. B. (1997), Mass transfer limitation of biotransformation: Quantifying bioavailability, *Environ. Sci. Technol.* **31**, 248–252.

BOSSERT, I., KACHEL, M. W., BARTHA, R. (1984), Fate of hydrocarbons during oily sludge disposal in soil, *Appl. Environ. Microbiol.* **47**, 763–767.

BOURG, A. C. M. (1988), Metals in aquatic and terrestrial systems: sorption, speciation and mobilization, in: *Chemistry and Biology of Solid Waste* (SALOMONS, W., FÖRSTNER, U., Eds.), pp. 3–32. Berlin: Springer-Verlag.

BOYD, S. A., SUN, S. (1990), Residual petroleum and polychlorobiphenyl oils as sorptive phases for organic contaminants in soils. *Environ. Sci. Technol.* **24**, 142–144.

BRUSSEAU, M. L., WANG, X., HU, Q. (1994), Enhanced transport of low-polarity organic compounds through soil by cyclodextrin, *Environ. Sci. Technol.* **28**, 952–956.

CARMICHAEL, L. M., CHRISTMAN, R. F., PFAENDER, F. K. (1997), Desorption and mineralization kinetics of phenanthrene and chrysene in contaminated soils, *Environ. Sci. Technol.* **31**, 126–132.

CHIOU, C. T. (1989), Theoretical considerations of the partition uptake of non-ionic organic compounds by soil organic matter, in: *Reactions and Movement of Organic Chemicals in Soils* (SAWHNEY, B. L., BROWN, K., Eds.), Soil Science Society of America (SSSA), Special Publication No. 22., pp. 1–30. Madison, WI: SSSA.

DAGASTINO, L., GOODMAN, A. E., MARSHALL, K. C.

(1991), Physiological responses induced in bacteria adhering to surfaces, *Biofouling* **4**, 113-119.

DAMKÖHLER, G. (1937), Influence of diffusion, fluid flow and heat transport on the yield in chemical reactors, *Der Chemie Ingenieur* **3**, 359–485 [translated version in: Int. Chem. Eng. **28**, 132-198 (1988)].

DAVIES, D. G., CHAKRABARTY, A. M., GEESEY, G. G. (1993), Exopolysaccharide production in biofilms: substratum activation of alginate gene expression by *Pseudomonas aeruginosa, Appl. Environ. Microbiol.* **59**, 1181–1186.

DECHEMA (1992), *Preprints* Int. Symp. "Soil Decontamination Using Biological Processes". Karlsruhe, Dec. 6–9 1992. Frankfurt: DECHEMA.

DESCHENES, L., LAFRANCE, P., VILLENEUVE, J. P., SAMSON, R. (1996), Adding sodium dodecyl sulfate and *Pseudomonas aeruginosa* UG2 biosurfactants inhibits polycyclic aromatic hydrocarbon biodegradation in a weathered creosote-contaminated soil, *Appl. Microbiol. Biotechnol.* **46**, 638–646.

DEZIEL, E., PAQUETTE, G., VILLEMUR, R., LÉPINE, F., BISAILLON, J. G. (1996), Biosurfactant production by a soil *Pseudomonas* strain growing on polycyclic aromatic hydrocarbons, *Appl. Environ. Microbiol.* **62**, 1908–1912.

DFG (1998), Pesticide bound residues in soil (Senate Commission for the Assessment of Chemicals Used in Agriculture, Ed.), *Report* 2. Weinheim: Wiley-VCH.

DOTT, W., HUND, K. (1995) Biologische Testmethoden für Böden, in: *DECHEMA Fachgespräche Umweltschutz* (KREYSA, G., WIESNER, J., Eds.), 4. Bericht des interdisziplinären Arbeitskreises Umweltbiotechnologie – Boden. Frankfurt: DECHEMA.

EFROYMSON, R. A., ALEXANDER, M. (1991), Biodegradation by an *Arthrobacter* species of hydrocarbons partitioned into an organic solvent, *Appl. Environ. Microbiol.* **57**, 1441–1447.

EFROYMSON, R. A., ALEXANDER, M. (1995), Reduced mineralization of low concentrations of phenanthrene because of sequestering in nonaqueous-phase liquids, *Environ. Sci. Technol.* **29**, 515–521.

ENGEBRETSON, R. R., AMOS, T., VON WANDRUSZKA, R. (1996), Quantitative approach to humic acid associations, *Environ. Sci. Technol.* **30**, 990–997.

ERICKSON, D. C., LOEHR, R. C., NEUHAUSER, E. F. (1993), PAH loss during bioremediation of manufactured gas plant site soils, *Water Res.* **27**, 911–919.

ESCHENBACH, A., KÄSTNER, M., BIERL, R., SCHAEFER, G., MAHRO, B. (1994), Evaluation of a new and more effective method to extract polycyclic aromatic hydrocarbons from soil samples, *Chemosphere* **28**, 683–692.

ESCHENBACH, A., WIENBERG, R., MAHRO, B. (1998),

Fate and stability of non-extractable residues of [^{14}C]PAH in contaminated soils under environmental stress conditions, *Environ. Sci. Technol.* **32**, 2585–2590.

FEITKENHAUER, H., HEBENBROCK, S., TERSTEGEN, L., SCHNICKE, S., SCHÖB, T. et al. (1996), Bodenreinigung mit thermophilen Mikroorganismen, in: *Neue Techniken der Bodenreinigung* (STEGMANN, R., Ed.), Hamburger Berichte 10, pp. 361–372. Bonn: Economica Verlag.

FOGHT, J. M., GUTNICK, D. L., WESTLAKE, D. W. S. (1989), Effect of Emulsan on biodegradation of crude oils by pure and mixed bacterial cultures, *Appl. Environ. Microbiol.* **55**, 36–42.

FÖRSTNER, U. (1997), Ingenieurgeochemische Konzepte für Altlasten, in: *Chemie und Biologie der Altlasten* (Fachgruppe Wasserchemie in der GDCh, Ed.), pp. 140–154. Weinheim: VCH.

FRITSCHE, W. (1998), *Umweltmikrobiologie.* Jena: Gustav Fischer Verlag.

FZK/TNO (1998), Contaminated soil '98, *Proc. 6th Int. FZK/TNO Conf. on Contaminated Soil*, 17–21 May 1998, Edinburgh, UK. London: Thomas Telford Publishing.

GEERDINK, M., VAN LOOSDRECHT, M., LUYBEN, K. C. A. M. (1996), Model for microbial degradation of non-polar organic contaminants in a soil slurry reactor, *Environ. Sci. Technol.* **30**, 779–786.

GEERDINK, M., HARDEVELT, E., SCHOUTEN, I., VAN LOOSDRECHT, M., LUYBEN, K. C. A. M. (1993), Microbial treatment of contaminated soils in a slurry reactor, in: *Contaminated Soils '93* (ARENDT, F., ANNOKKEE, G. J., BOSMAN, R., VAN DEN BRINK, W. J., Eds.), pp. 487–488. Dordrecht: Kluwer Academic Publishers.

GHOSHAL, S., LUTHY, R. G. (1996), Bioavailability of hydrophobic organic compounds from non-aqueous-phase liquids: The biodegradation of naphthalene from coal tar, *Environ. Toxicol. Chem.* **15**, 1894–1900.

GORDON, A. S., MILLERO, F. J. (1985), Adsorption mediated decrease in the biodegradation rate of organic compounds, *Microb. Ecol.* **11**, 289–298.

GRIMBERG, S. J., STRINGFELLOW, W. T., AITKEN, M. D. (1996), Quantifying the biodegradation of phenanthrene by *Pseudomonas stutzeri* P16 in the presence of a non-ionic surfactant, *Appl. Environ. Microbiol.* **62**, 2387–2392.

GUERIN, W. F., BOYD, S. A. (1992), Differential bioavailability of soil-sorbed naphthalene to two bacterial species, *Appl. Environ. Microbiol.* **58**, 1142–1152.

GUHA, S., JAFFÉ, P. R. (1996a), Bioavailability of hydrophobic compounds partitioned into the micellar phase of non-ionic surfactants, *Environ. Sci. Technol.* **30**, 1382–1391.

GUHA, S., JAFFÉ, P. R. (1996b), Biodegradation kinetics of phenanthrene partitioned into the micellar phase of non-ionic surfactants, *Environ. Sci. Technol.* **30**, 605–611.

HARMS, H. (1996), Bacterial growth on distant naphthalene diffusing through water, air, and water-saturated and non-saturated porous media, *Appl. Environ. Microbiol.* **62**, 2286–2293.

HARMS, H., ZEHNDER, A. J. B. (1994), Influence of substrate diffusion on biodegradation of dibenzofuran and 3-chlorodibenzofuran by attached and suspended bacteria, *Appl. Environ. Microbiol.* **60**, 2736–2745.

HARMS, H., ZEHNDER, A. J. B. (1995), Bioavailability of sorbed 3-chlorodibenzofuran, *Appl. Environ. Microbiol.* **61**, 27–33.

HASSETT, J. J., BANWART, W. L. (1989), The sorption of non-polar organics by soils and sediments, in: *Reactions and Movement of Organic Chemicals in Soils* (SAWHNEY, B. L., BROWN, K., Eds.), Soil Science Society of America (SSSA), Special Publication No. 22., pp. 31–44. Madison, WI: SSSA.

HEITZER, A., MALACHOWSKY, K., THONNARD, J. E., BIENKOWSKI, P. R., WHITE, D. C., SAYLER, G. S. (1994), Optical biosensor for environmental on-line monitoring of naphthalene and salicylate bioavailability with an immobilized bioluminescent catabolic reporter bacterium, *Appl. Environ. Microbiol.* **60**, 1487–1494.

HERMAN, D. C., ZHANG, Y., MILLER, R. M. (1997), Rhamnolipid (biosurfactant) effects on cell aggregation and biodegradation of residual hexadecane under saturated flow conditions, *Appl. Environ. Microbiol.* **63**, 3622–3627.

JIMENEZ, I., BARTHA, R. (1996), Solvent-augmented mineralization of pyrene by a *Mycobacterium* sp., *Appl. Environ. Microbiol.* **62**, 2311–2316.

JOHNSON, W. P., AMY, G. L. (1995), Facilitated transport and enhanced desorption of polycyclic aromatic hydrocarbons by natural organic matter in aquifer sediments, *Environ. Sci. Technol.* **29**, 807–817.

KÄPPELI, O., WALTHER, P., MUELLER, M., FIECHTER, A. (1984), Structure of the cell surface of the yeast *Candida tropicalis* and its relation to hydrocarbon transport, *Arch. Microbiol.* **138**, 279–292.

KELSEY, J. W., ALEXANDER, M. (1997), Declining bioavailability and inappropriate estimation of risk of persistent compounds, *Environ. Toxicol. Chem.* **16**, 582–585.

KELSEY, J. W., KOTTLER, B. D., ALEXANDER, M. (1997), Selective chemical extractants to predict bioavailability of soil-aged organic chemicals, *Environ. Sci. Technol.* **31**, 214–217.

KING, J. M. H., DiGRAZIA, P. M., APPLEGATE, B., BURLAGE, R., SANSEVERINO, J. et al. (1990), Rapid, sensitive bioluminescence reporter technology for naphthalene exposure and biodegradation, *Science* **249**, 778–781.

KLEIN, W., KÖRDEL, W. (1984), Evaluation of the bio-availability of chemical substances in relation to the effects of these substances on living organisms, in: *EEC Report No. 84-B-6602* (EEC, Ed.). Bruxelles: EEC.

KNAEBEL, D. B., FEDERLE, T. W., MCAVOY, D. C., VESTAL, J. R. (1994), Effect of mineral and organic soil constituents on microbial mineralization of organic compounds in a natural soil, *Appl. Environ. Microbiol.* **60**, 4500–4508.

KOCH, A. L. (1982), Multistep kinetics: choice of models for the growth of bacteria, *J. Theor. Biol.* **98**, 401–417.

KOCH, A. L. (1990), Diffusion: the crucial process in many aspects of the biology of bacteria, *Adv. Microb. Ecol.* **11**, 37–70.

KÖHLER, A., BRYNIOK, D., EICHLER, B., SCHÜTTOFF, M., KNACKMUSS, H. J. (1992), Mikrobieller Abbau polycyclischer aromatischer Kohlenwasserstoffe. II: Ein einfaches mathematisches Modell zum Abbau von kristallinem Phenanthren und seine Anwendung zur Steigerung des Abbaus in einem 2-Phasen-System, in: *Mikrobiologische Reinigung von Böden: Beiträge des 9. DECHEMA-Fachgesprächs Umweltschutz* (BEHRENS, D., WIESNER, J., Eds.), pp. 279–283, am 27.–28. Feb. Frankfurt: DECHEMA.

KÖRDEL, W., HUND, K. (1998), Soil extraction methods for the assessment of the ecotoxicological risk of soils, in: Contaminated Soil '98, *Proc. 6th Int. FZK/TNO Conf.* (FZK/TNO, Eds.), pp. 241–250. London: Thomas Telford Publishing.

KÖRDEL, W., DASSENAKIS, M., LINTELMANN, J., PADBERG, S. (1997), The importance of natural organic material for environmental processes in waters and soils (International Union of Pure and Applied Chemistry: Commission on Soil and Water Chemistry, Ed.), *Pure Appl. Chem.* **69**, 1571–1600.

LAHA, S., LUTHY, R. G. (1991), Inhibition of phenanthrene mineralization by non-ionic surfactants in soil-water systems, *Environ. Sci. Technol.* **25**, 1920–1930.

LANE, W. F., LOEHR, R. C. (1992), Estimating the equilibrium aqueous concentrations of polynuclear aromatic hydrocarbons in complex mixtures, *Environ. Sci. Technol.* **26**, 983–990.

LEE, L. S., HAGWELL, M., DELFINO, J. J., RAO, S. C. (1992a), Partitioning of polycyclic aromatic hydrocarbons from diesel fuel into water, *Environ. Sci. Technol.* **26**, 2104–2110.

LEE, L. S., RAO, S. C., OKUDA, I. (1992b), Equilibrium partitioning of polycyclic aromatic hydrocarbons from coal tar into water, *Environ. Sci. Technol.* **26**, 2110–2115.

LIU, Z., JACOBSON, A. M., LUTHY, R. G. (1995), Biodegradation of naphthalene in aqueous non-ionic surfactant systems, *Appl. Environ. Microbiol.* **61**, 145–151.

LUTHY, R. G., RAMASWAMI, A., GHOSHAL, S., MER-

KEL, W. (1993), Interfacial films in coal tar non-aqueous-phase liquid–water systems, *Environ. Sci. Technol.* **27**, 2914–2918.

LUTHY, R. G., DZOMBAK, D. A., PETERS, C. A., ROY, S. B., RAMASWAMI, A. et al. (1994), Remediating tar-contaminated soils at manufactured gas plant sites, *Environ. Sci. Technol.* **28**, 266–276.

LUTHY, R. G., AIKEN, G. R., BRUSSEAU, M. L., CUNNINGHAM, S. D., GSCHWEND, P. M. et al. (1997), Sequestration of hydrophobic contaminants by geosorbents, *Environ. Sci. Technol.* **31**, 3341–3347.

MAHRO, B., KÄSTNER, M. (1993), Der mikrobielle Abbau polyzyklischer aromatischer Kohlenwasserstoffe (PAK) in Böden und Sedimenten: Mineralisierung, Metabolitenbildung und Entstehung gebundener Rückstände, *Bioengineering* **9**, 50–58.

MAHRO, B., SCHAEFER, G. (1998), Bioverfügbarkeit als limitierender Faktor des mikrobiellen Abbaus von PAK im Boden – Ursachen des Problems und Lösungsstrategien, *Altlastenspektrum* **7**, 127–134.

MAHRO, B., PETERMANN-HERMANN, A., WALENTA, S., MÜLLER-KLIESER, W., KASCHE, V. (1992), Non-extractive localization and imaging of luminescent bacteria in liquid and soil samples by luminescence microscopy, *Microb. Releases* **1**, 79–85.

MAHRO, B., KÄSTNER, M., SCHAEFER, G. (1994), Pathways of microbial degradation of polycyclic aromatic hydrocarbons in soil, in: *Bioremediation of Chlorinated and Polycyclic Aromatic Hydrocarbon Compounds* (HINCHEE, R. E., LEESON, A., SEMPRINI, L., ONG, S. K., Eds.), pp. 203–217. Boca Raton, FL: Lewis Publishers.

MAHRO, B., ESCHENBACH, A., SCHAEFER, G., KÄSTNER, M. (1996), Possibilities and limitations for the microbial degradation of polycyclic aromatic hydrocarbons (PAH) in soil, in: *Wider Application and Diffusion of Bioremediation Technologies*, The Amsterdam '95 Workshop (OECD, Ed.), pp. 297–307. Paris: OECD.

MAHRO, B., SCHMIDT, L., ESCHENBACH, A. (1999), Möglichkeiten und Grenzen mikrobiologischer Verfahren bei der Sanierung kontaminierter Böden, in: *Biotechnologie im Umweltschutz. Bioremediation: Entwicklungsstand, Anwendungen, Perspektiven* (HEIDEN, S., ERB, R., WARRRELMANN, J., DIERSTEIN, R., Eds.), pp. 99–107. Berlin: E. Schmidt-Verlag.

MANILAL, V. B., ALEXANDER, M. (1991), Factors affecting the microbial degradation of phenanthrene in soil, *Appl. Microbiol. Biotechnol.* **35**, 401–405.

MANN, V. G., PFEIFER, F., SINDER, C., KLEIN, J. (1996), Reinigung PAK-kontaminierter, feinkörniger Böden in Suspensionsreaktoren, in: *Schriftenreihe Biologische Abwasserreinigung*, Vol. 7 (TU Berlin, Ed.), pp. 141–158. Berlin: Technical University Berlin, Germany.

MARIN, M., PEDREGOSA, A., LABORDA, F. (1996),

Emulsifier production and microscopical study of emulsions and biofilms formed by the hydrocarbon-utilizing bacteria *Acinetobacter calcoaceticus* MM5, *Appl. Microbiol. Biotechnol.* **44**, 660–667.

MARTEL, R., GELINAS, P. J., DESNOYERS, J. E., MASSON, A. (1993), Phase diagrams to optimize surfactant solutions for oil and DNAPL recovery in aquifers, *Ground Water* **31**, 789–800.

MEANS, J. C., WOOD, S. G., HASSETT, J. J., BANWART, W. L. (1980), Sorption of polynuclear hydrocarbons by sediments and soils, *Environ. Sci. Technol.* **14**, 1525–1528.

MIETHE, D., RIIS, V., BABEL, W. (1996), Zum Problem der Restkonzentration beim mikrobiellen Abbau von Mineralölen, in: *Neue Techniken der Bodenreinigung*, Hamburger Berichte Nr. 10 (STEGMANN, R., Ed.), pp. 289–302. Bonn: Economica Verlag.

MIHELCIC, J. R., LUEKING, R., MITZELL, R. J., STAPLETON, J. M. (1993), Bioavailability of sorbed and separate phase chemicals, *Biodegradation* **4**, 141–153.

MUELLER, J. G., LANTZ, S. E., BLATTMAN, B. O., CHAPMAN, P. J. (1991a), Bench-scale evaluation of alternative biological treatment processes for the remediation of pentachlorophenol- and creosote-contaminated materials: solid phase bioremediation, *Environ. Sci. Technol.* **25**, 1045–1055.

MUELLER, J. G., LANTZ, S. E., BLATTMAN, B. O., CHAPMAN, P. J. (1991b), Bench-scale evaluation of alternative biological treatment processes for the remediation of pentachlorophenol- and creosote-contaminated materials: slurry phase bioremediation, *Environ. Sci. Technol.* **25**, 1055–1061.

MULDER, H., BREURE, A. M., RULKENS, W. H. (1998), Bioremediation potential as influenced by the physical states of PAH pollutants, in: Contaminated Soil '98, *Proc. 6th Int. FZK/TNO Conf.* (FZK/TNO, Eds.), pp. 133–142. London: Thomas Telford Publishing.

NAM, K., ALEXANDER, M. (1998), Role of nanoporosity and hydrophobicity in sequestration and bioavailability: tests with model solids, *Environ. Sci. Technol.* **32**, 71–74.

NELSON, E. C., GHOSHAL, S., EDWARDS, J. C., MARSH, G. X., LUTHY, R. G. (1996), Chemical characterization of coal tar–water interfacial films, *Environ. Sci. Technol.* **30**, 1014–1022.

NELSON, E. C., WALTER, M. V., BOSSERT, I. D., MARTIN, D. G. (1996), Enhancing biodegradation of petroleum hydrocarbons with guanidium fatty acids, *Environ. Sci. Technol.* **30**, 2406–2411.

OECD (1996), Wider Application and Diffusion of Bioremediation Technologies, The Amsterdam '95 Workshop. Paris: OECD Documents.

OGRAM, A. V., JESSUP, L. T., OU, L. T., RAO, P. S. C. (1985), Effects of sorption on biological degradation rates of (2,4-dichlorophenoxy)acetic acid in soils, *Appl. Environ. Microbiol.* **49**, 582–587.

OUYANG, Y., MANSELL, R. S., RHUE, R. D. (1995), Emulsion-mediated transport of non-aqueous-phase liquid in porous media: a review, *Crit. Rev. Environ. Sci. Technol.* **25**, 269–290.

PEITZSCH, N., EBERZ, G., NIES, D. H. (1998), *Alcaligenes eutrophus* as a bacterial chromate sensor, *Appl. Environ. Microbiol.* **64**, 453–458.

PETERMANN, A., SCHLOBOHM, I., MAHRO, B. (1993), Evaluation of different techniques to monitor bioluminescent microorganisms in soil or non-marine water samples, in: *Contaminated Soils '93* (ARENDT, F., ANNOKKEE, G. J., BOSMAN, R., VAN DEN BRINK, W. J., Eds.), pp. 965–966. Dordrecht: Kluwer Academic Publishers.

PETERS, R. W., MONTEMAGNO, C. D., SHEM, L. (1992), Surfactant screening of diesel-contaminated soil, *Hazardous Waste Hazardous Materials* **9**, 113–136.

PIGNATELLO, J. J., XING, B. (1996), Mechanisms of slow sorption of organic chemicals to natural particles, *Environ. Sci. Technol.* **30**, 1–11.

PRESTON, S., BARBOSA-JEFFERSON, V. L., ZHANG, H., TYE, A., CROUT, N. et al. (1998), Assessment of the bioavailability of soil pollutants using lux-based biosensors: an interdisciplinary approach, in: Contaminated Soil '98, *Proc. 6th Int. FZK/TNO Conf.* (FZK/TNO, Eds.), pp. 123–132. London: Thomas Telford Publishing.

PROSSER, J. I. (1994), Molecular marker systems for detecting of genetically engineered microorganisms in the environment, *Microbiology* **140**, 5–17.

PROSSER, J. I., KILLHAM, K., GLOVER, L. A., RATTRAY, E. A. S. (1996), Luminescence-based systems for detection of bacteria in the environment, *Crit. Rev. Biotechnol.* **16**, 157–183.

PROVIDENTI, M. A., FLEMMING, C. A., LEE, H., TREVORS, J. T. (1995), Effect of addition of rhamnolipid biosurfactants or rhamnolipid-producing *Pseudomonas aeruginosa* on phenanthrene mineralization in soil slurries, *FEMS Microbiol. Lett.* **17**, 15–26.

RIEGER, P. G., KNACKMUSS, H. J. (1995), Basic knowledge and perspectives on biodegradation of 2,4,6-trinitrotoluene and related nitroaromatic compounds in contaminated soil, in: *Biodegradation of Nitroaromatic Compounds* (SPAIN, J., Ed.), pp. 1–18. New York: Plenum Press.

ROBERTSON, B. K., ALEXANDER, M. (1996), Mitigating toxicity to permit bioremediation of constituents of non-aqueous-phase liquids, *Environ. Sci. Technol.* **30**, 2066–2070.

ROSENBERG, M., ROSENBERG, E. (1981), Role of adherence in growth of *Acinetobacter calcoaceticus* RAG-1 on hexadecane, *J. Bacteriol.* **148**, 51-57.

ROSENBERG, M., BEYER, E. A., DELAREA, J., ROSENBERG, E. (1982), Role of thin fimbriae in adherence and growth of *Acinetobacter calcoaceticus*

RAG-1 on hexadecane, *Appl. Environ. Microbiol.* **44**, 929–937.

ROUSE, J. D., SABATINI, D. A., SUFLITA, J. M., HARWELL, J. H. (1994), Influence of surfactants on microbial degradation of organic compounds, *Crit. Rev. Environ. Sci. Technol.* **24**, 325–370.

SAWHNEY, B. L., BROWN, K., Eds. (1989), *Reactions and Movement of Organic Chemicals in Soils*, Soil Science Society of America (SSSA), Special Publication No. 22. Madison, WI: SSSA.

SCHACHTSCHABEL, P., BLUME, H. P., BRÜMMER, G., HARTGE, K. H., SCHWERTMANN, U. (1992), *Lehrbuch der Bodenkunde*. Stuttgart: Ferdinand Enke Verlag.

SCHAEFER, G., HATTWIG, S., UNTERSTE-WILMS, M., HUPE, K., HEERENKLAGE, J. et al. (1995), PAH-degradation in soil: microbial activation or inoculation. A comparative evaluation with different supplements and soil materials, in: *Contaminated Soil '95* (VAN DEN BRINK, W. J., BOSMAN, R., ARENDT, F., Eds.), pp. 415–416. Dordrecht: Kluwer Academic Publ.

SCHWARTZ, A., BAR, R. (1995), Cyclodextrin-enhanced degradation of toluene and p-toluic acid by *Pseudomonas putida*, *Appl. Environ. Microbiol.* **61**, 2727–2731.

SCOTT, C. C. L., FINNERTY, W. R. (1976), A comparative analysis of the ultrastructure of hydrocarbon-oxidizing microorganisms, *J. Gen. Microbiol.* **94**, 342–350.

SEITINGER, P., BAUMGARTNER, A., SCHINDLBAUER, H. (1994), Die Ausbreitung von Mineralölkontaminationen im Untergrund, *Erdöl Erdgas Kohle* **110**, 211–215.

SELIFONOVA, O., BURLAGE, R., BARKAY, T. (1993), Bioluminescent sensors for detection of bioavailable Hg(II) in the environment, *Appl. Environ. Microbiol.* **59**, 3083–3090.

SIKKEMA, J., DE BONT, J. A. M., POOLMAN, B. (1995), Mechanisms of membrane toxicity of hydrocarbons, *Microbiol. Rev.* **59**, 201–222.

SOEDER, C. J., PAPADEROS, A., KLEESPIES, M., KNEIFEL, H., HAEGEL, F. H., WEBB, L. (1996), Influence of phytogenic surfactants (quillaya saponin and soya lecithin) on bioelimination of phenanthrene and fluoranthene by three bacteria, *Appl. Microbiol. Biotechnol.* **44**, 654-659.

SOUTHWORTH, G. R., HERBES, S. E., ALLEN, C. P. (1983), Evaluation of a mass transfer model for the dissolution of organics from oil films into water, *Water Res.* **17**, 1647–1651.

STEILEN, N., HEINKELE, T. H., REINEKE, W., NECKER, U., ODENSASS, M., WILLERSHAUSEN, K. H. (1993), Ergebnisse von Feldversuchen zur Behandlung eines PAK-belasteten Gaswerksbodens mit verschiedenen mikrobiologischen Mietenverfahren, *Altlastenspektrum* **3**, 152–163.

STICHER, P., JASPERS, M. C. M., STEMMLER, K.,

HARMS, H., ZEHNDER, A., VAN DER MEER, J. R. (1997), Development and characterization of a whole-cell bioluminescent sensor for bioavailable middle chain alkanes in contaminated groundwater samples, *Appl. Environ. Microblol.* **63**, 4053–4060.

STIEBER, M., TIEHM, A., WERNER, P., FRIMMEL, F. H. (1994), PAK-Abbau durch eine Bakterien-Mischkultur in Bodensäulen, in: *Schriftenreihe Biologische Abwasserreinigung*, Vol. 4 (TU Berlin, Ed.), pp. 69–89. Berlin: Technical University Berlin, Germany.

STUCKI, G., ALEXANDER, M. (1987), Role of dissolution rate and solubility in biodegradation of aromatic compounds, *Appl. Environ. Microbiol.* **53**, 292–297.

SUN, S., BOYD, S. A. (1993), Sorption of non-ionic organic compounds in soil–water systems containing petroleum sulfonate–oil surfactants, *Environ. Sci. Technol.* **27**, 1340–1346.

SUN, S., INSKEEP, W. P., BOYD, S. A. (1995), Sorption of non-ionic organic compounds in soil–water systems containing a micelle-forming surfactant, *Environ. Sci. Technol.* **29**, 903–913.

TABAK, H. H., GOVIND, R., FU, C., YAN, X., GAO, C., PFANSTIEL, S. (1997), Development of bioavailability and biokinetics determination methods for organic pollutants in soil to enhance *in situ* and on site bioremediation, *Biotechnol. Prog.* **13**, 43–52.

TAURIANEN, S., KARP, M., CHANG, W., VIRTA, M. (1997), Recombinant luminescent bacteria for measuring bioavailable arsenite and antimonite, *Appl. Environ. Microbiol.* **63**, 4456–4461.

THIBAULT, S. L., ANDERSON, M., FRANKENBERGER, W. T. (1996), Influence of surfactants on pyrene desorption and degradation in soils, *Appl. Environ. Microbiol.* **62**, 283–287.

THOMAS, J. M., YORDY, J. R., AMADOR, J. A., ALEXANDER, M. (1986), Rates of dissolution and biodegradation of water-insoluble organic compounds, *Appl. Environ. Microbiol.* **52**, 290–296.

TIEHM, A. (1994), Degradation of polycyclic aromatic hydrocarbons in the presence of synthetic surfactants, *Appl. Environ. Microbiol.* **60**, 258–263.

Umweltbundesamt (1999), *Tagungsbericht zum 1. Statusseminar des BMBF-Verbundvorhabens "Langzeitstabilität und Remobilisierungsverhalten von Schadstoffen".* Berlin: Umweltbundesamt.

VAN DER SLOOT, H. A., DE GROOT, G. J. (1988), Mobility of trace elements derived from combustion residues and products containing these residues in soil and ground water, in: *Report of the Commission of the EC*, Directorate General for Science, Research and Development XII/E6. Pet-

ten/Bruxelles: EC-Print.

VOLKERING, F., BREURE, A. M., VAN ANDEL, J. G., RULKENS, W. H. (1995), Influence of non-ionic surfactants on bioavailability and biodegradation of polycyclic aromatic hydrocarbons, *Appl. Environ. Microbiol.* **61**, 1699–1705.

VOLKERING, F., BREURE, A. M., STERKENBURG, A., VAN ANDEL, J. G. (1992), Microbial degradation of polycyclic aromatic hydrocarbons: effect of substrate availability on bacterial growth kinetics, *Appl. Microbiol. Biotechnol.* **36**, 548–552.

VOLKERING, F., QUIST, J. J., VAN VELSEN, A. F. M., THOMASSEN, P. H. G., OLIJVE, M. (1998), A rapid method for predicting the residual concentration after biological treatment of oil-polluted soil, in: Contaminated Soil '98. *Proc. 6th Int. FZK/TNO Conf.* (FZK/TNO, Eds.), pp. 251–259. London: Thomas Telford Publishing.

WEISSENFELS, W. D., KLEWER, H. J., LANGHOFF, J. (1992), Adsorption of polycyclic aromatic hydrocarbons (PAHs) by soil particles: influence on biodegradability and biotoxicity, *Appl. Microbiol. Biotechnol.* **36**, 689–696.

WHITE, J. C., KELSEY, J. W., HATZINGER, P. B., ALEXANDER, M. (1997), Factors affecting sequestration and bioavailability of phenanthrene in soils, *Environ. Toxicol. Chem.* **16**, 2040–2045.

WILLUMSEN, P. A., ARVIN, E. (in press), Kinetics of degradation of surfactant-solubilized fluoranthene by a *Sphingomonas paucimobilis*, *Environ. Sci. Technol.*

WILLUMSEN, P., KARLSON, U., PRITCHARD, P. H. (1998), Response of fluoranthene-degrading bacteria to surfactants, *Appl. Microbiol. Biotechnol.* **50**, 475–483.

WODZINSKI, R. S., BERTOLINI, D. (1972), Physical state in which naphthalene and dibenzyl are utilized by bacteria, *Appl. Microbiol.* **23**, 1077–1081.

WODZINSKI, R. S., COYLE, J. E. (1974), Physical state of phenanthrene for utilization by bacteria, *Appl. Microbiol.* **27**, 1081–1084.

WODZINSKI, R. S., LAROCCA, D. (1977), Bacterial growth kinetics on diphenylmethane and naphthalene-heptamethylnonane mixtures, *Appl. Environ. Microbiol.* **33**, 660–665.

YEOM, I. T., GHOSH, M. M., COX, C. D. (1996), Kinetic aspects of surfactant solubilization of soil-bound polycyclic aromatic hydrocarbons, *Environ. Sci. Technol.* **30**, 1589–1595.

ZHANG, Y., MILLER, R. M. (1995), Effect of rhamnolipid (biosurfactant) structure on solubilization and biodegradation of *n*-alkanes, *Appl. Environ. Microbiol.* **61**, 2247–2251.

ZIECHMANN, W., MÜLLER-WEGENER, U. (1990), *Bodenchemie*. Mannheim: Wissenschaftsverlag.

4 "Humification" Process or Formation of Refractory Soil Organic Matter

MATTHIAS KÄSTNER

Leipzig, Germany

List of Abbreviations

2,4-D   2,4-dichlorophenoxy acetic acid
CAC   cation exchange capacity
IUPAC   International Union of Pure and Applied Chemistry
PAH   polycyclic aromatic hydrocarbons

1 Introduction

Over the past few years, the humification of xenobiotic compounds (anthropogenic or naturally occurring compounds with environmental impacts) has been discussed and sometimes applied as a technique for the elimination and immobilization of xenobiotics in contaminated soils (BOLLAG and LOLL, 1983; BERRY and BOYD, 1985; BOLLAG, 1992; VERSTRAETE and DEVLIEGHER, 1996). Other authors argued that such processes in the organic matrix of soil (humic substances or refractory soil organic matter) may result in a sink for the contaminants containing a potential risk of the remobilization of hazardous compounds, especially hardly degradable compounds such as polycyclic aromatic hydrocarbons (PAH) (HÜTTERMANN et al., 1988; GOßEL and PÜTTMANN, 1990; PÜTTMANN, 1990; STIEBER et al., 1990). Protagonists of humification assume the covalent binding of the compounds to the organic soil matrix after transformation or metabolization by microorganisms. However, critical authors imply sorption and retention of the parent compounds or toxic metabolites. In fact, both processes occur in soil, necessitating precise analysis. The question of whether the incorporation of metabolites into the organic soil matrix is actually a process of humification could not be answered due to the inadequate knowledge about the humification processes. Consequently, the propagated "controlled" humification of xenobiotics in soils is still not available as a reliable remediation technique. However, humification processes are generally thought to immobilize and detoxify the hazardous potential of xenobiotics (BOLLAG and LOLL, 1983; BOLLAG et al., 1988; PARK et al., 1988; WANG et al., 1990). If we are able to shed light on this issue, the reactions which cause the fixation of xenobiotic compounds or the metabolites with the formation of residues in soil are of crucial importance for understanding the carbon flux and microbial metabolism of xenobiotics in soil.

Xenobiotic compounds that enter soil systems are subject to a great variety of interactions with the soil matrix which depend on the properties of the compounds and of the different soil components. Bioavailability and microbial degradation of xenobiotics are substantially affected by these interactions. Besides clay minerals, the organic soil matrix derived from postmortal biopolymers (fragments of plants, animals, and microorganisms) and humic substances play an important role in these interactions. The microbial degradation of plant material in soil is essential for the global carbon cycle and allows mineralization and the recycling of carbon incorporated into biopolymers by fixation of carbon dioxide from the primary production of plants. Degradation by microorganisms leads to the intermediate formation of refractory soil organic matter, which constitutes a considerable sink of organic carbon on earth. This carbon depot, in which dead but not fossilie organic compounds are fixed in soils and sediments, is estimated at $3.7 \cdot 10^{11}$ t and represents more than five times the amount of carbon dioxide in the atmosphere (ATLAS and BARTHA, 1997). The tremendous stock of carbon in humic matter is necessary for primary production by plants in soils, and would not exist without the degrading activity of microorganisms. More than 75% of the carbon from primary production $(1.3 \cdot 10^{11}$ t a$^{-1})$ is metabolized by microbial decay, the formation of humic matter, and subsequent humus turnover in soil. Humic substances and the reactions of humus genesis play an important role during the incorporation of xenobiotics into the soil organic matrix. However, humic substances have not been considered in most experiments on the microbial degradation of organic compounds in soil. Thus, the processes of carbon fixation in humic matter during microbial degradation in soil and the effect of soil organic matter on the activity of microorganisms are still not understood. Although humic substances are also formed in aquatic systems, soils differ from aquatic systems by producing a carbon depot, whereas in aquatic systems mostly a particular precipitation and a C depletion of the water phase are observed. Hence, aquifers and sediments should be considered as underwater soils. The following considerations on humic substances are mainly focused on soil systems and will review current knowledge of humic matter and its effects on the microbial degradation of xenobiotic compounds.

2 Soil and Soil Components

Soil is a highly variable and complex medium with a large number of interaction sites for xenobiotic organic compounds. The mode and extent of interactions depend on the properties of the compounds and on the accompanying matrices (i.e., tar oils or coal and coke particles at sites contaminated by PAH, byproducts of synthesis of explosives like 2,4,6-trinitrotoluol, or others). Depending on the history of contamination, these matrices may vary greatly. In the case of contamination by mineral or tar oils, a fourth-phase (oil) is introduced into a three-phase soil system (gas, water, and solid matter), which alters the interaction conditions of the other soil constituents. The composition, texture, and water content affect the physicochemical behavior of the soil, the turnover of organic matter, and interactions with xenobiotic compounds. Contamination with liquid mineral oils may result in different distributions in the soil system in the event of spillage on dry or wet soil. In the case of dry soil, the particles will be wetted by the oil phase. Sequestering the oil components into the soil pores is facilitated, which supports ageing and immobilization of the oil. However, the oil matrix is much more mobile and bioavailable after oil spillage into wet soil. Hydrophobic compounds like PAH may also be sorbed to dissolved organic matter, which mediate the transport of those compounds within the water phase of the soil (MAGEE et al., 1991; KÖGEL-KNABNER and KNABNER, 1991; DESCHAUER et al., 1994; FRIMMEL et al., 1994; NANNY et al., 1997).

2.1 Soil Components

Soil (pedosphere) as the living upper part of the solid surface on earth (lithosphere) consists of a system of mineral components (weathered rock), organic components (humic matter including living and dead biomass), *gas* within the pore space, and water coating particles and capillary space. The pore water contains dissolved salts and nutrients and is a solvent for the microorganisms. The water contents of soil vary over a wide range, which fundamentally affects microbial activity in soil. The great variety of soil types differ in terms of the level of sand, silt, clay, and humic matter specific to the respective soil type. Mineral components and humic matter are arranged in a specific manner in three dimensions and create a porous structure (Fig. 1). The microcosms within the pore space of particles and aggregates may be colonized by microorganisms.

Natural soil organic matter has complex structures and exists in different kinds in soil. The organic matter can be divided into living biomass (bacteria, fungi, algae, higher animals, and plant roots) and non-living organic substances. The latter consist mainly of fragments or degradation products of biomass and humic substances. Fragments of biomass, low molecular weight products of microbial degradation and existing humic substances *rearrange* in soil to form refractory soil organic matter. During the genesis of refractory soil organic matter, the resistance to microbial attack increases by secondary synthesis reactions of the molecules or by aggregation processes. The formation of humic molecules generally takes place in stabilization and aggregation reactions. In addition, soils may contain black carbon in the form of charcoal-like material as a result of biomass burning or as soot particles from immisions of combustion of fossil fuels. The carbon pools are difficult to characterize and quantify in soils and sediments. The interaction of black carbon fractions with hydrophobic xenobiotics may affect their bioavailability in soils and sediments (GUSTAFSSON and GSCHWEND, 1998). Black carbon in particular

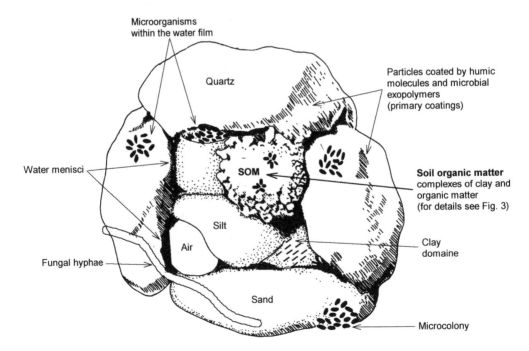

Fig. 1. Aggregation of soil constituents (modified after ATLAS and BARTHA, 1997).

as a component of the xenobiotic matrix-like coal particles within tar oil matrices has a major impact on the bioavailability of hydrophobic xenobiotic compounds (WEIßENFELS et al., 1992).

Non-living fragments of biomass and humic substances are essential for soil fertility and act as a source of fertilizing and growing substrates for plants and other soil organisms. Humic substances have impacts on water holding and ion exchanging capacity, water and gas permeability, and the degree of aggregation of soil particles and colonization by microorganisms (BOLLAG and LOLL, 1983). Humic substances are effective sorbents for both polar cationic organic molecules and lipophilic substances, and cause the retention of such compounds against the water phase of the soil (HASSETT and BANWART, 1989). Although the organic matter fraction is believed mainly to be responsible for the interactions with hydrophobic organic xenobiotics, clay minerals also play an important role. In many cases, both soil constituents are involved (CALDERBANK, 1989; STEVENSON, 1994). The interaction of pesticides and organic xenobiotics with clay minerals has been reviewed several times (WHITE, 1976; THENG, 1982; MORTLAND, 1986; ZIELKE et al., 1989). Reactivity and catalytic properties of clays and metal constituents in soils may generate several oxidative coupling processes of organic molecules to the humic matrix, as discussed in Sects. 4.3 and 4.4 and reviewed by several authors (RUGGIERO, 1998; PAL et al., 1994; SHINDO, 1994). Clays may function as electron donors and acceptors and may be involved in the following reactions:

- transaminations,
- decarboxylations,
- polymerization of phenols, quinones, and aromatic amines,
- synthesis and polymerization of amino acids (BURNS, 1989).

In particular, clay and organic colloids have a large active surface and high cation exchange capacity (CAC) affecting the physical and chemical properties of soils (pH, redox potential, ion concentration). Expanding lattice clays such as montmorillinite and vermiculite provide large external and internal surfaces to interact with organic molecules. They may adsorb organic molecules in their swelling interlayers. The surfaces and cation exchange capacity of other non-expanding clays such as illite and kaolinite are, in general, much smaller (Tab. 1). Due to the porous structure and the amount of clay, the water content varies in soils from a few percent to $\geq 45\%$. In natural soil systems not saturated with water, all particles are usually coated by water layers of different thickness. Water menisci cause the adherence of particles.

2.2 Structures and Aggregation

Soil organic matter interacts with clay to form stable clay organic complexes and aggregates. This interaction has important consequences for the physical, chemical, and biological properties of the soil matrix. In many cases the organic matter contents in soils are closely related to the amount of clay, which

Tab. 1. Physical Properties of Soil Components (modified after RICHNOW et al., 1998; GÖTZ, personal communication)

Soil Constituent	Cation Exchange Capacity [meq kg^{-1}]	Surface Area [m^2 g^{-1}]
Humic substances	180–300	800–1,000
Vermiculite	150–200	600–700
Smectite	70–130	600–800
Illite	20–200	50–200
Koalinite	10–40	25–40
Oxides and hydroxides	20–60	100–800
Sandy soil (<2 mm)	<30	<0.1
Washed sand (<2 mm)	<5	≪0.1

tends to stabilize organic matter. Organic matter can serve to bridge soil particles by forming stable aggregates. Humic substances are bound to clay minerals

(1) by attachment forming stable clay–organic complexes through cation or anion exchange, bridging by polyvalent metal cations or other interactions and

(2) by penetration into expanding interlayers of clay minerals.

Usually, 50% of the organic matter in natural soils or more are fixed within clay–organic complexes (STEVENSON, 1994). More than 95% of the total amount of soil organic matter is connected to inorganic particles; with more than 40% in the silt fraction (2–50 µm) and nearly 60% in the clay fraction (< 0.2–2 µm) (HAIDER, 1998b). Because soil organic matter is closely bound to clays, it is difficult to distinguish between the individual contribution of clay and organic matter to the "humification" processes and the interaction with xenobiotics. Organic substances form coatings on other mineral particles and modify the surface properties of the particles with respect to the sorption of xenobiotics (CALDERBANK, 1989; STEVENSON, 1994). In addition, the presence of reactive functional groups and hydrophobic moieties in humic substances allows chemical and physical associations with hydrophobic and hydrophilic xenobiotics. Soil particles like sand or quartz particles usually have primary coatings of organic molecules either by humic molecules or by microbial exopolymers (Fig. 1).

The refractory organic matter also coats fragments of biomass and may thus stabilize otherwise degradable biomolecules. Fragments and particles of plant biomass are also encrusted by inorganic material like clay particles. Soil particles of 100–200 µm have been shown to release partly degraded plant debris after ultrasonic disruption of the aggregates (OADES, 1993). Macrostructures with biomass fragments or humic clay complexes formed by such mechanisms protect their inner compounds, which become resistant to microbial attack. The type of clay minerals and the size of aggregates affect the bioavailability of degradable organic compounds. In addition, living microorganisms supplied with suitable substrates (i.e., labile plant material) stabilize higher macroaggregates (> 250 µm) by excreting adhesive exopolymers, and fungal hyphae on roots stabilize macroaggregates in the form of "sticky string bags". There is a positive correlation in soil between the organic matter content and aggregation. These results lead to the development of concepts of aggregate hierarchy (OADES, 1993). Clay microstructures, fragments of biomass, and humic particles arrange to form higher, macroscopic aggregation systems with diameters up to several millimeters. At this point, organisms feeding on the soil such as termites or earthworms may have a selective nutritional advantage in being able to break up such aggregates during gut passage. Stabilized labile organic molecules will become exclusively accessible to these organisms, as is discussed by BRUNE (1998). The soil aggregates may be destroyed on a technical scale by disruption through wet and drying processes, or by mixing and tilling procedures (mechanical stress), which cause the increased bioavailability of stabilized labile organic compounds. Disaggregation occurs in a stepwise fashion with increasing energy input (OADES, 1993). However, it was assumed that organic molecules entrapped within pores with a diameter < 100 nm are not available to microorganisms (HAIDER, 1998a).

The aggregate size is also affected by the water content of the soil. The transpiration of plants leads to continuous water transport from the roots to the leaves. The water is transferred from soil particles to the rhizosphere, which causes the continuous drying of soil particles coated and adhered to by water films. This water transfer is accompanied by a chromatography of mobile molecules in the water phase against solid soil matter. Numerous fine roots may dry the soil at countless sites, thus creating many non-oriented cracks in the soil texture which give rise to the granular, crumby structure of soils. The greater the clay content, the greater the shrink–swell capacity and the more vigorous the structural formation during dry–wet cycles (OADES, 1993). In addition, the rhizosphere is influenced by the exudates of plants and by syntrophic interactions with microorganisms (mycorrhiza, root nodules).

2.3 Microorganisms

In line with the dimensions of microorganisms of a few µm, the pores between particles of the soil structure (Fig. 1) represent separate microcosms with different conditions. Within the niches and habitats of these microcosms, various microorganisms compete for C substrates, nutrients, electron donors and acceptors. Easily degradable C or N substrates are usually limited or of limited bioavailability in soil environments with the exception of occasional inputs of dead plant material. However, hardly degradable or refractory organic compounds are available in abundance. Soil microorganisms able to grow on such compounds with low but constant activity are characterized as autochthonous microflora. By contrast, the zymogenic microflora is unable to use humic substances, but is characterized by the ability to grow at high growth rates and activities if easily degradable substrates become available, i.e., by the occasional input of dead plant material. Both types of microorganisms represent the indigenous microflora of a soil (ATLAS and BARTHA, 1997). Organisms of the zymogenic microflora with intermittent activities inevitably have to develop dormant or starvation stages or other resting forms. Besides the spore formation of fungi, actinomycetes, clostridia and bacilli, there are several other adaptations of the respective cells to starvation such as dwarf cells or dormant stages, that are based on considerably lowered metabolic activity (ROSZAK and COLWELL, 1987). On average, microbial biomass amounts to 2–4% of the organic C of a soil (JENKINSON and LADD, 1981), and the numbers of microorganisms in soil are much higher than those found in aquatic habitats. Normally, 10^6–10^9 bacteria per gram of soil are found by cultivating methods. Most types of bacteria and fungi can be detected in soil and wide differences in the relative proportions of individual bacterial genera are encountered in samples from different sites (ALEXANDER, 1977). Fungi are usually found with numbers of colony forming units by two orders of magnitude lower than bacteria. However, fungi represent a high proportion of the microbial biomass in soil (DOMSCH et al., 1980) and may exhibit greater motility than bacteria. Hyphae of basidiomy-

cetes may switch from slow-diffuse to fast-diffuse growth in search of suitable substrates (DOWSON et al., 1988).

When composting plant material, the C–N ratio of the organic matter decreases from >30 at the beginning to values of about 10 after 6 months, indicating that about 65% of the carbon is lost. Due to bacterial C–N ratios of about 5, only a small fraction (max. 16%) of the carbon can be converted into biomass during the degradation of organic matter. The composting process can only be explained by assuming a few consecutive microbial populations during humification each deteriorating at the end of their growth phase and supplying their nitrogen to the next population, as is discussed in detail by KUTZNER (1999). In soil systems, the much slower degradation of plant material and litter in soil systems generate humic substances with a C–N ratio of 10 as well. The C–N ratio of soil usually decreases with decreasing particle size. The particle fraction <0.1 µm of several soils showed C–N ratios of 5, which are close to the C–N ratio of bacteria (STEMMER et al., 1998). These considerations enabled us to estimate that the soil microflora multiplies only by a factor of 1–2 per year. Similar values were calculated by other authors reflecting the annual litter production and the total biomass in a forest soil of a temperate climate (SCHINK, 1998).

Microorganisms that are detected in soil include archaea, eubacteria, actinomycetes and fungi, as well as algae, protozoa, and viruses. Soils are extremely heterogeneous habitats and individual soils may favor populations of microorganisms with particular types of metabolism. All the properties of the particular soil have substantial effects on the carbon distribution, the activity of microorganisms and the possible functions of the soil from a specific site. The soil system is very difficult to characterize because of the different nutritional requirements of various groups of microorganisms. Only 1–10% of the microscopically visible soil bacteria were estimated to be actually culturable by artificial media (PICKUP, 1991; SAUNDERS and SAUNDERS, 1993; LEADBETTER, 1997). However, the question of whether this phenomenon is an artefact of available culture methods or is an inherent quality of the organisms has still to be evaluated (DIXON, 1998). In addition, many organisms are detectable in soils using gene probes, whereas the whole organisms cannot be detected by their activity or by cultivation methods. This shows that many parts of the soil system are not fully understood (Torsvik et al., 1990; Ward et al., 1992; Stahl, 1997).

3 Microbial Degradation of Organic Compounds in Soil

3.1 Microbial Activity in Soil

With plenty of microorganisms in soil, most of the natural and anthropogenic xenobiotic compounds are subject to microbial attack, assuming the compounds are biodegradable at all. The metabolism starts immediately after introduction of easily degradable compounds. After introduction of large amounts of such compounds, degradation starts with a phase of adaptation which is similar to that after the introduction of readily degradable compounds. Hardly degradable compounds such as several xenobiotics or lignin compounds of wood are only degraded under specific conditions and even then only at very slow rates. Thus, the relative amount of such compounds increases over time in the organic matrix of soil. In general, biodegradation categories such as easily, readily, or hardly degradable do not only depend on the quality of the respective compound, but on the sum of qualities of the whole environmental system. We can distinguish between three types of microbial degradation in complex biocenoses. Usually, easily degradable substrates are metabolized immediately upon formation of biomass and mineralization products (complete metabolization and mineralization). Readily and hardly degradable compounds are metabolized at first at slow rates or at higher rates after protracted phases of adaptation. Many such compounds are metabolized upon the release of metabolites and without the formation of biomass (cometabolic transformation). Other substances, mostly

random polymers like lignin or melanins, are degraded very slowly. In most cases the degradation of such compounds occurs only by enzymatically generated, non-specific radical oxidation mechanisms (e.g., by lignin peroxidases) which do not provide energy for the organisms (non-specific oxidation by radicals). These types of bio-oxidation are not yet understood in detail (see Chapter 6, this volume).

The persistence of organic substance in the environment is caused by qualities of a compound such as low water solubility and high molecular weights, or whether the microflora is able to develop metabolic pathways for degradation or whether any pathways are available at all. Microbial degradation may in turn be hindered by factors depending on the environmental system such as limited bioavailability and sorption or sequestering into the soil matrix. In general, the presence of solid matter in the water phase of axenic cultures of bacteria leads to the formation of additional carbon residues within the system. Increased amounts of sorption sites lead to the decreasing availability of bacteria, i.e., *Pseudomonas* sp. to mineralize easily degradable substrates such as glutamate (Alexander and Scow, 1989). Similar effects were confirmed for the mineralization of several aromatic compounds like benzoic acid in the presence of humic substances (Armador and Alexander, 1988). In contrast to experiments in liquid cultures, glucose was depleted in soil systems before mineralization started (Coody et al., 1986).

In many cases xenobiotics like dichlorobenzenes (van der Meer, et al., 1987) did not completely disappear after microbial degradation and remain at trace level concentrations of a few $\mu g \ kg^{-1}$ soil. The reasons for such threshold concentrations may be that residual amounts are insufficient

(1) to maintain growth and thus the degradation potential dies out in the population,
(2) to keep the enzymes of a degrading pathway induced, or
(3) exceed the sequestration or sorption capacities of the specific soil and become bioavailable.

The decreased bioavailability of phenanthrene was shown after ageing of the compound in sterilized humic soil (HATZINGER and ALEXANDER, 1995). However, the effect of ageing was only significant in soil macroaggregates of a size of 0.125 up to 4.0 mm. With a particle size <0.125 mm, the mineralization rate was reduced immediately after spiking the soil to levels observed in aged samples. These results imply that ageing behavior and the bioavailability of the compounds depend on the aggregate size of soil particles. Wet–dry cycles may increase the amount sequestered in soil (WHITE et al., 1997).

The water content of soils varies from a few percent up to ≥45%. However, the availability of water expressed as water potential is the controlling factor of the metabolic activity and numbers of microorganisms in a particular soil. Water potentials of <0.05 MPa are usually sufficient to support microbial activity, whereas at high potentials >10 MPa even fungi with higher tolerance to limited water availability are completely inhibited. Generally, bacteria and fungi show optima of activity at water potentials of 0.01–0.05 MPa, which approximately correspond to the water potential of a clay soil at 40–45% water holding capacity. C oxidation reaches its maximum at 60% of water filled soil porosity. Bacterial activity decreases rapidly >0.5 MPa and fungal activity >4.0 MPa (HAIDER, 1998b). Periodical processes like wet–dry cycles or changes of temperature have additional effects on the properties of soils. Wet and dry cycles are shown to increase the bioavailability of the labile pool of organic matter from macroaggregates in soils (VAN GESTEL et al., 1991). However, other research showed that the microbial mineralization of soil organic matter was not accelerated after repeated wet–dry cycles in a sandy loam (DEGENS and SPARLING, 1995). This indicates that soil organic matter may be very resistant to microbial degradation even after the disruption of the aggregates.

The availability of oxygen in the gas phase of soil is another factor controlling soil microbial activity. Partial anaerobiosis considerably decreases the degradation and turnover of organic matter in soil. The content of CO_2 is usually higher in soils with active microflora and may reach values up to 1% (HAIDER, 1996).

Water and oxygen availability control the redox conditions and the microbial activity in soil and thus the rates of humus formation and turnover. The core of greater soil aggregates (>4 mm) is virtually free of oxygen in otherwise well aerated soils. This corresponds to a declining microbial colonization of the inner volume of the aggregates with increasing diameter (HAIDER, 1998b). In addition, new results show that humic acids may be used as terminal electron acceptors by bacteria in order to oxidize appropriate electron donors (i.e., chlorinated hydrocarbons) under anoxic conditions (CURTIS and REINHARD, 1994; BRADLEY et al., 1998). Thus, humic acids may support microbial growth conditions. This phenomenon should actually be classified as a respiration process (LOVLEY et al., 1996). Due to the redox conditions, the process fits between nitrate and sulfate reduction and is able to shuttle the electron transfer onto Fe^{3+} ions. Humic substances contain considerable amounts of active Fe in ferroheme or porphyrine systems derived from microbial enzymes like cytochromes or peroxidases or from chlorophyll type compounds after the exchange of Mg by Fe (STEVENSON, 1994). A lot of reductive processes in anaerobic soils or anaerobic micro-environments, especially the dechlorination of halogenated pesticides and xenobiotics in the presence of humic substances, have already been described (KUHN and SUFLITA, 1989). Humic substances are also able to catalyze the reduction of aromatic nitro groups in the presence of reduced iron or sulfur compounds (DUNNIVANT and SCHWARZENBACH, 1992). The often observed reduction of 2,4,6-trinitrotoluene in contaminated soils (RIEGER and KNACKMUSS, 1995) may also result from such processes.

Most biogenic organic molecules that enter soil (cellulose, hemicellulose, pectin, starch, lignin, chitin, lipids, proteins, microbial exopolysaccharides, etc.) are macromolecules and, therefore, not suitable for uptake into bacterial cells until they have been depolymerized. In general, intracellular degradation only occurs at mol weights of <600 Da (HAIDER, 1998b). Thus, all biogenic macromolecules have to be cut up by mechanical actions and depolymerized by chemical actions or by exoenzmyes before the resulting monomeric building blocks and molecule fragments become available to other cells. After depolymerization, these fragments are susceptible to the uptake by microbial cells or to interactions with the soil matrix. However, due to mechanistic and spatial considerations (i.e., limited amount, activity, and small size of bacteria, abundance of humic matter, limited water availability), the effect of the released macromolecule fragments to a specific organism should be considered to be extremely restricted to the nearby surroundings, whereas the contribution to the formation of humic substances ought to be much greater (BURNS, 1989). This might be one of the reasons for the immanent ability of soil systems to form humic aggregates and to stabilze otherwise labile biomolecules.

Most *free enzymes* added to soil survive only briefly due to inactivation by denaturation or proteolysis. Thus, the undoubted capacities of the soil microflora to degrade most macromolecular substrates are difficult to understand (BURNS, 1989). However, with the exception of ligninolytic enzymes, depolymerizing exoenzymes form enzyme–substrate complexes in the case of a surplus of polymeric substrates. Thus, the enzymes will be bound and stabilized within enzyme–substrate complexes. The depolymerizing exoenzyme activity of xylanases was found to be mainly associated with fractions of plant organic matter in the coarse particle fraction >63 μm, whereas the activity of invertases was mainly associated with the fraction >2 μm to <63 μm containing the highest content of microorganisms. In addition, the bulk soil activity of xylanase was much lower than the summarized activity of the different particle fractions after gentle disruption of the soil macroaggregates (STEMMER et al., 1998). This fact indicates that enzymes associated with fragments of plant organic matter are protected and stabilized by aggregation. The production and distribution of xylanase and invertase during the decomposition of maize straw in various soils indicate that even in the bulk soil enzyme activities remained on a higher level for longer periods (STEMMER et al., 1999).

The substantial metabolic activity of soils depends on enzymes that are not associated with cells. After the death and lysis of the microorganisms, fragments of the cells begin to react with the organic and inorganic matrix of the soils. If proteins and enzymes escape from inactivation by proteases during this process, the biomolecules may be stabilized in the long run. Thus, these enzymes contribute to the total reactivity of the soil matrix. Besides extracellular enzymes, several intracellular enzymes such as oxidoreductases, transferases, hydrolases, and lyases were actually observed in cell free humic substances or clay mineral complexes (BURNS, 1989). In addition, enzymes that are attached to the cell wall may be stabilized by cell debris. For extracellular ligninolytic enzymes such as peroxidases and laccases, stabilization by sorption to humic substances or clay minerals has already been proven. This stabilization leads to the protection of the enzymes against heat or proteolytic inactivation, which, however, is often accompanied by decreased enzyme activity (SERBAN and NISSENBAUM, 1986; BURNS, 1989; RUGIERO and RADOGNA, 1988; SARKAR et al., 1989; GIANFREDA and VIOLANTE, 1995). Some authors even reported a 50% increase in the activity of laccases of *Trametes versicolor* by bentonite (FILIP and CLAUS, 1995). Humic acids and to a greater extent fulvic acids are reported to act as competitive inhibitors for peroxidases and laccases (SARKAR and BOLLAG, 1987). These enzymes are able to catalyze oxidations of fulvic and humic acids and thus generate oxidative coupling reactions of organic xenobiotic compounds or of their metabolites. Such coupling activity is present in most soils and the substances responsible for this activity can be extracted by aqueous solutions. This activity can be inactivated by heat or by radical scavengers and is thought to be of biogenic origin (SUFLITA and BOLLAG, 1980; BERRY and BOYD, 1985). Soils receive a persistent catalytic capacity by stabilized enzymes which is independent of the existing living microflora and may represent a kind of biochemical memory of the foregoing microflora and decaying processes. Of course, the supply of necessary cofactors for some enzymes is limited, since redox cycles will be exhausted apart from the cells. However, the cycles may be maintained for a certain amount of time by clay minerals or humic substances. The presence of externally stabilized enzymes fixed to the humic matrix also benefits the supply of nutrients for the microorganisms, as is assumed for the supply of ammonia by ureases (BURNS, 1989). Presumably, several advantages for the organisms are provided by a matrix of humic substances which are able to detoxify toxic metabolites by coupling reactions or by scavenging radicals. In addition, microorganisms are often surrounded by clay particles or clay closely associated with particulate organic matter. Clay minerals and microbial cells bear dense negative charges at their surface and concentrate cations between the clay particle and cells, which provides nutritional benefits for the cells (BURNS, 1989).

3.2 Plant Components – Modification of Organic Matter

Humic substances or refractory soil organic matter are built of fragments from decaying postmortal biomass (plants, animals and mi-

croorganisms). The precursor molecules are derived from plant material (carbohydrates, proteins, lipids, cellulose, hemicelluloses, lignins, cutins, suberins, waxes), components of the soil fauna, material from the microflora (especially from cell walls of bacteria), fungi and algae (mureines, chitins, melanines) as well as from secondary metabolites of all groups of microorganisms. The majority of humic substances in soil is derived from fragments of plant material. After the death of the plants, the macroscopic structures of the matter are decayed by organisms of the soil fauna and microflora. Processes of mechanical breakup of the macrostructures by organisms of the soil fauna are essential for the preparation of further microbial degradation. The fragments are either mineralized to CO_2 and H_2O with the formation of microbial biomass, or metabolized to low molecular weight compounds, or converted in secondary reactions to humic substances (see Sect. 4.3 and Fig. 2,

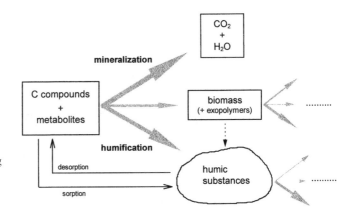

Fig. 2. Scheme of carbon flow during microbial degradation of organic compounds in soil (modified after KÄSTNER et al., 1993).

Tab. 2. Carbon Partitioning of Plant Matter after Microbial Degradation in Soil (modified after STOTT et al., 1983a; STEVENSON, 1994)

Components	[%]			
	of Plant Biomass	Mineralized[a]	Converted to Humic Substances[a]	of Humic Substances
Carbohydrates	>50	>60	<40	(47)
N compounds	<20	>50	<50	(12)
Lignins	10–40	<25	>75	(9–36)
Cutines, Suberines, Waxes	<5	<10	>90	(4.5)

[a] Referred to plant biomass, column 1.

Tab. 2). Even complex fragments of the plant biomass may be included in such reactions. Whenever dead plant material comes into contact with soil, rapid degradation with the formation of CO_2 from the easily degradable compounds such as polysaccharides and proteins begins. The degrees to which the compounds are degradable determine the succession of organisms during the decaying process. A portion of the carbon is converted into microbial biomass which increases in amount during the first few weeks after the entry of the material. Polysaccharides incorporated into lignocelluloses are degraded more slowly than free cellulose, but faster than lignin molecules by orders of magnitude (HAIDER, 1996). Lignin is formed by terrestrial plants, giving wood mechanical strength and resistance to microbial attack. Lignin is a substance of spheric, macromolecular heteropolymers of molecular weights up to 100,000 Da or more. The polymers consist of phenyl propanoic monomers which are connected via $C-C$ and $C-O-C$ bonds generated by enzyme catalyzed radical reactions (random polymers). The microbial degradation of lignin is likewise performed by enzyme catalyzed radical reactions. The ability to degrade lignin is generally restricted to ligninolytic basidiomycetes with the exception of some fungi imperfecti and deuteromycetes as well as individual species of bacteria and actinomycetes. The degradation of lignin is carried out by the *cleavage* of the linkages of the side chains between $C\alpha$ and $C\beta$ atoms and by the cleavage of the aromatic ring systems. With continued degradation, increasing amounts of low molecular weight fragments are formed, as is described in detail in other reviews (HAIDER, 1988; BOOMINATHAN and REDDY, 1992; WOOD, 1994; SCHOEMAKER et al., 1994; SHEVCHENKO and BAILEY, 1996). In the course of lignin degradation, many carboxylic groups are generated at the molecules. This effect gives rise to signals assigned to polysaccharides in NMR spectroscopic analyses and may contribute to the overestimation of residual amounts of polysaccharides in humic substances. Other natural or xenobiotic compounds can be included in the non-specific oxidation processes by radicals in lignin macromolecule degradation (BUMPUS et al., 1985; EATON, 1985; FIELD et al., 1993; AUST, 1990; RAJ et al., 1992;

BARR and AUST, 1994). The decaying velocity of plant matter in soil essentially depends on the physical and chemical composition of the plants. Up to 70% of the carbon input from plants into soils are mineralized to CO_2 within one year (HAIDER, 1990). With increasing complexity and lignin content, the possible microbial degradation of the plant material decelerates. Finally, soil organic matter is formed from non-degraded residual material (especially lignin), products of secondary polymerzation reactions of metabolites, fragments of the plant biomass, and *de novo* formed components of microbial biomass.

During the microbial degradation of any organic compounds in soils, many degradation and conversion reactions take place, leading to a distribution of carbon in different compartments. Besides metabolization, mineralization, and formation of biomass, a certain amount of carbon is generally incorporated into refractory soil organic matter. Humic substances, metabolites and biomass are likewise parent materials for renewed degradation and conversion reactions (Fig. 2). The stimulation of microbial activity through the addition of composts or other organic materials to soils usually leads to accelerated turnover in all parts of the processes. The sum of these processes during the degradation of natural compounds is described as humification. Xenobiotic compounds or their metabolites are included in such reaction sequences in an analogous manner.

The degradation processes of dead plant material in humic rich soils can be described by a two-stage exponential function, with each stage being represented by first order kinetics. For example, 70% of ^{14}C labeled straw from rye grass were degraded with a half-life time of 0.25 a, whereas the remaining 30% were degraded with a half-life time of 8 a (JENKINSON, 1977). After more than 10 a, the degradation processes of the residual material have to slow down considerably, since age analyses of humic substances revealed residence times of carbon of several 100 a (HAIDER, 1996). Thus, humic substances with higher ages must be enriched within soils, which may indicate even a three-stage exponential function of degradation. To look at the overall soil system, an accumulation of humic substances in soil is only possible if the velocity of humification is higher than the velocity of other elimination processes such as mineralization [$V_{humification} > V_{mineralization}$] (ZIECHMANN, 1994).

Comparative examination of the degradation and humification rates of plant material in soil from different climates has shown temperature to have an essential impact on the processes of the formation of soil organic matter. A temperature increase of 10 °C leads to an increase of degradation rates by a factor of 2–3. Thus, soils in warmer climates need much higher inputs of parent plant matter to form similar amounts of humic substances (HAIDER, 1996; ZECH et al., 1997). In most cases, only slight mineralization occurs under anaerobic conditions, resulting in the formation of peat (FRITSCHE, 1990; HAIDER, 1998b). In the climax stage of a soil, the *de novo* formation and the degradation of humic substances are balanced and the amount of carbon is usually constant. After changes to external factors (exploitation, clearing by wood fire, climatic conditions) or internal conditions (acidification, moistening), the soil system reacts with rapid changes in the content of humic matter (GISI, 1997). During such changes, refractory soil organic matter may be metabolized at increased rates. Humic acids were shown to be degraded by lignin degrading white-rot fungi, i.e., Phanerochaete chrysosoporium and Trametes versicolor. In addition, humic acids are inductors for the expression of lignin degrading peroxidase enzymes, which are shown to play an important role during the degradation of humic acids (HAIDER and MARTIN, 1988; DEHORTER and BLONDEAU, 1992).

3.3 Xenobiotics and Mineral Oil Components

Similar to dead plant material and microbial biomass, xenobiotic and mineral oil compounds are subject to microbial degradation and humification processes after introduction into the soil system. The microbial degradation of most of these compounds is delayed, which is due to the properties of the molecules such as toxicity and low water solubility or attributable to effects of the xenobiotic or soil matrices. Details of the microbial metabolism of the

various groups of xenobiotic and mineral oil components are described in Chapters 6, 7, this volume.

The slower the microbial degradation of a given compound proceeds, the more time is available for abiotic interactions with the soil matrix and for migrating into aggregate structures and molecules (sequestration). The bioavailability of a compound (see Chapter 3, this volume) may be reduced substantially over time by sequestration. Thus, compounds that are otherwise easily biodegradable may persist under certain conditions in soil. Metabolites of compounds which are released from the cells into the environment or metabolites generated by exoenzymes are preferentially subject to interactions with the soil matrix. However, even metabolites of mineralizing intracellular pathways of bacteria are found to be covalently linked to the organic soil matrix by ester bonds, as was demonstrated for the PAH compounds anthracene and pyrene (RICHNOW et al., 1994; RICHNOW, unpublished data). Such metabolic pathways normally facilitate the complete metabolization of the PAH compounds with the formation of biomass, CO_2, and H_2O. The results indicate that generally all metabolites which undergo intermediate accumulation in the cells as well as components of the biomass itself are subject to similar interactions with the soil matrix and contribute to the formation of carbon residues. Thus, theoretical concepts that only metabolites of cometabolism or dead end pathways and products of extracellular non-specific oxidation reactions such as ligninolysis are potential precursors of humification reactions (KÄSTNER et al., 1993; FRITSCHE et al., 1994; BOLLAG, 1996) were too restricted. The potential of organic compounds to interact with the soil matrix is determined by functional groups of the molecule. This potential is often increased by the activity of microorganisms, which in the case of PAH introduce reactive hydoxylic or carboxylic groups into the molecules. Hydroxylated metabolites of PAH or hydroxynaphthoic acids are shown to be able to undergo autoxidative polymerization reactions to form macromolecules which resemble humic acids (KÄSTNER et al., 1999). The properties of such metabolites may explain why PAH contamination decreases rapidly in nature after forest fires. Presumably, an initial photooxidation which primes microbial degradation (MILLER et al., 1988) is followed by humification reactions.

The input of degradable carbon sources such as plant material or composts leads to accelerated microbial degradation and transformation into the organic soil matrix. The increased microbial activity causes the additional elimination of xenobiotic substances, if the compounds are at all degradable. Hardly degradable or non-degradable compounds may be stabilized by adding further carbon sources, which may decrease the bioavailability of the compounds, especially for low concentrations of contaminants (< ppm range). Sequestered phenanthrene showed decreased bioavailability and uptake by the earthworm *Eisenia foetida* (WHITE et al., 1997). Sequestration or incorporation into the soil matrix are mostly accompanied by decreasing bioavailability of toxic compounds (BOLLAG, 1992; VERSTRAETE and DEVLIEGHER et al., 1996). During the bioremediation of soils, the poor bioavailability of a compound after sequestering into soil aggregates may be overcome by the mechanical cracking of the soil texture, e.g., by extensive mixing, which usually leads to more metabolism and less stabilization. However, repeated drying and rewetting during such treatment may be counterproductive (WHITE et al., 1997).

4 Humic Substances and the "Humification" Process

4.1 Humic Substances

The complex mixture of naturally occurring, hardly biodegradable, yellow to brownish colored, colloidal organic substances in soils is usually called humic matter (AIKEN, 1985). Humic substances are formed postmortally from decaying matter of plants, animals and microorganisms, and show molecular weights of 500–20,000 Da and in some cases up to 100,000 Da. (To date, no experimental methods are available which give reliable data about the true molecular weights of humic

substances). Humic substances may occur in soils in amounts of a few thousands up to 30%. Although humic molecules have relatively defined chemical and physical properties, they do not show any uniform, reproducible chemical structure (ZIECHMANN, 1994). The molecules cover mineral particles with primary coatings (see Sect. 4.2) or arrange with fragments of the biomass and components of the inorganic soil matrix (clay minerals) to form microscopic aggregates. Humic substances are commonly classified into humins, humic and fulvic acids by acid and alkaline solubility. This classification was made operationally by superficial criteria, which do not indicate the chemical behavior or structures of the respective molecules (SWIFT, 1985; MÜLLER-WEGENER, 1990):

(1) *fulvic acids*, which are soluble in acids and bases (molecular weights: 500–2,100 Da);
(2) *humic acids*, which are soluble in bases and precipitate in acids (molecular weights: 1,400–100,000 Da);
(3) *humins*, which are insoluble in acids and bases (molecular weights assumed to be similar to humic acids, more complex structures).

Different fractions of humic substances are assumed to have similar molecular structures, which differ in the degree of cross-linking of the macromolecules, the content of oxygen and the amount of hydroxyl, carbonyl and carboxyl groups. Tab. 3 shows the elementary composition of humic substances in comparison to microbial biomass and plants. The molecular weights of humic acids and the degree of cross-linking increase with increasing age of the molecules. This usually leads to increasing biological half-life times and resistance to microbial attack. Biologically inert humic molecules show ages of as much as over a 1,000 a, as was measured by the radiocarbon content (HAIDER, 1996). Organic matter composition varies with the soil type. In comparison to soils with high clay contents, sandy soils contain more alkyl C and less carbohydrates or proteins (CAPRIEL et al., 1995). The soil organic matter of these soils is more hydrophobic, which might be due to the lack of stabilization of labile biomolecules by clay minerals.

Tab. 3. Elementary Compositions of Humic Substances and Averages of Microbial and Plant Biomass (adapted from HAIDER, 1996; SCHLEGEL, 1992; SCHRÖDER, 1984)

	C	H	O	N	S [% dw]	C−N Ratio
Humins	55.4–56.3	5.5–6.0	31.8–33.8	4.6–5.1	0.7–0.8	11
Humic acids	53.8–60.4	3.7–5.8	31.9–36.8	1.6–4.1	0.4–1.1	20
Fulvic acids	42.5–50.9	3.3–5.9	44.8–47.3	0.7–2.8	0.3–1.7	30
Biomass (bacteria)	~50	~8	~20	~14	~1	~4–5
Biomass (plants)	~47	~7	~44	< 2	<1	25–90

The application of modern analytical methods such as ^{13}C NMR spectroscopy and pyrolysis GC-MS revealed different structural compositions of humic substances in comparison to older models. The amounts of aliphatic groups were usually underestimated and the amounts of aromatic structural units were overestimated (STEVENSON, 1994; SCHNITZER and SCHULTEN, 1995). The carbon distribution of humic substances from different soils is presented in Tab. 4. The data were obtained from 8 different soils and represent the range of carbon within the different types of bonds. The content of carbon within *N*-alkyl, methoxyl, or aromatic units is much smaller in fulvic acids and the amount of carbon hydrates is much higher in comparison to humic acids. In addition, various distributions of aromatic and aliphatic structural units were observed in soil fractions of different particle size. Higher amounts of aliphatic units were observed in coatings of sand particles and in the fraction <2 µm, whereas in the fraction of 2–50 µm higher levels of aromatic structures were found (HAIDER, 1998b).

Tab. 4. Distribution of Carbon Atoms in Different Structural Units and Binding Types of Humic Substances (modified after STEVENSON, 1994)

C content [%]	Fulvic Acids	Humic Acids
Aliphatic	12–24	16–30
N-alkyl (methoxyl)	3–6	5–10
Carbohydrate	24–55	13–17
Aromatic	8–25	26–43
Carboxylic	12–20	13–17
Carbonyl	4–7	5–7

Nitrogen occurs in humic substances in mean amounts of up to 3.5%. It comprises NH_3-N (20–35%), amino acid and sugar N (30–50%), whereas 25–55% of N have to be classified as non-acid soluble or acid soluble unknown N compounds (STEVENSON, 1994). The majority of the unknown N is thought to be derived from purine and pyrimidine bases. Humic substances contain considerably smaller amounts of C and H in comparison to the parent plant biomass. Since N is especially enriched in humins and humic acids, C and H have to be eliminated by microbial metabolism in soils, whereas N remains within the organic matter. However, humic substances contain less N in comparison to bacterial biomass. Thus, it must be concluded that bioavailable N is repeatedly recycled by successive populations of bacteria and the N originating from parent plant biomass is preserved in soil (see also Sect. 2.3).

4.2 Models of Humic Substances

Although the heterogeneity of humic substances does not allow precise chemical definitions to be given of the overall structure, various models and concepts have been discussed (for a review, see HAYES et al., 1989). Humic substances are suggested to be complex soil colloids with micellar like structures (PAULI, 1967; WERSCHAW, 1993) and hierarchically formed aggregates (OADES, 1993). The building blocks of humic molecules consist of fragments of biomolecules that are dominantly linked by certain types of covalent bonds and may contain quinoid, hydrophobic aromatic and aliphatic structures. Humic substances car-

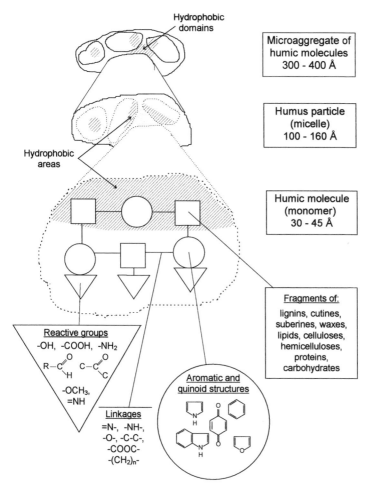

Fig. 3. Humic molecules and aggregation model (according to PAULI, 1967; ZIECHMANN, 1988; RICHNOW et al., 1998); the humic micelles and particles may be encrusted with inorganic components.

ry a number of reactive functional groups with a hydrophilic character (Fig. 3). Due to the hydrophobic and hydrophilic areas of fragments of plant material and humic molecules, they may exhibit an amphiphilic, tensid-like character in water solutions (WERSHAW, 1986). Hydroxyl and carboxylic functional groups enable fulvic and humic acid molecules to form complex multivalent metal cations and to create intermolecular networks and bind to inorganic soil components, especially to clay minerals. Molecules and particles combined in the process in which hydrophobic areas con-

gregate side by side to form hydrophobic domains (pseudomicellar structures). Both hydrophobic domains and hydrophilic functional groups determine the ability of humic substances to interact with xenobiotic compounds. Hydrophobic domains are important for the sorption of hydrophobic compounds. Reactive functional groups may act as binding sites for chemical reactions with xenobiotics. With increasing concentrations of metal cations or increasing temperatures, these aggregates undergo rearrangements to become more compact aggregate structures with cage-like cavities

(ENGEBRETSON and WANDRUSZKA, 1994). At neutral to acidic pH, the microaggregated particles show sponge-like micropores in electron micrographs. The aggregates incorporate other molecules and, therefore, stabilize labile compounds (STEVENSON, 1994). The particles stick together by sorptive or covalent bindings, by adhesive mineral substances, or by microbial exopolymers (HAIDER, 1996). Again, such particles arrange to form larger aggregates and clusters (macroaggregates >250 µm) and incorporate microorganisms, fragments of biomass, clay compounds and xenobiotic compounds in the aggregation process. These self-aggregating and stabilizing properties are amplified by connecting the clusters with roots and hyphae of fungi and actinomycetes or by coating fragments of plant material and microorganisms with humic substances or clay minerals (TISDALL and OADES, 1982; see also Fig. 1). However, such aggregates are easily destroyed by mechanical effects, eliminating the stabilization of labile organic substances and followed by increased microbial degradation activity.

Humic particles were shown to exhibit diameters of up to 100 Å or more (FLAIG et al., 1975). These diameters correspond to molecular weights of up to 20,000 Da and are in the range of the particle size of colloids, which is usually regarded as $10^{-6}-10^{-2}$ µm (STEVENSON, 1994). The particles contain hydrophobic inner areas and hydrophilic surfaces. Hydrophobic compounds and xenobiotics tend to migrate into the inner areas of such particles causing residues with limited bioavailability of otherwise degradable compounds (NAM and ALEXANDER, 1997). The phenomena observed in natural humic substances may also be explained by the model assumption that humic substances or aggregates may vary in consistence from "glassy" to elastic "rubber-like" areas. Lipophilic xenobiotics may be dissolved within the rubber-like areas, enabling the compounds to diffuse into internal microvoids of the glassy areas of the humic molecule. The compounds are then entrapped by hole filling mechanisms (XING and PIGNATELLO, 1997). This model was supported by the observation that about half of the non-extractable residues derived from a radiolabeled fungicide (Cyprodinil®) were liberated after the solubilization of the soil humin fraction by silylation procedures. The majority of the released radioactivity was only sequestered within the humin fraction and was not covalently bound (DEC et al., 1997a, b). In general, the stability of organic compounds in soil may be caused by intrinsic effects of the chemical molecule structure (i.e., of melanins, polyphenols, and black carbon components) or by interactions with the organic or inorganic solid phase (clay–humus complexes, aggregation processes; see Figs. 1 and 2). Moreover, stabilization may simply be caused by the physical separation of the molecules from the metabolizing microflora (by sequestration into micropores not accessible to microorganisms) or by the limitation of biodegradation (by anaerobic conditions within water saturated aggregates). For residues derived from xenobiotic compounds, bioavailability and toxicity are usually assessed to be much smaller in the case of stabilized, unavailable components compared to free compounds (ALEXANDER, 1995; KELSEY and ALEXANDER, 1997).

Several structural models of humic acids have been presented, which are only useful for visual instruction but do not have any general validity. Models which refer to recent results of pyrolysis GC-MS analyses indicate molecular weights of ~6,600 Da for humic acids (SCHNITZER and SCHULTEN, 1995). The proposed empirical formula $(C_{342}H_{388}O_{124}N_{12})$ can be roughly simplified to a basic formula of $[C_3H_3O]$, which contains less H and O than the basic formula of $[CH_2O]$ for parent biomass. However, N is obviously underestimated in these models in comparison to the elementary analysis of real humic acids (see also Tab. 3).

Humic substances show electron conductivity which is in the range between isolators and semiconductors (ZIECHMANN, 1994). Electrons are liberated in humic substances or at least in parts of the molecular network and thus negative charges (electrons) and positive charges (sites of liberated electrons) are movable within the macromolecule. This behavior enables a bipolar charge transport, causing increased reactivity. Recent experiments showed that the molecular size of dissolved molecules of humic acids is significantly smaller under anoxic conditions than in the presence of oxygen (VON DER KAMMER and FÖRSTNER, 1998). Using electron spin resonance spectroscopy, significant amounts of *stable* free radicals were observed in humic substances. For instance, *S*-triazines with quinoid structures have been shown to generate semiquinone radicals during the formation of electron acceptor–donator–complexes. Reactive benzo- and naphthoquinones were observed in humic substances in amounts up to 10–15% (STEVENSON, 1994). Humic substances absorb ultraviolet light. The molecules become excited (triplet state) in the process and transfer energy to ground state oxygen ($^3O_2$) to form singulett oxygen ($^1O_2$), which, in turn may generate humic acid cation radicals (SENESI, 1993). Both clay minerals and humic substances contain Fe and Mn ions, which are able to generate oxidative coupling reactions of phenolic compounds to form humic substances (BOLLAG and LOLL, 1983). Radical structures are observed in the process on phenolic compounds such as syringic acid, ferulic acid and guaiacol, which are formed during the biodegradation of lignin (HAIDER et al., 1975). Radical scavenging reactions lead to coupling reactions with other molecules. Hence the formation of humic macromolecules in the presence of phenolic degradation products of lignins may be initiated perpetually. Radicals that are generally formed more frequently during the degradation of plant biomass may also contribute to the genesis of humic substances.

4.3 Genesis of Humic Substances

The formation of humic substances from the dead biomass of plants is a common feature of soils. Several new substances are formed and aggregate with fragments of biomass to form complex macromolecules in the process. The sum of the reactions is called humification, although the individual processes are not understood in detail. The mass turnover of carbon during humification is in a range similar to that of turnover by photosynthesis. Nearly all substances which are formed by photosynthesis are subject to microbial degradation with subsequent humification in soil during the carbon cycle. Thus, humic substances should be assessed as a specific state of biological matter and carbon, respectively. Despite extensive re-

search efforts, the genesis of humic substances is still obscure. It is very likely that all known formation reactions may contribute to the genesis of humic substances (STEVENSON, 1994). Several models are discussed in which degradative and aggregative processes can be distinguished (HEDGES, 1988). Degradative models of genesis are deduced from the condensation of partly degraded biogenic macromolecules such as lignins, melanins, and other hardly degradable compounds. For a long time, the theory of WAKSMAN (1932) was most generally accepted in which modified lignins condense with N compounds (degradation products of proteins) to form humic substances. Aggregative models were favored later on, which postulate the formation of humic substances by condensation and repolymerization from reactive aromatic compounds of low molecular weight (LARSON and WEBER, 1994). Such phenolic precursor molecules originate from the complete depolymerization of biogenic macromolecules such as lignin and were observed in biologically active soils (STEVENSON, 1994). The aggregative models propose that humic macromolecules are formed after abiotically or enzymatically generated radical reactions of precursor molecules, leading to polyphenolic and quinoid structures (MARTIN and HAIDER, 1971). Microorganisms are also able to synthesize phenolic precursor molecules *de novo* (HAIDER et al., 1975). It is assumed that the formation of humic substances by condensation and polymerization reactions is initiated by an electrophilic attack on aromatic precursor molecules and following radical reactions (phase I). These radicals react with each other or with other precursor molecules that show similar reactivity, and condense to macromolecules with covalent bonds. These macromolecules age to humic acid molecules by intramolecular reactions during the further course of genesis (phase II). Humic acid molecules aggregate to form larger clusters and molecular networks, which are sometimes insoluble and called humins, by intermolecular interactions such as ionic bonds, charge transfer complexes, and stabilization by clay minerals (phase III) (ZIECHMANN, 1994). Since aromatic compounds are often coupled by oxygen bridges in molecules, it was assumed that phenolic compounds which originate from the biodegradation of lignins play a substantial role during the formation and metabolism of humic substances. However, new results show that the cleavage of dimeric lignin model compounds by lignin and manganese peroxidases from ligninolytic fungi mostly occurs between the $C\alpha$ and $C\beta$ atom of side chains in the molecule and thus phenolic compounds are not the primary products (ODIER and ARTAUD, 1992; HAIDER, 1996). The latest results also show similar cleavage by laccase enzymes in the presence of specific mediator substances (EG-GERT et al., 1996, 1997). NMR spectra of lignins show a decrease of phenolic and methoxyl C in the course of humification (KÖGEL-KNABNER et al., 1991).

When considering the results, older models about the genesis of humic substances from partly degraded biogenic macromolecules are becoming more important again. These hypotheses assume the functionalization and rearrangement of the macromolecules instead of the complete degradation and depolymerization of lignin during biotransformation to humic and fulvic acids in the environment (SHEV-CHENKO and BAILEY, 1996). NMR spectroscopic data show similarities between native lignins and humic or fulvic acids that would hardly be expected if humic acids were generated by the secondary condensation of low molecular weight phenols. These results suggest that the plant components have not undergone major alteration during humification, except for the extensive transformation of oxygen functional groups of lignin to form carboxyl groups (SHEVCHENKO and BAILEY, 1996). Analyses of cattle manured compost at various stages of decomposition showed that the amount of humic acids, fulvic acids and humins doubled during composting and reached stable values after 90 d. However, only slight compositional and structural changes were observed in the humic acids by ^{13}C-NMR spectra and chemical analysis. No significant changes in elemental analysis and only a slight increase in carboxyl (12%) and aromatic (8%) signal areas and a decrease of methoxyl (-16%) and alkyl (-8%) signal areas were observed. This led to the conclusion that humic substances are composed of partially degraded constituents of plant tissues which still retain their chemical structures to a certain extent (INBAR et al., 1990). During the

composting of ^{13}C and ^{15}N enriched rye grass (*Lolium perenne*) and wheat (*Triticum sativum*) for two years, the continuous, simultaneous depletion of all signal areas in ^{13}C-NMR spectra were observed compared to the parent plant material. The equal degradation of all classes of carbon compounds obviously occurred in the process. Only intermediate carboxylic C (160–220 ppm) increases occurred within the first 100 d (KNICKER and LÜDEMANN, 1995). Such results indicate degradative and modifying processes of biopolymers during the genesis of humic substances in the composting experiment. Assuming aggregative and condensation models from reactive, low molecular weight compounds, significant changes to the aromatic signals ought to have been expected (KNICKER and LÜDEMANN, 1995). However, quinoid structures that are formed during the oxidative coupling of aromatic precursor molecules such as guaiacol can be stabilized by nucleophilic addition reactions (DAWEL et al., 1997). Such reactions lead to the depletion of 6-π electron systems and to an increase of 4-π electron systems which show no aromatic signals. Thus, the aromatic structures and signals are eliminated in the composting process by such reactions so that nearly constant signal ratios in the ^{13}C-NMR spectra do not necessarily rule out the aggregative models of polymerization of low molecular weight aromatic compounds.

During the composting of rye grass and wheat, the corresponding ^{15}N-NMR spectra showed no significant changes of the ^{15}N signals of the proteins (KNICKER and LÜDEMANN, 1995). The continued existence of non-modified ^{15}N enriched plant material within amide bonds and peptide structures after composting was not expected at first, since proteins were assessed as being easily degradable by microorganisms (KNICKER and LÜDEMANN, 1995). Non-modified fragments of biocomponents are obviously included in the formation of humic aggregates in the process. Thus, mechanisms have to exist which lead to the stabilization and protection of biodegradable components (TEGELAAR et al., 1989; KÖGEL-KNABNER and GUGGENBERGER, 1995). Such components are stabilized by hardly degradable humic substances or by clay minerals, and become an integral part of the refractory soil organic matter. This hypothesis is enforced by the findings that proteinogenic material of algae stabilized by refractory biopolymers was observed in 4,000 a old sediments (KNICKER et al., 1996). The breakdown of such aggregates and the resulting bioavailability of easily degradable substances like proteins or carbon hydrates may be the key to understand the nutritional requirements of soil feeding organisms like earthworms or termites. The conditions in termite guts, i.e., that lead to the resolution of the aggregates, are reviewed in other pulications (BRUNE, 1998).

Another model of the formation of humic substances is based on the amphiphilic properties of molecules and fragments that are formed during the degradation of plant material. Such molecules aggregate spontaneously in water solutions to form double membrane-like structures, vesicles, and micelles. These aggregates develop hydrophobic areas on their inner surface that are able to incorporate lipophilic compounds, whereas hydrophilic areas are located on the outside unable to coat the surface of mineral particles or to take up clay minerals. Such organo-mineral associates substantially contribute to the stabilization of the aggregates of organic molecules (WERSHAW, 1989, 1993).

Besides modifying processes on biopolymers and self-assembling reactions, the condensation reaction of low molecular weight compounds is generally considered to be involved in the genesis of humic substances. However, such processes may have been overestimated over the past few decades. To initiate such condensation processes, several model reactions can be given that generate brown macromolecules with properties of humic substances (synthetic humic substances) from polyphenols and quinones by autoxidation processes or by enzymatic oxidation (laccases and peroxidases). Radicals are formed initially by the electrophilic attack of an electron acceptor (oxygen, metal oxides, and clay minerals) to phenolic compounds, generating reactions to form macromolecular networks of such compounds coupled by covalent $C-C$ or $C-O-C$ bonds with high binding enthalpy. Such reactions generate macromolecules to which less reactive compounds such as carbon hydrates or peptides can bind. Thus, the reac-

tive aromatic macromolecules are transformed into stable products that are able to incorporate residues of xenobiotic compounds or their metabolites, as was shown, e.g., with PAH (RICHNOW et al., 1994; KÄSTNER et al., 1999).

The incorporation of nitrogen into the formation of humic substances is considered to occur via Maillard-like model reactions from amino acids or amino sugars. In the process, carbon hydrates and amino acids react with the formation of polyfunctional intermediates, which then react with further carbon hydrates (ZIECHMANN, 1994). The polymerization of quinones and amino acids (RINDERKNECHT and JURT, 1958; HACKMAN and TODD, 1953) and the reaction of phenols and ammonia with following autopolymerization are additional reactions that are assessed as essential for the incorporation of nitrogen (FLAIG et al., 1975). Recent findings show that the incorporation of N from aromatic amines (N containing xenobiotics such as aniline, reduced metabolites of 2,4,6-trinitrotoluene and others) via nucleophilic addition of the amino group in *ortho* position to the carbonyl group of humic substances (WEBER et al., 1996; THORN et al., 1996; DAWEL et al., 1997) has to be considered as a specific fixation reaction of nitrogen. Due to the high $C-N$ ratio of bacteria, we can assume that bioavailable N in soil will be assimilated at first by microorganisms and then fixed to the humic matrix by the incorporation of biopolymers from microbial biomass (see also Sect. 2.3 and Tab. 3).

The formation of high orders of aggregates from humic molecules (derived either from condensation reactions of low molecular weight compounds or from the modification of biopolymers) depends on intermolecular interactions of the humic molecules such as ion bonds, H bridges and electron donator–acceptor complexes with low binding enthalpy. Charge transfer complexes facilitate intermolecular mesomeric stabilization. Humification has to be assessed as a series of natural processes, which stabilize molecules intramolecularly and higher networks intermolecularly and lead to the overall stabilization of the humifying matter with decreasing reactivity. Humic substances may be a kind of intermittence of the disorder of chemical molecules in soil. Both lignins and humic acids are fractal objects with specific similarities and disorder, which has to be considered in any characterzation scale used to examine them (SHEVCHENKO and BAILEY, 1996). Beside cellular biotic systems, humification can be assessed as a simple way of the self-organization of organic molecules. As expected, brown macromolecular substances with the character of humic acids were also observed in experiments that repeated the classic experiment of the formation of amino acids (MILLER, 1955) under the conditions of the earth's early atmosphere (ZIECHMANN, 1994).

4.4 Residues within the Organic Soil Matrix

The formation of non-extractable residues is a process which is well-known from the application of pesticides and herbicides. Such residues have been shown to occur in animals, plants, sediments and soils. Their significance, occurrence and formation in the environment have been the subject of several reviews (BAILEY and WHITE, 1964; KHAN, 1982; FÜHR et al., 1985; FÜHR, 1987; CALDERBANK, 1989; MANSOUR et al., 1993). Generally it was observed that after extractions with various solvents, a certain amount of the active compound remained in the soils. In order to focus the problem of assessment based on the diverse results, the commission of the International Union of Pure and Applied Chemistry (IUPAC) specified in 1981 the definition of non-extractable bound residues of crop protection agents in soil and plants. *Non-extractable residues, (sometimes also referred to as bound residues, or non-extracted residues) in plants and soils are defined as chemical species originating from pesticides and herbicides (parent compounds, metabolites, and fragments) used in agriculture practice, that are unextractable by methods which do not significantly change the chemical nature of these residues. These non-extractable residues are considered to exclude fragments recycled through metabolic pathways leading to natural products* (ROBERTS, 1984). The formation of bound residues can be analyzed by isotopic labels and in trace amounts only by the use of radiolabeled com-

pounds. The radiotracer technique allows the transformation to be balanced and the residues formed to be quantified. However, the technique does not provide any information about the structure of the residues. Thus, the residues may result from the parent compound, from the biogenic or abiotic formation of metabolites, or from fixed carbon dioxide after the microbial mineralization of the compound (KÄSTNER et al., 1999). In most cases nothing is known about the chemical structure and especially about the binding types of the residues, so that we can only refer to non-extractable bound residues according to this definition in individual cases. Thus the term residues will be used here synonymously for non-extractable or bound residues.

Many crop protection agents (above all older ones) tend to form bound residues in terms of the IUPAC definition, especially if the compounds are not easily biodegradable. The extent of residue formation in soil amounted to 7–90% of the applied doses (BAILEY and WHITE, 1964; MEANS et al., 1980; KHAN, 1982; BOLLAG and LOLL, 1983; FÜHR et al., 1985; FÜHR, 1987; CALDERBANK, 1989). The formation of residues seems to be more abundant in the range of low concentrations of xenobiotic compounds in soils ($\ll 200$ mg kg^{-1}) in comparison to higher concentrations. Beside sorption and sequestration processes, the chemical reactivity of the parent compound or the metabolic products effects the formation of bound residues. Factors controlling the chemical reactivity of the molecule are functional groups such as carboxyl, hydroxyl, phosphate, nitro and amino groups that are able to react in condensation processes with natural organic matter. Chlorinated and cyclodiene compounds only exhibit a low potential to form bound residues, whereas carbamate, triazine, and organic phosphate compounds have a much higher potential. Nitro groups of anthropogenic chemicals are often reduced to amino groups before significant residue formation occurs. The amino derivative of certain pesticides contributes to relatively high amounts of residues and causes fast formation rates (ROBERTS, 1984). Reactions catalyzed by fungal ligninolytic exoenzymes were shown to generate oxidative cross-linking to typical humic structure moieties, which causes the formation of amide bonds (THORN et al., 1996; DAWEL et al., 1997).

Many plant protection agents are partly degraded and the residues are formed by the fixation of the metabolites. Several years ago, the degradation of phenylamide herbicides was shown to release chloroaniline compounds, of which more than 90% are fixed within the non-extractable residues (HSU and BARTHA, 1976). Addition and condensation reactions of aromatic amines with carbonyl groups of humic acids were demonstrated in principle in the early 1980s (PARRIS, 1980). The formation of hybrid oligomers with the humic components protocatechuic acid and syringic acid by laccases of the fungus *Rhizoctonia practicola* was shown for the compounds 4-chloroaniline, 3,4-dichloroaniline, and 2,6-dichloroaniline (BOLLAG et al., 1983). A few years later, the oligomerization of these compounds was demonstrated with horseradish peroxidase and laccase of *Trametes versicolor* (HOFF et al., 1985). However, conclusive proof of covalent binding of anilines to natural fulvic and humic acids only became feasible over the past few years using ^{15}N labeled aniline and the application of ^{15}N-NMR spectroscopy. Aniline hydroquinone, aniline quinone, anilide and imine bonds as well as nitrogen within heterocyclic structures are formed in the process (THORN et al., 1996). The formation of heterocyclic rings around the amino N, which amounted to more than 50% of the applied nitrogen, leads to the nearly irreversible fixation of the nitrogen. Such incorporation of N into the organic soil matrix has also to be considered for other amino aromatic compounds.

The herbicide 2,4-D (2,4-dichlorophenoxy acetic acid) was mineralized in soil to 73–94% within 1 a (STOTT et al., 1983b). However, dihydroxylated metabolites such as catechol, 4-chlorocatechol, and 4,5-dichlorocatechol were only mineralized to 38–50%, 22–62%, and 22–63%, respectively, which indicates the fixation of such metabolites from 2,4-D within the organic soil matrix. The formation of hybrid oligomers with humic components such as orcinol, syringic acid and vanillinic acid was demonstrated with 2,4-dichlorophenol by the laccase and tyrosinase of *Rhizoctonia practicola* as well (BOLLAG et al., 1980). The coupling of 2,4-dichlorophenol to natural fulvic acids

was shown with peroxidases of *R. practicola* and *Trametes versicolor* (SARKAR et al., 1989). Using such coupling reactions enabled the detoxification (BOLLAG et al., 1988) and the dehalogenation (DEC and BOLLAG, 1995) of the parent compounds to be demonstrated. The ultimate proof of the covalent bonds of 2,4-dichlorophenol to natural humic acids was likewise successfully presented for these compounds after the application of ^{13}C labeled compounds and NMR spectroscopy. About 50% of the applied compound remained within the residues and were either dimerized or bound to the organic matrix via ester, ether, and C—C bonds at various positions of the aromatic ring (HATCHER et al., 1993). It was shown for the residue forming fungicide anilazin that the active compound was only slightly degraded and was specifically incorporated into the humic matrix via dialkoxyl bonds after hydroxylation (WAIS et al., 1995). The proof of these reactions succeeded after the depletion of the natural ^{13}C amount of humic acids by the composting of ^{13}C depleted *Zea mays* straw in the soil and with a high dosage of fungicide (200 mg kg^{-1}). Experiments with ^{13}C and ^{14}C labeled fungicide Cyprodinil and a dosage of 500 mg kg^{-1} soil revealed the formation of 18% residues from these compounds. After 6 months of incubation, changes of the ^{13}C-NMR signals of humic acid indicated the cleavage of Cyprodinil between the aromatic rings and specific coupling of the phenyl and pyrimidyl fragments to the organic matrix. 14% of these residues were observed in fulvic acids, 23% in humic acids, and 26% in insoluble humins, whereas 23% of the amount were fixed in the fulvic acids as non-extractable, sequestered original or only slightly altered active compound (DEC et al., 1997a, b).

The formation of bound residues from xenobiotic compounds was similarly observed in experiments conducted to assess the half-life time in soil using radiolabeled compounds (HERBES and SCHWALL, 1978; HOSLER et al., 1988; PARK et al., 1988; LAMMAR and DIETRICH, 1990; KÄSTNER et al., 1993, 1995; RICHNOW et al., 1994). The residues amounted to from 20–40% during the microbial degradation of anthracene (KÄSTNER et al., 1995, 1999) up to 57-84% with 2,4,6-trinitrotoluol (DRZYZGA et al., 1998). From pyrene, the higher formation of bound residues within the soil organic matter fraction was observed in biologically active soil (49%) in comparison to soil poisoned with NaN_3 (14%). Residues in the soil with degrading microorganisms were associated with humic and fulvic acids (GUTHRIE and PFAENDER, 1998). The residue formation is usually accompanied by microbial degradation processes. However, so far only single indications concerning the molecular structures of the residues are available (DAUN et al., 1998; DRZYZGA et al., 1998; RICHNOW et al. 1994, 1998; KÄSTNER et al., 1999; KNICKER et al., 1999). The formation of residues of compounds without reactive groups such as PAH or mineral oil hydrocarbons within soil organic matter requires the *modification* of the molecules by chemical or microbial reactions in order to introduce functional groups into the molecules. The possible types of binding of the metabolites thus depend on the respective functional groups. The covalent binding of 2,4,6-trinitrotoluene only occurs after the reduction of the nitro to the corresponding amino groups, whereas a microbially catalyzed oxidation is required for PAH that leads to hydroxylated or quinoid compounds. The further degradation of the PAH proceeds via the stepwise elimination of the aromatic rings and the respective *o*-hydroxyl carbon acids which are formed. Each of the intermediate metabolites potentially contributes to the formation of bound residues. In addition, hydroxylated metabolites and hydroxynaphthoic acid are able to undergo oxidative polymerization to macromolecules with properties of humic acids after oxidation (KÄSTNER et al., 1999). Such abilities may explain why contamination by PAH is eliminated soon after forest fires in natural systems. Presumably, photooxidation enables the initiation of microbial degradation processes, which are accompanied by humification as was described for benzo(a)pyrene (MILLER et al., 1988).

During the microbial degradation of organic compounds in soil, the partitioning of carbon occurs in principle into the mineralization products CO_2 and H_2O, biomass, and the residue forming part fixed to soil organic matter (see Sect. 3.2). The incorporation of carbon into the biomass occurs in most cases intermediately within the main degradation pro-

cess in soil. After this phase, a certain amount of the residues is created by the fixation of fragments of dead microbial biomass which incorporated the carbon after the degradation of the parent compound. The higher the degree of metabolization of a given compound, the more the formation of residues should be regarded as processes of humification. The formation of residues in soil is not an unusual process. Even after the degradation of natural compounds such as plant material or glucose in soil, up to 35% of residues are formed from the carbon (STOTT et al., 1983a; BALDOCK et al., 1989). These results are of fundamental significance and can be generalized for every microbial degradation of organic compounds. Even after the degradation of ^{14}C-glucose, more than 30% of the radiolabel remained in the soil after 34 d and after one year 10% were still found to be incorporated within the organic matrix of the soil (BALDOCK et al., 1989; HAIDER, 1990). After incubation, 65% of the initial ^{13}C signals of glucose in the NMR spectra were still found within the original *O*-alkyl area, whereas 25% were shifted to the alkyl and 8% to the carboxylic areas (BALDOCK et al., 1989).

The general modes of residue formation from ^{14}C labeled compounds depend on the compounds, the association process, and the binding matrix (Tab. 5). These modes are validated for residues of xenobiotics and may be valid as well for the formation of refractory soil organic matter from fragments of biomass. Three possible binding modes with increasing stability and decreasing bioavailability can be classified: sorption, physical entrapment into cavities of macromolecules, and covalent bonds. Sorption is considered as a reversible process which leads to relatively labile associations of xenobiotics with the organic matrix. Van der Waals forces, hydrogen bonds, dipole–dipole and hydrophobic interactions, electrostatic Coulomb forces (ion bonds), ligand exchange and charge transfer complexes are the forces of interaction (HASSETT and BANWART, 1989; SENESI, 1993). These forces are additive. In addition, non-equilibrium sorption processes or entrapment may cause the adsorption–desorption hysteresis of PAH in organic soil matrices, in which 30–50% of the parent compounds were not desorbed

Tab. 5. Formation of Non-Extractable Residues from ^{14}C Labeled Xenobiotic Compounds in Soil

Origin	Modes of Association	Binding Matrix
Parent compound	sorption	xenobiotic matrix (tar oil, coal particles)
↓		
Metabolites	inclusion (physical entrapment)	*inorganic matrix (clay minerals)*
↓		
Biomass/(CO₂)	*covalent bonds*	*organic matrix (humic substances)*

Italics indicate humification reactions

after several weeks (KAN et al., 1994). Hydrophobic sorption is the partitioning of non-polar organic compounds out of the polar aqueous phase onto hydrophobic areas of organic molecules, and is relevant for the partitioning of non-polar xenobiotics into hydrophobic domains of humic aggregates in soil–water systems (GUSTAFSSON and GSCHWEND, 1998). The primary sorption forces are entropy changes resulting from the removal of the solute from the solution. Entropy changes are due to the destruction of the structured water shell surrounding the hydrophobic compound in the solution (HASSETT and BANWART, 1989). In addition, charge transfer complexes can initiate sorption reactions of compounds in which an electron donor and an acceptor molecule with partial overlapping of molecule orbitals and partial exchange of electrons are involved. For example, 6-π electron systems of dihydroxybenzenes are able to create such associations with 4-π electron systems of quinoid structures of humic substances (SENESI, 1992). Physical entrapment into cavities of macromolecules occurs after sequestering of the compounds into hydrophobic domains or sponge-like micropores of humic molecules or aggregates. The compounds can be entrapped after changes to the structure of the molecules or aggregates by altering the conditions. More stable associations result from the formation of covalent bonds (ester, ether, C—C or C—N bonds) of xenobiotic molecules or their metabolites to soil organic matter. Covalent bonds are formed by chemical reactions and display much higher reaction enthalpy (HASSETT and BANWART, 1989). As a consequence of the formation of stable chemical linkages, the xenobiotic molecule looses its chemical identity. Once bound, the xenobiotic fragment

is suggested to be an integral part of the humic material (BOLLAG and LOLL, 1983). Covalent binding requires the specific reactivity of the involved compounds or the activation by enzymes or chemical oxidation reactions. The mode of binding depends on the functional groups of compounds or on the metabolites usually formed by microbial oxidations in soil.

In the case of residues of compounds within lipophilic oil matrices, i.e., PAH in tar oil, no complete transformation occurs. A certain amount of the parent compounds remains extractable in soil. Bound residues in such systems may be created by strongly delayed desorption or physical entrapment. The residues can be caused by the parent compounds as well as by primary oxidation products of the PAH. However, the residues may be extractable under altered conditions. Such processes may cause the increase in PAH concentrations at sites contaminated by tar oil after bioremediation actions, as has sometimes been observed. Sequestered compounds (sorbed or entrapped) can be distinguished analytically from compounds actually bound within the residues by pyrolysis GC-MS techniques (RICHNOW et al., 1995) or by methods involving the silylation of the organic matrix, by which the aggregates of molecules break down and most of the substances are solubilized (DEC et al. 1997b; HAIDER et al., 1999). Subsequent mass spectroscopic or NMR analyses facilitate the identification of the residue structures.

The formation of covalent ester, ether, C—N, or C—C bonds of the xenobiotic compounds or the metabolites to humic substances only takes place by specific enzyme reactions with a low probability, since the supply of the specific enzymes, coenzymes or the cycling of redox equivalents outside the microbial cells is very limited. Due to the size of humic molecules and aggregates, the substances cannot be expected to access the active sites of the enzymes. The covalent coupling of xenobiotic compounds or of the metabolites is much more likely, assuming chemically or enzymatically generated non-specific oxidation or radical reactions. These reactions require either oxidation agents such as metal cations or reactive oxygen species derived from enzymes like laccases and peroxidases. The reactive oxygen species generate in turn reactive, low molecular weight compounds which mediate the oxidative power. Such mediator compounds include phenolic compounds like veratryl alcohol or Mn^{3+} ions, as was demonstrated for peroxidases of ligninolytic white-rot fungi (ODIER and ARTAUD, 1992; SCHOEMAKER et al., 1994; WOOD, 1994). Oxidative coupling reactions of aromatic xenobiotic compounds that are generated by non-specific oxidizing enzymes depend on the extent and type of substituents and lead to covalent $C-C$, $C-N$ or $C-O-C$ bonds. This type of coupling indicates reaction sequences that are only initiated by enzymatic action and continue chemically. Such a reaction sequence was found for the coupling of reduced metabolites of 2,4,6-trinitrotoluene to guaiacol. A nucleophilic addition of 2,4-diamino-6-nitrotoluene in *o*-position to the $C=O$ function of a diphenoquinone was observed, which was formed after the initial oxidation of guaiacol by laccases of *Pyricularia oryzae* and *Trametes versicolor* (DAWEL et al., 1997). Using laccases, the coupling of chloroanilines to dimers of ferulic acid was also observed (TATSUMI et al., 1994). A reaction mechanism via free radicals was proven for the dehalogenation of chlorophenols during the coupling and formation of oligomers by horseradish peroxidase and laccase of *Trametes versicolor* (DEC and BOLLAG, 1995).

Brown, macromolecular compounds are also formed from phenolic compounds by *abiotic processes* in the presence of *clay minerals, iron* or *manganese oxides*, and *oxygen* (BOLLAG and LOLL, 1983; RUGGIERO, 1998). Co-oligomerization of 4-chloroanilin and guaiacol in the presence of manganese dioxide as well as tyrosinases, peroxidases, and laccases leads to similar product patterns (SIMMONS et al.,

1989). Both clay minerals (birnesite) and enzymes (laccases and tyrosinase) are able to oxidize phenolic compounds. However, whereas the oxidation capacity of the manganese containing birnesite was quickly exhausted, the enzymatic activity remained constant with the consumption of oxygen even after the repeated addition of catechol. Thus, oxygen has to be regarded as an electron acceptor in the case of enzymes, and manganese is the electron acceptor in the case of clay minerals (PAL et al., 1994). Besides Mn(IV) oxides, Fe(III) oxides were also described to catalyze the formation of humic-like compounds from several di- and trihydroxybenzenes and -benzoic acids (SHINDO and HUANG, 1982; SHINDO, 1994). A cyclic reaction mechanism in the presence of MnO_2, H_2O, and O_2 for the formation of polymers from dihydroxynaphthalenes was proven. Mn^{4+} ions generate 2 naphthalene semiquinone radicals and are reduced to Mn^{2+}, which is oxidized with oxygen uptake at a MnO_2 surface to regenerate Mn^{4+} (WHELAN et al., 1995). Since humic substance within the organic soil matrix contains clay minerals, Fe and Mn ions, both abiotic and enzymatic factors might be responsible for the generation of oxidative coupling reactions (BOLLAG et al., 1995). Oxidative coupling activity to bind xenobiotic compounds to humic constituents is generally present in soils and they can be extracted (BERRY and BOYD, 1985). Thus, the ability to form bound residues from xenobiotic compounds is an immanent property of the soil system, which does not exclusively depend on the activity of microorganisms. The ability is a complex function of the respective soil site:

(1) activity of microorganisms,
(2) fixed enzymes,
(3) clay minerals,
(4) humic substances,
(5) Fe and Mn ions,
(6) input of organic matter (plant material, compost, etc.),
(7) physical and chemical parameters (water content, temperature, O_2 content, soil texture etc.).

However, without any evidence of carbon transformation from xenobiotic to natural compounds, differentiation between bound residues and humification products is only a theoretical concept. Hence, further attempts to elucidate the molecular structures are required. In general, bound residues were assessed to be less toxic, less bioavailable, and less mobile in comparison to the free parent compounds (BOLLAG and LOLL, 1983; BOLLAG et al., 1988; WANG and BARTHA, 1990; SIMS et al., 1990). Some authors argued that the enhanced formation of non-extractable residues from xenobiotics may in general be used as a technique to lower the toxic potential and bioavailability at contaminated sites (BOLLAG and LOLL, 1983; BERRY and BOYD, 1985; BOLLAG, 1992; VERSTRAETE and DEVLIEGHER, 1996). However, for a critical assessment of such techniques, further investigations of the maximum residue remobilization under environmental conditions and of long-term stability are necessary.

4.5 Turnover of Residues and Humic Substances

Humic substances have various ages, which range from a few months to several thousand years (HAIDER, 1996). Based on the different degradation rates of plant material (which are caused by varying contents of lignins), the carbon turnover and the content of humic matter was calculated for several soils (VAN VEEN et al., 1984; PARTON et al., 1987). In addition to easily degradable fragments of plants and soil biomass, there are at least three more carbon pools with different turnover rates:

(1) the active soil, i.e., the C content with rapid turnover rates (residence times of 1–5 a) by which the easily degradable C of plant material (labile fraction) is metabolized,
(2) the slowly degradable amount (residence time < 20–40 a) by which hardly degradable plant residues (stable fraction) are metabolized and which represents the largest share, and
(3) the passive carbon, which may have turnover times of more than 100 a.

These carbon pools are not separate compartments, but should be considered as mixed res-

ervoirs with floating borders and overlapping residence times. Mass transfer between these pools exists with varying turnover rates. Due to microbial activity, this turnover leads generally to the release of parts of the carbon as CO_2 and overall represents the mineralization activity of the soil. The existence of carbon pools with different turnover rates was proven later by the analysis of the turnover of stable C isotopes [$^{12}C/^{13}C$] (LICHTFOUSE et al., 1995). The majority of the CO_2 released from the soil is derived from the active soil which, however, only represents 2–12% of the C in soils (HAIDER, 1996).

Several authors assumed that, after partial degradation non-extractable residues from plant protection agents and chemicals of environmental relevance behave like natural plant components and contribute to the formation of the humic matrix (FÜHR et al., 1985; FÜHR, 1987; BOLLAG, 1992). Thus, the fate of the residue carbon should be determined by the turnover of humic substances. However, this hypothesis has not yet been proven since the structure of the residues is unknown and residues can be represented by more or less modified parent compounds, by metabolites, or by C that has been converted to biomass (see Tab. 5). Especially in the case of residue formation which is accompanied by the mineralizing degradation of the parent compounds, it should be considered carefully whether residues are actually bound residues in terms of the IUPAC definition. To assess humification techniques and composting processes in order to eliminate and detoxify components of contaminated sites, the long-term behavior and the stability of the residues should be evaluated by way of example for each compound.

Older results with residues derived from the herbicide 3,4-dichloroaniline revealed that the amounts which can be mobilized by hydrolysis was 36% after 18 d of incubation in soil. With increasing incubation time, this amount decreased to 24% after 190 d and shifted to the non-hydrolyzable fraction (HSU and BARTHA, 1976). Comparable results were observed with residues from 2,4,6-trinitrotoluolene (THORNE and LEGGETT, 1997). The question, however, of whether the transformation products of the xenobiotic compounds may be released from the organic soil matrix in the long run has still not been answered. The radiolabeled residues of herbicides derived from dichloroaniline were found to be mineralized in soils to 2–8% in 100 d. The mineralization rates of the residues depended on the strength of binding, but were always slightly higher than the turnover of soil humic matter (2–5%), so that the extensive accumulation of the residues even after repeated application of the herbicide was not to be expected (SAXENA and BARTHA, 1983). Bound residues derived from catechol or chlorophenol (synthetic humic acids) were mineralized to CO_2 within 13 weeks to an extent of 1.2–10% of the initially bound radioactivity and similar amounts (0.4–12.4%) were mostly mobilized unaltered. The majority of the radioactivity remained bound to the synthetic humic acid (DEC and BOLLAG, 1988). Experiments with the white-rot fungus *Phanerochaete chrysosporium* showed that peroxidase generated residues from ^{14}C labeled chloro and amino aromatic compounds at synthetic humic acids were mineralized in some cases to a higher degree than the free parent compounds under similar conditions (HAIDER and MARTIN, 1988). The release of CO_2 was mostly found in the range of the mineralization of the humic acids themselves. Even with chloroaniline conjugates of plants, much higher mineralization by the activity of *Phanerochaete chrysosporium* was found in comparison to the parent compounds. The mineralization of the conjugates was similar to that of the non-conjugated lignins. Although metabolites of the chloroanilines were released during these experiments, the parent compounds were not observed (ARJMAND and SANDERMANN, 1985).

Remobilization experiments with residues derived from microbial degradation of PAH showed no *mobilization* of the parent compounds (RICHNOW et al., 1995, 1998; KÄSTNER, unpublished data). After the complete depletion of ^{14}C labeled anthracene in soil, only a slight continuous mineralization of the labeled carbon from the residues was observed, which should be considered as biogenic residues. To assess the residue stability, the potential maximum turnover was evaluated in experiments after different mechanical treatments of the soil and after inoculation with white-rot fungi, which are able to depolymerize humic acids. The climatic effects of several years were sim-

ulated by freezing and thawing of the soil for 5 cycles. A total breakdown of soil texture and aggregates was achieved after milling the soil in a ball mill. Independent of the pretreatment and the addition of white-rot fungi, the residue radioactivity was diminished by mineralization in all soil cultures (2–17%, depending on the soil charge and the age of residues; see Fig. 4). However, the amount of activity that was mobilized by extraction and alkaline hydrolysis remained nearly constant. Neither mechanical nor temperature treatment nor the addition of white-rot fungi led to significantly increased mobilization or residues. In contrast to experiments with residues from chloroaniline–lignin conjugates, white-rot fungi were evidently unable to attack and mobilize residues from the microbial degradation of PAH in soil, whereas the parent PAH was transformed under these conditions. Even the addition of compost, which stimulated the metabolism of the parent anthracene (KÄSTNER and MAHRO, 1996; KÄSTNER et al., 1999) did not affect the turnover of the residues. Freeze and thaw treatments revealed a slight, but nevertheless significant increase in extractable residue activity, whereas the breakup of soil texture did not alter the extractability (unpublished data). Only the addition of fungi, which are able to depolymerize humic acids from low rank coal (HOFRICHTER and FRITSCHE, 1997), showed a slight increase in mineralization (5–6%) in contrast to the control (1%) and the increased mobilization of the residue activity into the water phase (3% ↔ 16%; unpublished data). However, the release of the parent compound or known specific metabolites was not observed in any of these cultures. The mineralization of the residues may be explained by the cleavage of labeled marginal carboxylic groups within the organic soil matrix. This hypothesis is also supported by the fact that the mineralized amount decreased with the increasing age of the residues. Microbially mediated oxidation of marginal structures was also observed for resins of crude oil (JENISCH et al., 1997), which represent macromolecules of similar complexity. To summarize these findings, it can be estimated that bound residues derived from microbially metabolized PAH display high stability to mechanical and microbial mobilization.

Residues derived from the microbial degradation of PAH in soils contaminated by tar oil similarly showed no significant changes to mobilization and mineralization after the addition of humus degrading white-rot fungi, or mechanical stress like freeze and thaw treatments (NORDLOHNE et al., 1995). However, a slight but significant increase (2–9%) in the release of ^{14}C activity from the non-extractable residues was observed after treatment with EDTA (ESCHENBACH et al., 1998). This treatment destabilizes metal organic complexes, causing the decline of aggregation followed by a release of entrapped molecules and the solubilization of humic molecules. Release was not due to the parent PAH compounds or to metabolites. More than 50% of the mobilized amount were precipitated with the humic acid fraction, and could not be extracted by organic solvents. However, not only the residues from the degradation of PAH [^{14}C naphthalene, ^{14}C anthracene, ^{14}C pyrene, and ^{14}C benzo(a)pyrene] but also the parent compounds and the residues that are associated with the tar–oil matrix were present in this soil material. The metabolism of the residues thus occurred simultaneously with the degradation of the parent compound which was still extractable, and the total mineralization was higher in comparison to the degradation of residues derived from completely metabolized single compounds. Therefore, the remobilization of residues in soils containing oil matrices cannot

be usefully examined until the parent compounds are completely transformed and no longer extractable.

The age of formation of biogenically formed residues seems to play a much more important role for the metabolization and mineralization of the residues than mechanical treatment and inoculated microorganisms. Similar experiments with increasing residue ages reveal a substantial decrease in the mineralization by the autochthonous microflora, whereas the extractability remained nearly constant (unpublished data). The decrease in mineralization was accompanied by the increasing inertness of the residues (Fig. 4). According to present knowledge, these residues can be assessed as mostly stable with only low emissions into the water path. After a residue age of 250–300 d, mineralization decreased to the range of the mean turnover of humic matter of 2–7% per year (ALEXANDER, 1977; FÜHR et al., 1985). However, this has to be ascertained for each soil. Two phases are to be observed during the microbial turnover of ^{14}C labeled anthracene in soil, which can be divided into a fast phase of the metabolism of the parent compound and a slow one of residue metabolism. After 89 d the parent compound could not be detected any more (<1%). Assuming first order kinetics for the degradation of anthracene or the residues according to the equation $C_t = C_o \cdot e^{-at}$, phase I showed a half-life time of approximately 55 d and phase II of 720 d

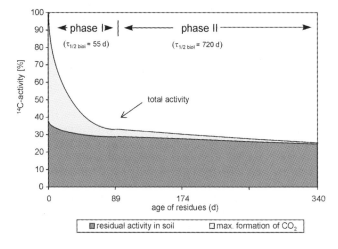

Fig. 4. Maximal mineralization (CO_2) and residual radioactivity after microbial degradation of ^{14}C anthracene at d 0 and of bound residues from labeled anthracene starting with residue ages of 89, 174, and 340 d.

(unpublished data). Similar biphasic degradation was also observed for the degradation of glucose in soil. However, the turnover of this compound amounted to 8.5 d in phase I and 60.8 d in phase II (BALDOCK et al., 1989). Thus, the observed half-life times of anthracene were higher by one order of magnitude. The fact that the half-life times of the residues of anthracene and glucose were 10-fold higher than those of the parent compounds may indicate a biological principle (HAIDER, 1996; BALDOCK et al., 1989). These two phases of degradation of radiolabeled PAH cannot be clearly distinguished in soils contaminated by tar oils matrices, since the parent compound remains extractable and no distinct transitions from the metabolism of the parent compound to the parallel metabolism of the residues can be observed (NORDLOHNE et al., 1995; ESCHENBACH et al., 1998).

No remobilization or uptake of radioactivity from the residues derived from microbial degradation of anthracene into plants were observed in germination and growth experiments with mono- and dicotylous plants (*Avena sativa* and *Lepidium sativum*). Only 0.002% and 0.006%, respectively, of the residue activity available in the root zone were taken up by the plants, and no significant mobilization was observed in this soil (unpublished data). Thus, the relevant mobilization, uptake, and transfer of the radioactivity within the residues into plants are not very likely. In experiments with rye and soy beans other authors found an uptake of 0.01–0.08% and 0.7–0.2%, respectively, from labeld anthracene into the topsoil parts of plants (GOODIN and WEBER, 1995). Similar results were obtained with tomato plants and soil contaminated by fluoranthene. Only 0.7–1% of the extractable activity were accumulated within the plants (KOLB et al., 1996). However, the majority of the applied radioactivity from anthracene or fluor-anthene in the latter experiments was still detectable as parent compounds or primary metabolites. This does not indicate a transfer from the residues, but rather the transfer of the parent compounds, which is relatively low due to the low water solubility of the compounds. The more intensive the metabolization and humification of a given compound, the lower the transfer into the plants. A transfer of 0.1% of the resid-

ual activity has to be considered from residues after microbial degradation of PAH, whereas in the presence of the parent compound within a tar oil matrix transfer may account for up to 1%. Older results concerning the transfer of labeled pesticides have been summarized by CALDERBANK (1989) such that generally small amounts of the residual activity were transferred into the plants, amounting to values <1%.

5 Summary and Concluding Remarks

The formation of refractory soil organic matter usually characterized as humification is a series of complex processes in which biomass from plants and microorganisms is post-mortally metabolized by microorganisms, while metabolites and fragments of the biomass rearrange to form macromolecular networks (molecular aggregates). In addition, new molecules are created by chemically and enzymatically generated reactions from reactive, mostly aromatic metabolites. The molecules grow with the incorporation of less reactive metabolites to form macromolecules that are characterized as humins, humic and fulvic acids, depending on solubility. Such macromolecules rearrange with each other, with fragments of the biomass and with the inorganic matrix of the soil (especially with clay minerals) to form larger aggregates. During the process, xenobiotic compounds or their metabolites were either included by interactions with clay minerals, with single macromolecules or by incorporation into the aggregation reactions. The microbial degradation of organic compounds in soil is generally accompanied by the distribution of the carbon to form mineralization products (CO_2 and H_2O), biomass, and the part which is incorporated into the organic matrix of soil. Tracing the degradation processes in soil with ^{14}C labeled compounds, the formation of non-extractable non-mobile residues within the organic matrix can be observed. In general, the possible modes of residue formation from organic compounds and xenobiotics depend on the compounds (parent molecule, metabolites, assimilated carbon in microbial biomass, or CO_2) and on the binding matrix (organic, inorganic soil matrix, or xenobiotic matrix). Three possible binding modes of xenobiotics or metabolites can be classified: sorption, physical entrapment, and covalent bonds. The formation of covalent bonds of xenobiotic compounds or their metabolites ($C-O-C, C-C, C-N$ or multivalent bonds) is very likely via non-specific, radical oxidation reactions that are abiotically generated or by enzymes. These reactions are initiated by the oxygen or Fe and Mn catalyzed abstraction of hydrogen from the respective aromatic compounds (electrophilic attack with $1-e^-$ oxidation). By this formation of radicals, the nonspecific nucleophilic addition reactions of potential reactants were generated which led to scavenging reactions of the radicals and to coupling products with a higher degree of stabilization. Covalent bonds show the high stability of the products (residues) and low probability of the remobilization of the educts. The ability to initiate oxidative coupling reactions is an immanent property of soil, which does not depend exclusively on the activity of soil microorganisms. This ability is a complex function of the respective soil site: [activity of microorganisms, fixed enzymes, clay minerals, humic substances, Fe and Mn ions, input of organic matter (plant material, compost, etc.), physical and chemical parameters (water content, temperature, O_2 content, soil texture, etc.)].

Processes of residue formation have to be differentiated into bound residue formation in terms of the IUPAC definition of pesticides that are derived from parent xenobiotic compounds or from toxicologically relevant metabolites, and biogenic residue formation. Biogenic residues have to be assessed as results of humification processes because they are derived from microbial products of metabolism (non-toxic, central metabolites of microbial metabolism) or from carbon which has been incorporated into the biomass by assimilation processes. Therefore, the term "bound carbon" instead of bound residues should be used for this type of contribution to the soil organic matter. Once incorporated into the organic soil matrix, the bioavailability of the residue

components is decreased by orders of magnitude and only minimal transfer into plants was observed. Compounds which are incorporated by covalent bonds can no longer be identified and have lost their chemical identity. Thus it can be concluded that the general turnover of humic matter is the essential factor controlling the further fate of the carbon from the residues. Biphasic degradation kinetics are observed during the microbial metabolism of organic compounds in soil that are first dominated by the degradation of the parent component. In the second phase the fixation products and residues within the organic matrix are metabolized with 10^{-1} of the turnover rates of initial degradation. With the increasing age of the residues and especially with the biogenic residues, the molecules become more and more inert, leading to humic substances with high long-term stability.

6 References

AIKEN, G. R. (1985), *Humic Substances in Soil Sediment and Water*. New York: John Wiley & Sons.

ALEXANDER, M. (1977), *Introduction into Soil Microbiology*. New York: John Wiley & Sons.

ALEXANDER, M. (1995), How toxic are toxic chemicals in soil, *Environ. Sci. Technol.* **29**, 2713–2717.

ALEXANDER, M., SCOW, K. M. (1989), Kinetics of biodegradation in soil, in: *Reactions and Movement of Organic Chemicals in Soils* (SAHWNEY, B. L., BROWN, K., Eds.), pp. 243–270. SSSA Special Publication No. 22. Soil Science Society of America, Inc.; American Society of Agronomy, Inc., Madison, WI, USA.

ARJMAND, M., SANDERMANN, H. (1985), Mineralization of chloroaniline/lignin conjugates and of free chloroanilines by the white-rot fungus *Phanerochaete chrysosporium*, *J. Agric. Food Chem.* **33**, 1055–1060.

ARMADOR, J. A., ALEXANDER, M. (1988), Effect of humic acids on the mineralization of low concentrations of organic compounds, *Soil Biol. Biochem.* **20**, 185–191.

ATLAS, R. M., BARTHA, R. (1997), *Microbial Ecology, Fundamentals and Applications* 4th Edn. Redwood City, CA: The Benjamin/Cummings Publishing Company.

AUST, S. D. (1990), Degradation of environmental pollutants by *Phanaerochaete chrysosporium*, *Microb. Ecol.* **20**, 197–209.

BAILEY, G. W., WHITE, J. L. (1964), Review of adsorption and desorption of organic pesticides by soil colloids with implication concerning pesticide bioactivity, *J. Agric. Food Chem.* **12**, 324–332.

BALDOCK, J. A., OADES, J. M., VASSALLO, A. M., WILSON, M. A. (1989), Incorporation of uniformly labeled ^{13}C glucose carbon into the organic fraction of a soil. Carbon balance and CP/MAS-^{13}C-NMR measurements, *Soil Biol. Biochem.* **27**, 725–746.

BARR, P. D., AUST, S. D. (1994), Mechanisms white-rot fungi use to degrade pollutants, *Environ. Sci. Technol.* **28**, 79A–87A.

BERRY, D. F., BOYD, S. A. (1985), Decontamination of soil through enhanced formation of bound residues, *Environ. Sci. Technol.* **19**, 1132–1133.

BOLLAG, J.-M. (1992), Decontaminating soil with enzyme, *Environ. Sci. Technol.* **26**, 1876–1881.

BOLLAG, J.-M. (1996), Decontamination of soil through immobilization of pollutants, in: *Neue Techniken der Bodenreinigung* (STEGMANN, R., Ed.), pp. 243–255. Hamburger Berichte Abfallwirtschaft Vol. 10. Bonn: Economica Verlag.

BOLLAG, J.-M., LOLL, M. J. (1983), Incorporation of xenobiotics into soil humus, *Experientia* **39**, 1221–1231.

BOLLAG, J.-M., LIU, S. Y., MINARD, R. D. (1980), Cross-coupling of phenolic humus contituents and 2,4-dichlorophenol, *Soil Sci. Soc. Am. J.* **44**, 52–57.

BOLLAG, J.-M., MINARD, R. D., LIU, S. Y. (1983), Cross-linkage between anilines and phenolic humus constituents, *Environ. Sci. Technol.* **17**, 72–80.

BOLLAG, J.-M., SHUTTLEWORTH, K. L., ANDERSON, D. H. (1988), Laccase-mediated detoxification of phenolic compounds, *Appl. Environ. Microbiol.* **54**, 3086–3091.

BOLLAG, J.-M., MYERS, C., PAL, S., HUANG, P. M. (1995), The role of abiotic and biotic catalysts in the transformation of phenolic compounds, in: *Environmental Impact of Soil Component Interactions* (HUANG, P. M., BERTHELIN, J., BOLLAG, J.-M., McGHILL, W. B., PAGE, A. L., Eds.), pp. 299–309. Boca Raton, FL: CRC Lewis Publishers.

BOOMINATHAN, K., REDDY, A. (1992), Fungal degradation of lignin: biotechnological applications, in: *Handbook of Applied Mycology, Fungal Biotechnology* Vol. 4 (ARORA, D. K., ELANDER, R. P., MUKERJI, K. G., Eds.). New York: Marcel Dekker.

BRADLEY, P. M., CHAPELLE, F. H., LOVLEY, D. R. (1998), Humic acids as electron acceptors for anaerobic microbial oxidation of vinyl chloride and dichloroethene, *Appl. Environ. Microbiol.* **64**, 3102–3105.

BRUNE, A. (1998), Termite guts: the world's smallest bioreactors, *TIBTECH* **16**, 16–21.

BUMPUS, J. A., TIEN, M., WRIGHT, D., AUST, S. D. (1985), Oxidation of persistent environmental pollutants by a white-rot fungus, *Science* **228**, 1434–1436.

BURNS, R. G. (1989), Microbial and enzymatic activities in soil biofilms, in: *Structure and Function of Biofilms* (CHARAKLIS, W. G., WILDERER, P. A., Eds.), pp. 333–349. New York: John Wiley & Sons.

CALDERBANK, A. (1989), The occurrence and significance of bound pesticide residues in soil, *Rev. Environ. Contam. Toxicol.* **108**, 71–102.

CAPRIEL, P., BECK, P., BORCHERT, H., GRONHOLZ, J., ZACHMANN, G. (1995), Hydrophobicity of the organic matter in arable soils, *Soil Biol. Biochem.* **27**, 1453–1458.

COODY, P. N., SOMMERS, L. E., NELSON, D. W. (1986), Kinetics by glucose uptake by soil microorganisms, *Soil Biol. Biochem.* **18**, 283–289.

CURTIS, G. P., REINHARD, M. (1994). Reductive dehalogenation of hexachloroethane, carbon tetrachloride, and bromoform by anthrahydroquinone disulfonate and humic acid, *Environ. Sci. Technol.* **28**, 2393–2401.

DAUN, G., LENKE, H., REUSS, M., KNACKMUSS, H.-J. (1998), Biological treatment of TNT-contaminated soil. 1. Anaerobic cometabolic reduction and interaction of TNT and metabolites with soil components, *Environ. Sci. Technol.* **32**, 1956–1963.

DAWEL, G., KÄSTNER, M., MICHELS, J., POPPITZ, W., GÜNTHER, W., FRITSCHE, W. (1997), Structure of a laccase-mediated product of coupling of 2,4-diamino-6-nitrotoluene to guiacol, a model for coupling of 2,4,6-trinitrotoluene metabolites to a humic organic soil matrix, *Appl. Environ. Microbiol.* **63**, 2560–2565.

DEC, J., BOLLAG, J.-M. (1988), Microbial release and degradation of catechol and chlorophenols bound to synthetic humic acid, *Soil Sci. Soc. Am. J.* **52**, 1366–1371.

DEC, J., BOLLAG, J.-M. (1995), Effect of various factors on dehalogenation of chlorinated phenols and anilines during oxidative coupling, *Environ. Sci. Technol.* **29**, 657–663.

DEC, J., HAIDER, K., BENESI, A., RANGASWAMY, V., SCHÄFFER, A. et al. (1997a), Analysis of soil-bound residues of ^{13}C labeled fungicide Cyprodinil by NMR spectroscopy, *Environ. Sci. Technol.* **31**, 1128–1135.

DEC, J., HAIDER, K., SCHÄFFER, A., FERNANDES, F., BOLLAG. J.-M. (1997b), Use of silylation procedure and ^{13}C-NMR spectroscopy to characterize bound and sequestered residues of Cyprodinil in soil, *Environ. Sci. Technol.* **31**, 2991–2997.

DEGENS, B. P., SPARLING, G. P. (1995), Repeated wet and dry cycles do not accelerate the mineralization of organic C involved in the macro-aggregation of a sandy loam, *Plant Soil* **175**, 197–203.

DEHORTER, B., BLONDEAU, R. (1992), Extracellular enzyme activities during humic acid degradation by the white-rot fungi *Phanerochaete chrysosporium* and *Trametes versicolor*, *FEMS Microbiol.*

Lett. **94**, 209–216.

DESCHAUER, H., HARTMANN, R., KÖGEL-KNABNER, I., ZECH, W. (1994), The influence of dissolved organic matter (DOM) on the transport of polycyclic aromatic hydrocarbons (PAHs) in a forest soil under *Pinus sylvestris*, in: *Humic Substances in the Global Environment and Implication on Human Health* (SENESI, N., MIANO, T. M., Eds.), pp. 1063–1069. Amsterdam: Elsevier.

DIXON, B. (1998), Viable but not culturable? *ASM News* **64**, 372–373.

DOMSCH, K. H., GAMS, W., ANDERSON, T. H. (1980), *Compendium of Soil Fungi*. New York: Academic Press.

DOWSON, C. G., RAYNER, A. D. M., BODDY, L. (1988), Foreaging patterns of *Phallus impudicus, Phanerochaete laevis*, and *Steccherinum fimbriatum* between discontinuous resource units in soil, *FEMS Microbiol. Ecol.* **53**, 291–298.

DUNNIVANT, F. M., SCHWARZENBACH, R. (1992), Reduction of substituted nitrobenzenes in aqueous solutions containing natural organic matter, *Environ. Sci. Technol.* **26**, 2133–2141.

DRZYZGA, O., BRUNS-NAGEL, D., GORONTZY, T., BLOTEVOGEL, K.-H., GEMSA, D., VON LÖW, E. (1998), Incorporation of ^{14}C-labeled 2,4,6-trinitrotoluene metabolites into different soil fractions after anaerobic and anaerobic/aerobic treatment of soil–molasses mixtures, *Environ. Sci. Technol.* **32**, 3529–3535.

EATON, D. C. (1985), Mineralization of polychlorinated biphenyls by *Phanerochaete chrysosporium:* a ligninolytic fungus, *Enzyme Microb. Technol.* **7**, 194–196.

EGGERT, C., TEMP, U., DEAN, J. F. D., ERIKSSON, K. E. L. (1996), A fungal metabolite mediates degradation of non-phenolic lignin structures and synthetic lignin by laccase, *FEBS Lett.* **391**, 144–148.

EGGERT, C., TEMP, U., ERIKSSON, K. E. L. (1997), Laccase is essential for lignin biodegradation by the white-rot fungus *Pycnoporus cinnabarinus, FEBS Lett.* **407**, 89–92.

ENGEBRETSON, R. R., WANDRUSZKA, R. (1994), Microorganization of dissolved humic acids, *Environ. Sci. Technol.* **28**, 1934–1941.

ESCHENBACH, A., WIENBERG, R., MAHRO, B. (1998), Fate and stability of nonextractable residues of [^{14}C] PAH in contaminated soils under environmental stress conditions, *Environ. Sci. Technol.* **32**, 2585–2590.

FIELD, J. A., DE JONG, E., COSTA, G. F., DEBONT, J. A. M. (1993), Screening for ligninolytic fungi applicable to the biodegradation of xenobiotics, *Trends Biotechnol.* **11**, 44–49.

FILIP, Z. K., CLAUS, H. (1995), Effects of soil minerals on the microbial formation of enzymes and their possible use in remediation of chemically polluted sites, in: *Environmental Impact of Soil Component Interactions* Vol. I (HUANG, P. M., BERTHELIN, J., BOLLAG, J.-M., McGILL, W. B., PAGE, A., Eds.), pp. 409–421. Boca-Raton, FL: CRC Lewis Publishers.

FLAIG, W., BEUTELSPACHER, H., RIETZ, E. (1975), Chemical composition and physical properties of humic substances, in: *Soil Components* Vol. 1 (GIESEKING, J. E., Ed.), pp. 1–211. New York: Springer-Verlag.

FRIMMEL, F. H., ABBT-BRAUN, G., HAMBSCH, B., HUBER S., SCHECK, C., SCHMIEDEL, U. (1994), Behavior and functions of freshwater humic substances – Some biological physical and chemical aspects, in: *Humic Substances in the Global Environment and Implication on Human Health* (SENESI, N., MIANO, T. M., Eds.), pp. 735–755. Amsterdam: Elsevier.

FRITSCHE, W. (1990), *Mikrobiologie*. Jena: Gustav Fischer Verlag.

FRITSCHE, W., GÜNTHER, T., HOFRICHTER, M., SACK, U. (1994), Metabolismus von polyzyklischen aromatischen Kohlenwasserstoffen durch Pilze verschiedener ökologischer Gruppen, in: *Biologischer Abbau von polyzyklischen aromatischen Kohlenwasserstoffen*, Schriftenreihe Biologische Abwasserreinigung 4, SFB 193 (WEIGERT, B., Ed.), pp. 167–182. Berlin: Technical University.

FÜHR, F. (1987), Nonextractable pesticide residues in soil, in: *Pesticide Science and Biotechnology* (GREENHALGH, R., ROBERTS, T. R., Eds.), pp. 381–389. Oxford: Blackwell Scientific Publications.

FÜHR, F., KLOSKOWSKI, R., BURAUEL, P. W. (1985), Bedeutung der gebundenen Rückstände, in: *Pflanzenschutzmittel im Boden*. Zeitschrift für Agrarpolitik und Landwirtschaft, 198. Sonderheft (Bundesminister für Ernährung, Landwirtschaft und Forsten, Bonn, Ed.). Hamburg, Berlin: Verlag Paul Parey.

GIANFREDA, L., VIOLANTE, A. (1995), Activity, Stability, and kinetic properties of enzymes immobilized on clay minerals and organomineral complexes, in: *Environmental Impact of Soil Component Interactions* Vol. I (HUANG, P. M., BERTHELIN, J., BOLLAG, J.-M., McGILL, W. B., PAGE, A. L., Eds.), pp. 201–209. Boca Raton, FL: CRC Lewis Publishers.

GISI, U. (1997), *Bodenökologie*. Stuttgart: Thieme Verlag.

GOODIN, J. D., WEBER, M. D. (1995), Persistence and fate of anthracene and benzo(a)pyrene in municipal sludge treated soil, *J. Environ. Qual.* **24**, 271–278.

GOBEL, W., PÜTTMANN, W. (1990), Untersuchungen zum mikrobiologischen Abbau von Kohlenwasserstoffen im Grundwasserkontaktbereich,

GWF – Wasser Abwasser **132**, 126–131.

GUSTAFSSON, Ö., GSCHWEND, P. (1998), Phase distributions of hydrophobic chemicals in the aquatic environment, in: *Bioavailability of Organic Xenobiotics in the Environment* (BLOCK, J. C., BAVEYE, P., GONCHARUK, V. V., Eds.). NATO ASI Series, Vol. XX. Dordrecht: Kluwer Academic Publishers.

GUTHRIE, E. A., PFAENDER, F. K. (1998), Reduced pyrene availability in microbially active soil, *Environ. Sci. Technol.* **32**, 501–508.

HACKMANN, R. H., TODD, A. R. (1953), Some observations on the reaction of catechol derivatives with amines and amino acids in the presence of oxidizing agents, *Biochem. J.* **55**, 631–637.

HAIDER, K. (1988), Der mikrobielle Abbau des Lignins und seine Bedeutung für den Kreislauf des Kohlenstoffes, *Forum Mikrobiol.* **11**, 477–483.

HAIDER, K. (1990), Humus and its significance as a bioactive soil component, *Proc. Int. Symp. Adv. in Bioactive Natural Product Chemistry*, 19/20. Oct. 1990, Seoul, Korea.

HAIDER, K. (1996), *Biochemie des Bodens*. Stuttgart: Enke.

HAIDER, K. (1998a), Microbe–soil–organic contaminant interactions, in: *Bioremediation of Contaminated Soils*, pp. 33–51. Agronomy Monorgraphs No. 37. ASA, CSSA, SSSA, Madison, WI.

HAIDER, K. (1998b), Physical and chemical stabilization mechanisms of RSOM, *Mitt. Dtsch. Bodenkundl. Ges.* **87**, 119–132.

HAIDER, K. M., MARTIN, J. P. (1988), Mineralization of ^{14}C-labeled humic acids and of humic-acid bond ^{14}C xenobiotics by *Phanerochaete chrysosporium, Soil Biol. Biochem.* **20**, 425–429.

HAIDER, K., MARTIN, J. P., FILIP, Z. (1975), Humus biochemistry, in: *Soil Biochemistry* Vol. 4 (PAUL, E. A., McLAREN, A. D., Eds.), pp. 195–244. New York: Marcel Dekker.

HAIDER, K., SPITELLER, M., DEC, J., SCHÄFFER, A. (1999), Silylation of organic matter: extraction of humic compounds and soil bound residues, in: *Soil Biochemistry* Vol. 10. (BOLLAG, J.-M., STOTZKY, G., Eds.). New York: Marcel Dekker.

HASSETT, J. J., BANWART, W. L. (1989), The sorption of non-polar organics by soils and sediments, in: *Reactions and Movement of Organic Chemicals in Soils* (SAWHNEY, B. L., BROWN, K., Eds.), pp. 31–44. SSSA Special Publication 22.

HATCHER, P. G., BORTIATYNSKI, J. M., MINARD, R. D., DEC, J., BOLLAG, J.-M. (1993), Use of high-resolution ^{13}C-NMR to examine the enzymatic covalent binding of ^{13}C labeled 2,4-dichlorophenol to humic substance, *Environ. Sci. Technol.* **27**, 2098–2103.

HATZINGER, P. B., ALEXANDER, M. (1995), Effect of aging of chemicals in soil on their biodegradability and extractability, *Environ. Sci. Technol.* **29**,

537–545.

HAYES, M. H. B., MacCARTHY, P., MALCOM, R. L., SWIFT, R. S. (Eds.) (1989), *Humic Substances II – In Search of Structure*. Chichester: John Wiley & Sons.

HEDGES, J. I. (1988), Polymerization of humic substances in natural environment, in: *Humic Substances and Their Role in the Environment* (FRIMMEL, F. H., CHRISTMAN, R. F., Eds.), pp. 45–48. Chichester: John Wiley & Sons.

HERBES, S. E., SCHWALL, L. R. (1978), Microbial transformation of polycyclic aromatic hydrocarbons in pristine and petroleum-contaminated sediments, *Appl. Environ. Microbiol.* **35**, 306–316.

HOFF, T., LIU, S. Y., BOLLAG, J.-M. (1985), Transformation of halogen-, alkyl-, and alkoxy-substituted anilines by a laccase of *Trametes versicolor, Appl. Environ. Microbiol.* **49**, 1040–1045.

HOFRICHTER, M., FRITSCHE, W. (1997), Depolymerization of low-rank coal by extracellular fungal enzyme systems. II. The ligninolytic enzymes of the coal–humic acid depolymerizing fungus *Nematoloma frowardii* b19, *Appl. Microbiol. Biotechnol.* **47**, 419–424.

HOSLER, K. R., BULMAN, T. L., FOWLIE, P. J. A. (1988), Der Verbleib von Naphtalin, Anthracen und Benz(a)pyren im Boden bei einem für die Behandlung von Raffinerieabfällen genutztem Gelände, in: *Altlastensanierung '88* (WOLF, K., VAN DEN BRINK, W. J., COLON, F. J., Eds.), pp. 111–113. Dordrecht: Kluwer.

HSU, T. S., BARTHA, R. (1976), Hydrolyzeable and non-hydrolyzeable 3,4-dichloroaniline humus complexes and their respective rate of biodegradation, *J. Agric. Food Chem.* **24**, 118–122.

HÜTTERMANN, A., LOSKE, D., MAJCHERZCYK, A. (1988), Der Einsatz von Weißfäulepilzen bei der Sanierung von besonders problematischen Altlasten, in: *Altlasten 2* (THOME-KOZMIENSKY, K. J., Ed.), pp. 1–14. Berlin: EF-Verlag.

INBAR, Y., CHEN, Y., HADAR, Y. (1990), Humic substances formed during the composting of organic matter, *Soil Sci. Soc. Am. J.* **54**, 1316–1323.

JENKINSON, D. S. (1977), Studies on the decomposition of plant material in soil. V. The effect of plant cover and soil type on the loss of carbon from ^{14}C labeled ryegrass decomposing under field conditions, *J. Soil. Sci.* **28**, 424–434.

JENKINSON, D. S., LADD, J. N. (1981), Microbial biomass in soil: Measurement and turnover, in: *Soil Biochemistry* Vol. 5 (PAUL, E. A., LADD, J. N., Eds.), pp. 415–471. New York: Marcel Dekker.

JENISCH, A., ADAM, P. P., ALBRECHT, P., MICHAELIS, W. (1997), Degradation of macromolecular crude oil fractions by bacteria, *Terra Nova* **9**, 662.

KAN, A. T., FU, G., TOMSON, M. B. (1994), Adsorption/desorption hysteresis in organic pollutant and soil/sediment interaction, *Environ. Sci. Tech-*

nol. **28**, 859–867.

KÄSTNER, M., MAHRO, B. (1996), Microbial degradation of polycyclic aromatic hydrocarbons in soils affected by the organic matrix of compost, *Appl. Microbiol. Biotechnol.* **44**, 668–675.

KÄSTNER, M., MAHRO, B., WIENBERG, R. (1993), Biologischer Schadstoffabbau in kontaminierten Böden (unter besonderer Berücksichtigung der Polyzyklischen Aromatischen Kohlenwasserstoffe PAK), *Hamburger Berichte* Vol. 5. Bonn: Economica Verlag.

KÄSTNER, M., LOTTER, S., HEERENKLAGE, J., BREUER-JAMMALI, M., STEGMANN, R., MAHRO, B. (1995), Fate of ^{14}C labeled anthracene and hexadecane in compost manured soil, *Appl. Microbiol. Biotechnol.* **43**, 1128–1135.

KÄSTNER, M., STREIBICH, S., BEYRER, M., RICHNOW, H. H., FRITSCHE, W. (1999), Formation of bound residues during microbial degradation of [^{14}C]-anthracene in soil, *Appl. Environ. Microbiol.* **65**, 1834–1842.

KELSEY, J. W., ALEXANDER, M. (1997), Declining bioavailability and inappropriate estimation of risk of persistent compounds, *Environ. Tox. Chem.* **16**, 582–585.

KHAN, S. U. (1982), Bound pesticide residues in soil and plants, *Residue Rev.* **84**, 1–25.

KNICKER, H., LÜDEMANN, H.-D. (1995), N-15 and C-13 CPMAS and solution NMR studies of N-15 enriched plant material during 600 days of microbial degradation, *Org. Geochem.* **23**, 329–341.

KNICKER, H., SCARONI, A. W., HATCHER, P. G. (1996), ^{13}C and ^{15}N NMR spectroscopic investigation on the formation of fossil algal residues, *Org. Geochem.* **24**, 661–669.

KNICKER, H., BRUNS-NAGEL, D., DRZYZGA, O., VON LÖW, E., STEINBACH, K. (1999), Characterization of ^{15}N-TNT residues after an anaerobic/aerobic treatment of soil/molasses mixtures by solid state ^{15}N-NMR spectroscopy. 1. Determination and optimization of relevant NMR spectroscopic parameters, *Environ. Sci. Technol.* **33**, 343–349.

KÖGEL-KNABNER, I., KNABNER, P. (1991), Einfluß von gelöstem Kohlenstoff auf die Verlagerung organischer Umweltchemikalien, *Mitt. Dtsch. Bodenkundl. Ges.* **63**, 119–122.

KÖGEL-KNABNER, I., GUGGENBERGER, G. (1995), Stabilisierungsprozesse der organischen Substanz in Böden, *Mitt. Dtsch. Bodenkundl. Ges.* **76**, 843–846.

KÖGEL-KNABNER, I., HATCHER, P. G., ZECH, W. (1991), Chemical and structural studies of forest soil humic acids: Aromatic carbon fraction, *Soil Sci. Soc. Am. J.* **55**, 241–247.

KOLB, M., BOCK, C., HARMS, H. (1996), Bioakkumulation und Persistenz organischer Schadstoffe aus Bioabfallkomposten in Pflanzen, in: *Neue Techniken der Kompostierung*, Hamburger Berichte Abfallwirtschaft 11 (STEGMANN, R., Ed.), pp. 345–360. Bonn: Economica Verlag.

KUHN, E. P., SUFLITA, J. M. (1989), Dehalogenation of pesticides by anaerobic microorganisms in soil and groundwater – a review, in: *Reactions and Movement of Organic Chemicals in Soils* (SAHWNEY, B., BROWN, K. L., Eds.), pp. 111–180. SSSA Special Publication No. 22. Madison, WI: Soil Science Society of America, Inc.; American Society of Agronomy, Inc.

KUTZNER, H. J. (1999), The microbiology of composting, in: *Biotechnology* 2nd Edn., Vol. 11a. (REHM, H.-J., REED, G., PÜHLER, A., STADLER, P., Eds.). Weinheim: Wiley-VCH.

LAMMAR, R. T., DIETRICH, D. M. (1990), *In situ* depletion of pentachlorophenol from contaminated soil by *Phanerochaete chrysosporium* sp., *Appl. Environ. Microbiol.* **56**, 3093–3100.

LARSON, R. A., WEBER, E. J. (1994), *Reaction Mechanisms in Environmental Organic Chemistry*. Boca Raton, FL: Lewis Publishers.

LEADBETTER, E. R. (1997), Prokaryotic diversity: form, ecophysiology, and habitat, in: *Manual of Environmental Microbiology* (HURST, C. J., KNUDSEN, G. R., MCINERNEY, M. J., STETZENBACH, L. D., WALTER, M. V., Eds.), pp. 14–24. Washington, DC: ASM Press.

LICHTFOUSE, E., DOU, S., HOUOT, S., BARRIUSO, E. (1995), Isotope evidence for soil organic carbon pools with distinct turnover rates. II. Humic substances, *Org. Geochem.* **23**, 845–847.

LOVLEY, D. R., COATES, J. D., BLUNT-HARRIS, E. L., PHILLIPS, E. J. P., WOODWARD, J. C. (1996), Humic substances as electron acceptors for microbial respiration, *Science* **382**, 445–448.

MAGEE, B. R., LION, L. W., LEMLEY, A. T. (1991), Transport of dissolved organic macromolecules and their effect on the transport of phenanthrene in porous media, *Environ. Sci. Technol.* **25**, 323–331.

MANSOUR, M., SCHEUNERT, I., KORTE, F. (1993), Fate of persistent organic compounds in soil and water, in: *Migration and Fate of Pollutants in Soils and Subsoils* (PETRUZELLI, D., HELFFERICH, F. G., Eds.), pp. 111–139. NATO ASI Series Vol. 32, Berlin: Springer-Verlag.

MARTIN, J. P., HAIDER, K. (1971). Microbial activity in relation to soil humus formation, *Soil Sci.* **111**, 54–63.

MEANS, J. G., WOOD, S. G., HASSETT, J. J., BANWART, W. L. (1980), Sorption of polynuclear aromatic hydrocarbons by sediments and soils, *Environ. Sci. Technol.* **14**, 1524-1528.

MILLER, S. L. (1955), Production of some organic compounds under possible primitive earth conditions, *J. Am. Chem. Soc.* **77**, 2352–2362.

MILLER, R. M., SINGER, G. M., ROSEN, R. D., BARTHA, R. (1988), Photolysis primes biodegradation of benzo(a)pyrene, *Appl. Environ. Microbiol.* **54**, 1724–1730.

MORTLAND, M. M. (1986) Mechanisms of adsorption of non-humic organic species by clays, in: *Interaction of Soil Minerals with Natural Organics and Microbes*. (HUANG, P. M., SCHNITZER, M., Eds.), pp. 59–75. SSSA Special Publication 17.

MÜLLER-WEGENER, U. (1990), *Bodenchemie*. Mannheim, Wien, Zürich: BI-Wiss. Verlag.

NAM, K., ALEXANDER, M. (1997), Role of nanoporosity and hydrophobicity in sequestration and bioavailability: Test with model solids, *Environ. Sci. Technol.* **32**, 71–74.

NANNY, M. A., BORTIATYNSKY, J. M., HATCHER, P. G. (1997), Non-covalent interactions between acenaphthenone and dissolved fulvic acid as determined by ^{13}C NMR T_1 relaxation measurements, *Environ. Sci. Technol.* **31**, 530–534.

NORDLOHNE, L., ESCHENBACH, A., WIENBERG, R., MAHRO, B., KÄSTNER, M. (1995), Versuche in Kleinreaktoren und Batchversuche mit Zudotierung ^{14}C-markierter Schadstoffe. Teilprojekt 4, Wissenschafliches Untersuchungsprogramm "Veringstraße" (Verbundvorhaben): Sanierungsbegleitende Untersuchung zur Stoffbilanz und zur Metabolitenbildung bei der Durchführung eines Weißfäule-Mietenverfahrens zur Reinigung des PAK-kontaminierten Bodens von dem Schadensfall "Veringstraße 2", *Abschlußberichte* an die Umweltbehörde der Hansestadt Hamburg, February, 1995.

OADES, J. M. (1993), The role of biology in the formation, stabilization and degradation of soil structure, *Geoderma* **56**, 377–400.

ODIER, E., ARTAUD, I. (1992), Degradation of lignin, in: *Microbial Degradation of Natural Products* (WINKELMANN, G., Ed.), pp. 161–191. Weinheim: VCH.

PAL, S., BOLLAG, J.-M., HUANG, P. M. (1994), Role of abiotic and biotic catalysts in the transformation of phenolic compounds through oxidative coupling reactions, *Soil Biol. Biochem.* **26**, 813–820.

PARK, K. S., SIMS, R. C., DOUCETTE, W. J., MATTHEWS, J. E. (1988), Biological transformation and detoxification of 7,12-dimethyl benz(a)anthracene in soil systems, *J. Water Pollut. Control Fed.* **60**, 1822–1825.

PARTON, W. J., SCHIMEL, D. S., COLE, C. V., OJIMA, D. S. (1987), Analysis of factors controlling soil organic matter level in Great Plains Grasslands, *Soil Sci. Soc. Am. J.* **51**, 1173–1179.

PAULI, F. W. (1967), *Soil Fertility*. London: Adam Hilger.

PARRIS, G. E. (1980), Covalent bilding of aromatic amines to humates. 1. Reactions with carbonyls and quinones, *Environ. Sci. Technol.* **14**, 1099–1106.

PICKUP, R. W. (1991), Development of molecular methods for the detection of specific bacteria in the environment, *J. Gen. Microbiol.* **137**, 1009–1019.

PÜTTMANN, W. (1990), Kriterien zur Beurteilung von Sanierungsverfahren auf mikrobiologischer Basis, in: *Bodenschutz – Loseblattsammlung* (ROSENKRANZ, D., EINSELE, G., HARREß, H.-M., Eds.), pp. 1–25. 5. Lieferung V/90. Berlin: Erich Schmidt Verlag.

RAJ, H. G., SAXENA, M., ALLAMEH, A., MURKERJI, K. G. (1992), Metabolism of foreign compounds by fungi, in: *Handbook of Applied Mycology* Vol. 4 (ARORA, D. K., ELANDER, R. P., MUKERJI, K. G., Eds.), pp. 881-904. New York: Marcel Dekker.

RICHNOW, H. H., SEIFERT, R., HEFTER, J., KÄSTNER, M., MAHRO, B., MICHAELIS, W. (1994), Metabolites of xenobiotics and mineral oil constituents linked to macromolecular organic matter in polluted environments, *Adv. Org. Geochem.* **22**, 671–681.

RICHNOW, H. H., SEIFERT, R., KÄSTNER, M., MAHRO, B., HORSFIELD, B. et al. (1995), Rapid screening of PAH residues in bioremediated soils, *Chemosphere* **31**, 3991–3999.

RICHNOW, H. H., ANNWEILER, E., FRITSCHE, W., KÄSTNER, M. (1998), Organic pollutants associated with macromolecular soil organic matter and the formation of bound residues, in: *Xenobiotics in the Environment* (BLOCK, J. C., BAVEYE, P., GONCHARUK, V. V., Eds.) NATO ASI Series, Vol. XX, Dordrecht: Kluwer Academic Publishers.

RIEGER, P. G., KNACKMUSS, H.-J. (1995), Basic knowledge and perpectives on biodegradation of 2,4,6,-trinitrotoluene and related nitroaromatic compounds in contaminated soil, in: *Biodegradation of Nitroaromatic Compounds* (SPAIN, J. C., Ed.), pp. 1–18. New York: Plenum Press.

RINDERKNECHT, H., JURD, L. (1958), A novel non-enzymatic browning reaction, *Nature* **181**, 1268–1269.

ROBERTS, T. R. (1984), Nonextractable pesticide residues in soil and plants. IUPAC Reports on pesticides (17), *Pure Appl. Chem.* **56**, 945–956.

ROSZAK, D. B., COLWELL, R. R. (1987), Survival strategies of bacteria in natural environment, *Microbiol. Rev.* **51**, 365–379.

RUGGIERO, P. (1998), Abiotic transformations of organic xenobiotic in soils: A compounding factor in the assessment of bioavailability, in: *Bioavailability of Organic Xenobiotics in the Environment* (BLOCK, J. C., BAVEYE, P., GONCHARUK, V. V., Eds.). NATO ASI Series, Vol. XX. Dordrecht: Kluwer Academic Publishers.

RUGGIERO, P., RADOGNA, V. M. (1988), Humic acids–tyrosinase interactions as a model of soil humic-enzyme complexes, *Soil. Biol. Biochem.* **20**, 350–353.

SARKAR, J. M., BOLLAG, J. M. (1987), Inhibitory effect of humic and fulvic acids on oxidoreductases as measured by the coupling of 2,4-dichlorophenol to humic substances, *Sci. Tot. Environ.* **62**, 367–377.

SARKAR, J. M., LEONOWICZ, A., BOLLAG, J. M. (1989), Immobilization of enzymes on clays and soils, *Soil Biol. Biochem.* **21**, 223–230.

SAUNDERS, J. R., SAUNDERS, V. A. (1993), Genotypic and phenotypic methods for the detection of specific released microorganisms, in: *Monitoring Genetically Manipulated Microorganisms in the Environment* (EDWARDS, C., Ed.), pp. 27–29. Chichester: John Wiley & Sons.

SAXENA, A., BARTHA, R. (1983), Microbial mineralization of humic acid-3,4-dichloroaniline complexes, *Soil Biol. Biochem.* **15**, 59–62.

SCHINK, B. (1998), Habitats of prokaryotes, in: *Biology of the Prokaryotes* (LENGELER, J. W., DREWS, G., SCHLEGEL, H. G., Eds.), pp. 763–801. Stuttgart, New York: Thieme Verlag.

SCHLEGEL, H. G. (1992), *Allgemeine Mikrobiologie.* Stuttgart, New York: Thieme Verlag.

SCHNITZER, M., SCHULTEN, H.-R. (1995), Analysis of organic matter by pyrolysis–mass spectrometry, *Adv. Agronomy* **55**, 167–217.

SCHRÖDER, D. (1984), *Bodenkunde in Stichworten.* Unterägeri, CH: Verlag Ferdinand Hirt.

SCHOEMAKER, H. E., KUNDELL, T. K., HATTAKA, A., PIONTEK, K. (1994): The oxidation of veratryl alcohol, dimeric lignin models and lignin by peroxidase: The redox cycle revisited, *FEMS Microbiol Rev.* **13**, 321–332.

SENESI, N. (1992), Binding mechanisms of pesticides to soil humic substances, *Sci. Tot. Environ.* **123/124**, 63–76.

SENESI, N. (1993), Organic pollutant migration in soils as affected by soil organic matter. Molecular and mechanistic aspects, in: *Migration and Fate of Pollutants in Soils and Subsoils* (PETRUZELLI, D., HELFFERICH, F., G., Eds.), pp. 47–74. NATO ASI Series Vol. 32. Berlin: Springer-Verlag.

SERBAN, A., NISSENBAUM, A. (1986), Humic acid association with peroxidase and catalase, *Soil. Biol. Biochem.* **18**, 41–44.

SHEVCHENKO, S. M., BAILEY, G. W. (1996), Life after death: Lignin–humic relationships re-examined, *Crit. Rev. Environ. Sci. Technol.* **26**, 95–153.

SHINDO, H. (1994), Significance of Mn(IV) and Fe(III) oxides in the synthesis of humic acids from phenolic compounds, in: *Humic Substances in the Global Environment and Implications on Human Health* (SENESI, N., MIANO, T. M., Eds.), pp. 361–366. Amsterdam: Elsevier.

SHINDO, H., HUANG, P. M. (1982), Role of Mn(IV) oxide in abiotic formation of humic substances in the environment, *Nature* **298**, 363.

SIMMONS, K. E., MINARD, R. D., BOLLAG, J.-M. (1989), Oxidative co-oligomerization of guajacol and 4-chloroaniline, *Environ. Sci. Technol.* **23**, 115–121.

SIMS, J., SIMS, R. C., MATTHEW, J. E. (1990), Approach to bioremediation of contaminated soil, *Haz. Waste Haz. Mat.* **7**, 117–149.

STAHL, D. A. (1997), Molecular approaches for the measurement of density, diversity, and phylogeny, in: *Manual of Environmental Microbiology* (HURST, C. J., KNUDSEN, G. R., MCINERNEY, M. J., STETZENBACH, L. D., WALTER, M. V., Eds.), pp. 102–114. Washington, DC: ASM Press.

STEMMER, M., GERZABEK, M. H., KANDELER, E. (1998), Organic matter and enzyme activity in particle-size fractions of soils obtained after low-energy sonication, *Soil Biol. Biochem.* **30**, 9–17.

STEMMER, M., GERZABEK, M. H., KANDELER, E. (1999), Invertase and xylanase activity of bulk soil and particle-size fractions during maize straw decomposition, *Soil Biol. Biochem.* **31**, 9–18.

STEVENSON, F. J. (1994), *Humus Chemistry: Genesis, Composition, Reactions,* 2nd Edn. New York: John Wiley & Sons.

STIEBER, M., BÖCKLE, K., WERNER, P., FRIMMEL, F. H. (1990), Abbauverhalten von polyzyklischen aromatischen Kohlenwasserstoffen (PAK) im Untergrund, in: *Altlastensanierung '90* (ARENDT, F., HINSENVELD, M., VAN DEN BRINK, W. J., Eds.), pp. 551–557. Dordrecht: Kluwer Academic Publisher.

STOTT, D. E., KASSIM, G., JARRELL, W. M. J., MARTIN, J. P., HAIDER, K. (1983a), Stabilization and incorporation into biomass of specific plant carbons during biodegradation in soil, *Plant Soil* **70**, 15–26.

STOTT, D. E., MARTIN, J. P., FOCHT, D. D., HAIDER, K. (1983b), Biodegradation, stabilization in humus, and incorporation into soil biomass of 2,4-D and chlorocatechol carbons, *Soil Sci. Soc. Am. J.* **47**, 66–70.

SUFLITA, J. M., BOLLAG, J.-M. (1980), Oxidative coupling activity in soil extracts, *Soil. Biol. Biochem.* **12**, 177–183.

SWIFT, R. S. (1985), Fractionation of soil humic substances, in: *Humic Substances in Soil Sediment and Water* (AIKEN, G. R., Ed.), pp. 387–409. New York: John Wiley & Sons.

TATSUMI, K., FREYER, A., MINARD, R. D., BOLLAG, J.-M. (1994), Enzyme-mediated coupling of 3,4-dichloroaniline and ferulic acid: a model for pollutant binding to humic materials, *Environ. Sci. Technol.* **28**, 210–215.

TEGELAAR, E. W., DELEEUW, DERENNE, S., LARGEAU, C. (1989), A reappraisal of kerogen formation, *Geochim. Cosmochim. Acta* **53**, 3103–3106.

THORNE, P. G., LEGGETT, D. C. (1997), Hydrolytic release of bound residues from composted soil contaminated with 2,4,6-trinitrotoluene, *Environ. Tox. Chem.* **16**, 1132–1134.

THENG, B. K. G. (1982), Clay activated organic reactions, *Dev. Sedimentol.* **35**, 197–238.

THORN, K. A., PETTIGREW, P. J., GOLDENBERG, W. S., WEBER, E. J. (1996), Covalent binding of aniline to humic substances. 2. ^{15}NMR studies on nucleophilic addition reactions, *Environ. Sci. Technol.* **30**, 2764–2774.

TISDALL, J. M., OADES, J. M. (1982), Organic matter and water-stable aggregates in soils, *J. Soil Sci.* **33**, 141–163.

TORSVIK, V., GOKSOYR, J., DAAE, F. L. (1990), High diversity in DNA of soil bacteria, *Appl. Environ. Microbiol.* **56**, 782–787.

VAN DER MEER, J. R., ROELOFSEN, SCHRAA, G., ZEHNDER, A. J. B. (1987), Degradation of low concentrations of dichlorobenzenes and 1,2,4-trichlorobenzene by *Pseudomonas* sp. strain P51 in nonsterile soil columns, *FEMS Microb. Ecol.* **45**, 333–341.

VAN GESTEL, M., LADD, J. N., AMATO, M. (1991), Carbon and nitrogen mineralization from two soils of contrasting texture and micro-aggregate stability: influence of sequential fumigation, drying and storage, *Soil Biol. Biochem.* **22**, 817–823.

VAN VEEN, J. A., LADD, J. N., FRISSEL, M. J. (1984), Modelling C and N turnover through the microbial biomass in soil, *Plant Soil* **76**, 257–274.

VERSTRAETE, W., DEVLIEGHER, W. (1996), Formation of non-bioavailable organic residues in soil: Perspectives for site remediation, *Biodegradation* **7**, 471–485.

VON DER KAMMER, F., FÖRSTNER, U. (1998), Natural colloid characterization using flow-field-flow-fractionation followed by multi-detector analysis, *Water Sci. Technol.* **37**, 173–180.

WAIS, A., HAIDER, K., SPITELLER, M., deGRAF, A. A., BURAUEL, P., FÜHR, F. (1995), Using ^{13}C-NMR spectroscopy to evaluate the binding mechanism of bound pesticide residues in soils, *J. Environ. Sci. Health* **B30**, 1–24.

WANG, X., YU, X., BARTHA, R. (1990), Effect of bioremediation on polycyclic aromatic hydrocarbon residues in soil, *Environ. Sci. Technol.* **24**, 1086–1089.

WARD, D. M., BATESON, M. M., WELLER, R., RUFF-ROBERTS, A. L. (1992), Ribosomal RNA analysis of microorganisms as they occur in nature, in: *Advances in Microbial Ecology.* Vol. 12. (MARSCHALL, K. C., Ed.). New York: Plenum Press.

WAKSMAN, S. A. (1932), *Humus.* Baltimore, MD: Williams and Wilkins.

WEBER, E. J., SPIDLE, D. L., THORN, K. A. (1996), Covalent binding of aniline to humic substances. 1. Kinetic studies, *Environ. Sci. Technol.* **30**, 2755–2763.

WEIßENFELS, W. D., KLEWER, H. J., LANGHOFF, J. (1992), Adsorption of polycyclic aromatic hydrocarbons (PAHs) by soil particles: influence on biodegradability and biotoxicity, *Appl. Microbiol. Biotechnol.* **36**, 689–696.

WERSHAW, R. L. (1986), Application of a membrane model to the sorptive interactions of humic substances, *Environ. Health Persp.* **83**, 191–203.

WERSHAW, R. L. (1989), Application of a membrane model to the sorptive interactions of humic substances, *Environ. Health Persp.* **83**, 191–293.

WERSHAW, R. L. (1993), Model for humus, *Environ. Sci. Technol.* **27**, 814–816.

WHELAN, C., SIMS, R. C., MURARKA, I. P. (1995), Interactions between manganese oxides and multiple-ringed aromatic compounds, in: *Environmental Impact of Soil Component Interactions* Vol. I (HUANG, P. M., BERTHELIN, J., BOLLAG, J.-M., McGILL, W. B., PAGE, A. L., Eds.), pp. 345–361. Boca Raton, FL: CRC Lewis Publishers.

WHITE, J. L. (1976), Clay–pesticide interaction, in: *Bound and Conjugated Pesticide Residues* (KAUFMANN, D. D., STILL, G.G., PAULSON, G. D., BANDAL, S. K., Eds.), pp. 208–218. ACS Symposiums Series 29.

WHITE, J. C., KELSEY, J. W., HATZINGER, P. B., ALEXANDER, M. (1997), Factors affecting sequestration and bioavailability of phenanthrene in soils, *Environ. Tox. Chem.* **16**, 2040–2045.

WOOD, P. M. (1994), Pathways for production of Fenton's reagent by wood-rotting fungi, *FEMS Microbiol. Rev.* **13**, 313–320.

XING, B., PIGNATELLO, J. J. (1997), Dual-mode sorption of low-polarity compounds in glassy poly(vinyl chloride) and soil organic matter, *Environ. Sci. Technol.* **31**, 792–799.

ZECH, W., SENESI, N., GUGGENBERGER, G., KAISER, K., LEHMANN, J. et al. (1997), Factors controlling humification and mineralization of soil organic matter in the tropics, *Geoderma* **79**, 117–161.

ZIELKE, R. C., PINNAVAIA, T. J., MORTLAND, M. M. (1989), Adsorption reactions of selected organic molecules on clay mineral surfaces, in: *Reactions and Movement of Organic Chemicals in Soils* (SAWHNEY, B. L., BROWN, B. K., Eds.). SSSA Special Publication **22**, 81–97.

ZIECHMANN, W. (1988), Evolution of structural models from consideration of physical and chemical properties, in: *Humic-Substances and their Role in the Environment* (FRIMMEL, F. H., CHRISTMAN, R. F., RUSSEL, F., Eds.), pp. 113–150. *Life Science Research Report 41.* Chichester, UK: John Wiley & Sons.

ZIECHMANN, W. (1994), *Humic Substances.* Mannheim, Wien, Zürich: BI-Wiss. Verlag.

5 Ecotoxicological Assessment

ADOLF EISENTRÄGER

Aachen, Germany

KERSTIN HUND

Schmallenberg, Germany

List of Abbreviations

COD   chemical oxygen demand
DOC   dissolved organic oxygen
EC   effect concentration
FNU   formazine nephelometric units
G_A algae test – smallest reciprocal dilution factor of the test sample which results in a reduction in fluorescence of $<20\%$ under test conditions
G_D *Daphnia* test – smallest reciprocal dilution factor of the test sample which results in a reduction in immobilization of $<10\%$ under test conditions
G_L luminescent bacteria test – smallest reciprocal dilution factor of the test sample which results in a reduction in luminescence of $<20\%$ under test conditions
G_{LW} growth test with luminescent bacteria – smallest reciprocal dilution factor of the test sample which results in a reduction in growth of $<20\%$ under test conditions
NOEC   no observed effect concentration
PAH   polyaromatic hydrocarbons
PNEC   predicted no effect concentration
TU   toxic unit

1 Introduction

The ecotoxic potential of soil samples from contaminated sites is assessed, e.g., in the scope of a hazard assessment or as a control of the success of a remediation measure. At present, the evaluation of these sites is based mainly on data from substance-specific chemical analyses. During these examinations usually the whole amount of specific substances which are expected from the history of the site is extracted and quantified. Within this frequently used approach chemical data are compared either with generic guideline values or with quality criteria derived from ecotoxicity data (FERGUSON et al., 1998; SHEPPARD et al. 1992).

These ecotoxicity data are obtained in standardized ecotoxicological single species tests with pure substances in the laboratory previously performed. For risk assessment, e.g., ecotoxicological protection levels are estimated from LC_{50} values or no observed effect toxicity data (NOEC values) (KOOIJMAN, 1987; VAN STRAALEN and DENNEMAN, 1989; WAGNER and

LØKKE, 1991; ALDENBERG and SLOB, 1993), and no effect concentrations are predicted (PNEC values).

These approaches are widely used in many countries (FERGUSON et al., 1998). However, since they are based on the chemical quantification of defined substances they are not suitable to detect ecotoxic effects of unknown substances or synergistic effects of the pollutants. Additionally, the contaminants entering the soil are subject to biotic and abiotic influences. In addition to transport and accumulation, especially the metabolization of a substance is of importance, since the compounds formed by metabolic processes may possess a higher toxic potential than the parent compounds. Metabolites are not routinely determined by chemical analyses (VAAM-Arbeitsgruppe Ökosystem Boden/Fachgruppe Umweltbiologie, 1994). Furthermore, the bioavailability of hazardous substances in soil is strongly dependent on site-specific soil characteristics.

Biological test procedures can be applied to overcome this lack of information within the ecotoxicological characterization of contaminated soil samples. With these bioassays soil samples are tested directly, and the toxicological effects of hazardous soil-bound substances are detected. They can be applied supplementary to chemical analysis. Synergistic, antagonistic, and additive effects in the soil as well as the bioavailability of pollutants are considered by these ecotoxicological examinations (DECHEMA, 1995). First approaches to compare chemical and ecotoxicological results and to detect, respectively quantify combined effects of hazardous contaminants in environmental samples are based on the calculation of toxic units (TU) (for the determination of TU, see VAN DER GAAG, 1992; PETERSON, 1994; for the application for soil assessment, see HUND and KÖRDEL, 1996).

Most bioassays originally had been developed for the assessment of chemical substances and were then applied for assessing environmental samples as, e.g., soil samples. For this reason there is a lot of experience with the individual test systems available, but for several methodological reasons the methods cannot be transferred directly to the field of soil assessment. Although some approaches are already available, no standardized and validated strategy for the testing of contaminated soil samples exists so far.

In analogy to chemical analyses a suitable investigation program has to be set up for the application of bioassays in the field of risk assessment of contaminated sites. In addition to the selection of biological test procedures the investigation program should include instructions on the preparation of the samples, on the combination of different biological test procedures, on the interpretation of the test results, and it should include some information on the relevance of these ecotoxicological results in relation to the results of chemical analyses.

Some of these aspects are shown in this chapter, and some approaches for the implementation of biological test procedures into the ecotoxicological assessment of contaminated soils are described. The fundamental concepts of ecotoxicological risk assessment including the above mentioned approach of extrapolating data from single-species toxicity testing to real-world situations in the field are reviewed by FERGUSON et al. (1998).

2 Principles of the Existing Methodological Approaches

The existing methodological approaches which are described in this chapter cannot be directly compared. Whereas within the DECHEMA approach a test battery has been set up and even threshold values have been recommended, others just consist of a collection of different test systems.

KEDDY et al. (1995) recommend the application of an acute earthworm test and a germination test for screening the contaminated soil. An aqueous soil extract is investigated using the algal growth test. The test program can be extended by tests with arthropods and bacteria. A more comprehensive investigation of the soil comprises the acute earthworm test, the germination test, and the growth test with algae. The extended test battery would include reproduction tests with lumbricidae, arthropods, and other soil organisms such as nematodes. Moreover, performance of the germination test using further species is expected, a test with bacteria using the extract and the soil as well as the growth test with algae. Concepts for interpretation of the measuring data and for a detailed assessment scheme have not yet been proposed by the authors.

In Sweden, a national research program (MATS) has recently resulted in the development of a number of guidelines for single tests designed for the testing of single substances as well as contaminated samples, e.g., soil and sewage sludge (TORSTENSSON, 1993). The tests which are based on respiration, denitrification, nitrogen fixation, ammonium oxidation, and phosphatase acitvity reflect different stages of development. Whereas some tests already can be recommended for use others should be regarded as the first step towards novel tests. A test battery for the terrestrial environment has not been built up within this study.

A scheme for the ecotoxicological assessment of contaminated soils depending on the (intended) soil use and on the soil functions to be considered was established by the DECHEMA *ad hoc* working group "*Methods for the Toxicological/Ecotoxicological Assess-*

ment of Soils" (DECHEMA, 1995). In the concept different soil functions are considered and the following test procedures are proposed:

(1) retention function of the soil (toxic/ ecotoxic potential reaching the environment via the water path), which can be investigated by means of aquatic test systems with daphnids, algae, and the luminescence test;

(2) habitat function of the soil for soil biocenoses (general biological/ecological soil functions), comprising tests for soil respiration and nitrification and tests with earthworms (mortality);

(3) habitat function of the soil as a site for plant growth (ecotoxic potential influencing plant growth), which can be investigated by means of plant tests (*Avena sativa, Brassica rapa, Phaseolus aureus*, and *Lepidium sativum*).

Further to test systems proposed for an assessment of soil functions for which threshold values have been agreed upon, optional tests without indication of threshold values are compiled in a list.

The selection of the test systems depends on the intended use of the soil. Except for sealed areas a hazard for ground water cannot principally be excluded. An investigation of the retention function is recommended for commercially or industrially used areas which are not sealed. Horticulturally and/or agriculturally used areas should be investigated not only for their retention capacity, but also for their habitat function as a site for plant growth and soil biocenoses. This system is presently improved by the ISO TC Soil Quality in form of the *"Guidance on the Ecotoxicological Characterization of Soil and Soil Materials"* (ISO CD 15,799).

3 Examples of Methodological Approaches

3.1 Test Procedures with Aquatic Organisms

3.1.1 Preparation of the Extract

An assessment of the ecotoxic potential of contaminated sites via the water path requires the preparation of aqueous extracts. However, a commonly accepted procedure for sample preparation does not exist so far. Existing procedures show variations for all parameters such as type of extract, ratio of extract to solid matter, elution method and time, and procedure for the separation of soil and extract. Some examples are given in Tab. 1.

DOMBROWSKI et al. (1996) selected some methods representative for the numerous existing procedures to conduct comparative tests with three environmental samples. As expected, a strong influence of single parameters on the toxicities determined in the extracts was found.

The applied extracts have to meet the following requirements to obtain information on the retention function of soils:

(1) simulation of the pollutant concentration in the soil pore water reflecting natural conditions as far as possible,

(2) no or only slight modification of the soil properties during the preparation of the extract,

(3) suitability for ecotoxicological testing with respect to salt and nutrient concentrations as well as pH value,

(4) feasible and well standardizable procedure which is suitable for a great variety of soils and soil substrates,

(5) sufficient amount of aqueous soil extract for tests with a test battery,

(6) adequate separation steps to obtain a sufficiently clear solution without elimination of contaminants (e.g., unacceptably high adsorption to filter materials).

Tab. 1. Examples for Preparation of Soil Extracts for Aquatic Ecotoxicity Test Systems

Extraction Solution	Elution Method	Ratio of Soil to Extraction Solution	Time	Separation Soil/Water	Authors
CO_2-saturated distilled water	agitating at 30 rpm in a rotary soil extraction	1:4 (weight:weight)	48 h	allowed to settle for 2 h, then filtered (0.45 μm)	SIMINI et al. (1995)
Distilled water	shaken overhead	1:2 (weight:weight)	2 h	filtration (0.2 μm)	RÖNNPAGEL et al. (1995)
EPA elution water	stirred (4 h), untouched overnight, stirred (1 h)	1:4 (weight:volume)	see "elution method"	filtration (0.22 μm, Millipore filter)	BIERKENS et al. (1998)
Deionized water	shaken	1:2.5 (soil dry weight:weight)	24 h	centrifugation (20,000 g, 20 min)	LENKE et al. (1998)

A first proposal for a commonly accepted procedure refers to the application of an aqueous solution, use of one soil sample (a 1:2 ratio of dry matter to water), shaking of the soil suspension for 24 h, and the separation of water and soil by means of centrifugation (DECHEMA, 1995). The result of comprehensive investigations on the preparation of extracts (WAHLE and KÖRDEL, 1997; KÖRDEL and HUND, 1998) was the proposal illustrated in Fig. 1 (KÖRDEL and HUND, 1998).

A soil–water ratio of 1:2 does not exactly correspond to the natural conditions in the soil pore water (higher dilution in the extract). On the other hand soil aggregates are disturbed as a result of shaking leading to an increase of the pollutant concentration in the extract. These oppositely directed processes allow to achieve a pollutant concentration in the extract which approximately corresponds to or slightly exceeds the concentrations in the soil saturation extract (KÖRDEL and HUND, 1998).

In a comparison of ecotoxicological and chemical analytical data a turbidity of the extract reaching about 50 FNU is acceptable for determination of the ecotoxicological data, whereas the chemical analysis has to be carried out in an extract with a turbidity below 10 FNU (formazine nephelometric units). This approach facilitates the preparation of the extracts, since time consuming cleaning steps are

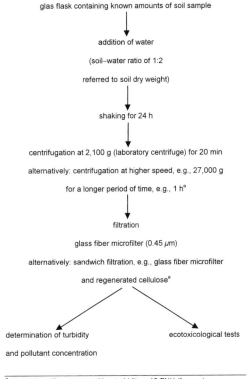

glas flask containing known amounts of soil sample

↓

addition of water

(soil–water ratio of 1:2

referred to soil dry weight)

↓

shaking for 24 h

↓

centrifugation at 2,100 g (laboratory centrifuge) for 20 min

alternatively: centrifugation at higher speed, e.g., 27,000 g

for a longer period of time, e.g., 1 h[a]

↓

filtration

glass fiber microfilter (0.45 μm)

alternatively: sandwich filtration, e.g., glass fiber microfilter

and regenerated cellulose[a]

determination of turbidity   ecotoxicological tests

and pollutant concentration

---

[a]alternatives, if no extract with a turbidity ≤ 10 FNU (formazine nephelometric units) can be achieved

Fig. 1. Proposed procedure for the preparation of water extracts from soil samples.

required only for a relatively small amount of samples (KÖRDEL and HUND, unpublished data).

3.1.2 Aquatic Test Systems for Ecotoxicological Assessment of Contaminated Soil Samples

Since the test organisms are used as an analytical instrument and a direct comparison with natural soil conditions is not required, principally any organism is suitable for the determination of toxic pollutants. Frequently, test systems are applied for which national and international guidelines from chemicals or wastewater testing exist (Tab. 2).

Further quite different test organisms and parameters have been described in the literature referring to tests with aqueous samples. Examples are the reproduction rates of nematodes (DEBUS and NIEMANN, 1994; DONKIN and DUSENBERRY, 1993) and protozoa (FORGE et al., 1993), nodulation of legumes (WETZEL and WERNER, 1995), the growth of duckweed (JENNER and JANSSEN-MOMMEN, 1993), and root growth of *Allium* (FRISKEJÖ, 1993). In the overview given here only a few authors are referred to in order to give some examples. Slightly modified test systems and test systems using different organisms are described in numerous other publications. However, not all of these test systems have been applied so far for tests with soil extracts. Some have been applied for tests with sediments, waste materials, or mineral media to which test substances were added or for testing with aqueous environmental samples (e.g., landfill leachate). In these cases an adaptation of the test systems to the requirements of soil extracts seems to be possible.

3.1.3 Expression and Evaluation of Test Results

In order to investigate the effects of the different concentrations on the test organism dilution series with the soil extracts are prepared. For tests performed in the context of soil assessments a dilution factor of 2 has proved to be adequate. Quantification of the toxicity can be done in different ways. One possibility is the calculation of the EC_X value (e.g., EC_{20}, EC_{50}) according to chemicals testing. In this case the undiluted extract is set at 100%. The EC value can be calculated either by graphic presentation or by statistic evaluation.

A further, very simple way of quantification is the indication of G values. This method was originally applied in wastewater testing in Germany and has been extended on tests with soil extracts (DECHEMA, 1995).

The G value represents the smallest reciprocal dilution factor of the test sample which gives a reduction of the test parameter of $<20\%$ (exception: acute test with daphnids, 10%). An example for the determination is given in Fig. 2.

Aquatic test systems are a means of obtaining information on whether the retention capacity of soils is reduced or whether migration of the pollutant into deeper soil layers or the ground water is possible. In some cases a response of test organisms upon exposure to soil extracts has been observed for which a contamination with organic pollutants and/or inorganic pollutants like heavy metals could be excluded. This seems to indicate that the organisms can also respond to natural soil components. Therefore, a background toxicity was considered acceptable to avoid false-positive statements for tests with algae, daphnids, and luminescent bacteria (DECHEMA, 1995).

Tab. 2. Aquatic Test Systems for the Assessment of Chemicals According to International Guidelines (Selection) which are Frequently Used for Soil Assessments

Test Organism	Parameter	Duration	Guideline
Daphnids (*Daphnia magna*)	immobilization	48 h	OECD 202, ISO 6341
Algae (e.g., *Scenedesmus subspicatus*)	growth	72 h	OECD 201; EN 28692
Microorganisms (*Pseudomonas putida*)	growth	16 h	ISO 10712
Microorganisms (*Vibrio fischeri*)	light emission	30 min	ISO/FDIS 11348-1 (-2; -3)

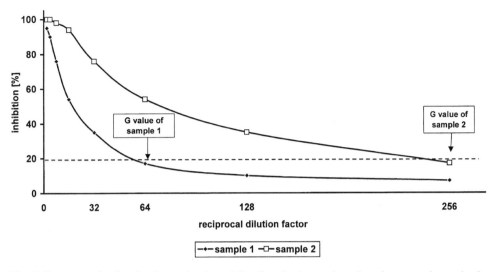

Fig. 2. Two examples for the determination of G values in the testing of environmental samples in aquatic test systems with a dual dilution series.

Upon application of a soil–water ratio of 1:2 toxic contaminants entering the environment via the water path are expected for the following dilutions of the soil extract (G values): $(G_A > 4, G_D > 4, G_L > 8, G_{LW} > 2$ with G_A: test with algae; G_D: test with daphnids; G_L: luminescence test; G_{LW}: growth test with luminescent bacteria).

If one of these "threshold values" has been exceeded, it has to be stated that there is a risk that the pollutant enters the environment; for values below the threshold danger is estimated low. Proposals for assessments with further test organisms do not exist so far.

3.1.4 Factors Influencing the Test Results

Aqueous extracts may considerably vary in their chemical-physical properties as, e.g., nutrient content, pH value, and coloring caused by eluted humic substances. Every test organism has specific nutrient requirements. These can be adequately considered when investigating pure substances by using test media which are adopted optimally to the specific requirements of the organism. As the nutrient compo-

sition of soil extracts may considerably differ from optimal nutrient supply and the concentration gradients obtained in dilution series using fixed solutions lead to a variation of the chemical composition, potential disturbing factors have to be known for all test systems and organisms to avoid misinterpretations.

Tab. 3 presents frequently observed interference factors, examples for the respective test parameters, consequences and examples for possible corrections.

3.1.5 Sensitivity and Selection of Test Systems

Ecotoxic effects of contaminants in soil can be detected with aquatic test systems if the concentrations in the aqueous soil extract are exceeding the toxicity thresholds which depend on the contaminant and on the test system. In Tab. 4 some examples for the sensitivity of ecotoxicological test systems are listed.

For most of the contaminants the sensitivity is in the range of mg L^{-1}. As ecotoxicological test systems indicate cumulative effects and soils are usually polluted with mixed contami-

Tab. 3. Interference in Aquatic Ecotoxicological Test Systems

Interference Factor	Test Parameter/ Organism (Example)	Consequences	Correction
High nutrient content compared to the optimal nutrient medium for the considered test organism	growth, reproduction, e.g., algae	stimulation of growth by increased nutrient contents and resulting interference of toxicity, therefore, underestimation of toxicity	• diluting solution also containing increased nutrient contents • test organism for which the prescribed test medium contains higher nutrient contents (e.g., duckweeds) • no reproduction tests
Extreme pH value	principally all test systems	overestimation of toxicity	• pH tolerant test organism (more detailed investigations and recommendations are still lacking) • pH correction (may, however, cause modification of bioavailability of contaminants, e.g., heavy metals)
CSB contents > approximately 1,000 mg L^{-1} (an indicator for this are high DOC contents)[a]	luminescence test (inhibition of luminescence)	overestimation of toxicity	• no correction possible
Coloring	• reproduction of organisms depending on light intensity in the colored solution (e.g., algae) • reduced luminescence signal due to absorption (luminescent bacteria) • photometric determination of enzymatic reactions (e.g., dehydrogenase activity)	overestimation of toxicity	• test sets with smaller layer thickness • commercially available color correction bulbs (luminescence test) • organisms reproducing themselves while swimming on the colored solution (e.g., duckweed)

[a] High contents of natural organic and easily degradable soil components (high CSB or DOC contents) may cause an inhibition of luminescence, which in this case is not to be interpreted as an indicator of toxic contaminants; if a sample is considered toxic contaminants should always be detectable by chemical analysis.

nations frequently showing additive or synergistic effects, single concentrations detectable in the soil may be lower. Investigations of hazardous waste sites or potentially polluted sites aim to clarify whether or not harmful soil changes have occurred. For this purpose the sensitivity of ecotoxicological test systems is sufficient.

From testing of chemicals it is well known that the sensitivity of the test systems is strongly dependent on the pollutant. Therefore, the application of a test battery is recommended. Different reactions of single organisms were also observed during ecotoxicological investigations in the course of a PAH remediation measure (HUND and TRAUNSPURGER, 1994).

Tab. 4. Ecotoxicological Analytical Data Obtained from Testing of Chemicals

	Algae, Growth Inhibition EC_{50} [mg L^{-1}]	Daphnids, Immobilization EC_{50} [mg L^{-1}]	Luminescent Bacteria, Inhibition of Luminescence EC_{50} [mg L^{-1}]
Hexogen	6.8[a]	>100[b]	**1.2**[a]
Trinitrotoluene	**1.8**[a]	12[c]	2.3[a]
4A-2,6-Dinitrotoluene	**4.4**[a]	5.2[b]	27[a]
2A-4,6-Dinitrotoluene	**2.3**[a]	4.5[b]	>10[a]
2,4-DNT	2.5[a]	26[d]	**2.0**[a]
Nitrobenzene	43[e]	**27**[e]	46[e]
Phenol	150[f]	**7.7**[g]	36[h]
Pentachlorophenol	**0.09**[i]	0.3[k]	0.4[e]

For each test substance the lowest measured values are printed bold.
[a] HUND (unpublished data)
[b] PEARSON et al. (1979)
[c] BENTLEY et al. (1977)
[d] RANDALL and KNOPP (1980)
[e] GOTTSCHALK et al. (1986)
[f] SHIGEOKA et al. (1988)
[g] PHIPPS et al. (1984)
[h] KAISER and RIBO (1985)
[i] KORTE and FREITAG (1984)
[k] HERMENS et al. (1984)

On the one hand it is frequently recommended to include organisms from different trophic levels into a test battery. In the approach of DECHEMA (1995) different trophic levels were considered (tests with destruents: bacteria; tests with consumers: daphnids; tests with phototrophic organisms: algae). But on the other hand it is well known that the sensitivities of the test systems differ in a wide range even within one trophic level (SCHMITZ et al., 1998). For this reason a test battery should primarily consist of several test systems with complementing sensitivities, and several test systems should be applied for a comprehensive ecotoxicological assessment.

3.1.6 Contact Assays

Besides the testing of aquatic extracts using aquatic organisms to characterize the retention function tests with suspended soil material are performed, e.g., using a test system developed by RÖNNPAGEL et al. (1995) or the Mi-

crotox™ solid-phase test with *Vibrio fischeri*. These test systems provide information on the bioavailability of contaminants for the respective organisms. They are, therefore, a link between aquatic and terrestrial test systems.

3.2 Test Systems with Terrestrial Test Organisms

3.2.1 Selection of Test Organisms

Terrestrial test systems are applied to obtain information on impaired habitat functions of soils due to the pollutant impact. As the soil is a habitat for a wide variety of plants and organisms, an impact on the habitat function of soils can only be determined by tests using representative organisms. Criteria for the selection of test organisms were compiled by DUNGER (1982) and SAMSOE-PETERSEN and PEDERSEN (1994). Soil organisms do not live independently, but form a complex food web.

Therefore, the elimination of a target organism due to the pollutant impact can have indirect effects on other, not directly affected organisms. A further criterium to be considered for the selection of organisms is their specific habitat which depends, e.g., on size and living behavior of a species (organism) and leads to a different exposure to a contaminant. Soft-bodied organisms living in the soil pore water or in close contact to the soil are predominantly affected by water soluble contaminants, whereas fairly soluble sorbed contaminants can be taken up by organisms for which organic material is the main nutrient source. For organisms residing in soil pores filled with air (e.g., collemboles) the gas phase will be the major exposure route, and for some organisms the major path of exposure may change in the course of the individual development (e.g., larvae, adults).

Therefore, a test battery should consider not only the sensitivity of organisms towards a chemical substance, but should also take into account the specific living behavior of an organism.

Test systems with terrestrial organisms can only be successfully applied in soil assessments, if the obtained results are interpretable (see Sect. 3.2.2). Tab. 5 presents a list of test methods for which a proposal for evaluation already exists or is being prepared. It is evident from the scheme that the test battery still needs considerable improvement with respect to simulating the complexity of soil life. Test systems which are already standardized or which are presently in process of standardization are systems with collembola, gamasida, staphylinidae, lumbricidae, oribatidae, nematoda, enchytreidae, carabidae, ichneumonidae, *Rhizobium* as well as test systems based on direct measurement of enzymatic microbial activities. The establishment of respective assessment criteria could contribute to closing the existing gaps in a terrestrial test battery in a short period of time.

3.2.2 Setup and Assessment of the Test Results

Organisms already present in the investigated soil (e.g., microorganisms) as well as organisms added to the soil sample for the specific test (e.g., earthworms, plants) are used for the tests.

Growth, reproduction, and activity of organisms residing in soil are influenced by the physical-chemical soil characteristics. For a soil assessment it is, therefore, necessary that the results obtained with the applied test systems can be interpreted with respect to potential adverse effects on the habitat function of the soil. However, uncontaminated reference soils with chemical-physical properties comparable to the test soils frequently are not available. To overcome this problem several strategies exist. Minimal values regarding microbial respiration, nitrification, plant growth, and survival rate of the earthworms were proposed by DE-CHEMA (1995). If higher values are determined, it can be assumed that there are no adverse effects on nutrient cycles. From this, however, it cannot be concluded that contaminants do not occur in the soils.

For microorganisms and plants some approaches are already available to obtain information on a potential presence of toxic contaminants. One refers to microbial respiration (TORSTENSSON, 1993). For this test system the quotient from basal and substrate-induced res-

Tab. 5. Terrestrial Test Systems for which Proposals for a Results Evaluation with Respect to the Assessment of the "Habitat" Function of the Soil are Already Available

Test Organism	Parameter	Test Duration	Guideline
Microorganisms	respiration	at least 6 h	ISO 14240-1
Microorganisms	nitrification	6 h	ISO NP 15685
Earthworms	mortality	14 d	ISO 11268-1
Plants	emergence, growth	14 d	ISO 11269-2

piration and the lag phase after addition of glucose until reproduction of the organisms turned out to be different for contaminated and uncontaminated soils. In contaminated soils quotients of basal and substrate-induced respiration activity >0.3 and lag times >20 h are obtained (WILKE et al., 1998). These values are lower in uncontaminated soil samples.

For the plant test and for the determination of nitrification it has been proposed to test a control soil, the soil sample, and a 50:50 mixture of both soils. According to the results obtained so far with the plant test the value for the mixture of soil sample and control soil does not correspond to the calculated mean value of biomass obtained for the soil sample and the control soil (WILKE et al., 1998).

4 Examples of the Ecotoxicological Characterization of Contaminated Soils

The ecotoxic potential from contaminated soil samples reaching the environment via the water path has been assessed by MAXAM et al.

(in press). Several soil samples containing different toxicants have been characterized by means of chemical analysis and ecotoxicity tests in parallel. Soil samples LMKW1 and SPMKW1 are contaminated with mineral oil. Samples HTNT1 and CTNT1a were taken from military production plants. Sample HTNT1 contains large amounts of nitroaromatics (16,700 mg kg^{-1}), polycyclic aromatic hydrocarbons and lead. Sample CTNT1a contains 1,130 mg kg^{-1} nitroaromatics. Sample SPAK1a is contaminated with PAH and chromium (960 mg kg^{-1} PAH, 885 mg kg^{-1} Cr^{2+}), and R1 contains several heavy metals (MAXAM et al., in press).

The results presented in Fig. 3 clearly demonstrate the different sensitivity of the tester strains on the one hand and the differing water extractability, respectively mobility of the contaminants on the other hand. There is a low risk of mobilization from the sites contaminated with mineral oils, whereas the samples taken from military sites that are contaminated with nitroaromatics show extremely high ecotoxicity in the aquatic test systems. Sample SPAK1a shows high *G* values in the algae test, the luminescence test and the *Daphnia* test. This ecotoxic potential is probably caused by the high contents of chromium, which was not considered during remediation of PAH. The high concentrations of organic matter of sample R1 reduce the mobility and the toxicity of

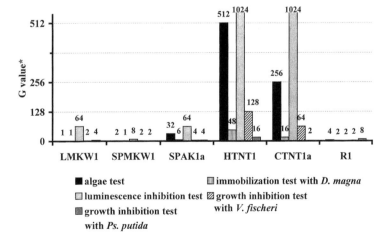

Fig. 3. Water extractable ecotoxicological potential of the soil samples contaminated with organic and inorganic substances (MAXAM et al., in press), *G value: smallest reciprocal dilution factor where the inhibition of the test factor is <20%; *Daphnia* test <10%.

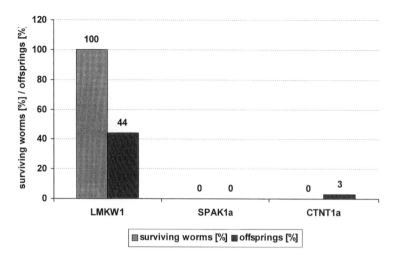

Fig. 4. Ecotoxicological test with earthworms for soil assessment.

heavy metals. The results show that there is a low risk of transfer of heavy metals from sample R1 (MAXAM et al., in press).

Results of ecotoxicity tests with earthworms are presented in Fig. 4. The soils were adjusted to 60% of the maximum water holding capacity and the tests were performed following the incubation conditions described in the guidelines ISO 11268-1 and 11268-2. With the tests a differentiation between the soils is possible. Whereas no mortality was obtained in LMKW1, 100% mortality was detected in SPAK1a and CTNT1a. According to DECHEMA (1995) only LMKW1 would be classified as *"habitat function for earthworms is given"*. With respect to the number of offsprings an effect could also be detected for LMKW1. Reproduction is a more sensitive indicator than mortality. However, as the test is much more time-consuming than the test with respect to mortality it is still under discussion whether the reproduction test should be proposed for routine soil assessment.

The results clearly show that aquatic and terrestrial test systems give different and complementary information: Sample LMKW1 has a low toxic potential in the aquatic test systems and no toxic potential in the earthworm tests. Samples SPAK1a and CTNT1a give different responses in the aquatic test systems, whereas toxicity on earthworms is in the same range.

5 Summary of the State of the Art and Future Aspects

A lot of powerful test systems for the ecotoxicological characterization of contaminated sites have been developed in recent years. Some of them are standardized, some are in the process of standardization right now. Although the degree of standardization of the test systems has been improved, there is still a great variety in the fields of sample preparation and sample storage on the one hand and in the field of interpretation of test results on the other hand. Due to these circumstances it can be recommended that the methodic aspects should be documented in detail in order to ensure comparability of test results. There should be a strong emphasis on quality assessment and quality control.

Despite of this lack of standardization ecotoxicological test systems can already be successfully applied before, during, and after remediation of contaminated sites. The obtained results provide important information on the biological and ecotoxicological status of the soil, and they should be considered during every process of soil remediation.

For the implementation of the exotoxicological characterization of contaminated sites into the legally binding assessment of contaminated sites the following aspects seem to be of great importance for future research:

(1) standardization of soil sampling, transport, sample preparation, sample storage, and test procedures;
(2) selection of test systems with respect to soil functions;
(3) harmonized definition of assessment criteria.

By the application of these test systems a path-specific assessment of contaminated sites considering the intended future land use is possible. It provides important additional information which is complementary to the information derived from chemical analyses.

6 References

ALDENBERG, T., SLOB, W. (1993), Confidence limits for hazardous concentrations based on logistically distributed NOEC data, *Ecotox. Environ. Safety* **25**, 48–63.

BENTLEY, R. E., DEAN, J. W., ELLIS, S. J., HOLLISTER, T. A., LeBLANC, G. A. et al. (1977), Laboratory evaluation of the toxicity of cyclo trimethylene trinitramine (RDX) to aquatic organisms, U.S. Army Med. Res. Develop. Command. Frederick, mD *Government Report* Announce, Index 79 (7), U.S. NTIS AD-A061 730 (cited in ETNIER, 1986).

BIERKENS, J., KLEIN, G., CORBISIER, P., VAN DEN HEUVEL, R., VERSCHAEVE, L. et al. (1998), Comparative sensitivity of 20 bioassays for soil quality, *Chemosphere* **37**, 2935–2347.

DEBUS, R., NIEMANN, R. (1994), Nematode test to estimate the hazard potential of solved contamination, *Chemosphere* **29**, 611–621.

DECHEMA (1995), Bioassays for soils, *4th Report of the Interdisciplinary DECHEMA Committee "Environmental Biotechnology – Soil"*, Ad hoc committee "Methods for Toxicological/Ecotoxicological Assessment of Soils", DECHEMA e.V., Frankfurt/Main. Schön & Wetzel.

DOMBROWSKI, E. C., GAUDET, I. D., FLORENCE, L. Z. (1996), A comparison of techniques used to extract solid samples prior to acute toxicity analysis using the Microtox test, *Environ. Toxicol. Water Qual.* **11**, 121–128.

DONKIN, S. G., DUSENBERRY, D. B. (1993), A soil toxicity test using the nematode *Caenorhabditis ele-*

gans and an effective method of recovery, *Arch. Environ. Contam. Toxicol.* **25**, 145–151.

DUNGER, W. (1982), Die Tiere des Bodens als Leitformen für anthropogene Umweltveränderungen, *Decheniana Beiheft* **26**, 151–157.

EN 28692 (1993), Water quality; fresh water algal growth inhibition test with *Scenedesmus subspicatus* and *Selenastrum capricornutum*, European Committee for Standardization (CEN) (Ed.).

FERGUSON, C., DARMENDRAIL, D., FREIER, K., JENSEN, B. K., JENSEN J. et al. (Eds.) (1998), Risk assessment for contaminated sites in Europe, Vol. 1, Scientific Basic. Nottingham: LQM Press.

FORGE, T. A., BERRWO, M. L., DARBYSHIRE, J. F., WARREN, A. (1993), Protozoan bioassays of soil amended with sewage sludge and heavy metals, using the common soil ciliate *Colpoda steinii*, *Biol. Fertil. Soils* **16**, 282–286.

FRISKEJÖ, G. (1993), *Allium* test I: A 2–3 plant test for toxicity assessemt by measuring the mean root growth of onions (*Allium cepa* L.), *Environ. Toxicol. Water Qual.* **8**, 461–470.

GOTTSCHALK, C., MARKARD, C., HELLMANN, H., KREBS, F., HANSEN, P. D., KÜHN, R. (1986), Beitrag zur Beruteilung von 19 gefährlichen Stoffen in oberirdischen Gewässern (Umweltbundesamt, Ed.). Berlin.

HERMENS, J., CANTON, H., STEYGER, N., WEGMAN, R. (1984), Joint effects of a mixture of 14 chemicals on mortality and inhibition of reproduction of *Daphnia magna, Aquat. Toxicol.* **5**, 315–322.

HUND, K., TRAUNSPURGER, W. (1994), Ecotox evaluation strategy for soil bioremediation exemplified for a PAH-contaminated site, *Chemosphere* **29**, 371–390.

HUND, K., KÖRDEL, W. (1996), Erfassung der Grundwassergefährdung durch aquatische Testsysteme, in: *Hamburger Berichte 10*, Neue Techniken der Bodenreinigung (STEGMANN, R., Ed.), pp. 207–218. Economica Verlag, Bonn.

ISO 11268-1 (1993), Soil quality – Effects of pollutants on earthworms (*Eisenia fetida*) – Part 1: Determination of acute toxicity using artificial soil substrate.

ISO 10712 (1995), Water quality – *Pseudomonas putida* growth inhibition test (*Pseudomonas* cell multiplication inhibition test).

ISO 11269-2 (1995), Soil quality – Determination of the effects of pollutants on soil flora – Part 2: Effects of chemicals on the emergence and growth of higher plants.

ISO 6341 (1996), Water quality – Determination of the inhibition of the mobility of *Daphnia magna* Straus (*Cladocera, Crustacea*) – Acute toxicity test.

ISO 14240-1 (1997), Soil quality – Determination of soil microbial biomass – Part 1: Substrate-induced respiration method.

ISO 11268-2 (1998), Soil quality – Effects of pollutants on earthworms (*Eisenia fetida*) – Part 2: Determination of effects on reproduction.

ISO CD 15799 (1998), Guidance on the ecotoxicological characterization of soil and soil materials.

ISO/FDIS 11348-1 (1998), Water quality – Determination of the inhibitory effect of water samples on the light emission of *Vibrio fischeri* (luminescent bacteria test) – Part 1: Method using feshly prepared bacteria.

ISO/FDIS 11348-2 (1998), Water quality – Determination of the inhibitory effect of water samples on the light emission of *Vibrio fischeri* (luminescent bacteria test) – Part 1: Method using liquid-dried bacteria.

ISO/FDIS 11348-3 (1998), Water quality – Determination of the inhibitory effect of water samples on the light emission of *Vibrio fischeri* (luminescent bacteria test) – Part 1: Method using freeze-dried bacteria.

ISO NP 15685 (1998), Ammonium oxidation – a rapid method to test potential nitrification in soil.

JENNER, H. A., JANSSEN-MOMMEN, J. P. M. (1993), Duckweed *Lemna minor* as tool for testing toxicity of coal residues and polluted sediments, *Arch. Environ. Contam.* **25**, 3–11.

KAISER, K. L. E., RIBO, J. M. (1985), QSAR of toxicity of chlorinated aromatic compounds, in: *QSAR in Toxicology and Xenobiochemistry* (TICHY, M., Ed.), pp. 27–38. Amsterdam: Elsevier.

KEDDY, C. J., GREENE, J. C., BONNELL, M. A. (1995), Review of whole-organism bioassays: soil, freshwater sediment, and freshwater assessment in Canada, *Ecotox. Environ. Safety* **30**, 221–251.

KOOIJMAN, S. A. L. M. (1987), A safety factor for LC50 values allowing for differences in sensitivity among species, *Water Res.* **21**, 269–276.

KÖRDEL, W., HUND, K. (1998), Soil extraction methods for the assessment of the ecotoxicological risk of soils, *Proc. Contaminated Soil '98, 6th Int. FZK/TNO Conf. ConSoil '98*, pp. 241–250. London: Thomas Telford.

KORTE, F., FREITAG, D. (1984), Überprüfung der Durchführbarkeit von Prüfungsvorschriften und der Aussagekraft der Stufe I und II des Entwurfs des Chemikaliengesetzes (Bericht der Gesellschaft für Strahlen- und Umweltforschung mbH, Neuherberg, Institut für Ökologische Chemie und Institut für Biochemie und Toxikologie, Abt. Toxikologie, an das Umweltbundesamt, Berlin), *Forschungsbericht Nr. 106 04 011/02.*

LENKE, H., WARRELMANN, J., DAUN, G., HUND, K., SEIGLEN, U. et al. (1998), Biological treatment of TNT-contamianted soil. 2. Biologically induced immobilization of the contaminants and full-scale application, *Environ. Sci. Technol.* **32**, 1964–1971.

MAXAM, G., RILA, J.-P., DOTT, W., EISENTRAEGER, A. (in press), Use of bioassays for the assessment of the water extractable ecotoxic potential of soils, *Ecotox. Environ Safety*.

OECD 201 (1984), *OECD Guideline for Testing of Chemicals 201* (adopted 7 June, 1984), *Alga Growth Inhibition Test*. Paris: OECD, 1981, 4. Vol. 1.

OECD 202 (1984), *OECD Guideline for Testing of Chemicals 202* (adopted April 4, 1984), *Daphnia* sp., *Acute Immobilization Test and Reproduction Test*. Paris: OECD, 1981, 4. Vol. 1.

PEARSON, J. G., GLENNON, J. P., BARKLEY, J. J., HIGHFILL, J. W. (1979), An approach to the toxicological evaluation of a complex industrial wastewater, in: *Aquatic Toxicology* (MARKING, L. L., KIMMERLE, R. A., Eds.), ASTM STP 667, pp. 284–301. Philadelphia.

PETERSON, D. R. (1994), Calculating the aquatic toxicity of hydrocarbon mixtures, *Chemosphere* **29**, 2493–2506.

PHIPPS, G. L., HARDEN, M. J., LEONARD, E. N., ROUSH, T. H., SPEHAR, D. L. et al. (1984), Effects of pollution on freshwater organisms, *J. Water Pollut. Control Fed.* **46**, 725.

RANDALL, T. L., KNOPP, P. V. (1980), Detoxification of specific organic substances by wet oxidation, *J. Water Pollut. Control Fed.* **52**, 2117–2130.

RÖNNPAGEL, K., LIß, W., AHLF, W. (1995), Microbial bioassays to assess the toxicity of solid-associated contaminants, *Ecotox. Environ. Safety* **31**, 99–103.

SAMSOE-PETERSEN, L., PEDERSEN, F. (1994), Discussion paper regarding guidance for terrestrial effect assessment, Water Quality Institute, Horsholm (Denmark), *Draft Final Report*, August 1994.

SCHMITZ, R. P. H., EISENTRÄGER, A., DOTT, W. (1998): Miniaturized kinetic growth inhibition assays with *Vibrio fischeri* and *Pseudomonas putida* (application, validation and comparison), *J. Microbiol. Methods* **31**, 159–166.

SHEPPARD, S. C., GAUDET, C., SHEPPARD, M. I., CURETON, P. M., WONG, M. P. (1992), The development of assessment and remediation guidelines for contaminated soils, a review of the science, *Can. J. Soil Sci.* **72**, 359–394.

SHIGEOKA, T., SATO, Y., TAKEDA, Y., YOSHIDA, K., YAMAUCHI, F. (1988), Acute toxicity of chlorophenols to green algae, *Selenastrum capricomutum* and *Chlorella vulgaris*, and quantitative structure–activity relationships, *Environ. Toxicol. Chem.* **7**, 847–854.

SIMINI, M., WENTSEL, R. S., CHECKAI, R. T., PHILLIPS, C. T., CHESTER, N. A. et al. (1995), Evaluation of soil toxicity at Joliet army ammunition plant, *Environ. Toxicol. Chem.* **14**, 623–630.

TORSTENSSON L. (Ed.) (1993), Guidelines – Soil biological variables in environmental hazard assessment, Swedish Environmental Protection Agency, *Report No. 4262*. Uppsala: GEO tryckeriet.

VAAM-Arbeitsgruppe Ökosystem Boden/Fachgruppe Umweltmikrobiologie (1994), Regeneration mikrobieller Aktivität in Böden nach natürlichen Streßsituationen – Bewertungskriterium für die Bodenqualität, *Bioengineering* **6**, 38–41.

van der Gaag, M. (1992), Combined effects of chemicals: an essential element in risk extrapolation for aquatic ecosystems, *Water Sci. Technol.* **25**, 441–447.

van Straalen, N. M., Dennemann, C. A. J. (1989), Ecotoxicological evaluation of soil quality criteria, *Ecotox. Environ. Safety* **18**, 241–251.

Wagner, C., Løkke, H. (1991), Estimation of eco-toxicological protection levels from NOEC toxicity data, *Water Res.* **25**, 1237–1242.

Wahle, U., Koerdel, W. (1997), Development of analytical methods for the assessment of ecotoxicological relevant soil contamination, *Chemosphere* **35**, 223–237.

Wetzel, A., Werner, D. (1995), Ecotoxicological evaluation of contaminated soil using the legume root nodule symbiosis as effect parameter, *Environ. Toxicol. Water Qual.* **10**, 127–133.

Wilke, B.-M., Winkel, B., Fleischmann, S., Gong, P. (1998), Higher plant growth and microbial toxicity tests for the evaluation of the ecotoxic potential of soils, *Proc. Contaminated Soil '98, 6th Int. FZK/TNO Conf. ConSoil '98*, pp. 345–354. London: Thomas Telford.

II Microbiological Aspects

6 Aerobic Degradation by Microorganisms

WOLFGANG FRITSCHE
MARTIN HOFRICHTER

Jena, Germany

List of Abbreviations

AsO	arsenic-containing organic compounds
BTX	benzene, toluene, xylenes
DBDs	dibenzo-*p*-dioxines
DBFs	dibenzofurans
DCA	3,4-dichloroaniline
DCP	2,4-dichlorophenol
DDT	1,1,1-trichloro-2,2′-bis(4-chlorophenyl)ethane
DNT	2,4-dinitrotoluene
KCN	potassium cyanide
LiP	lignin peroxidase
MnP	manganese peroxidase
PAHs	polycyclic aromatic hydrocarbons
PCBs	polychlorinated biphenyls
PCP	pentachlorophenol
TCC	tricarboxylic acid cycle
TCE	trichloroethene
TNT	2,4,6-trinitrotoluene

1 Introduction: Characteristics of Aerobic Microorganisms Capable of Degrading Organic Pollutants

The most important classes of organic pollutants in the environment are mineral oil constituents and halogenated products of petrochemicals. Therefore, the capacities of aerobic microorganisms are of particular relevance for the biodegradation of such compounds and are exemplarily described with reference to the degradation of aliphatic and aromatic hydrocarbons as well as their chlorinated derivatives. The most rapid and complete degradation of the majority of pollutants is brought about under aerobic conditions (RISER-ROBERTS, 1998).

The essential characteristics of aerobic microorganisms degrading organic pollutants are (Fig. 1):

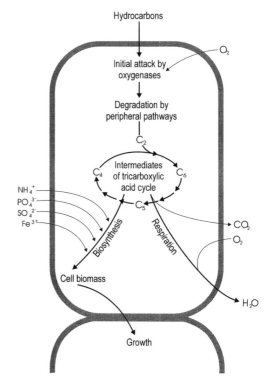

Fig. 1. Main principle of aerobic degradation of hydrocarbons: growth associated processes.

(1) Metabolic processes for optimizing the contact between the microbial cells and the organic pollutants. The chemicals must be accessible to the organisms having biodegrading activities. For example, hydrocarbons are water-insoluble and their degradation requires the production of biosurfactants.

(2) The initial intracellular attack of organic pollutants is an oxidative process, the activation and incorporation of oxygen is the enzymatic key reaction catalyzed by oxygenases and peroxidases.

(3) Peripheral degradation pathways convert organic pollutants step by step into intermediates of the central intermediary metabolism, e.g., the tricarboxylic acid cycle.

(4) Biosynthesis of cell biomass from the central precursor metabolites, e.g., acetyl-CoA, succinate, pyruvate. Sugars required for various biosyntheses and growth must be synthesized by gluconeogenesis.

A huge number of bacterial and fungal genera possess the capability to degrade organic pollutants. Biodegradation is defined as the biologically catalyzed reduction in complexity of chemical compounds (ALEXANDER, 1994). It is based on two processes: growth and cometabolism. In the case of growth, organic pollutants are used as sole source of carbon and energy. This process results in a complete degradation (mineralization) of organic pollutants as demonstrated in Sect. 2.2. Cometabolism is defined as the metabolism of an organic compound in the presence of a growth substrate which is used as the primary carbon and energy source. The principle is explained in Sect. 2.4.

Enzymatic key reactions of aerobic biodegradation are oxidations catalyzed by oxygenases and peroxidases. Oxygenases are oxidoreductases that use O_2 to incorporate oxygen into the substrate. Degradative organisms need oxygen at two metabolic sites, at the initial attack of the substrate and at the end of the respiratory chain (Fig. 1). Distinct higher fungi have developed a unique oxidative system for the degradation of lignin based on extracellular ligninolytic peroxidases and laccases. This enzymatic system possesses increasing significance for the cometabolic degradation of persistent organopollutants. Thus it is in particular the domain of basidiomycetous fungi, which requires a deeper insight and an extensive consideration. Therefore, this chapter has been divided in two sections: bacterial and fungal degradation capacities.

2 Principles of Bacterial Degradation

2.1 Typical Aerobic Degrading Bacteria

The predominant degraders of organopollutants in the oxic zone of contaminated areas are chemo-organotrophic species able to use a huge number of natural and xenobiotic compounds as carbon sources and electron donors for the generation of energy. Although many bacteria are able to metabolize organic pollutants, a single bacterium does not possess the enzymatic capability to degrade all or even most of the organic compounds in a polluted soil. Mixed microbial communities have the most powerful biodegradative potential because the genetic information of more than one organism is necessary to degrade the complex mixtures of organic compounds present in contaminated areas. The genetic potential and certain environmental factors such as temperature, pH, and available nitrogen and phosphorus sources, therefore, seem to determine the rate and the extent of degradation.

The predominant bacteria of polluted soils belong to a spectrum of genera and species listed in Tab. 1. The composition of this list of bacteria is determined by the fact whether they can be cultured on nutrient-rich media. We have to consider that the majority of bacteria present in soils cannot be cultured in the laboratory yet.

Pseudomonads, aerobic gram-negative rods that never show fermentative activities, seem to have the highest degradative potential, e.g., *Pseudomonas putida* and *P. fluorescens*. Further important degraders of organic pollutants can be found within the genera *Comamonas,*

Tab. 1. Predominant Bacteria in Soil Samples Polluted with Aliphatic and Aromatic Hydrocarbons, Polycyclic Aromatic Hydrocarbons, and Chlorinated Compounds[a]

Gram-Negative Bacteria	Gram-Positive Bacteria
Pseudomonas spp.	*Nocardia* spp.
Acinetobacter spp.	*Mycobacterium* spp.
Alcaligenes sp.	*Corynebacterium* spp.
Flavobacterium/ Cytophaga group	*Arthrobacter* spp.
Xanthomonas spp.	*Bacillus* spp.

[a] The reclassification of bacteria has been based on phylogenetic markers resulting in changes of some genera and species. This is why the names of species are not mentioned.

Burkholderia, and *Xanthomonas.* Some species utilize > 100 different organic compounds as carbon sources. The immense potential of the pseudomonas does not solely depend on the catabolic enzymes, but also on their capability of metabolic regulation (HOUGHTON and SHANLEY, 1994). A second important group of degrading bacteria are the gram-positive rhodococci and coryneform bacteria. Many species, now classified as *Rhodococcus* spp. had originally been described as *Nocardia* spp., *Mycobacterium* spp., and *Corynebacterium* spp. Rhodococci are aerobic actinomycetes showing considerable morphological diversity. A certain group of these bacteria possess mycolic acids at the external surface of the cell. These compounds are unusual long-chain alcohols and fatty acids, esterified to the peptidoglycan of the cell wall. Probably, these lipophilic cell structures have a significance for the affinity of rhodococci to lipophilic pollutants. In general, rhodococci have high and diverse metabolic activities and are able to synthesize biosurfactants.

2.2 Growth-Associated Degradation of Aliphatics

The aerobic initial attack of aliphatic and cycloaliphatic hydrocarbons requires molecular oxygen. Fig. 2 shows both types of enzymatic reactions involved in these processes. It depends on the nature of the substrate and the enzymatic equipment of the involved microorganisms, what kind of enzymatic reaction is realized. *n*-Alkanes are the main constituents of mineral oil contaminations (HINCHEE et al., 1994). Long-chain *n*-alkanes ($C_{10}-C_{24}$) are degraded most rapidly by mechanisms demonstrated in Fig. 3. Short-chain alkanes (less than C_9) are toxic to many microorganisms, but they evaporate rapidly from petroleum contaminated sites. Oxidation of alkanes is classified as being terminal or diterminal. The monoterminal oxidation is the main pathway. It proceeds via the formation of the corresponding alcohol, aldehyde, and fatty acid. β-Oxidation of the fatty acids results in the formation of acetyl-CoA. *n*-Alkanes with an uneven number of carbon atoms are degraded to propionyl-CoA, which is in turn carboxylated to methylmalonyl-CoA and further converted to succinyl-CoA. Fatty acids of a physiological chain length may be directly incorporated into membrane lipids, but the majority of degradation products is introduced into the tricarboxylic acid cycle. The subterminal oxidation occurs with lower (C_3-C_6) and longer alkanes with the formation of a secondary alcohol and subsequent ketone. Unsaturated 1-alkenes are oxidized at the saturated end of the chains. A minor pathway has been shown to proceed via an epoxide, which is converted to a fatty acid. Branching, in general, reduces the rate of biodegradation. Methyl side groups do not drastically decrease the biodegradability, whereas complex branching chains, e.g., the tertiary butyl group, hinder the action of the degradative enzymes.

Cyclic alkanes representing minor components of mineral oil are relatively resistant to microbial attack. The absence of an exposed terminal methyl group complicates the primary attack. A few species are able to use cyclohexane as sole carbon source; more common is its cometabolism by mixed cultures. The mechanism of cyclohexane degradation is shown in Fig. 4. In general, alkyl side chains of cycloalkanes facilitate the degradation.

Aliphatic hydrocarbons become less water soluble with increasing chain length. Hydrocarbons with a chain length of C_{12} and above are virtually water insoluble. Two mechanisms

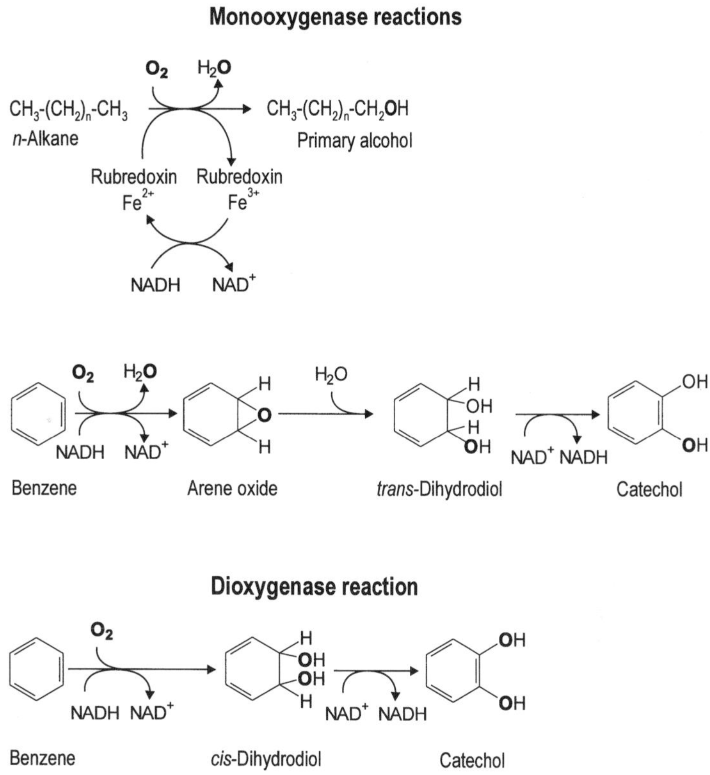

Monooxygenase reactions

Fig. 2. Initial attack on xenobiotics by oxygenases. Monooxygenases incorporate one atom of oxygen of O_2 into the substrate, the second atom is reduced to H_2O. Dioxygenases incorporate both atoms into the substrate.

Dioxygenase reaction

are involved in the uptake of these lipophilic substrates: the attachment of microbial cells at oil droplets and the production of biosurfactants (HOMMEL, 1990). The uptake mechanism linked to attachment of the cells is still unknown, whereas the effect of biosurfactants has been studied well (Fig. 5). Biosurfactants are molecules consisting of a hydrophilic and a lipophilic moiety. They act as emulsifying agents, by decreasing the surface tension and by forming micelles. The microdroplets may be encapsulated in the hydrophobic microbial cell surface. The products of hydrocarbon degradation, introduced to the central tricarboxylic acid cycle, have a dual function. They are substrates of the energy metabolism and building blocks for biosynthesis of cell biomass and growth (Fig. 1). The synthesis of amino acids and proteins needs a nitrogen and sulfur source, that of nucleotides and nucleic acids a phosphorus source. The biosynthesis of the

bacterial cell wall requires activated sugars synthesized by gluconeogenesis.

Products of growth-associated degradation are CO_2, H_2O, and cell biomass. The cells act as the complex biocatalysts of degradation. In addition, cell biomass may be mineralized after exhaustion of the degradable pollutants in a contaminated site.

2.3 Diversity of Aromatic Compounds – Unity of Catabolic Processes

Aromatic hydrocarbons, e.g., benzene, toluene, ethylbenzene and xylenes (BTEX compounds), and naphthalene belong to the large volume petrochemicals, widely used as fuels and industrial solvents. Phenols and chlorophenols are released into the environment as

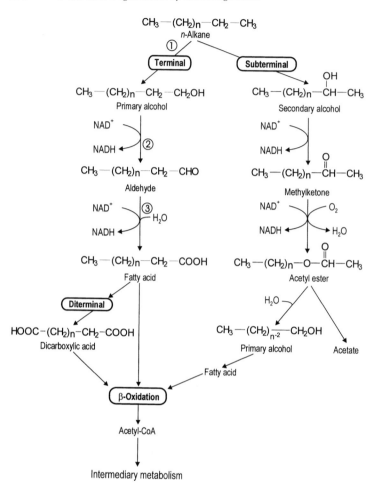

Fig. 3. Peripheral pathways of alkane degradation. The main pathway is the terminal oxidation to fatty acids catalyzed by ① n-alkane monoxygenase, ② alcohol dehydrogenase and ③ aldehyde dehydrogenase.

products and waste materials from industry. Aromatic compounds are formed in large amounts by all organisms, e.g., as aromatic amino acids, phenols, or quinones. Thus, it is not surprising that many microorganisms have evolved catabolic pathways to degrade aromatic compounds. In general, man-made organic chemicals (xenobiotics) can be degraded by microorganisms, when the respective molecules are similar to natural compounds. The diversity of man-made aromatics shown in Fig. 6 can be converted enzymatically to natural intermediates of the degradation: catechol and protecatechuate. In general, benzene and related compounds are characterized by a higher thermodynamic stability than aliphatics are.

Only few reports on bacteria capable of attacking benzene have been published (SMITH, 1990). The first step of benzene oxidation is a hydroxylation catalyzed by a dioxygenase (Fig. 2). The product, a diol, is then converted to catechol by a dehydrogenase. These initial reactions, hydroxylation and dehydrogenation, are also common to pathways of degradation of other aromatic hydrocarbons.

The introduction of a substituent group onto the benzene ring renders alternative mechanisms possible to attack side chains or to oxidize the aromatic ring. The versatility and adaptability of bacteria is based on the existence of catabolic plasmids. Catabolic plasmids have been found to encode enzymes degrading

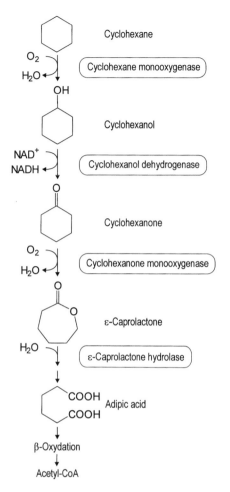

O₂ ⌐
H₂O ◄┘

Fig. 4. Peripheric metabolic pathway of cycloaliphatic compounds (cycloparaffins).

lated derivatives are then oxidized by toluate dioxygenase and decarboxylated to catechol (SMITH, 1990).

The oxygenolytic cleavage of the aromatic ring occurs via *o*- or *m*-cleavage. The significance of the diversity of degradative pathways and of the few key intermediates is still under discussion. Both pathways may be present in one bacterial species. *"Whenever an alternative mechanism for the dissimilation of any compound becomes available (ortho- versus meta-cleavage of ring structures, for example) control of each outcome must be imposed"* (HOUGHTON and SHANLEY, 1994). The metabolism of a wide spectrum of aromatic compounds by one species requires the metabolic isolation of intermediates into distinct pathways. This kind of metabolic compartmentation seems to be realized by metabolic regulation. The key enzymes of the degradation of aromatic substrates are induced and synthesized in appreciable amounts only when the substrate or structurally related compounds are present. Enzyme induction depends on the concentration of the inducing molecules. The substrate specific concentrations represent the threshold of utilization and growth and are in the magnitude of µM. A recent report on the regulation of TOL catabolic pathways has been published by RAMOS et al. (1997).

Fig. 7 shows the pathways of the oxygenolytic ring cleavage to intermediates of the central metabolism. At the branchpoint catechol either is oxidized by the intradiol *o*-cleavage, or the extradiol *m*-cleavage. Both ring cleavage reactions are catalyzed by specific dioxygenases. The product of the *o*-cleavage – *cis,cis*-muconate – is transferred to the instable enol-lactone, which is in turn hydrolyzed to oxoadipate. This dicarboxylic acid is activated by transfer to CoA, followed by the thiolytic cleavage to acetyl-CoA and succinate. Protocatechuate is metabolized by a homologous set of enzymes. The additional carboxylic group is decarboxylated and, simultaneously, the double bond is shifted to form oxoadipate enol-lactone. The oxygenolytic *m*-cleavage yields 2-hydroxymuconic semialdehyde, which is metabolized by the hydrolytic enzymes to formate, acetaldehyde, and pyruvate. These are then utilized in the central metabolism. In general, a wealth of aromatic substrates is degrad-

naturally occurring aromatics such as camphor, naphthalene, and salicylate. Most of the catabolic plasmids are self-transmissible and have a broad host range. The majority of gram-negative soil bacteria isolated from polluted areas possess degradative plasmids, mainly the so called TOL plasmids. These pseudomonads are able to grow on toluene, *m*- and *p*-xylene, and *m*-ethyltoluene. The main reaction involved in the oxidation of toluene and related arenes is the methyl group hydroxylation. The methyl group of toluene is oxidized stepwise to the corresponding alcohol, aldehyde, and carboxylic group. Benzoate formed or its alky-

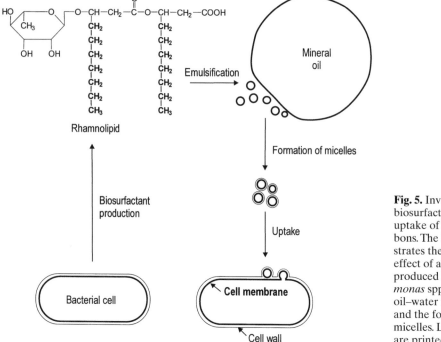

Fig. 5. Involvement of biosurfactants in the uptake of hydrocarbons. The figure demonstrates the emulsifying effect of a rhamnolipid produced by *Pseudomonas* spp. within the oil–water interphase and the formation of micelles. Lipid phases are printed in bold.

ed by a limited number of reactions: hydroxylation, oxygenolytic ring cleavage, isomerization, hydrolysis. The inducible nature of the enzymes and their substrate specificity enable bacteria with a high degradation potential, e.g., pseudomonads and rhodococci, to adapt their metabolism to the effective utilization of substrate mixtures in polluted soils and to grow at a high rate.

2.4 Extension of Degradative Capacities

2.4.1 Cometabolic Degradation of Organopollutants

Cometabolism, the transformation of a substance without nutritional benefit in the presence of a growth substrate, is a common phenomenon of microbial activities. It is the basis of biotransformations (bioconversions) used in biotechnology to convert a substance to a chemically modified form. Microorganisms growing on a particular substrate gratuitously oxidize a second substrate (cosubstrate). The cosubstrate is not assimilated, but the product may be available as substrate for other organisms of a mixed culture.

The prerequisites of cometabolic transformations are the enzymes of the growing cells and the synthesis of cofactors necessary for enzymatic reactions, e.g., of hydrogen donors (reducing equivalents, NADH) for oxygenases. The principle is shown in Fig. 8. The example demonstrated in Fig. 8 has been used in field experiments for the elimination of trichloroethylene (THOMAS and WARD, 1989). Methanotrophic bacteria used in this experiment can utilize methane and other C_1 compounds as sole sources of carbon and energy. They oxidize methane to CO_2 via methanol, formaldehyde, and formate. The assimilation requires special pathways, and formaldehyde is the intermediate assimilated. The first step of

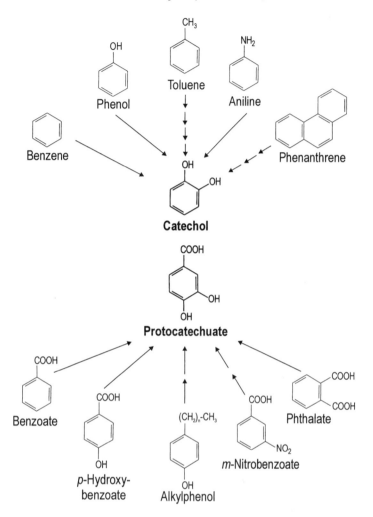

Fig. 6. Degradation of a broad spectrum of aromatic natural and xenobiotic compounds into two central intermediates: catechol and protocatechuate.

methane oxidation is catalyzed by methane monooxygenase, which attacks the inert CH_4. It is an unspecific enzyme that also oxidizes various other compounds, e.g., alkanes, aromatic compounds, and trichloroethylene (TCE). The proposed mechanism of TCE transformation according to HENRY and GRBIĆ-GALLIĆ (1994) is shown in Fig. 8. TCE is oxidized to an epoxide excreted from the cell. The unstable oxidation product breaks down to compounds, which may be used by other microorganisms. Methanotrophic bacteria are aerobic indigenous bacteria, in soil and aqui-

fers, but methane has to be added as growth substrate and inducer for the development of methanotrophic biomass. The addition of methane as substrate limits the application for bioremediation.

Cometabolism of chloroaromatics is a widespread activity of bacteria in mixtures of industrial pollutants. KNACKMUSS (1997) demonstrated that the cometabolic transformation of 2-chlorophenol gives rise to dead end metabolites, e.g., 3-chlorocatechol. This reaction product may be auto-oxidized or polymerized in soil to humic-like structures. Irreversible

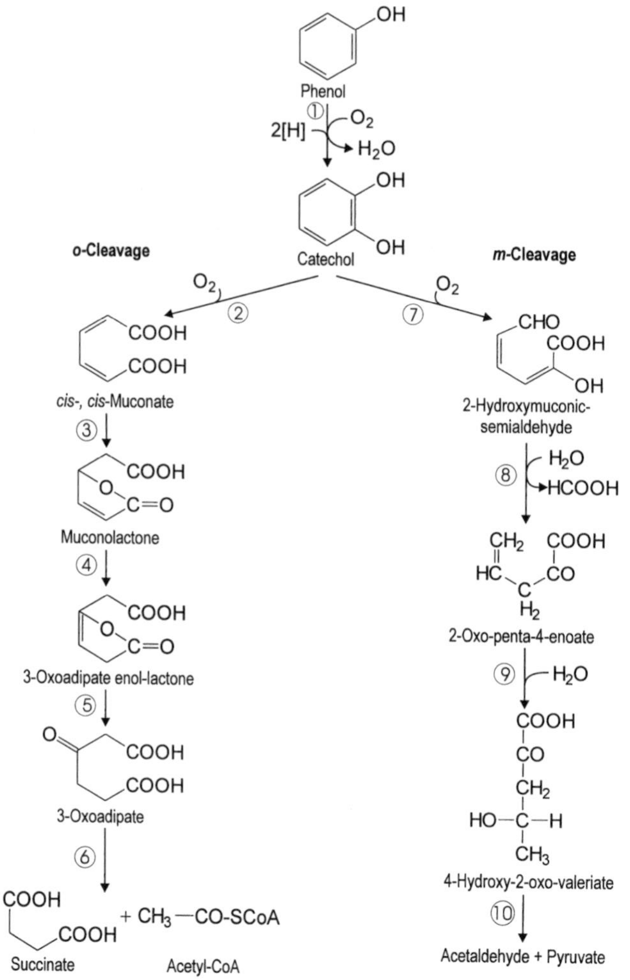

Fig. 7. The two alternative pathways of aerobic degradation of aromatic compounds: o- and m-cleavage, ① phenol monooxygenase, ② catechol 1,2-dioxygenase, ③ muconate lactonizing enzyme, ④ muconolactone isomerase, ⑤ oxoadipate enol-lactone hydrolase, ⑥ oxoadipate succinyl-CoA transferase, ⑦ catechol 2,3-dioxygenase, ⑧ hydroxymuconic semialdehyde hydrolase, ⑨ 2-oxopent-4-enoic acid hydrolase, ⑩ 4-hydroxy-2-oxovalerate aldolase.

binding of dead end metabolites may fulfill the function of detoxification. The accumulation of dead end products within microbial communities under selection pressure is the basis for the evolution of new catabolic traits (REINECKE, 1994).

2.4.2 Overcoming the Persistence by Cooperation of Anaerobic and Aerobic Bacteria

As a rule, recalcitrance of organic pollutants increases with increasing halogenation. Substi-

tution of halogen as well as nitro and sulfo groups at the aromatic ring is accomplished by an increasing electrophilicity of the molecule. These compounds resist the electrophilic attack by oxygenases of aerobic bacteria. Compounds that persist under oxic conditions are, e.g., PCBs (polychlorinated biphenyls), chlorinated dioxins, some pesticides, e.g., DDT.

To overcome the relatively high persistence of halogenated xenobiotics, reductive attack of anaerobic bacteria is of significance. The degradation of environmental pollutants by anaerobic bacteria is the subject of Chapter 7, this volume. Because of the significance of reductive dehalogenation for the first step in the

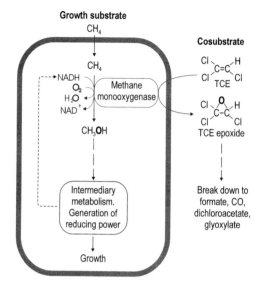

Growth substrate
CH₄

Cosubstrate

Fig. 8. Cometabolic degradation of trichloromethane by the methane monooxygenase system of methanotrophic bacteria.

degradation of higher halogenated compounds, this process has been announced. Reductive dehalogenation effected by anaerobic bacteria is either a gratuitous reaction or a new type of anaerobic respiration (ZEHNDER, 1988). The process reduces the degree of chlorination and, therefore, makes the product more accessible to mineralization by aerobic bacteria.

The potential of a sequence of anaerobic and aerobic bacterial activities for the mineralization of chlorinated xenobiotics is described in Fig. 9. PCBs, which are selected as an example for degradation of halogenated compounds, are well-studied objects (TIEDJE et al. 1993; SYLVESTRE and SANDOSSI, 1994; BEDARD and QUENSEN, 1995). The scheme demonstrates the principle of enzymatic dehalogenation mechanisms. The realization of the reactions depends on the structure of the chemical compounds as well as on the microorganisms and conditions in a polluted ecosystem. We have to distinguish between the general degradation potential and the actual conditions necessary for its realization. Reductive dehalogenation, the first step of PCB degradation, requires anaerobic conditions and organic substrates acting as electron donors. The PCBs

have the function of an electron acceptor to allow the anaerobic bacteria to transfer electrons to these compounds. Anaerobic bacteria capable of catalyzing reductive dehalogenation seem to be relatively ubiquitous in nature. Most dechlorinating cultures are mixed cultures (consortia). Aanaerobic dechlorination is always incomplete, products are di- and monochlorinated biphenyls. These products can be metabolized further by aerobic microorganisms. The substantial reduction of PCBs by sequential anaerobic and aerobic treatment has been demonstrated in the laboratory (AB-RAMOWICZ, 1990).

The principle of aerobic microbial dehalogenation reactions of chloroaromatics are described in Fig. 9. Hydrolytic dechlorination has been elucidated using 4-chlorobenzoate as substrate for *Pseudomonas* and *Nocardia* spp. A halidohydrolase is capable of replacing the halogen substituent by a hydroxy group originating from water. This type of reaction seems to be restricted to halobenzoates substituted in the *p*-position. Dechlorination after ring cleavage is a common reaction of the *o*-pathway of chlorocatechols catalyzed by catechol 1,2-dioxygenases to produce chloromuconates. The oxygenolytic dechlorination is a rare fortuitous reaction catalyzed by mono- and dioxygenases. During this reaction, the halogen substituent is replaced by oxygen of O_2.

Higher chlorinated phenols, e.g., pentachlorophenol, have been widely used as biocides. Several aerobic bacteria that degrade chlorophenols have been isolated (*Flavobacterium, Rhodococcus*). The degradation mechanism has been elucidated in some cases (McALLIS-TER et al., 1996). Thus, *Rhodococcus chlorophenolicus* degrades pentachlorophenol through a hydrolytic dechlorination and three reductive dechlorinations, producing trihydroxybenzene (APAJALAKTI and SALKINOJA-SALONEN, 1987). The potential of these bacteria is limited to some specialists and specific conditions. Therefore, the use of polychlorinated phenols has been banned in many countries.

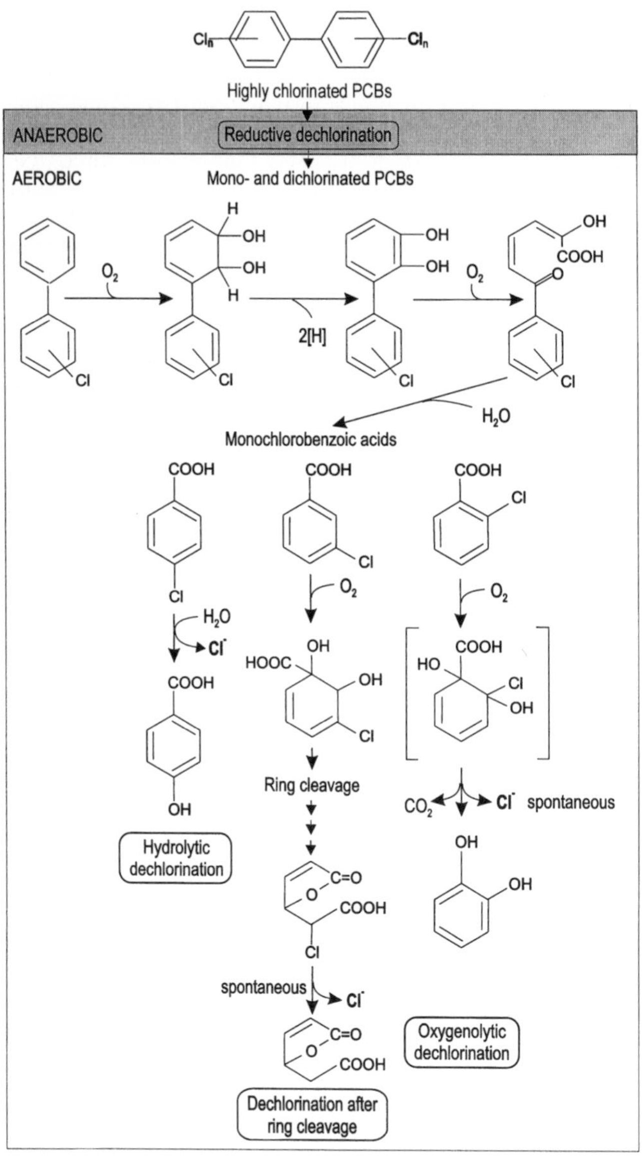

Fig. 9. Principle of dehalogenation: degradation of PCBs by a sequence of anaerobic and aerobic bacterial processes.

3 Degradative Capacities of Fungi

As saprophytic decomposers, symbionts, and parasites, members of the fungal kingdom permeate the living scene. There is an enormous number of fungi, and it has been estimated that there may be as many as 1.5 million fungal species. This magnitude is only comparable with the huge biodiversity of insects. Fungi exist in a wide range of habitats: in fresh water and the sea, in soil, litter, decaying remainders of plants and animals, in dung, and in living organisms. It was estimated that fungi account for more than 60% of the living microbial biomass in certain habitats (e.g., forest and moor soils; DIX and WEBSTER, 1995). Last but not least, fungi possess important degradative capacities and capabilities which are involved in the elimination of hazardous wastes from the environment. Below, some aspects of fungal degradation of organopollutants are discussed.

3.1 Metabolism of Organopollutants by Microfungi

Yeasts and molds can be grouped together into the microfungi (GRAVESEN et al., 1994). Taxonomically they belong to the ascomycetous, deuteromycetous and zygomycetous fungi. In addition, it has been discussed whether the so called "black yeasts" are descended from the basidiomycetes. Yeasts preferentially grow as single cells or form pseudomycelia, whereas molds typically grow as hyphae-forming real mycelia (GRAVESEN et al., 1994). Microfungi colonize many habitats and some genera are typical, ubiquitously distributed soil microorganisms (for soil fungi, see DOMSCH et al., 1993).

3.1.1 Aliphatic Hydrocarbons

The degradation of aliphatic hydrocarbons occurring in crude oil and petroleum products has been investigated well, especially for yeasts. *n*-Alkanes are the most widely and readily utilized hydrocarbons, with those between C_{10} and C_{20} being most suitable as substrates for microfungi (BARTHA, 1986), but also the biodegradation of *n*-alkanes with molecular weights up to *n*-C_{24} has been demonstrated (BLASIG et al., 1988, HOFRICHTER and FRITSCHE, 1996). Typical representatives of alkane-utilizing yeasts are *Candida lipolytica, C. tropicalis, Rhodoturula rubra*, and *Aureobasidium (Trichosporon) pullulans*, examples for molds using *n*-alkanes as growth substrates are *Cunninghamella blakesleeana, Aspergillus niger*, and *Penicillium frequentans* (WATKINSON and MORGAN, 1990). Yeasts and molds preferentially oxidize long-chain *n*-alkanes; short-chain liquid alkanes (*n*-C_5–C_9) are considered to be toxic (disorganization of the cytoplasmic membrane). The toxic effects of certain short-chain alkanes can be reversed by adding a non-toxic long-chain hydrocarbon. Thus, *Candida* spp. could grow on *n*-octane if 10% pristane was present (BRITTON, 1984). The soil fungus *Penicillium frequentans* is even capable of utilizing mono-halogenated *n*-alkanes (e.g., 1-fluorotetradecane) and dehalogenates them completely; higher halogenated alkanes, however, are resistant to fungal attack (WUNDERWALD, 1998). Since all aliphatic hydrocarbons are nearly insoluble in water, fungi produce surface-active compounds (biosurfactants) which disperse the substrates into oil-in-water emulsions, thus increasing the interfacial area and thereby enhancing the bioavailability of hydrocarbons (HOMMEL, 1990).

In microfungi, alkanes are mostly terminal-oxidized to their corresponding primary alcohols (*n*-alkan-1-ols) by a monooxygenase enzyme complex containing cytochrome P-450 and NAD(P)H cytochrome P-450 reductase (BRITTON, 1984).

$$R-CH_2-CH_3+O_2+NAD(P)H_2 \rightarrow \\ R-CH_2-CH_2OH+NAD(P)+H_2O \quad (1)$$

Subterminal oxidation yielding different secondary alcohols is a rarer type of primary attack on *n*-alkanes and has been documented for certain molds (*Aspergillus* spp., *Fusarium* spp.; REHM and REIFF, 1982). Peroxisomal enzymes carry out the remaining degradation of alkanols to intermediates that are transferable to the mitochondria (BRITTON, 1984). Following terminal oxidation, the produced alcohol normally is oxidized to the corresponding aldehyde and fatty acid by means of pyridine nucleotide-linked dehydrogenases. In some *Candida* spp. and several molds, alcohol oxidases have been shown to be present in place of dehydrogenases (BLASIG et al., 1988). Secondary alcohols resulting from subterminal oxidation systems are oxidized to the corresponding ester and are hydrolytically cleaved to acetic acid and an alcohol which is subsequently also converted into a fatty acid (REHM et al., 1983). In all cases, the fatty acids produced are further metabolized by β-oxidation and finally to CO_2 via the tricarboxylic acid cycle (MORGAN and WATKINSON, 1994).

Less is known about the pathways of *n*-alkene degradation by fungi, but as much is certain, both yeasts (e.g., *Candida lipolytica*) and molds (e.g., *Penicillium frequentans*) can grow with *n*-alkenes (e.g., 1-hexadecene) as sole carbon source (HORNICK et al., 1983; HOFRICHTER and FRITSCHE, 1996). In case of *Candida lipolytica*, the double bond in *n*-alkenes is attacked resulting in the formation of alkane-1,2-diols which are further metabolized to α-hydroxy fatty acids (HORNICK et al., 1983). In contrast to *n*-alkanes, fungi do not seem to be able to utilize branched alkanes or cycloaliphatic compounds as sole source of carbon and energy, because alkyl branching hinders normal β-oxidation and the destruction of alicyclic rings requires special cleaving enzymes which, in contrast to bacteria, lack in fungi (BRITTON, 1984; MORGAN and WATKINSON, 1994).

3.1.2 Aromatic Compounds

A number of yeasts and molds utilize aromatic compounds as growth substrates (sole sources of carbon and energy; Tab. 2), but most important is their ability to degrade aromatic substances cometabolically. Cometabolic degradation means the transformation of a substrate that cannot be utilized as carbon source in the presence of a second utilizible substrate. Fig. 10 describes this principle by the example of 3,4-dichlorophenol transformation in the presence of phenol as growth substrate by the soil inhabiting mold *Penicillium frequentans* (HOFRICHTER et al., 1994). Whereas phenol is completely converted into biomass, carbon dioxide and water, the chlorinated phenol is only transformed to the corresponding catechol which cannot be degraded further and remains as a dead-end product.

Tab. 2 is a partial list of microfungal species that have been shown to utilize aromatic compounds as growth substrates. Microfungi cleave aromatic rings via the *o*-pathway (Fig. 11; NEUJAHR and VARGA, 1970; HOFRICHTER et al., 1994). For this purpose, it is necessary to insert activating hydroxyl groups into the aromatic ring. Phenol hydroxylase, benzoate-4-hydroxylase, and 4-hydroxybenzoate-3-hydroxylase are examples for such enzymes hydroxylating the aromatic ring in *o*- or *p*-position. These enzymes are NADPH$_2$-dependent monooygenases and have been described for

yeasts and molds (e.g., *Trichosporon cutaneum, P. frequentans;* WRIGHT, 1993; HOFRICHTER et al., 1994). After activation, dioxygenases cleave the aromatic rings to form *cis,cis*-muconic acids. The latter are lactonized, isomerized and hydrolyzed resulting in the formation of β-ketoadipate which is degraded to CO_2 via the tricarboxylic acid cycle (TCC). The whole degradation process occurs intracellularly and microfungi have developed specific, energy-dependent uptake systems for aromatic substrates (e.g., the phenol uptake system in *T. cutaneum*; MÖRTBERG and NEUJAHR, 1985).

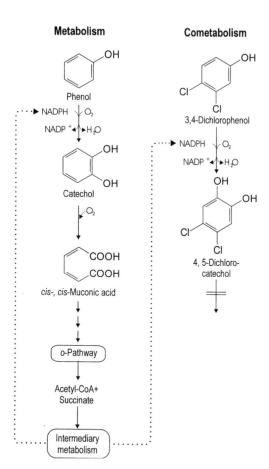

Fig. 10. Principle of cometabolic transformation of organopollutants by microfungi: hydroxylation of 3,4-dichlorophenol by the mold *Penicillium frequentans* growing on phenol.

Tab. 2. Some Species of Yeasts and Molds that Utilize Various Aromatic Compounds as Growth Substrates (Sole Sources of Carbon and Energy)

Species	Growth Substrates	Reference
Yeasts		
Aureobasidium pullulans	phenol, *o*-cresol, *p*-cresol, benzoic acid	TAKAHASHI et al. (1981)
Candida maltosa	phenol, catechol, benzoic acid	POLNISCH et al. (1992)
Exophiala jeanselmei	phenol, styrene, benzoic acid, acetophenone	COX et al. (1993)
Rhodotorula glutinis	phenol, *m*-cresol, benzoic acid	WALKER (1972), KATAYA-MA-HIRAYAMA et al. (1994)
Trichosporon cutaneum	*phenol, p*-cresol, benzoic acid, salicylic acid	NEUJAHR and VARGA (1970), HASEGAWA et al. (1990)
Molds		
Aspergillus niger	2,4-dichloro-phenoxy acetic acid, benzoic acid, salicylic acid, monochlorobenzoic acids	SHAILUBHAI et al. (1982, 1983)
Aspergillus fumigatus	phenol, *p*-cresol, 4-ethylphenol, phenylacetic acid	JONES et al. (1993, 1995)
Fusarium flocciferum	phenol, resorcinol	ANSELMO and NOVAIS (1984)
Penicillium frequentans	phenol, *p*-cresol, resorcinol, phloroglucinol, anisole, benzyl alcohol, benzoic acid, salicylic acid, gallic acid, phenylacetic acid, 1-phenyl-ethanol acetophenone	HOFRICHTER et al. (1994, 1995), HOFRICHTER and FRITSCHE (1996)
Penicillium simplicissimum	phenol, phloroglucinol, monofluorophenols	PATEL et al. (1990), MARR et al. (1996)

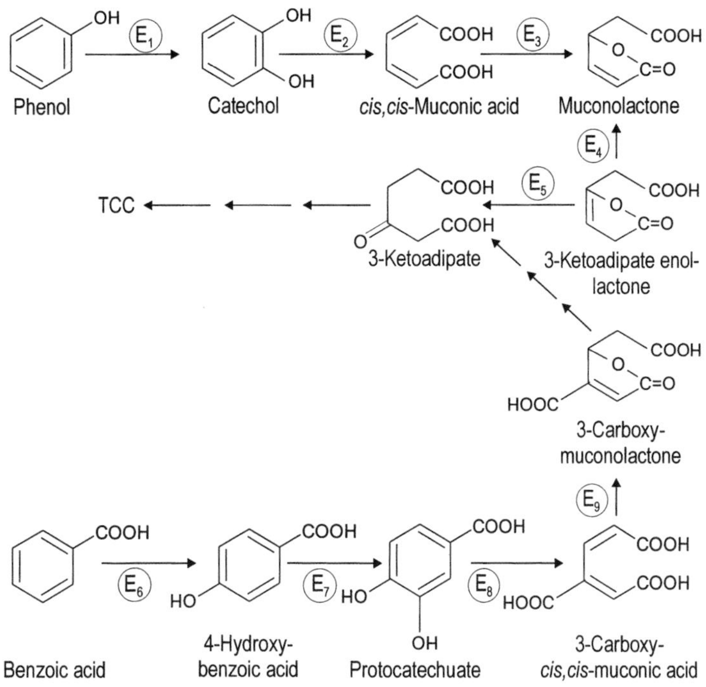

Fig. 11. The *o*-ring cleavage pathways for phenol and benzoic acid in microfungi. E$_1$: phenol hydroxylase, E$_2$: catechol-1,2-dioxygenase, E$_3$: muconate cycloisomerase, E$_4$: muconolactone isomerase, E$_5$: 3-ketoadipate enol-lactone hydrolase, E$_6$: benzoate-4-hydroxylase, E$_7$: 4-hydroxybenzoate-3-hydroxylase, E$_8$: protocatechuate-3,4-dioxygenase, E$_9$: carboxymuconate lactonizing enzyme.

Both aromatic ring hydroxylating and ring cleaving enzymes of yeasts and molds are relatively unspecific and also convert differently substituted derivatives including halo- and nitroaromatics (NEUJAHR and VARGA, 1970; POLNISCH et al., 1992; HOFRICHTER et al., 1994). Thus, in dependence of the subtitution pattern, fluoro- and chlorophenols are converted into the corresponding catechols, cleaved to halogenated muconic acids, or dehalogenated. The alkane and phenol utilizing mold *P. frequentans* has been shown to degrade a mixture of phenol, *p*-cresol, and two chlorophenols without additional growth substrate in soil (HOFRICHTER et al., 1995). In addition to halogenated aromatic compounds, microfungi transform numerous other aromatic organopollutants cometabolically, among others polycyclic aromatic hydrocarbons (PAHs) and biphenyls, dibenzofurans, nitroaromatics, various pesticides, and plasticizers (BERK et al., 1957; BOLLAG, 1972; SMITH et al., 1980; GIBSON and SUBRAMANIAN, 1984; HAMMER et al., 1998). Typical fungal transformations are glycosylations, hydroxylations, and ring cleavages, methoxylations or the reduction of nitro groups to amino groups (Tab. 3). Hydroxylation reactions are particularly important for the elimination of organopollu-

tants in soils, because they increase the reactivity of the molecules and make their subsequent covalent coupling to humus possible (humification, see Chapter 4, this volume). The introduction of hydroxyl groups is often catalyzed by relatively unspecific cytochrome P-450 containing monooxygenases which, e.g., oxidize benzo(a)pyrene in *Aspergillus ochraceus* (DUTTA et al., 1983).

In summary, the degradative capacities of yeasts and molds play an important role in the carbon cycle and for the self-cleaning potential of soils. Their application in bioremediation technologies might be promising, in particular with respect to their high competitive potential.

3.2 Degradative Capabilities of Basidiomycetous Fungi

Basidiomycetous fungi form characteristic macroscopic fruiting bodies and belong, like certain ascomycetes, to the macrofungi (DIX and WEBSTER, 1995). Colloquially, the edible fungi of this group are also called mushrooms. Basidiomycetes preferentially colonize the litter and upper soil layers in woodlands and pas-

Tab. 3. Selected Organopollutants and Their Metabolites that are Cometabolically Formed by Yeasts and Molds

Organopollutants	Metabolites	Reference
Mono- and dichlorophenols	chlorocatechols[a–c], chloroguaiacols[c], chlorinated muconic acids[c], chloride[c]	WALKER (1972), HOFRICHTER et al. (1994)
Pentachlorophenol	pentachloroanisole[f]	CSERJESI (1967)
Fluorophenols	fluorocatechols[c], fluorinated muconic acids[c], fluoride[c]	WUNDERWALD et al. (1998)
2,4-Dinitrophenol	monoaminonitrophenols[e]	MADHOSINGH (1961)
2,4,6-Trinitrotoluene	aminodinitrotoluenes[c–e] hydroxylamino dinitrotoluenes[c–e]	SCHEIBNER et al. (1997)
Pyrene	1-pyrenol[c–e], 1-methoxypyrene[d], dihydroxypyrenes[c]	SACK et al. (1997a)
Benzo(a)pyrene	hydroxybenzo(a)pyrenes[d]	DUTTA et al. (1983)
Dibenzofurane	2,3-dihydroxybenzofuran[b], ring cleavage products[b]	HAMMER et al. (1998)
Biphenyl	4,4′-dihydroxybiphenyl[d]	GIBSON and SUBRAMANIAN (1984)
Atrazine	*N*-dealkylated products[c–f]	BOLLAG (1972)

[a] *Rhodotorula* spp., [b] *Trichosporon* spp., [c] *Penicillium* spp., [d] *Aspergillus* spp., [e] *Fusarium* spp., [f] *Trichoderma* spp.

tures as well as lignocellulosic materials (wood, straw). From the ecophysiological point of view, they can be broadly classified into litter-decomposing, wood-decaying, and myccorhizal fungi, although there is, inevitably, some overlap of roles. Basidiomycetes can form huge mycelial clones, and it has been estimated that they can extend over an area of several ha, while reaching an age of 1,500 years and a mycelial biomass of 100 t (DIX and WEBSTER, 1995).

3.2.1 The Ligninolytic Enzyme System

Lignin, like cellulose, is a major component of plant material and the most abundant form of aromatic carbon in the biosphere providing rigidity to all higher plants by acting as a glue between the cellulose fibers. In addition, lignin forms a barrier against microbial destruction and protects the readily degradable carbohydrates (cellulose, hemicellulose). From the chemical point of view, lignin is a heterogeneous, optically inactive polymer consisting of phenylpropanoid interunits, which are linked by several covalent bonds (aryl–ether, aryl–aryl, carbon–carbon bonds; FENGEL and WEGENER, 1989). The polymer arises from enzyme initiated polymerization of phenolic precursors (coniferyl, sinapyl, *p*-cumaryl alcohol) via the radical coupling of their corresponding phenoxy radicals. Due to the types of bonds and their heterogenity, lignin cannot be degraded by hydrolytic mechanisms as most other natural polymeres can. Within the course of evolution, only certain basidiomycetous fungi have developed an efficient enzyme system to degrade lignin so that these microorganisms play an important role in maintaining the global carbon cycle (GRIFFIN, 1994). Due to their ability to selectively remove lignin, while leaving white cellulose fibers, these fungi are also called white-rot fungi. Typical representatives of these fungi are the wood degraders *Trametes versicolor*, *Phanerochaete chrysosporium*, *Pleurotus ostreatus*, and *Nematoloma frowardii* as well as the litter decayers *Stropharia rugosoannulata*, *Mycena galopus*, and *Agrocybe praecox*. Lignin degradation does

not provide a primary source of carbon and energy for fungal growth and is, therefore, a general cometabolic process (mostly hemicelluloses are used as growth substrate; ERIKSSON et al., 1990).

Lignin degradation is very probably brought about by the synergistic cooperation of different ligninolytic enzymes. Three phenol-oxidizing enzymes – manganese peroxidase (MnP), lignin peroxidase (LiP) and laccase – are generally thought to be responsible for the unspecific attack of lignin (HATAKKA, 1994). Both peroxidases are ferric iron containing heme proteins requiring peroxides (e.g., H_2O_2) for function, and laccase belongs to the copper containing blue oxidases which use molecular oxygen (O_2). The enzymes have the ability in common to catalyze one-electron oxidations resulting in the formation of reactive free radical species inside of the lignin polymer (FERNANDO and AUST, 1990). Subsequently, the radicals undergo spontaneous reactions leading to the incorporation of oxygen (O_2), bond cleavages, and finally to the breakdown of the lignin molecule (KIRK and FARRELL, 1987). LiP and laccase react directly with aromatic substrates (e.g., aromatic lignin structures), whereas MnP takes effect via chelated Mn^{3+} ions acting as low-molecular weight redox mediator. Thus, the function of MnP is the generation of Mn^{3+} from Mn^{2+} which is the actual substrate of the enzyme. Similar to the catalytic cycle of other peroxidases including LiP, that of MnP involves the formation of peroxidase compounds I and II. The latter are highly reactive and abstract one electron from Mn^{2+} to form Mn^{3+} (Fig. 12). MnP catalysis has an absolute requirement for Mn chelat-ing, organic acids (e.g., oxalate, malate, malonate) which increase the affinity of Mn^{2+} to the enzyme and stabilize the extremely reactive Mn^{3+} to high redox potentials (GLENN et al., 1986; KISHI et al., 1994). In contrast to the relatively large enzyme molecules, chelated Mn^{3+} is small enough to diffuse into the compact lignocellulosic complex, where it preferentially reacts with phenolic lignin structures. Therefore, it is supposed that in most fungi the primary attack on lignin is brought about by the MnP system (WARIISHI et al., 1992; HATAKKA, 1994). An exception is the wood-degrading fungus *Pycnoporus cinnabarinus* which has developed an al-

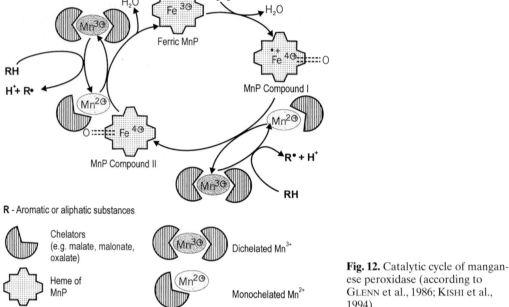

Fig. 12. Catalytic cycle of manganese peroxidase (according to GLENN et al., 1986; KISHI et al., 1994).

ternative mediator system for the primary attack on lignin based on laccase and the fungal metabolite 3-hydroxy anthranilic acid (EGGERT et al., 1996). The latter is first transformed by laccase into a reactive radical which then can act in a similar way as chelated Mn^{3+} inside the lignocellulosic complex. Another widespread fungal metabolite, veratryl alcohol, has been thought of for a long time as a redox mediator of LiP. Latest findings, however, have demonstrated that the lifespan of the aryl cation radical derived from veratryl alcohol is too short to act as an electron mediator (AKHTAR et al., 1997).

As the result of MnP or laccase catalyzed primary attack, water soluble lignin fragments are formed which are now accessible for the further conversion through ligninolytic enzymes. Laccase oxidizes phenolic structures, whereas LiP preferentially cleaves recalcitrant non-phenolic lignin moieties. MnP is also involved in the further degradation of the formed lignin fragments. Thus, there are indications that the MnP system mineralizes aromatic lignin structures directly. It is the only enzyme – as far as known – that converts aromat-

ic compounds including lignin partly to carbon dioxide (CO_2). This means, lignin can be mineralized – at least in part – outside the fungal hyphae (HOFRICHTER et al., 1999). The reactions underlying the MnP catalyzed mineralization are still not completely understood, but as much is certain, decarboxylation reactions caused by Mn^{3+} are the final step in this process. Furthermore, certain radicals (carbon-centered radicals, peroxyl radicals, superoxide) deriving from the autocatalytic decomposition of organic acids in the presence of Mn^{3+} may also be involved in the mineralization process (HOFRICHTER et al., 1998b). The oxidative strength of the MnP system can be considerably enhanced in the presence of additional mediating agents such as unsaturated fatty acids or organic thiols. Both groups of substances are converted by Mn^{3+} to highly reactive radical species (peroxyl and alkoxy radicals of fatty acids, thiyl radicals), which increase lignin mineralization (CO_2 formation) and render the cleavage of structures possible that are normally not attacked by the MnP system (e.g., non-phenolic aromatic aryl ethers; WARIISHI et al., 1989; JENSEN et al., 1996; HOFRICHTER et al., 1999). Following a concept of KIRK and FRARRELL (1987), the process of MnP catalyzed mineralization of lignin and other substances has been described as "enzymatic combustion" (Fig. 13; HOFRICHTER et al., 1998a).

3.2.2 Degradation of Organopollutants

The unique ligninolytic enzyme system of basidiomycetous fungi being based on a highly reactive free radical depolymerization mecha-

nism would be ideal for the biodegradation of organopollutants in the environment (FERNANDO and AUST, 1990). Compared to other potential bioremediation systems, the extracellular, non-specific, non-stereoselective lignin-degrading system of basidiomycetous fungi has the advantage of being applicable to a variety of recalcitrant and toxic chemicals. Examples for such chemicals and selected fungi responsible for the biodegradation are given in Tab. 4. Among these substances are hazardous xenobiotic compounds like polychlorinated dibenzodioxins and dibenzofurans, other chlorinated aromatics, nitroaromatic compounds (ex-

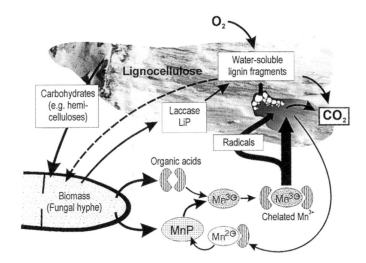

Fig. 13. "Enzymatic combustion" of lignin by manganese peroxidase producing basidiomycetes. The degradation of lignin is a generally aerobic, cometabolic process. Sugars deriving from hemicelluloses and cellulose serve as growth substrates.

Tab. 4. Selected Organopollutants that are Transformed or Mineralized by Ligninolytic Basidiomycetes and/or their Ligninolytic Enzymes (according to SACK et al., 1997b; RAJARATHNAM et al., 1998; HAAS et al., 1998; HOFRICHTER et al., 1998a)

Fungus	Organopollutants[a]
Bjerkandera adusta	benzo(a)pyrene, other PAHs, TNT, dyes
Nematoloma frowardii	benzo(a)pyrene, other PAHs, TNT, DCP, PCP, AsO
Phanerochaete chrysosporium	benzo(a)pyrene, other PAHs, BTX, DNT, TNT, DDT, DCP, PCP, PCBs, DCA, dyes, polystyrenes, KCN
Phanerochaete sordida	PAHs, polychlorinated DBDs and DBFs
Phlebia radiata	TNT, dyes
Pleurotus ostreatus	benzo(a)pyrene, other PAHs, dibenzothiophene, TNT
Stropharia rugosoannulata	DCP, PCP, TNT
Trametes versicolor	benzo(a)pyrene, other PAHs, DCA, DCP, PCP, dyes

[a] AsO: arsenic-containing organic compounds, BTX: benzene, toluene, xylenes, DCA: 3,4-dichloroaniline, DCP: 2,4-dichlorophenol, DBDs: dibenzo-*p*-dioxins, DBF: dibenzofurans, DNT: 2,4-dinitrotoluene, DDT: 1,1,1-trichloro-2,2′-bis(4-chlorophenyl)ethane, KCN: potassium cyanide, PAHs: polycyclic aromatic hydrocarbons, PCBs: polychlorinated biphenyls, PCP: pentachlorophenol, TNT: 2,4,6-trinitrotoluene.

plosives), and cancerogenic organopollutants belonging to the polycyclic aromatic hydrocarbons. Both MnP, LiP and laccase have been shown to be involved in the transformation of organopollutants (BOLLAG, et al., 1988; BARR and AUST, 1994; HOFRICHTER et al., 1998a). As the result of the attack of organopollutants by ligninolytic enzymes, various metabolites are formed which either can be further degraded intracellularly, coupled to the humus, or mineralized by the MnP system. The latter has been shown to mineralize various organopollutants directly in cell-free systems indicating that an extracellular "enzymatic combustion" of hazardous chemicals is possible in principle (HOFRICHTER et al., 1998a).

Over the last decade, efforts have been made to use ligninolytic basidiomycetes in bioremediation technologies. Laboratory experiments have shown that the degradation of certain organopollutants (i.e., PAHs, PCP, TNT) is stimulated by wood-inhabiting white-rot fungi in contaminated soils (*Phanerochaete chrysosporium, Lentinus edodes, Kuehneromyces mutabilis;* SACK and FRITSCHE, 1997; RAJARATHNAM et al., 1998). However, a disadvantage of these fungi is their small competitive potential in soil. Especially, the best investigated and highly active white-rot fungus *P. chrysosporium* only survives in soil for a very short time and cannot compete with the authochtonous microorganisms. Therefore, research has recently been focused on litter-decaying basidiomycetes which naturally colonize the upper soil layers. *Stropharia rugosoannulata* is an example of such a fungus which has already been successfully tested for TNT decontamination of soils (HERRE et al., 1997). Future investigations should screen for novel litter-decomposing strains that can be used in remediation technologies.

principle that we absolutely depend on microbial activities for the replenishment of our environment and the maintenance of the global carbon cycle (RATLEDGE, 1991). Among the substances which can be degraded or transformed by microorganisms are a huge number of synthetic compounds (xenobiotics) and other chemicals with environmental relevance (e.g., mineral oil components). However, it has to be considered that this statement concerns potential degradabilities which – in most cases – were estimated in the laboratory by using pure cultures and ideal growth conditions. Under natural conditions in soil, the actual degradability of organopollutants is in any case lower, due to a whole string of factors: competition with other microorganisms, insufficient supply with essential substrates (C, N, P, S sources), unfavorable external conditions (O_2, H_2O, pH, temperature), and low bioavailability of the pollutant that is to be degraded. Thus, it is an important assignment of environmental biotechnology to tackle and solve these problems in order to permit the use of microorganisms in bioremediation technologies. For this purpose, it is necessary to support the activities of the indigenous microorganisms in polluted soils and to enhance their degradative potential by bioaugmentation. The former measure particularly applies to bacteria, whereas bioaugmentation concerns basidiomycetous fungi above all (e.g., soil- and litter-colonizing fungi). In this connection, it has to be pointed out that neither bacteria nor fungi are "the better degraders". In nature, both groups of microorganisms work together, of course, and complement one another in their degradative capabilities.

4 Conclusions

It has been said that without microorganisms and their degradative capabilities animal life including humans on the planet would cease to exist within about 5 years. Whether or not this is an exaggeration in time scale, it is true in

5 References

ABRAMOWICZ, D. A. (1990), Aerobic and anaerobic biodegradation of PCBs: A review, *Crit. Rev. Biotechnol.* **10**, 241–251.

AKTHAR, M., BLANCHETTE, R. A., KIRK, T. K. (1997), Fungal delignification and biomechanical pulping of wood, in: *Biotechnology in the Pulp and Paper Industry* (ERIKSSON, K.-E. L., Ed.), pp. 159–195. Berlin: Springer-Verlag.

ALEXANDER, M. (1994), *Biodegradation and Biore-*

mediation. San Diego, CA: Academic Press.

ANSELMO, A. M., NOVAIS, J. M. (1984), Isolation and selection of phenol-degrading microorganisms from an industrial effluent, *Biotechnol. Lett.* **6**, 601–606.

APAJALAHTI, J. H. A., SALKINOJA-SALONEN, M. S. (1987), Complete dechlorination of terachloro hydroquinone by cell extracts of pentachlorophe-nol-induced *Rhodococcus chlorophenolicus*, *J. Bacteriol.* **169**, 5125–5130.

BARR, D. P., AUST, S. D. (1994), Mechanisms of white-rot fungi to degrade pollutants, *Environ. Sci. Technol.* **2**, 78A–87A.

BARTHA, R. (1986), Biotechnology of petroleum pollutant degradation, *Microb. Ecol.* **12**, 155–172.

BEDARD, D. L. QUENSEN, J. F. (1995), Microbial reductive dechlorination of polychlorinated biphenyls, in: *Microbial Transformation and Degradation of Toxic Organic Chemicals* (YOUNG, L. Y., CERNIGLIA, C. E., Eds.), pp. 127–216. New York: Wiley-Liss.

BERK, S., EBERT, H., TEITELL, L. (1957), Utilization of plasticizers and related organic compounds by fungi, *Ind. Eng. Chem.* **49**, 1115–1124.

BLASIG, R., MAUERSBERGER, S., RIEGE, P., SCHNUCK, W.-H., FRANKE, P., MÜLLER, H.-G. (1988), Degradation of long-chain *n*-alkanes by the yeast *Candida maltosa*. II. Oxidation of *n*-alkanes and intermediates using microsomal membrane fractions, *Appl. Microbiol. Biotechnol.* **28**, 589–597.

BOLLAG, J.-M. (1972), Biochemical transformation of pesticides by soil fungi, *CRC Crit. Rev. Microbiol.* **2**, 35–58.

BOLLAG, J.-M., SHUTTLEWORTH, K. L., ANDERSON, D. H. (1988), Laccase-mediated detoxification of phenolic compounds, *Appl. Environ. Microbiol.* **54**, 3086-3091.

BRITTON, L. N. (1984), Microbial degradation of aliphatic hydrocarbons, in: *Microbial Degradation of Organic Compounds* (GIBSON, D. T., Ed.), pp. 89–129. New York: Marcel Dekker.

COX, H. H. J., HOUTMAN, J. H. M., DODDEMA, H. J., HARDER, W. (1993), Growth of the black yeast *Exophiala jeanselmei* on styrene and styrene related compounds, *Appl. Microbiol. Biotechnol.* **39**, 372–376.

CSERJESI, A. J. (1967), The adaptation of fungi to pentachlorophenol and its biodegradation, *Can. J. Microbiol.* **13**, 1243–1249.

DIX, N. J., WEBSTER, J. (1995), *Fungal Ecology*. London: Chapman & Hall.

DOMSCH, K. H., GAMS, W., ANDERSON, T.-H. (1993), *Compendium of Soil Fungi*, Vol. I. Eching: IHW-Verlag.

DUTTA, D., GHOSH, D. K., MISHRA, A. K., SAMANTA, T. B. (1983), Induction of benzo(a)pyrene hydroxylase in *Aspergillus ochraceus* TS: evidences of multiple forms of cytochrome P-450, *Biochem.*

Biophys. Res. Commun. **115**, 692-699.

EGGERT, C., TEMP, U., DEAN, J. F. D., ERIKSSON, K.-L. E. (1996), A fungal metabolite mediates degradation of non-phenolic lignin structures and synthetic lignin by laccase, *FEBS Lett.* **391**, 144–148.

ERIKSSON, K.-E., BLANCHETTE, R. A., ANDER, P. (1990), *Microbial and Enzymatic Degradation of Wood and Wood Components*. Berlin: Springer-Verlag.

FENGEL, D., WEGENER, G. (1989), *Wood: Chemistry, Ultrastructure, Reactions*. Berlin: Walter de Gruyter.

FERNANDO, T., AUST, S. D. (1990), Biodegradation of toxic chemicals by white rot fungi, in: *Biological Degradation and Bioremediation of Toxic Chemicals* (CHAUDHRY, G. R., Ed.), pp. 386–402. London: Chapman & Hall.

GIBSON, D. T., SUBRAMANIAN, V. (1984), Microbial degradation of aromatic hydrocarbons, in: *Microbial Degradation of Organic Compounds* (GIBSON, D. T., Ed.) pp. 181–252. New York: Marcel Dekker.

GLENN, J. K., AKILESWARAN, L., GOLD, M. H. (1986), Mn(II) oxidation is the principal function of the extracellular Mn peroxidase from *Phanerochaete chrysosporium*, *Arch. Biochem. Biophys.* **251**, 688–696.

GRAVESEN, S., FRISVAD, J. C., SAMSON, R. A. (1994), *Microfungi*. Copenhagen: Munksgaard.

GRIFFIN, D. H. (1994), *Fungal Physiology*. New York: Wiley-Liss.

HAAS, R., SCHEIBNER, K., HOFRICHTER, M., KRIPPEN-DORF, A. (1998), Enzymatic degradation of arsenic-containing chemical warfare agents, in: *Proc. Conf. Bioremediation of Contaminated Soils* (registered project of the EXPO 2000) V23/1–V23/15, Munster: North Rhine-Westphalian Ministry of Science and Culture (in German).

HAMMER, E., KROWAS, D., SCHÄFER, A., SPECHT, M., FRANCKE, W., SCHAUER, F. (1998), Isolation and characterization of a dibenzofuran-degrading yeast: identification of oxidation and ring cleavage products, *Appl. Environ. Microbiol.* **64**, 2215–2219.

HASEGAWA, Y., OKAMOTO, T., OBATA, H., TOKUYAMA, T. (1990), Utilization of aromatic compounds by *Trichosporon cutaneum* KUY-6A, *J. Ferm. Bioeng.* **2**, 122–124.

HATAKKA, A. (1994), Lignin-modifying enzymes from selected white-rot fungi: production and role in lignin degradation, *FEMS Microbiol. Rev.* **13**, 125–135.

HENRY, S. M., GRBIĆ-GALLIĆ, D. (1994), Biodegradation of trichloroethylene in methanotrophic systems and implications for process applications, in: *Biological Degradation and Bioremediation of Toxic Chemicals* (CHAUDHRY, G. R., Ed.), pp.

314–344. London: Chapman & Hall.

HERRE, A., MICHELS, J., SCHEIBNER, K., FRITSCHE, W. (1997), Bioremediation of 2,4,6-trinitrotoluene (TNT) contaminated soil by a litter decaying fungus,. in: *In situ and On-Site Bioremediation*, Vol. 1 (ALLEMAN, B. C., LEESON BATTELLE, A., Eds.), pp. 493–498. Columbus, OH: Batelle Press.

HINCHEE, R. E., ALLEMAN, B. C., HOEPPEL, R. E., MILLER, R. N. (Eds.) (1994), *Hydrocarbon Bioremediation*. Boca Raton, FL: Lewis Publisher.

HOFRICHTER, M., FRITSCHE, W. (1996) Degradation of aromatic hydrocarbons by the mold *Penicillium frequentans*, *Z. Wasser-Abwasser* **137**, 199–204.

HOFRICHTER, M., BUBLITZ, F., FRITSCHE, W. (1994), Unspecific degradation of halogenated phenols by the soil fungus *Penicillium frequentans* Bi 7/2, *J. Basic Microbiol.* **34**, 163–172.

HOFRICHTER, M., SCHEIBNER, K., FRITSCHE, W. (1995), The role of soil fungi for the degradation of organopollutants, *TerraTech* **3**, 69–71 (in German).

HOFRICHTER, M., SCHEIBNER, K., SCHNEEGAß, I., FRITSCHE, W. (1998a), Enzymatic combustion of aromatic and aliphatic compounds by manganese peroxidase from *Nematoloma frowardii*, *Appl. Environ. Microbiol.* **64**, 399–404.

HOFRICHTER, M., ZIEGENHAGEN, D., VARES, T., FRIEDRICH, M., JÄGER, M. G. et al. (1998b), Oxidative decomposition of malonic acid as basis for the action of managnese peroxidase in the absence of hydrogen peroxide, *FEBS Lett.* **434**, 362–366.

HOFRICHTER, M., VARES, T., GALKIN, S., SIPILÄ, J., HATAKKA, A. (1999), Mineralization and solubilization of synthetic lignin (DHP) by manganese peroxidases from *Nematoloma frowardii* and *Phlebia radiala*, *J. Biotechnol.* **67**, 217–228.

HOFRICHTER, M., SCHEIBNER, K., BUBLITZ, F., SCHNEEGAß, I., ZIEGENHAGEN, D. et al. (1999), Depolymerization of straw lignin by manganese peroxidase is accompanied by release of carbon dioxide, *Holzforsch.* **53**, 161–166

HOMMEL, R. K. (1990), Formation and physiological role of biosurfactants produced by hydrocarbon-utilizing microorganisms, *Biodegradation* **1**, 107–109.

HORNICK, S. B., FISHER, R. H., PAOLINI, P. A. (1983), Petroleum wastes, in: *Land Treatment of Hazardous Wastes* (PARR, J. F., MARSH, P. B., KLA, J. M., Eds.), pp. 321–337. Park Ridge, NJ: Noyes Data Corp.

HOUGHTON, J. E., SHANLEY, M. S. (1994), Catabolic potential of pseudomonads: A regulatory perspective, in: *Biological Degradation and Bioremediation of Toxic Chemicals* (CHAUDHRY, G. R., Ed.), pp. 11–32. London: Chapman & Hall.

JENSEN, J. A., JR., BAO, W., KAWAI, S., SREBOTNIK, E., HAMMEL, K. E. (1996), Manganese-dependent cleavage of non-phenolic lignin structures by *Ceriporiopsis subvermispora* in the absence of lignin peroxidase, *Appl. Environ. Microbiol.* **62**, 3679–3686.

JONES, K. H., TRUGDILL, P. W., HOPPER, D. J. (1995), Evidence of two pathways for the metabolism of phenol by *Aspergillus fumigatus, Arch. Microbiol.* **163**, 176–181.

JONES, K. H., TRUGDILL, W., HOPPER, D. J. (1993), Metabolism of p-cresol by the fungus *Aspergillus fumigatus, Appl. Environ. Microbiol.* **59**, 1125–1130.

KATAYAMA-HIRAYAMA, K., TOBITA, S., HIRAYAMA, K. (1994), Biodegradation of phenol and monochlorophenols by yeast *Rhodotorula glutinis, Water Sci. Technol.* **30**, 59–66.

KIRK, T. K., FARRELL, R. L. (1987), Enzymatic "combustion": the microbial degradation of lignin, *Ann. Rev. Microbiol.* **41**, 465–505.

KISHI, K., WARIISHI, H., MARQUEZ, L., DUNFORD, H. B., GOLD, M. H. (1994), Mechanism of manganese peroxidase compound II reduction. Effect of organic acid chelators and pH, *Biochemistry* **33**, 8694–8701.

KNACKMUSS, H. J. (1997), Abbau von Natur- und Fremdstoffen, in: *Umweltbiotechnologie* (OTTOW, C. G., BILLINGMAIER, W., Eds.), pp. 39–80. Stuttgart: Gustav Fischer.

MADHOSINGH, C. (1961), The metabolic detoxification of 2,4-dinitrophenol by *Fusarium oxysporum, Can. J. Microbiol.* **7**, 553–567.

MARR, J., KREMER, S., STERNER, O., ANKE, H. (1996), Transformation of halophenols by *Penicillium simplicissimum* SK9117, *Biodegradation* **7**, 165–171.

MCALLISTER, K. A., LEE, H., TREVORS, J. T. (1996), Microbial degradation of pentachlorophenol, *Biodegradation* **7**, 1–40.

MORGAN, P., WATKINSON, R. J. (1994), Biodegradation of components of petroleum, in: *Biochemistry of Microbial Degradation* (RATLEDGE, C., Ed.), pp. 1–31. Dordrecht: Kluwer Academic Publishers.

MÖRTBERG, M., NEUJAHR, H. Y. (1985), Uptake of phenol in Trichosporon cutaneum, *J. Bacteriol.* **161**, 615-619.

NEUJAHR, H. Y., VARGA, J. M. (1970), Degradation of phenols by intact cells and cell-free preparations of Trichosporon cutaneum, *Eur. J. Biochem.* **13**, 37–44.

PATEL, T. R., HAMEED, N., MARTIN, A. M. (1990), Initial steps of phloroglucinol metabolism in *Penicillium simplicissimum, Arch. Microbiol.* **153**, 438–443.

POLNISCH, E., KNEIFEL, H., FRANZKE, H., HOFFMANN, H. H. (1992), Degradation and dehalogenation of monochlorophenols by the phenol assimilating yeast *Candida maltosa, Biodegradation*

2, 193–199.

RAJARATHNAM, S., SHASHIREKHA, N. U., BANO, Z. (1998), Biodegradative and biosynthetic capacities of mushrooms: present and future stategies, *Crit. Rev. Biotechnol.* **18**, 91–236.

RAMOS, J. L., MARQUES, S., TIMMIS, K. N. (1997), Transcriptional control of the *Pseudomonas* TOL plasmid catabolic operons is achieved through an interplay of host factors and plasmid-encoded regulators, *Annu. Rev. Microbiol.* 51, 341–373.

RATLEDGE, C. (1991), Editorial, in: *Physiology of Biodegradative Microorganisms* (RATLEDGE, C., Ed.), pp. VII-VIII. Dordrecht: Kluwer Academic Publishers.

REHM, H.-J., HORTMANN, L., REIFF, I. (1983), Regulation of the formation of fatty acids during microbial oxidation of alkanes, *Acta Biotechnol.* **3**, 279–288 (in German).

REHM, H. J., REIFF, I. (1982), Regulation of microbial alkane oxidation with respect to the formation of products, *Acta Biotechnol.* **3**, 279–288 (in German).

REINECKE, W. (1994), Degradation of chlorinated aromatic compounds by bacteria: strains development, in: *Biological Degradation and Bioremediation of Toxic Chemicals* (CHAUDHRY, G. R., Ed.), pp. 416–454. London: Chapman & Hall.

RISER-ROBERTS, E. (1998), *Remediation of Petroleum Contaminated Soils*. Boca Raton, FL: Lewis Publishers.

SACK, U., FRITSCHE, W. (1997), Enhancement of pyrene mineralization in soil by wood-decaying fungi, *FEMS Microbiol. Ecol.* **22**, 77–83.

SACK, U., HEINZE, T. M., DECK, J., CERNIGLIA, C. E., CAZAU, M. C., FRITSCHE, W. (1997a), Novel metabolites in phenanthrene and pyrene transformation by *Aspergillus niger*, *Appl. Environ. Microbiol.* **63**, 2606–2609.

SACK, U., HOFRICHTER, M., FRITSCHE, W. (1997b), Degradation of polycyclic aromatic hydrocarbons by manganese peroxidase of *Nematoloma frowardii*, *FEMS Lett.* **152**, 227–234.

SCHEIBNER, K., HOFRICHTER, M., HERRE, A., MICHELS, J., FRITSCHE, W. (1997), Screening for fungi intensively mineralizing 2,4,6-trinitrotoluene, *Appl. Microbiol. Biotechnol.* **47**, 452–457.

SHAILUBHAI, K., RAO, N. N., MODI, V. V. (1982), Degradation of benzoate and salicylate by *Aspergillus niger*, *Ind. J. Exp. Biol.* **20**, 166–168.

SHAILUBHAI, K., SAHASRAUDHE, S. R., VORA, K. A., MODI, V.V. (1983), Degradation of chlorinated derivatives of phenoxyacetic acid and benzoic acid by *Aspergillus niger*, *Appl. Microbiol. Biotechnol.* **21**, 365–367.

SMITH, M. R. (1990), The biodegradation of aromatic hydrocarbons by bacteria, *Biodegradation* **1**, 191–206.

SMITH, R. V., DAVIS, P. J., CLARK, A. M., GLOVER-MILTON, S. (1980), Hydroxylations of biphenyl by fungi, *J. Appl. Bacteriol.* **49**, 65–73.

SYLVESTRE, M., SONDOSSI, M. (1994), Selection of enhanced polychlorinated biphenyl-degrading bacterial strains for bioremediation: Consideration of branching pathways, in: *Biological Degradation and Bioremediation of Toxic Chemicals* (CHAUDHRY, G. R., Ed.), pp. 314–344. London: Chapman & Hall.

TAKAHASHI, S., ITOH, M., KANEKO, Y. (1981), Treatment of phenolic wastes by *Aureobasidium pullulans* adhered to the fibrous supports, *Eur. J. Appl. Microbiol. Biotechnol.* **13**, 175–178.

THOMAS, J., WARD, C. (1989), *In situ* biorestoration of organic contaminants in the subsurface, *Environ. Sci. Technol.* **23**, 760–766.

TIEDJE, J. M., QUENSEN III, J. F., CHEE-SANFORD, J., SCHIMEL, J. P., BOYD, S. A. (1993), Microbial reductive dechlorination of PCBs, *Biodegradation* **4**, 231–240.

WALKER, N. (1972), Metabolism of chlorophenols by *Rhodotorula glutinis*, *Soil Biol. Biochem.* **5**, 525–530.

WARIISHI, H., VALLI, K., GOLD, M. H. (1992), Manganese(II) oxidation by manganese peroxidase from the basidiomycete *Phanerochaete chrysosporium*. Kinetic mechanism and role of chelators, *J. Biol. Chem.* **267**, 23688–23695.

WARIISHI, H., VALLI, K., RENGANATHAN, V., GOLD, M. H. (1989), Thiol-mediated oxidation of nonphenolic lignin model compounds by manganese peroxidase of *Phanerochaete chrysosporium*, *J. Biol. Chem.* **264**, 14185–14191.

WATKINSON, R. J., MORGAN, P. (1990), Physiology of aliphatic hydrocarbon-degrading microorganisms, *Biodegradation* **1**, 79–92.

WRIGHT, J. D. (1993), Fungal degradation of benzoic acids and related compounds, *World J. Microbiol. Biotechnol.* **9**, 9–16.

WUNDERWALD, U. (1998), Investigations into the metabolism of fluorinated organic compounds by fungi, *Thesis*, University of Jena, Germany (in German).

WUNDERWALD, U., HOFRICHTER, M., KREISEL, G., FRITSCHE, W. (1998), Transformation of difluorinated phenols by *Penicillium frequentans* Bi 7/2, *Biodegradation* **8**, 379–385.

ZEHNDER, A. J. B. (Ed.) (1988), *Biology of Anaerobic Bacteria*. New York: John Wiley & Sons.

7 Principles of Anaerobic Degradation of Organic Compounds

BERNHARD SCHINK

Konstanz, Germany

1 General Aspects of Anaerobic Degradation Processes

Our atmosphere contains about 21% O_2, and this high amount of a bioavailable oxidant allows the vast majority of organic compounds produced in nature or through human manufacture to be degraded aerobically, with molecular oxygen as terminal electron acceptor. As long as oxygen is available, it is the preferred electron acceptor for microbial degradation processes in nature.

Anaerobic degradation processes have always been considered as being inferior in their kinetics and capacities, in comparison to aerobic degradation. They are thought to be slow and inefficient, especially with certain comparably stable types of substrates. Nonetheless, there are certain anoxic environments such as the cow's rumen where the turnover of, e.g., cellulose is much faster than in the presence of oxygen, with average half-life times in the range of one day. Fermentative degradation of fibers in the rumen finds its limits with plant tissues rich in lignin which largely withstands degradation in the absence of oxygen.

Also in waste treatment, especially with high loads of easy-to-degrade organic material (wastewaters of the sugar industry, slaughterhouses, food industry, paper industry, etc.), anaerobic processes have proved to be very efficient and far less expensive than aerobic treatment: they require only small amounts of energy input, different from treatment in aeration basins, and produce most often a mixture of methane and CO_2 ("biogas"), which can be applied efficiently for energy generation. This holds true for most waste materials that are easily accessible to degradation without the participation of oxygen, such as polysaccharides, proteins, fats, nucleic acids, etc. These polymers are hydrolyzed through specific extracellular enzymes, and the oligo- and monomers can be degraded inside the cell through enzyme reactions similar to those known from aerobic metabolism. The activities of such enzymes in anaerobic cultures are in the same range (0.1–1 µmol substrate per min and mg cell protein) as those of growing aerobic bacteria, and thus the transformation rates per unit biomass should be equivalent.

Nonetheless, anaerobic bacteria obtain typically far less energy from substrate turnover than their aerobic counterparts. Whereas aerobic oxidation of hexose to 6 CO_2 yields 2,870 kJ per mol, dismutation of hexose to $3 CH_4$ and $3 CO_2$ yields only 390 kJ per mol, about 15% of the aerobic process, and this small amount of energy has to be shared by at least three different metabolic groups of bacteria (see SCHINK, 1997). As a consequence, they can produce far less biomass per substrate molecule. Their growth yields are low, and most often growth is also substantially slower than that of aerobes. Therefore, the development of efficient reactors for methanogenic treatment of wastewaters, etc., aimed at retaining the microbial biomass inside the reactor, either by binding it to surfaces (fixed bed, fluidized bed reactors) or by stimulating aggregation of cells with particles that are maintained in the reactor through sedimentation (Upflow Anaerobic Sludge Blanket reactors). This overcomes the problem of low and slow biomass production in anaerobic degradation, and largely uncouples substrate turnover from biomass increase (for an overview, see, e.g., SCHINK, 1988). These systems allow to make anaerobic wastewater treatment nearly as efficient and less expensive than the aerobic process, with methane as a useful product, but the microbial communities in these advanced anaerobic reactors still are comparably sluggish in reacting to changes in substrate composition or in their re-establishment after accidental population losses due to toxic ingredients in the feeding waste.

In nature as well as in technical reactors, degradation of organic matter in the absence of oxygen can be coupled to the reduction of alternative electron acceptors. These alternative processes follow a certain sequence that appears to be determined roughly by the redox potential of the respective acceptor systems. As long as molecular oxygen (O_2/H_2O $E_h = +810$ mV; E_h values calculated for pH 7.0) is available it acts as the preferred electron acceptor, followed by nitrate (NO_3^-/NO_2^- $E_h = +430$ mV), manganese(IV) oxide (MnO_2/Mn^{2+} $E_h = +400$ mV), iron(III) hydroxides ($FeOOH/Fe^{2+}$ $E_h = +150$ mV), sulfate (SO_4^{2-}/HS^- $E_h = -218$ mV), and finally CO_2 (CO_2/CH_4 $E_h = -244$ mV), with the release of nitrite, ammonia, dinitrogen, manganese(II) and iron(II) carbonates, sulfides, and finally methane as products (STUMM and MORGAN, 1981; ZEHNDER and STUMM, 1988). Reduction of these acceptors with electrons from organic matter (average redox potential of glucose $\rightarrow 6\,CO_2$: -0.434 V; calculated after data from THAUER et al., 1977) provides metabolic energy in the mentioned sequence. Thus, the energy yields of the various anaerobes mentioned differ, and we will see later that the availability of either high or low potential electron acceptors may also influence the biochemistry of anaerobic degradation processes.

Limits of anaerobic degradation become obvious if we look at those organic compounds that accumulate in anoxic sediments, or that persist in anoxic soil compartments that have been contaminated with spills of mineral oil or other rather recalcitrant compounds. Mineral oil consists mainly of aliphatic and aromatic hydrocarbons which in the presence of molecular oxygen are attacked biochemically through so-called oxygenase reactions which introduce molecular oxygen into the respective molecule (SCHLEGEL, 1992; LENGELER et al., 1999). Monooxygenases introduce one oxygen atom into, e.g., aliphatics and certain aromatics to form the corresponding alcohols. Dioxygenases introduce both atoms of an oxygen molecule into an aromatic compound, transforming, e.g., benzene to catechol. A further oxygenase reaction cleaves catechol to form an open-chain carboxylic acid. Oxygenase reactions cannot be employed in the absence of oxygen, and especially the compounds that require oxygenases for aerobic breakdown might resist degradation under anoxic conditions. We will see below that in most cases there are alternatives in the anoxic world that also allow oxygen-independent degradation of such compounds.

Oxygen is not always advantageous in degradation processes. Oxygenases introduce hydroxyl groups into aromatics, and further oxygen may cause formation of phenol radicals that initiate uncontrolled polymerization and condensation to polymeric derivatives, similar to humic compounds in soil, that are very difficult to degrade further, either anaerobically or

aerobically. Therefore, anaerobic degradation processes may be applied for treatment of specific wastewaters rich in phenolic compounds, e.g., from the chemical industry, to avoid formation of unwanted side products, such as condensed polyphenols. In other cases, aerobic treatment may cause technical problems, e.g., by extensive foam formation during aerobic degradation of surface-active compounds, tensides, etc. Thus, knowledge of the limits and principles of anaerobic degradation processes under the various conditions prevailing in natural habitats might help to design suitable alternative techniques for cleanup of contaminated soils or for treatment of specific wastewaters that have so far been applied only insufficiently.

The following survey will try to give an overview of our present knowledge of the limits and principles of anaerobic degradation of organic compounds. The focus will be on those compounds which were for long times considered to be stable in the absence of oxygen.

2 Key Reactions in Anaerobic Degradation of Certain Organic Compounds

2.1 Degradation of Hydrocarbons

Saturated aliphatic hydrocarbons are only slowly attacked in the absence of oxygen, and the first reliable proof of such a process was provided only about 8 years ago for a culture of sulfate-reducing bacteria (AECKERSBERG et al., 1991). Growth of this culture with hexadecane was very slow, with doubling times of more than one week under optimal conditions. In the meantime, several strains of alkane-oxidizing anaerobes were isolated (AECKERSBERG et al., 1998; RUETER et al., 1994) which are specialized either on long-chain (C_{12}–C_{20}) or on medium-chain (C_{6-16}) alkanes and use either sulfate or nitrate as electron acceptor. No anaerobic degradation of short-chain alkanes ($<C_6$) has been observed so far. The biochemistry of alkane activation in the absence of oxygen is still enigmatic and awaits elucidation. Analysis of the lipid fatty acid composition of certain sulfate reducers after growth with even- or odd-chained alkanes suggested that a one-carbon unit was added to the substrate molecule in the initial attack, perhaps through a carbonylation reaction, however, other sulfate reducers or nitrate reducers gave no indication of such chain elongations (AECKERSBERG et al., 1998). Thus, there might even be more than one way of alkane activation in the absence of oxygen, or alternatives that have not been thought of so far. In any case, anaerobic hydrocarbon degradation is very slow, and will hardly be applicable to, e.g., biological clean-up of soil sites polluted with petroleum or diesel fuel.

A special case, although not of technical interest, is the anaerobic degradation of methane, e.g., with sulfate as electron acceptor, a process which is of major importance in global carbon transformations. No bacterium has been isolated so far which catalyzes this reaction although it is thermodynamically feasible (see SCHINK, 1997). There is indirect evidence that methanogenic bacteria themselves can reverse the methane formation reaction (ZEHNDER and BROCK, 1980; HARDER, 1997) but they do so only to a small extent, concomitant with methane formation. Thus, this concept cannot account for the net methane oxidation processes observed in marine sediments. Perhaps a cooperation of methanogens with sulfate reducers could explain this phenomenon, but experimental evidence is lacking so far, and the biochemistry of oxygen-independent methane oxidation remains an open question.

Unsaturated long-chain hydrocarbons with terminal double bonds can be hydrated, obviously to the corresponding primary alcohols (although against the Markownikoff rule!), and completely degraded through β-oxidation (SCHINK, 1985a). Branched-chain unsaturated hydrocarbons can be attacked as well. Squalene degradation has been documented in methanogenic enrichment cultures (SCHINK, 1985a), however, degradation was incomplete, probably due to the formation of saturated branched derivatives. Other unsaturated isoprene derivatives such as terpenes have been shown recently to be completely degraded with nitrate as electron acceptor (HARDER and PROBIAN, 1995; FOß and HARDER, 1998; FOß et al., 1998). Although the structures of terpenes differ substantially with respect to the way of possible attack, an amazingly broad variety of different terpenes was completely degraded. Some concepts of the biochemistry of degradation of these compounds have been developed (HYLEMON and HARDER, 1999) but experimental evidence is missing so far. Acetylene as a highly unsaturated hydrocarbon is fermented comparably fast to acetate and ethanol through acetaldehyde (SCHINK, 1985b), which is formed by a hydratase enzyme (ROSNER and SCHINK, 1995). No anaerobic degradation was documented so far with ethylene, propylene, propine, and higher homologs up to 6 carbon atoms.

2.2 Degradation of Ether Compounds and Non-Ionic Surfactants

Ether linkages are rather stable, and chemical cleavage requires rough conditions, e.g., boiling at strongly alkaline or acidic pH. Biological ether cleavage in the presence of oxygen employs oxygen as cosubstrate in an oxygenase reaction which transforms the ether to an unstable hemiacetal compound (BERNHARDT et al., 1970). Methyl groups of lignin monomers are thus released as formaldehyde, not as methanol.

Anaerobic demethylation of lignin monomers by the homoacetogen *Acetobacterium woodii* was first described by BACHE and PFENNIG (1981) and later repeatedly for several other homoacetogens. The mechanism of this phenyl methyl ether cleavage could only recently be elucidated. Studies with the homoacetogen *Holophaga foetida* showed that the methyl group is first transferred as a methyl cation to a fully reduced cob(I)alamin carrier which later methylates the coenzyme tetrahydrofolate (Fig. 1; KREFT and SCHINK, 1993, 1994). Similar studies with *Acetobacterium woodii*, *Sporomusa ovata*, or the homoacetogenic strain MC revealed that also in these cases the methyl group is transferred as a methyl

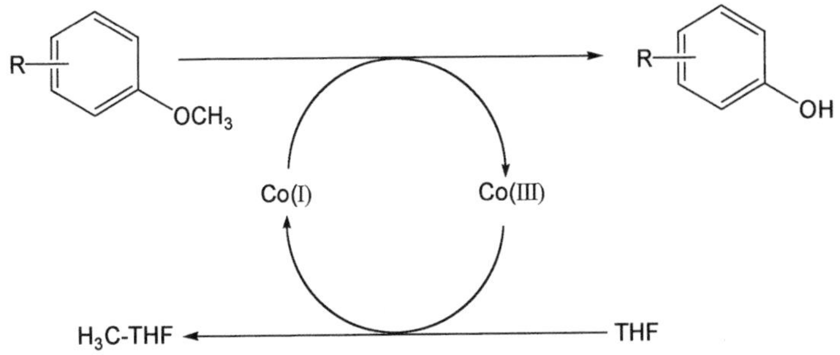

Fig. 1. Anaerobic demethylation of phenyl methyl ethers. Co(I), Co(III), cobalamine at different redox states; THF, tetrahydrofolate.

cation, but that the details of further methyl transfer to the coenzymes may differ with the respective strains studied (BERMAN and FRAZER, 1992; STUPPERICH and KONLE, 1993; KAUFMANN et al., 1997).

A different type of anaerobic ether cleavage was observed with the synthetic polyether polyethylene glycol (PEG). Formation of acetaldehyde as first cleavage product, extreme oxygen sensitivity of the ether-cleaving enzyme in cell-free extracts, and interference with cobalamines strongly suggest that the first step in this degradation is a cobalamine-dependent shift of the terminal hydroxyl group to the subterminal C-atom, analogous to a diol dehydratase reaction (Fig. 2; SCHRAMM and SCHINK, 1991; FRINGS et al., 1992). This reaction again transforms the ether to a hemiacetal derivative which decomposes easily on its own. Since the ether-cleaving enzyme is located in the cytoplasmic space the polymeric PEG (with molecular masses up to 40,000 Da!) has to cross the cell membrane(s) before it is cleaved inside, and the same is true for PEG-containing non-ionic surfactants. Since the bacterial strains studied so far are specialized only on the degradation of the PEG chain, the lipophilic residues of the surfactants have to cross the membrane(s) again on their way back out. It is obvious that these transport steps considerably limit the applicability of such degradation capacities for treatment of, e.g., wastewaters rich in such surfactants. Nonetheless, the applicability of anaerobic fixed bed reactors for treatment of non-ionic surfactants of various types to methane, CO_2 and mixtures of fatty acids has been demonstrated (WAGNER and SCHINK, 1987, 1988).

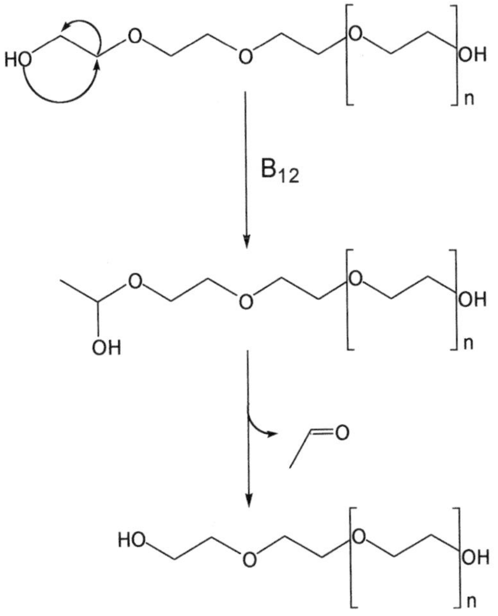

Fig. 2. Anaerobic degradation of polyethylene glycol by fermenting bacteria. B_{12}, coenzyme B_{12}.

The same strategy of ether cleavage is also used obviously in the degradation of the industrial solvents phenoxyethanol or methoxyethanol by homoacetogenic bacteria. With these substrates, a terminal free hydroxy substituent can also be shifted to the subterminal carbon atom, releasing acetaldehyde (FRINGS and SCHINK, 1994). This strategy cannot be applied in degradation of phenoxyacetate, the mother substance of 2,4-D and 2,4,5-T. Also phenoxyacetate is degraded in the absence of oxygen, but this process is extremely slow and nothing is known so far about its biochemistry (FRINGS and SCHINK, unpublished data).

The anaerobic ether cleavage reactions described here only proceed inside the bacterial cell. With this, they differ from the corresponding reactions reported for certain basidiomycetes (KEREM et al., 1998) which apply these cleavage capacities, e.g., in lignin degradation. It is not surprising, therefore, that highly polymeric, condensed ether compounds such as lignin are not degraded in the absence of molecular oxygen to any significant extent.

2.3 Degradation of *N*-Alkyl Compounds and NTA

Among the natural *N*-alkyl compounds are, besides amino acids, several methylated amines such as trimethylamine, which is formed during the initial decay of fish tissue through reduction of trimethylamine-*N*-oxide by several enterobacteria and others. Under strictly anoxic conditions, methanogenic Archaea have been found to efficiently demethylate trimethylamine via dimethylamine to monomethylamine, and ferment the methyl moieties by dismutation to methane and CO_2 (HIPPE et al., 1979). The cleavage between the nitrogen and the neighboring carbon atom is accomplished by a nucleophilic attack of a cob(I)alamin, analogous to demethylation of phenyl methyl ethers by homoacetogenic bacteria (see above). So far, only methanogens have been found to demethylate methylamines whereas homoacetogens appear to be specialists for demethylation of phenyl methyl ethers which, in turn, are not attacked by methanogens. It should be emphasized at this point that the strategy of methyl cation removal by cob(I)alamin derivatives is applicable only with these one-carbon compounds. Ethyl or higher homologs cannot be cleaved this way.

N-alkyl compounds of technical and environmental concern are ethylene diamine tetraacetate (EDTA) and nitrilotriacetate (NTA); the latter has largely replaced polyphosphates as calcium chelator in most commercial washing detergents. The main problem in EDTA degradation is the formation of strong complexes of EDTA with metal ions which render this substrate very difficult to attack. Nonetheless, microbial EDTA degradation in the presence of oxygen could be documented (NÖRTEMANN, 1992), but no reports exist on a possible anaerobic degradation of EDTA. NTA is degraded aerobically through oxygenase-dependent hydroxylation of one methylene carbon. The resulting hydroxy compound is unstable and releases glyoxylic acid. Removal of a further carboxymethylene residue produces glycine as coproduct. Anaerobic degradation of NTA is possible with nitrate as electron acceptor. The first degradation step is a dehydrogenation to an unsaturated iminium derivative which, upon hydration, could release again glyoxylic acid to form the iminodiacetate derivative (EGLI et al., 1990). Since the oxidation of NTA to the unsaturated derivative has a rather high redox potential ($> +100$ mV) it is not surprising that so far only nitrate reducers have been found to be able to degrade NTA. We have to assume that under strictly reducing conditions, i.e., in deeper sediment layers, NTA is stable towards microbial attack because neither sulfate-reducing nor fermenting bacteria can release electrons arising at this high redox potential.

2.4 Degradation of *S*-Alkyl Compounds

Dimethylsulfoniopropionate is an osmoprotectant of several green algae and seaweeds. Its cleavage by anaerobic bacteria leads to acrylate and dimethylsulfide which can escape into the atmosphere. The thioether dimethylsulfide can also be degraded anaerobically by methanogenic bacteria (KIENE et al., 1986; OREMLAND et al., 1989). The carbon–sulfur linkage is cleaved by methyl cation removal, analogous to the demethylation reactions described above (KELTJENS and VOGELS, 1993).

2.5 Degradation of Ketones

Aerobic degradation of ketones appeared to be well-established. Indications of an oxygenase-catalyzed hydroxylation of acetone to acetol by aerobic bacteria were provided early (LUKINS and FOSTER, 1963; TAYLOR et al., 1980), but this type of reaction was never confirmed unequivocally. Anaerobic degradation of acetone, by either nitrate-reducing, sulfate-reducing, or fermenting bacteria cooperating with methanogenic partners uses a carboxylation reaction as primary step of activation, leading to an acetoacetyl derivative which undergoes subsequent cleavage to two acetate moieties (Fig. 3; PLATEN and SCHINK, 1987, 1989; BONNET-SMITS et al., 1988; PLATEN et al., 1990). Unfortunately, the primary carboxylation reaction could never be proved convincingly with these cultures. Very little activity of acetone carboxylation was provided by enzyme assays (PLATEN and SCHINK, 1991) or by radiotracer experiments with suspensions of intact cells (JANSSEN and SCHINK, 1995). Acetone carboxylation activity in the phototrophic anaerobe *Rhodobacter capsulatus* was very weak *in vitro* as well (BIRKS and KELLY, 1997). An acetone-carboxylating enzyme complex of good activity has recently been found in an aerobic *Xanthobacter* strain (SLUIS et al., 1996) and was purified and characterized (SLUIS and ENSIGN, 1997). The reaction requires as energy source for the acetone carboxylation reaction one ATP which is hydrolyzed to AMP plus two P_i. Whether sulfate-reducing or fermenting bacteria provide the necessary energy for carboxylation in different ways has still to be examined. Especially the fermenting bacteria cannot afford to spend the equivalent of two ATP into this carboxylation reaction. To our surprise, all aerobic bacteria enriched with acetone used as well a carboxylation step rather than an oxygenase reaction in acetone activation, even if CO_2 was trapped during the enrichment process (SCHINK, unpublished data). Thus, the original reports of oxygenase-dependent acetone activation may describe an ex-

ceptional situation with one single strain which is not representative for the majority of aerobic acetone degraders.

Higher homologs of acetone appear to be attacked in a similar manner by carboxylations. This applies also to acetophenone, the phenyl-substituted analog of acetone (SCHINK, unpublished data).

2.6 Degradation of Aromatic Compounds

In aerobic degradation of aromatic compounds, oxygenases activate the comparably stable oxygen molecule in such a way as to produce highly electrophilic species which add to a comparably inert aromatic compound, to form hydroxylated products such as catechol (DAGLEY, 1971; SCHLEGEL, 1992). A further oxygenase-dependent step opens the aromatic ring, either between or vicinal to the two hydroxyl groups of catechol, thus forming an unsaturated, open-chain carboxylic acid which undergoes further degradation, typically to an acetyl and a succinyl derivative.

That aromatic compounds can also be degraded anaerobically was documented as early as 1934. A broad variety of mononuclear aromatic compounds, such as benzoate, phenols, and several lignin monomers, was converted stoichiometrically to methane and CO_2 (TARVIN and BUSWELL, 1934). Later these observations were forgotten, and some textbooks maintained even into the 1980s the dogma that aromatics can be attacked only with oxygen as cosubstrate. During the 1970s, EVANS (1977) developed the concept that destabilization of the aromatic nucleus in the absence of oxygen could proceed through a reductive rather than an oxidative reaction. Today we know at least

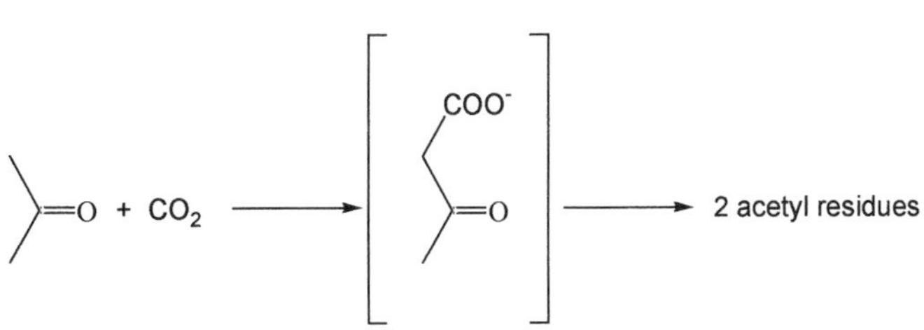

Fig. 3. Anaerobic degradation of acetone through carboxylation.

three different pathways of anaerobic degradation of mononuclear compounds, i.e., the benzoyl-CoA pathway, the resorcinol pathway, and the phloroglucinol pathway (EVANS and FUCHS, 1988; SCHINK et al., 1992; FUCHS et al., 1994; HEIDER and FUCHS, 1997). In all these cases, a 1,3-dioxo structure is formed through a reduction step, either inside the ring itself or in combination with a carboxyl coenzyme A moiety. This structure allows a nucleophilic attack on one of the ring ketone carbon atoms, and subsequent ring fission. Depending on the respective aromatic substrate, either a pimelic (C_7-dicarboxylic) residue bound to coenzyme A, or a partly oxidized caproic (C_6-monocarboxylic) acid is formed, and undergoes subsequent β-oxidation to three acetyl moieties. Formation of other products in fermentative benzoate degradation such as succinate or propionate, as claimed in earlier papers, could

never be reproduced with defined cultures and may have been due to uncontrolled side reactions, or misinterpretations of insufficient chemical analyses.

2.6.1 Benzoate and the Benzoyl-CoA Pathway

The benzoyl-CoA pathway appears to be the most important one in anaerobic degradation of aromatics because a broad variety of compounds enter this path, including phenol, various hydroxybenzoates, phenylacetate, aniline, certain cresols, and even the hydrocarbon toluene (Fig. 4; SCHINK et al., 1992; HEIDER and FUCHS, 1997; HARWOOD et al., 1999). Once benzoyl-CoA is formed, the stability of the aromatic ring structure is overcome by a

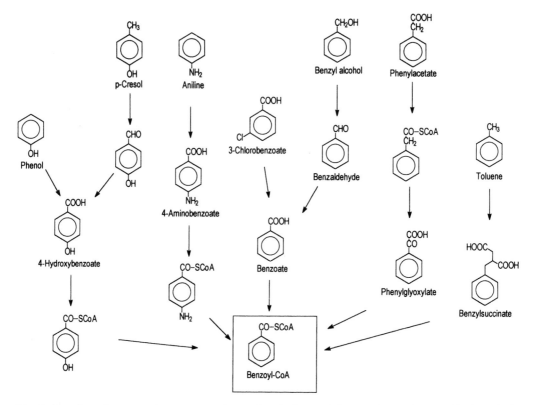

Fig. 4. Overview of mononuclear compounds entering the benzoyl-CoA pathway of anaerobic degradation (courtesy of Prof. G. FUCHS, Freiburg).

reductive step which introduces two single electrons and protons, probably through a radicalic intermediate, to form cyclohexadiene carboxyl-CoA as the first identifiable product (KOCH et al., 1993; BOLL and FUCHS, 1995). Since the reduction of the benzene ring to a cyclohexadiene derivative, even with electrons at the ferredoxin level, is endergonic it requires the investment of energy in the form of 2 ATP (BOLL and FUCHS, 1995; HARWOOD et al., 1999). Nitrate-reducing bacteria can recover this energy investment through the further breakdown of the C_7-dicarboxylic acid derivative produced upon ring cleavage, via β-oxidation to three acetyl-CoA residues that are finally oxidized in the citric acid cycle (Fig. 5). Fermenting bacteria and sulfate-reducing bacteria recover only little energy in the further breakdown of the open-chain intermediate. They may apply a different type of reaction for benzoyl-CoA dearomatization that leads to a cyclohexene carboxyl derivative (SCHÖCKE and SCHINK, 1999; HARWOOD et al., 1999). This reduction is exergonic, even with NADH electrons, and could thus be catalyzed without a net investment of ATP. However, the biochemistry of this new reaction still needs to be elucidated.

2.6.2 Phenol, Hydroxybenzoates, and Aniline

Aromatic compounds such as phenol or aniline which do not carry a carboxyl group are first carboxylated to a *p*-hydroxy or a *p*-amino benzoic acid residue which is subsequently activated with coenzyme A (Fig. 6; TSCHECH and FUCHS, 1989; LACK and FUCHS, 1992; SCHNELL and SCHINK, 1991). The carboxylation of phenol by a nitrate-reducing *Thauera aromatica* strain can be followed *in vitro* with phenyl phosphate as substrate which is carboxylated to 4-hydroxybenzoate and further degraded as such (see below). There is good evidence that phenyl phosphate is really the physiological substrate of the carboxylation reaction although the phosphate donor for phenol phosphorylation is still unknown (LACK and FUCHS, 1992, 1994). Whether sulfate-reducing bacteria or fermenting bacteria cooperating with methanogenic partners use the same pathway for phenol degradation still remains to be examined. A H/D-exchange at the carbon atom 4 of phenol by cell suspensions of a methanogenic phenol-degrading enrichment culture indicates that these cultures also activate phenol through a carboxylation at this position (GALLERT et al., 1991). Whether the carboxylation reaction in these bacteria is also initiated by phenol phosphorylation is still an open question. The overall energy budget of fermentative phenol degradation is very tight and hardly allows spending a full ATP equivalent or

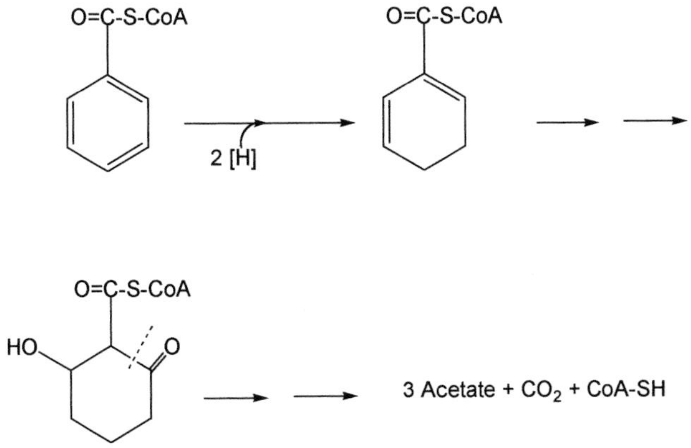

Fig. 5. Initial steps in anaerobic degradation of benzoate by the nitrate-reducing bacterium *Thauera aromatica*.

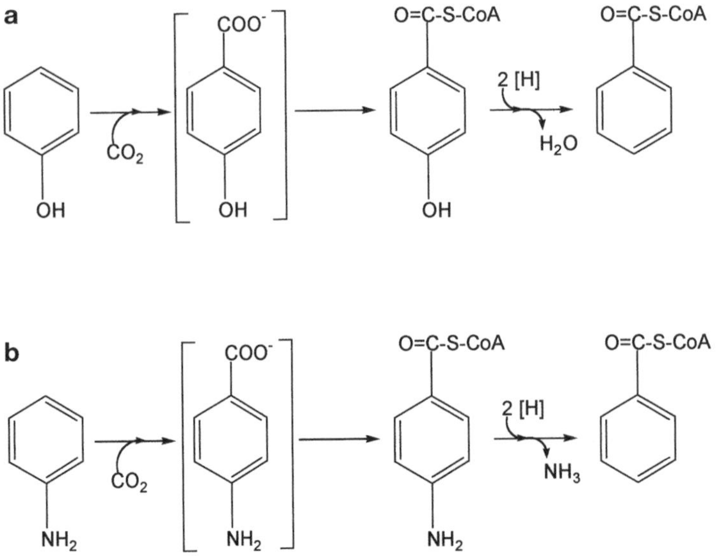

a

COO⁻   O=C-S-CoA   O=C-S-CoA

CO₂   2 [H]   H₂O

OH   OH   OH

b

COO⁻   O=C-S-CoA   O=C-S-CoA

CO₂   2 [H]   NH₃

NH₂   NH₂   NH₂

Fig. 6. Anaerobic degradation of **a** phenol and **b** aniline by anaerobic bacteria. The compounds in brackets have not been identified.

even more into this carboxylation reaction. The biochemistry of phenol degradation by sulfate-reducing bacteria has not been studied so far, but it is likely to proceed basically through the same pathway.

The 4-hydroxybenzoate formed is activated through a ligase reaction, analogous to benzoyl-CoA formation from benzoate, to form 4-hydroxybenzoyl-CoA which is subsequently reductively dehydroxylated to benzoyl-CoA (Fig. 6a).

4-Hydroxybenzoate may be degraded anaerobically through the same pathway. However, in cultures 4-hydroxybenzoate is not stable but is slowly decarboxylated to phenol, perhaps by an enzyme activity related to phenyl phosphate carboxylase.

3-Hydroxybenzoate is comparably stable and does not decarboxylate spontaneously. Instead, the hydroxyl group is reductively eliminated by fermenting bacteria to allow further degradation through the benzoyl-CoA pathway, as shown with *Sporotomaculum hydroxybenzoicum* (MÜLLER and SCHINK, unpublished data). An entirely different strategy in anaerobic degradation of 3-hydroxybenzoate is used by a nitrate-reducing bacterium, Strain BoNHB. This bacterium does not activate or reduce 3-hydroxybenzoate, but oxidizes it to gentisate (2,5-dihydroxybenzoate) and further

to hydroxyhydroquinone (MÜLLER and SCHINK, unpublished data). The further fate of hydroxyhydroquinone is discussed below.

Aniline is degraded anaerobically through a pathway analogous to phenol degradation. The initial activation is accomplished through carboxylation to 4-aminobenzoate which is subsequently activated to 4-aminobenzoyl-CoA and undergoes reductive deamination to benzoyl-CoA (Fig. 6b; SCHNELL and SCHINK, 1991). The initial carboxylation reaction has not been studied in cell-free extracts so far, and nothing is known about an activated intermediate to provide the carboxylation reaction with the necessary energy.

Aminobenzoates, diaminobenzenes, and aminohydroxybenzenes are degraded very slowly in anaerobic enrichment cultures, but nothing is known about the degradation pathways (SCHNELL and SCHINK, unpublished data).

2.6.3 Cresols

Cresols (methylphenols) are anaerobically degraded through three different pathways, depending on the type of substitution. *p*-Cresol is hydroxylated at the methyl group by an oxygen-independent reaction, probably through a quinomethide intermediate as suggested earlier for an aerobic *Pseudomonas* strain (Fig. 7a; HOPPER, 1978). The redox potential of this oxidation reaction is in the range of +100 mV (calculated after THAUER et al., 1977), and the reaction is, therefore, easy to do for a nitrate-reducing bacterium that couples this oxidation, e.g., with the reduction of a *c*-type cytochrome at +235 mV (HOPPER et al., 1991). Sulfate-reducing or fermenting bacteria, on the other hand, have difficulties in disposing of these electrons. *o*-Cresol can be carboxylated to 3-methyl-4-hydroxybenzoate and further degraded as such (Fig. 7b; BISAILLON et al., 1991; RUDOLPHI et al., 1991). An alternative pathway could lead through methyl group hydroxylation, analogous to *p*-cresol, to form salicylic acid as an intermediate (SUFLITA et al., 1989; SCHINK et al., 1992) but this pathway has only been hypothesized so far. The pathway of anaerobic, *m*-cresol degradation has been elucidated recently with the sulfate-

reducing bacterium *Desulfobacterium cetonicum*. This degradation follows a strategy analogous to anaerobic toluene degradation by nitrate-reducing bacteria (MÜLLER et al., 1999): the methyl group of *m*-cresol adds to fumarate to form 3-hydroxybenzylsuccinate. Activation and β-oxidation lead to succinyl-CoA and benzoyl-CoA (Fig. 7c). Thus, the new type of methyl group activation by addition to fumarate appears not to be restricted to the activation of aromatic hydrocarbons (see below).

2.6.4 Hydroquinone and Catechol

Hydroquinone is degraded by sulfate-reducing and fermenting bacteria. The degradation pathway has been studied with a sulfate-reducing *Desulfococcus* strain (GORNY and SCHINK, 1994a) and a fermenting bacterium that was later described as *Syntrophus gentianae*

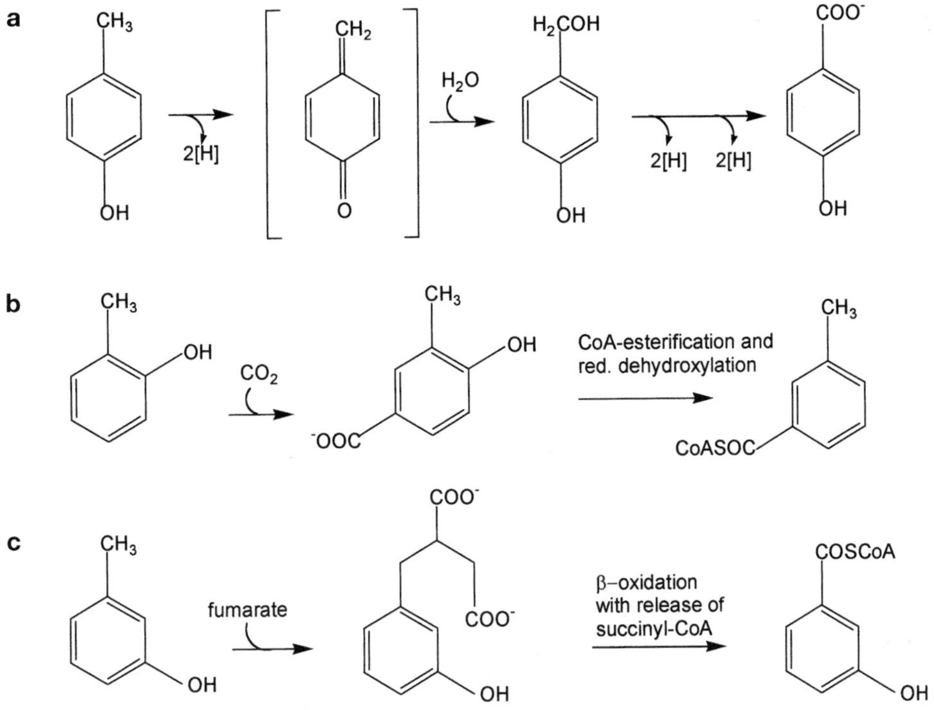

Fig. 7. Anaerobic degradation of cresols. **a** Degradation of *p*-cresol, **b** degradation of *o*-cresol, **c** degradation of *m*-cresol.

(GORNY and SCHINK, 1994b). In both cases, hydroxyhydroquinone is carboxylated to gentisic acid; again, this carboxylation could never be studied in cell-free extracts and the way of energetization of this reaction is unknown. Gentisate is activated to gentisyl-CoA through a CoA-ligase reaction, and reductively dehydroxylated to benzoyl-CoA which enters the modified benzoyl-CoA pathway. The dehydroxylation of both hydroxyl groups proceeds in one single step. Gentisic acid is utilized by the fermenting bacterium alternatively and probably enters the described degradation pathway at a later stage.

Catechol, the key intermediate of aerobic breakdown of aromatic compounds, is by far the slowest phenolic compound to be degraded under anoxic conditions. The biochemistry of catechol degradation has been studied only

with a sulfate-reducing *Desulfobacterium* strain, which carboxylates catechol to protocatechuate (GORNY and SCHINK, 1994c). Protocatechuate is activated to form protocatechuyl-CoA, which is subsequently dehydroxylated to benzoyl-CoA. Efforts to isolate nitrate-reducing or fermenting bacteria with hydroquinone or catechol as substrate have failed so far.

2.6.5 Resorcinol

An entirely different strategy is taken in the anaerobic degradation of resorcinol and its derivatives. The two hydroxyl groups in resorcinol are in positions to each other that allow tautomerization to a cyclohexene dione derivative with an isolated double bond (Fig. 8a).

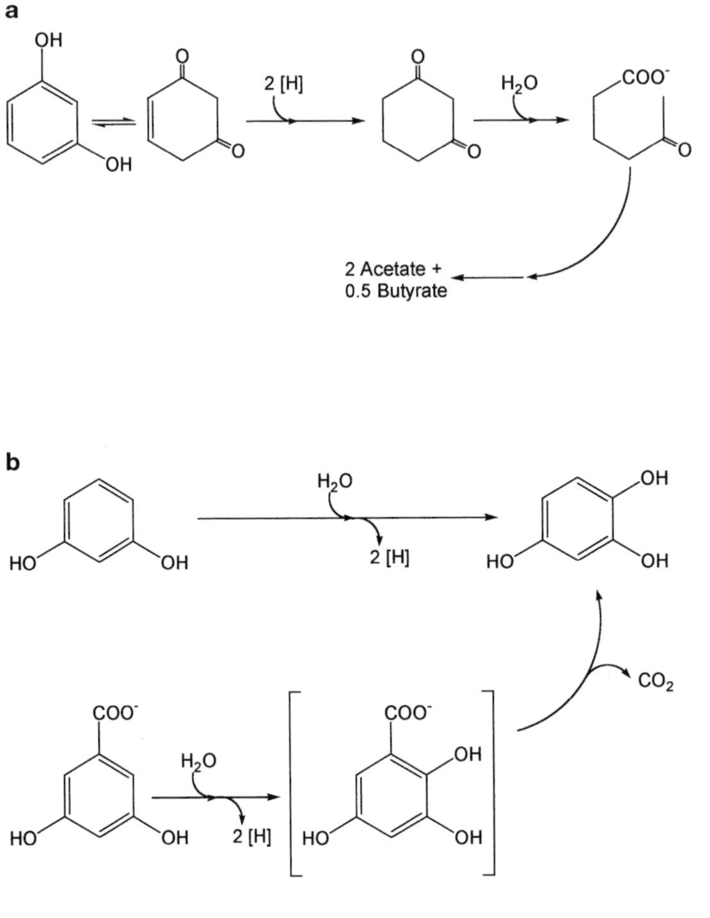

Fig. 8. Anaerobic degradation of resorcinol and α-resorcylate. **a** Resorcinol degradation by a fermenting bacterium, *Clostridium* strain KN245, **b** degradation of resorcinol and α-resorcylate by nitrate-reducing bacteria.

Cell-free extracts of a fermenting *Clostridium* strain convert resorcinol to dihydroresorcinol (cyclohexanedione; TSCHECH and SCHINK, 1985; KLUGE et al., 1990) which is further hydrolyzed to 5-oxohexanoate, probably through a nucleophilic attack on one of the carbonyl carbon atoms (Fig. 8a).

The resorcinol carboxylates β- and γ-resorcylate are degraded by the same fermenting bacterium after decarboxylation to resorcinol. These decarboxylations are chemically easy because in these cases the carboxylic group is located in *ortho*- or *para*-position to electron-withdrawing hydroxyl groups.

In cultures of nitrate-reducing bacteria growing with resorcinol as the sole substrate, no resorcinol-reducing activity could be identified (KLUGE et al., 1990; GORNY et al., 1992). Only recently we have found that the pattern of resorcinol degradation by nitrate reducers uses an entirely different chemistry. The nitrate reducer *Azoarcus anaerobius* does not cleave the ring reductively or hydrolytically, but destabilizes it by introduction of a further hydroxyl group to form hydroxyhydroquinone (Fig. 8b; PHILIPP and SCHINK, 1998). The enzyme activity involved is membrane-bound and the hydroxylation is coupled to nitrate reduction to nitrite. In a further oxidation step, hydroxyhydroquinone is oxidized to hydroxybenzoquinone (PHILIPP and SCHINK, 1998). The ring fission reaction is still not entirely clear: in the presence of NADH plus nitrate, cell-free extracts of *A. anaerobius* convert hydroxybenzoquinone slowly to succinate and acetate (PHILIPP and SCHINK, unpublished data).

1989; BRUNE and SCHINK, 1992), and the same strategy is followed by *Holophaga foetida* strain TMBS4 (KREFT and SCHINK, 1993). Hydrolytic ring cleavage leads to 3-hydroxy-5-oxohexanoic acid which is thiolytically cleaved to three acetate residues (BRUNE and SCHINK, 1992). This pathway is easy to conceive because the 1,3,5-arrangement oft he three hydroxyl groups on the aromatic ring allows tautomerization to 1,3,5-trioxocyclohexane to a degree which lends itself to a nucleophilic attack on the oxocarbon groups. The second trihydroxybenzene isomer, pyrogallol, cannot be hydrolyzed directly but is isomerized to phloroglucinol through a transhydroxylation reaction (Fig. 9; KRUMHOLZ and BRYANT, 1988; BRUNE and SCHINK, 1990). The reaction requires 1,3,4,5-tetrahydroxybenzene as a cosubstrate, and the enzyme transfers a hydroxyl group from the tetrahydroxybenzene to pyrogallol, thus releasing phloroglucinol as product and the tetrahydroxybenzene as coproduct (BRUNE and SCHINK, 1990; REICHENBECHER et al., 1996).

The third trihydroxybenzene isomer, hydroxyhydroquinone, is converted by the fermenting bacterium *Pelobacter massiliensis* to three acetates as well (SCHNELL et al., 1991), indicating that this pathway also leads through phloroglucinol. The isomerization to phloroglucinol requires three subsequent transhydroxylation reactions analogous to the pyrogallol–phloroglucinol transhydroxylation, and, indeed, phloroglucinol is the final aromatic compound that is reduced and cleaved hydrolytically (BRUNE et al., 1992).

2.6.6 Trihydroxybenzenes and Trihydroxybenzoates

Among the three trihydroxybenzene isomers, pyrogallol and phloroglucinol are degraded quickly by fermenting bacteria (SCHINK and PFENNIG, 1982). Phloroglucinol degradation has been studied in detail with *Eubacterium oxidoreducens* and *Pelobacter acidigallici*. Phloroglucinol is reduced by an NADPH-dependent reductase to dihydrophloroglucinol (Fig. 9; HADDOCK and FERRY,

2.6.7 Hydroxyhydroquinone, a New Important Intermediate

Besides the strategy of isomerization to phloroglucinol which is taken by all fermenting bacteria, we have recently found alternative pathways of hydroxyhydroquinone degradation with nitrate-reducing and sulfate-reducing bacteria. Hydroxyhydroquinone degradation by nitrate-reducing bacteria was mentioned above in the context of nitrate-dependent resorcinol degradation (PHILIPP and SCHINK, 1998). The reaction sequence leads to

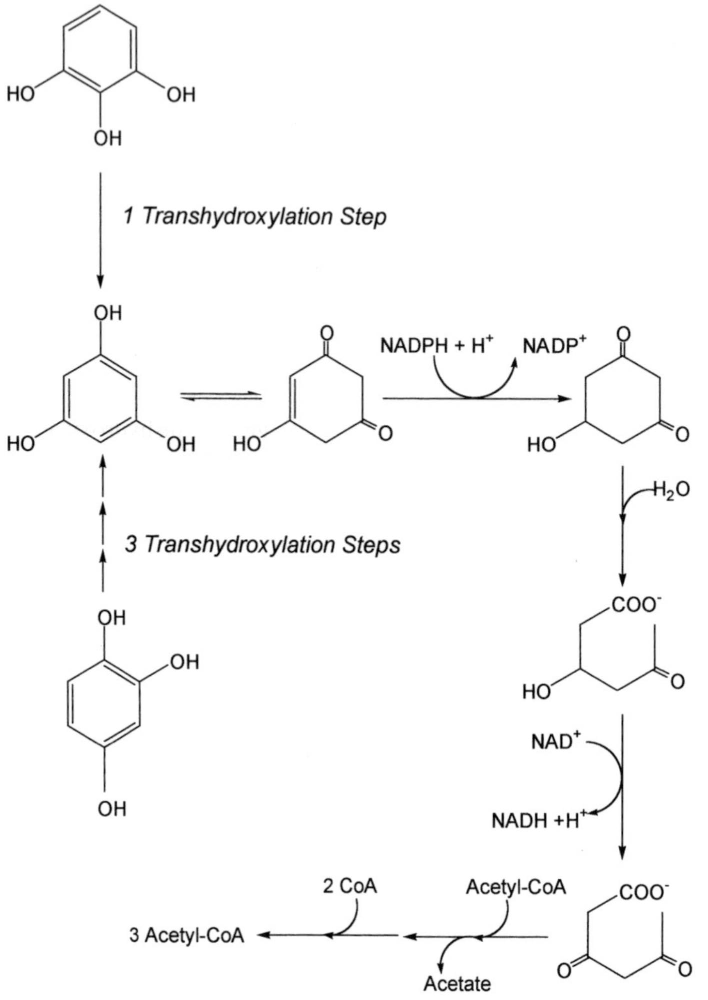

Fig. 9. Degradation of trihydroxybenzenes by fermenting bacteria.

an acetate and a succinate residue, suggesting that the hydroxyhydroquinone intermediate is cleaved between the carbon atoms 1 and 2 and 3 and 4 (Fig. 10). The cleavage products found can easily be oxidized to CO_2, with nitrate as electron acceptor. A further alternative of hydroxyhydroquinone degradation was found with the sulfate-reducing bacterium *Desulfovibrio inopinatus*. This bacterium destabilizes hydroxyhydroquinone by reduction to dihydrohydroxyhydroquinone, to form acetate and a so far unidentified 4-carbon derivative (REICHENBECHER et al., unpublished data). Since *D. inopinatus* is unable to oxidize ace-

tate, the final products are two acetate and two CO_2, and 1 mol of sulfate is reduced concomitantly to sulfide.

Hydroxyhydroquinone has gained additional interest recently because it was found to be an intermediate in the nitrate-dependent degradation of resorcinol, 3-hydroxybenzoate, 3,5-dihydroxybenzoate, and perhaps also gentisate and catechol, although these substrates have not been examined for degradation by nitrate-reducing bacteria yet. The strategy of oxidative destabilization of aromatic compounds through hydroxylation, that these nitrate-reducing bacteria apply, resembles to some ex-

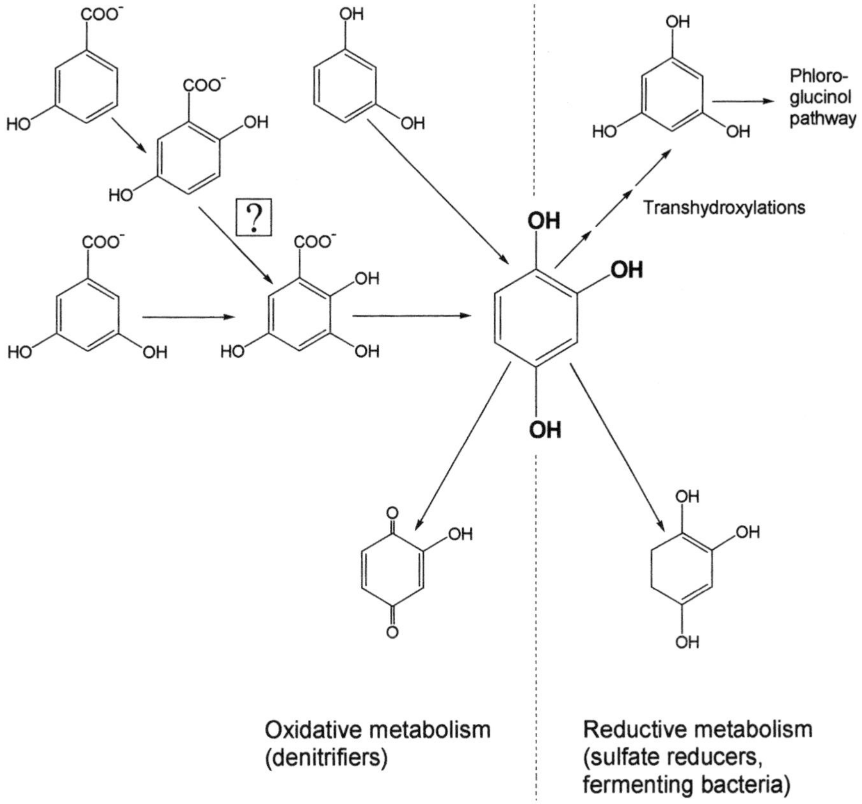

Fig. 10. Hydroxyhydroquinone as a new intermediate in anaerobic degradation of various aromatic compounds.

tent that taken by aerobic bacteria, and nitrate-dependent degradation of phenolic compounds thus follows a strategy that is somehow a mix between the oxidative aerobic strategy and the typical reductive strategy followed by strictly anaerobic bacteria.

2.6.8 Aromatic Hydrocarbons

Anaerobic degradation of aromatic hydrocarbons has been a matter of dispute for several years until reliable conversion balances with fast growing enrichment cultures or pure cultures were provided. Today, several pure cultures of nitrate-reducing or sulfate-reducing bacteria are available which oxidize toluene (methylbenzene) and have been characterized in detail (DOLFING et al., 1990; SCHOCHER et al., 1991; EVANS et al., 1991; ALTENSCHMIDT, and FUCHS, 1991; SEYFRIED et al., 1994; RABUS et al., 1993; RABUS and WIDDEL, 1995; BELLER et al., 1996). Initial experiments indicated that toluene degradation proceeded through oxidation of the methyl group (ALTENSCHMIDT and FUCHS, 1991), with benzoyl-CoA as central intermediate. Nonetheless, the anticipated methyl hydroxylation reaction transforming toluene to benzylalcohol could never be observed *in vitro*. Labeling experiments with intact cells provided evidence that toluene was activated by addition of fumarate, probably through a radicalic intermediate, to form benzyl succinate (Fig. 11a; BIEGERT et al., 1996), and this mechanism could be confirmed to be used by nitrate-reducing and by sulfate-reducing bacteria (BELLER and SPORMANN, 1997a, b, 1998).

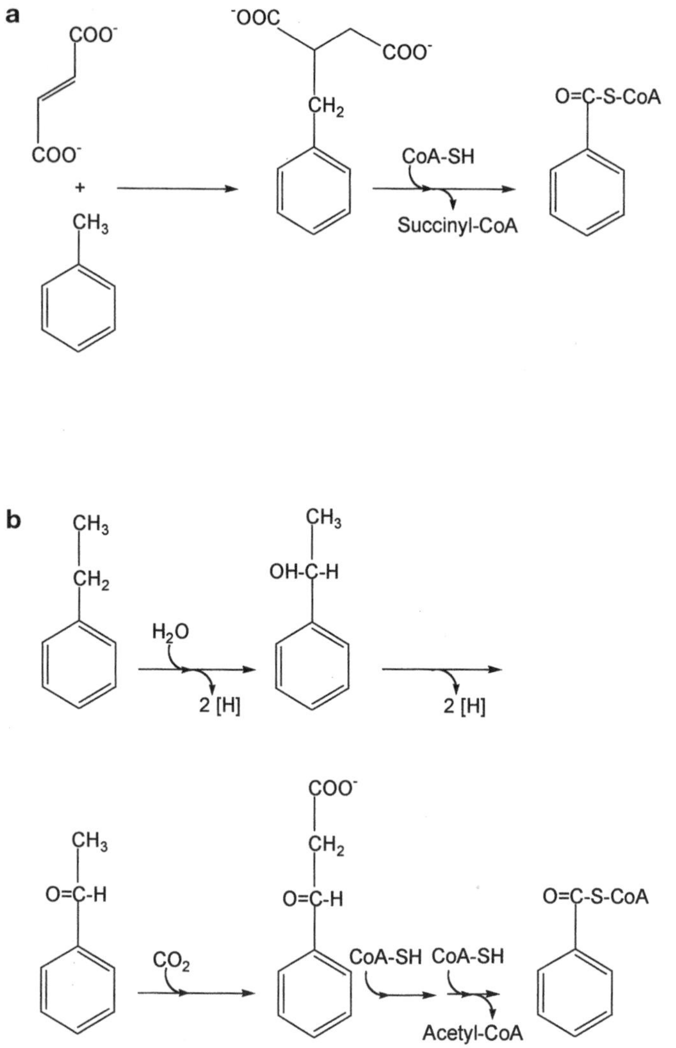

a

b

Fig. 11. Anaerobic degradation of **a** toluene and **b** ethylbenzene by nitrate-reducing or sulfate-reducing bacteria.

Benzyl succinate releases succinyl-CoA through β-oxidation, leading again to benzoyl-CoA as key intermediate.

This new type of methyl group activation is also employed in anaerobic degradation of *m*-cresol (see above).

Anaerobic degradation of *o*-, *m*-, and *p*-xylene has been documented mostly in tracer experiments with sediment samples or in enrichment cultures. Pure cultures of sulfate-reducing xylene degraders are now available (HARMS et al., 1999), and one might speculate that activation by addition of fumarate may al-

so be applied in the anaerobic oxidation of xylene isomers or of saturated alkanes for which so far no convincing degradation concepts exist.

Ethylbenzene is oxidized by denitrifying and sulfate-reducing bacteria (RABUS and WIDDEL, 1995; BALL et al., 1996; RABUS and HEIDER, 1998). Here the side chain is desaturated and water is added to form 1-phenylethanol as first oxidation product (Fig. 11b; BALL et al., 1996; RABUS and HEIDER, 1998). The further pathway leads via oxidation to acetophenone and carboxylation to benzoylacetate, and through thiolytic cleavage to an acetyl residue and benzoyl-CoA (HEIDER et al., 1999).

Much less is known about anaerobic benzene oxidation. Oxygen-independent hydroxylation and degradation of benzene was observed with methanogenic enrichment cultures (VOGEL and GRBIĆ-GALIĆ, 1986; EDWARDS and GRBIĆ-GALIĆ, 1994) and with sulfate-reducing (LOVLEY et al., 1995), nitrate-reducing (BURLAND and EDWARDS, 1999), or iron(III)-reducing enrichments (LOVLEY et al., 1996), but the bacteria acting in these processes could not yet be isolated. Thus, the biochemistry of benzene activation remains an open question.

Naphthalene is degraded anaerobically by nitrate-reducing (MIHELCIC and LUTHY, 1988) and also by pure cultures of sulfate-reducing bacteria (GALUSHKO et al., 1999). Preliminary experiments with an undefined enrichment culture indicate that a carboxylation may initiate anaerobic naphthalene degradation (ZHANG and YOUNG, 1997) but the further pathway of naphthalene breakdown is unknown so far.

2.7 Degradation of Halogenated Organics

Halogenated organics are widespread in nature and are formed especially as secondary metabolites by plants, marine algae, fungi, and certain bacteria at low rates (for a review, see HUTZINGER, 1982). It is not surprising, therefore, that a broad variety of bacteria and fungi can degrade such compounds, and that this capacity has grown in the past to comprise also the majority of synthetic halogenated compounds. Dehalogenation can proceed basically through an oxidative, a hydrolytic, or a reductive reaction (FETZNER and LINGENS, 1994; EL FANTROUSSI et al., 1998). Among anaerobic bacteria, the reductive elimination of halogen substituents is the most common type of reaction, and was first observed in enrichment cultures with 3,5-dihydroxybenzoate (SUFLITA et al., 1982). Later several other aliphatic and aromatic compounds were found to be dehalogenated (mainly: dechlorinated) in anoxic incu-

bation experiments (BOUWER and MCCARTY, 1983), and today nearly all chlorinated organics can be dehalogenated and further degraded in strictly anaerobic microbial cultures. As a rule, reductive dechlorination is the preferred process of degradation the higher the degree of halogenation is. The reaction is an excellent electron sink (E_h at $+250$ to $+580$ mV), and highly halogenated compounds are much more amenable to a nucleophilic attack on the respective carbon atom than to an oxidative reaction.

The overall reaction of reductive dehalogenation can be described as shown in Fig. 12 for a chlorinated compound: Electrons derived from molecular hydrogen, formate, or more complex organic compounds are transferred to the halogenated substrate to release the organic residue in a reduced form, together with chloride. It has been shown in several instances that this redox process can yield metabolic energy through a respiratory mechanism, which implies that the process establishes a net translocation of protons across the cytoplasmic membrane, and the proton gradient thus established drives ATP synthesis through a membrane-bound ATP synthase complex. The biochemistry of the dechlorination reaction has been studied best with bacteria converting tetrachloroethene via trichloroethene to dichloroethene (HOLLIGER et al., 1999). The dechlorination reaction employs a corrinoid as cofactor which attacks the carbon–chlorine linkage in its reduced Co(I) form, and is converted to the Co(III) form, probably through two subsequent one-electron transfer reactions. Whether this reaction mechanism (which resembles the reductive dehydroxylation of 3-hydroxybenzoyl CoA; see above) is used also by other dehalogenating bacteria has still to be elucidated. There appear to be numerous differences with respect to the electron carriers involved and the spatial arrangement of the enzyme components in the cytoplasmic membrane.

$$R\text{-Cl} \xrightarrow{\text{2 [H]}} R\text{-H} + Cl^- + H^+$$

Fig. 12. Reductive dehalogenation of a chloro-organic compound.

2.8 Degradation of Sulfonates

Sulfonated organics are rare in nature; only taurine, coenzyme M, cysteate, and a few secondary metabolites are known. Aerobic degradation of such compounds typically requires an oxygenase reaction which hydroxylates the neighboring carbon atom and releases sulfite (COOK et al., 1999). Some sulfonates can also be partly degraded in the absence of molecular oxygen, and can serve as sulfur sources under conditions of sulfur limitation (CHIEN et al., 1995; COOK et al., 1999). The biochemistry of sulfur release from the organic residue in these cases has not been studied. Dissimilative utilization of sulfonated organics as sources of carbon and energy or as electron acceptors has so far only been observed with taurine and related compounds which are metabolized through sulfoacetaldehyde as key intermediate (LAUE et al., 1997a, b). The sulfono group is released from sulfoacetaldehyde probably through a thiamine pyrophosphate-dependent reaction which forms sulfite and acetate, analogous to the reaction elucidated earlier with an aerobic bacterium (KONDO and ISHIMOTO, 1975). Since this desulfonation reaction requires an oxo group in β-position to the sulfur atom for linkage to the coenzyme, it remains questionable whether this concept can also be applied for desulfonation of commercial sulfonates such as alkyl sulfonates or alkylbenzene sulfonates.

2.9 Degradation of Nitroorganics

Among the nitroaromatic compounds, trinitrotoluene (TNT) is of major importance as a soil pollutant because it has accumulated at old ammunition factory sites over several decades. The electron-withdrawing effect of the nitro substituents renders an oxidative attack on nitroaromatics rather difficult. Therefore, aerobic bacteria also attack trinitrotoluene primarily through a reductive reaction which transforms nitroaromatics to the corresponding amino derivatives (NAUMOVA et al., 1989). The same approach is taken by strictly anaerobic bacteria, e.g., sulfate-reducing bacteria: Trinitrotoluene is converted via diaminonitrotoluene to triaminotoluene. Whereas the first step can also be catalyzed purely chemically without participation of microbial cells or enzymes, reduction of diaminonitrotoluene to triaminotoluene needs the participation of microbial cells or enzyme fractions (PREUSS et al., 1993). The further fate of triaminotoluene is unclear. It is partly utilized as a nitrogen source by a sulfate-reducing bacterium; the remnant product possibly polymerizes, especially in the presence of traces of oxygen (PREUSS et al., 1993). The same is likely to happen in contaminated soils; so far, there is no reliable proof of a complete degradation of TNT by anaerobic bacteria or by anaerobic and aerobic bacteria cooperating in a two-step process (see Chapter 11, this volume).

3 Concluding Remarks

Anaerobic degradation can be applied in technical devices for treatment of waste material, often leading to CH_4 and CO_2 as products, which can be exploited as energy source or as a basis for biosynthetic processes. Moreover, anaerobic degradation proceeds in many anoxic habitats such as intestinal tracts of humans and animals, sediments, and oxygen-deprived microenvironments in soil, sewage sludge, etc. Knowledge of the capacities, strategies, and limits of anaerobic degradation processes is, therefore, needed to assess the potential risk of synthetic compounds to health or to the environment, no matter whether such synthetics are released intentionally (as with plant protection agents), inadvertently through wastewater treatment, or unintendedly through accidental spills.

This overview shows that the degradative potential of anaerobic microbial communities is much greater than assumed only a few years ago: A broad variety of compounds can be subject to anaerobic degradation, most often down to methane and carbon dioxide as final products. Aliphatic hydrocarbons are degraded if they contain unsaturated bonds, preferentially if these are located terminally, but also saturated long-chain aliphatic hydrocarbons are anaerobically degradable. These processes are slow and can be applied only in long-term incubations, if at all. Ether compounds are degraded anaerobically either if they are methyl ethers, or if they can be transformed into hemiacetals through, e.g., hydroxyl shift reactions. In both cases, B_{12} coenzymes are directly involved in the ether cleavage reaction. In the anaerobic degradation of ketones, the primary activation reaction is a carboxylation rather than an oxidation step.

Mononuclear aromatic compounds can be degraded anaerobically rather efficiently if they carry at least one carboxy, hydroxy, methoxy, amino, or methyl substituent, and four major degradation pathways have been elucidated in the recent past which all differ basically from the well-known aerobic oxygenase-dependent pathways. The degradation kinetics differ considerably, depending on the sites of substitution.

Halogenated aliphatics and aromatics are reductively dehalogenated, most efficiently and better than by aerobes, the higher the degree of halogenation. Anaerobic degradation of sulfonates appears to be restricted to only few compounds whereas the majority of synthetic sulfonates (detergents) are degraded efficiently only in the presence of oxygen. Nitro-substituted compounds are preferentially attacked through reduction, and, therefore, anaerobic processes appear to be advantageous over aerobic ones. The same applies to azo compounds.

Several types of reactions were identified which activate or destabilize comparably inert substrates in the absence of oxygen. Among those are carboxylations, addition to fumarate, reductions and reductive eliminations, cobalamine-dependent rearrangements of aliphatics, cobalamine-dependent nucleophilic substitutions, and oxygen-independent hydroxylations. Some of the reactions described proceed through radicalic mechanisms, and the diversity of radical chemistry in the absence of oxygen appears to be considerably greater than in its presence.

Transformation of polymeric compounds is restricted in the anaerobic world to extracellular hydrolysis reactions unless the polymer can be taken up into the cell as in the case of polyethylene glycol. There is no equivalent in the anoxic world to the fungal lignin-degrading enzyme apparatus. Therefore, polynuclear aromatics (lignin, other polyphenols) remain comparably recalcitrant in anoxic environments and represent barriers to microbial attack in the absence of molecular oxygen.

Acknowledgements

Experimental work in the author's lab was funded mainly by the Deutsche Forschungsgemeinschaft, the Bundesministerium für Forschung und Technologie, the Fonds der Chemischen Industrie, and the University of Konstanz. BODO PHILIPP and JOCHEN MÜLLER provided unpublished results and designed the figures. Thanks are also due to the dedicated and talented graduate students who carried out the experimental work and contributed to the concepts with their own ideas. This overview is based on two review articles from our group (SCHINK, 1995; SCHINK et al., unpublished data) which cover special aspects of this survey.

4 References

AECKERSBERG, F., BAK, F., WIDDEL, F. (1991), Anaerobic oxidation of saturated hydrocarbons to CO_2 by a new type of sulfate-reducing bacterium, *Arch. Microbiol.* **156**, 5–14.

AECKERSBERG, F., RAINEY, F. A., WIDDEL, F. (1998), Growth, natural relationships, cell fatty acids and metabolic adaptation of sulfate-reducing bacteria utilizing long-chain alkanes under anoxic conditions, *Arch. Microbiol.* **170**, 361–369.

ALTENSCHMIDT, U., FUCHS, G. (1991), Anaerobic degradation of toluene in denitrifying *Pseudomonas* sp.: Indication of toluene methylhydroxylation and benzoyl-CoA as central aromatic intermediate, *Arch. Microbiol.* **156**, 152–158.

BACHE, R., PFENNIG, N. (1981), Selective isolation of *Acetobacterium woodii* on methoxylated aromatic acids and determination of growth yields, *Arch. Microbiol.* **130**, 255–261.

BALL, H. D., JOHNSON, H. A., REINHARD, M., SPORMANN, A. M. (1996), Initial reactions in anaerobic ethylbenzene oxidation by a denitrifying bacterium, strain EB1, *J. Bacteriol.* **178**, 5755–5761.

BELLER, H. R., SPORMANN, A. M. (1997a), Anaerobic activation of toluene and *o*-xylene by addition to fumarate in denitrifying strain T, *J. Bacteriol.* **179**, 670–676.

BELLER, H. R., SPORMANN, A. M. (1997b), Benzylsuccinate formation as a means of anaerobic toluene activation by sulfate-reducing strain PRTOL1, *Appl. Environ. Microbiol.* **63**, 3729–3731.

BELLER, H. R., SPORMANN, A. M. (1998), Analysis of the novel benzyl succinate synthase reaction for anaerobic toluene activation based on structural studies of the product, *J. Bacteriol.* **180**, 5454–5457.

BELLER, H. R., SPORMANN, A. M., SHARMA, P. K., COLE, J. R., REINHARD, M. (1996), Isolation and characterization of a novel toluene-degrading sulfate-reducing bacterium, *Appl. Environ. Microbiol.* **62**, 1188–1196.

BERMAN, M. H., FRAZER, A. Z. (1992), Importance of tetrahydrofolate and ATP in the anaerobic *O*-demethylation reaction for phenylmethylethers, *Appl. Environ. Microbiol.* **58**, 925–931.

BERNHARDT, F. H., STAUDINGER, H., ULLRICH, V. (1970), Eigenschaften einer *p*-Anisat-*O*-Demethylase im zellfreien Extrakt von *Pseudomonas* species, *Hoppe-Seyler's Z. Physiol. Chem.* **351**, 467–478.

BIEGERT, T., FUCHS, G., HEIDER, J. (1996), Evidence that oxidation of toluene in the denitrifying bacterium *Thauera aromatica* is initiated by formation of benzylsuccinate from toluene and fumarate, *Eur. J. Biochem.* **238**, 661–668.

BIRKS, S. J., KELLY, D. J. (1997), Assay and properties of acetone carboxylase, a novel enzyme involved in acetone-dependent growth and CO_2 fixation in *Rhodobacter capsulatus* and other photosynthetic and denitrifying bacteria, *Microbiology* **143**, 755–766.

BISAILLON, J. G., LÉPINE, F., BEAUDET, R., SYLVESTRE, M. (1991), Carboxylation of *o*-cresol by an anaerobic consortium under methanogenic conditions, *Appl. Environ. Microbiol.* **57**, 2131–2134.

BOLL, M., FUCHS, G. (1995), Benzoyl-CoA reductase (dearomatizing), a key enzyme of anaerobic aromatic metabolism. ATP dependence of the reaction, purification and some properties of the enzyme from *Thauera aromatica* strain K172, *Eur. J. Biochem.* **234**, 921–933.

BONNET-SMITS, E. M., ROBERTSON, L. A., VAN DIJKEN, J. P., SENIOR, E., KUENEN, J. G. (1988), Carbon dioxide fixation as the initial step in the metabolism of acetone by *Thiosphaera pantotropha*, *J. Gen. Microbiol.* **134**, 2281–2289.

BOUWER, E. J., MCCARTY, P. L. (1983), Transformation of halogenated organic compounds under denitrification conditions, *Appl. Environ. Microbiol.* **45**, 1295–1299.

BRUNE, A., SCHINK, B. (1990), Pyrogallol-to-phloroglucinol conversion and other hydroxyl-transfer reactions catalyzed by cell extracts of *Pelobacter acidigallici*, *J. Bacteriol.* **172**, 1070–1076.

BRUNE, A., SCHINK, B. (1992), Phloroglucinol pathway in the strictly anaerobic *Pelobacter acidigallici:* Fermentation of trihydroxybenzenes to acetate via triacetic acid, *Arch. Microbiol.* **157**, 417–424.

BRUNE, A., SCHNELL, S., SCHINK, B. (1992), Sequential transhydroxylations converting hydroxy hydroquinone to phloroglucinol in the strictly anaerobic fermenting bacterium, *Pelobacter massiliensis*, *Appl. Environ. Microbiol.* **58**, 1861–1868.

BURLAND, S. M., EDWARDS, E. A. (1999), Anaerobic benzene biodegradation linked to nitrate reduction, *Appl. Environ. Microbiol.* **65**, 529–533.

CHIEN, C.-C., LEADBETTER, E. R., GODCHAUX III, W. (1995), Sulfonate sulfur can be assimilated for fermentative growth, *FEMS Microbiol. Lett.* **129**, 189–194.

COOK, A. M., LAUE, H., JUNKER, F. (1999), Microbial desulfonation, *FEMS Microbiol. Rev.* **22**, 399–419.

DAGLEY, S. (1971), Catabolism of aromatic compounds by microorganisms, *Adv. Microb. Physiol.* **6**, 1–46.

DOLFING, J., ZEYER, P., BINDER-EICHER, P., SCHWARZENBACH, R. P. (1990), Isolation and characterization of a bacterium that mineralizes toluene in the absence of molecular oxygen, *Arch. Microbiol.* **154**, 336–341.

EDWARDS, E. A., GRBIĆ-GALIĆ, D. (1994), Anaerobic degradation of toluene and *o*-xylene by a methanogenic consortium, *Appl. Environ. Microbiol.* **60**, 313–322.

EGLI, T., BALLY, M., UETZ, T. (1990), Microbial degradation of chelating agents used in detergents with special reference to nitrilotriacetic acid (NTA), *Biodegradation* **1**, 121–132.

EL FANTROUSSI, S., NAVEAU, H., AGATHOS, S. N. (1998), Anaerobic dechlorinating bacteria, *Biotechnol. Prog.* **14**, 167–188.

EVANS, W. C. (1977), Biochemistry of the bacterial catabolism of aromatic compounds in anaerobic environments, *Nature* **270**, 17–22.

EVANS, W. C., FUCHS, G. (1988), Anaerobic degradation of aromatic compounds, *Annu. Rev. Microbiol.* **42**, 289–317.

EVANS, P. J., MANG, D. T., KIM, K. S., YOUNG, L. Y. (1991), Anaerobic degradation of toluene by a denitrifying bacterium, *Appl. Environ. Microbiol.* **57**, 1139–1145.

FETZNER, S., LINGENS, F. (1994), Bacterial dehalogenases: biochemistry, genetics, and biotechnological applications, *Microbiol. Rev.* **58**, 641–685.

FOß, S., HARDER, J. (1998), *Thauera linaloolentis* sp. nov. and *Thauera terpenica* sp. nov., isolated on oxygen-containing monoterpenes (linalool, menthol, and eucalyptol) and nitrate, *Syst. Appl. Microbiol.* **21**, 365–373.

FOß, S., HEYEN, U., HARDER, J. (1998), *Alcaligenes defragrans* sp. nov., description of four strains isolated on alkenoic monoterpenes ((+)-menthene, α-pinene, 2-carene and α-phellandrene) and nitrate, *Syst. Appl. Microbiol.* **21**, 237–244.

FRINGS, J., SCHINK, B. (1994), Fermentation of phenoxyethanol to phenol and acetate by a homoacetogenic bacterium, *Arch. Microbiol.* **162**, 199–204.

FRINGS, J., SCHRAMM, E., SCHINK, B. (1992), Enzymes involved in anaerobic polyethylene glycol degradation by *Pelobacter venetianus* and *Bacteroides* strain PG1, *Appl. Environ. Microbiol.* **58**, 2164–2167.

FUCHS, G., EL-SAID MOHAMED, M., ALTENSCHMIDT, U., KOCH, J., LACK, A. et al. (1994), Biochemistry of anaerobic biodegradation of aromatic compounds, in: *Biochemistry of Microbial Degradation* (RATLEDGE, C., Ed.), pp. 513–553. Dordrecht: Kluwer Academic Publ.

GALLERT, C., KNOLL, G., WINTER, J. (1991), Anaerobic carboxylation of phenol to benzoate: use of deuterated phenols revealed carboxylation exclusively in the C4 position, *Appl. Microbiol. Biotechnol.* **36**, 124–129.

GALUSHKO, A., MINZ, D., SCHINK, B., WIDDEL, F. (1999), Anaerobic degradation of naphthalene by a pure culture of a novel type of marine sulfate-reducing bacterium, *Environ. Microbiol.* **1**, 415–420.

GORNY, N., SCHINK, B. (1994a), Hydroquinone degradation via reductive dehydroxylation of gentisyl-CoA by a strictly anaerobic fermenting bacterium, *Arch. Microbiol.* **161**, 25–32.

GORNY, N., SCHINK, B. (1994b), Complete anaerobic oxidation of hydroquinone by *Desulfococcus* sp. strain Hy5: Indications of hydroquinone carboxylation to gentisate, *Arch. Microbiol.* **162**, 131–135.

GORNY, N., SCHINK, B. (1994c), Anaerobic degradation of catechol by *Desulfobacterium* sp. strain Cat2 proceeds via carboxylation to protocatechuate, *Appl. Environ. Microbiol.* **60**, 3396–3340.

GORNY, N., WAHL, G., BRUNE, A., SCHINK, B. (1992), A strictly anaerobic nitrate-reducing bacterium growing with resorcinol and other aromatic compounds, *Arch. Microbiol.* **158**, 48–53.

HADDOCK, J. D., FERRY, J. G. (1989), Purification and properties of phloroglucinol reductase from *Eubacterium oxidoreducens* G-41, *J. Biol. Chem.* **264**, 4423–4427.

HARDER, J. (1997), Anaerobic methane oxidation by bacteria employing ¹⁴C-methane uncontaminated with ¹⁴C-carbon monoxide, *Mar. Geol.* **137**, 13–23.

HARDER, M., PROBIAN, C. (1995), Microbial degradation of monoterpenes in the absence of molecular oxygen, *Appl. Environ. Microbiol.* **61**, 3804–3808.

HARMS, G., ZENGLER, K., RABUS, R., AECKERSBERG, F., MINZ, D. et al. (1999), Anaerobic oxidation of *o*-xylene, *m*-xylene, and homologous alkyl

benzenes by new types of sulfate-reducing bacteria, *Appl. Environ. Microbiol.* **65**, 999–1004.

HARWOOD, C. S., BURCHHARDT, G., HERRMANN, H., FUCHS, G. (1999), Anaerobic metabolism of aromatic compounds via the benzoyl-CoA pathway, *FEMS Microbiol. Rev.* **22**, 439–458.

HEIDER, J., FUCHS, G. (1997), Anaerobic metabolism of aromatic compounds, *Eur. J. Biochem.* **243**, 577–596.

HEIDER, J., SPORMANN, A. M., BELLER, H. R., WIDDEL, F. (1999), Anaerobic bacterial metabolism of hydrocarbons, *FEMS Microbiol. Rev.* **22**, 459–473.

HIPPE, H., CASPARI, D., FIEBIG, K., GOTTSCHALK, G. (1979), Utilization of trimethylamine and other *N*-methyl compounds for growth and methane formation by *Methanosarcina barkeri*, *Proc. Natl. Acad. Sci. USA* **76**, 494–498.

HOLLIGER, C., WOHLFARTH, G., DIEKERT, G. (1999), Reductive dechlorination in the energy metabolism of anaerobic bacteria, *FEMS Microbiol. Rev.* **22**, 383–398.

HOPPER, D. J. (1978), Incorporation of [¹⁸O] water in the formation of *p*-hydroxybenzyl alcohol by the *p*-cresol methylhydroxylase from *Pseudomonas putida*, *Biochem. J.* **175**, 345–347.

HOPPER, D. J., BOSSERT, I. D., RHODES-ROBERTS, M. E. (1991), *p*-Cresol methylhydroxylase from a denitrifying bacterium involved in anaerobic degradation of *p*-cresol, *J. Bacteriol.* **173**, 1298–1301.

HUTZINGER, O. (1982), *The Handbook of Enironmental Chemistry*. Berlin: Springer-Verlag.

HYLEMON, P. B., HARDER, J. (1999), Biotransformation of monoterpenes, bile acids, and other isoprenoids in anaerobic ecosystems, *FEMS Microbiol. Rev.* **22**, 475–488.

JANSSEN, P. H., SCHINK, B. (1995), ¹⁴CO₂ exchange with acetoacetate catalyzed by dialyzed cell-free extracts of the bacterial strain BunN grown with acetone and nitrate, *Eur. J. Biochem.* **228**, 677–682.

KAUFMANN, F., WOHLFAHRT, G., DIEKERT, G. (1997), Isolation of *O*-demethylase, an ether-cleaving enzyme system of the homoacetogenic strain MC, *Arch. Microbiol.* **168**, 136–142.

KELTJENS, J. T., VOGELS, G. D. (1993), Conversion of methanol and methylamines to methane and carbon dioxide, in: *Methanogenesis* (FERRY, J. G., Ed.), pp. 253–303. New York, London: Chapman & Hall.

KEREM, Z., BAO, W., HAMMEL, K. E. (1998), Rapid polyether cleavage via extracellular one-electron oxidation by a brown-rot basidiomycete, *Proc. Natl. Acad. Sci. USA* **95**, 10373–10377.

KIENE, R. P., OREMLAND, R. S., CATENA, A., MILLER, L. G., CAPONE, D. G. (1986), Metabolism of reduced methylated sulfur compounds in anaerobic sediments and by a pure culture of an estuarine methanogen, *Appl. Environ. Microbiol.* **52**, 1037–1045.

KLUGE, C., TSCHECH, A., FUCHS, G. (1990), Anaerobic metabolism of resorcylic acids (*m*-dihydroxybenzoic acids) and resorcinol (1,3-benzenediol) in a fermenting and in a denitrifying bacterium, *Arch. Microbiol.* **155**, 68–74.

KOCH, J., EISENREICH, W., BACHER, A., FUCHS, G. (1993), Products of enzymatic reduction of benzoyl-CoA, a key reaction in anaerobic aromatic metabolism, *Eur. J. Biochem.* **211**, 649–661.

KONDO, H., ISHIMOTO, M. (1975), Purification and properties of sulfoacetaldehyde sulfolyase, a thiamine pyrophosphate-dependent enzyme forming sulfite and acetate, *J. Biochem.* **78**, 317–325.

KREFT, J.-U., SCHINK, B. (1993), Demethylation and further degradation of phenyl methylethers by the sulfide-methylating homoacetogenic bacterium strain TMBS4, *Arch. Microbiol.* **159**, 308–315.

KREFT, J., SCHINK, B. (1994), *O*-Demethylation by the homoacetogenic anaerobe Holophaga foetida studied by a new photometric methylation assay using electrochemically produced cob(I)alamin, *Eur. J. Biochem.* **226**, 945–951.

KRUMHOLZ, L. R., BRYANT, M. P. (1988), Characterization of the pyrogallol–phloroglucinol isomerase of *Eubacterium oxidoreducens*, *J. Bacteriol.* **170**, 2472–2479.

LACK, A., FUCHS, G. (1992), Carboxylation of phenylphosphate by phenol caboxylase, an enzyme system of anaerobic phenol metabolism in a denitrifying *Pseudomonas* sp., *J. Bacteriol.* **174**, 3629–3636.

LACK, A., FUCHS, G. (1994), Evidence that phenol phosphorylation to phenylphosphate is the first step in anaerobic phenol metabolism in a denitrifying *Pseudomonas* sp., *Arch. Microbiol.* **161**, 306–311.

LAUE, H., DENGER, K., COOK, A. (1997a) Taurine reduction in anaerobic respiration of *Bilophila wadsworthia* RZA-TAU, *Appl. Environ. Microbiol.* **63**, 2016–2021.

LAUE, H., DENGER, K., COOK, A. (1997b) Fermentation of cysteate by a sulfate-reducing bacterium, *Arch. Microbiol.* **168**, 210–214.

LENGELER, J. W., DREWS, G., SCHLEGEL, H. G. (Eds.) (1999), *Biology of the Prokaryotes*, Stuttgart: Georg Thieme Verlag.

LOVLEY, D. R., COATES, J. D., WOODWARD, J. C., PHILLIPS, E. J. P. (1995), Benzene oxidation coupled to sulfate reduction, *Appl. Environ. Microbiol.* **61**, 953–958.

LOVLEY, D. R., WOODWARD, J. C., CHAPELLE, F. H. (1996), Rapid anaerobic benzene oxidation with a variety of chelated Fe(III) forms, *Appl. Environ. Microbiol.* **62**, 288–291.

LUKINS, H. B., FOSTER, J. W. (1963), Methylketone metabolism in hydrocarbon utilizing mycobacteria, *J. Bacteriol.* **85**, 1074–1087.

MIHELCIC, J. R., LUTHY, R. G. (1988), Degradation of

polycyclic aromatic hydrocarbon compounds under various redox conditions in soil-water systems, *Appl. Environ. Microbiol.* **54**, 1188–1198.

MÜLLER, J. A., GALUSHKO, A. S., KAPPLER, A., SCHINK, B. (1999), Anaerobic degradation of *m*-cresol by *Desulfobacterium cetonicum* is initiated by formation of 3-hydroxybenzylsuccinate, *Arch. Microbiol.* **172**, 287–294.

NAUMOVA, R. P., SELIVANOVSKAYA, S. Y., CHEREPNEVA, I. E. (1989), Conversion of 2,4,6-trinitrotoluene under conditions of oxygen and nitrate respiration of *Pseudomonas fluorescens*, *Appl. Biochem. Microbiol.* **24**, 409–413.

NÖRTEMANN, B. (1992), Total degradation of EDTA by mixed cultures and a bacterial isolate, *Appl. Environ. Microbiol.* **58**, 671–676.

OREMLAND, R. S., KIENE, R. P., MATHRANI, I., WHITICAR, M. J., BOONE, D. R. (1989), Description of an estuarine methylotrophic methanogen which grows on dimethylsulfide, *Appl. Environ. Microbiol.* **55**, 994–1002.

PHILIPP, B., SCHINK, B. (1998), Evidence of two oxidative reaction steps initiating anaerobic degradation of resorcinol (1,3-dihydroxybenzene) by the denitrifying bacterium *Azoarcus anaerobius*, *J. Bacteriol.* **180**, 3644–3649.

PLATEN, H., SCHINK, B. (1987), Methanogenic degradation of acetone by an enrichment culture, *Arch. Microbiol.* **149**, 136–141.

PLATEN, H., SCHINK, B. (1989), Anaerobic degradation of acetone and higher ketones via carboxylation by newly isolated denitrizing bacteria, *J. Gen. Microbiol.* **135**, 883–891.

PLATEN, H., SCHINK, B. (1991), Enzymes involved in anaerobic degradation of acetone by a denitrifying bacterium, *Biodegradation* **1**, 243–251.

PLATEN, H., TEMMES, A., SCHINK, B. (1990), Anaerobic degradation of acetone by *Desulfococcus biacutus* sp. nov., *Arch. Microbiol.* **154**, 355–361.

PREUSS, A., FIMPEL, J., DIEKERT, G. (1993), Anaerobic transformation of 2,4,6-trinitrotoluene (TNT), *Arch. Microbiol.* **159**, 345–355.

RABUS, R., HEIDER, J. (1998), Initial reactions of anaerobic metabolism of alkyl benzenes in denitrifying and sulfate-reducing bacteria, *Arch. Microbiol.* **170**, 337–384.

RABUS, R., WIDDEL, F. (1995), Anaerobic degradation of ethyl benzene and other aromatic hydrocarbons by new denitrifying bacteria, *Arch. Microbiol.* **163**, 96–103.

RABUS, R., NORDHAUS, R., LUDWIG, W., WIDDEL, F. (1993), Complete oxidation of toluene under strictly anaerobic conditions by a new sulfate-reducing bacterium, *Appl. Environ. Microbiol.* **59**, 1444–1451.

REICHENBECHER, W., BRUNE, A., SCHINK, B. (1994), Transhydroxylase of *Pelobacter acidigallici*: A molybdoenzyme catalyzing the conversion of pyrogallol to phloroglucinol, *Biochim. Biophys. Acta* **1204**, 217–224.

REICHENBECHER, W., RÜDIGER, A., KRONECK, P. M. H., SCHINK, B. (1996), One molecule of molybdopterin guanine dinucleotide is associated with each subunit of the heterodimeric Mo-Fe-S protein transhydroxylase of *Pelobacter acidigallici* as determined by SDS/PAGE and mass spectrometry, *Eur. J. Biochem.* **237**, 406–413.

ROSNER, B., SCHINK, B. (1995), Purification and characterization of acetylene hydratase of *Pelobacter acetylenicus*, a tungsten iron-sulfur protein, *J. Bacteriol.* **177**, 5767–5772.

RUDOLPHI, A., TSCHECH, A. FUCHS, G. (1991), Anaerobic degradation of cresols by denitrifying bacteria, *Arch. Microbiol.* **155**, 238–248.

RUETER, P., RABUS, R., WILKES, H., AECKERSBERG, F., RAINEY, F. A. et al. (1994), Anaerobic oxidation of hydrocarbons in crude oil by denitrifying bacteria, *Nature* **372**, 445–458.

SCHINK, B. (1985a), Degradation of unsaturated hydrocarbons by methanogenic enrichment cultures, *FEMS Microbiol. Ecol.* **31**, 69–77.

SCHINK, B. (1985b), Fermentation of acetylene by an obligate anaerobe, *Pelobacter acetylenicus* sp. nov., *Arch. Microbiol.* **142**, 295–301.

SCHINK, B. (1988), Principles and limits of anaerobic degradation – environmental and technological aspects, in: *Biology of Anaerobic Microorganisms* (ZEHNDER, A. J. B., Ed.), pp. 771–846. New York: John Wiley & Sons.

SCHINK, B. (1995), Chances and limits of anaerobic degradation of organic compounds, in: *Mikrobielle Eliminierung chlororganischer Verbindungen* (CUNO, M., Ed.), Schriftenreihe Biologische Abwasserreinigung Vol. 6, pp. 57–67. Berlin: Technische Universität.

SCHINK, B. (1997), Energetics of syntrophic cooperations in methanogenic degradation, *Microbiol. Mol. Biol. Rev.* **61**, 262–280.

SCHINK, B., PFENNIG, N. (1982), Fermentation of trihydroxybenzenes by *Pelobacter acidigallici* gen. nov. sp. nov., a new strictly anaerobic non-spore-forming bacterium, *Arch. Microbiol.* **133**, 195–201.

SCHINK, B., BRUNE, A., SCHNELL, S. (1992), Anaerobic degradation of aromatic compounds, in: *Microbial Degradation of Natural Compounds* (WINKELMANN, G., Ed.), pp. 219–242. Weinheim: VCH.

SCHLEGEL, H. G. (1992), *Allgemeine Mikrobiologie*. Stuttgart: Georg Thieme Verlag.

SCHNELL, S., SCHINK, B. (1991), Anaerobic aniline degradation via reductive deamination of 4-aminobenzoyl CoA in *Desulfobacterium anilini*, *Arch. Microbiol.* **155**, 183–190.

SCHNELL, S., BRUNE, A., SCHINK, B. (1991), Degradation of hydroxy hydroquinone by the strictly an-

aerobic fermenting bacterium *Pelobacter massiliensis* sp. nov., *Arch. Microbiol.* **155**, 511–516.

SCHOCHER, R. J., SEYFRIED, B., VAZQUEZ, F., ZEYER, J. (1991), Anaerobic degradation of toluene by pure cultures of denitrifying bacteria, *Arch. Microbiol.* **157**, 7–12.

SCHÖCKE, L., SCHINK, B. (1999), Biochemistry and energetics of fermentative benzoate degradation by *Syntrophus gentianae*, *Arch. Microbiol.* **171**, 331–337.

SCHRAMM, E., SCHINK, B. (1991), Ether-cleaving enzyme and diol dehydratase involved in anaerobic polyethylene glycol degradation by an *Acetobacterium* sp., *Biodegradation* **2**, 71–79.

SEYFRIED, B., GLOD, G., SCHOCHER, R., TSCHECH, A., ZEYER, J. (1994), Anaerobic degradation of toluene by pure cultures of denitrifying bacteria, *Appl. Environ. Microbiol.* **60**, 4047–4052.

SLUIS, M. K., ENSIGN, S. A. (1997), Purification and characterization of acetone carboxylase in a CO_2-dependent pathway of acetone metabolism by *Xanthobacter* strain Py2, *Proc. Natl. Acad. Sci. USA* **94**, 8456–8461.

SLUIS, M. K., SMALL, F. J., ALLEN, J. R., ENSIGN, S. A. (1996), Involvement of an ATP-dependent carboxylase in a CO_2-dependent pathway of acetone metabolism by *Xanthobacter* strain Py2, *J. Bacteriol.* **178**, 4020–4026.

STUMM, W., MORGAN, J. J. (1981), *Aquatic Chemistry*, 2nd Edn. New York: John Wiley & Sons.

STUPPERICH, E., KONLE, R. (1993), Corrinoid-dependent methyl transfer reactions are involved in methanol and 3,4-dimethoxybenzoate metabolism by *Sporomusa ovata*, *Appl. Environ. Microbiol.* **59**, 3110–3116.

SUFLITA, J. M., HOROWITZ, A., SHELTON, D. R., TIEDJE, J. M. (1982), Dehalogenation: A novel pathway for the anaerobic biodegradation of haloaromatic compounds, *Science* **218**, 1115–1117.

SUFLITA, J. M., LIANG, L.-N., SAXENA, A. (1989), The anaerobic degradation of *o*-, *m*-, and *p*-cresol by sulfate-reducing bacterial enrichment cultures obtained from a shallow anoxic aquifer, *J. Ind. Microbiol.* **4**, 255–266.

TARVIN, D., BUSWELL, A. M. (1934), The methane fermentation of organic acids and carbohydrates, *J. Am. Chem. Soc.* **56**, 1751–1755.

TAYLOR, D. G., TRUDGILL, P. W., GRIPPS, R. E., HARRIS, P. R. (1980), The microbial metabolism of acetone, *J. Gen. Microbiol.* **118**, 159–170.

THAUER, R. K., JUNGERMANN, K., DECKE, K. (1977), Energy conservation in chemotrophic anaerobic bacteria, *Bacteriol. Rev.* **41**, 100–180.

TSCHECH, A., FUCHS, G. (1989), Anaerobic degradation of phenol via carboxylation to 4-hydroxybenzoate: *in vitro* study of isotope exchange between $^{14}CO_2$ and 4-hydroxybenzoate, *Arch. Microbiol.* **152**, 594–599.

TSCHECH, A., SCHINK, B. (1985), Fermentative degradation of resorcinol and resorcylic acids, *Arch. Microbiol.* **143**, 52–59.

VOGEL, T. M., GRBIĆ-GALIĆ, D. (1986), Incorporation of oxygen from water into toluene and benzene during anaerobic fermentative transformation, *Appl. Environ. Microbiol.* **51**, 200–202.

WAGENER, S., SCHINK, B. (1987), Anaerobic degradation of non-ionic and anionic surfactants in enrichment cultures and fixed-bed reactors, *Water Res.* **21**, 615–622.

WAGENER, S., SCHINK, B. (1988), Fermentative degradation of non-ionic surfactants and polyethylene glycol by enrichment cultures and by pure cultures of homoacetogenic and propionate-forming bacteria, *Appl. Environ. Microbiol.* **54**, 561–565.

ZEHNDER, A. J. B., BROCK, T. D. (1980), Anaerobic methane oxidation: occurrence and ecology, *Appl. Environ. Microbiol.* **39**, 194–204.

ZEHNDER, A. J. B., STUMM, W. (1988), Geochemistry and biogeochemistry of anaerobic habitats, in: *Biology of Anaerobic Microorganisms* (ZEHNDER, A. J. B., Ed.), pp. 1–38. New York: John Wiley & Sons.

ZHANG, X., YOUNG, L. Y. (1997), Carboxylation as an initial reaction in the anaerobic metabolism of naphthalene and phenanthrene by sulfidogenic consortia, *Appl. Environ. Microbiol.* **63**, 4759–4764.

8 Bacterial Degradation of Aliphatic Hydrocarbons

Johann E. T. van Hylckama Vlieg
Dick B. Janssen

Groningen, The Netherlands

List of Abbreviations

AlkB	alkane hydroxylase
AlkG	rubredoxin
AlkH	aldehyde dehydrogenase
AlkJ	alcohol dehydrogenase
AlkK	acyl-CoA synthase
AlkT	rubredoxin reductase
MTBE	methyl *tert*-butyl ether
NAPL	non-aqueous phase liquids
pMMO	particulate methane monooxygenase
sMMO	soluble methane monooxygenase

1 Bacterial Degradation of Aliphatic Compounds

1.1 Sources and Environmental Relevance of Aliphatic Hydrocarbons and Oxo-Compounds

A wide variety of aliphatic compounds are used in industry and manufacture as solvents, intermediates, fuels, cleaning agents, and for other applications. This includes numerous alkanes, alkenes, alcohols, ketones, ethers, epoxides, carboxylic acids, and esters. The production and use of these chemicals may be accompanied by exposure or emission, which may have various adverse effects. Volatile aliphatic hydrocarbons may escape into the air and then play an important role in atmospheric chemistry, whereas many alkanes and oxo-derivatives are found as soil and groundwater pollutants. Alkanes, ketones, and ethers can have diverse toxic effects, causing neurological disorders and liver damage. Aldehydes and epoxides are electrophilic compounds that may react with amine functions or other nucleophilic groups in biomolecules, such as DNA, which also leads to toxic effects in higher organisms and carcinogenicity in some cases. Hence, there has been a continuing interest in the biological fate of industrially produced aliphatic compounds, and in the potential of microbial techniques for their removal from the environment and waste streams.

Most aliphatic compounds which are produced from petrochemicals and are used in industry cannot be regarded as xenobiotics, since they also occur in a wide variety of naturally occurring materials, such as fermentation products, plant secondary metabolites, metabolic intermediates, and oil. Some non-oxygenated aliphatic hydrocarbons are also produced in large quantities from biotic sources. For example, methane is generated in many anaerobic environments from where it escapes into the atmosphere. It is thought to play an important role in global warming. Data on the atmospheric relevance of methane have been reviewed extensively (CICERONE and OREMLAND, 1988; LELIEVELD et al., 1993).

Isoprene is emitted from plants as a reaction to conditions of thermal stress, and its production is closely linked to photosynthesis. Plants may emit 2% of the carbon that is assimilated as isoprene, but values as high as 15% have been reported for oak trees (MONSON and FALL, 1989; SHARKEY and SINSGAAS, 1995; SHARKEY, 1996). It is well known that thermal damage rapidly reduces photosynthesis rates. In photosynthetic membranes isoprene may function as a stabilizer of the lipid bilayer, or protect protein–membrane and protein–protein interactions. Isoprene emission is not limited to plants, but it has also been described for bacteria (KUZMA et al., 1995) and animals (SHARKEY, 1996). The role of isoprene in atmospheric chemistry has drawn considerable attention. Isoprene is quite reactive due to the presence of the two unsaturated bonds and has a half-life of a few hours in the atmosphere. In a series of photochemical reactions, it is involved in the production of tropospheric ozone (TRAINER et al., 1987; THOMPSON, 1992). Other important naturally produced aliphatic hydrocarbons are released from (decaying) plants. These include the plant hormone ethylene, medium and long-chain alkanes, and isoprenoids.

Saturated alkanes with chain lengths up to 30 carbon atoms are the major constituents of mineral oil. Oil spillage has frequently caused contamination of soil, groundwater, beaches and shores. Long-chain alkanes are poorly soluble in water which limits their bioavailability, and at sites that are contaminated with high concentrations these compounds are often present as non-aqueous phase liquids (NAPL) (ATLAS, 1981, 1984). In some cases bacteria produce biosurfactants that can increase the aqueous dispersion of poorly soluble long-chain alkanes to facilitate their utilization as growth substrate (ZHANG and MILLER, 1994; ROSENBERG, 1986). In field situations, biodegradation is often limited by nutrient availability and by oxygen supply, and, therefore, treatment strategies have usually focused on the removal of these barriers by ensuring improved oxygenation and adding fertilizers, rather than on stimulating specific organisms.

Finally, oxygenated aliphatics enter the environment from both natural and anthropogenic sources. Most alcohols, carboxylic acids, and esters are of little environmental concern since they are rapidly biodegraded, and toxicity is very limited. A special case is represented by oxygenates that are added to gasoline to reduce emission of carbon monoxide and volatile organic compounds in exhaust gases. These compounds recently came into use because of the substitution of lead-based antiknock agents in automotive fuel. Methyl *tert*-butyl ether (MTBE) is the most common gasoline additive, and as a result of spillage and leaking tanks many sites have been contaminated with MTBE (SQUILLACE et al., 1996). Ether linkages are also present in polyglycols, which are components of various detergents (CAIN, 1981), and in ethers used as solvents such as tetrahydrofuran.

Aldehydes and epoxides are not expected to accumulate in the environment because of their chemical reactivity. However, they are frequently observed in industrial waste streams since their properties make them attractive building blocks for application in synthetic chemical processes.

As many aliphatic hydrocarbons and oxo-compounds that occur as pollutants are also produced in nature, it is not surprising that these compounds can be assimilated by a wide variety of organisms (ATLAS, 1981; BRITTON, 1984; WATKINSON and MORGAN, 1990). Studies on the occurrence of hydrocarbon utilizing organisms have indicated that the order of degradability is aromatics > linear alkanes > branched alkanes, but even a compound such as 2,2,4-trimethylpentane was found to serve as a growth substrate for several organisms (RIDGWAY et al., 1990). Over the last decades considerable efforts have been dedicated to elucidate the physiology, biochemistry, and genetics of the microorganisms involved.

In this chapter some of the reaction types and the corresponding enzymes that are involved in the degradation of aliphatics will be discussed. Most of the work on microbiology and physiology of organisms involved in the degradation of aliphatic compounds has been reviewed (see various contributions in GIBSON, 1984; RATLEDGE, 1990; JANSSEN, 1994; VAN AGTEREN et al., 1998, FUKUI and TANAKA, 1981; REHM and REIFF, 1981; EISELE, 1983), and will only briefly be summarized here. The discussion will focus on key reactions and pathways for the aerobic microbial metabolism of environmentally relevant aliphatic compounds which have recently drawn more attention.

2 Reaction Types

2.1 Oxygenase Reactions for Oxidation of Alkanes

Microorganisms that degrade simple *n*-alkanes, alkenes, and cycloalkanes under aerobic conditions can easily be isolated from soil and water and have been known for many years. The degradation of these compounds is usually initiated by the action of monooxygenases, which convert an alkane to a corresponding alcohol or an alkene to an epoxide. The first step can be regarded as a critical bioactivation reaction which converts the chemically inert alkane to a hydroxylated product which is more susceptible to further biochemical conversion. The reaction barrier is energetically taken by coupling it to the oxidation of a strong reductant, usually NADH, by molecular oxygen. This critical step explains that alkanes are generally recalcitrant to anaerobic metabolism. The oxygen atom which is incorporated by the monooxygenase is derived from molecular oxygen, of which the other oxygen atom is reduced to H_2O, with the use of NADH as electron donor. The monooxygenases from methane utilizing bacteria (Fig. 1) and from the *n*-alkane utilizer *Pseudomonas oleovorans* (Fig. 2) have been characterized in most detail (LANGE and QUE, 1998). These unrelated proteins are composed of multiple polypeptides. One contains an active site with two Fe atoms and is involved in the hydroxylation of the substrate, and at least two play a role in transfer-

ring electrons from NADH to the monooxygenase.

The only alkane monooxygenases for which X-ray structures are known are the soluble methane monooxygenases (sMMO) from *M. trichosporium* and *M. capsulatus* Bath. The sMMOs of these organisms are very similar and the results of many biochemical and biophysical studies with these enzymes have been reviewed (LIPSCOMB, 1994). Of both organisms the genes encoding the sMMO have been cloned and sequenced (MURRELL, 1994). The hydroxylase component of these enzymes is composed of 6 subunits ($\alpha_2\beta_2\gamma_2$; with polypeptides of 60, 45 and 20 kDa). The second component is a 38 kDa NADH-dependent reductase which contains an $[Fe_2S_2]$-cluster and FAD, and is involved in electron transfer from NADH to the hydroxylase. The third component is a 15 kDa protein with a regulatory function. The active site is located in the α-subunit of the hydroxylase and contains a non-heme di-Fe-cluster of which the Fe atoms [Fe(II)] are bound to glutamic acid, aspartic acid, and histidine side chains. The enzyme reaction starts with reduction of the enzyme by NADH. Subsequently, oxygen is bound and cleaved, and the resulting highly reactive Fe-bound oxo-species abstracts a hydrogen atom from the substrate. Product is formed after recombination of the methyl radical with the active site oxygen (Fig. 1). This mechanism is similar to that of ribonucleotide reductase, as already inferred from sequence similarity around the metal-binding residues (MURRELL, 1994).

From sequence analysis, a similar mechanism and structure were recently predicted for the hydroxylase components of the alkene monooxygenase from *Rhodococcus rodochrous* B-276 (GALLAGHER et al., 1998), and alkene monooxygenase from *Xanthobacter* Py2 (ENSIGN et al., 1998). The residues that coordinate the active site Fe atoms are present in two conserved sequence motifs along the polypeptide chain: GXXH.

A different well studied hydroxylase is the membrane localized alkane hydroxylase from *Pseudomonas oleovorans*, which also possesses a non-heme Fe site. The same was found for a hexadecane utilizing *Acinetobacter* strain, and it was proposed that these enzymes consti-

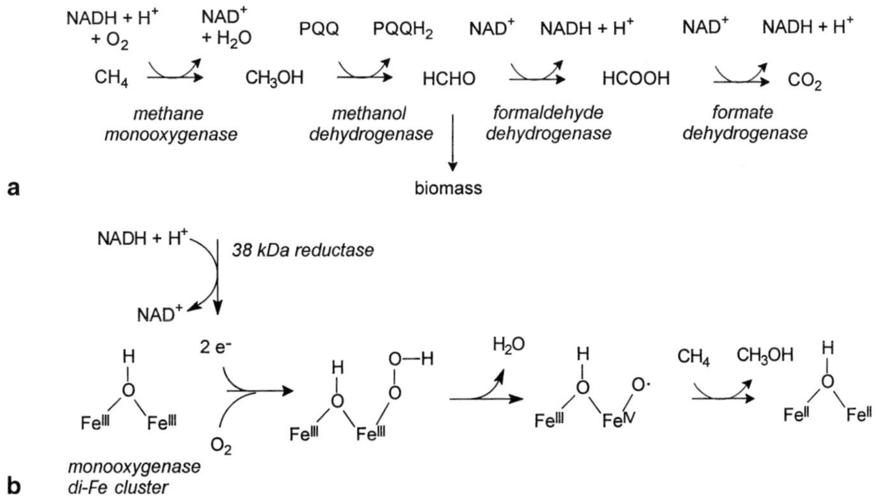

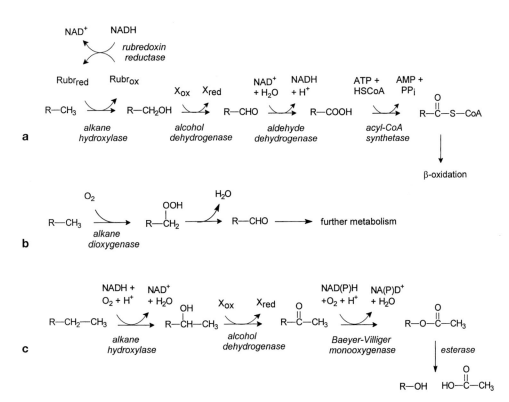

Fig. 1a, b. Pathway of methane oxidation (**a**) in methanotrophic bacteria (ANTHONY, 1986; LIPSCOMB, 1994) and catalytic mechanism of soluble methane monooxygenase (**b**).

Fig. 2. a–c. Pathways for the oxidation of *n*-alkanes. (**a**) Hydroxylation and alkanol conversion in *Pseudomonas oleovorans*; (**b**) proposed dioxygenation in hexadecane oxidizing *Acinetobacter*; (**c**) subterminal oxidation and chain cleavage by Baeyer–Villiger oxidation.

tute a separate class of hydroxylases to which also xylene monooxygenase (side chain oxidizing) of *Pseudomonas putida* belongs (RATAJC-ZAK et al., 1998; LANGE and QUE, 1998). No structures are known, but up to 8 conserved histidine residues in conserved sequence boxes (HXXXH, HXXXHH), which are also found in fatty acid desaturase, may be used for identifying members of this class. Since four histidines may form a structure somewhat similar to a tetrapyrrole ring, it is well possible that the enzyme also contains two Fe atoms in the active site (VAN BEILEN et al., 1994).

Another type of monooxygenase is cytochrome P450, an enzyme isolated from a camphor utilizing *Pseudomonas putida* strain (TRUDGILL, 1984). The X-ray structure of the enzyme (P450$_{cam}$) is known, and it is now used for modeling the structures of cytochrome P450 isoenzymes from eukaryotic organisms, which play an important role in the metabolism of xenobiotic and endogenous compounds. The course of the reaction catalyzed by this enzyme is very similar to that of methane monooxygenase. It also involves generation of a strongly oxidizing Fe–O species, which abstracts a hydrogen atom from the substrate. Recombination of the radical with oxygen yields the hydroxylated product.

There is some evidence for a possible role of dioxygenases in alkane biodegradation. MAENG et al. (1996) described an *Acinetobacter* strain capable of growth on *n*-alkanes from C_{13}–C_{44}, and found that the organism produces an *n*-alkane oxidizing enzyme which does not require reducing cofactors for activity and contains FAD and copper as prosthetic groups. However, only a minor fraction of the hexadecane consumed was recovered as hydroperoxide due to instability of the product. Likely, the hydroperoxide would rapidly decompose to the corresponding aldehyde, which then can be metabolized by oxygen-independent pathways (Fig. 2).

Most alkane monooxygenases initiate degradation of alkanes by terminal attack, but there are also examples of subterminal attack (BRITTON, 1984). For example, alkanes may be oxidized to secondary alcohols and subsequently ketones. This may be followed by production of an ester, according to a Baeyer–Villiger type of reaction, after which an esterase

cleavage gives an alcohol and a carboxylic acid. This last part of the pathway has been demonstrated for tridecanone degradation by a *Pseudomonas* and a *Nocardia* strain (BRITTON, 1984) (Fig. 2).

2.2 Ring Cleavage of Cycloalkanes

The most commonly used cycloalkanes are cyclohexane and cyclopentane. Several microorganisms that degrade these compounds are known (TRUDGILL, 1984). The bacterial degradation of cycloalkanes and cyclic terpenoids starts with conversion to the hydroxy derivatives, which is carried out by a cytochrome P450 type of enzyme or by a non-heme monooxygenase (see Sect. 2.1) (TOWER et al., 1985). The cyclic alkanols are subsequently oxidized to cycloketones, and ring cleavage then occurs by Baeyer–Villiger monooxygenase type reactions, such as the cyclohexanone monooxygenase of *Acinetobacter calcoaceticus*, *Rhodococcus* or *Xanthobacter* strains (Fig. 3).

The enzymes that carry out this latter step are flavin-dependent and use either NADH or NADPH as electron donor. Such Baeyer–Villiger enzymes occur in the metabolism of alkanes, steroids, cycloalkanes, and terpenoids. The Baeyer–Villiger monooxygenases have raised significant interest because of their potential use in the production of chiral compounds, such as the lactones that are generated from ketones or the sulfoxides that are generated from sulfides (WILLETTS, 1997). Several enzymes have been purified and characterized. Only the sequence from the *Acinetobacter* enzyme has been determined, and it revealed the presence of a FAD-binding site close to the *N*-terminus in this 59 kDa monomeric protein. On the basis of kinetic data and comparison of the *N*-terminal amino acid sequences Baeyer–Villiger monooxygenases have been divided into two distinct classes. Type 1 enzymes are FAD-binding and NADPH-dependent while the type 2 enzymes are FMN-binding and NADH-dependent. Active-site models and an enzyme mechanism have been proposed on the basis of biotransformation data (WILLETTS, 1997).

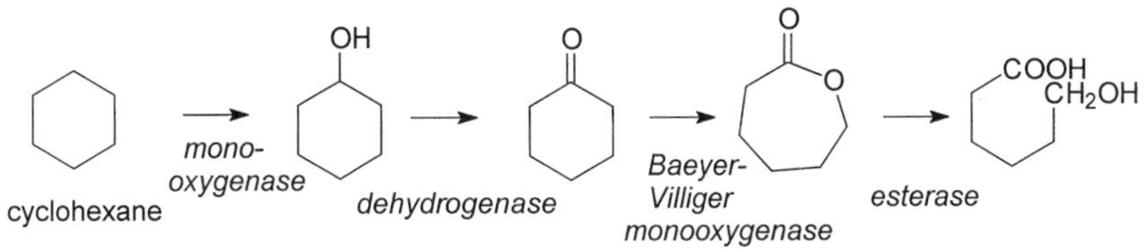

Fig. 3. Involvement of a monooxygenase in the hydroxylation of cyclohexane. Oxidation of cyclohexanol by a Baeyer–Villiger monooxygenase results in the formation of an ester, which is subsequently cleaved by an esterase (TOWER et al., 1985).

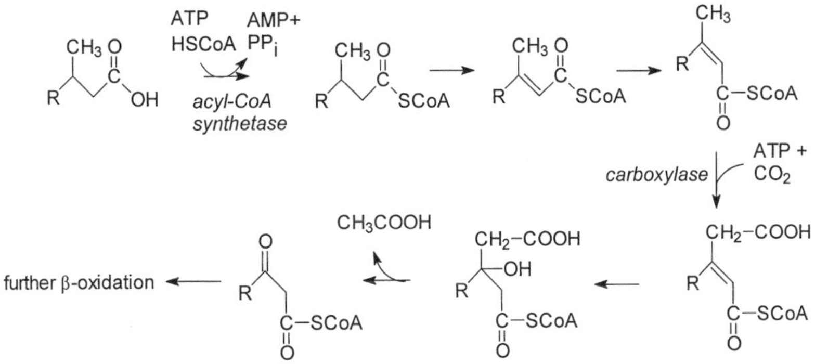

Fig. 4. Mechanism for the debranching of a methyl substituted alcohol. Carboxylation is the key step, and preceeds removal of the methyl group as acetate (BRITTON, 1984).

2.3 Debranching of Branched Alkanes

In general, a high degree of branching of alkyl groups inhibits degradation of aliphatics (BRITTON, 1984). This holds for alkanes, ethers, and alkylbenzene sulfonates surfactants. Both the restricted substrate range of initial monooxygenases and the limited activity of debranching mechanisms may be responsible for the reduced degradation of branched alkanes.

In the case of methyl groups as branching substituents, removal mainly occurs after carboxylation. For this, a methyl group containing aliphatic compound is first converted to a fatty acid–CoA derivative. Then, the methyl group is carboxylated in an ATP-dependent reaction, after which β-decarboxymethylation gives a linear acyl-CoA and releases acetic acid

(Fig. 4). This mechanism has been found in citronellol degrading *Pseudomonas citronellolis*, an organism which produces multiple carboxylases for debranching reactions. The combined capacity of terminal oxidation and carboxylation/decarboxymethylation allowed the utilization of methyl branched alkanes by strains which were obtained by conjugation of the OCT plasmid, harboring alkane oxidation genes, to *P. citronellolis* (FALL et al., 1979).

Long-chain branched alkanes, such as pristane, are very recalcitrant to biodegradation, although a pathway has been proposed based on the detection of some intermediates (BRITTON, 1984). However, degradation in the environment is usually so slow that pristane can be used as a conservative non-degradable marker during bioremediation of petroleum contaminated sites.

2.4 Conversion of Alcohols, Aldehydes, Ketones, and Carboxylic Acids

Oxo-derivatives of aliphatic hydrocarbons are usually more susceptible to biodegradation than the hydrocarbons themselves, which is in agreement with a general correlation between polarity and biodegradability (GIBSON, 1984). Furthermore, whereas degradation of alkanes occurs only rapidly in the presence of molecular oxygen, most alcohols, aldehydes, ketones, and similar compounds may also be utilized by bacteria which use other electron acceptors or by fermentative organisms. For example, alcohols are generally good substrates for nitrate reducing organisms.

Alcohol and aldehyde dehydrogenases, which may be coupled either to protein electron acceptors via quinoproteins, or directly to NAD, convert alcohols to carboxylic acids. These are further oxidized in C-1, C-2 or C-3 assimilation pathways, or first shortened by β-oxidation. A large diversity of alcohol and aldehyde oxidizing enzymes exist, but they will not be discussed here since they appear not to be involved in the critical steps in the biodegradation pathways. The same holds for esters, which are usually rapidly cleaved by microorganisms, in agreement with the observation that many organisms contain esterase activity.

2.5 Cleavage of Ethers

Ether linkages occur both in xenobiotic and natural hydrocarbons. They occur in lignin, which is one of the most abundant organic polymers, and the presence of these ether bonds is one of the main reasons for its recalcitrance. Ether bonds also occur in synthetic compounds such as polyethoxylates which are used in detergents and in oxygenates which are used as solvents or gasoline additives. Little detailed work has been done on the biochemistry of bacterial ether cleaving enzymes, but at least 8 different mechanisms have been proposed based on measurements with whole cells, crude extracts, and purified enzymes. Data on this subject have been reviewed extensively (WHITE et al., 1996; CAIN, 1981; KAWAI, 1995).

In most of the ether cleaving reactions the key step is the formation of a hemiacetal after which the oxygen–carbon bond is cleaved (Fig. 5). Under aerobic conditions ether bond cleavage may be catalyzed by

(1) non-heme oxygenases or
(2) cytochrome P450 as for instance in the metabolism of vanillate or 3-ethoxybenzoate,
(3) dehydrogenases which are involved in the metabolism of polyethylene glycol,
(4) hydrolases which cleave the ether bond in isochorismic acid, or
(5) lyases which are involved in carboxy methyloxy succinate metabolism.
(6) A reductive glutathione dependent enzyme has been reported that cleaves ether bonds in lignin-like substrates.
(7) A vitamin B_{12}-dependent enzyme catalyzes the hydroxy shift to the carbon atom adjacent to the ether bond which is subsequently non-enzymatically cleaved.
(8) Finally, an enzymatic activity cleaving the ether bond in the anaerobic metabolism of methoxyaromatics has been reported. It was postulated that this reaction involves a two-step transfer of the methyl group from the substrate to tetrahydrofolate via a corrinoid.

Interestingly, oxygenation of a carbon atom which is flanking an ether oxygen is well known to lead to dealkylation, and the O-dealkylation by the microsomal cytochrome P450 monooxygenase reaction is a well established conversion. This conversion also explains the demethylation of MTBE and the ring cleavage of cyclic ethers such as dioxane and furan that will be discussed in detail in Sect. 3.5. Other examples of ethers for which a (partial) degradation pathway has been established are 1,8-cineole (WILLIAMS et al., 1989), dialkylethers, and diphenylethers.

2.6 Conversion of Epoxides

Besides occurring as contaminants of wastes, epoxides are important intermediates in the bacterial oxidation of alkenes, which is

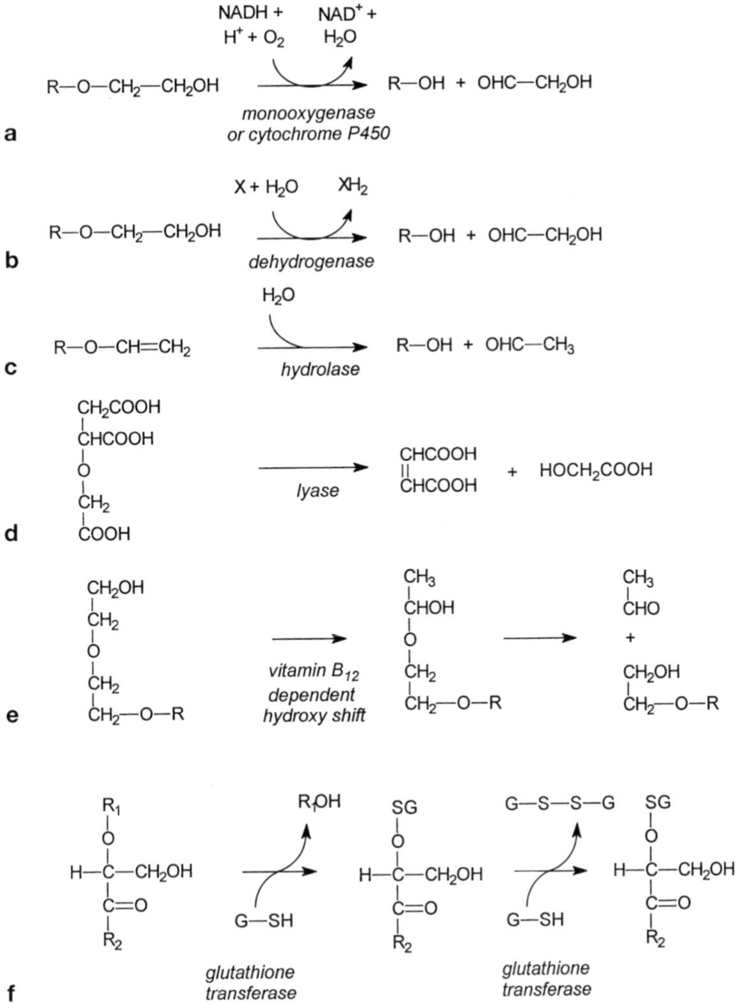

NADH +   NAD⁺ +
H⁺ + O₂   H₂O

R—O—CH₂—CH₂OH ⟶ R—OH + OHC—CH₂OH

*monooxygenase
or cytochrome P450*

a

X + H₂O   XH₂

R—O—CH₂—CH₂OH ⟶ R—OH + OHC—CH₂OH

b   *dehydrogenase*

H₂O

R—O—CH=CH₂ ⟶ R—OH + OHC—CH₃

c   *hydrolase*

CH₂COOH
|
CHCOOH
|
O   CHCOOH
| ⟶   ‖   + HOCH₂COOH
CH₂   CHCOOH
| *lyase*
d COOH

CH₂OH   CH₃   CH₃
| | |
CH₂   CHOH   CHO
| | +
O   ⟶   O   ⟶
| | CH₂OH
CH₂   *vitamin B₁₂* CH₂ |
| *dependent* | CH₂—O—R
e CH₂—O—R *hydroxy shift* CH₂—O—R

R₁   R₁OH SG   G—S—S—G SG
| | |
O   O   O
| | |
H—C—CH₂OH ⟶   H—C—CH₂OH ⟶   H—C—CH₂OH
| | |
C=O   G—SH   C=O   G—SH   C=O
| | |
R₂   R₂   R₂

*glutathione   glutathione
transferase   transferase*
f

Fig. 5a–f. Reactions leading to cleavage of ether bonds. (**a**) oxygenation; (**b**) dehydrogenation; (**c**) hydrolysis; (**d**) lyase reaction; (**e**) hydroxy shift; (**f**) glutathione-dependent reduction; for details, see WHITE et al. (1996), CAIN (1981), TIDSWELL et al. (1996), and KAWAI (1995).

initiated by monooxygenases with similar properties as the alkane monooxygenases (HARTMANS et al., 1989). A wide range of enzymes are involved in the further metabolism of epoxides. These enzymes include epoxide isomerases, carboxylases, dehydrogenases, hydrolases, reductases, and lyases (Fig. 6) (SWAVING and DE BONT, 1998; WEIJERS et al., 1988). Because of the reactivity of epoxides, epoxide converting enzymes are important for the biodegradation of various pollutants. Furthermore, they may also be valuable tools in the production of chiral synthons that can be used in organic synthesis (KASAI et al., 1998).

Epoxide hydrolases catalyze the hydration of the epoxide ring to the corresponding diol. Examples of bacterial epoxide hydrolases are enzymes involved in the degradation of epi-

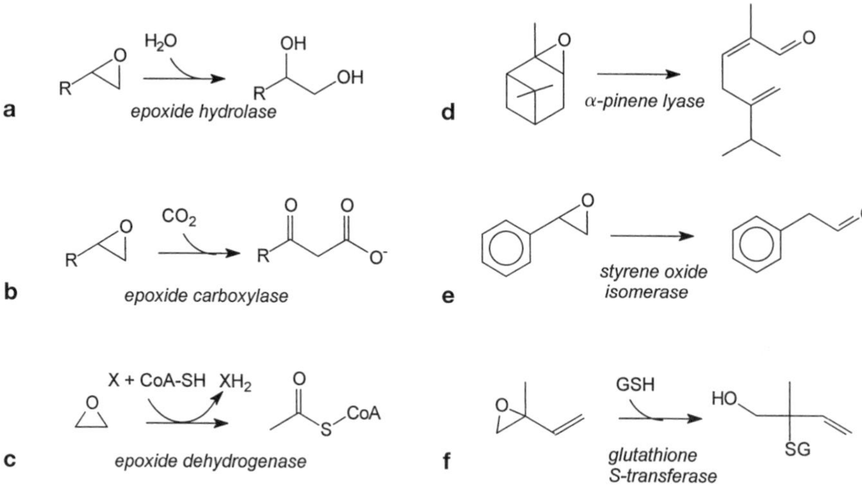

Fig. 6a–f. Enzymes involved in the metabolism of epoxides. (**a**) hydrolase; (**b**) carboxylase; (**c**) dehydrogenase; (**d**) lyase; (**e**) isomerase; (**f**) glutathione S-transferase.

chlorohydrin (3-chloroepoxypropane) and cyclohexene (MISAWA et al., 1998; RINK et al., 1997). The gene encoding the epoxide hydrolase of the epichlorohydrin utilizing bacterium *Agrobacterium radiobacter* AD1 is also active with many non-chlorinated epoxides (RINK et al., 1997). Evidence was presented that the enzyme belongs to the α/β hydrolase fold family, and the mechanism and catalytic triad were shown to be similar to that of haloalkane dehalogenase (RINK et al., 1997). A similar enzyme appeared to be involved in the cleavage of cyclohexene epoxide by a gram-positive organism (MISAWA et al., 1998).

The metabolism of epoxypropane in *Xanthobacter* sp. Py2 has been the subject of several detailed studies (ENSIGN et al., 1998; SWAVING et al., 1996). Epoxides are carboxylated to β-ketoacids with a concomitant transhydrogenation reaction during which NADPH is oxidized and NAD$^+$ is reduced (Fig. 6). Epoxide carboxylase, the enzyme catalyzing this reaction, is composed of four protein components that are all required for carboxylase activity. To each component a function could be assigned, and a correlation with the gene sequences determined by SWAVING et al. (1995) was established with N-terminal sequence

analysis of the purified proteins. Recently, a highly similar epoxide carboxylase has been detected in *Rhodococcus rhodochrous* B-276 (previously identified as *Nocardia corallina*), which indicates that carboxylation of epoxides may be common among phylogenetically distinct bacteria (ALLEN and ENSIGN, 1998).

In ethene utilizing bacteria, ethene is oxidized to epoxyethane (HARTMANS et al., 1989), which is further converted to acetyl-CoA by epoxide dehydrogenase in a reaction that is dependent on NAD$^+$, FAD, CoA, and a fourth unknown cofactor (Fig. 6) (HARTMANS et al., 1989).

GRIFFITHS et al. (1987a,b) described an epoxide lyase that is involved in the degradation of the bicyclic monoterpene α-pinene. Upon oxidation of α-pinene, an epoxide is formed that is converted by α-pinene oxide lyase. The remarkable reaction that is catalyzed by this enzyme involves the cleavage of two carbon–carbon bonds which results in the opening of both the 4- and 6-membered ring system. The enzyme was purified to homogeneity from a *Nocardia* strain and was found to be highly specific for α-pinene oxide (Fig. 6) (TRUDGILL, 1990).

3 Examples of Catabolic Pathways

3.1 Methane Utilization by *Methylosinus trichosporium*

Methanotrophic bacteria, which can utilize methane as the sole source of carbon and energy, are ubiquitous in the environment. A detailed review on their taxonomy, physiology, and ecological role in the global carbon cycle was published by HANSON and HANSON (1996). Methane is oxidized in four consecutive steps (Fig. 1) to methanol, formaldehyde, formate, and carbon dioxide, respectively. Methanotrophs are divided into three distinct classes that are characterized by the pathway that is used for formaldehyde assimilation. Type I methanotrophs utilize the ribulose monophosphate pathway, whereas type II methanotrophs use the serine pathway as the primary route for formaldehyde utilization. Type X organisms possess enzymes of both pathways. The primary step in methane metabolism is the oxidation of methane to methanol which is catalyzed by the methane monooxygenase.

Two very different forms of methane monooxygenases have been found in methanotrophic bacteria. The soluble form (sMMO), described above, derives electrons from NADH and is only found in a limited number of methanotrophs, mainly type II and type X bacteria. This enzyme is expressed under copper limitation (ANTHONY, 1986). Although methane is the only sMMO substrate that can support rapid growth of methanotrophs, the enzyme also catalyzes the fortuitous oxidation of many saturated, unsaturated, linear, branched, (poly)cyclic, and halogenated hydrocarbons (LIPSCOMB, 1994).

A particulate or membrane-bound methane monooxygenase (pMMO) is produced by most methanotrophs if copper is present in sufficient amounts. Thus, it appears to be the preferred enzyme for methane assimilation, which is in agreement with the higher affinity for methane displayed by methanotrophs which produce the pMMO. It is, therefore, probably of greater environmental relevance with respect to methane oxidation than sMMO. The pMMO is a copper containing enzyme that is bound to cytoplasmic membranes and is easily inactivated in extracts, which have been major obstacles for biochemical studies. The enzyme consists of three subunits of 45, 26, and 23 kDa. It was recently shown that pMMO is very sensitive to oxygen and optimization of the isolation procedure has resulted in an almost completely pure pMMO preparation which may facilitate future research (NGUYEN et al., 1998). Sequence analysis has revealed that pMMO is similar to the copper containing ammonium monooxygenase involved in ammonium oxidation by *Nitrosomonas*, which makes it even more interesting to obtain further mechanistic and structural insight.

3.2 Alkane Utilization by *Pseudomonas oleovorans*

Degradation of *n*-alkanes is widespread among bacteria, and such organisms can easily be enriched from soil. Most detailed studies have been done on the degradation of alkanes by *Pseudononas oleovorans* Gpo1. This includes physiological, genetic, enzymological, and biocatalytic studies. Alkane oxidation is initiated by a plasmid encoded alkane hydroxylase (AlkB), which is a membrane-bound non-heme monooxygenase of 46 kDa, containing Fe in the active site. After conversion of the alkane to the alcohol and further oxidation to a carboxylic acid, coupling to a CoA moiety takes place, which is followed by β-oxidation. Genes for a cofactor-independent flavin containing alcohol dehydrogenase (AlkJ, 58 kDa), a non-essential NADP-dependent aldehyde dehydrogenase (AlkH, 49 kDa), and an ATP-dependent acyl-CoA synthase (AlkK, 59 kDa) are also encoded on the plasmid. Thus, a complete pathway for alkane oxidation to the β-oxidation intermediate is located on a plasmid, although complementary chromosomally located genes seem to be present as well.

The alkane oxidation system obains electrons from NADH via a rubredoxin (AlkG, 18 kDa) and a rubredoxin reductase (*alk*T, 48 kDa), which is located in another operon together with a regulatory gene. There is another intriguing open reading frame in the *alk* system designated AlkL, which seems to encode an outer membrane protein, but its function in alkane oxidation or uptake is still unkown.

3.3 Isoprene Degradation by *Rhodococcus*

Despite the important role of isoprene in atmospheric chemistry, little work has been done on the microbial degradation of this compound. Recently, CLEVELAND and YAVITT (1998) have shown that microorganisms consume isoprene even at trace-level concentrations and that soil microorganisms may provide a significant biological sink for atmospheric isoprene. Information about the physiology of isoprene degradation is scarce. VAN GINKEL et al. (1987) have shown that isoprene degradation in *Nocardia* sp. starts with a monooxygenase. When isoprene was added to cell suspensions where epoxide metabolism was inhibited by excess 1,2-epoxybutane, they observed an accumulation of 1,2-epoxy-2-methyl-3-butene and the diepoxide (1,2–3,4-diepoxybutane). EWERS et al. (1991) have reported a glutathione-dependent activity towards 1,2-epoxy-2-methyl-3-butene in cell-free extracts of an isoprene utilizing *Rhodococcus* sp., but the enzyme catalyzing this reaction was not further characterized.

Isoprene metabolism in *Rhodococcus* sp. strain AD45, that was isolated by VAN HYLCKAMA VLIEG et al. (1998), starts with the oxidation of the sterically most hindered double bond, resulting in the formation of 1,2-epoxy-2-methyl-3-butene (Fig. 7). The monooxygenase of strain AD45 favors mainly the genera-

tion of the *R*-enantiomer of isoprene monoxide. A glutathione *S*-transferase then catalyzes the conjugation of glutathione and opening of the epoxide ring yielding 1-hydroxy-2-gluta-thionyl-2-methyl-3-butene (VAN HYLCKAMA VLIEG et al., 1999). This compound is further converted by a NAD$^+$-dependent dehydrogenase in two sequential oxidation steps to yield 2-glutathionyl-2-methyl-3-butenoic acid with the concomitant reduction of NAD$^+$ to NADH.

The genes involved in isoprene utilization in *Rhodococcus* sp. strain AD45 have recently been cloned and sequenced, which showed the presence of genes encoding the glutathione *S*-transferase with activity towards 1,2-epoxy-2-methyl-3-butene and the 1-hydroxy-2-gluta-thionyl-2-methyl-3-butene dehydrogenase. Furthermore, a gene encoding a second glutathione *S*-transferase was identified. Downstream of the glutathione transferase genes, in a separate operon, six genes were found that showed high homology to components of the toluene-4-monooxygenase from *Pseudomonas mendocina* KR1 and multicomponent monooxygenases of the methane monooxygenase class (see Sect. 2.1).

The results are in agreement with a catabolic route for isoprene involving epoxidation by a monooxygenase, conjugation to glutathione and oxidation of the hydroxyl group to a carboxylate. Further metabolism may proceed by fatty acid oxidation after removal of glutathione by a still unknown mechanism. Interestingly, some glutathione *S*-transferases catalyze the reversible Michael reaction of glutathione with α,β-unsaturated carbonyl compounds (CHEN and ARMSTRONG, 1995). In strain AD45, the second glutathione transferase could possibly be involved in the (reversible) elimination of the glutathione moiety from

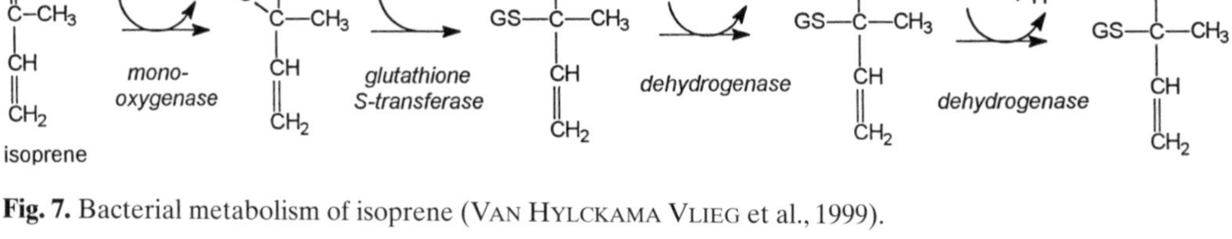

Fig. 7. Bacterial metabolism of isoprene (VAN HYLCKAMA VLIEG et al., 1999).

2-glutathionyl-1-oxo-2-methyl-3-butene or 2-glutathionyl-2-methyl-3-butenoic acid, which would result in the generation of 2-methylene-3-butene-1-al or 2-methylene-3-butenoic acid. The metabolic route for isoprene degradation is one of the first examples where a glutathione transferase is involved in a catabolic pathway of an epoxide.

3.4 Bacterial Degradation of Methyl *tert*-Butyl Ether

Few studies on the biodegradation of methyl *tert*-butyl ether have been carried out, although this compound has in a short period become one of the most important groundwater contaminants due to its widespread use as a gasoline additive. It can serve as a carbon source for some slow growing cultures under aerobic conditions. SALANITRO et al. (1994) isolated a mixed culture that was able to utilize MTBE as a growth substrate under aerobic conditions but the low growth rate is illustrated by the fact that wash-out occurred in a continuous culture at retention times below 50 d. In degradation experiments with cell suspensions it was found that during the consumption of MTBE, the primary metabolite *t*-butyl alcohol accumulated. Low growth rates were also observed by MO et al. (1997), who described several pure isolates that were obtained from a mixed MTBE utilizing mixed culture.

Cometabolic degradation by propane utilizing bacteria was recently studied, using propane utilizing bacterial cultures (STEFFAN et al., 1997). With [^{14}C]-MTBE, it was found that the methoxy group was oxidized to CO_2 and the *t*-butyl alcohol was oxidized to 2-methyl-1,2-propanediol and 2-hydroxybutyric acid (Fig. 8a). Especially *Mycobacterium vaccae* JOB5 was efficient in MTBE degradation, while the organism also degraded related ethers such as ethyl *tert*-butyl ether and methyl *tert*-amyl ether. It was concluded from spectral data that the initial oxidation of the methoxy group was catalyzed by a cytochrome P450.

3.5 Bacterial Degradation of Furan and Dioxan

Some *Rhodococcus* spp. have been isolated that utilize tetrahydrofuran as a growth substrate (BERNHARDT and DIEKMANN, 1991; BOCK et al., 1996). One of these strains, *Rhodococcus ruber* strain 219, also grew on dioxane. Although few biochemical studies have been carried out, a pathway was postulated in which the hydroxylation of the C-2 carbon was followed by oxidation to a ketone and ester cleavage (Fig. 8b). This degradation pathway for tetrahydrofuran was proposed on the basis of intermediates that accumulated during the metabolism of tetrahydrofuran and 2,5-dimethyl tetrahydrofuran. This route seems plausible also in view of the fact that a similar pathway via 2-oxygenation occurs in higher organisms.

4 Future Outlook

Although a large amount of work on the degradation of haloaliphatic compounds has been done, several important issues remain unsolved. In the last part of this review, we indicate some areas which are still relatively unexploited.

First very little work has yet been done on the anaerobic degradation of alkanes (HOLLIGER and ZEHNDER, 1996). Although anaerobic metabolism of alkanes coupled to denitrification or sulfate reduction is very well feasible from a thermodynamic point of view, few indications exist that such reactions are indeed important and take place on a larger scale although there is convincing evidence for their occurrence under some conditions (RUETER et al., 1994). Furthermore, the pathway of anaerobic alkane utilization, which has for a long time assumed to be non-existent, is still poorly understood and insight into the mechanisms of the enzymes involved is completely lacking.

The diversity of alkane hydroxylation reactions is also still not well established. The role and distribution of dioxygenases and copper containing enzymes is hardly known, although,

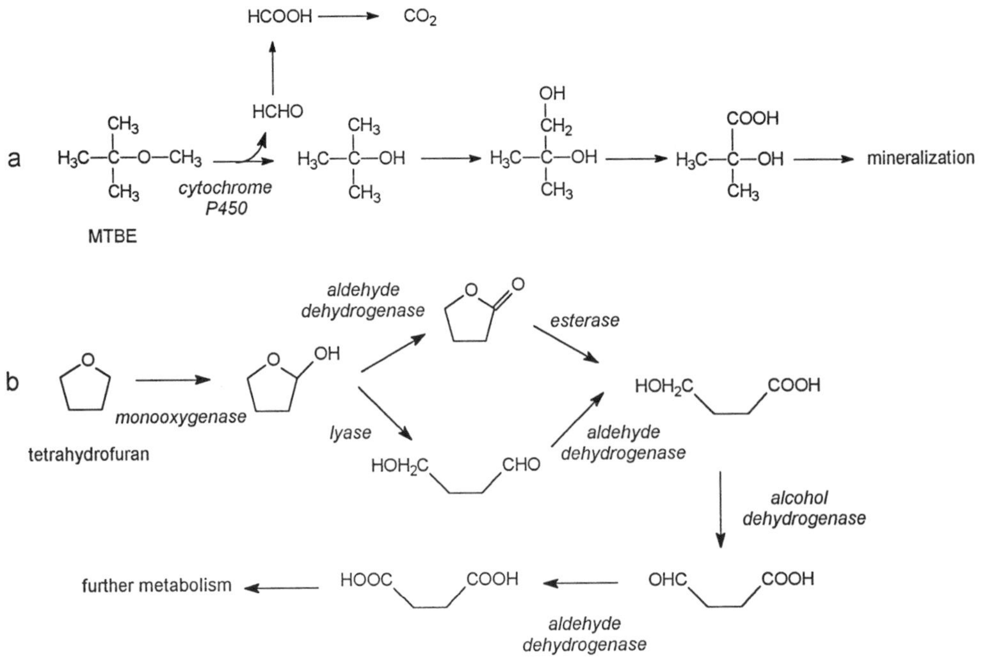

Fig. 8a, b. Degradation of MTBE by gram-positive microorganisms (**a**) (STEFFAN et al., 1997) and tetrahydrofuran (**b**) by a *Rhodococcus* sp. (BERNARDT and DIEKMANN, 1991).

in the case of methanotrophs, methane oxidation via the copper containing oxygenase is preferred over the use of the much better understood di-Fe monooxygenase.

Long-chain alkanes are poorly soluble in water, which may limit their rate of biodegradation. Various microorganisms produce biosurfactants which facilitate emulsification of the organic phase and accelerate biodegradation. The interaction between alkanes enclosed in microdroplets or in micelles with the cell and the exact role of biosurfactants is in many cases still unclear, although this process is very important for the degradation of many important soil contaminants.

Little is also known about ether-cleaving enzymes. We foresee that interest in this topic will strongly increase in the near future. As over the last years the recalcitrance of gasoline oxygenates such as MTBE has drawn a lot of attention, it can be expected that establishing more detailed insight in the fission of ether bonds will become more important.

5 References

ALLEN, J., ENSIGN, S. A. (1981), Identification and characterization of epoxide carboxylase activity in cell extracts of *Nocardia corallina* B276, *J. Bacteriol.* **180**, 2072–2078.

ANTHONY, C. (1986), Bacterial oxidation of methane and methanol, *Adv. Microb. Physiol.* **27**, 113–210.

ATLAS, R. M. (1984), *Petroleum Microbiology*. New York: Macmillan.

ATLAS, R. M. (1981), Microbial degradation of petroleum hydrocarbons: an environmental perspective, *Microbiol. Rev.* **45**, 180–209.

BERNHARDT, D., DIEKMANN, H. (1991), Degradation of dioxane, tetrahydrofuran and other cyclic ethers by an environmental *Rhodococcus* strain, *Appl. Microbiol. Biotechnol.* **36**, 120–123.

BOCK, C., KROPPENSTEDT, R. M., DIEKMANN, H. (1996), Degradation and bioconversion of aliphatic and aromatic hydrocarbons by *Rhodococcus ruber* 219, *Appl. Microbiol. Biotechnol.* **45**, 408–410.

BRITTON, L. N. (1984), Microbial degradation of aliphatic hydrocarbons, in: *Microbial Degradation* of Organic Compounds (GIBSON, D. T., Ed.), pp. 89–129. New York: Marcel Dekker.

CAIN, R. B. (1981), Microbial degradation of surfactants "builder" components, in: Microbial Degradation of Xenobiotics and Recalcitrant Compounds (LEISINGER, T., COOK, A. M., HUTTER, R., NUESCH, J., Eds.), pp. 325–370. London: Academic Press.

CHEN, J., ARMSTRONG, R. N. (1995), Stereoselective catalysis of a retro-Michael reaction by class mu glutathione transferases. Consequences for the internal distribution of products in the active site, Chem. Res. Toxicol. 8, 580–585.

CICERONE, R. J., OREMLAND, R. S. (1988), Biogeochemical aspects of atmospheric methane, Global Biochem. Cycles 2, 299–327.

CLEVELAND, C. C., YAVITT, J. B. (1998), Microbial consumption of atmospheric isoprene in a temperature forest soil, Appl. Environ. Microbiol. 64, 172–177.

EISELE, A. (1983), Biomass from higher n-alkanes, in: Biotechnology 1st Edn., Vol. 3 (REHM, H.-J., REED, G., Eds.), pp. 43–81. Weinheim: VCH.

ENSIGN, S. A., SMALL, F. J., ALLEN, J. R., SLUIS, M. K. (1998), New roles for CO_2 in the microbial metabolism of aliphatic epoxides and ketones, Arch. Microbiol. 169, 179–187.

EWERS, J., CLEMENS, W., KNACKMUSS, H.-J. (1991), Biodegradation of chloroethenes using isoprene as cosubstrate, in: Proc. Int. Symp. Environ. Biotechnol., Oostende (VERACHTERT, H., VERSTRAETE, W., Eds.), pp. 77–83. Royal Flemish Society of Engineers, Belgium.

FALL, R. R., BROWN, J. L., SCHAEFFER, T. L. (1979), Enzyme recruitment allows the biodegradation of recalcitrant branched hydrocarbons by Pseudomonas citronellolis, Appl. Environ. Microbiol. 38, 715–722.

FUKUI, S., TANAKA, A. (1981), Metabolism of alkanes by yeasts, Adv. Biochem. Eng. 19, 217–237.

GALLAGHER, S. C., GEORGE, A., DALTON, H. (1998), Sequence alignment modeling and molecular docking studies of the epoxygenase component of alkene monooxygenase from Nocardia corallina B-276, Eur. J. Biochem. 254, 480–489.

GIBSON, D. T. (Ed.) (1984), Microbial Degradation of Organic Compounds. New York: Marcel Dekker.

GRIFFITHS, E. T., BOCIEK, S. M., HARRIES, P. C., JEFFCOAT, R., SISSONS, D. J., TRUDGILL, P. W. (1987a), Bacterial metabolism of α-pinene: Pathway from α-pinene oxide lyase to acyclic metabolites in Nocardia sp. strain P18.3, J. Bacteriol. 169, 4972–4979.

GRIFFITHS, E. T., HARRIES, P. C., JEFFCOAT, R., TRUDGILL, P. W. (1987b), Purification and properties of α-pinene oxide lyase from Nocardia sp. strain P18.3, J. Bacteriol. 169, 4980–4983.

HANSON, R. S., HANSON, T. E. (1996), Methanotrophic bacteria, Microbiol. Rev. 60, 439–471.

HARTMANS, S., DE BONT, J. A. M., HARDER, W. (1989), Microbial metabolism of short-chain unsaturated hydrocarbons, FEMS Microbiol. Rev. 63, 235–264.

HOLLIGER, C., ZEHNDER, A. J. (1996), Anaerobic biodegradation of hydrocarbons, Curr. Opin. Biotechnol. 7, 326–330.

JANSSEN, D. B. (Ed.) (1994), Genetics of Biodegradation of Synthetic Compounds, Biodegradation Vol. 5 (3–4). Delft: Kluwer Academic Publishers.

JOHN, D. M., WHITE, G. F. (1998), Mechanism for biotransformation of nonylphenol polyethoxylates to xenoestrogens in Pseudomonas putida, J. Bacteriol. 180, 4332–4338.

KASAI, N., SUZUKI, T., FURUKAWA, Y. (1998), Chiral C3 epoxides and halohydrins: their preparation and synthetic application, J. Mol. Catal. B 4, 237–252.

KAWAI, F. (1995), Breakdown of plastics and polymers by microorganisms, Adv. Biochem. Eng. Biotechnol. 52, 151–194.

KUZMA, J., NEMECEK-MARSHALL, M., POLLOCK, W. H., FALL, R. (1995), Bacteria produce the volatile hydrocarbon isoprene, Curr. Microbiol. 30, 97–103.

LANGE, S. J., QUE, L., JR. (1998), Oxygen activating nonheme iron enzymes, Curr. Opin. Chem. Biol. 2, 159–172.

LELIEVELD, J., CRUTZEN, P. J., BRUHL, C. (1993), Climate effects of atmospheric methane, Chemosphere 26, 739–768.

LIPSCOMB, J. D. (1994), Biochemistry of the soluble methane monooxygenase, Annu. Rev. Microbiol. 48, 371–399.

MAENG, J. H., SAKAI, Y., TANI, Y., KATO, N. (1996), Isolation and characterization of a novel oxygenase that catalyzes the first step of n-alkane oxidation in Acinetobacter sp. strain M-1, J. Bacteriol. 178, 3695–3700.

MISAWA, E., CHAN KWO CHION, C. K., ARCHER, I. V., WOODLAND, M. P., ZHOU, N. Y. et al. (1998), Characterization of a catabolic epoxide hydrolase from a Corynebacterium sp., Eur. J. Biochem. 253, 173–183.

MO, K., LORA, C. O., WANKEN, A. E., JAVANMARDIAN, M., YANG, X., KULPA, C. F. (1997), Biodegradation of methyl t-butyl ether by pure bacterial cultures, Appl. Microbiol. Biotechnol. 47, 69–72.

MONSON, R. K., FALL, R. (1989), Isoprene emission from aspen leaves. Influence of environment and relation to photosynthesis and respiration, Plant Physiol. 90, 267–274.

MURELL, J. C. (1994), Molecular genetics of methane oxidation, Biodegradation 5, 145–159.

NGUYEN, H.-H., ELLIOTT, S. J., YIP, J. H., CHAN, S. I.

(1998), The particulate methane monooxygenase from *Methylococcus capsulatus* (Bath) is a novel copper-containing three-subunit enzyme. Isolation and characterization, *J. Biol. Chem.* **273**, 7957–7966.

RATAJCZAK, A., GEISSDORFER, W., HILLEN, W. (1998), Expression of alkane hydroxylase from *Acinetobacter* sp. strain ADP1 is induced by a broad range of *n*-alkanes and requires the transcriptional activator AlkR, *J. Bacteriol.* **180**, 5822–5827.

RATLEDGE, C. (Ed.) (1990), *Physiology of Biodegradative Organisms*, Biodegradation Vol. 1 (2–3). Delft: Kluwer Academic Publishers.

REHM, H.-J., REIFF, I. (1981), Mechanism and occurrence of microbial oxidation of long-chain alkanes, *Adv. Biochem. Eng.* **19**, 175–215.

RIDGWAY, H. F., SAFARIK, J., PHIPPS, D., CLARK, C. P. (1990), Identification and catabolic activity of well-derived gasoline-degrading bacteria from a contaminated aquifer, *Appl. Environ. Microbiol.* **56**, 3565–3575.

RINK, R., FENNEMA, M., SMIDS, M., DEHMEL, U., JANSSEN, D. B. (1997), Primary structure and catalytic mechanism of the epoxide hydrolase from *Agrobacterium radiobacter* AD1, *J. Biol. Chem.* **272**, 14650–14657.

ROSENBERG, E. (1986), Microbial surfactant, *CRC Rev. Biotechnol.* **3**, 109–131.

RUETER P., RABUS, R., WILKES, H., AECKERSBERG, F., RAINEY, F. A., JANNASCH, H. W., WIDDEL, F. (1994), Anaerobic oxidation of hydrocarbons in crude oil by new types of sulfate-reducing bacteria, *Nature* **372**, 455–458.

SALANITRO, J. P., DIAZ, L. A., WILLIAMS, M. P., WISNIEWSKI, H. L. (1994), Isolation of a bacterial culture that degrades methyl *t*-butyl ether, *Appl. Environ. Microbiol.* **60**, 2593–2596.

SHARKEY, T. D. (1996), Isoprene synthesis by plants and animals, *Endeavour* **20**, 74–78.

SHARKEY, T. D., SINGSAAS, E. L. (1995), Why plants emit isoprene, *Nature* **374**, 769.

SQUILLACE, P. J., ZOGARSKI, J. S., WILBER, W. G., PRICE, C. V. (1996), Preliminary assessment of the occurrence and possible sources of MTBE in groundwater in the United States, *Environ. Sci. Technol.* **27**, 15–24.

STEFFAN, R. J., McCLAY, K., VAINBERG, S., CONDEE, C. W., ZHANG, D. (1997), Biodegradation of the gasoline oxygenates methyl *tert*-butyl ether by propane-oxidizing bacteria, *Appl. Environ. Microbiol.* **63**, 4216–4222.

SWAVING, J., DE BONT, J. A. M. (1998), Microbial transformation of epoxides, *Enzyme Microb. Technol.* **22**, 19–26.

SWAVING, J., WEIJERS, C. A., VAN OOYEN, A. J., DE BONT, J. A. (1995), Complementation of *Xantho-*

bacter Py2 mutants defective in epoxyalkane degradation, and expression and nucleotide sequence of the complementing DNA fragment, *Microbiology* **141**, 477–484.

SWAVING, J., DE BONT, J. A. M., WESTPHAL, A., DE KOK, A. (1996), A novel type of pyridine nucleotide disulfide oxidoreductase is essential for NAD$^+$ and NADPH-depentent degradation of epoxyalkanes by *Xanthobacter* strain Py2, *J. Bacteriol.* **178**, 6644–6646.

THOMPSON, A. M. (1992), The oxidizing capacity of the earth's atmosphere; probable past and future changes, *Science* **256**, 1157–1165.

TIDSWELL, E. G. (1996), Ether-bond scission in the biodegradation of alcohol ethoxylate non-ionic surfactants by *Pseudomonas* sp. strain SC25A, *Microbiology* **142**, 1123–1131.

TOWER, M. K., BUCKLAND, R. M., HIGGINS, R., GRIFFIN, M. (1985), Isolation of a cyclohexane-metabolizing *Xanthobacter* sp., *Appl. Environ. Microbiol.* **49**, 1282–1289.

TRAINER, M., WILLIAMS, E. J., PARRISH, D. D., BUHR, M. P., ALLWINE, E. J. et al. (1987), Models and observations of the impact of natural hydrocarbons on rural ozone, *Nature* **329**, 705–707.

TRUDGILL, P. W. (1984), Microbial degradation of the alicyclic ring. Structural relationships and metabolic pathways, in: *Microbial Degradation of Organic Compounds* (GIBSON, D. T., Ed.), pp. 131–180. New York: Marcel Dekker.

TRUDGILL, P. W. (1990), Cyclopentanone 1,2-monooxygenase from *Pseudomonas* NCIMB 9872, *Methods Enzymol.* **188**, 77–81.

VAN AGTEREN, M. H., KEUNING, S., JANSSEN, D. B. (1998), *Handbook on Biodegradation and Biological Treatment of Harzardous Organic Compounds*. Dordrecht: Kluwer Academic Publishers.

VAN BEILEN, J. B., WUBBOLTS, M. G., WITHOLT, B. (1994), Genetics of alkane oxidation by *Pseudomonas oleovorans*, *Biodegradation* **5**, 161–174.

VAN GINKEL, C. G., DE JONG, E., TILANUS, J. W. R., DE BONT, J. A. M. (1987), Microbial oxidation of isoprene, a biogenic foliage volatile and of 1,3-butadiene, an anthropogenic gas, *FEMS Microbiol. Ecol.* **45**, 275–279.

VAN HYLCKAMA VLIEG, J. E. T., KINGMA, J., VAN DEN WIJNGAARD, A. J., JANSSEN, D. B. (1998), A glutathione *S*-transferase with activity towards *cis*-1,2-dichloroepoxyethane is involved in isoprene utilization by *Rhodococcus* sp. strain AD45, *Appl. Environ. Microbiol.* **64**, 2800–2805.

VAN HYLCKAMA VLIEG, J. E. T., KINGMA, J., KRUIZINGA, W., JANSSEN, D. B. (1999), Purification of a glutathione *S*-transferase and a glutathione conjugate dehydrogenase involved in isoprene metabolism in *Rhodococcus* sp. strain AD45, *J. Bacteri-*

ol. **181**, 2094– 2101.

WATKINSON, R. J., MORGAN, P. (1990), Physiology of aliphatic hydrocarbon-degrading microorganisms, *Biodegradation* **1**, 79–92.

WEIJERS, C. A. G. M., DE HAAN, A., DE BONT, J. A. M. (1988), Microbial production and metabolism of epoxides, *Microb. Sci.* **5**, 156–159.

WHITE, G. F., RUSSEL, N. J., TIDSWELL, E. C. (1996), Bacterial scission of ether bonds, *Microbiol. Rev.* **60**, 216–232.

WILLETTS, A. (1997), Structural studies and synthetic applications of Baeyer–Villiger monooxygenases, *Trends Biotechnol.* **15**, 55–62.

WILLIAMS, S. T., TRUDGILL, P. W., TAYLOR, D. G. (1989), Metabolism of 1,8-cineole by a *Rhodococcus* species – ring cleavage reactions, *J. Gen. Microbiol.* **135**, 1957–1967.

ZHANG, Y., MILLER, R. M. (1994), Effect of a *Pseudomonas* biosurfactant on cell hydrophobicity and biodegradation of octadecane, *Appl. Environ. Microbiol.* **60**, 2101–2102.

9 Degradation of Aromatic and Polyaromatic Compounds

Matthias Kästner

Leipzig, Germany

List of Abbreviations

ABTS   2,2'-amino (3-ethylene benzothiazoline-6-sulfonate)
BTX   benzene, toluene, xylene
EPA   Environmental Protection Agency
NAPL   non-aqueous phase liquid
PAH   polycyclic aromatic hydrocarbons

1 Aromatic Compounds

Aromatic compounds are ubiquitous in the environment and are derived from both man-made and natural sources. Natural sources are aromatic compounds of biomass such as aromatic amino acids (phenylalanine, tryptophan, tyrosine), lignin compounds (polymers of phenyl propanoic units) and components of fossil mineral oil and coals which are also of biosynthetic origin. However, the latter compounds normally are not available to the biosphere. Thus, the amount of such compounds present in the biosphere is due to human action. Other man-made sources are emissions of pyrolysis of organic compounds and chemical synthesis. Due to the abundance of aromatic compounds in the environment microorganisms are able to degrade and to mineralize aromatic compounds. In this chapter, the degradation pathways of aromatic compounds are presented with respect to the simple monoaromatic hydrocarbons. There is a great concern about the occurrence of benzene, toluene, ethylbenzene, styrene, and xylenes in the environment which are amongst the 50 largest volume industrial chemicals produced in the order of millions of tons per year. The compounds are mostly used as fuels and industrial solvents (SMITH, 1990).

1.1 Benzene

Benzene and related compounds are characterized by large negative resonance energy resulting in thermodynamic stability. Thus, the chemical properties are very different from those observed for aliphatic compounds. Microbial biodegradation of aromatic compounds requires specific enzymatic systems to cleave the aromatic ring which is necessary for the mineralization of the carbon. The pathways of degradation by soil microorganisms (mostly *Pseudomonas* sp.) has extensively been reviewed (GIBSON and SUBRAMANIAN, 1984; CERNIGLIA, 1984; DAGLEY, 1986; SMITH, 1990).

Only a few reports about the aerobic microbial degradation pathways of benzene are available in the last years (anaerobic degradation, see chapter 7, this volume). The main contributions to elucidate the pathways (metabolites and enzyme systems) in bacteria (mostly soil Pseudomonads) were carried out in the last decades (MARR and STONE, 1961; GIBSON et al., 1968, 1970; HÖGN and JAENICKE, 1972; AXELL and GEARY, 1975) and were excellently reviewed by GIBSON and SUBRAMANIAN (1984). Most of the described new bacterial strains able to utilize benzene use the same degradation routes. Fig. 1 shows the degradation pathway of benzene by bacteria. The reaction sequence is initialized by oxidation with a dioxygenase to *cis*-benzene dihydrodiol which is then re-aromatized to catechol (1,2-dihydroxybenzene). The adjacent hydroxyl groups are prerequisites for the ring cleavage which proceeds either by the catechol 1,2-dioxygenase (*o*- or intradiol cleavage) or by the less

common catechol 2,3-dioxygenase (*m*- or extradiol cleavage). The non-aromatic ring fission products are further metabolized to compounds of the central metabolism either via the characteristic metabolite 3-oxoadipate (oxoadipate pathway) to succinate and acetate (acetyl CoA) or via the other route to pyruvate and acetaldehyde (FUCHS, 1999).

1.2 Alkylated Aromatic Compounds

Substituted aromatic compounds undergo modification of the peripherical groups to central ring cleavage metabolites such as catechol, protocatechuate (3,4-dihydroxybenzoate), or gentisate (3,6-dihydroxybenzoate). In addition, alternative modes of degradation such as side chain attack or ring attack are opened up, i.e., by alkyl substituents. Longer chain alkyl-

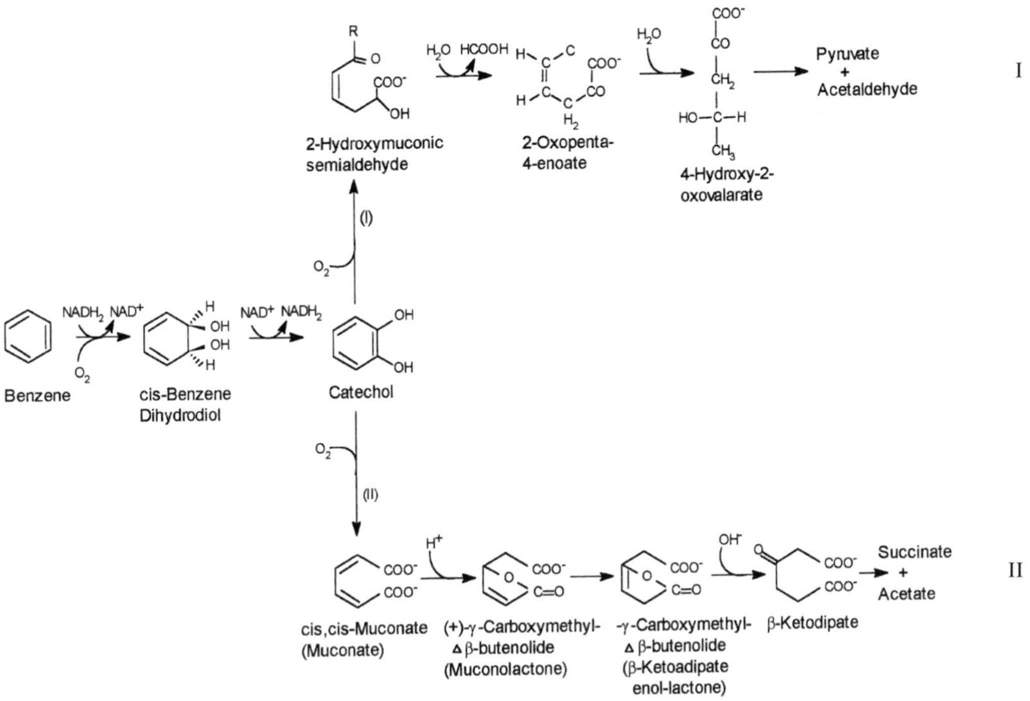

Fig. 1. Microbial degradation of benzene: (I) *m*-cleavage, (II) *o*-cleavage and formation of central metabolites (modified after SMITH, 1990).

benzenes provide enough energy during oxidation of the side chain and enable the growth of the organisms that might not be able to cleave the aromatic ring. Such compounds may be regarded as substituted alkanes rather than substituted benzenes (SMITH, 1990). Toluene, the simplest of the substituted benzenes is metabolized by both ring attack with dioxygenases and methyl group hydroxylation. The alternative pathways were reviewed in detail by HOOPER (1978) and are presented in Figs. 2 and 3. Recent findings show that *Burkholderia* (*Pseudomonas*) sp. uses multiple pathways (one dioxygenase and two different monooxygenases) for the metabolism of toluene (JOHNSON and OLSON, 1997). Direct ring cleavage of several higher alkylbenzenes has also been demonstrated (SMITH, 1990). A general pathway for the complete metabolism of mono-

alkylbenzenes is given in Fig. 4. When the alkyl chain length exceeds C7 the preferred route seems to be the attack of the alkyl chain proceeded by ω- and β-oxidation to homogentisinic acid (2,5-dihydroxyphenylacetic acid) whith subsequent ring cleavage (SMITH, 1990).

Microbial degradation of dialkylbenzenes, i.e., of xylenes is mostly restricted to the metabolisms of *m*- and *p*-isomers by pseudomonads. The degradation proceeds by initial oxidation of one of the methyl groups to the corresponding methyl benzylalcohols, tolualdehydes, toluic acids, to methyl catechol as the central metabolites for ring cleavage (GIBSON and SUBRAMANIAN, 1984). The resulting different methyl catechols (3-methylcatechol from *m*-xylene and 4-methylcatechol from *p*-xylene) undergo *m*-cleavage and the ring fission products are further catabolized by different

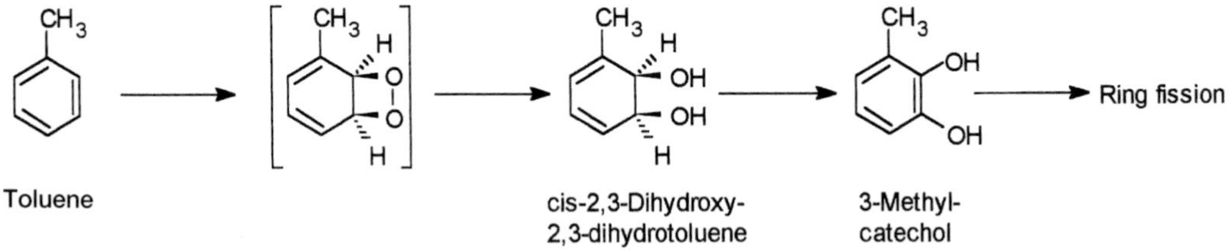

Fig. 2. Direct ring cleavage of toluene.

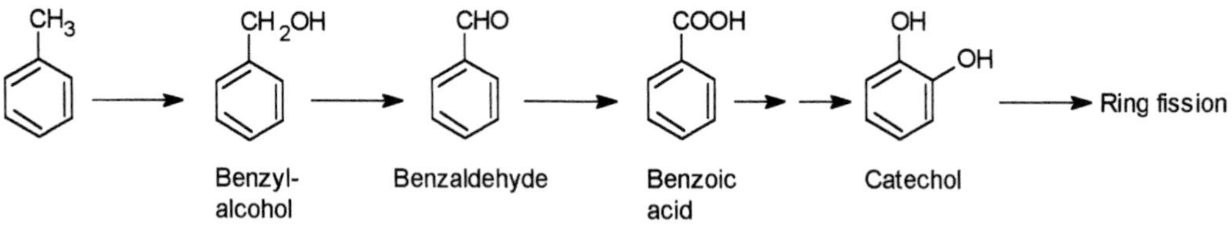

Toluene → cis-2,3-Dihydroxy-2,3-dihydrotoluene → 3-Methylcatechol → Ring fission

Fig. 3. Side chain attack of toluene.

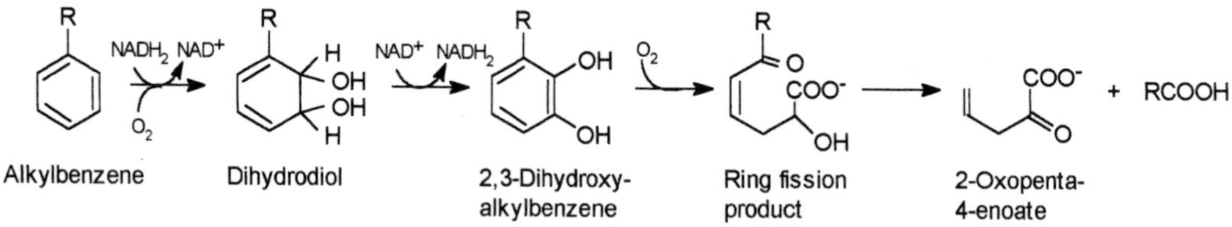

Benzylalcohol → Benzaldehyde → Benzoic acid → Catechol → Ring fission

Fig. 4. Microbial degradation of alkylbenzenes (C1–C7).

Alkylbenzene → Dihydrodiol → 2,3-Dihydroxyalkylbenzene → Ring fission product → 2-Oxopenta-4-enoate + RCOOH

enzyme systems (DUGGLEBY and WILLIAMS, 1986). The alternative mode of attack of *m*- and *p*-xylene via direct dioxygenase reaction leads to the corresponding dimethylcatechols. However, these compounds are mostly not further degraded and this route should be regarded as a cometabolic reaction. The degradation of *o*-xylene by *Pseudomonas* and *Corynebacterium* sp. via 3,4-methylcatechol with subsequent ring fission was discovered several years later (BAGGI et al., 1987; SCHRAA et al., 1987). However, a decade later the question why none of the strains able to degrade *m*- and *p*-xylene can attack *o*-xylene and vice versa can still not be answered.

1.3 Mixed Substrates

Biodegradation of aromatic hydrocarbons can be affected by the co-occurrence of mixed substrates. Such interactions usually occur in the environment since contaminations with single compounds are not very likely. It is well established that certain mixtures are more rapidly degraded than single present compounds (McCARTY et al., 1984). However, these studies are mostly based on work with non-growth supporting substrates. Little is known about substrate interactions among biodegradable aromatic hydrocarbons present in growth supporting concentrations (SMITH, 1990). The presence of naphthalene enhanced the biodegradation of phenanthrene in marine sediments but not the one of anthracene (BAUER and CAPONE, 1988). The presence of toluene or xylene also stimulated the degradation of benzene whereas antagonistic effects on the utilization of carbon by benzene were found (ARVIN et al., 1989). However, no differences in the mineralization of BTX compounds were observed in soils, regardless of whether they were presented singly or in combinations (TSAO et al., 1998). Other authors found competitive inibitions by mixed substrates, i.e., of phenanthrene or fluorene in the presence of naphthalene (SHUTTLEWORTH and CERNIGLIA, 1996; STRINGFELLOW and AIKEN, 1995).

Such phenomena cannot be explained on a molecular basis, since the amounts of active biocatalysts have mostly not been estimated in these experiments. More recent findings show that the ability to degrade naphthalene, phenanthrene, and anthracene are encoded on the same plasmids of *Pseudomonas* sp. isolated from a contaminated site (SANSEVERINO et al., 1993).

2 Polycyclic Aromatic Hydrocarbons

Polycyclic aromatic hydrocarbons (PAH) comprise a group of compounds the molecular structure of which consists of two or more condensed benzene rings. Several 100 compounds of this type are known. More than 70 isomers are possible of merely the 4- to 6-ring PAH. Heterocyclic polyaromatic hydrocarbons containing N, S and O atoms are also considered to be PAH in the broader sense as they exhibit comparable characteristics (SIMS and OVERCASH, 1983). Owing to their hydrophobic properties, PAH show a strong tendency to absorb on surfaces in aqueous systems and to undergo no or only slow decomposition in the environment (CERNIGLIA and HEITKAMP, 1989). Because of their toxic effects and their frequent occurrence in the environment, the US Environmental Protection Agency (EPA) numbers PAH among the "priority pollutants". In view of the complexity of this group, the EPA has selected 16 non-substituted, well analyzable PAH compounds as a model standard for analysis for PAH (16 EPA standard) which have come to be globally accepted as an analysis standard (Fig. 5).

Varying concentrations of PAH are contained in petroleum and oil-based fuels. They occur during the pyrolysis and incomplete combustion of biological material and organic compounds above 800 °C (RAMDAHL, 1985). At lower temperatures, in addition to unsubstituted PAH, large quantities of alkyl substituted, mainly methylated PAH originate. The quantities formed depend on the structure of the combusted material. PAH also occur naturally during processes such as volcanic activity, forest fires, and all technical combustion processes. In addition, trace amounts of PAH may also be formed through biogeneous synthesis

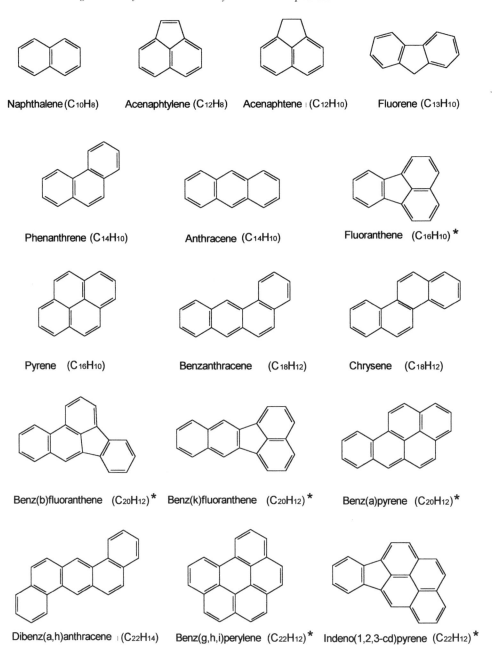

Naphthalene (C₁₀H₈)   Acenaphtylene (C₁₂H₈)   Acenaphtene ₁ (C₁₂H₁₀)   Fluorene (C₁₃H₁₀)

Phenanthrene (C₁₄H₁₀)   Anthracene (C₁₄H₁₀)   Fluoranthene (C₁₆H₁₀) *

Pyrene (C₁₆H₁₀)   Benzanthracene (C₁₈H₁₂)   Chrysene (C₁₈H₁₂)

Benz(b)fluoranthene (C₂₀H₁₂)* Benz(k)fluoranthene (C₂₀H₁₂)*   Benz(a)pyrene (C₂₀H₁₂)*

Dibenz(a,h)anthracene ₁(C₂₂H₁₄)   Benz(g,h,i)perylene (C₂₂H₁₂)*   Indeno(1,2,3-cd)pyrene (C₂₂H₁₂)*

Fig. 5. Structures of several PAH (16 EPA PAH).

in microorganisms (SUESS, 1976). The main sources of the anthropogenic formation and emission of PAH are the combustion of fossil fuels in vehicles and for power generation (BJØRSETH and RAMDAHL, 1985). Due to their hydrophobic properties, the PAH released during combustion are adsorbed on dust and soot particles, and are evenly distributed in the environment by air circulation and subsequent atmospheric washout and deposition. As a result of these processes, background PAH pollution levels of 1–5 mg total PAH kg^{-1} soil nowadays must be expected in urban districts, while levels as high as 10 mg total PAH kg^{-1} soil have been measured in the vicinity of busy roads. Furthermore, additional contamination must be anticipated on agricultural areas wherever sewage sludge is used as fertilizer. Average concentrations in such areas were, e.g., 1–5 mg total PAH kg^{-1} soil in Great Britain (BECK et al., 1995). All in all, the tendency of general PAH pollution in the environment is rising. General soil contamination at reference sites in Great Britain has quadrupled since 1880. Concentration factors exceeding 10 have been found for the 5-ring PAH, especially benz[a]pyrene (JONES et al., 1983a, b). Assuming that 2- to 4-ring PAH undergo limited decomposition in the soil, the actual increase in immissions is probably on this latter scale – and hence approximately corresponds to the rise in the consumption of fossil fuels.

There are basically two ways in which PAH penetrate the soil:

(1) immission via soot particles from combustion processes, and
(2) direct entry via products containing PAH – especially tar oils, which arise during the incomplete combustion of wood and coal.

Historically, the main quantities of PAH were formed during the production of coke for steel production, as well as in the manufacture of coal gas and fuel gas from hard coal, during which the tar oils were left over as distillation residues (WIESMANN, 1994). As the toxic and biocidal effects of tar oils were known since in the 18th century, such distillation residues were used on early for the conservation of wood and rope. During gas production and paint manufacture in the tar processing industry, elements which could no longer be used and superfluous quantities were generally disposed of in pits on the plant premises. Consequently, gasworks sites and impregnation plants for the production of railway sleepers are nowadays regarded as contaminated locations (WIESMANN, 1994).

2.1 Physico-Chemical Properties and Bioavailability of PAH

Describing the problems during the microbial degradation of PAH in soil entails knowledge of main physico-chemical properties of PAH (Tab. 1). At room temperature, PAH are solid crystalline substances, whose boiling points rise with increasing molecular weight and the growing number of condensed rings. By contrast, their melting points are not closely correlated to the number of rings or their molecular weight, although they are nearly all much higher than $100\,^{\circ}C$. The vapor pressure of the compounds is not very high and except for PAH with 2 and 3 rings up to phenanthrene the compounds do not have the tendency to volatilize. Therefore, their solubility in water decreases from naphthalene to dibenz(a,h)anthracene by more than 5 orders of magnitude and is very low in most PAH, with values below 1 mg L^{-1}. In addition, water solubility greatly depends on temperature and salt content of the water (McELROY et al., 1989). An increase in temperature from 10–$20\,^{\circ}C$ almost doubles water solubility – a significant factor concerning, i.e., the operating temperature of remediation piles.

The solubility characteristics of PAH, which decline with decreasing molecular weights, and their distribution behavior in multiphase systems are decisive for the microbial degradation of PAH. Their distribution behavior is usually expressed by the logarithmic distribution coefficients k between two phases. The coefficients of PAH rise with increasing molecular weights by 3–4 orders of magnitude. Inversely, the possible concentrations in the water phase, therefore, decline. The 1-octanol water distribution coefficient k_{ow} is a standardized parameter which indicates the ratio between the concentrations of a PAH in phases of 1-octanol and water in direct contact with each other. The concentration of the respective PAH in equilibrium is below the physical water solubility. By contrast, the distribution coefficient k_{oc} describes the concentration distribution within a system consisting of an organic soil matrix and the water phase. This value is standardized to the organic C content of the respective soil matrix. It represents an em-

Tab. 1. Physical and Chemical Properties of the 16 EPA-PAK

Compound	MW	Melting Point [°C]	Boiling Point [°C]	Vapor Pressure [mPa[a]]	Water Solubility [mgL^{-1}][b]	log k_{ow}	log k_{oc}
Naphthalene	128	80	218	$10.8 \cdot 10^3$	30.00	3.37	3.1
Acenaphthylene	152	92	265	–	16.1	4.07	?
Acenaphthene	154	96	279	$1.16 \cdot 10^3$	3.47	4.33	3.8
Fluorene	166	116	298	$4.5 \cdot 10^2$	1.8	4.18	3.9
Phenanthrene	178	101	340	$9.3 \cdot 10^1$	1.29	4.46	4.1
Anthracene	178	218	342	$1.1 \cdot 10^1$	0.073	4.45	4.3
Fluoranthene	202	110	375	$2.4 \cdot 10^2$	0.26	5.33	4.3
Pyrene	202	150	404	$1.6 \cdot 10^0$	0.135	5.32	4.8
Benzo(a)anthracene	228	159	435	$1.0 \cdot 10^{-1}$	0.014	5.61	4.8
Chrysene	228	256	448	$1.5 \cdot 10^{-3}$	0.0006	5.86	4.9
Benzo(b)fluoranthene	252	168	–	$2.9 \cdot 10^{-2}$	0.0012	6.57	6.2
Benzo(k)fluoranthene	252	217	480	$1.8 \cdot 10^{-2}$	0.00055	6.84	5.6
Benzo(a)pyrene	252	179	495	$3.8 \cdot 10^{-3}$	0.0038	6.04	5.3
Dibenzo(a,h)anthracene	278	267	524	$6.7 \cdot 10^{-6}$	0.0005	6.75	6.3
Benzo(g,h,i)perylene	276	278	–	$1.8 \cdot 10^{-4}$	0.00026	7.23	–
Indeno(1,2,3-cd)pyrene	276	162	–	–	0.062	7.66	6.2
Reference:	(5)	(5)	(1,5)	(6)	(1,3,4)	(1,2,3,5,6)	(1,2,3,5,6)

Adapted from: (1) KOCH and WAGNER, 1989; (2) MEANS et al., 1980; (3) RICHARDSON and GANGOLLI, 1993; (4) RIPPEN, 1994; (5) SIMS and OVERCASH, 1983; (6) BECK et al., 1995
[a] 25 °C
[b] 20 °C

pirically determined value, which in the case of PAH is always smaller than k_{ow}, and which can be used to estimate the behavior of compounds in the soil system. Sorption processes also play a major role in this respect. Sorption is considered to be a reversible process and leads to the formation of a relatively labile association between xenobiotics and natural organic matter in soils. The sorption of PAH, i.e., to organic soil compounds, is driven by van der Waals forces ($419 \cdot 10^3$–$8,38 \cdot 10^3$ J mol^{-1}), and electrostatic Coulomb forces (HASSETT and BANWART, 1989). Moreover, charge transfer complexes consisting of an electron donor and an electron acceptor molecule with a partial overlap of orbitals and a partial exchange of electron densities are responsible for sorption reactions. In particular, π-bonds from the overlap of π-electron systems may be relevant for the association of alkenes and aromatic compounds with soil organic matter.

Hydrophobic sorption is the partitioning of non-polar organic compounds out of the polar aqueous phase onto a hydrophobic surface in soil. In this case, the primary sorption forces are entropy changes resulting from the removal of the solute from the solution. Entropy changes are due to the destruction of the structured water shell surrounding the hydrophobic compound in the solution (HASSETT and BANWART, 1989). Hydrophobic sorption is a weak interaction between a hydrophobic solute and a polar solvent, and is relevant for the partitioning of non-polar xenobiotics in soil–water systems. The sorption process causes a decrease in the concentration of dissolved hydrophobic xenobiotics in the water phase of a soil to a certain equilibrium. The partitioning coefficient depends on the solubility of the substance in the aqueous phase and on suitable soil surfaces, whereas soil organic matter concentration is the major factor controlling sorption in soils (HASSETT and BANWART, 1989). Furthermore, crossover effects in liquid–liquid phase partitioning from non-dissolved and dissolved NAPL (non-aqueous phase liquid) phases may enhance the solubility of hydrophobic chemicals in the water phase, and thus influence their partition in soil–water systems. Therefore, the organic matter concentration has to be considered in partitioning models to predict the behavior of xenobiotics in soils (KARICKHOFF, 1981; KARICKHOFF et al., 1979). Details of interactions between xenobiotic compounds and organic matrices of soils are described in Chapter 3, this volume.

Moreover, PAH are of toxicological relevance for humans, as the compounds are partly regarded as carcinogenic and mutagenic (CERNIGLIA and HEITKAMP, 1989; LaVOIE and RICE, 1988; WISLOCKI and LU, 1988). Toxicologically relevant contact with PAH by humans occur

(1) via inhalation of soot particles, coal dust and smoke (including cigarette smoke),
(2) via the oral intake of contaminated food, and
(3) via skin resorption upon contact with tar, soot particles or mineral oil products.

When assessing the toxic impact of PAH on humans, a distinction must be drawn between acute toxic effects and the potentially carcinogenic or mutagenic effect of PAH. PAH have a relatively low acute toxic potential with LD_{50} values of 250–700 ppm in mice (CERNIGLIA and HEITKAMP, 1989). The carcinogenic, mutagenic and teratogenic potentials of PAH are of much greater toxicological significance and are thus important for assessing PAH and their residues.

2.2 Microbial Degradation of PAH

Microbial degradation pathways of PAH are necessary to understand and to control biodegradation in biotechnological systems. Furthermore, the underlying reactions and possible metabolites are largely responsible for the distribution of carbon in soils and sediment and the formation of residues (see Chapter 4, this volume). PAH are metabolized not only by cytochrome P-450 monooxygenases in mammalian cells, but also by a large number of enzymes in bacteria, fungi and algae. The ability to degrade PAH is not limited to individual

species, but occurs in various groups and even in thermophilic microorganisms (FEITKEN-HAUER et al., 1996). The microbial degradation of PAH as a whole was well documented in numerous reviews (CERNIGLIA, 1884; CERNIGLIA and HEITKAMP, 1989; CERNIGLIA, 1992, 1993; CERNIGLIA et al., 1992; POTHULURI and CERNIGLIA, 1994; HAMMEL, 1995; SUTHERLAND et al., 1995). Fig. 6 shows the degradation pathways of various groups of organisms. PAH are oxidized by the activity of cytochrome P-450 monooxygenases in eukaryotes and according to recent findings also in prokaryotes to form arene oxides. These compounds are subsequently hydrolyzed by enzymes into *trans*-dihydrodiols or non-enzymatically converted into phenols, which can form conjugates with cell components.

Bacteria and some algae are also able to oxidize PAH with specific dioxygenases to form *cis*-dihydrodiols. This degradation pathway, which permits complete metabolization and mineralization, was found with 2- and 3-ring PAH and proceeds via

(1) formation of a *cis*-dihydrodiol,
(2) dehydrogenation to form dihydroxy PAH,
(3) extradiol ring cleavage,
(4) release of C3 or C2 and C 1 compounds (elimination of the first ring),
(5) decarboxylation of the hydroxy naphthoic acid,
(6) extradiol ring cleavage and degradation of the (second) ring, and so on.

However, the exclusive assignment of dioxygenase reactions to bacteria and of monooxygenase reactions to eukaryotes described in many reviews must, however, be qualified. It has been shown that various bacteria also express generally lower activities of monooxygenases besides the dioxygenase activity (SCHOCKEN and GIBSON, 1984; HEITKAMP et al., 1988b; TROWER et al., 1988; SUTHERLAND et al., 1990; KELLEY et al., 1990; SUTHERLAND et al., 1995).

Another possibility of PAH degradation comprises non-specific radical oxidation by ligninolytic enzymes of white-rot fungi, which use their ability to degrade plant material containing lignocellulose. This type of degra-

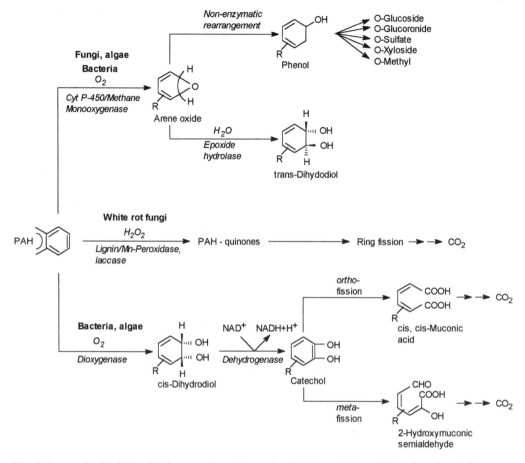

Fig. 6. Summarized initial oxidation reaction during microbial degradation of PAH (modified after CERNI-GLIA, 1993).

dation primarily leads to reactive compounds and to PAH quinones, which in some fungi are subsequently cleaved or even are mineralized to CO_2. Recently, PAH were shown to be directly mineralized (to an extent of 3–7%) by the activity of manganese-dependent peroxidase (SACK et al., 1997a). However, more recent findings also indicate that white-rot fungi can similarly oxidize PAH by cytochrome P-450 monooxygenases (SUTHERLAND et al., 1991; BEZALEL et al., 1996a, b; MASAPHY et al., 1996; SACK et al., 1997b). Saprophytic fungi and other soil fungi are also able to produce PAH quinones (LAMBERT et al., 1994; LAUNEN et al., 1995). The metabolic pathways which a few years ago were strictly assigned to differ-

ent groups of organisms, therefore, nowadays can only be regarded as main metabolic routes.

In addition to classifying PAH degradation by various pathways of initial oxidation, pollutant degradation of xenobiotics can more generally be distinguished by physiological aspects (KÄSTNER et al., 1993):

(1) complete metabolization and mineralization: intracellular, complete degradation of the ring structure, theoretically no accumulation of metabolites, degradation with the formation of biomass (usage as carbon and energy source), CO_2 and H_2O are the main products;

(2) cometabolic transformation: mainly intracellular; partial oxidation of the ring structure, generally accumulation of partly oxidized metabolites, unproductive degradation without the formation of biomass; CO_2 and H_2O are possible products;

(3) non-specific oxidation by radicals: mainly extracellular, initial oxidation by radical formation, unspecific secondary reaction of the primary oxidation products, reaction chains up to CO_2 possible.

2.2.1 Complete Metabolization and Mineralization

The type of complete metabolization and mineralization includes the almost total degradation of xenobiotics and for PAH has hitherto only been described for bacteria and compounds up to 4 rings. The substance concerned is used by the microorganisms as a source of carbon and energy to build up biomass, while the rest is mineralized to form CO_2 and water. The term "complete mineralization" frequently used for this type of degradation no longer ought to be applied since it suggests that the carbon from the substance is found as CO_2. Of course, this can never be the case, because a certain amount of the carbon will generally be utilized to maintain and build up biomass. However, during complete metabolization of a PAH compound, the accumulation of small amounts of metabolites may temporarily take place whenever reaction bottlenecks occur in the pathways. Under certain conditions, the metabolites are also found outside the cells in the surrounding medium, and have even been detected in soils and surface water. These metabolites have been described for naphthalene, phenanthrene, anthracene and benz(a)anthracene, and mostly represent the initially identified metabolites from metabolic pathways. Such compounds are o-hydroxy carboxylic acids like salicylic acid, 1-hydroxy-2-naphthoic acid, 2-hydroxy-3-naphthoic acid, and three possible compounds from benz(a)anthracene (2-hydroxy-3-phenanthroic acid, 3-hydroxy-2-phenanthroic acid and 1-hydroxy-2-anthronic acid) remaining after degradation of a certain

ring system (GUERIN and JONES, 1988a; MAHAFFEY et al., 1988) (Fig. 7–11). For fluoranthene, a comparable compound (9-fluorenon-1-carboxylic acid) has also been found, which, however, owing to the central 5-membered ring in the 9-position bears a keto group instead of a hydroxyl group (KELLEY et al., 1991). The extent of the accumulation of these metabolites largely depends on the initial concentration of the original PAH compound. For example, the corresponding hydroxyl acid (1-hydroxy-2-naphthoic acid) accumulated in marine sediments with high concentrations of phenanthrene. This compound only underwent further degradation when the initial PAH quantities decreased and limitations occurred in metabolism. The metabolite was used by the bacteria as secondary substrate and the degradation proceeded by a 2-stage conversion process (GUERIN and JONES, 1988b).

The cis-dihydrodiols occurring during complete metabolization of PAH by dioxygenase reactions are not regarded as toxicologically relevant. They only occur as metabolization intermediates within the cells. However, the trans-dihydrodiol configuration may be associated with a carcinogenic effect produced by cytochrome P-450 monooxygenase reaction with formation of epoxides and subsequent hydrolysis (CERNIGLIA, 1984).

2.2.1.1 Naphthalene

The degradation pathway of naphthalene is the pathway which has been most thoroughly investigated. The oxidation of naphthalene in the presence of *Pseudomonas* sp., *Acinetobacter calcoaceticus*, *Mycobacterium* sp. and *Rhodococcus* sp. is initiated by the naphthalene dioxygenase multienzyme complex and leads to the formation of cis-1,2-dihydroxy-1,2-dihydro naphthalene (Fig. 7). This compound is dehydrogenated to 1,2-dihydroxy naphthalene by a dehydrogenase. The ring then undergoes extradiol cleavage by another dioxygenase to create cis-2-hydroxy benzalpyruvate.

An aldolase eliminates the first ring fragments and produces salicylic aldehyde, with the release of pyruvate. Salicylic aldehyde is further oxidized to form salicylic acid, which accumulates as intermediate in some organ-

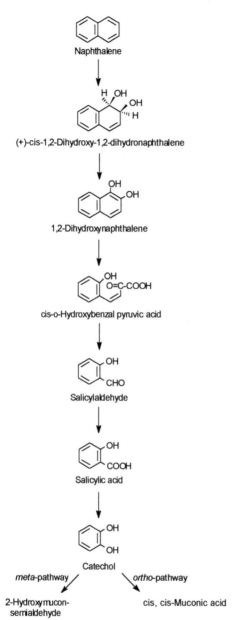

Naphthalene

(+)-cis-1,2-Dihydroxy-1,2-dihydronaphthalene

1,2-Dihydroxynaphthalene

cis-o-Hydroxybenzal pyruvic acid

Salicylaldehyde

Salicylic acid

Catechol

meta-pathway   *ortho*-pathway

2-Hydroxymucon-   cis, cis-Muconic acid
semialdehyde

Fig. 7. Naphthalene degradation by bacteria (modified after CERNIGLIA and HEITKAMP, 1989).

isms. Salicylic acid is oxidized by monooxygenase to form catechol, which then depending on the organism undergoes either *o*- or *m*-cleavage. The non-cyclic cleavage products passed

the degradation pathway for monoaromatic compounds (see Fig. 1) in order to feed the carbon and energy metabolism (CERNIGLIA and HEITKAMP, 1989).

2.2.1.2 Anthracene

The bacterial degradation pathways have also been elucidated for the 3-ring PAH anthracene (Fig. 8) and phenanthrene (Fig. 9), which in many respects exhibit analogs to the degradation pathway of naphthalene. Various *Pseudomonas* sp. and *Beijerinckia* strains oxidize anthracene by a dioxygenase in 1,2-position. This *cis*-dihydrodiol compound is re-aromatized to form 1,2-diol, which in turn (like naphthalene) undergoes extradiol cleavage by a dioxygenase. Once a C3 body has split off, this results in 2-hydroxy-3-naphthoic acid, which can intermediately accumulate. The naphthoic acid is decarboxlyated to form 2,3-dihydroxy naphthalene, which reacts with the enzymes of the naphthalene pathway via salicylic acid and catechol before being transferred to the central metabolism (CERNIGLIA, 1984; CERNIGLIA and HEITKAMP, 1989). This suggests that decomposition takes place via the degradation pathway of naphthalene. However, at this point a problem occurs as 2,3-dihydroxy naphthalene results in different ring cleavage products than 1,2-dihydroxy naphthalene. Hence further degradation cannot take place via the naphthalene pathway.

The discovery of *o*-phthalic acid in the degradation pathway of anthracene by *Sphingomonas paucimobilis* BA 2 (RICHNOW, unpublished data) as well as in other autochthonous soil microorganisms indicates that degradation does not occur via the metabolic pathway of naphthalene (via 1,2-dihydroxy naphthalene) to form salicylic acid. The original work on the degradation of anthracene by *Pseudomonas* sp. leaves this problem unsettled, as merely the occurrence of salicylic acid and catechol in cell extracts incubated with 2-hydroxy-3-naphthenic acid has been demonstrated (EVANS et al., 1965). As this part of the anthracene degradation pathway has not yet been elucidated, the data presented close this gap. The degradation of anthracene must take place directly via the (extradiol) *m*-cleavage of 2,3-hydroxy naphthalene formed by decarboxylation of 2-hydroxy-3-naphthoic acid. After cleavage, only a C2 compound (instead of pyruvate in case of naphthalene) is eliminated, and phthalic acid originates as the product. In addition, there may also be an alternative pathway during the degradation of the second ring, since in cultures of *S. paucimobilis* BA 2 both 2-hydroxy-3-naphthoic acid and 2,3-dicarboxy naphthalene are found as degradation products of the first ring, indicating both intradiol (*o*-) and extradiol (*m*-) ring cleavage mechanisms (RICH-NOW, unpublished data). Phthalic acid is subsequently decarboxylated to form salicylic acid and enters the central metabolic pathways via catechol. Despite the formation of considerable quantities of biomass, the majority of the marked carbon from 9-anthracene was found as CO_2 in these experiments. This is an additional indication that the C atom in the 9-position of anthracene is released into the metabolic pathway from either the phthalic acid or the salicylic acid as CO_2. Hence, carbon balances with 9-labeled anthracene are overvalued during mineralization and undervalued during incorporation into the biomass. Previously, *o*-phthalic acid has only been found in the alternative metabolic pathway for the degradation of phenanthrene starting from 1,2-hydroxy naphthoic acid (Kiyohara pathway, Fig. 9). Phthalic acid can also be formed via the same degradation sequence from pyrene (Fig. 11). After non-specific oxidation of anthracene by *Phanerochaete chrysosporium* 9,10-anthraquinone and *o*-phthalic acid also occur (HAMMEL et al., 1991; see Sect. 2.2.3). This metabolism takes place via lipid peroxidation catalyzed by Mn peroxidase (MOEN and HAMMEL, 1994; HAMMEL, 1995).

2.2.1.3 Phenanthrene

Phenanthrene can be converted via various metabolic pathways and is initially oxidized by bacteria in 1,2′ or mainly in 3,4-position to form the corresponding *cis*-dihydrodiols (Fig. 9). *Pseudomonas* and *Nocardia* sp. further oxidize this compound into 3,4-dihydroxy-phenanthrene, which is cleaved and further converted to 1-hydroxy-2-naphthoic acid. After decarboxylation to form 1,2-dihydroxy naph-

thalene, this compound can directly enter the metabolic pathway of naphthalene (Evans pathway, EVANS et al., 1965). *Aeromonas, Vibrio, Alcaligenes* and *Micrococcus* sp. use an alternative degradation pathway during further oxidation of 1-hydroxy-2-naphthoic acid (Kiyohara pathway, KIYOHARA et al., 1976), which leads by intradiol cleavage directly to a dicarboxylic acid intermediate, and during further course of degradation to phthalic and protocatechoic acid.

2.2.1.4 Fluorene

Recently, indications have also suggested that fluorene, to which only minor attention was paid in the past, also undergoes complete metabolization by *Arthrobacter* sp. strain F101, which is able to grow on this compound as a sole source of carbon and energy (GRIFOLL et al., 1992). The strain possesses three different degradation pathways. One of them is non-productive and leads to the accumulation of metabolites like 4-hydroxy-9-fluorenon (CASELLAS et al., 1997). 34% of the carbon from fluorene are converted to biomass while 7.5% remain in the medium as metabolites. In the productive degradation pathways of fluorene 1-formyl-2- and 2-formyl-1-indanon accumulate as degradation products of the first ring. This compound is, therefore, analogous to the hydroxy naphthoic acids in the degradation pathways of phenanthrene and anthracene. Due to the cleavage of the central C5 ring to form biphenyl, an alternative degradation pathway (which is similar to the degradation of dibenzofuran) for fluorene arises with different species of bacteria (SELIFONOV et al., 1993; GRIFOLL et al., 1994; TRENZ et al., 1994). In addition, other bacteria (*Pseudomonas cepacia* strain F297) exist which, although they can perform the initial dioxygenase reaction, stop at the stage of the degradation product of the first ring (GRIFOLL et al., 1995).

2.2.1.5 Fluoranthene

As far as PAH with more than 3 rings are concerned, degradation pathways have only been shown for fluoranthene and pyrene. How-

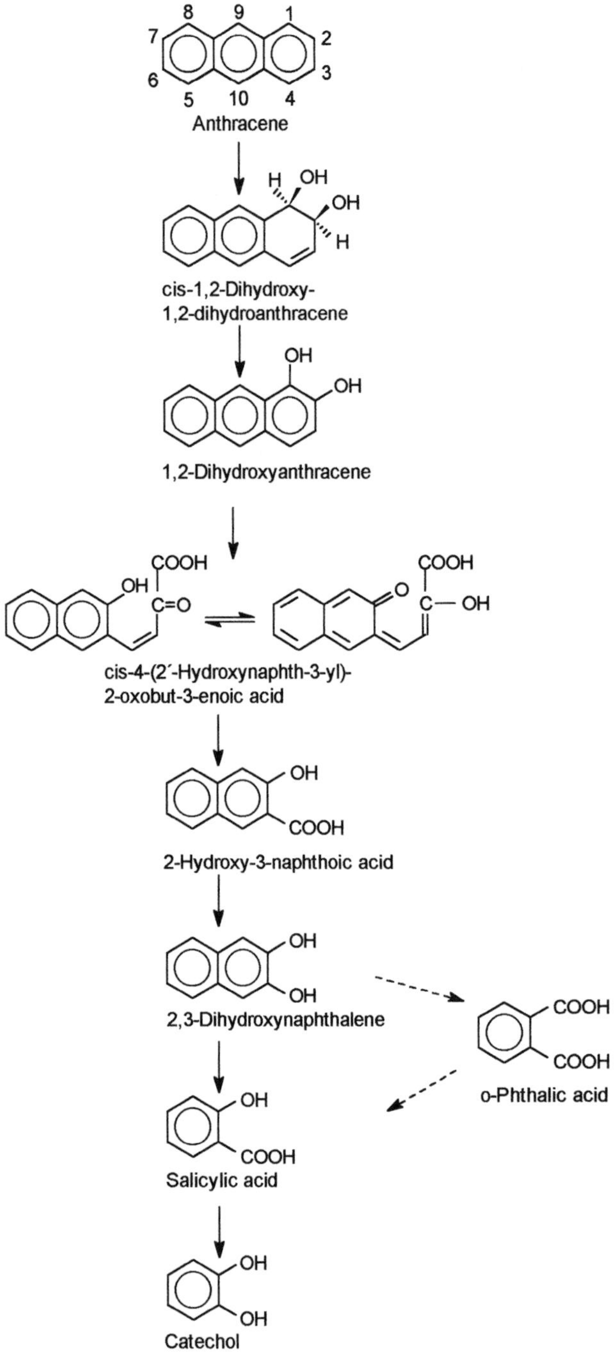

Fig. 8. Anthracene degradation by bacteria (modified after CER-NIGLIA and HEITKAMP, 1989).

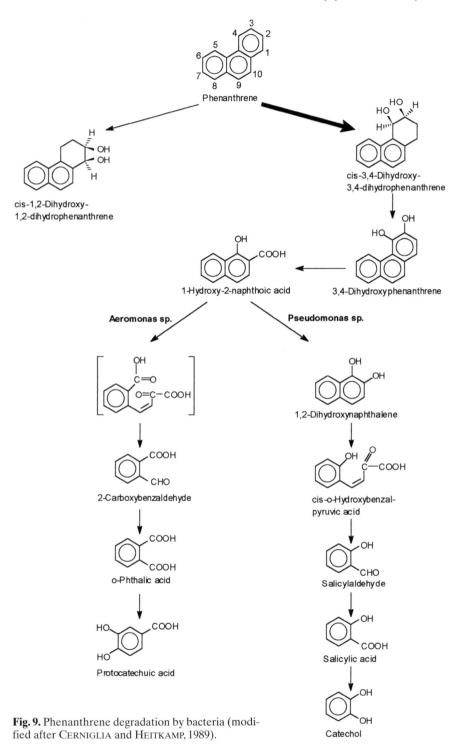

Fig. 9. Phenanthrene degradation by bacteria (modified after CERNIGLIA and HEITKAMP, 1989).

ever, it has not yet been established for all organisms whether degradation actually takes place by utilizing the compound as sole source of carbon and energy. But it has been demonstrated that the ability to grow on fluoranthene and pyrene as the sole source of carbon and energy in contaminated soils is more widespread than had been previously supposed (KÄSTNER et al., 1994). Some organisms are able to grow with fluoranthene and/or pyrene as the sole carbon source (MUELLER et al., 1989, 1990; WEISSENFELS et al., 1990, 1991; WALTER et al., 1991; KÄSTNER et al., 1994; JIMENEZ and BARTHA, 1996), whereas other organisms such as the *Mycobacterium* strain PYR-1 and RJGII-35, which utilize pyrene, also need complex carbon sources for growth (HEITKAMP et al., 1988a, b; KELLEY et al., 1991; GROSSER et al., 1991; SCHNEIDER et al., 1996; BOLDRIN et al., 1993). Although final metabolization is hence formally cometabolic, it actually corresponds more closely to the criteria of complete metabolization and mineralization, as both processes take place. The degradation pathways of fluoranthene (Fig. 10) and pyrene (Fig. 11) suggested for *Mycobacterium* sp. PYR-1, which largely coincide with the identified metabolites of the other organisms, are, therefore, presented at this point. However, during this metabolization, the formation of metabolites such as 4-hydroxy perinaphthenon (HEITKAMP et al., 1988b) or 7-methoxy-8-hydroxy fluoranthene, which cannot be further degraded, also partly occurs (KELLEY et al., 1991).

The initial attack of fluoranthene is conducted by a dioxygenase, leading after the rearomatization of the intermediate products to dihydroxylated compounds in the 7,8' the 9,10', and in particular the 1,2-position (Fig. 10). This diol compound undergoes extradiol cleavage and is then oxidized to 9-fluorenon-1-carboxyl acid, with glyoxylate being split off (KELLEY et al., 1991). Depending on the pH, this substance is in a state of equilibrium with the corresponding hydroxylated compound (CERNIGLIA, 1992). These two intermediately accumulating compounds are the product of elimination of the first ring, and are analogous to the hydroxy naphthoic acids from the degradation pathways of the anthracene or phenanthrene. They are subsequently decarboxylated and

then transferred to the central metabolism via other, as yet unidentified degradation pathways, which may be related to the degradation of fluorene.

2.2.1.6 Pyrene

As shown for fluoranthene, the initial attack of pyrene by a dioxygenase occurs at various points of the molecule and leads after re-aromatization to 1,2- or 4,5-pyrene-diol (Fig. 11). In addition, *trans*-4,5-pyrene-dihydrodiol which is not further converted is formed to a lesser extent by a cytochrome P-450-dependent monooxygenase (HEITKAMP et al., 1988b). The 1,2-diol is oxidized to form 4-hydroxy perinaphenon, which also is not further metabolized. In contrast to other degradation pathways, 4,5-pyrene-diol undergoes intradiol cleavage, which leads after decarboxylation to 4-phenanthroic acid as the product of the first ring elimination. This compound accumulates as well and has also been detected in sediments (HEITKAMP and CERNIGLIA, 1988). After further dioxygenation in 4,5-position and subsequent decarboxylation, the 1-hydroxy-2-naphthoic acid arising is further broken down via the two degradation pathways of phenanthrene (see also Fig. 9). One of the degradation pathways resembles the Kiyohara pathway, whereas in the other pathway the next step initially takes place via the Evans pathway (*Aeromonas* sp.) form 1,2-dihydroxy naphthalene. This compound, however, is then further degraded after intradiol cleavage (CERNIGLIA, 1992). By contrast, pyrene degradation in *Rhodococcus* sp. occurs via other routes, because after the at first equivalent initial oxidation products, other metabolites emerge from the extradiol ring cleavage of the 1,2- and 4,5-dihydroxy pyrene (small arrows at 1,2- and 4,5-dihydroxy pyrene in Fig. 11), leading to *cis*-2-hydroxy-3-(perinaphthenon-9-yl)-propionic acid and 2-hydroxy-2-(phenanthrene-5-on-4-enyl)-acetic acid, and the corresponding cyclic lactones (WALTER et al., 1991).

2.2.2 Cometabolic Transformation

Cometabolic degradation of PAH has been described for bacteria and also for a variety of

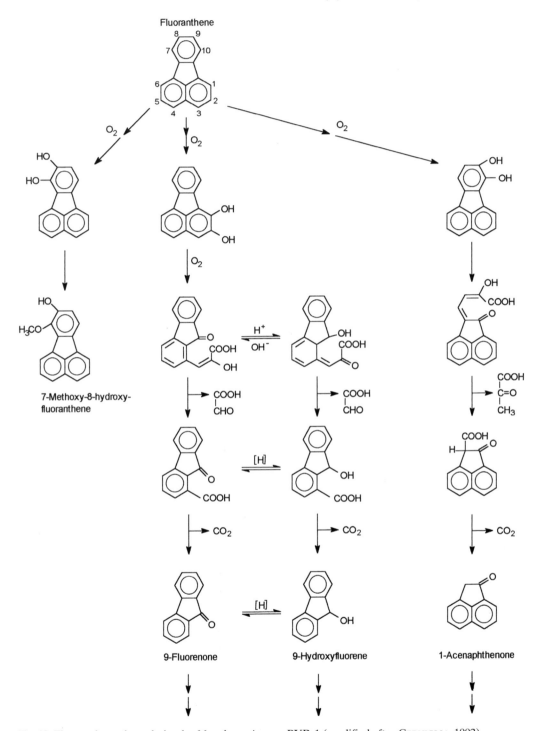

Fig. 10. Fluoranthene degradation by *Mycobacterium* sp. PYR-1 (modified after CERNIGLIA, 1992).

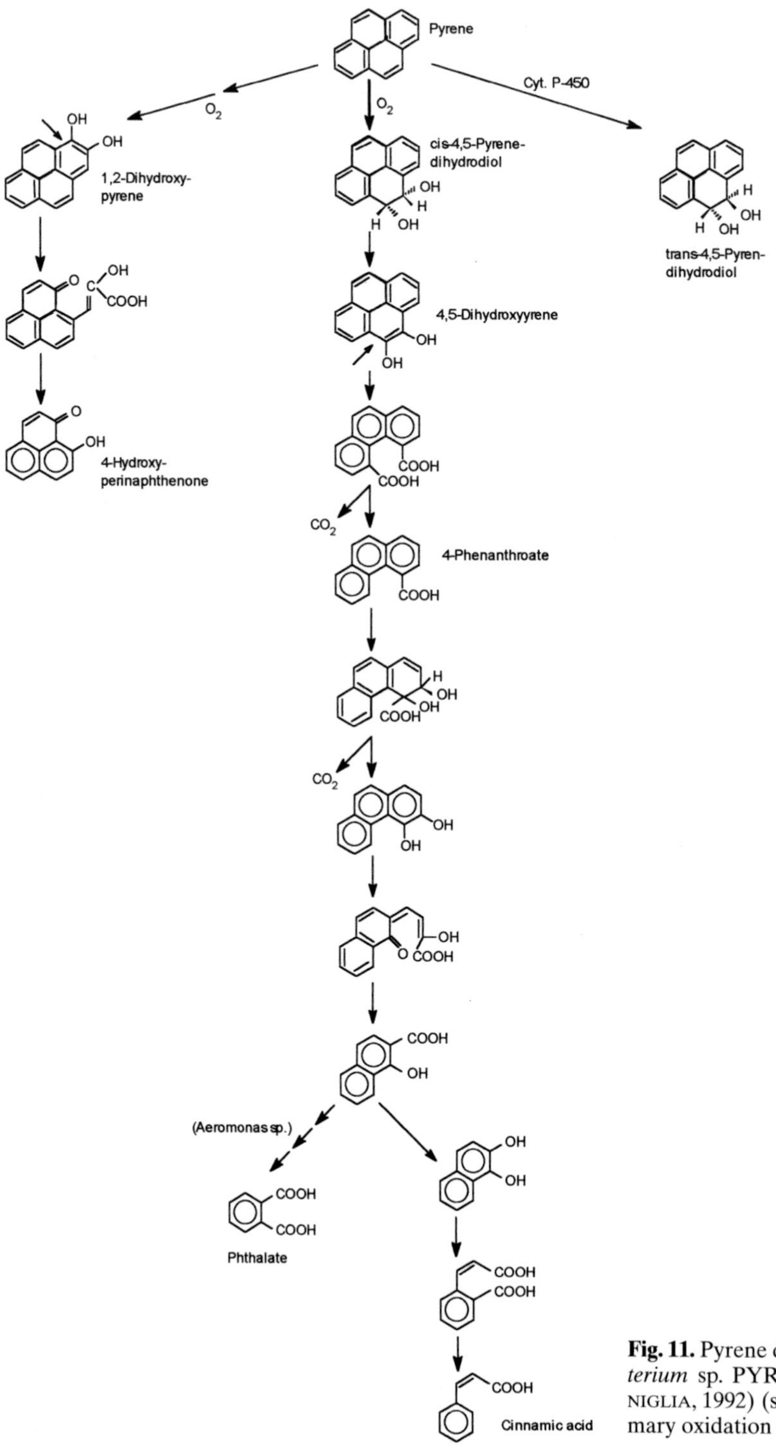

Fig. 11. Pyrene degradation by *Mycobacterium* sp. PYR-1 (modified after CERNIGLIA, 1992) (small arrows indicate primary oxidation by *Rhodococcus* sp.).

fungi (CERNIGLIA et al., 1992). The main criterion for this type of degradation is that the corresponding microorganisms cannot use the compounds as sole source of carbon and energy. This type of degradation is, apart from a detoxification function, of no identifiable benefit or growth advantage for the organisms and is not restricted to PAH compounds with 2- to 4-ring systems. Therefore, the organisms depend on additional carbon and energy sources in order to maintain their metabolism or to grow. Practically all the degradation sequences presented (see Figs. 7–11) can be considered for this type of degradation, as long as they do not lead to growth. Usually, the primary oxidation products of PAH or the following products of the first ring degradation arise as metabolites which are not further transformed. More recent findings show, that PAH with high molecular weights previously thought to be highly resistant to microbial degradation were biodegraded to an extent of up to 95% (SCHNEIDER et al., 1996; KANALY et al., 1997). Compounds such as benzo(a)pyrene, benzo(b)fluoranthene, and dibenz(a,h)anthracene are metabolized and mineralized in parts by *Mycobacterium* and *Sphingomonas* sp. (SCHNEIDER et al., 1996; YE et al., 1996). 4,5-chrysene dicarboxylic acid, 7,8-dihydro pyrene-7' and -8-carboxylic acids were observed as the products of the first ring degradation (SCHNEIDER et al., 1996). However, these experiments were conducted with very high amounts of biomass in the presence of a few mg L^{-1} of the PAH and the kinetics are very slow.

In addition, all direct oxidation products and derived compounds of cytochrome P-450-dependent monooxygenases can be classified under this type of metabolization (see Fig. 6). In particular with higher condensed PAH, degradation comes to a standstill after initial oxidation, causing hydroxylated or quinoid derivatives to accumulate. This means that abundance of metabolites is to be expected from cometabolic metabolism, as has been thoroughly documented by CERNIGLIA et al. (1992). The formation of tetraols previously only described for mammalian cells has also been demonstrated with the formation of 8,9,10,11-tetrahydro tetradiol from benz(a)anthracene by *Cuninghamella elegans* (CERNIGLIA et al., 1994). In addition, more recent findings show that methoxy derivatives such as 1-methoxy pyrene and 1,6-dimethoxy pyrene (WUNDER et al., 1997) as well as 1-methoxy phenanthrene and 1-methoxy pyrene (SACK et al., 1997c) are also formed by some species of fungi from the hydroxylated, primary oxidation products. It has been shown as well that non-ligninolytic fungi can form quinones such as 1,6- and 1,8-pyrene quinone (LAMBERT et al., 1994) or 1,6- and 1,8-benz(a)pyrene quinone (LAUNEN et al., 1995). Furthermore, the conversion of anthracene, fluorene, and pyrene by cytochrome P-450-dependent monooxygenases has also been described for ligninolytic fungi (*Pleurotus ostreatus*), leading to metabolites which have previously been found only with non-ligninolytic fungi (BEZALEL et al., 1996b). During the conversion of phenanthrene with *P. ostreatus*, the formation of *trans*-9,10-phenanthrene dihydrodiol and 2,2'-dicarboxy biphenyl was observed. The formation of these compounds could be suppressed by cytochrome P-450 inhibitors (BEZALEL et al., 1996a). Metabolites of this degradation have so far only been described for conversion by ligninolytic enzymes (see Sect. 2.2.3). Inversely, it may be possible to conclude from this that cytochrome P-450-dependent monooxygenases also play a greater role in lignin degradation than previously thought.

After cometabolic transformation of PAH, a toxicologically relevant potential has to be expected as accumulating metabolites are usually released into the environment by the microorganisms. Special concern should arise for the *trans*-dihydrodiol metabolites of cytochrome P-450-dependent monooxygenase reactions. However, the exact regiospecific and stereospecific characteristics of metabolites have been found to be of decisive importance for the actual mutagenic and carcinogenic potential of the compounds. By examining the mutagenicity of metabolically active extracts of the fungus *Cuninghamella elegans* in the Ames test, it was demonstrated that the fungal extract (in contrast to an S9-liver cell extract) was unable to metabolically activate the PAH. On the contrary, there is even a direct correlation between the conversion of benzo(a)pyrene and 7,12-dimethyl-benz(a)anthracene, and the reduction of toxicity (CERNIGLIA et al.,

1985). 8*R*,9*R*-Dihydrodiol-benz(a)-anthracene and its other oxidation products formed from mammalian cells of benzanthracene were highly mutagenic, whereas 8*S*,9*S*-dihydrodiol and other products formed from *Cuninghamella elegans* do not have any mutagenic effect (YANG, 1988). A fundamental difference between the effect of fungal cell extracts and mammalian cell extracts of PAH is that the stereochemistry of the two PAH transformations is different. Whereas mammalian cells mainly convert PAH into the potentially mutagenic *R*,*R*-dihydrodiols, fungi usually produce dihydrodiols with *S*,*S*-configuration (CERNIGLIA et al., 1990, 1992). Other detoxification mechanisms of fungi for PAH and their metabolites take place via formation of water soluble sulfate, glucuronide and glycoside conjugates (CERNIGLIA and HEITKAMP, 1989; LANGE et al., 1994; POTHULURI et al., 1990, 1992, 1995; SUTHERLAND et al., 1991; WUNDER et al., 1997). Water soluble conjugates can be precipitated during the course of further metabolization. The total metabolization of PAH in fungi is evidently directed more towards detoxification than to metabolic activation to form mutagens (CERNIGLIA et al., 1990, 1992).

2.2.3 Non-Specific Oxidation by Radicals

Another possibility for the degradation of PAH is non-specific oxidation by radicals. This type depends on the properties of ligninolytic enzyme systems of white-rot fungi. Coating cellulose and hemicellulose, water insoluble lignins create the firmness and durability component of wood in higher plants. Lignin is an irregular, three-dimensional, aromatic heteropolymer. Depending on the plants, lignin is formed from various substituted cinnamyl alcohols in a random polymerization reaction initiated by peroxidases (KIRK, 1984). Owing to the macromolecular structure and water insolubility of lignin, degradation can only occur by extracellular enzymes. With these enzyme systems [lignin, Mn peroxidases (LiP, Mn) and laccases], whose reaction mechanisms have not yet been elucidated in detail, the fungi are able to depolymerize lignin via radical catalyzed cleavage reactions. According to current mod-

els (BOOMINATHAN and REDDY, 1992), radical cations are initially produced by the enzymes such as veratryl alcohol in the case of LiP or chelated Mn^{3+} in the case of MnP. The radicals diffuse from the enzymes into the macromolecular structure and catalyze the cleavage reactions there. Thus, such compounds mediate between the enzymes and the substrate to be degraded, which no longer comes into contact with the enzyme. Recent publications show that in the presence of mediators laccases also can depolymerize lignin (BOURBONNAIS et al., 1995; YOUN et al., 1995). The non-specific oxidation reactions are ideal tools for biological oxidative attack of persistent environmental pollutants such as PAH. Using the white-rot fungus *Phanerochaete chrysosporium*, it was shown that under ligninolytic conditions tetrachlorodibenzodioxin, benzo(a)pyrene and polychlorinated biphenyls can be metabolized and mineralized in parts (BUMPUS et al., 1985; EATON, 1985; AUST, 1990; RAJ et al., 1992; FIELD et al., 1993; BARR and AUST, 1994). More recently, this non-specific oxidation has also been found for other environmental contaminants from various groups of pollutants (FIELD et al., 1993). *Phanerochaete chrysosporium* expresses the greatest activity of LiP and MnP under conditions of N-deficiency and ought not to express any laccase activity (HATAKKA, 1994). Owing to its easy cultivation and rapid growth, this organism has become a model organism for investigations into the metabolism of lignin and environmental chemicals by white-rot fungi. This may have led to a restricted understanding of the ligninolytic system and the metabolization of environmental pollutants. For example, recent findings show that other species of white-rot fungi such as *Bjerkandera adusta* are also ligninolytically active when there is a sufficient N-supply, and sometimes even develop a higher activity (FIELD et al., 1992; KAAL et al., 1993, 1995). In addition, high activities of laccase have also been found for *Phanerochaete chrysosporium* under altered cultivation conditions (SRINIVASAN et al., 1995). However, further investigations are required to understand the reaction processes of ligninolytic enzyme systems (which can be regarded as "radical transfer systems") in detail.

Various publications have shown the conversion of PAH *in vitro* with purified lignin peroxidases, LiP (HAEMMERLI et al., 1986; SANGLARD et al., 1986; HAMMEL et al., 1986, 1991, 1992). An important criterion for the oxidation of PAH by LiP is the ionization potential. It has been shown for PAH that LiP of *P. chrysosporium* can oxidize PAH up to a potential of 7.55 eV, thereby exceeding the abilities of other peroxidases. Compounds with a higher potential (phenanthrene, chrysene and benz(e)pyrene) were not oxidized (HAMMEL et al., 1986). However, the living organism is easily able to oxidize and mineralize phenanthrene even under non-ligninolytic conditions (DHAWALE et al., 1992). Therefore, other mechanisms for the oxidation of PAH have to be assumed in white-rot fungi which are not based on the activity of lignin peroxidases (SUTHERLAND et al., 1991, 1993; HAMMEL et al., 1992). Using *Pleurotus ostreatus* it was even shown that cytochrome P-450-dependent monooxygenases are responsible for this reaction (BEZALEL et al., 1996a), whereas in the case of *P. chrysosporium* MnP-dependent lipid peroxidation has been identified as the catalyzing mechanism (MOEN and HAMMEL, 1994). Recently, the oxidation of fluorene via this mechanism has also been demonstrated (BOGAN et al., 1996). Furthermore, phenanthrene was metabolized by manganese-dependent peroxidases of *Nematoloma frowardii* after addition of a suitable secondary mediator, and phenanthrene was even mineralized to a greater extent directly by the activity of this enzyme (SACK et al., 1997a). Much higher ionization potentials may also be reached by varying the mediators in the reactions catalyzed by ligninolytic enzymes. The oxidation of anthracene and benz(a)pyrene was also enabled with laccases in the presence of mediators such as ABTS [2,2'-amino(3-phthylene benzothiazoline-6-sulfonate)] or hydroxy benzotriazole (COLLINS et al., 1996; JOHANNES et al., 1996). This shows that under suitable conditions all ligninolytic enzymes are able to oxidize PAH. Hence, most of the abilities previously attributed only to lignin peroxidases should be nowadays attributed to all ligninolytic enzymes.

The proposed oxidation mechanism of benz(a)pyrene by LiP is shown in Fig. 12. The benz(a)pyrene radical cations arising from the reaction with the veratryl alcohol radical cations produced by LiP via 1-e$^-$ transitions can be stabilized by mesomerism, and lead via quenching reactions with H_2O to quinones (HAEMMERLI et al., 1986). Comparable mechanisms have also been proposed for the oxidation of other environmental pollutants. As quenching reactions may lead to addition products with other compounds, such aromatic radicals stabilized by mesomerism are regarded as possible initial reactions in the genesis of humic substances and the formation of bound residues (see Chapter 4, this volume).

The primary oxidation products of several PAH found after action of ligninolytic enzymes of *P. chrysosporium* are summarized in Fig. 13. As a rule of thumb, these products are mostly quinoid compounds. In addition, ring cleavage and further degradation also occur in the case of anthracene and phenanthrene. By contrast, *Bjerkandera* sp. metabolize anthracene to form mainly 9,10-anthraquinone – a metabolite which does not occur at all with *Trametes* sp. (FIELD et al., 1992). The degradation of PAH by ligninolytic enzymes initially leads to the formation of primary oxidation products in the medium, which either accumulate or immediately react in additional reaction sequences depending on the enzymes and mediators of the respective fungi. However, the direct formation of CO_2 via unspecific oxidations (SACK et al., 1997a; HOFRICHTER et al., 1998) remains the exception. Therefore, in most degradation experiments, far more PAH were metabolized than were actually converted into CO_2. More than 95% of ^{14}C-marked benz(a)pyrene were metabolized in cultures of *P. chrysosporium*, whereas only 19% were found as $^{14}C-CO_2$ (SANGLARD et al., 1986). Also in the case of phenanthrene mineralization was about 8% while metabolization was around 61–75% (BUMPUS, 1989).

Decomposition by white-rot fungi was also observed if PAH were present in complex matrices of anthracene oil. Apart from two exceptions, PAH components from a mixture were degraded to residual amounts of less than 20% (BUMPUS, 1989). During remediation actions of soil contaminated by tar oil, the addition of white-rot fungi on straw led to the elimination of more than 50% of the PAH, whereas the mineralization of added ^{14}C anthracene amounted just to 2.5% (MAJCHERCZYK et al., 1993). However, more recent findings show

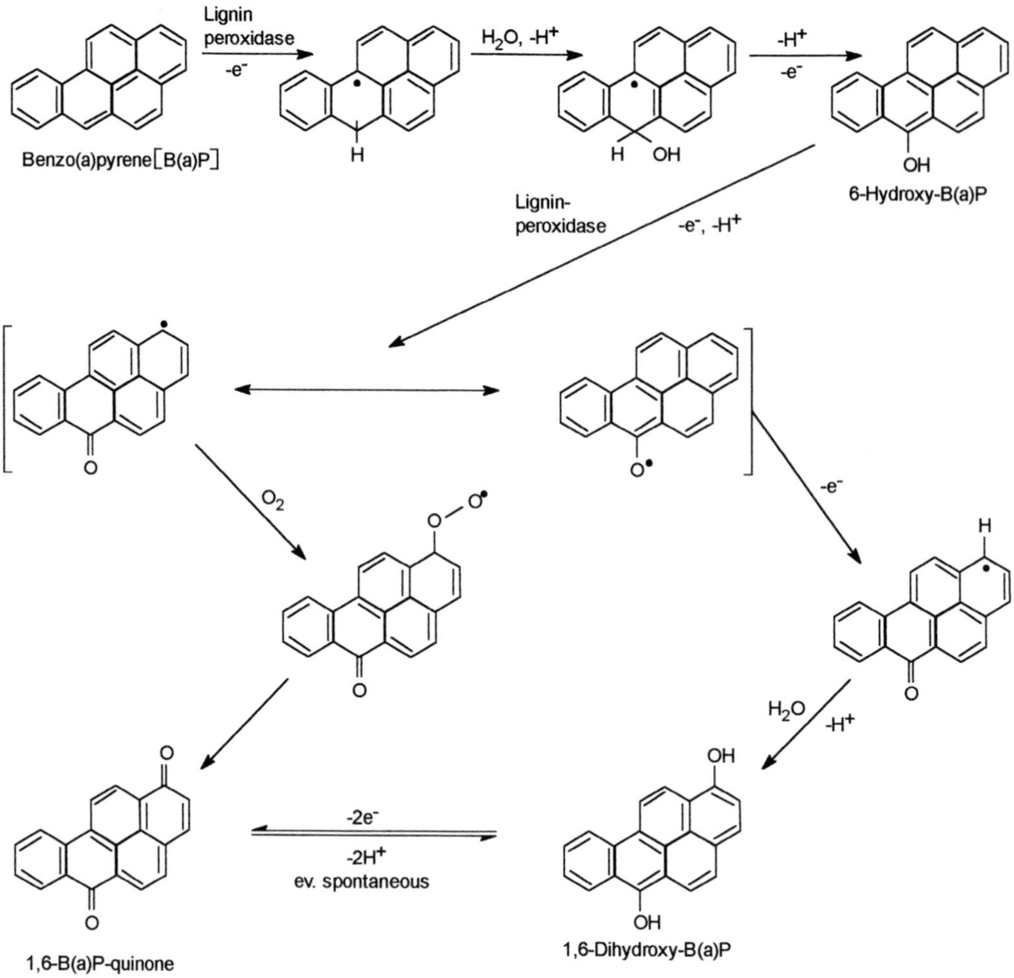

Fig. 12. Proposed mechanism of benzo(a)pyrene oxidation lignin peroxidase generated radicals (modified after HAEMMERLI et al., 1986).

that ligninolytic fungi may act synergistically together with the autochthonous soil microflora. The mineralization in the soil mixture with the fungus is significantly higher than found only with the fungus or the soil alone (IN DER WIESCHE et al., 1996; SACK and FRITSCHE, 1997). Evidently, a higher degradation potential can be mobilized by synergistic effects of white-rot fungi and indigenous microorganisms, which is in accordance with the ecological function of white-rot fungi in the natural systems. Moreover, the activity of ligninolytic enzyme systems appears to be much lower without the presence of lignin compounds. In comparison to liquid cultures, the amount of mineralized PAH is considerably higher when the components are adsorbed on straw (SACK et al., 1997b).

Concerning the toxicological effect by metabolites from non-specific oxidation by radicals, the same basical considerations are valid as for metabolites of cometabolic degradation. Moreover, the metabolites stemming from radical mechanisms may exert toxic effects

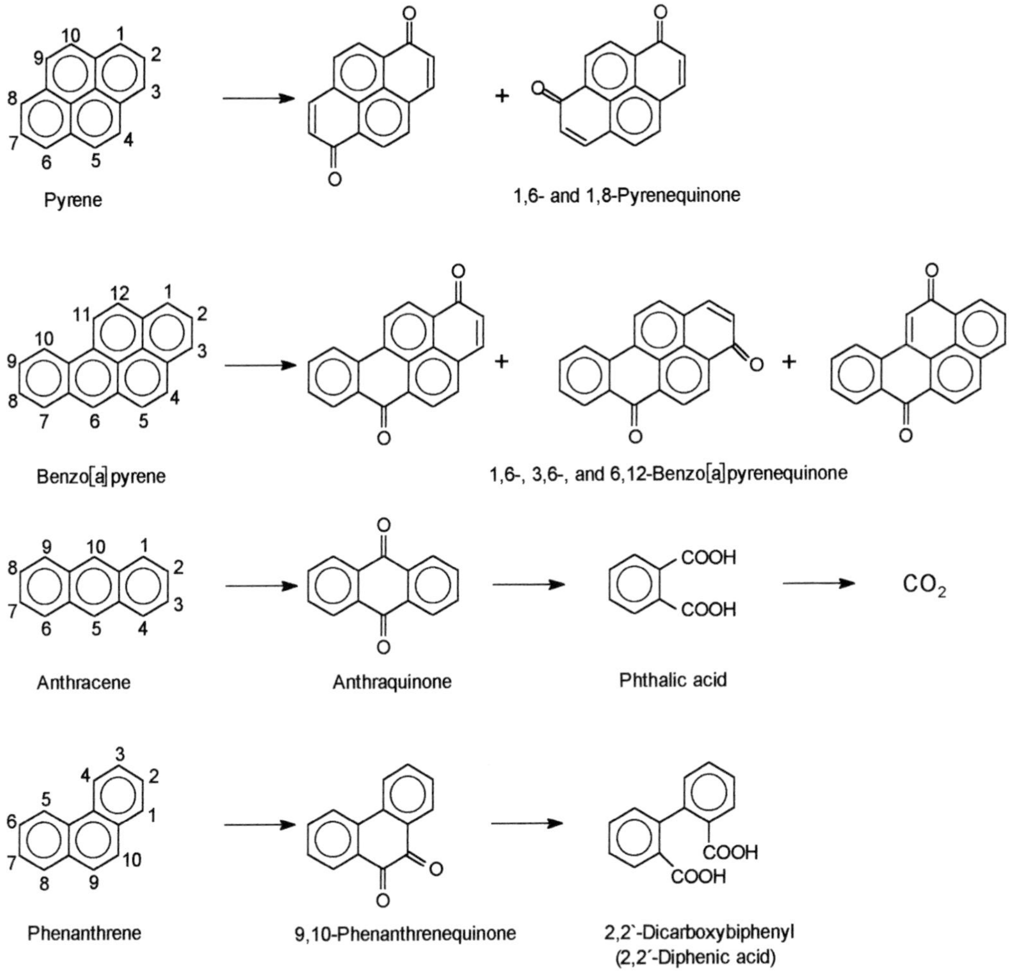

Fig. 13. Oxidation of several PAH by the white-rot fungus *Phanerochaete chrysosporium* (modified from SUTHERLAND et al., 1995).

during the process of formation. Furthermore, quinones as the main metabolites of non-specific oxidations are also able to develop mutagenic effects during the Ames test (CHESIS et al., 1984), and hence are also of ecotoxicological relevance. Plant microsomes are also able to activate benz(a)pyrene metabolically to form benz(a)pyrene quinone (VON DER TRENK and SANDERMANN, 1980; SANDERMANN, 1988). Older publications show that benz(a)pyrene quinone and other environmental pollutants

can copolymerize during the formation of synthetic lignins (VON DER TRENK and SANDER-MANN, 1981; SANDERMANN et al., 1983). These metabolites can, therefore, be observed as residues in cell walls of plants. These findings show that quinones can be incorporated during radically initiated polymerization reactions of aromatic alcohols. Owing to the activity of ligninolytic enzymes in soils, PAH may also undergo addition reactions to lignin fragments or humic substances.

3 Summary and Concluding Remarks

Monocyclic aromatic hydrocarbons are metabolized aerobically by a great variety of microorganisms and several classes of enzymes involved. The general principles of microbial attack on aromatic structures are:

(1) extensive modification or elimination of the ring substituents,
(2) introduction of molecular oxygen to form adjacent hydroxyl groups,
(3) cleavage of the aromatic ring,
(4) further degradation of the non-cyclic ring fission products to intermediates of central metabolic pathways.

Depending on the patterns of substituents most aromatic compounds are completely metabolized.

Polycyclic aromatic hydrocarbons are metabolized by a large number of enzymes in bacteria, fungi, algae and cytochrome P-450 monooxygenases in mammalian cells. The ability to degrade PAH occurs in various groups and even in thermophilic and anaerobic microorganisms. The microbial degradation of PAH can be classified under physiological aspects of the organisms. Complete metabolization and mineralization leads to the formation of biomass and mineralization products (CO_2 and H_2O). Bacteria and some algae are also able to utilize PAH (complete metabolization) with specific dioxygenases to form *cis*-dihydrodiols. This degradation proceeds via formation of a *cis*-dihydrodiol, dehydrogenation to form dihydroxy PAH, extradiol ring cleavage, release of C3 or C2 and C1 compounds, decarboxylation of the residual aromatic hydroxy carbon acids to form *o*-dihydroxy compounds, extradiol ring cleavage and degradation of the second ring, and so on. Cometabolic transformation leads only to the formation of metabolites that may accumulate in the environment. PAH are cometabolically oxidized by monooxygenases in prokaryotes and eukaryotes to form arene oxides which are enzymatically hydrolyzed into *trans*-dihydrodiols or non-enzymatically converted into phenols. Such metabolites can form conjugates with cell

components and other organic molecules in the environment. Non-specific oxidation by radicals due to the activity of ligninolytic enzymes of white-rot fungi leads primarily to the formation of reactive metabolites which tend to couple to other surrounding organic molecules. The type of degradation determines the carbon flux, the formation of metabolites and bound residues in the environment (see Chapter 4, this volume).

4 References

ARVIN, E., JENSEN, B. K., GUNDERSEN, A. T. (1989), Substrate interactions during the aerobic degradation of benzene, *Appl. Environ. Microbiol.* **55**, 3221–3225.

AUST, S. D. (1990), Degradation of environmental pollutants by *Phanerochaete chrysosporium*, *Microb. Ecol.* **20**, 197–209.

AXELL, B. C., GEARY, P. J. (1975), The metabolism of benzene by bacteria, *Biochem. J.* **136**, 927–934.

BAGGI, G., BARBIERI, P., GALLI, E., TOLLARI, S. (1987), Isolation of a *Pseudomonas stutzeri* strain that degrades *o*-xylene, *Appl. Environ. Microbiol.* **53**, 2129–2132.

BARR, D. P., AUST, S. D. (1994), Mechanisms white-rot fungi use to degrade pollutants, *Environ. Sci. Technol.* **28**, 79A–87A.

BAUER, J. E., CAPONE, D. G. (1988), Effects of co-occurring aromatic hydrocarbons on the degradation of individual polycyclic aromatic hydrocarbons in marine sediment slurries, *Appl. Environ. Microbiol.* **54**, 1649–1655.

BECK, A. J., ALCOCK, R. E., WILSON, S. C., WANG, M. J., WILD, S. R. et al. (1995), Long-term persistence of organic chemicals in sewage sludge-amended agricultural land: A soil quality perspective, *Adv. Agron.* **55**, 345–391.

BEZALEL, L., HADAR, Y., FU, P., FREEMAN, J. P., CERNIGLIA, C. E. (1996a), Metabolism of phenanthrene by the white-rot fungus *Pleurotus ostreatus*, *Appl. Environ. Microbiol.* **62**, 2547–2553.

BEZALEL, L., HADAR, Y., FU, P., FREEMAN, J. P., CERNIGLIA, C. E. (1996b), Initial oxidation products in the metabolism of pyrene, anthracene, fluorene, and dibenzothiophene by the white-rot fungus *Pleurotus ostreatus*, *Appl. Environ. Microbiol.* **62**, 2554–2559.

BJØRSETH, A., RAMDAHL, T. (1985), Sources and emissions of PAH, in: *Handbook of Polycyclic Aromatic Hydrocarbons* (BJØRSETH, A., RAM-

DAHL, T., Eds.), pp. 1–20. New York: Marcel Dekker.

BOGAN, B. W., LAMAR, R. T., HAMMEL, K. E. (1996), Fluorene oxidation *in vivo* by *Phanerochaete chrysosporium* and *in vitro* during manganese peroxidase-dependent lipid peroxidation, *Appl. Environ. Microbiol.* **62**, 1788–1792.

BOLDRIN, B., THIEM, A., FRITSCHE, C. (1993), Degradation of phenanhtrene, fluorene, fluoranthene, and pyrene by a *Mycobacterium* sp., *Appl. Environ. Microbiol.* **59**, 1927–1930.

BOOMINATHAN, K., REDDY, A. (1992), Fungal degradation of lignin: biotechnological applications, in: *Handbook of Applied Mycology* Vol. 4 (ARORA, D. K., ELANDER, R. P., MUKERJI, K. G., Eds.), pp. 763–822. New York: Marcel Dekker.

BOURBONNAIS, R., PAICE, M. G., REID, I. D., LANTHIER, P., YAGUCHI, M. (1995), Lignin oxidation by laccase isoenzymes from *Trametes versicolor* and role of the mediator 2,2′-azinobis(3-ethylebenzothiazoline-6-sulfonate) in kraft lignin depolymerization, *Appl. Environ. Microbiol.* **61**, 1876–1880.

BUMPUS, J. A. (1989), Biodegradation of polycyclic aromatic hydrocarbons by *Phanerochaete chrysosporium*, *Appl. Environ. Microbiol.* **55**, 154–158.

BUMPUS, J. A., TIEN, M., WRIGHT, D., AUST, S. D. (1985), Oxidation of persistent environmental pollutants by a white-rot fungus, *Science* **228**, 1434–1436.

CASELLAS, M., GRIFOLL, M., BAYONA, J. M., SOLANAS, A. M. (1997), New metabolites in the degradation of fluorene by *Arthrobacter* sp. strain F101, *Appl. Environ. Microbial.* **63**, 819–826.

CERNIGLIA, C. E. (1984), Microbial degradation of polycyclic aromatic hydrocarbons, *Adv. Appl. Microbiol.* **30**, 31–71.

CERNIGLIA, C. E. (1992), Biodegradation of polycyclic aromatic hydrocarbons, *Biodegradation* **3**, 351–368.

CERNIGLIA, C. E. (1993), Biodegradation of polycyclic aromatic hydrocarbons, *Curr. Opin. Biotechnol.* **4**, 331–338.

CERNIGLIA, C. E., HEITKAMP, M. A. (1989), Microbial metabolism of polycyclic aromatic hydrocarbons (PAH) in the aquatic environment, in: *Metabolism of Polycyclic Aromatic Hydrocarbons in the Aquatic Environment* (VARANASI, U., Ed.), pp. 41–68. Boca Raton, FL: CRC Press.

CERNIGLIA, C. E., WHITE, G. L., HEFLICH, R. H. (1985), Fungal metabolism and detoxification of polycyclic aromatic hydrocarbons, *Arch. Microbiol.* **143**, 105–110.

CERNIGLIA, C. E., CAMPBELL, W. L., FU, P. P., FREEMANN, J. P., EVANS, F. E. (1990), Stereoselective fungal metabolism of methylated anthracenes, *Appl. Environ. Microbiol.* **56**, 661-668.

CERNIGLIA, C. E., SUTHERLAND, J. B. CROW., S. A. (1992), Fungal metabolism of aromatic hydrocarbons, in: *Microbial Degradation of Natural Products* (WINKELMANN, G., Ed.), pp. 193–217. Weinheim: VCH.

CERNIGLIA, C. E., GIBSON, D. T., DODGE, R. H. (1994), Metabolism of benz(a)anthracene by the filamentous fungus *Cuninghamella elegans, Appl. Environ. Microbiol.* **60**, 3931–3938.

CHESIS, P. L., LEVIN., D. E., SMITH, M. T., ERNSTER, L., AMES, B. N. (1984), Mutagenicity of quinones: Pathways of metabolic activation and detoxification, *Proc. Natl. Acad. Sci. USA* **81**, 1696–1700.

COLLINS, P. J., KOTTERMANN, M. J., FIELD, J. A., DOBSON, A. D. W. (1996), Oxidation of anthracene and benzo(a)pyrene by laccases from *Trametes versicolor, Appl. Environ. Microbiol.* **62**, 4563–4567.

DAGLEY, S. (1986), Biochemistry of aromatic hydrocarbon degradation in Pseudomonads, in: *The Bacteria* Vol. 10 (SOKATCH, J. R., Ed.), pp. 527–555. New York, London: Academic Press.

DHAWALE, S. W., DHAWALE, S. S., DEAN-ROSS D. (1992), Degradation of phenanthrene by *Phanerochaete chrysosporium* occurs under ligninolytic as well as non-ligninolytic conditions, *Appl. Environ. Microbiol.* **58**, 3000–3006.

DUGGLEBY, C. J., WILLIAMS, P. A. (1986), Purification and some properties of the 2-hydroxy-6-oxo-hepta-2,4-dienoate hydrolase (2-hydroxymuconic semialdehyde hydrolase) encoded by the TOL plasmid pWWO from *Pseudomonas putida* mt 2, *J. Gen. Microbiol.* **132**, 717–726.

EATON, D. C. (1985), Mineralization of polychlorinated biphenyls by *Phanerochaete chrysosporium:* a ligninolytic fungus, *Enzyme Microb. Technol.* **7**, 194–196.

EVANS, W. C., FERNLEY, H. N., GRIFFITHS, E. (1965), Oxidative metabolism of phenanthrene and anthracene by soil pseudomonads, *Biochem. J.* **95**, 819–831.

FEITKENHAUER, H., HEBENBROOK, S., TERSTEGEN, L., SCHNICKE, S., SCHÖB, T. et al. (1996), Bodenreinigung mit thermophilen Mikroorganismen, in: *Neue Techniken der Bodenreinigung, Hamburger Berichte Abfallwirtschaft* Vol. 10 (STEGMANN, R., Ed.), pp. 361–376. Bonn: Economica Verlag.

FIELD, J. A., DE JONG, E., COSTA, G. F., DEBONT, J. A. M. (1992), Biodegradation of polycyclic aromatic hydrocarbons by new isolates of white-rot fungi, *Appl. Environ. Microbiol.* **58**, 2219–2226.

FIELD, J. A., DE JONG, E., COSTA, G. F., DEBONT, J. A. M. (1993), Screening for ligninolytic fungi applicable to the biodegradation of xenobiotics, *Trends Biotechnol.* **11**, 44–49.

FUCHS, G. (1999), Oxidation of organic compounds, in: *Biology of the Prokaryotes* (LENGLER, J. W., DREWS, G., SCHLEGEL, H. G., Eds.), pp. 187–232. Stuttgart: Thieme Verlag.

GIBSON, D. T., SUBRAMANIAN, V. (1984), Microbial degradation of aromatic hydrocarbons, in: *Micro-*

bial Degradation of Organic Compounds (GIBSON, D. T., Ed.), pp. 361–369. New York: Marcel Dekker.

GIBSON, D. T., KOCH, J. R., KALLIO, R. E. (1968), Oxidative degradation of aromatic hydrocarbons by microorganisms. I. Enzymatic formation of catechol from benzene, *Biochemistry* **7**, 2643–2656.

GIBSON, D. T., CARDINI, G. E., MAESELS, F. C., KALLIO, R. E. (1970), Incorporation of 18O into benzene, *Biochemistry* **9**, 1631–1636.

GRIFOLL, M., CASELLAS, M., BAYONY, J. M., SOLANAS, A. M. (1992), Isolation and characterization of a fluorene-degrading bacterium: identification of ring oxidation and ring fission products, *Appl. Environ. Microbiol.* **58**, 2910–2917.

GRIFOLL, M., SELIFONOV, S. A., CHAPMAN, P. J. (1994), Evidence for a novel pathway in the degradation of fluorene by *Pseudomonas* sp. strain F274, *Appl. Environ. Microbiol.* **60**, 2438–2449.

GRIFOLL, M., SELIFONOV, S. A., GATLIN, C. V., CHAPMAN, P. J. (1995), Action of a versatile fluorene-degrading bacterial isolate on polycyclic aromatic hydrocarbons, *Appl. Environ. Microbiol.* **61**, 3711–3723.

GROSSER, R. J., WARSHAWSKY, D., VESTAL, R. (1991), Indigeous and enhanced mineralization of pyrene, benzo[a]pyrene, and carbazole in soils, *Appl. Environ. Microbiol.* **57**, 3462–3469.

GUERIN, W. F., JONES, G. E. (1988a), Two stage mineralization of phenanthrene by estuarine enrichment cultures, *Appl. Environ. Microbiol.* **54**, 929–936.

GUERIN, W. F., JONES, G. E. (1988b), Mineralization of phenanthrene by a *Mycobacterium* sp., *Appl. Environ. Microbiol.* **54**, 937–944.

HAEMMERLI, S. D., LEISOLA, M. S. A., SANGLARD, D., FIECHTER, A. (1986), Oxidation of benzo(a)pyrene by extracellular ligninases of *Phanerochaete chrysosporium, J. Biol. Chem.* **261**, 6900–6903.

HAMMEL, K. E. (1995), Organopollutant degradation by ligninolytic fungi, in: *Microbial Transformation and Degradation of Toxic Organic Chemicals, Wiley Series in Ecological and Applied Microbiology* (YOUNG, L. Y., CERNIGLIA, K. E., Eds.), pp. 331–346. New York: Wiley Liss.

HAMMEL, K. E., KALYANARAMAN, B., KIRK, T. K. (1986), Oxidation of polycyclic aromatic hydrocarbons and dibenzo[p]dioxins by *Phanerochaete chrysosporium* ligninase, *J. Biol. Chem.* **261**, 16948–16952.

HAMMEL, K. E., GREEN, B., GAI, W. Z. (1991), Ring fission of anthracene by an eukaryote, *Proc. Natl. Acad. Sci. USA* **88**, 10605–10608.

HAMMEL, K E., GAI, W. Z., GREEN, B., MOEN, M. (1992), Oxidative degradation of phenanthrene by the ligninolytic fungus *Phanerochaete chrysosporium, Appl. Environ. Microbiol.* **58**, 1832–1838.

HASSETT, H. J., BANWART, W. L. (1989), The sorption of nonpolar organics by soils and sediments, in: *Reactions and Movement of Organic Chemicals in Soils. SSSA Special Publication No. 22, Soil Science Society of America, Inc.* (SAHWNEY, B. L., BROWN, K., Eds.), pp. 31–44. Madison, WI: American Society of Agronomy, Inc.

HATAKKA, A. (1994), Lignin-modifying enzymes from selected white-rot fungi: production and role in lignin degradation, *FEMS Microbiol. Rev.* **13**, 125–135.

HEITKAMP, M. A., CERNIGLIA, C. E. (1988), Polycyclic aromatic hydrocarbon degradation by a *Mycobacterium* sp., in microcosms containing sediment and water from a pristine ecosystem, *Appl. Environ. Microbiol.* **33**, 1968–1973.

HEITKAMP, M. A., FRANKLIN, W., CERNIGLIA, C. E. (1988a), Microbial metabolism of polycyclic aromatic hydrocarbons: Isolation and characterization of a pyrene-degrading bacterium, *Appl. Environ. Microbiol.* **54**, 1612–1614.

HEITKAMP, M. A., FREEMAN, J. P., MILLER, D. W., CERNIGLIA, C. E. (1988b), Pyrene degradation by a *Mycobacterium* sp: Identification of ring oxidation and ring fission products, *Appl. Environ. Microbiol.* **54**, 2556–2565.

HÖGN, T., JAENICKE, L. (1972), Benzene metabolism of *Moraxella* sp., *Eur. J. Biochem.* **30**, 369–375.

HOFRICHTER, M., SCHEIBNER, K., SCHNEEGAß, Y., FRITSCHE, W. (1998), Enzymatic combustion of aromatie and aliphatic compounds by manganese peroxidase from *Nematoloma frowardii, Appl. Environ. Microbiol.* **64**, 399–404.

HOOPER, D. J. (1978), Microbial degradation of aromatic hydrocarbons, in: *Developments in Biodegradation of Hydrocarbons* (WATKINSON, R. J., Ed.), pp. 85–112. London: Science Publishers.

JIMENEZ, I. Y., BARTHA, R. (1996), Solvent-augmented mineralization of pyrene by a *Mycobacterium* sp., *Appl. Environ. Microbiol.* **62**, 2311–2316.

JOHANNES, C., MAJCHEREZYK, A., HÜTTERMANN, A. (1996), Degradation of anthracene by laccase of *Trametes versicolor* in the presence of different mediator compounds, *Appl. Biotechnol. Microbiol.* **46**, 313–317.

JOHNSON, G. R., OLSEN, R. H. (1997), Multiple pathways for toluene degradation in *Burkholderia* sp. strain JS150, *Appl. Environ. Microbiol.* **63**, 4047–4052.

JONES, K. C., STRATTFORD, J. A., WATERHOUSE, K. S., FURLONG, E. T., GIGER, W. et al. (1989a) Increases in the polynuclear aromatic hydrocarbon content of an agricultural soil over the last century, *Environ. Sci. Technol.* **23**, 95–101.

JONES, K. C., STRATTFORD, J. A., WATERHOUSE, K. S., VOGT, N. B. (1989b), Organic contaminants in welsh soils: polynuclear aromatic hydrocarbons, *Environ. Sci. Technol.* **23**, 540–550.

KAAL, E. E, DEJONG, E., FIELD, J. A. (1993), Stimulation of ligninolytic peroxidase activity by nitrogen nutrients in the white rot fungus *Bjerkandera adusta* sp. strain BOS55, *Appl. Environ. Microbiol.* **59**, 4031–4036.

KAAL, E. E., FIELD, J. A., JOYCE, T. W. (1995), Increasing ligninolytic activities in several white-rot basidiomycetes by nitrogen-sufficient media, *Biores. Technol.* **53**, 133–139.

KANALY, R., BARTHA, R., FOGEL, S., FINDLAY, M. (1997), Biodegradation of [^{14}C] benzo[a]pyrene added in crude oil to uncontaminated soil, *Appl. Environ. Microbiol.* **63**, 4511–4515.

KARICKHOFF, S. (1981), Semi-empirical estimation of sorption of hydrophobic pollutants on natural sediments and soils, *Chemosphere* **10**, 833–846.

KARICKHOFF, S. W., BROWN, D. S., SCOTT, T. A. (1979), Sorption of hydrophobic pollutants of natural sediments, *Water Res.* **13**, 241–248.

KÄSTNER, M., MAHRO, B., WIENBERG, R. (Eds.) (1993), *Biologischer Schadstoffabbau in kontaminierten Böden (unter besonderer Berücksichtigung der Polyzyklischen Aromatischen Kohlenwasserstoffe PAK), Hamburger Berichte*, Vol. 5. Bonn: Economica Verlag.

KÄSTNER, M., BREUER-JAMMALI, M., MAHRO, B. (1994), Enumeration and characterization of the soil microflora from hydrocarbon-contaminated soil sites able to mineralize polycyclic aromatic hydrocarbons, *Appl. Microbiol. Biotechnol.* **41**, 267–273.

KELLEY, I., FEEMAN, J. P., CERNIGLIA, C. E. (1990), Identification of metabolites from degradation of naphthalene by a *Mycobacterium* sp., *Biodegradation* **1**, 283–290.

KELLEY, I., FREEMAN, J. P., EVANS, F. E., CERNIGLIA, C. E. (1991), Identification of a carboxylic acid metabolite from the catabolism of fluoranthene by a *Mycobacterium* sp., *Appl. Environ. Microbiol.* **57**, 636–641.

KIYOHARA, H., NAGAO, K., NOMI, R. (1976), Degradation of phenanthrene through *o*-phthalate by an *Aeromonas* sp., *Agric. Biol. Chem.* **40**, 1075–1082.

KIRK, T. K. (1984), Degradation of lignin, in: *Microbial Degradation of Organic Compounds* (GIBSON, D. T., Ed.), pp. 399–437. New York: Marcel Dekker.

KOCH, R., WAGNER, B. O. (Eds.) (1989), *Umweltchemikalien*. Weinheim: VCH.

LAMBERT, M., KREMER, S., STERNER, O., ANKE, H. (1994), Metabolim of pyrene by the basidiomycete *Crinipellis stipitaria* and identification of pyrenequinones and their hydroxylated prcursors in strain JK375, *Appl. Environ. Microbiol.* **60**, 3597–3601.

LANGE, B., KREMER, S., STERNER, O., ANKE, H. (1994), Pyrene metabolism in *Crinipellis stipitaria*: identification of *trans*-4,5-dihydro-4,5-dihy-

droxypyrene and 1-pyrenylsulfate in strain JK364, *Appl. Environ. Microbiol.* **60**, 3602–3607.

LAUNEN, L., PINTO, L., WIEBE, C., HIEHLMANN, E., MOORE, M. (1995), The oxidation of pyrene and benzo(a)pyrene by nonbasidiomycete soil fungi, *Can. J. Microbiol.* **41**, 477–488.

LA VOIE, E. J., RICE, J. E. (1988), Structure–acitivity relationships among tricyclic polynuclear hydrocarbons, in: *Polycyclic Aromatic Hydrocarbon Carcinogenesis: Structure–Activity Relationships* Vol. I (YANG, S. K., SILVERMAN, B. D., Eds.), pp. 151–176. Boca Raton, FL: CRC Press.

MAHAFFEY, W. R., GIBSON, D. T., CERNIGLIA, C. E. (1988), Bacterial oxidation of chemical carcinogens: formation of polycyclic aromatic acids from benz(a)anthracene. *Appl. Environ. Microbiol.* **54**, 2415–2423.

MAJCHERCZYK, A., ZEDDEL, A., KELSCHEBACH, M., LOSKE, D., HÜTTERMANN, A. (1993), Abbau von Polyzyklischen Aromatischen Kohlenwasserstoffen durch Weißfäulepilze, *Bioengineering* **9**, 27–31.

MARR, E. K., STONE, R. W. (1961), Bacterial oxidation of benzene, *J. Bacteriol.* **85**, 425–430.

MASAPHY, S., LEVANON, D., HENIS, Y., VENKATESWARLU, K., KELLY, S. L. (1996), Evidence for cytochrome P-450 and P-450 mediated benzo(a)pyrene hydroxylation in the white-rot fungus *Phanerochaete chrysosporium*, *FEMS Microbiol. Lett.* **135**, 51–55

MCCARTY, P. L., RITTMAN, B. E., BOUWER, E. J. (1984), Microbial processes affecting chemical transformations in groundwater, in: *Groundwater Pollution Microbiology* (BITTON, G., GERBA, C. P., Eds.), pp. 89–115. New York: John Wiley & Sons.

MCELROY, A. E., FARRINGTON, J. W., TEAL, J. M. (1989), Bioavailability of polycyclic aromatic hydrocarbons in the aquatic environment, in: *Metabolism of Polycyclic Aromatic Hydrocarbons in the Aquatic Environment* (VARANASI, U., Ed.), pp. 1–39. Boca Raton, FL: CRC Press.

MEANS, J. G., WOOD, S. G., HASSETT, J. J., BANWART, W. L. (1980), Sorption of polynuclear aromatic hydrocarbons by sediments and soils, *Environ. Sci. Technol.* **14**, 1524–1528.

MOEN, M. A., HAMMEL, K. E. (1994), Lipid peroxidation by the manganese peroxidase of *Phanerochaete chrysosporium* is the basis for phenanthrene oxidation by the intact fungus, *Appl. Environ. Microbiol.* **60**, 1956-1961.

MUELLER, J. G., CHAPMAN, P. J., PRITCHARD, P. H. (1989), Creosote contaminated sites: their potential for bioremediation, *Environ. Sci. Technol.* **23**, 1197–1201.

MUELLER, J. G., CHAPMAN, P. J., BLATTMAN, B. O., PRITCHARD, P. H. (1990), Isolation and characterization of a fluoranthene-utilizing strain of *Pseudomonas paucimobilis, Appl. Environ. Microbiol.*

56, 1079-1086.

POTHULURI, J. V., CERNIGLIA, C. E. (1994), Microbial metabolism of polycyclic aromatic hydrocarbons, in: *Biological Degradation of Toxic Chemicals* (CHAUDHRY, G. R., Ed.), pp. 92–123. Portland, OR: Dioscorides Press.

POTHULURI, J. V., FREEMAN, J. P., EVANS, F. E., CERNIGLIA, C. E. (1990), Fungal transformation of fluoranthene, *Appl. Environ. Microbiol.* **56**, 2974–2983.

POTHULURI, J. V., FU, P. P., CERNIGLIA, C. E. (1992), Fungal metabolism and detoxification of fluoranthene, *Appl. Environ. Microbiol.* **58**, 937–941.

POTHULURI, J. V., SELBY, A., EVANS, F. E., FREEMAN, J. P., CERNIGLIA, C. E. (1995), Transformation of chrysene and other polycyclic aromatic hydrocarbon mixtures by the fungus *Cuninghamella elegans, Can. J. Bot.* **73**, 1025–1033.

RAJ, H. G., SAXENA, M., ALLAMEH, A. (1992), Metabolism of foreign compounds by fungi, in: *Handbook of Applied Mycology* Vol. 4 (ARORA, D. K., ELANDER, R. P., MUKERJI, K. G., Eds.), pp. 881–904. New York: Marcel Dekker.

RAMDAHL, T. (1985), PAH emissions from combustion of biomass, in: *Handbook of Polycyclic Aromatic Hydrocarbons* (BJØRSETH, A., RAMDAHL, T., Eds.), pp. 61–85. New York: Marcel Dekker.

RICHARDSON, M. I., GANGOLLI, D. (1993), The dictionary of substances and their effects, 3rd Edn. London: Royal Society of Chemistry.

RIPPEN, G. (Ed.) (1994), *Handbuch Umweltchemikalien* Vols. 5, 6 (Loseblattsammlung). Landsberg: Ecomed-Verlag.

SACK, U., FRITSCHE, W. (1997), Enhancement of pyrene mineralization in soil by wood-decaying fungi, *FEMS Microbiol. Ecol.* **22**, 77–83.

SACK, U., HOFRICHTER, M., FRITSCHE, W. (1997a), Degradation of polycyclic aromatic hydrocarbons by manganese peroxidase of *Nematoloma frowardii, FEMS Microbiol. Lett.* **152**, 227–234.

SACK, U., HOFRICHTER, M., FRITSCHE, W. (1997b), Degradation of phenanthrene and pyrene by *Nematoloma forwardii, J. Bascic Microbiol.* **37**, 287–293.

SACK, U., HEINZE, T. M., DECK, J., CERNIGLIA, C. E., CAZAU, M. C., FRITSCHE, W. (1997c), Novel metabolites in phenanthrene and pyrene transformation by *Aspergillus niger, Appl. Environ. Microbiol.* **63**, 2906–2909.

SANDERMANN, H. (1988), Mutagenic activation of xenobiotics by plant enzymes, *Mutat. Res.* **197**, 183–194.

SANDERMANN, H., SCHEEL, D., VON DER TRENK, T. (1983), Metabolism of environmental chemicals by plants – copolymerization into lignin, *J. Appl. Polymer Sci.* **37**, 407–420.

SANGLARD, D., LEISOLA, M. S. A., FIECHTER, A. (1986), Role of extracellular ligninases in bio-

degradation of benzo(a)pyrene by *Phanerochaete chrysosporium, Enzyme Microb. Technol.* **8**, 209–212.

SANSEVERINO, J., APPLEGATE, B. M., KING, J. M. H., SAYLER, G. S. (1993), Plasmid-mediated mineralization of naphthalene, phenanthrene, and anthracene, *Appl. Environ. Microbiol.* **59**, 1931–1937.

SCHNEIDER, J., GROSSER, R., JAYASIMHULU, K., XUE, W., WARSHAWSKY, W. (1996), Degradation of pyrene, benz(a)anthracene, and benzo(a)pyrene by a *Mycobacterium* sp. strain RJGII-135, isolated from a former coal gasification site, *Appl. Environ. Microbiol.* **62**, 13–19.

SCHOCKEN, M. J., GIBSON, D. T. (1984), Bacterial oxidation of the polycyclic aromatic hydrocarbons acenaphthene and acenaphthylene, *Appl. Environ. Microbiol.* **48**, 10–16.

SCHRAA, G., BETHE, B. M., VAN NEERVEN, A. R. W., VAN DEN TWEL, W. J. J., VAN DER WENDE, E., ZEHNDER, A. J. B. (1987), Degradation of 1,2-dimethylbenzene by *Corynebacterium* strain C125, *Antonie van Leuwenhoek* **53**, 159–170.

SELIFONOV, S. A., GRIFOLL, M., GURST, J. E., CHAPMAN. P. J. (1993), Isolation and characterization of (+)-1,1α-dihydroxy-1-hydrofluoren-9-one formed by angular dioxygenation in the bacterial catabolism of fluorene, *Biochem. Biophys. Res. Commun.* **193**, 67–76.

SHUTTLEWORTH, K. L., CERNIGLIA, C. E. (1996), Bacterial degradation of low concentrations of phenanthrene and inhibition by naphthalene, *Microb. Ecol.* **31**, 305–317.

SIMS, R. C., OVERCASH, M. R. (1983), Fate of polynuclear aromatic compounds (PNAs) in soil plant systems, *Residue Rev.* **88**, 1–68.

SMITH, M. R. (1990), The biodegradation of aromatic hydrocarbons by bacteria, *Biodegradation* **1**, 191–206.

SRINIVASAN, C., D'SOUZA, T. M., BOOMINATHAN, K., REDDY, C. A. (1995), Demonstration of laccase in the white-rot basidiomycete *Phanerochaete chrysosporium* BKM-F1767, *Appl. Environ. Microbiol.* **61**, 4274–4277.

STRINGFELLOW, M. T., AITKEN, M. D. (1995), Competitive metabolism of naphthalene, methylnaphthalene and fluorene by phenanthrene-degrading *Pseudomonas, Appl. Environ. Microbiol.* **61**, 357–362.

SUESS, M. J. (1976), The environmental load and cycle of polycyclic aromatic hydrocarbons, *Sci. Total Environ.* **6**, 239–250.

SUTHERLAND, G. B., FREEMAN, J. P., SELBY, A. L., FU, P. P., MILLER, D. W., CERNIGLIA, C. E. (1990), Stereoselective formation of a K-region dihydrodiol from phenanthrene by *Streptomyces flavovirens, Arch. Microbiol.* **154**, 260–266.

SUTHERLAND, G. B., SELBY, A. L., FREEMAN, J. P.,

EVANS, F. E., CERNIGLIA, C. E. (1991), Metabolism of phenanthrene by *Phanerochaete chrysosporium, Arch. Microbiol.* **154**, 260–266.

SUTHERLAND, G. B., FU, P. P., YANG, S. K., VON TUNGELN, L. S., CASILLAS, R.P .et al. (1993), Enatiomeric composition of the trans-dihydrodiols produced from phenanthrene by fungi, *Appl. Environ. Microbiol.* **59**, 2145–2149.

SUTHERLAND, J. B., RAFII, F., KHAN, A. A., CERNIGLIA, C. E. (1995), Mechanisms of polycyclic aromatic hydrocarbon degradation, in: *Microbial Transformation and Degradation of Toxic Organic Chemicals* (YOUNG, L. Y., CERNIGLIA, C. E., Eds.), pp. 269–306. New York: Wiley-Liss.

VON DER TRENK, T., SANDERMANN, H. (1980), Oxigenation of benzo(a)pyrene by plant microsomal fractions, *FEBS Lett.* **124**, 227–231.

VON DER TRENK, T., SANDERMANN, H. (1981), Incorporation of benzo(a)pyrene quinones into lignin, *FEBS Lett.* **125**, 72–76.

TRENZ, S. P., ENGESSER, K.-H., FISCHER, P., KNACKMUSS, H.-J. (1994), Degradation of fluorene by *Brevibacterium* sp. strain DPQ 1361: a novel C—C bond cleavage mechanism via 1,10-dihydroxyfluoren-9-one, *J. Bacteriol.* **176**, 789–795.

TSAO, C. W., SONG, H. G., BARTHA, R. (1998), Metabolism of benzene, toluene, and xylene hydrocarbons in soil, *Appl. Environ. Microbiol.* **64**, 4924–4929.

TROWER, M. K., SARIASLANI, F. S., KITSON, F. G. (1988), Xenobiotic oxidation by cytochrome P-450-enriched extracts of *Streptomyces griseus*, *Biochem. Biophys. Res. Commun.* **157**, 1417–1422.

WALTER, U., BEYER, M., KLEIN, J., REHM, H.-J. (1991), Degradation of pyrene by *Rhodococcus* sp. UW 1, *Appl. Microbiol. Biotechnol.* **34**, 671–676.

WEISSENFELS, W. D., BEYER, M., KLEIN, J. (1990), Degradation of phenanthrene, fluorene and fluoranthene by pure bacterial cultures, *Appl. Microbiol. Biotechnol.* **32**, 479–484.

WEISSENFELS, W. D., BEYER, M., KLEIN, J., REHM, H.-J. (1991), Microbial metabolism of fluoranthene: isolation and identification of ring fission products, *Appl. Microbiol. Biotechnol.* **34**, 528–535.

WIESCHE, C. IN DER, MARTENS, R., ZADRAZIL, F. (1996), Two-step degradation of pyrene by whiterot fungi and soil microorganisms, *Appl. Microbiol. Biotechnol.* **46**, 653–659.

WIESMANN, U. (1994), Der Steinkohleteer und seine Destillationsprodukte – Ein Beitrag zur Geschichte der Technik und der Bodenverschmutzung, in: *Biologischer Abbau von polyzyklischen aromatischen Kohlenwasserstoffen, Schriftenreihe Biologische Abwasserreinigung* **4**, SFB 193. (WEIGERT, B., Ed.), pp. 3–18. Technical University, Berlin.

WISLOCKI, P. G., LU, A. Y. H. (1988), Carcinogenicity and mutagenicity of proximate and ultimate carcinogens of polycyclic aromatic hydrocarbons, in: *Polycyclic Aromatic Hydrocarbon Carginogenesis: Structure–Activity Relationships* Vol. I (YANG, S. K., SILVERMAN, B. D., Eds.), pp. 1–30. Boca Raton, FL: CRC Press.

WUNDER, T., MARR, J., KREMER, S., STERNER, O., ANKE, H. (1997), 1-Methoxypyrene and 1,6-dimethoxypyrene: two novel metabolites in fungal metabolism of polycyclic aromatic hydrocarbons, *Arch. Microbiol.* **167**, 310–316.

YANG, S. K. (1988), Metabolism and activation of benz(a)anthracene and methyl benz(a)anthracenes, in: *Polycyclic Aromatic Hydrocarbon Carcinogenesis: Structure–Activity Relationship* Vol. I (YANG, S. K., SILVERMAN, B. D., Eds.), pp. 129–150. Boca Raton, FL: CRC Press.

YE, D., SIDDIQI, M. A., MACCUBBIN, A. E., KUMAR, S., SIKKA, H. C. (1996), Degradation of polynuclear aromatic hydrocarbons by *Sphingomonas paucimobilis, Environ. Sci. Technol.* **30**, 136–142.

YOUN, H.-D., HAH, Y. CH., KANG, S.-O. (1995), Role of laccase in lignin degradation by white-rot fungi, *FEMS Microbiol. Lett.* **132**, 183–188.

10 Degradation of Chlorinated Compounds

CATRIN WISCHNAK
RUDOLF MÜLLER

Hamburg, Germany

List of Abbreviations

2,4,5-T	2,4,5-trichloro phenoxyacetic acid
2,4-D	2,4-dichloro phenoxyacetic acid
CB	chlorobenzene
DCA	dichloroethane
DCM	dichloromethane
HCH	hexachloro cyclohexane
MCBP	4-(4-Cl-2-methylphenoxy)butyric acid
MCPA	4-Cl-2-methylphenoxy acetic acid
MCPP	2-(4-Cl-2-methylphenoxy)propionic acid
ODP	ozone depletion potential
PCB	polychlorinated biphenyl
PCDD	polychlorinated dibenzodioxin
PCDF	polychlorinated dibenzofuran
PCE, PER	tetrachloroethene
PCP	pentachlorophenol
TCE	trichloroethene
TETRA	tetrachloromethane
TRI	1,1,1-trichloroethane

1 Introduction

1.1 Properties of Halogenated Compounds

The halogens constitute the 7th main group within the periodic system of elements and are found in about 0.2 weight % of the Earth's crust. The main halogen is chlorine followed by fluorine, bromine, and iodine. The relative abundance of these elements is 3,000:100:10:1 (Cl:F:Br:I). In nature, these elements usually occur as metal salts and as ions in aqueous solution. While fluorine and iodine are pure isotopes, chlorine and bromine are isotopic mixtures (75.5% ^{35}Cl and 24.5% ^{37}Cl; 50.5% ^{79}Br and 49.5% ^{81}Br). Chemically, the halogens are characterized by their stable bonds with carbon atoms and by their strong electronegativity. The strength of the carbon halogen bond and its polarization markedly decrease with increasing atomic radius of the halogen. In correlation to the electronegativity the enthalpy of the reaction between methane and halogen X (F, Cl, Br, I) to form halomethane and HX varies from exotherm to endotherm (-430, -106, -31, $+53$ kJ mol^{-1}). The enthalpy ΔH^0 of C–Cl (339 kJ mol^{-1}) is similar to C–C (348 kJ mol^{-1}) and C–O (358 kJ mol^{-1}) but less than C–H (413 kJ mol^{-1}) and C–F (489 kJ mol^{-1}). The length of the chemical bond between C and Cl (178 pm) is rather comparable to the C–C (154 pm) than to the C–O (134 pm) or the C–H (110 pm) bonds. Regarding steric effects, the van der Waals radii of the chlorine (180 pm) substituent are most similar in size to CH$_3$ (200 pm), Br (195 pm), and S (185 pm). Oxygen (140 pm) is much smaller than chlorine, but the hydroxyl group is of similar size.

The reactivity of the C–X bond is strongly affected by the type of carbon atom to which the halogen is attached, e.g., the enthalpy ΔH^0 for chloride liberation decreases from a primary to a tertiary C–C bond.

In the past, chemists have developed chlorinated hydrocarbons within all groups of chemical substances. Most of these new compounds are not inflammable and/or durable solvents or materials. This advantage, i.e., the applica-

tion of stable materials, competes with the risk of persistence, environmental pollution, and accumulation in the food chain. Several mechanisms for the persistence of chemicals have been discussed (ALEXANDER, 1973). Research has been conducted until now in order to circumvent this recalcitrance and the environmental and physiological limitations and to develop environmentally convenient compounds.

One of the main problems in the degradation of halogenated compounds is that either microorganisms with degradative abilities or an appropriate environment for these microorganisms are not present everywhere. Another important problem is the low bioavailability of the highly substituted halocarbons which are hardly water-soluble and tend to adsorb to surface materials, or they are highly volatile and, therefore, not available to microorganisms. Transport into the cell may be a barrier, if active transport is involved. Inducible enzymes need an effective inductor at adequate concentrations below the toxicity level of the pure substance and of mixtures in order to permit biological degradation.

The physical behavior of halocarbons determines their distribution in the different environmental compartments. The melting and boiling point of a substance as well as its water solubility play an important role in the environmental behavior of a halogenated substance. The ratio between vapor pressure and water solubility of a substance is called Henry constant [Pa m^3 mol^{-1}] and is influenced by temperature and strongly depends on the method used for determination. High values such as 100 Pa m^3 mol^{-1} represent volatile, and low values of about 1 Pa m^3 mol^{-1} less volatile compounds. 1,1,1-trichloroethane, e.g., is volatile and diffuses from water to the atmosphere (Tab. 1). The lipophilic nature of a compound is described by its distribution coefficient between *n*-octanol and water (log P_{ow}). For highly lipophilic substances (log P_{ow} >3.5) bioaccumulation is very likely. Such compounds accumulate in the soil and in the food chain. Substances with a low log P_{ow} are commonly found in the aqueous compartment. Tab. 1 summarizes the physico-chemical data of some representative and frequently used chlorinated compounds.

Tab. 1. Physicochemical Data of Selected Halogenated Hydrocarbons (Gesellschaft Deutscher Chemiker, 1985–1997)

Chemical Substance	Formula	MW [g mol⁻¹]	mp [°C]	bp [°C]	Water Solubility at 25 °C [mg L⁻¹]	log P_{ow}	Henry Constant [Pa m³ mol⁻¹]	Bacterial Growth Inhibition [mg L⁻¹]	Estimated Global[d] Production
Chloromethane	CH_3Cl	50.49	− 97	− 24	6,000[a]	0.9	5,000[a]	500[EC3]	5,000,000 t a⁻¹
Tetrachloromethane (TETRA)	CCl_4	153.84	− 22	+ 77	800	2.6	2,300[a,c]	750[EC3]	1,000,000 t a⁻¹ 442,000 t[EU, 1990]
1,2-Dichloroethane	C_2H_5Cl	98.96	− 35[a]	+ 84[b]	8,600[c]	1.5	130[a]	135[EC3]	2,000,000 t[1993, Ger]
1,1,1-Trichloroethane	$C_2H_3Cl_3$	133.41	− 32[a]	+ 74[b]	1,300	2.5	1,479[a,c]	93	726,000 t[1990]
Chloroethene (Vinyl chloride)	C_2H_3Cl	62.50	−153	− 14	9,500[15°C]	1.6	2,820	>63	14,000,000 t[1985]
1,1-Dichloroethene	$C_2H_2Cl_2$	96.95	−122[b]	+ 32[b]	2,500[c]	1.7[a]	2,300[a,c]	2,000	100,000 t a⁻¹[EU]
Trichloroethene	C_2HCl_3	131.4	− 86[a]	+ 87[b]	1,100	2.7[a]	9,000[a,c]	65	432,000 t[1988]
Tetrachloroethene (PER)	C_2Cl_4	165.83	− 21[a]	+121[b]	150	2.7	2,015[a,c]	11[EC10]	550,000 t[EU, 1990]
2-Chloro-1,3-butadiene	C_4H_5Cl	88.54	−130	+ 59	300	1.9[a]	7,950	>250	600,000 t[1989]
Chlorobenzene	C_5H_5Cl	112.56	− 45	+132[b]	500	2.8	370[a]	17[EC3]	220,000 t[1988]
1,2,4,5-Tetrachlorobenzene	$C_6H_2Cl_4$	215.9	+141	+245	0.5	4.6	15,290[c]	no data	4,000 t[EU, 1978]
Hexachlorobenzene	C_6Cl_6	284.78	+230	+323[a,b]	0.005	5.8[a]	170	no data	3,000 t[EU, 1981]
2,4-Dichlorophenol	$C_6H_4Cl_2O$	163	+ 44	+208	4,500[c]	3.2	0.3	3	59,000 t[EU, 1985]
Pentachlorophenol	C_6HCl_5O	266.4	+189	+309[b]	2,000[pH7]	3[pH7]	no data	no data	40,000 t[1985]

[a] Average value.
[b] At 1,013 hPa.
[c] At 20 °C.
[d] Global view by means of the Western hemisphere, growth inhibition test with *P. putida* (16 h) determines the toxic limit concentration (EC 3).

1.2 Natural Occurrence of Halogenated Compounds

In the biosphere approximately 1,500 different halogenated substances have been detected so far. They are ubiquitously produced and released by plants, animals, microorganisms, and natural combustion processes, e.g., in volcanism, oceans, and forest fires. In some cases, the concentrations of these compounds exceed their anthropogenic level. While the global emission of chloromethane is mainly produced by marine organisms (about 5 mio t a^{-1}), the man-made amount of this chemical is about 26,000 t a^{-1}. Like chloromethane, iodomethane (4 mio t a^{-1}) and fluoromethane (650,000 t a^{-1}) mainly come from natural sources (SIUDA and DEBERNARDIS, 1973).

Although fluoroacetic acid is highly toxic by blocking the Krebs cycle, fluoro fatty acids are found in some higher terrestrical plants. About 1% of the dry weight of a marine sponge are mixed bromofluorodiphenyl ethers believed to be xenobiotics. Nevertheless, fluorinated compounds are rather rare in nature.

Simple chlorophenols such as 2,5-dichlorophenol and 2,6-dichlorophenol are secreted by grashoppers as a repellent to ants and by female ticks as a sex pheromone. The thyroid gland hormone thyroxine is a tetraiodo tyrosine derivative. Other biologically produced halocarbons show antibiotic, antiviral, antifungal, herbicidal, insecticidal, and anticancer activity. Haloperoxidases, lactoperoxidases, or vanadium peroxidases are involved in the biosynthesis of organohalogens (GRIBBLE, 1992).

2 General Principles of the Degradation of Chlorinated Hydrocarbons

2.1 Biotic

Although chlorinated compounds are considered to be recalcitrant, many microorganisms have been discovered which are able to degrade these compounds and cleave the carbon halogen bond. During studies on the microbial degradation of chlorinated compounds several principal mechanisms have been discovered which have been extensively reviewed during the last years (CASTRO, 1998; CHAUDHRY and CHAPALAMADUGU, 1991; FETZNER and LINGENS, 1994; HARDMAN, 1991; JANSSEN et al., 1994; LEISINGER and BADER, 1993; SLATER et al., 1995). For a better understanding of the degradation of single substances the basic principles of dehalogenation are outlined in Sects. 2.1.1–2.1.7.

The reactions described in Sects. 2.1.1–2.1.7 exist to cleave the carbon halogen bond.

2.1.1 Reductive Dehalogenation

The carbon–chlorine bond can be reduced by hydrogen or electrons to form the hydrocarbon and hydrochloric acid. This is called reductive dehalogenation. Reductive dehalogenation has been reviewed by BEDARD and QUENSEN (1995), EL FANTROUSSI et al. (1998) and by MOHN and TIEDJE (1992). Two reaction mechanisms are known, which both require an electron donor or reductant. The first process is hydrogenolysis: A halogen substituent is replaced by a hydrogen atom in alkyl or aryl hydrocarbons (Fig. 1a). Examples for this reaction are the stepwise reductive dehalogenation

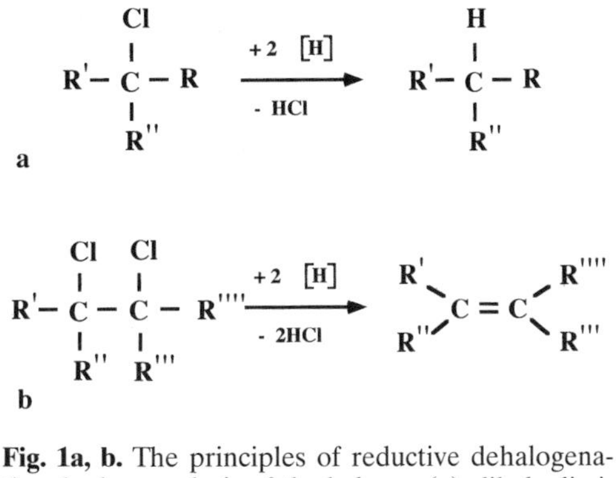

Fig. 1a, b. The principles of reductive dehalogenation, hydrogenolysis of the halogen (**a**), dihaloelimination (**b**).

of tetrachloroethene or the conversion of 3-chlorobenzoate to benzoate.

The second mechamsm is a vicinal reduction or dihaloelimination. Two adjacent halogens are removed with the formation of a carbon double bond in alkyl hydrocarbons (Fig. 1b).

The reductive dechlorination depends on the redox potential and on the presence of a reductant. In a few cases the bacteria able to perform the reductive dehalogenation have been isolated: molecular hydrogen is the initial reductant, in other cases organic hydrogen sources like pyruvate, lactate, and formiate are used. The redox chain, which ends in the transfer of electrons to the chlorinated hydrocarbon is not yet fully understood. The involvement of cytochromes and menaquinones is discussed.

KHINDARIA et al. (1995) have described the mineralization of TCE and CCl$_4$ via reductive dehalogenation through lignin peroxidase of *Phanerochaete chrysosporium*. In this reaction EDTA or oxalate served as electron donors. In reductive dehalogenations higher chlorinated hydrocarbons (e.g., tetrachloromethane and perchloroethylene) are preferred to less chlorinated hydrocarbons. In Fig. 2 the redox potentials of the various chlorinated ethanes and ethenes are given. The redox potential of the chlorinated hydrocarbons between 266 mV for the reduction of chlorobenzene to benzene

and 478 mV for the reduction of tetrachloroethene to trichloroethene are in the same range as the reduction of nitrate to nitrite with 433 mV. Therefore, it is not surprising that some microorganisms can gain energy from the reduction of chlorinated compounds. Under suitable conditions the chlorinated hydrocarbon can serve as terminal electron acceptor in an energy-yielding respiratiory process. This process was termed "halorespiration". The respiration produces an energy profit from the difference between the reduction and the coupled oxidation process. The halocarbon itself serves as the oxidant. Theoretically, the transformation of tetrachloroethene to dichloroethene could yield 5 ATP. Fig. 3 shows a model as it is discussed for halorespiration.

Higher chlorinated aromatic compounds are also reductively dechlorinated under anaerobic conditions. In theory, chlorobenzenes and chlorophenols can be completely dehalogenated to benzene or phenol, but – because of the preference of higher chlorinated compounds – the less chlorinated hydrocarbons mostly remain persistent under anaerobic conditions. COLE et al. (1994) have shown the dechlorination of monochlorophenol to phenol. For pentachlorophenol a complete, reductive dechlorination to phenol has been described. ADRIAN et al. (1998) have shown the reduction of trichlorobenzene to monochlorobenzene

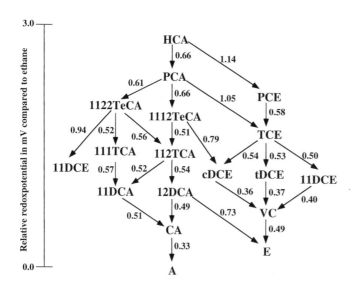

Fig. 2. Redox potentials of chloroalkanes and chloroalkenes.
HCA: hexachloroethane,
PCA: pentachloroethane,
TeCA: tetrachloroethane,
DCA: dichloroethane,
CA: chloroethane, A: ethane,
PCE: tetrachloroethene,
TCA: trichloroethane,
TCE: trichloroethene,
DCE: dichloroethene,
VC: vinylchloride, E: ethene,
A: ethane.

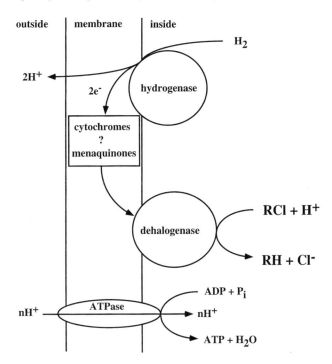

Fig. 3. Proposed reaction scheme for halorespiration.

via dichlorobenzene. For hexachlorobenzene, perchloroethylene or highly chlorinated biphenyls substitution of the halogen by hydrogen is the sole possibility for dechlorination.

Several bacterial pure cultures catalyzing reductive dehalogenation have been isolated in recent years (Tab. 2). From some of these bacteria the dehalogenating enzymes have been

Tab. 2. Recently Isolated Anaerobic Microorganisms Reductively Dehalogenating Halogenated Hydrocarbons

Microorganism/Electron Acceptors	Reference
PCE, TCE	
Dehalobacter restrictus	HOLLIGER et al. (1998)
Dehalococcoides ethenogenes	MAYMO-GATELL et al. (1997)
Desulfitobacterium sp. strain PCE1	GERRITSE et al. (1996)
Strain TT4B	KRUMHOLZ et al. (1996)
Strain MS-1	SHARMA and MCCARTY (1996)
Strain TEA	WILD et al. (1996)
Dehalospirillum multivorans	SCHOLZ-MURAMATSU et al. (1995)
Chlorinated Aromatic Compounds	
Desulfitobacterium chlororespirans	SANFORD et al. (1996)
Desulfitobacterium frappieri	BOUCHARD et al. (1996)
Desulfitobacterium hafniense	CHRISTIANSEN and AHRING (1996)
Desulfitobacterium dehalogenans	UTKIN et al. (1994)
Strain 2CP-1	COLE et al. (1994)
Desulfomonile tiedjei	DEWEERD et al. (1990)

purified. All of these enzymes contain 8Fe8S centers and a corrinoid. In all studies synthetic strong reductants like methyl viologen were used. The compound from which these electrons originate in the organism is still unclear.

In addition to the respiratory dehalogenations a number of processes have been described in which reduced cofactors in whole cells and in cell extracts can reduce chlorinated compounds without energy gain.

For example nickel-containing coenzyme F_{430}, cobalt-containing vitamin B_{12}, and iron porphyrins (hematin) are redox-active cofactors biosynthesized by diverse species (WACKETT, 1995). The reduced cofactors can catalyze an overall 2-electron transfer to form C—H bonds, yielding free halide ions from, e.g., tetrachloromethane or tetrachloroethene.

Non-physiological reductive dehalogenation reactions are also known to be catalyzed by cytochrome P_{450}CAM monooxygenase. The physiological reaction of this enzyme is the oxidation of camphor with oxygen. Obviously, hexachloroethane binds more tightly to Fe(III)-P_{450}CAM than the physiological substrate camphor (LI and WACKETT, 1993).

Although most of these reductive dehalogenations occur under anaerobic conditions, reductive dehalogenation under aerobic conditions has also been described. In the degradation of 2,4-dichlorobenzoate (VAN DEN TWEEL et al., 1987) as well as in the aerobic degradation of pentachlorophenol reductive dehalogenation steps were found (APAJALAHTI and SALKINOJA-SALONEN, 1987; XUN et al., 1992).

2.1.2 Oxygenolytic Dehalogenation

Oxygenolytic dehalogenations are catalyzed by monooxygenases or dioxygenases which incorporate one or both oxygen atoms of molecular oxygen into the substrate. The chlorine-binding carbon atom is oxidized resulting in unstable chlorohydroxy compounds that spontaneously release chloride. Long-chain haloalkanes and chloroaromatic compounds like 2-chlorobenzoate and 4-chlorophenylacetate, e.g., are dehalogenated oxygenolytically (Fig. 4a, b). The enzymes involved in these dehalogenations have been purified and characterized. They are multi-component enzyme systems consisting of a reductase component containing a flavin cluster and an oxidase component containing an iron–sulfur cluster. This composition is the same as described for other oxygenases catalyzing the oxidation of non-chlorinated substrates like benzene dioxygenase or naphthaline dioxygenase. Therefore, it may be speculated that the dehalogenating oxygenases are derived from their non-dehalogenating analogs by extending their substrate range (HARAYAMA et al., 1992).

2.1.3 Hydrolytic Dehalogenation

Hydrolytic dehalogenations are an exchange of chlorine in chlorinated hydrocarbons by a hydroxyl group from water. Halidohydrolases mediate the nucleophilic substitution reaction, especially with aliphatic com-

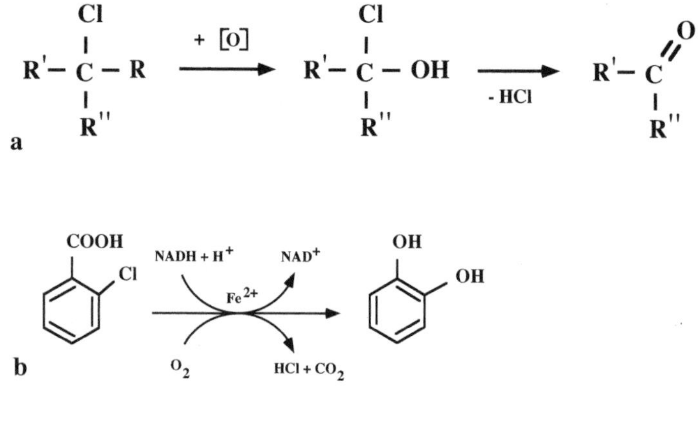

Fig. 4. Oxygenolytic dehalogenation by monooxygenases (**a**) or by dioxygenases (**b**).

pounds. Typically, short-chain haloalkanes with 2–8 carbon atoms (JANSSEN et al., 1985; SCHOLTZ et al., 1987) or 2-chloroalkanoic acids (MÖRSBERGER et al., 1991; SCHNEIDER et al., 1991; SCHWARZE et al., 1997; SLATER et al., 1979) are dehalogenated hydrolytically (Fig. 5). The enzymes involved in hydrolytic dehalogenation have been studied in great detail. Many of these enzymes have been purified, cloned and sequenced, and a lot of information on the dehalogenation mechanism on a molecular level is available (see Sect. 3.1.4).

For aromatic compounds a hydrolytic dehalogenation was observed so far only in the dehalogenation of 4-chlorobenzoate. However, in this case the carbon–halogen bond has to be activated by the formation of a thioester before hydrolytic dehalogenation is possible (LÖFFLER et al., 1991).

2.1.4 Thiolytic Dehalogenation

Thiolytic dehalogenations have an *S*-chloromethyl glutathione conjugate as intermediate. In the dehalogenation reaction chlorine is substituted by the sulfur atom of glutathione. Such a reaction was observed in the degradation of dichloromethane (KOHLER-STAUB and LEISINGER, 1985; STUCKI et al., 1981). The resulting thioether is hydrolyzed to form the hydroxylated product, and glutathione is regenerated (Fig. 6).

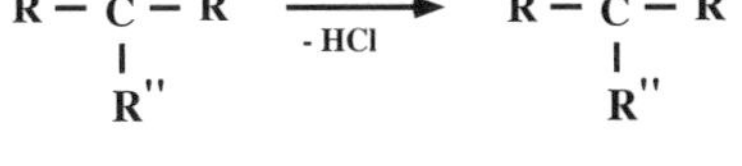

2.1.5 Dehalogenation by Intramolecular Substitution

In the case of vicinal haloalcohols an intramolecular substitution of the halogen by the neighboring hydroxyl group was observed. The first product of such a reaction is an epoxide. In some bacteria this dehalogenation is stereospecific. This reaction has been used to produce chiral C3 epoxides and halohydrins from racemic chloropropanols at a commercial scale. This process and the reactions (Fig. 7) included have been summarized recently (KASAI et al., 1998).

2.1.6 Dehydrodehalogenation

Dehydrodehalogenation involves the elimination of a chlorine together with a hydrogen atom. A C=C double bond is produced (Fig. 8). Such a reaction is only possible with aliphatic substances carrying a hydrogen atom at the carbon next to the one carrying the halogen substituent, e.g., hexachlorocyclohexane. Although this reaction has been known for some time, the enzymes involved in this reaction have not been described yet (MIYAUCHI et al., 1998; NAGATA et al., 1993).

2.1.7 Hydrodehalogenation

Finally in the case of 3-chloroacrylic acid (HARTMANS et al., 1991) the halogen is removed by spontaneous decomposition of the product derived from the addition of water to the double bond in a hydratase-catalyzed reaction also called "hydrodehalogenation" (Fig. 9). The resulting molecule carrying a chlorine and a hydroxyl group at the same carbon atom is probably not stable and looses HCl to form the corresponding aldehyde.

Fig. 5. The principle of hydrolytic dehalogenation.

Fig. 6. The principle of thiolytic dehalogenation. GSH: free glutathione, GS: bound glutathione.

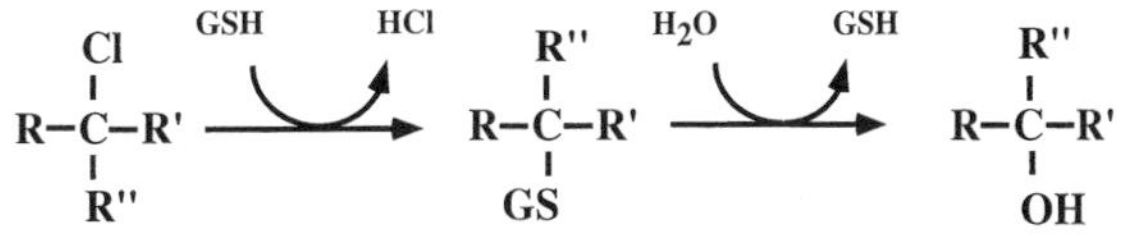

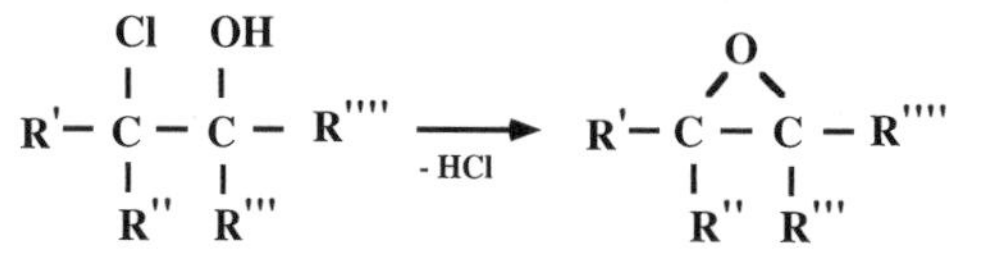

$$\begin{array}{ccc} Cl & OH \\ | & | \\ R'-C-C-R'''' \\ | & | \\ R'' & R''' \end{array} \xrightarrow[-\,HCl]{} \begin{array}{c} O \\ \diagup \backslash \\ R'-C-C-R'''' \\ | & | \\ R'' & R''' \end{array}$$

Fig. 7. The principle of dehalogenation by intramolecular substitution.

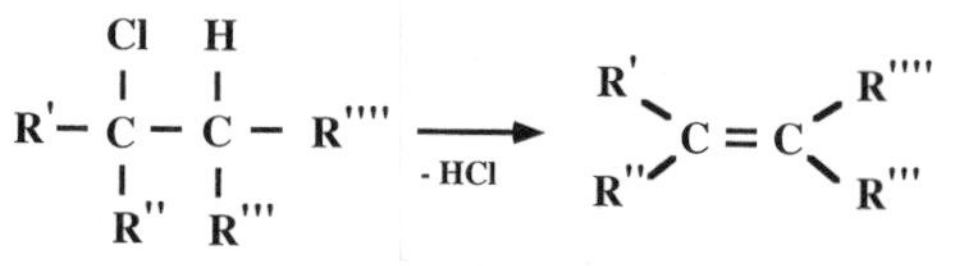

$$\begin{array}{ccc} Cl & H \\ | & | \\ R'-C-C-R'''' \\ | & | \\ R'' & R''' \end{array} \xrightarrow[-\,HCl]{} \begin{array}{c} R' \\ \diagdown \\ R'' \end{array} C = C \begin{array}{c} R'''' \\ \diagup \\ R''' \end{array}$$

Fig. 8. The principle of dehydrodehalogenation.

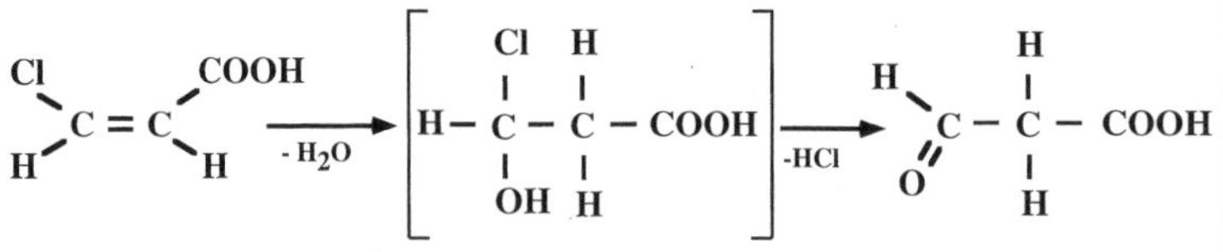

$$\begin{array}{c} Cl \\ \diagdown \\ H \end{array} C = C \begin{array}{c} COOH \\ \diagup \\ H \end{array} \xrightarrow[-\,H_2O]{} \left[\begin{array}{cc} Cl & H \\ | & | \\ H-C-C-COOH \\ | & | \\ OH & H \end{array} \right] \xrightarrow[-HCl]{} \begin{array}{c} H \\ \diagdown \\ O \diagup\diagup \end{array} C - \begin{array}{c} H \\ | \\ C \\ | \\ H \end{array} - COOH$$

Fig. 9. The principle of hydrodehalogenation (from HARTMANS et al., 1991).

2.2 Abiotic

For the abiotic degradation of halogenated compounds three major processes are known.

Hydrolysis takes place in the ubiquitous presence of water in all compartments: soil, water, and air. In a nucleophilic attack of a hydroxide ion from water the halide is substituted at suitable ionic strength, pH, and surface conditions. The photohydrolysis is activated by irradiation with sunlight. Hydrolysis predominates other abiotic reactions in the atmosphere and in surface waters.

Abiotic reduction of halogenated hydrocarbons can be catalyzed by metal ions in the laboratory. In nature metals exist in non-catalyzing ores such as oxides, sulfides, or silicates. Some sediments are capable of reductive reaction (PERLINGER, 1994) using, e.g., iron pyrite (FeS_2), Cr(II), or other low-valent complexes of iron, cobalt, and nickel can play a role in biochemical reactions. Nevertheless, the reaction of chlorinated hydrocarbons by metals can be used in the treatment of contaminated groundwater by the insertion of so-called reactive walls containing metallic iron (JOHNSON et al., 1998).

Oxidation is the main reaction of volatile halogenated hydrocarbons in the upper atmosphere. There the ultraviolet irradiation cleaves the carbon–halogen bond to radicals that react with O_2, O_3, and NO_x species resulting in oxygenated products. Certain ores of iron, TiO_2, and ZnO are thought to generate hydrogen peroxide or organic peroxides oxidizing halocarbons (CASTRO, 1998).

The major compounds responsible for the chlorine content in the stratosphere (16–64 km from the Earth's surface) are trichlorofluoromethane, dichlorofluoromethane, tetrachloromethane, chloromethane, and 1,1,1-trichloroethane. In the Northern hemisphere atmospheric emission from the use of 1,1,1-trichloroethane (TRI) is about 80% of the consumed TRI. The stability of the substance in the troposphere (0–16 km) facilitates its vertical diffusion into the stratosphere. The stratospheric radiation contains high-energy UV light which supports photodissociation. Photolysis results in chloro atoms and chloro oxide which catalytically destroy ozone. Substances absorbing UV light can be damaged directly. For example, TRI is not photolyzed in the troposphere because it does not absorb wavelengths >290 nm. TRI can be photodegraded indirectly by OH radicals to chloro radicals, chloride, phosgene, HCl, trichloroacetic acid, trichloroacetyl chloride, acetyl chloride, acetic acid, formaldehyde, CO, CO_2, and H_2O. Completely halogenated compounds are exclusively degraded in the stratosphere and show high ozone depletion potential (ODP). They have a higher potential to destroy the ozone layer than partially halogenated substances.

In aquatic systems the abiotic hydrolysis of TRI depends on temperature, pH value, and the catalysis by metal ions. Products are HCl, acetic acid, and 1,1-dichloroethane. The half-life of TRI decreases from 74 a at 0 °C to 4 h at 80 °C. The main compartment of TRI is the atmosphere. In the troposphere and stratosphere the half-life of TRI is estimated to be an average of 8 a. The atmospheric half-life of TRI seems to depend on the OH radical concentration. Due to legislative restrictions on the use of chlorinated solvents, the production of these compounds has steadily decreased during the last years.

3 Degradation of Environmentally Relevant Chlorinated Compounds

3.1 Aliphatic Compounds

The biodegradation of chlorinated aliphatic compounds has been reviewed recently (SLATER et al., 1997). This substance class can be divided into short-chain (C1–C8) and long-chain (>C9) halocarbons.

The degradation of short-chain compounds, namely chloromethane, dichloromethane (DCM), chloroform, carbon tetrachloride (TETRA), dichloroethene (DCE), trichloroethene (TCE), and tetrachloroethene (PER or PCE) has been summarized by LEISINGER (1996) and by VOGEL et al. (1987). Short-chain haloalkanes are mainly used as solvents in dry cleaning, degreasing machinery, manufacturing electronic components, as pesticides, and they are products of water chlorination. Due to the extensive use of chlorinated alkanes these substances have become major contaminants in soil. In consequence of their water solubilities their concentrations in groundwater range between 150 and 13,000 mg L^{-1} (HOWARD, 1990). The concentration limits in drinking water range between 2 and 2,000 µg L^{-1} (Bundesamt für Gesundheitswesen, 1995).

Halomethanes are volatile (low boiling point around 40 °C) and important in the destruction of stratospheric ozone, especially bromide more than chloride species (MONTZKA et al., 1996). Methyl bromide is used as fumigant in crop protection and a commodity preservation world wide and can be consumed aerobically by soil bacteria (HINES et al., 1998), methane monooxygenase (COLBY et al., 1977), ammonium monooxygenase (RASCHE et al.,

1990), or by anaerobic sediments or in agricultural soil (HANCOCK et al., 1998).

Halogenated C1 and C2 compounds are often degraded by methanotrophs which can utilize methane as the sole source of carbon and energy in the presence of oxygen, e.g., *Methylosinus trichosporium* and *Methylococcus capsulatus* Bath OB3 (LONTOH and SEMRAU, 1998).

3.1.1 Chloromethane

Chloromethane is formed during thermal decomposition of chloride-containing plastics and, e.g., tobacco. However, the bigger portion (3 mio t a^{-1}) of this compound comes from natural sites when methyl iodide from seaweed is chlorinated, volatilized, and reaches the atmosphere as a sink. Chloromethane can be metabolized as growth substrate or by co-metabolism. Three theoretical mechanisms have been described for the biodegradation of chloromethane (Fig. 10). Either an initial dechlorination by a hydrolase (KEUNING et al., 1985) or by a monooxygenase occurs before the formaldehyde produced is further oxidized via formic acid to CO_2 (VANNELLI et al., 1998). Finally, strictly anaerobic growth on chloromethane was described for the homoacetogenic bacterium strain MC which possesses a methyltransferase and a dehydrogenase (MESSMER et al., 1996).

3.1.2 Dichloromethane (DCM)

Dichloromethane is distributed globally because of its open usage and its volatilization from water to the atmosphere within a few days. DCM serves as the growth substrate for a number of aerobic methylotrophs. LEISINGER and coworkers isolated several bacteria which can grow on dichloromethane as the sole carbon and energy source (BRUNNER et al., 1980; KOHLER-STAUB et al., 1986). Recently, *Methylobacterium* sp. strain DM4 and *Methylophilus* sp. strain DM11 have been described (GISI et al., 1998). All these strains dehalogenate DCM to formaldehyde. The involved dehalogenases are homologous and gluthatione-dependent (Fig. 11). They have similar substrate ranges

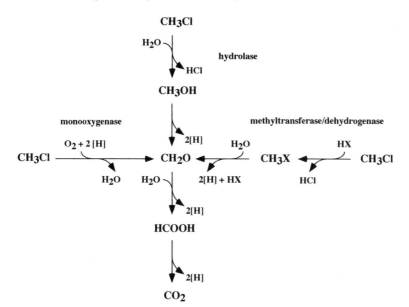

Fig. 10. Possible reaction mechanisms for degradation of chloromethane (from VANNELLI et al., 1988).

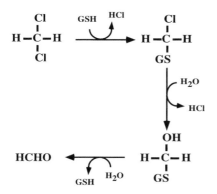

Fig. 11. Glutathione-dependent dehalogenation of dichloromethane (from BADER and LEISINGER, 1994). GSH: free glutathione, GS: bound glutathione.

(dihalomethanes), subunit compositions, and molecular masses. Differences in kinetic and immunological properties and the *N*-termini of the amino acid sequences divide the enzymes involved in the dehalogenation of dichloromethane into two groups (BADER and LEISINGER, 1994).

Under anaerobic conditions an anaerobic acetogen growing on dichloromethane has been reported (MÄGLI et al., 1998). *Dehalobacterium formicoaceticum* dehalogenates

dichloromethane via methylene tetrahydrofolate to formate and acetate (acetyl CoA pathway) in a molar ratio of 2:1 (Fig. 12). As a cofactor a Co(I) corrinoid is thought to be involved in the dehalogenating reaction. The formation of formate and acetate in the 2:1 ratio enables this organism to balance hydrogen and electrons without the involvement of exogenous electron donors. However, carbon dioxide is necessary for acetate formation.

3.1.3 Tetrachloromethane (TETRA)

In water tetrachloromethane is very resistant against chemical degradation, its half-life is about 1,000 a. The atmospheric half-life of TETRA is about 36 a (GOLOBEK and PRINN, 1986). Because of its long persistence in the troposphere it can pass to the stratosphere where short UV waves can destroy TETRA. The formed chloro radicals attack the ozone molecules and start radical chain reactions that only stop at binding atomic hydrogen.

For the biological degradation of TETRA two ways are described: Either the chlorides are removed in a stepwise reduction via chloroform, dichloromethane, and chloromethane to methane, or the chlorines are removed hydrolytically and carbon dioxide is formed. In *Acetobacterium woodii* both pathways were found to occur simultaneously (EGLI et al., 1990) (Fig. 13).

3.1.4 Dichloroethane (DCA)

The degradation of dichloroethane has been studied in great detail with *Xanthobacter autotrophicus* GJ10 (JANSSEN et al., 1989). This organism dehalogenates the substrate in a hydrolytic step to form chloroethanol which is then oxidized via the aldehyde to chloroacetic acid. Again, as a next step in a hydrolytic dehalogenation glycolic acid is formed which enters the central metabolism (Fig. 14). The two dehalogenases involved in the degradation are not identical.

The dehalogenase catalyzing the initial dehalogenation (class 1L 2HAA) has a molecular weight of 36 kDa and is composed of one subunit. It is the only dehalogenase the crystal structure of which has been resolved (ROZE-BOOM et al., 1988; VERSCHUEREN et al., 1993). It has been concluded from this structure that the enzyme possesses three structurally different domains where the active site is located between two domains. The reaction mechanism of this dehalogenase has also been elucidated in detail. By crystallization of the enzyme carrying the substrate in the transient state, it was shown, that the chlorine is first replaced by the free carboxylic group of an aspartate. The resulting ester is then cleaved hydrolytically (VERSCHUEREN et al., 1993) (Fig. 15). A similar mechanism has been proposed for the second dehalogenation of chloroacetic acid, where aspartate was also involved in the dehalogenation (LIU et al., 1995; LIU et al., 1994; SCHNEIDER et al., 1993). Other haloal-

kane dehalogenases are closely related to that of *Xanthobacter autotrophicus* GJ10 (JANSSEN et al., 1994).

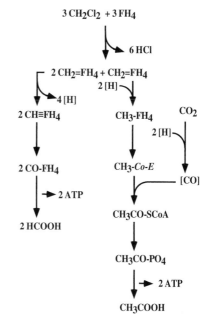

Fig. 12. Anaerobic degradation of dichloromethane (according to MÄGLI et al., 1998).

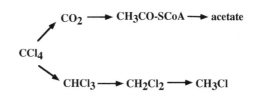

Fig. 13. The two degradation pathways for tetrachloromethane by *Acetobacterium woodii* (from EGLI et al., 1990).

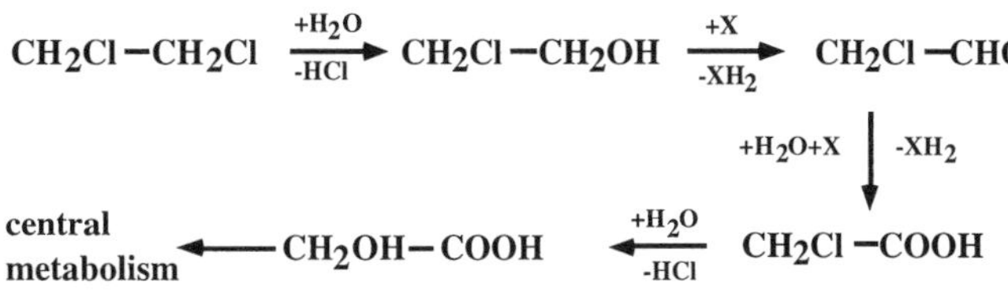

$$CH_2Cl-CH_2Cl \xrightarrow[-HCl]{+H_2O} CH_2Cl-CH_2OH \xrightarrow[-XH_2]{+X} CH_2Cl-CHO$$

$$\downarrow {+H_2O+X \atop -XH_2}$$

central metabolism $\longleftarrow CH_2OH-COOH \xleftarrow[-HCl]{+H_2O} CH_2Cl-COOH$

Fig. 14. The degradation of dichloroethane by *Xanthobacter autotrophicus* GJ10 (from JANSSEN et al., 1985).

3.1.5 Trichloroethene (TCE)

Trichloroethene is the most ubiquitous halo-alkane and is frequently encountered in groundwater. To date no bacterium is known which can utilize TCE as the sole carbon and energy source. The cometabolic dechlorination was summarized by HANSON and HANSON (1996) (Fig. 16). In this cometabolic transformation TCE is attacked by oxygenases which normally catalyze other reactions. These enzymes attack the ethene double bond, and an epoxide is formed. This is not stable and decomposes spontaneously to various products. TCE degradation can be mediated by alkanes or alkenes, by aromatic hydrocarbons (e.g., toluene, phenol, benzene, alkylbenzene, and biphenyl), or by ammonium. Recent studies demonstrate that TCE has a negative effect on the competition between TCE cometabolizing toluene degraders (MARS et al., 1998). One main problem in this reaction is the high reactivity of the epoxide formed. This epoxide rapidly destroys the enzymes forming the epoxide. Therefore, continuous production of the enzymes is required in order to obtain reasonable conversions. To circumvent this problem strains have been constructed in which these enzymes are constitutively expressed (BERENDES et al., 1998). In this case the original substrate is not necessary as an inductor, and higher conversion rates were obtained. Under psychrotrophic (12 °C) conditions ammonia oxidizers and methanotrophs degrade TCE at a significant rate of 145 nmol L^{-1} d^{-1} in groundwater microcosms, compared to mesophilic (24 °C) conditions with 775 and 900 nmol L^{-1} d^{-1}, respectively (MORAN and HICKEY, 1997). Under anaerobic conditions TCE can be dechlorinated reductively. The processes are essentially the same as for tetrachloroethylene. All perchloroethylene-degrading bacteria also degrade TCE. Therefore, this degradation will be included in the next paragraph (Sect. 3.1.6).

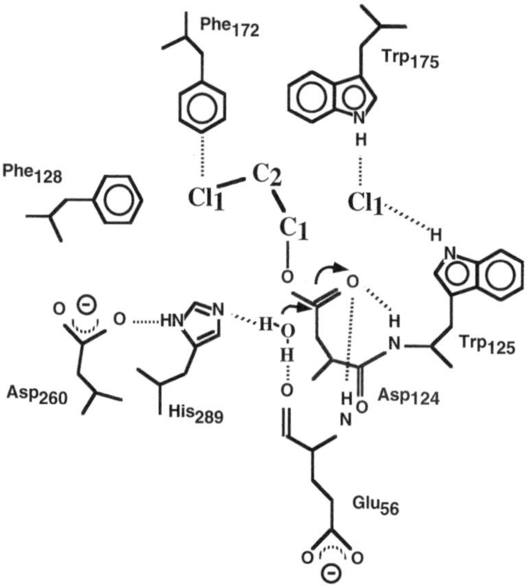

Fig. 15. The structure of the active site of dichloroethane dehalogenase with the intermediate ester formed with aspartate 124 during dehalogenation (from VERSCHUEREN et al., 1993).

3.1.6 Tetrachloroethene (PER or PCE)

Tetrachloroeth(yl)ene or perchloroeth(yl)ene (PER or PCE) is not degraded by aerobic bacteria. Under anaerobic conditions the reductive dehalogenation by non-enzymatic reactions involving coenzymes and transition

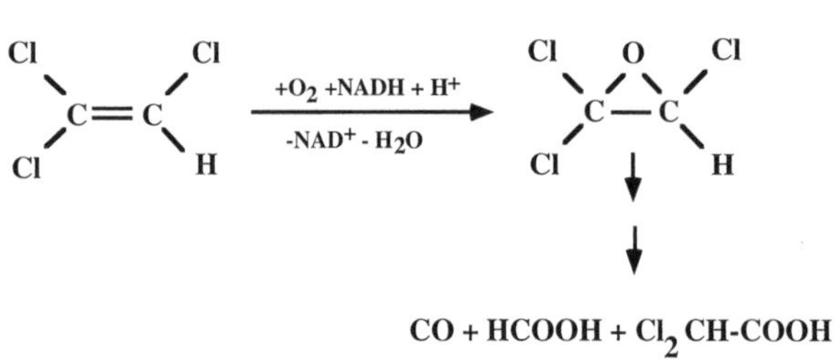

$$CO + HCOOH + Cl_2CH\text{-}COOH$$

Fig. 16. Cometabolic degradation of trichloroethylene by oxygenases.

metal complexes in anaerobic bacteria have been described (CASTRO, 1998; HANSON and HANSON, 1996). In addition, numerous anaerobic microorganisms dehalogenating PCE via TCE to dichloroethylene, vinyl chloride, ethylene or even to ethane have been reported (Tab. 1, Fig. 17). However, in many cases the reaction stops at the level of dichloroethene.

For this microbial reductive dehalogenation cosubstrates are necessary serving as electron donors. Several substances such as methanol, hydrogen, complex substrates from sediment, benzoate, sucrose, or acetate have been shown to serve as electron donors. The bacteria involved in these reductive dehalogenations and their properties have been reviewed recently (EL FANTROUSSI et al., 1998).

3.1.7 Di- and Trichloropropanes

Di- and trichloropropanes are not produced directly, but are formed as by-products in the chemical synthesis of epichlorohydrin, vinyl chloride, and 1,2-dichloroethane. In this way about 80,000 t were produced in Germany in 1992. The major part was used as solvents for oil, gum, waxes, etc., small amounts were recycled for thermal and material use, exported, temporarily stored in a waste landfill, and combusted. In some countries chlorinated propanes are used as nematocides.

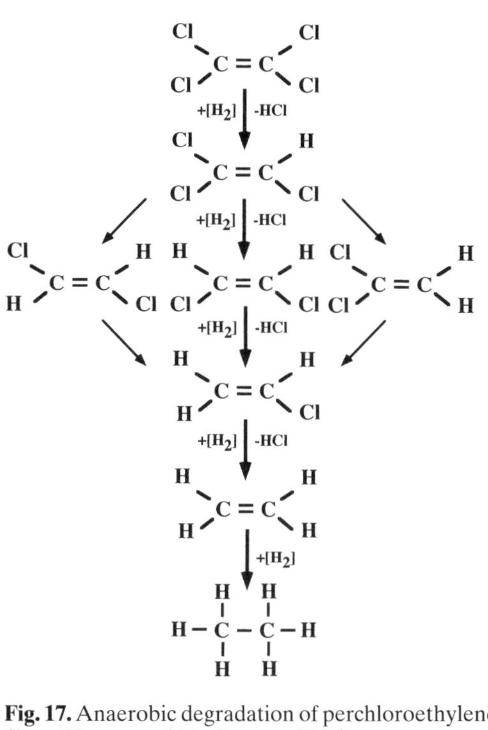

Chlorinated propanes are not readily degraded. Although oxidation of the methyl group in dichloropropane seems possible, such a reaction has never been found. However, some methane-oxidizing bacteria show very low activity towards dichloropropane (OLDENHUIS et al., 1989). Under anaerobic conditions reductive dehalogenation of chloropropanes by a mixed culture has been described recently. The mixed culture required an electron donor like hydrogen, lactate, fumarate, pyruvate, glycerol, methanol, ethanol, mannitol, sorbitol, glucose, fructose, or yeast extract. Propene was the end product of the dehalogenation (LÖFFLER et al., 1997) (Fig. 18). The organisms responsible for this dehalogenation have not been identified yet, but methanogens were not involved.

3.1.8 Long-Chain Chloroalkanes

Long-chain chloroalkanes and paraffins are used as plasticizers in paints, rubbers, and plastic materials. Despite their low water solubility they can be detected in river sediments.

These compounds can be dechlorinated by oxidative dehalogenation. In the initial oxidation a compound is formed carrying a chlorine and a hydroxy group at the same carbon atom. This intermediate is not stable and loses the halogen spontaneously to form an aldehyde (YOKOTA et al., 1987) which is then oxidized to the fatty acid and enters the general metabolism (Fig. 19).

Recently, a bacterium has been isolated, which degrades α,ω-dichloroalkanes in 2-phase systems. In this organism a broad-spectrum alkane monooxygenase converts the chlorinated alkanes (WISCHNAK et al., 1998). This enzyme is also induced by chlorinated alkanes in contrast to the enzymes from the known alkane degraders (ARMFIELD et al., 1995).

Fig. 17. Anaerobic degradation of perchloroethylene (from VOGEL and MCCARTHY, 1985).

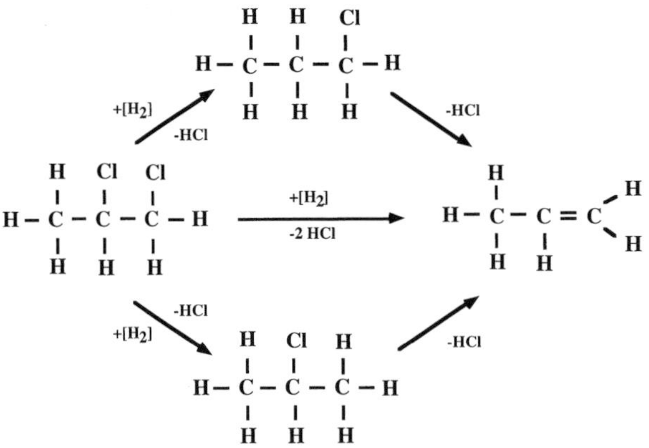

$$H-\overset{\overset{\displaystyle H}{|}}{\underset{\underset{\displaystyle H}{|}}{C}}-\overset{\overset{\displaystyle H}{|}}{\underset{\underset{\displaystyle H}{|}}{C}}-\overset{\overset{\displaystyle Cl}{|}}{\underset{\underset{\displaystyle H}{|}}{C}}-H$$

+[H₂] ↗ -HCl   -HCl ↘

$$H-\overset{\overset{\displaystyle H}{|}}{\underset{\underset{\displaystyle H}{|}}{C}}-\overset{\overset{\displaystyle Cl}{|}}{\underset{\underset{\displaystyle H}{|}}{C}}-\overset{\overset{\displaystyle Cl}{|}}{\underset{\underset{\displaystyle H}{|}}{C}}-H \quad\xrightarrow[\text{-2 HCl}]{\text{+[H}_2\text{]}}\quad H-\overset{\overset{\displaystyle H}{|}}{\underset{\underset{\displaystyle H}{|}}{C}}-\overset{\overset{\displaystyle H}{|}}{\underset{\underset{\displaystyle H}{|}}{C}}=C\overset{H}{\underset{H}{\diagdown}}$$

+[H₂] ↘ -HCl   -HCl ↗

$$H-\overset{\overset{\displaystyle H}{|}}{\underset{\underset{\displaystyle H}{|}}{C}}-\overset{\overset{\displaystyle Cl}{|}}{\underset{\underset{\displaystyle H}{|}}{C}}-\overset{\overset{\displaystyle H}{|}}{\underset{\underset{\displaystyle H}{|}}{C}}-H$$

Fig. 18. Anaerobic degradation of 1,2-dichloropropane (from LÖFFLER et al., 1997).

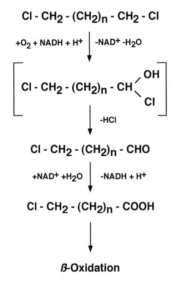

Fig. 19. Aerobic degradation of long-chain chloroalkanes.

separated from the mixture and marketed as "lindane". Today the pure substance is used in industrialized countries instead of the technical-grade insecticide still used in developing countries. HCH did not only affect insects, but all kinds of animals possessing a central nervous system where the release of synaptic neurotransmitters is enhanced by HCH. Furthermore, from experiments with animals α-HCH is a suspected carcinogen. Due to the application in agriculture and to inappropriate waste disposal large amounts of HCH were released into the environment and still persist there.

A number of aerobic microorganisms growing on α-, γ- or δ-HCH have been isolated. While α- and γ-HCH are readily degradable under aerobic conditions (Fig. 20), β-HCH is the most persistent isomer and has been shown to be hardly mineralized (SAHU et al., 1995). The isomerization of γ-HCH to α-HCH can be catalyzed by *Pseudomonas* sp. in aquatic sediments or abiotically (BENZET and MATSUMURA, 1973; DEO et al., 1994).

3.1.9 Hexachlorocyclohexane (HCH)

Hexachlorocyclohexane was widely used as an insecticide, especially after the restriction posed on DTT. In 1983 about 150,000 t were produced by chlorination of benzene. The technical quality consists of a mixture of the stereoisomers α-HCH (60–80%, aaeeee), β-HCH 5–10%, eeeeee), γ-HCH (10–16%, aaaeee), and δ-HCH (5–10%, aeeeee). The most effective isomer was γ-HCH which was

3.1.10 Alkanoic Acids

The haloalkanoic acid (HAA) dehalogenation has been studied and reviewed by SLATER et al. (1997). One widely used compound of this substance class is the herbicide Dalapon (2,2-dichloropropionic acid) which is readily degraded in soil. 2-HAAs are dehalogenated by hydrolytic dehalogenases which are historically grouped into 4 classes based on their substrate specifity against L-, D-, or both isomers

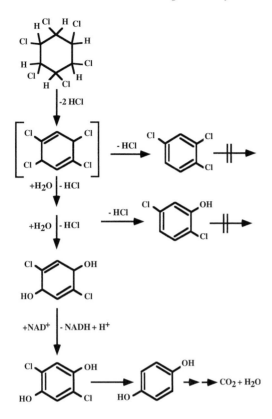

Fig. 20. The aerobic degradation of γ-hexachlorocyclohexane (HCH) by *Pseudomonas putida* (from MIYAUCHI et al., 1998).

combined with retention or inversion of the product configuration and with the sensitivity against sulfhydryl-blocking reagents. The enzymes involved have been described in Sect. 3.1.4.

Some microorganisms can grow on 3-haloalkanoic acids, but the detailed dehalogenation mechanism has not yet been elucidated.

3.2 Haloaromatic Compounds

The degradation pathways of haloaromatics were summarized by HÄGGBLOM (1992), MIDDELDORP (1997), and REINEKE and KNACKMUSS (1988). In general, the aromatic halocarbons that are stabilized by resonance

energy have to be activated for further reaction.

Under aerobic conditions several pathways are possible in principle. Either the halogen is removed before the aromatic ring is attacked, or the ring is attacked first, and then the halogen is removed in a later step. In the first case a halogen is removed during the initial attack of a dioxygenase leading to a non-chlorinated or lower chlorinated catechol as in the degradation of 2-chlorobenzoate (FETZNER et al., 1991), 4-chlorophenylacetate (MARKUS et al., 1984; MARKUS et al., 1986), or 1,2,4,5-tetrachlorobenzene (SANDER et al., 1991). This catechol is then degraded via the *ortho*-cleavage pathway. In the second case the aromatic ring is attacked at free ring positions leading to the formation of the chlorinated catechol. This attack can occur by dioxygenases as in the degradation of halobenzenes or halotoluenes or by monooxygenases as in the degradation of chlorophenols.

After the intradiol cleavage of the resulting catechol by another dioxygenase (HARAYAMA et al., 1992) the chlorine substituents are eliminated during lactonization of chloro-*cis,cis*-muconate. Another pathway for the degradation of chlorophenols involved the formation of chlorohydroquinones, which are then reductively dehalogenated step by step, before aromatic structure is lost.

In special cases as in the degradation of 4-chlorobenzoate special enzymes catalyze the hydrolytic removal of the halogen from the aromatic ring (LÖFFLER et al., 1991). Under anaerobic conditions the halogen is generally removed by reductive dehalogenation, before the ring is attacked.

As a rule of thumb, the higher chlorinated a compound is the more important is the reductive pathway of dehalogenation. Monohalogenated compounds are degraded preferentially aerobically.

In Sects. 3.2.1–3.2.10 the degradation pathways for the most relevant chlorinated aromatic compounds are described in detail.

3.2.1 Chlorobenzenes (CB)

The water solubility of chlorobenzenes significantly decreases with the number of chlorine substituents from 500–0.005 mg L^{-1} (IUPAC, 1989). All 12 isomers are highly volatile, lipophilic, toxic, and tend to accumulate in biota and in the environment.

In general, the aerobic biodegradation of chlorobenzenes involves the initial reaction of a dioxygenase to form the corresponding *cis*-dihydrodihydroxy benzene and a dioldehydrogenase to form the corresponding chlorocatechol which subsequently undergoes an *ortho*-cleavage. The chlorine substituents are either lost in further steps or they are not removed and dead end products are formed which cannot be metabolized. As an example, Fig. 21 shows the degradation pathway proposed for 1,2,4,5-tetrachlorobenzene (SANDER et al., 1991). Although in most cases chlorocatechols are cleaved in *ortho*-position, and although it was postulated that *meta*-cleavage results in suicide inactivation by reaction of the highly reactive acyl chloride formed, the *meta*-cleavage of chlorocatechol has been demonstrated recently (KASCHABEK et al., 1998).

Studies on the anaerobic reductive dehalogenation of chlorobenzenes are restricted to bacterial consortia so far, because high toxicity and low water solubility of chlorobenzenes prevented successful isolation of pure bacterial species using chlorobenzenes as electron acceptors (ADRIAN et al., 1998). Nevertheless, reductive dehalogenation of chlorobenzenes does occur under anaerobic conditions and the bacteria responsible for this dehalogenation will certainly be isolated in the near future.

Hexachlorobenzene is found as a contaminant in about 30 industrial compounds and to up to 14% in pesticides, e.g., poly-chloro-chlorophthalic acid dimethyl ester. It is degradad anaerobically via penta-, tetra- and trichlorobenzene to dichlorobenzene. The reduction reaction decreases from methanogenic via sulfate-reducing to denitrifying conditions (CHANG et al., 1998).

3.2.2 Dichloroanilines

While 2,4-dichloroaniline is completely degraded under anaerobic, but incompletely transformed under aerobic conditions, 3,4-dichloroaniline, used in crop protection, can be mineralized aerobically (YOU and BARTHA, 1982; ZEYER et al., 1985). The first attack is an oxidative desamination to form a dichlorocatechol which is then cleaved in *ortho*-position. The chloride substituents are successively released from the aliphatic product. In anaerobic estuarine sediment chloroanilines are reductively degraded in the *para*- and *ortho*-positions (SUSARLA et al. 1997).

3.2.3 Chloromethyl Anilines

Chloromethyl anilines (chlorotoluidines) are starting materials for the manufacture of herbicides, dyes, pigments, and pharmaceutical products. They are aerobically, but not anaerobically degradable by oxygenation of the

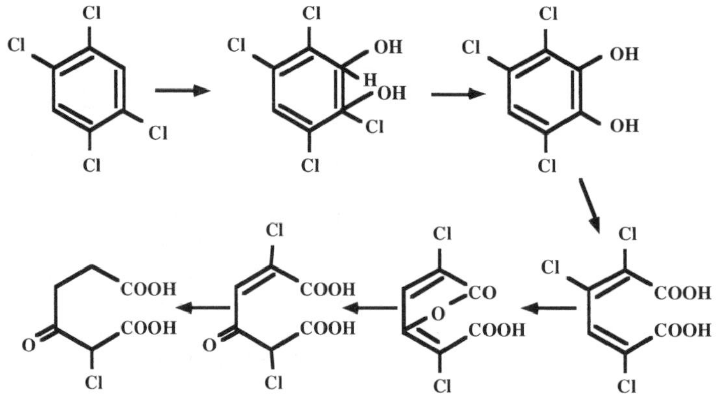

Fig. 21. The aerobic degradation of 1,2,4,5-tetrachlorobenzene by *Pseudomonas* strains PS12 and PS14 (from SANDER et al., 1991).

aromatic ring, elimination of the amino group and production of a methyl chlorocatechol, which is then degraded as described in Sect. 3.2.1.

3.2.4 Chloronitrobenzenes

Chloronitrobenzenes are used as starting materials for, e.g., dyes, pesticides, and rubber chemicals. More than 100,000 t are produced annually. Chlorinated nitrobenzenes are usually first modified in the nitro function and are not dechlorinated. They are converted by reduction of the nitro group to an amino function, and the resulting chloroanilines are degraded as described above.

3.2.5 Chlorophenols

Apart from the chloroethenes, the degradation of chlorophenols consisting of 19 different isomers is best studied in this area. In summary, several chlorophenol-degrading microorganisms have been isolated and their properties have been carefully investigated in the laboratory (for review see: ALLARD and NEILSON, 1997; HÄGGBLOM, 1992; HÄGGBLOM and VALO, 1995; MCALLISTER et al., 1996; MOHN and TIEDJE, 1992; VAN DER MEER et al., 1992).

The mineralization of chlorophenols may proceed aerobically (APAJALAHTI and SALKI-NOJA-SALONEN, 1986; SEECH et al., 1994; CRAWFORD and MOHN, 1985; HÄGGBLOM et al., 1988) or via reductive dehalogenation (HÄGG-BLOM et al., 1993; MIKESELL and BOYD, 1986). However, reductive dehalogenation does not always lead to complete mineralization, but to the accumulation of metabolic products (COLE et al., 1994; SMITH and WOOD, 1994).

Less chlorinated phenols like mono- and dichlorophenols are degraded via hydroxylation to the corresponding chlorocatechols, which are then degraded via *ortho*-cleavage as described for chlorobenzenes (HÄGGBLOM, 1992). In higher chlorinated phenols a hydroxylation occurs in *para*-position. From the resulting hydroquinone a chlorine can be replaced by a hydroxy group to form the hydroxy hydroquinone from which additional chlorines are reductively removed (see also Sect. 3.2.6).

In fungi another strategy for detoxification was observed. After oxidation of TCP or PCP to tri- or pentachlorobenzoqinone the quinone was reduced and methylated (JOSHI and GOLD, 1993; KREMER et al., 1992).

3.2.6 Pentachlorophenol (PCP)

PCP and its sodium salt have been widely used as wood and leather preservatives against fungi and bacteria or outside of Europe as insecticides and herbicides. The technical-grade PCP contained 87–89% PCP, other chlorinated phenols (around 6%), phenoxyphenols (around 3%) and traces (mg kg^{-1}) of chlorobenzenes, chlorodibenzodioxins and chlorodibenzofurans. During incomplete combustion of PCP-treated material polychlorinated dibenzodioxins and dibenzofurans can be formed. In consequence, the production of PCP stopped in Germany in 1985, and its use has been restricted since 1990.

The fate of PCP in the environment depends on its acid–base level, its volatility, lipophilicity, and its tendency to chemical reactions. In neutral to alkaline water the phenolate dominates. It is soluble and can be degraded photochemically. At acidic conditions the non-dissociated PCP ($pKa = 4.7$) predominates. It is less than two orders of magnitude less soluble, but it can vaporize from the solution or adsorb to soil or sediments. This example of substance-specific properties also emphasizes the influence of the water content on sorption or bioavailability of a compound. Dry alkaline soil retains the PCP sodium salt. Increasing water content of this soil mobilizes phenolate which can migrate into the depth until it reaches the groundwater.

A lot of work has been done concerning the aerobic degradation of PCP. Many gram-negative bacteria share a common pathway for the degradation of PCP and other highly chlorinated phenols (Fig. 22). The pathway involves an initial oxygenolytic dechlorination of PCP to tetrachloro hydroquinone by PCP monooxygenase. It has been suggested that pentachlorophenol monooxygenase has evolved from a previously existing flavin monooxygenase by gaining the ability to accommodate PCP (CO-

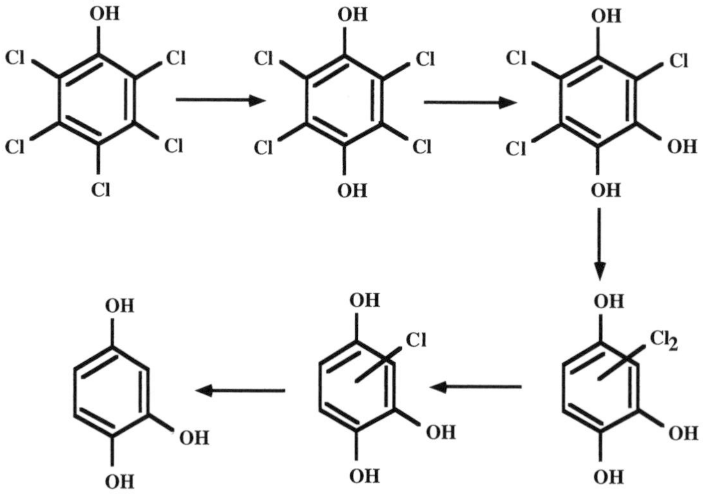

Fig. 22. The aerobic degradation of pentachlorophenol (from Häggblom et al., 1989).

PLEY, 1997). This dehalogenase of monooxygenase type and the corresponding *pcpB* gene share homologies among *Sphingomonas chlorophenolica* ATCC 39723 (ORSER and LANGE, 1994), *Sphingomonas* sp. (RADEHAUS and SCHMIDT, 1992), *Arthrobacter* sp. ATCC 33790 (SCHENK et al., 1989), and *Pseudomonas* sp. strain SR3 (RESNIK and CHAPMAN, 1994), as well as with other *Pseudomonas* spp. (LEUNG et al., 1997). According to studies of EDERER et al. (1997), the homology of *pcpB* among the above mentioned four PCP-degrading microorganisms was exceptionally high, and due to additional culture characteristics, it was suggested that these strains should be reclassified into the genus *Sphingomonas* of α-proteobacteria (KARLSON et al., 1995; NOHYNEK et al., 1995). In conclusion, most of the gram-negative bacteria able to degrade PCP appear to belong to the genus *Sphingomonas*. Chlorinated hydroquinones are the initial metabolites during aerobic degradation of PCP and other polychlorinated phenols by gram-positive chlorophenol-mineralizing strains of *Mycobacterium chlorophenolicum* (APAJALAHTI and SALKINOJA-SALONEN, 1987; HÄGGBLOM et al., 1994), *Mycobacterium fortuitum* (HÄGGBLOM et al., 1988; NOHYNEK et al., 1993), and *Streptomyces rochei* 303 (GOLOVLEVA et al., 1992). In contrast to gram-negative bacteria, these gram-positive bacteria have different PCP-hydroxylating enzymes (HÄGGBLOM and VALO,

1995; KARLSON et al., 1995) and the further degradation of chlorinated hydroquinones follows a different pathway (HÄGGBLOM, 1992). The genetic basis behind these reactions is still unknown.

Pentachlorophenol hydroxylase was purified from two bacteria. PCP-4-hydroxylase from *Flavobacterium* sp. strain ATCC 39723 is a broad-spectrum soluble monomeric flavoenzyme using 2 mol NADPH as the reductant per mol PCP (XUN and ORSER, 1991). The second *para*-hydroxylating enzyme from *Mycobacterium chlorophenolicum* is a membrane bound cytochrome P-450 containing monooxygenase that requires FAD and NADPH (UOTILA et al., 1991). This enzyme preferably incorporates oxygen from water rather than from molecular oxygen (UOTILA et al., 1992). The *ortho*-hydroxylating dehalogenase is a soluble halohydroquinone hydrolase. The *meta*- and the second *ortho*-chlorinated positions are reductively dechlorinated in both bacteria.

3.2.7 Phenoxyalkanoic Acids

Phenoxyalkanoic acids influence the growth of plants in the same way as the naturally occurring plant hormone β-indole acetic acid. Monocotyledona such as grain and rice are highly tolerant against the selective herbicides

4-Cl-2-methylphenoxyacetic acid (MCPA), 2-(4-Cl-2-methylphenoxy)propionic acid (MCPP), and 4-(4-Cl-2-methylphenoxy)butyric acid (MCPB, forbidden in Germany since 1987). The readily biodegradable 2,4-dichlorophenoxyacetic acid (2,4-D) has been used extensively as a herbicide. Its aerobic degradation is well studied and starts with the ether cleavage into 2,4-dichlorophenol and glycolate (Fig. 23). In the degradation of 2,4,5-trichlorophenoxyacetic acid (2,4,5-T) a hydroxylase removes the side chain resulting in 2,4,5-trichlorophenol and glyoxylate in a similar way. The resulting chlorophenols are then degraded via the chlorinated catechols as described above (PIEPER et al., 1988).

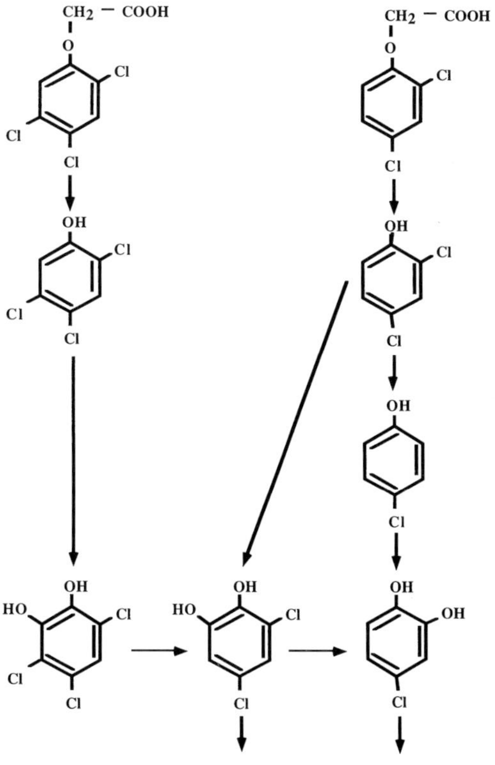

3.2.8 Halobenzoates

Halobenzoic acids are not of environmental concern, but because of their high water solubility and low toxicity these substances were used as model compounds in many studies. They were found as metabolites in the degradation of polychlorinated biphenyls and of herbicides (MÜLLER and LINGENS, 1986). In the degradation of chlorobenzoates different dehalogenation mechanisms were found. For 2-chlorobenzoate the dehalogenation in the initial dioxygenation step was described, resulting in the formation of unsubstituted catechol (FETZNER et al., 1991). In the case of 3-chlorobenzoate under aerobic conditions dioxygenation leads to the formation of chlorocatechol, which is degraded via *ortho*-cleavage (DORN et al., 1974; HARTMAN et al., 1979), and the chlorine is removed after ring cleavage. The degradation of 3-chlorobenzoate under anaerobic conditions proceeds via reductive dehalogenation to unsubstituted benzoate. The organism responsible for this dehalogenation, *Desulfomonile tiedjei*, was the first pure culture for which a reductive dehalogenating activity was shown. The organism produces energy from this dehalogenation (DOLFING and TIEDJE, 1987; MOHN and TIEDJE, 1990), and the enzyme system has been purified and characterized (NI et al., 1995). 4-Chlorobenzoate is the only aromatic chlorinated compound for which a hydrolytic dehalogenation has been shown so far. Before dehalogenation can occur the molecule has to be activated by the formation of a coenzyme A ester in an ATP-requiring process (LÖFFLER et al., 1991). The thioester is then dehalogenated, before coenzyme A is removed again by a thioesterase (Fig. 24).

3.2.9 Polychlorinated Biphenyls (PCB)

Polychlorinated biphenyls have high chemical stability, low water solubility, low toxicity, and low volatility. Therefore, these compounds have been used in many applications such as insulating liquids for transformers, as grease for vacuum pumps and turbines, as coolants, in

Fig. 23. The aerobic degradation of 2,4-dichlorophenoxyacetate and 2,4,5-trichlorophenoxyacetate.

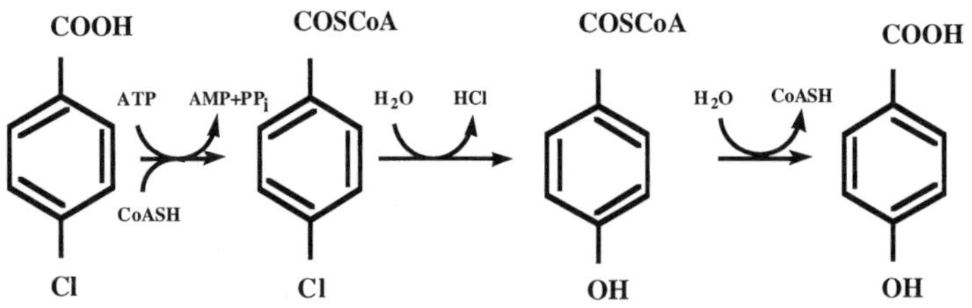

COOH   COSCoA   COSCoA   COOH

ATP AMP+PP$_i$ H$_2$O HCl   H$_2$O CoASH

CoASH

Cl   Cl   OH   OH

Fig. 24. Hydrolytic dehalogenation of 4-chlorobenzoate via formation of an intermediate coenzyme A ester. CoASH: free coenzyme A, CoAS: bound coenzyme A.

the production of wrapping paper, carbon paper, inks, paints, tires, and many other products. The calculated global burden is estimated to be 374,000 t PCB the majority of which is associated with seawater and coastal sediments (TANABE, 1988). PCB are a group of 209 different isomers. Commercial PCB are characterized by their C:Cl ratio, e.g., Arochlors® are substance mixtures containing up to 60 or 90 molecular species, so-called *congeners*. Oxidative, hydrolytic, and reductive dehalogenation of PCB has been reviewed by ABRAMOWICZ (1990) and FOCHT (1993). Anaerobic processes prefer the highly substituted congeners, while the less halogenated congeners are preferentially converted under aerobic conditions (COMMANDEUR et al., 1996). Several microorganisms have been isolated which degrade PCB under aerobic conditions. From these studies the following general principles have been concluded (DE VOOGT, 1996):

(1) less chlorinated PCB are degraded faster,
(2) dioxygenation occurs at the least substituted ring, and
(3) PCB with substituents on both rings are more recalcitrant than congeners containing an unsubstituted ring.

The 2,3-dioxygenation and subsequent ring cleavage result in the formation of chlorinated benzoates (Fig. 25). PCB are degraded anaerobically by reductive dehalogenation. Especially in river sediments the conversion of highly chlorinated PCB to lower chlorinated congeners has been demonstrated (BEDARD and QUENSEN, 1995). Generally, the substituents in *meta*- and *para*-position are removed resulting in the accumulation of *ortho*-substituted congeners which can be further degraded aerobi-

cally. *Ortho*-dehalogenation was only described in a few cases (WU et al., 1998).

Brominated biphenyls are also completely reductively dehalogenated by anaerobic sediment microorganisms (BEDARD and DORT, 1998).

3.2.10 Polychlorinated Dibenzodioxins (PCDD) and Dibenzofurans (PCDF)

Among the 75 possible PCDD and 135 PCDF there exists a wide variation in physicochemical properties, bioaccumulative tendencies, and toxicity. All these substances have extremely low water solubilities in the range of pg mL^{-1}. Polychlorinated dibenzodioxins and furans are often by-products in combustion or chemical synthesis.

Aerobic biodegradation of PCDDs appears to be restricted to congeners containing up to 4 chlorine atoms. In the initial dioxygenation the carbon–carbon bond adjacent to the ether linkage is oxidized, resulting in the formation of an unstable semiketal structure which rearranges to form a dihydroxyphenyl ether. This is then cleaved in *meta*-position (Fig. 26). In the case of the dibenzofurans hydroxylated biphenyls are formed instead of phenyl ethers.

The dioxin dioxygenase system of *Sphingomonas* sp. strain RW1 has been purified and characterized (BÜNZ and COOK, 1993). It is composed of a 2-subunit dioxygenase, a ferredoxin and a ferredoxin reductase.

The white-rot fungus *Phanaerochaete sordida* YK-624 degrades octachloro dibenzodioxin forming several less chlorinated dibenzodioxins, 4,5-dichlorocatechol, and tetrachlorocatechol (TAKADA et al., 1996). The exact mecha-

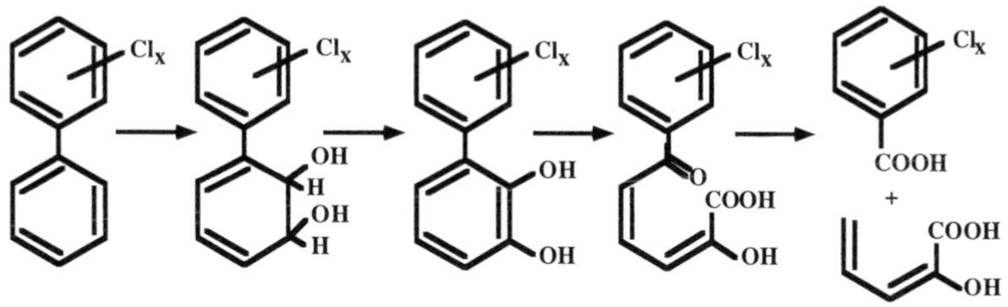

Fig. 25. Aerobic degradation pathway for polychlorinated biphenyls (from FuRUKAWA et al., 1979).

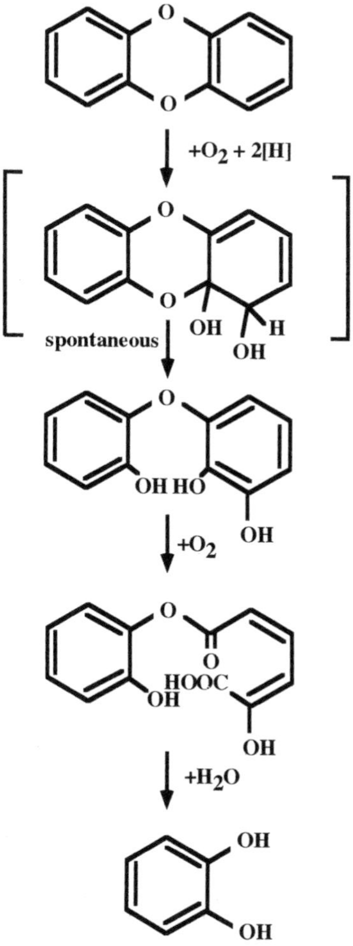

Fig. 26. Aerobic degradation of dibenzodioxin by *Sphingomonas* sp. RW1 (from WITTICH, 1998).

nated dibenzodioxins (BEURSKENS et al., 1995). BARKOVSKII and ADRIAENS (1996) proposed two dechlorination pathways for octachloro dibenzodioxin, the peri-dechlorination and the peri-lateral dechlorination pathway (Fig. 27). Another pathway is proposed for the lateral dechlorination of 1,2,3,4-tetrachloro dibenzodioxin to 2,3-dichloro dibenzodioxin in the Saale river sediment (BALLERSTEDT et al., 1997).

For the most toxic congener, 2,3,7,8-tetrachloro dibenzodioxin, the final proof for conversion by bacteria is still missing. This may be due to the properties of this molecule, but also to its high toxicity and the strict regulations on handling of this compound, which prevent an extensive search for microorganisms degrading it. Nevertheless, according to the above mentioned results, reductive dehalogenation of this compound seems possible.

4 Technical Applications and Future Aspects

The previous sections have shown that most of the chlorinated hydrocarbons can be degraded biologically. A lot of effort has been undertaken to use this potential of microorganisms for bioremediation. In the case of compounds which can be degraded easily under aerobic conditions, like the chlorophenols, several contaminated soils from wood treatment plants have been decontaminated successfully. In all cases supply of oxygen and nutrients was sufficient for successful remediation. In no case addition of exogenously grown microorganisms with the ability to degrade the chlorophenols was necessary. Addition of such

nisms underlying these conversions are not known.

Anaerobic microorganisms from river sediments reductively dechlorinated polychlori-

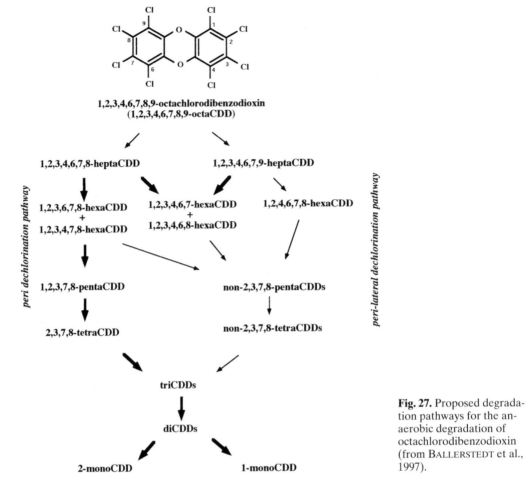

Fig. 27. Proposed degradation pathways for the anaerobic degradation of octachlorodibenzodioxin (from BALLERSTEDT et al., 1997).

organisms did not improve remediation, despite the fact that in the laboratory a positive effect of such microorganisms had been shown (LAINE and JØRGENSEN, 1996, 1997). The bacteria degrading dichloroethane have been used successfully on a technical scale to decontaminate groundwater contaminated with dichloroethane (STUCKI and THÜER, 1995).

From these results it can be concluded that compounds which can be degraded under aerobic conditions are generally degraded and do not cause environmental problems. If they persist in the environment oxygen usually is missing.

In case of the cometabolic conversion of chlorinated hydrocarbons, especially trichloro-

ethylene, several pilot-scale reactors have been set up. However, in order to achieve full-scale remediation, the problems associated with the rapid inactivation of the oxygenases by the epoxides formed have to be solved.

The degradation of chlorinated, especially highly chlorinated compounds like perchloroethene or highly chlorinated PCB by reductive dehalogenation under anaerobic conditions seems to have the most promising potential for technical applications. For reductive dehalogenation additional carbon and electron sources are required. Therefore, even if the conditions are anaerobic and the appropriate bacteria are present no degradation occurs, if the additional carbon and electron source is not present. This may explain why highly chlorinated compounds persist in anaerobic groundwater where no carbon source is available, and why some dehalogenation occurs in river sediments where low concentrations of organic degradable matter are present. As a consequence in *in situ* remediations where a highly chlorinated hydrocarbon is to be removed, a non-fermentable carbon source has to be added. At the moment it is still difficult to convince authorities that a site contaminated with a chlorinated hydrocarbon has to be contaminated with another chemical like methanol in order to achieve remediation. Nevertheless, successful *in situ* remediation of a site contaminated with perchloroethylene by using additional carbon sources on a technical scale has been reported (WILLERSHAUSEN, 1998). One major problem associated with reductive dechlorination is the fact that lower chlorinated compounds are only badly transformed, which may lead to an accumulation of these compounds. In the case of perchloroethene the reaction very often stopped at the level of dichloroethene. In the worst case highly toxic and carcinogenic vinyl chloride is formed. Our knowledge of the bacteria involved in reductive dehalogenation is still very limited. However, the more we learn about the needs of these organisms the more likely we will be able to control the extent of dehalogenation and to avoid formation of toxic chemicals and use them successfully on a technical scale.

5 References

ABRAMOWICZ, D. A. (1990), Aerobic and anaerobic biodegradation of PCBs: a review, *Biotechnology* **10**, 241–251.

ADRIAN, L., MANZ, W., SZEWZYK, U., GÖRISCH, H. (1998), Physiological characterization of a bacterial consortium reductively dechlorinating 1,2,3- and 1,2,4-trichlorobenzene, *Appl. Environ. Microbiol.* **64**, 496–503.

ALEXANDER, M. (1973), Biotechnology report: Nonbiodegradable and other recalcitrant molecules, *Biotechnol. Bioeng.* **15**, 611–647.

ALLARD, A. S., NEILSON, A. H. (1997), Bioremediation of organic waste sites: a critical review of microbiological aspects, *Int. Biodeterior. Biodegrad.* **39**, 253–285.

APAJALAHTI, J. H. A., SALKINOJA-SALONEN, M. S. (1986), Degradation of polychlorinated phenols by *Rhodococcus chlorophenolicus*, *Appl. Microbiol. Biotechnol.* **25**, 62–67.

APAJALAHTI, J. H. A., SALKINOJA-SALONEN, M. S. (1987), Dechlorination and *para*-hydroxylation of polychlorinated phenols by *Rhodococcus chlorophenolicus*, *J. Bacteriol.* **169**, 675–681.

ARMFIELD, S. J., SALLIS, P. J., BAKER, P. B., BULL, A. T., HARDMAN, D. J. (1995), Dehalogenation of haloalkanes by *Rhodococcus erythropolis* Y2, *Biodegradation* **6**, 237–246.

BADER, R., LEISINGER, T. (1994), Isolation and characterization of the *Methylophilus* sp. strain DM11 gene encoding dichloromethane dehalogenase/glutathione *S*-transferase, *J. Bacteriol.* **176**, 3466–3473.

BALLERSTEDT, H., KRAUS, A., LECHNER, U. (1997), Reductive dechlorination of 1,2,3,4-tetrachlorodibenzo-*p*-dioxin and its products by anaerobic mixed cultures from Saale river sediment, *Environ. Sci. Technol.* **31**, 1749–1753.

BARKOVSKII, A. L., ADRIAENS, P. (1996), Microbial dechlorination of historically present and freshly spiked chlorinated dioxins and diversity of dioxin-dechlorinating populations, *Appl. Environ. Microbiol.* **62**, 4556–4562.

BEDARD, D. L., QUENSEN III, J. F. (1995), Microbial reductive dechlorination of polychlorinated biphenyls, in: *Microbial Transformation and Degradation of Toxic Organic Chemicals* (YOUNG, L. Y., CERNIGLIA, C. E., Eds.), pp. 127–216. New York: Wiley-Liss.

BEDARD, D. L., DORT, H. M. V. (1998), Complete reductive dehalogenation of brominated biphenyls by anaerobic microorganisms in sediment, *Appl. Environ. Microbiol.* **64**, 940–947.

BENZET, H. J., MATSUMURA, F. (1973), Isomerization of γ-BHC to α-BHC in the environment, *Nature* (London) **243**, 480–481.

BERENDES, F., SABARTH, N., AVERHOFF, B., GOTTSCHALK, G. (1998), Construction and use of *ipb* DNA module to generate *Pseudomonas* strains with constitutive trichloroethene and isopropylbenzene oxidation activity, *Appl. Environ. Microbiol.* **64**, 2454–2462.

BEURSKENS, J. E. M., TOUSSAINT, M., DE WOLF, J., VAN DER STEEN, J. M. D., SLOT, P. C. et al. (1995), Dechlorination of chlorinated dioxins by an anaerobic microbial consortium from sediment, *Environ. Toxicol. Chem.* **14**, 939–943.

BOUCHARD, B., BEAUDET, R., VILLEMUR, R., MC SWEEN, G., LEPINE, F., BISAILLON, J.-G. (1996), Isolation and characterization of *Desulfitobacterium frapperi* sp. nov., an anaerobic bacterium which reductively dechlorinates pentachlorophenol to 3-chlorophenol, *Int. J. Syst. Bacteriol.* **46**, 1010–1015.

BRUNNER, W., STAUB, D., LEISINGER, T. (1980), Bacterial degradation of dichloromethane, *Appl. Environ. Microbiol.* **40**, 950–958.

Bundesamt für Gesundheitswesen (1995), *Fremd- und Inhaltsstoffverordnung* vom 26. Juni 1995, Schweiz, EDMZ, SR 817.021.23.

BÜNZ, P. V., COOK, A. M. (1993), Dibenzofuran 4,4α-dioxygenase from *Sphingomonas* sp. strain RW1: Angular dioxygenation by a three-component enzyme system, *J. Bacteriol.* **175**, 6467–6475.

CASTRO, C. E. (1998), Environmental dehalogenation: Chemistry and mechanism, *Rev. Environ. Contam. Toxicol.* **155**, 1–67.

CHANG, B.-V., SU, C.-J., YUAN, S.-Y. (1998), Microbial hexachlorobenzene dechlorination under three reducing conditions, *Chemosphere* **63**, 2721–2730.

CHAUDHRY, G. R., CHAPALAMADUGU, S. (1991), Biodegradation of halogenated organic compounds, *Microbiol. Rev.* **55**, 59–79.

CHRISTIANSEN, N., AHRING, B. K. (1996), *Desulfitobacterium hafniense* sp. nov., an anaerobic reductively dechlorinating bacterium, *Int. J. Syst. Bacteriol.* **46**, 1010–1015.

COLBY, J., STIRLING, D. I., DALTON, H. (1977), The soluble methane monooxygenase of *Methylococcus capsulatus* (Bath), *Biochem. J.* **165**, 395-402.

COLE, J. R., CASCARELLI, A. L., MOHN, W., TIEDJE, J. M. (1994), Isolation and characterization of a novel bacterium growing via reductive dehalogenation of 2-chlorophenol, *Appl. Environ. Microbiol.* **60**, 3536–3542.

COMMANDEUR, L. C. M., MAY, R. J., MOKROSS, H., BEDARD, D. L., REINEKE, W. et al. (1996), Aerobic degradation of polychlorinated biphenyls by *Alcaligenes* sp. JB1: metabolites and enzymes, *Biodegradation* **7**, 435–443.

COPLEY, S. D. (1997), Diverse mechanistic approaches to difficult chemical transformation: microbial dehalogenation of chlorinated aromatic compounds, *Appl. Environ. Microbiol.* **4**, 169–174.

CRAWFORD, R. L., MOHN, W. W. (1985), Microbiological removal of pentachlorophenol from soil using a *Flavobacterium, Enzyme Microb. Technol.* **7**, 617–620.

DE VOOGT, P. (1996), Ecotoxicology of chlorinated aromatic hydrocarbons, in: *Chlorinated Organic Micropollutants* (HESTER, R. E., HARRISON, R. M., Eds.), pp. 89–112. Cambridge: The Royal Society of Chemistry.

DEO, P. G., KARANTH, N. G., KARANTH, N. G. K. (1994), Biodegradation of hexachlorocyclohexane isomers in soil and food environment, *Crit. Rev. Microbiol.* **20**, 57–78.

DEWEERD, K. A., MANDELCO, L., TANNER, R. S.,

WOESE, C. R., SUFLITA, J. M. (1990), *Desulfomonile tiedje* gen. nov., a novel anaerobic, dehalogenating, sulfate-reducing bacterium, *Arch. Microbiol.* **154**, 23–30.

DOLFING, J., TIEDJE, J. M. (1987), Growth yield increase linked to reductive dechlorination in a defined 3-chlorobenzoate degrading methanogenic co-culture, *Arch. Microbiol.* **149**, 102–105.

DORN, E., HELLWIG, M., REINEKE, W., KNACKMUSS, H. J. (1974), Isolation and characterization of a 3-chlorobenzoate degrading pseudomonad, *Arch. Microbiol.* **99**, 61–70.

EDERER, M. M., CRAWFORD, R. L., HERWIG, R. P., ORSER, C. S. (1997), PCP degradation is mediated by closely related strains of the genus *Sphingomonas, Mol. Ecol.* **6**, 39–49.

EGLI, C., STROMEYER, S., COOK, A. M., LEISINGER, T. (1990), Transformation of tetra- and trichloromethane to CO_2 by anaerobic bacteria is a nonenzymic process, *FEMS Microbiol. Lett.* **68**, 207–212.

EL FANTROUSSI, S., NAVEAU, H., AGATHOS, S. N. (1998), Anaerobic dechlorinating bacteria, *Biotechnol. Prog.* **14**, 167–188.

FETZNER, S., LINGENS, F. (1994), Bacterial dehalogenases: Biochemistry, genetics, and biotechnological applications, *Microbiol. Rev.* **58**, 641–685.

FETZNER, S., MÜLLER, R., LINGENS, F. (1991), Purification and some properties of 2-halobenzoate 1,2-dioxygenase, a two-component enzyme system from *Pseudomonas cepacia* 2CBS, *J. Bacteriol.* **174**, 279–290.

FOCHT, D. D. (1993), Microbial degradation of chlorinated biphenyls, in: *Soil Biochemistry* (BOLLAG, J.-M., STROTZKY, G., Eds.), pp. 341–407. New York: Marcel Dekker.

FURUKAWA, K., TONOMURA, K., KAMIBAYASHI, A. (1979), Metabolism of 2,4,4′-trichlorobiphenyl by *Acinetobacter* sp. P6, *Agric. Biol. Chem.* **43**, 1577–1583.

GERRITSE, J., RENARD, V., PEDRO GOMES, T. M., LAWSON, P. A., COLLINS, M. D., GOTTSCHAL, J. C. (1996), *Desulfitobacterium* sp. strain PCE1, an anaerobic bacterium that can grow by reductive dechlorination of tetrachloroethene or *ortho*-chlorinated phenols, *Arch. Microbiol.* **165**, 132–140.

Gesellschaft Deutscher Chemiker (GDCh), Beratergremium für umweltrelevante Altstoffe (1985–1998), *BUA-Stoffberichte*. Weinheim: Wiley-VCH.

GISI, D., WILLI, L., TRABER, H., LEISINGER, T., VUILLEUMIER, S. (1998), Effects of bacterial host and dichloromethane dehalogenase on the competiveness of methylotrophic bacteria growing with dichloromethane, *Appl. Environ. Microbiol.* **64**, 1194–1202.

GOLOBEK, A., PRINN, R. G. (1986), A global three-dimensional model of the circulation and chemistry of CFCl₃, CF₂Cl₂, CH₃CCl, CCl₄, and N₂O, *J. Geophys. Res.* **91**, 3985–4001.

GOLOVLEVA, L.A., ZABORINA, O., PERTSOVA, R., BASKUNOV, B., SCHURUKHIN, Y., KUZMIN, S. (1992), Degradation of polychlorinated phenols by *Streptomyces rochei* 303, *Biodegradation* **2**, 201–208.

GRIBBLE, G. W. (1992), Naturally occurrring organohalogen compounds – a survey, *J. Nat. Prod.* (Lloydia) **55**, 1353–1395.

HÄGGBLOM, M. M. (1992), Microbial breakdown of halogenated aromatic pesticides and related compounds, *FEMS Microbiol. Rev.* **103**, 29–72.

HÄGGBLOM, M. M., JANKE, D., SALKINOJA-SALONEN, M. M. (1989), Hydroxylation and dechlorination of tetrachlorohydroquinone by *Rhodococcus* sp. strain CP-2 cell extracts, *Appl. Environ. Microbiol.* **55**, 516–519.

HÄGGBLOM, M. M., NOHYNEK, L. J., PALLERONI, N. J., KRONQVIST, K., NURMIAHO-LASSILA, E.-L. et al. (1994), Transfer of polychlorophenol-degrading *Rhodococcus chlorophenolicus* (APAJALATHI et al., 1986) to the genus *Mycobacterium* as *Mycobacterium chlorophenolicum* comb. nov., *Int. J. Syst. Bacteriol.* **44**, 485–493.

HÄGGBLOM, M. M., NOHYNEK, L. J., SALKINOJA-SALONEN, M. S. (1988), Degradation and O-methylation of chlorinated phenolic compounds by *Rhodococcus* and *Mycobacterium* strains, *Appl. Environ. Microbiol.* **54**, 3043–3052.

HÄGGBLOM, M. M., RIVERA, M. D., YOUNG, L. Y. (1993), Influence of alternative electron acceptors on the anaerobic biodegradability of chlorinated phenols and benzoic acids, *Appl. Environ. Microbiol.* **59**, 1162–1167.

HÄGGBLOM, M. M., VALO, R. J. (1995), Bioremediation of chlorophenol wastes, in: *Microbial Transformation and Degradation of Toxic Organic Chemicals* (YOUNG, L. Y., CERNIGLIA, C. E., Eds.), pp. 389–434. New York: Wiley-Liss

HANCOCK, T. L. C., COSTELLO, A. M., LIDSTROM, M. E., ORMELAND, R. S. (1998), Strain IMB-1, a novel bacterium for the removal of methyl bromide in fumigated agricultural soils, *Appl. Environ. Microbiol.* **64**, 2899–2905.

HANSON, R. S., HANSON, T. E. (1996), Methanotrophic bacteria, *Microbiol. Rev.* **60**, 439–471.

HARAYAMA, S., KOK, M., NEIDLE, E. L. (1992), Functional and evolutionary relationships among diverse oxygenases, *Annu. Rev. Microbiol.* **46**, 565–601.

HARDMAN, D. J. (1991), Biotransformation of halogenated compounds, *Crit. Rev. Biotechnol.* **11**, 1–40.

HARTMAN, J., REINEKE, W., KNACKMUSS, H. J. (1979), Metabolism of 3-chloro-4-chloro- and 3,5-dichlorobenzoate by a pseudomonad, *Appl. Environ. Microbiol.* **37**, 421–428.

HARTMANS, S., JANSEN, M. W., VAN DER WERF, M. J., DE BONT, J. A. (1991), Bacterial metabolism of 3-chloroacrylic acid, *J. Gen. Microbiol.* **137**, 2025–2032.

HINES, M. E., CRILL, P. M., VARNER, R. K., TALBOT, R. W., SHORTER et al. (1998), Rapid consumption of low concentrations of methyl bromide by soil bacteria, *Appl. Environ. Microbiol.* **64**, 1864–1870.

HOLLIGER, C., HAHN, D., HARMSEN, H., LUDWIG, W., SCHUMACHER, W. et al. (1998), *Dehalobacter restrictus* gen. nov. and sp. nov., a strictly anaerobic bacterium that reductively dechlorinates tetra- and trichloroethene in an anaerobic respiration, *Arch. Microbiol.* **169**, 313–321.

HOWARD, P. H. (1990), *Handbook of Environmental Fate and Exposure Data for Organic Chemicals.* Chelsea, MI: Lewis Publishers.

IUPAC (1989), Part II: *Hydrocarbons C8 to C36.* Oxford: Pergamon Press.

JANSSEN, D. B., SCHEPER, A., DIJKHUIZEN, L., WITHOLT, B. (1985), Degradation of halogenated aliphatic compounds by *Xanthobacter autotrophicus* GJ10, *Appl. Environ. Microblol.* **49**, 673–677.

JANSSEN, D. B., PRIES, F., VAN DER PLOEG, J., KAZEMIER, B., TERPSTRA, P., WITHOLT, B. (1989), Cloning of 1,2-dichloroethane degradation genes of *Xanthobacter autotrophicus* GJ10 and expression and sequencing of the *dhlA* gene, *J. Bacteriol.* **171**, 6791–6793.

JANSSEN, D. B., PRIES, F., VAN DER PLOEG, J. R. (1994), Genetics and biochemistry of dehalogenating enzymes, *Annu. Rev. Microbiol.* **48**, 163–191.

JOHNSON, T. L., FISH, W., GORBY, Y. A., TRATNYEK, P. G. (1998), Degradation of carbon tetrachloride by iron metal: Complexation effects on the oxide surface, *J. Contam. Hydrol.* **29**, 379–398.

JOSHI, D. K., GOLD, M. H. (1993), Degradation of 2,4,5-trichlorophenol by the lignin-degrading basidiomycete *Phanaerochaete chrysosporium*, *Appl. Environ. Microbiol.* **59**, 1779–1785.

KARLSON, U., ROJO, F., VAN ELSAS, J. D., MOORE, E. (1995), Genetic and serological evidence for the recognation of four pentachlorophenol-degrading bacterial strains as a species of the genus *Sphingomonas*, *Appl. Microbiol. Biotechnol.* **18**, 539–548.

KASAI, N., SUZUKI, T., FURUKAWA, Y. (1998), Chiral C3 epoxides and halohydrins: their preparation and synthetic application, *J. Mol. Catal. B: Enzymat.* **4**, 237–252.

KASCHABEK, S. R., KASBERG, T., MÜLLER, D., MARS, A. E., JANSSEN, D. B., REINEKE, W. (1998), Degradation of chloroaromatics: Purification and characterization of a novel type of chlorocatechol

2,3-dioxygenase of *Pseudomonas putida* GJ31, *J. Bacteriol.* **180**, 296–302.

KEUNING, S., JANSSEN, D. B., WITHOLT, B. (1985), Purification and characterization of hydrolytic haloalkane dehalogenase from *Xanthobacter autotrophicus* GJ10, *J. Bacteriol.* **163**, 635–639.

KHINDARIA, A., GROVER, T. A., AUST, S. D. (1995), Reductive dehalogenation of aliphatic halocarbons by lignin peroxidase of *Phanerochaete chrysosporium*, *Environ. Sci. Technol.* **29**, 719–725.

KOHLER-STAUB, D., HARTMANS, S., GÄLLI, R., SUTER, F., LEISINGER, T. (1986), Evidence for identical dichloromethane dehalogenases in different methylotrophic bacteria, *J. Gen. Microbiol.* **132**, 2837–2843.

KOHLER-STAUB, D., LEISINGER, T. (1985), Dichloromethane dehalogenase of *Hyphomicrobium* sp. strain DM2, *J. Bacteriol.* **162**, 676–681.

KREMER, S., STERNER, O., ANKE, H. (1992), Degradation of pentachlorophenol by *Mycena avenacea* TA 8480 – Identification of initial dechlorination metabolites, *Z. Naturforsch.* **47c**, 561-566.

KRUMHOLZ, L. R., SHARP, R., FISHBAIN, S. S. (1996), A freshwater anaerobe coupling acetate oxidation to tetrachloroethylene dehalogenation, *Appl. Environ. Microbiol.* **62**, 4108–4113.

LAINE, M. M., JØRGENSEN, K. S. (1996), Straw compost and bioremediated soil as inocula for the bioremediation of chlorophenol-contaminated soil, *Appl. Environ. Microbiol.* **62**, 1507–1513.

LAINE, M. M., JØRGENSEN, K. S. (1997), Effective and safe composting of chlorophenol-contaminated soil in pilot scale, *Environ. Sci. Technol.* **31**, 371–378.

LEISINGER, T. (1996), Biodegradation of chlorinated aliphatic compounds, *Curr. Opin. Biotechnol.* **7**, 295–300.

LEISINGER, T., BADER, R. (1993), Microbial dehalogenation of synthetic organohalogen compounds: Hydrolytic dehalogenases, *Chimia* **47**, 116–121.

LEUNG, K. T., CASSIDY, M. B., LEE, H., TREVORS, J. T., LOHMEIER-VOGEL, E. M., VOGEL, H. (1997), Pentachlorophenol biodegradation by *Pseudomonas* spp. UG25 and UG30, *World J. Microbiol. Biotechnol.* **13**, 305–313.

LI, S., WACKETT, L. P. (1993), Reductive dehalogenation by cytochrome P450$_{CAM}$: Substrate binding and catalysis, *Biochemistry* **32**, 9355–9361.

LIU, J. Q., KURIHARA, T., MIYAGI, M., ESAKI, N., SODA, K. (1995), Reaction mechanism of L-2-haloacid dehalogenase of *Pseudomonas* sp. YL, *Biol. Chem.* **270**, 18309–18312.

LIU, J. Q., KURIHARA, T., ESAKI, N., SODA, K. (1994), Reconsideration of the essential role of a histidine residue of L-2-halo acid dehalogenase, *J. Biochem.* **116**, 248–249.

LÖFFLER, F., MÜLLER, R., LINGENS, F. (1991), Dehalogenation of 4-chlorobenzoate by 4-chlorobenzoate dehalogenase from *Pseudomonas* sp. CBS3: an ATP/coenzyme A dependent reaction, *Biochem. Biophys. Res. Commun.* **176**, 1106–1111.

LÖFFLER, F. E., CHAMPINE, J. E., RITALAHTI, K. M., SPRAGUE, S. J., TIEDJE, J. M. (1997), Complete reductive dechlorination of 1,2-dichloropropane by anaerobic bacteria, *Appl. Environ. Microbiol.* **63**, 2870–2875.

LONTOH, S., SEMRAU, J. D. (1998), Methane and trichloroethylene degradation by *Methylosinus trichosporium* OB3b expressing particulate methane monooxygenase, *Appl. Environ. Microbiol.* **64**, 1106–1114.

MÄGLI, A., MESSMER, M., LEISINGER, T. (1998), Metabolism of dichloromethane by the strict anaerobe *Dehalobacterium formicoaceticum*, *Appl. Environ. Microbiol.* **64**, 646–650.

MARKUS, A., KLAGES, U., KRAUSS, S., LINGENS, F. (1984), Oxidation and dehalogenation of 4-chlorophenylacetate by a two-component system from *Pseudomonas* sp. strain CBS3, *J. Bacteriol.* **160**, 618–621.

MARKUS, A., KREKEL, D., LINGENS, F. (1986), Purification and some properties of component A of the 4-chlorophenylacetate 3,4-dioxygenase from a *Pseudomonas* species strain CBS, *J. Biol. Chem.* **261**, 12883–12888.

MARS, A. E., PRINS, G. T., WIETZES, P., DE KONING, W., JANSSEN, D. B. (1998), Effect of trichloroethylene on the competitive behavior of toluene-degrading bacteria, *Appl. Environ. Microbiol.* **64**, 208–215.

MAYMO-GATELL, X., CHIEN, Y., GOSSETT, J. M., ZINDER, S. H. (1997), Isolation of a bacterium that reductively dechlorines tetrachloroethene to ethene, *Science* **276**, 1568–1571.

MCALLISTER, K. A., LEE, H., TREVORS, J. T. (1996), Microbial degradation of pentachlorophenol, *Biodegradation* **7**, 1.

MESSMER, M., REINHARDT, S., WOHLFARTH, G., DIEKERT, G. (1996), Studies on methy chloride dehalogenase and *O*-demethylase in cell extracts of the homoacetogen strain MC based on a newly developed coupled enzyme assay, *Arch. Microbiol.* **165**, 18–25.

MIDDELDORP, P. J. M. (1997), Microbial transformation of highly persistent chlorinated pesticides and industrial chemicals, *Thesis*, Wageningen Agricultural University, The Netherlands.

MIKESELL, M. D., BOYD, S. A. (1986), Complete reductive dechlorination and mineralization of pentachlorophenol by anaerobic microorganisms, *Appl. Environ. Microbiol.* **52**, 861–865.

MIYAUCHI, K., SUH, S. K., NAGATA, Y., TAKAGI, M. (1998), Cloning and sequencing of a 2,5-dichloro-hydroquinone reductive dehydrogenase gene whose product is involved in degradation of γ-hexachlorocyclohexane by *Sphingomonas paucimobilis, J. Bacteriol.* **180**, 1354–1359.

MOHN, W. W., TIEDJE, J. M. (1990), Strain DCB-1 conserves energy for growth from reductive dechlorination coupled to formate oxidation, *Arch. Microbiol.* **153**, 267–271.

MOHN, W. W., TIEDJE, J. M. (1992), Microbial reductive dehalogenation, *Microbiol. Rev.* **56**, 482–507.

MONTZKA, S. A., BUTLER, J. H., MYERS, R. C. (1996), Decline in the tropospheric abundance of halogen from halocarbons: Implications for stratospheric ozone depletion, *Science* **272**, 1318–1322.

MORAN, B. N., HICKEY, W. J. (1997), Trichloroethylene biodegradation by mesophilic and psychrophilic ammonia oxidizers and methanotrophs in groundwater microcosms, *Appl. Environ. Microbiol.* **63**, 3866–3871.

MÖRSBERGER, F.-M., MÜLLER, R., OTTO, M. K., LINGENS, F., KULBE, K. D. (1991), Purification and characterization of 2-haloacid dehalogenase II from *Pseudomonas* spec. CBS3, *Biol. Chem. Hoppe Seyler* **372**, 915–922.

MÜLLER, R., LINGENS, F. (1986), Mikrobieller Abbau halogenierter Kohlenwasserstoffe: Ein Beitrag zur Lösung vieler Umweltprobleme, *Angew. Chem.* **98**, 778–787.

NAGATA, Y., NARIYA, T., OHTOMO, R., FUKUDA, M., YANO, K., TAKAGI, M. (1993), Cloning and sequencing of a dehalogenase gene encoding an enzyme with hydrolase activity involved in the degradation of γ-hexachlorocyclohexane in *Pseudomonas paucimobilis, J. Bacteriol.* **175**, 6403–6410.

NI, S., FREDRICKSON, J. K., XUN, L. (1995), Purification and characterization of a novel 3-chlorobenzoate-reductive dehalogenase from the cytoplasmic membrane of *Desulfomonile tiedjei* DCB-1, *J. Bacteriol.* **177**, 5135–5139.

NOHYNEK, L. J., SUHONEN, E. L., NURMIAHO-LASSILA, E.-L., HANTULA, J., SALKINOJA-SALONEN, M. S. (1995), Description of four pentachlorophenol-degrading bacterial strains as *Sphingomonas chlorophenolica* sp. nov., *Syst. Appl. Microbiol.* **18**, 527–538.

NOHYNEK, L. J., HÄGGBLOM, M. M., PALLERONI, N. J., KRONQVIST, K., NURMIAHO-LASSILA, E.-L., SALKINOJA-SALONEN, M. S. (1993), Characterization of a *Mycobacterium fortuicum* strain capable of degrading polychlorinated phenolic compounds, *Syst. Appl. Microbiol.* **16**, 126–134.

OLDENHUIS, R., VINK, R. L. J. M., JANSSEN, D. B., WITHOLT, B. (1989), Degradation of chlorinated aliphatic hydrocarbons by *Methylosinus trichosporium* OB3b expressing soluble methane monooxygenase, *Appl. Environ. Microbiol.* **55**, 2819–2826.

ORSER, C. S., LANGE, C. C. (1994), Molecular analysis of pentachlorophenol degradation, *Biodegradation* **5**, 277–288.

PERLINGER, J. A. (1994), Reduction of polyhalogenated alkanes by electron transfer mediators in aqueous solution, *Thesis*, Swiss Federal Institute of Technology, Zürich.

PIEPER, D. H., REINEKE, W., ENGESSER, K. H., KNACKMUSS, H. J. (1988), Metabolism of 2,4-dichlorophenoxyacetic acid 4-chloro-2-methyl-phenoxyacetic acid and 2-methylphenoxyacetic acid by *Alcaligenes eutrophus* JMP 134, *Arch. Microbiol.* **150**, 95–102.

RADEHAUS, P. M., SCHMIDT, S. K. (1992), Characterization of a novel *Pseudomonas* sp. that mineralizes high concentrations of pentachlorophenol, *Appl. Environ. Microbiol.* **58**, 2879–2885.

RASCHE, M. E., HICKS, R. E., HYMAN, M. R., ARP, D. J. (1990), Oxidation of monohalogenated ethanes and *n*-chlorinated alkanes by whole cells of *Nitrosomonas europaea, J. Bacteriol.* **172**, 5368–5373.

REINEKE, W., KNACKMUSS, H. J. (1988), Microbial degradation of haloaromatics, *Annu. Rev. Microbiol.* **42**, 263–287.

RESNIK, S. M., CHAPMAN, P. J. (1994), Physiological properties and substrate specifity of a pentachlorophenol degrading *Pseudomonas* species, *Biodegradation* **5**, 47–54.

ROZEBOOM, H. J., KINGMA, J., JANSSEN, D. B., DIJKSTRA, B. W. (1988), Crystallization of haloalkane dehalogenase from *Xanthobacter autotrophicus* GJ10, *J. Mol. Biol.* **200**, 611–612.

SAHU, S. K., PATNAIK, K. K., BHUYAN, S., SREEDHARAN, B., KURIHARA, N. et al. (1995), Mineralization of α-, γ- and β-isomers of hexachlorocyclohexane by a soil bacterium under aerobic conditions, *J. Agric. Food Chem.* **43**, 833–837.

SANDER, P., WITTICH, R. M., FORTNAGEL, P., WILKES, H., FRANCKE, W. (1991), Degradation of 1,2,4-trichloro- and 1,2,4,5-tetrachlorobenzene by *Pseudomonas* strains, *Appl. Environ. Microbiol.* **57**, 1430–1440.

SANFORD, R. A., COLE, J. R., LÖFFLER, F. E., TIEDJE, J. M. (1996), Characterization of *Desulfitobacterium chlororespirans* sp. now., which grows by coupling the oxidation of lactate to the reductive dechlorination of 3-chloro-4-hydroxybenzoate, *Appl. Environ. Microbiol.* **62**, 3800–3808.

SCHENK, T., MÜLLER, R., MÖRSBERGER, F., OTTO, K., LINGENS, F. (1989), Enzymatic dehalogenation of pentachlorophenol by extracts from *Arthrobacter* sp. strain ATCC 33790, *J. Bacteriol.* **171**, 5487–5491.

SCHNEIDER, B., MÜLLER, R., FRANK, R., LINGENS, F. (1993), Site-directed mutagenesis of the 2-haloalkanoic acid dehalogenase I gene from *Pseudomonas* sp. strain CBS3 and its effect on catalytic activity, *Biol. Chem. Hoppe Seyler* **374**, 489–496.

SCHNEIDER, B., MÜLLER, R., FRANK, R., LINGENS, F. (1991), Complete nucleotide sequences and comparison of the structural genes of two 2-haloalkanoic acid dehalogenases from *Pseudomonas* sp. CBS3, *J. Bacteriol.* **173**, 1530–1535.

SCHOLTZ, R., SCHMUCKLE, A., COOK, A. M., LEISINGER, T. (1987), Degradation of eighteen 1-monohaloalkanes by *Arthrobacter* sp. strain HA 1, *J. Gen. Microbiol.* **133**, 267–274.

SCHOLZ-MURAMATSU, H., NEUMANN, A., MESSMER, M., MOORE, E., DIEKERT, G. (1995), Isolation and characterization of *Dehalospirillum multivorans* gen. nov., sp. nov., a tetrachloroethene-utilizing, strictly anaerobic bacterium, *Arch. Microbiol.* **163**, 48–56.

SCHWARZE, R., BROKAMP, A., SCHMIDT, F. R. (1997), Isolation and characterization of dehalogenases from 2,2-dichloropropionate-degrading soil bacteria, *Curr. Microbiol.* **34**, 103–109.

SEECH, A. G., MARVAN, I. J., TREVORS, J. T. (1994), On-site/*ex-situ* bioremediation of industrial soils containing chlorinated phenols and polycyclic aromatic hydrocarbons, in: *Bioremediation of Chlorinated and Polycyclic Aromatic Hydrocarbon Compounds* (HINCHEE, A. L., Ed.), pp. 451–455. Boca Raton, FL: Lewis Publishers.

SHARMA, P. K., MCCARTY, P. L. (1996), Isolation and characterization of a facultatively aerobic bacterium that reductively dehalogenates tetrachloroethene to *cis*-1,2-dichloroethene, *Appl. Environ. Microbiol.* **62**, 761–765.

SIUDA, J. F., DEBERNARDIS, J. F. (1973), Naturally occurring halogenated organic compounds, *Lloydia* **36**, 107–143.

SLATER, J. H., BULL, A. T., HARDMAN, D. (1995), Microbial dehalogenation, *Biodegradation* **6**, 181–189.

SLATER, J. H., BULL, A. T., HARDMAN, D. J. (1997), Microbial dehalogenation of halogenated alkanoic acids, alcohols and alkanes, *Adv. Microb. Physiol.* **38**, 133–176.

SLATER, J. H., LOVATT, D., WEIGHTMAN, A. J., SENIOR, E., BULL, A. T. (1979), The growth of *Pseudomonas putida* on chlorinated aliphatic acids and its dehalogenase activity, *J. Gen. Microbiol.* **114**, 125–136.

SMITH, M. H., WOOD, S. L. (1994), Regiospecificity of chlorophenol reductive dechlorination by vitamin B_{12}, *Appl. Environ. Microbiol.* **60**, 4111–4115.

STUCKI, G., GÄLLI, R., EBERSOLD, H. R., LEISINGER, T. (1981), Dehalogenation of dichloromethane by cell extracts of *Hyphomicrobium* DM 2, *Arch. Microbiol.* **130**, 366–371.

STUCKI, G., THÜER, M. (1995), Experiences of a largescale application of 1,2-dichloroethane degrading microorganism for groundwater treat-ment, *Environ. Sci. Technol.* **29**, 2339–2345.

SUSARLA, S., YONEZAWA, Y., MASUNAGA, S. (1997), Reductive dehalogenation of chloroanilines in anaerobic estuarine sediment, *Environ. Technol.* **18**, 75–83.

TAKADA, S., NAKAMURA, M., MATSUEDA, T., KONDO, R., SAKAI, K. (1996), Degradation of polychlorinated dibenzo-*p*-dioxins and polychlorinated dibenzofurans by the white rot fungus *Phanerochaete sordida* YK-624, *Appl. Environ. Microbiol.* **62**, 4323–4328.

TANABE, S. (1988), PCB problems in the future: Foresight from current knowledge, *Environ. Pollut.* **50**, 5–28.

UOTILA, J. S., KITUNEN, V. H., APAJALAHTI, J. H. A., SALKINOJA-SALONEN, M. S. (1992), Environment-dependent mechanism of dehalogenation by *Rhodococcus chlorophenolicus* PCP-1, *Appl. Microbiol. Biotechnol.* **37**, 408–412.

UOTILA, J. S., SALKINOJA-SALONEN, M. S., APAJALAHTI, J. H. A. (1991), Dechlorination of pentachlorophenol by membrane bound enzymes of *Rhodococcus chlorophenolicus* PCP-1, *Biodegradation* **2**, 25–31.

UTKIN, I., WOESE, C., WIEGEL, J. (1994), Isolation and characterization of *Desulfitobacterium dehalogenans* gen. nov. sp., an anaerobic bacterium which reductively dechlorinates chlorophenolic compounds, *Int. J. Syst. Bacteriol.* **44**, 612–619.

VAN DEN TWEEL, W. J. J., KOK, J. B., DE BONT, J. A. M. (1987), Reductive dechlorination of 2,4-dichlorobenzoate to 4-chlorobenzoate and hydrolytic dehalogenation of 4-chloro-4-bromo- and 4-iodobenzoate by *Alcaligenes denitrificans* NTB-1, *Appl. Environ. Microbiol.* **53**, 810–815.

VAN DER MEER, J. R., DE VOS, W. M., HARAYAMA, S., ZEHNDER, A. J. B. (1992), Molecular mechanisms of genetic adaptation to xenobiotic compounds, *Microbiol. Rev.* **56**, 677–694.

VANNELLI, T., STUDER, A., KERTESZ, M., LEISINGER, T. (1998), Chlormethane metabolism by *Methylobacterium* sp. strain CM4, *Appl. Environ. Microbiol.* **64**, 1933–1936.

VERSCHUEREN, K. H. G., SELJEE, F., ROZEBOOM, H. J., KALK K. H., DIJKSTRA, B. W. (1993), Crystallographic analysis of the catalytic mechanism of haloalkane dehalogenase, *Nature* **363**, 693–698.

VOGEL, T. M., CRIDDLE, C. S., MCCARTY, P. L. (1987), Transformations of halogenated aliphatic compounds, *Environ. Sci. Technol.* **21**, 722–736.

VOGEL, T. M., MCCARTHY, P. L. (1985), Biotransformation of tetrachlorethylene to trichloroethylene, dichlorethylene, vinylchloride and carbon dioxide under methanogenic conditions, *Appl. Environ. Microbiol.* **49**, 1080–1083.

WACKETT, L. P. (1995), Bacterial co-metabolism of halogenated organic compounds, in: *Microbial*

Transformation and Degradation of Toxic Organic Chemicals (YOUNG, L. Y., CERNIGLIA, C. E., Eds.), pp. 217–241. New York: Wiley-Liss.

WILD, A., HERMANN, R., LEISINGER, T. (1996), Isolation of an anaerobic bacterium which reductively dechlorinates tetrachloroethene and trichloroethene, *Biodegradation* **7**, 507–511.

WILLERSHAUSEN, H. (1998), Sanierung des Geländes Dr. Freund, Sandhausen, in: *Workshop: Abbau chlorierter Kohlenwasserstoffe* (BRYNIOK, D., Ed.). Stuttgart: Fraunhofer IGB.

WISCHNAK, C., LÖFFLER, F. E., LI, J., URBANCE, J. W., MÜLLER, R. (1998), *Pseudomonas* sp. strain 273, an aerobic α,ω-dichloroalkane-degrading bacterium, *Appl. Environ. Microbiol.* **64**, 3507–3511.

WITTICH, R.-M. (1998), Aerobic degradation by bacteria of dibenzo-*p*-dioxins, dibenzofurans, diphenyl ethers and their halogenated derivatives, in: *Biodegradation of Dioxins and Furans* (WITTICH, R.-M., Ed.), pp. 1–28. Austin, TX: Landes Bioscience.

WU, Q., SOWERS, K. R., MAY, H. D. (1998), Microbial reductive dechlorination of Aroclor 1260 in anaerobic slurries of estuarine sediments, *Appl. Environ Microbiol.* **64**, 1052–1058

XUN, L., ORSER, C. S. (1991), Purification and properties of pentachlorophenol hydroxylase, a flavoprotein from *Flavobacterium* sp. strain ATCC 39723, *J. Bacteriol.* **173**, 4447–4453.

XUN, L., TOPP, E., ORSER, C. S. (1992), Purification and characterization of a tetrachloro-*p*-hydroquinone reductive dehalogenase from a *Flavobacterium* sp., *J. Bacteriol.* **174**, 8003–8007.

YOKOTA, T., OMORI, T., KODAMA, T. (1987), Purification and properties of haloalkane dehalogenase from *Corynebacterium* sp. strain m15-3, *J. Bacteriol.* **169**, 4049–4054.

YOU, I. S., BARTHA, R. (1982), Metabolism of 3,4-dichloroaniline by *Pseudomonas putida*, *J. Agric. Food Chem.* **30**, 274–277.

ZEYER, J., WASSERFALLEN, A., TIMMIS, K. N. (1985), Microbial mineralization of ring-substituted anilines through an *ortho*-cleavage pathway, *Appl. Environ. Microbiol.* **50**, 447–453.

11 Microbial Degradation of Compounds with Nitro Functions

KARL-HEINZ BLOTEVOGEL
THOMAS GORONTZY

Oldenburg, Germany

List of Abbreviations

ADNT	amino dinitrotoluene
2-ANT	2-amino nitrotoluene
2,4-DAHAT	2,4-diamino-6-hydroxylamino toluene
DANT	diamino nitrotoluene
Dinoseb	2-*sec*-butyl-4,6-dinitrophenol
2,4-DNT	2,4 dinitrotoluene
DPA	diphenylamine
EPA	Environmental Protection Agency
FAD	flavine adenine dinucleotide
FMN	flavine mononucleotide
GDN	glycerol dinitrate ester
GMN	glycerol mononitrate ester
HEXYL	hexanitro diphenylamine
NADH	nicotinamide adenine hydroxy dinucleotide
NADPH	nicotinamide adenine dinucleotide phosphate
PAH	polycyclic aromatic hydrocarbons
PETN	pentaerythritol tetranitrate
SABRE™	J. R. Simplot Anaerobic Bioremediation
TAT	2,4,6-triamino toluene
TNT	2,4,6-trinitro toluene

1 Introduction

The Earth is faced with the problem of water, soil, and even air pollution caused by several classes of chemicals. A large variety of chemical compounds have entered the environment as a result of our industrial but also daily household activities in the past and present. The improper discharge of wastes, accidental spills, or deliberate releases of these chemicals contribute to the contamination of our environment, and pollution will certainly continue into the future.

Microbial degradation of xenobiotics is a promising natural process which leads to the detoxification and hopefully to the elimination of such compounds from the environment. But most scientists and engineers believe in the thermal combustion of recalcitrant compounds. Because in response to the contamination the structure of the microbial community changes or even vanishes if the pollutant is toxic.

Only a few naturally occurring nitro-substituted organics have been identified to date that may be of importance concerning the observed recalcitrance and the biodegradability of these molecules. The reasons of persistence may be also a result of the relatively few catabolic pathways that characterize microbial cells (ALEXANDER, 1994). Therefore, it is not surprising that chemicals of xenobiotic character (i.e., not naturally occurring) will be metabolized or even mineralized to a certain extent only if biological systems exist that are able to catalyze their conversion into products which are intermediates or substrates of existing pathways (ALEXANDER, 1994). A main reason for the persistence of many organic compounds are their physiologically uncommon substitutes such as halogens, SO_3H, NO_2, $N=N-R_2$. It is their strong electron withdrawing effect which impedes oxygenolytic degradation processes (RIEGER and KNACKMUSS, 1995). On the other hand degradation is stimulated by the presence of hydroxyl, carboxyl, ester, or amino groups. In other words, the biodegradability of an organic molecule is determined by the identity of the added substituent.

Other possible effects of pollutants on microorganisms have to be considered. Factors which strongly influence toxicity and degradability are

(1) sorption/desorption
(2) bioavailability (often interpreted as water solubility)
(3) concentration
(4) chemical nature (natural or xenobiotic compound, toxic or not)
(5) abiotic transformations

(HADERLEIN and SCHWARZENBACH, 1993; ALEXANDER, 1994; HADERLEIN et al., 1996).

But nitroorganic molecules are not only recalcitrant, they also express a certain toxicity and even mutagenicity or cancerogenicity (RICKERT et al., 1986; SCHMEISSER and WIESSLER, 1995; HONEYCUTT et al., 1996; HASSPIELER et al., 1997; for review, see YINON, 1990).

Nitro compounds are released into the environment almost exclusively from anthropogenic sources. The main fraction of nitrated molecules is derived from dyes, pesticides, and especially explosives. Because nitro groups can be easily converted to other functional groups, they are useful as intermediate compounds in many chemical syntheses as mentioned above. Nitrobenzene, e.g., is the classical feedstock for aniline manufacture, although chlorobenzene and phenol are being used more and more (WEISSERMERL and ARPE, 1997). Nitrotoluenes are building blocks for explosives of the first rank for the production of 2,4,6-trinitrotoluene (TNT). A special group are the nitro musks which are used in the production of perfumes (IPPEN, 1994a, b).

But nitro compounds are synthesized not only in the chemical industry but also by various organisms. Best known are antibiotics such as chloramphenicol or azomycin, and plants, especially leguminosae produce nitro toxins such as 3-nitro-1-propionic acid and 3-nitro-1-propanol (BUTRUILLE and DOMINGUEZ, 1972; THALLER and TURNER, 1972; GLASBY, 1976; MIX et al., 1982; WILLIAMS and BARNEBY, 1977; SCHMEISSER and WIESLER, 1995). Some examples are given in Fig. 1.

Oxygenolytic attack of nitroaromatic compounds can be accomplished by monooxygenase and dioxygenase enzyme systems. But due to the strong electron withdrawing character of the nitro groups, nitroaromatic compounds are electron-deficient. It is the electron-deficiency of the aromatic ring which impedes possible electrophilic attack by oxygenases of aerobic microorganisms. Among the known reactions of the aerobic degradation of nitroaromatic compounds is the initial oxygenation of mononitroaromatic or to a limited degree also of dinitroaromatic compounds.

Therefore, it was not surprising when early research indicated that mineralization of chemically synthesized nitro compounds did neither occur under oxic nor under anoxic conditions. It must also be taken into account that naturally occurring nitro-substituted molecules generally only contain one nitro group, polysubstitution of biological molecules with nitro groups is not reported in the literature.

First reports on microbiological degradation of nitroaromatics were published in the 1950s, although the occurrence and synthesis of natural nitro compounds was found several years earlier (for review see SPAIN, 1995a; GORONTZY et al., 1994).

2 Microbial Catabolism of Nitrated Hydrocarbons

2.1 Microbial Degradation of Nitroalkanes

Although until today several studies have been done on the metabolism of nitroalkanes by fungi or streptomycetes relatively little is known about the biochemistry for catabolizing nitroalkanes (LITTLE, 1951; DHAWALE and HORNEMANN, 1979).

Recently a nitroalkane oxidase was isolated from *Fusarium oxysporum* which catalyzes the oxidation and denitration of various nitroalkanes to aldehydes, transferring the electrons to oxygen to form hydrogen peroxide (HEASLEY and FITZPATRICK, 1996).

3-nitro-1-propionic acid
Astragalus sp.
(WILLIAMS and BARNEBY, 1977)

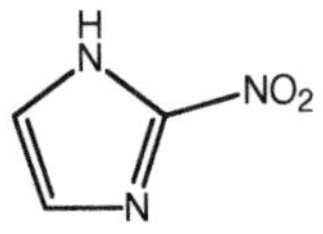

3-nitro-1-propanol

azomycin (5-nitroimidazole)
Pseudomonas fluorescens
(SHOJI, 1989)

4-nitroanisole
Lepista diemii Singer
(THALLER and TURNER, 1972)

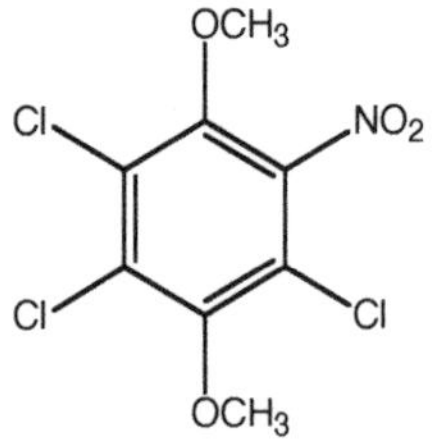

3,5,6-trichloro-2-nitro-1,4-dimethoxybenzene
Phellinus robiniae
(BUTRUILLE and DOMINGUEZ, 1972)

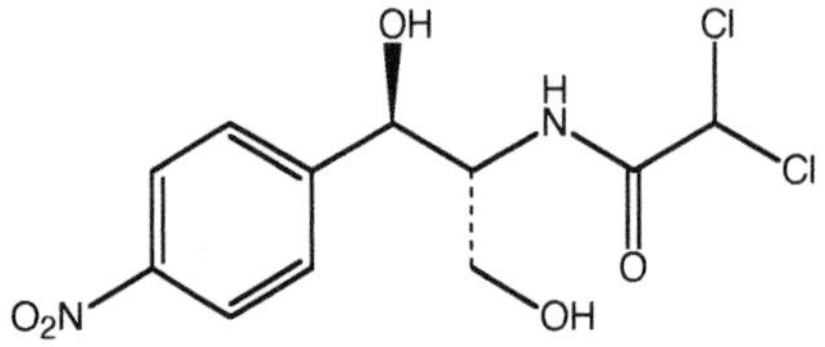

chloramphenicol
Streptomyces venezuelae
(GLASBY, 1976)

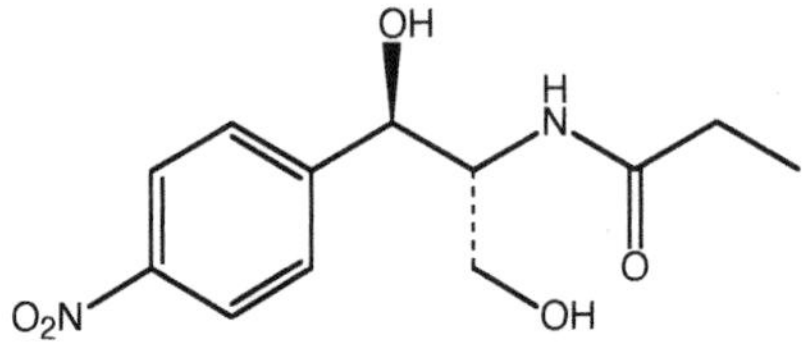

D-threo-1-*p*-nitrophenyl-2-propionyl-
aminopropane-1,3-diol
Arthrobacter oxameticus var. *propiophenolicus*
(GLASBY, 1976)

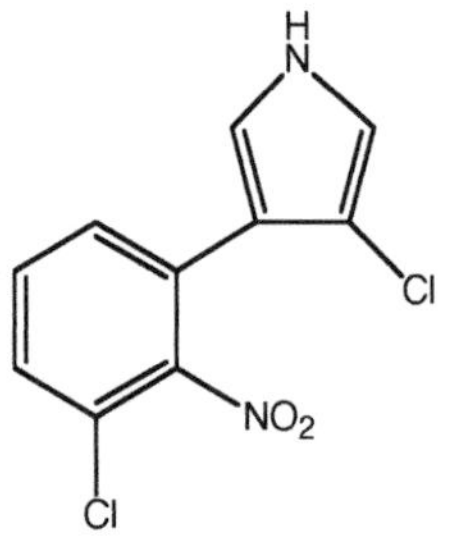

pyrrolnitrin
Pseudomonas pyrrocina / *Ps. acidula*
(GLASBY, 1976)

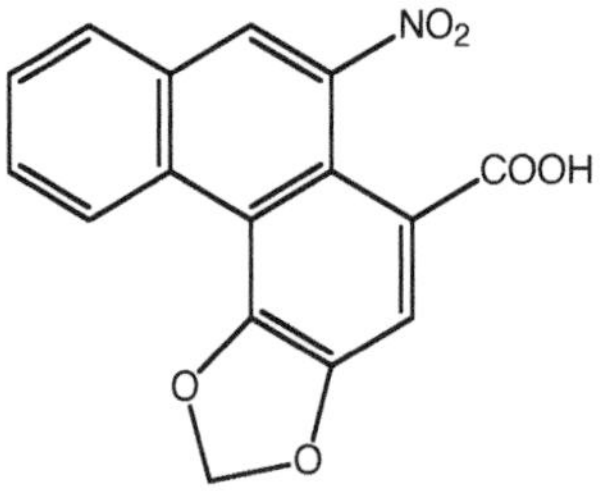

aristolochic acid
Aristolochia clematitis
(KARRER, 1985)

Fig. 1. Selected organics with nitro functions of biogenic origin.

The reaction mechanism of nitroalkane oxidase is proposed as follows:

$$R_1-CH(NO_2)-R_2+O_2+H_2O \rightarrow \\ R_1-CO-R_2+HNO_2+H_2O_2 \qquad (1)$$

Flavine adenine dinucleotide (FAD) was identified as the prosthetic group of nitroalkane oxidase. A notable feature of this enzyme is the modified identity of the FAD which is labile in the presence of oxygen when it is dissociated from the protein moiety. Recently it was identified as 5-isobutyl-FAD (GADDA et al., 1997).

Different kinds of flavoenzymes are responsible for the biochemical oxidation of nitroalkanes, but only for nitroalkane oxidase from *Fusarium oxysporum* could it be shown that this is obviously its physiological role.

2-nitropropane dioxygenase has been isolated from the yeast *Hansenula mrakii* which also contains FAD as a prosthetic group (KIDO et al., 1976). The reaction proceeds as follows:

$$2\,CH_3CH(NO_2)CH_3+O_2 \rightarrow \\ 2\,CH_3COCH_3+2\,HNO_2 \qquad (2)$$

This enzyme is unique in that the oxygen atoms of the dioxygen molecule are split and separately incorporated into two molecules of the substrates (KIDO et al., 1976). Normally dioxygenases incorporate both oxygen atoms into a single molecule of the substrate.

GORLATOVA et al. (1998) reported first on a flavine mononucleotide(FMN)-dependent 2-nitropropane dioxygenase from *Neurospora crassa* purified to homogeneity.

2.2 Microbial Metabolism of Nitrate Esters

Nitrate esters such as glycerol trinitrate (nitroglycerine) are well known as explosives or as vasodilators in pharmaceuticals (WHITE and SNAPE, 1993). It has not been well established until today whether $C-O-NO_2$ linkages occur naturally or not. Therefore, nitrate esters have been considered to be real xenobiotics (HALL et al., 1992). The bioavailability of nitrate esters has been disputed in the literature;

it was even concluded that some components are not degradable (WYMAN et al., 1984). Although it was shown later that nitro groups could be removed biologically (for review, see KAPLAN and WALKER, 1992), a half-life in the order of 2 years was assumed for some compounds such as diethylene glycol dinitrate (HAAG et al., 1991). But it was found in recent years that these often multi-nitrated molecules undergo sequential biological denitration by *Enterobacter* and *Bacillus* strains (MENG et al., 1995; SUN et al., 1996). Interestingly these studies indicated that nitrate esters are not required for denitration activity, even the regeneration of cofactors such as NAD(P)H or ATP seems not to be necessary for this activity. Therefore, it can be concluded that this is an additional unspecific function of certain constitutive enzymes during growth. Assimilation of nitrite was shown during degradation studies in pure cultures of *Pseudomonas* sp. and *Agrobacterium radiobacter*; the cells were able to denitrate glycerol tri- and dinitrate (GDN) esters to mononitrates (GMN), but not beyond (WHITE et al., 1996). Recently, an NADH-dependent glycerol trinitrate reductase enabling *Agrobacterium radiobacter* to utilize nitrate esters as sources of nitrogen was purified and characterized, and its gene was cloned and sequenced (SNAPE et al., 1997). Similar activities were found in cells of *Enterobacter cloacae*, except that this enzyme was dependent on NADPH (BINKS et al., 1996). These results are in contrast to the findings of MENG et al. (1995) in so far that cofactor dependence (i.e., NADH or NADPH) for bacterial denitration was clearly shown. But mineralization of the carbon skeleton did not occur. In a very recent study ACCASHIAN et al. (1998) were able to demonstrate for the first time that nitroglycerine can be used as the sole carbon, nitrogen, and energy source by a not yet defined mixed microbial culture under oxic conditions. The utilized pathway has been mentioned above with the exception that glycerol mononitrate (GMN) also disappeared during growth.

2.3 Aerobic Microbial Degradation of Nitroaromatic Compounds

2.3.1 General Apects of the Catabolism of Nitro-Substituted Aromatics by Aerobic Bacteria

Generally microbial mineralization of homo-cyclic nitroaromatics occurs by means of four presently known mechanisms:

(1) an initial oxygenation reaction yielding nitrite,
(2) an initial reduction yielding aromatic amines which may be further metabolized,
(3) a complete reductive elimination of the nitro group yielding nitrite,
(4) a partial reduction of the nitro group to a hydroxylamine which after a subsequent replacement is metabolized further (MARVIN-SIKKEMA and DE BONT, 1994).

The aromatic compounds resulting from these reactions are then degraded via known degradation patterns of homocyclic aromatics (KUHN et al., 1985; SMITH, 1990; FIELD et al., 1995). Nitro aromatic compounds can serve as nitrogen and as carbon source. But many microorganisms, preferably anaerobes, are only able to reduce such molecules to the corresponding amines. In this case it functions as an artificial electron acceptor, or it leads to its detoxification (DRZYZGA et al., 1995a).

Oxidative removal of nitro groups from the aromatic nucleus has been demonstrated with aerobic bacteria of different genera. Some of them use it as a nitrogen source only and leave the benzene ring untouched (MARVIN-SIKKEMA and DE BONT, 1994).

There are several reports on bacteria which were able to degrade nitrophenols and nitrobenzoates including *Pseudomonas* and *Nocardia (Rhodococcus)* strains (for review, see HIGSON, 1992; SPAIN, 1995b). But only in recent years light was brought to the biochemistry and enzymology of the catabolism of nitroaromatics mainly due to the works of the groups of DE BONT, GIBSON, KNACKMUSS, SPAIN, TIMMIS, and ZEYER.

About four different strategies for the removal or transformation of the nitro group by aerobic microorganisms are known today (SPAIN, 1995b). Firstly, there is a monooxygenase catalyzed reaction (Fig. 2) which converts 4-nitrophenol to catechol and nitrite via 1,4-benzoquinone as an intermediate (ZEYER and KEARNEY, 1984; ZEYER and KOCHER, 1988).

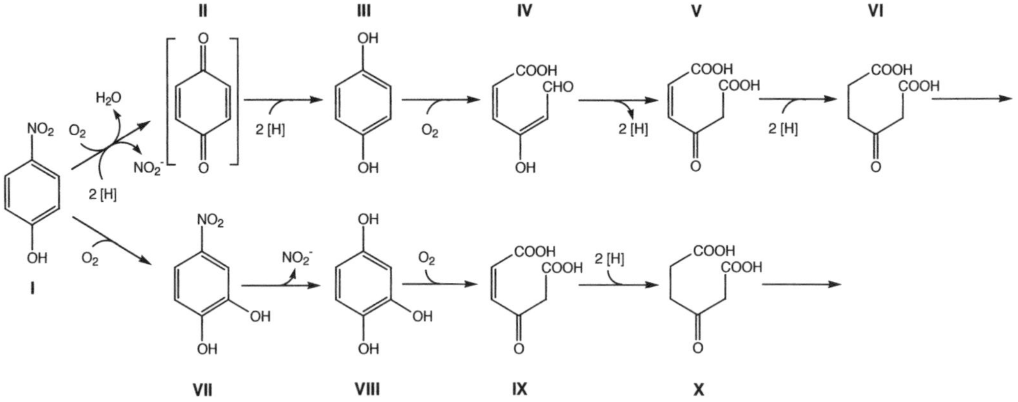

Fig. 2. Scheme of the degradation pathways for nitro-substituted aromatics initiated by monooxygenases as shown for *p*-nitrophenole. I: *p*-nitrophenol, II: benzochinone, III: 1,4-dihydroxy benzene, IV: γ-hydroxymuconic semialdehyde, V: maleyl acetate, VI: β-ketoadipate, VII: 4-nitrocatechol, VIII: 1,2,4-trihydroxy benzene, IX: maleyl acetate, X: β-ketoadipate.

Secondly, a dioxygenase catalyzes the insertion (Fig. 3) of two hydroxyl groups with the subsequent elimination of the nitro group as nitrite (SPAIN, 1995b). This mechanism was first observed with *Alcaligenes eutrophus* during transformation of 2,6-dinitrophenol (ECKER et al., 1992).

Thirdly, the nitro group is reduced to a hydroxylamine which is not catabolized further, but ammonium is liberated during this transformation yielding catechol. This mechanism was observed in studies with *Comamonas aci-*

dovorans while degrading 4-nitrobenzoate (GROENEWEGEN et al., 1992; GROENEWEGEN and DE BONT, 1992). Yet this pathway has also been demonstrated in two other *Pseudomonas* species (HAIGLER and SPAIN, 1993; MEULENBERG et al., 1996).

The fourth strategy is a partial reduction of the benzene ring forming a Meisenheimer complex (Fig. 4) by nucleophilic addition of a hydride ion and eliminating the nitro group as nitrite subsequently (LENKE et al., 1992; VORBECK et al., 1994; VORBECK et al., 1998).

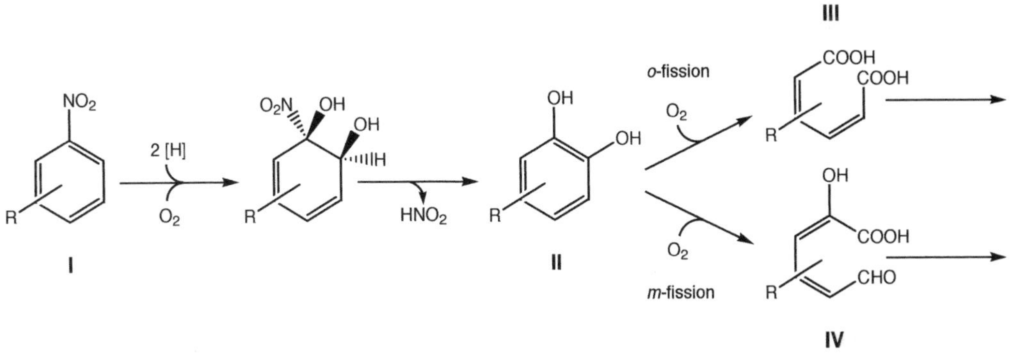

Fig. 3. Scheme of the degradation pathways for nitro-substituted aromatics initiated by dioxygenases. I: substituted nitrobenzene, II: substituted catechol, III: substituted muconic acid, IV: substituted hydroxymuconic semialdehyde.

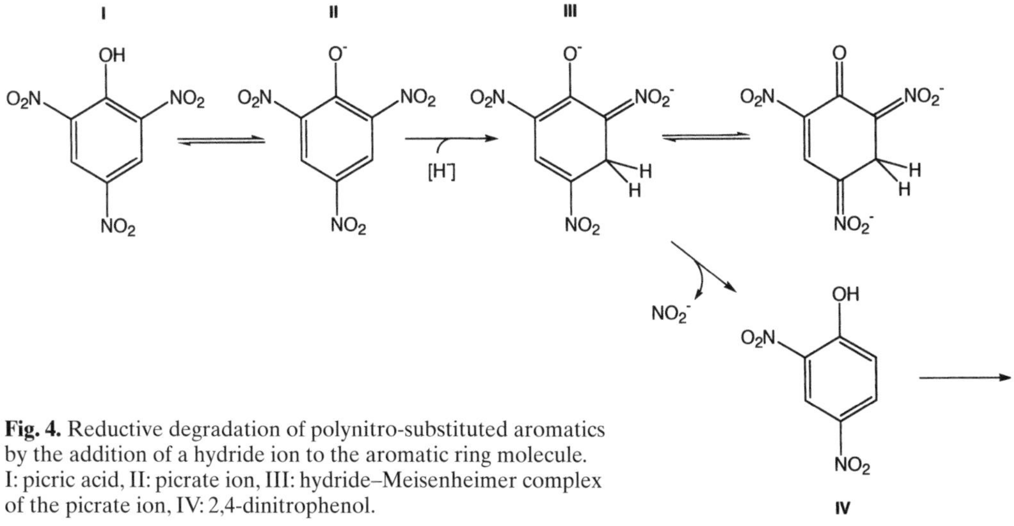

Fig. 4. Reductive degradation of polynitro-substituted aromatics by the addition of a hydride ion to the aromatic ring molecule. I: picric acid, II: picrate ion, III: hydride–Meisenheimer complex of the picrate ion, IV: 2,4-dinitrophenol.

2.3.2 Aerobic Microbial Degradation of Nitrobenzenes

The simplest catabolic pathway for nitrobenzenes is a direct dioxygenase attack forming catechol by a common pathway as described for a *Comamonas* strain (NISHINO and SPAIN, 1995). Another option would be a complete reduction of the nitro group forming an aniline which then undergoes a dioxygenase reaction also resulting in the formation of catechol (BACHOFER et al., 1975).

A more complex pathway is realized by *Pseudomonas pseudoalcaligenes*. This bacterium utilizes nitrobenzene as the sole source of nitrogen, carbon, and energy (HE and SPAIN, 1997). The catabolism of this substrate requires the participation of unusual enzymes (HE and SPAIN, 1997, 1998). First nitrobenzene is reduced by a NADPH-dependent nitrobenzene reductase to the corresponding hydroxylaminobenzene and not to the amine, while the latter is preferably a common reaction of bacterial nitroreductases. The hydroxyl compound is subjected to the rearrangement action of a mutase which results in the evolution of 2-aminophenol. This reaction is analogous to the so-called Bamberger rearrangement known in the chemistry of aromatics (CORBETT and CORBETT, 1995). The 2-aminophenol produced is further degraded to the ring fission product 2-aminomuconic semialdehyde. This is a rare example of a ring fission reaction in the absence of hydroxyl groups. But it seems that this pathway is the most common in contaminated soils (SPAIN, 1995c). In the next step a novel 2-aminomuconate deaminase converted 2-aminomuconate directly to 4-oxalocrotonate. This exhibits a close relationship to the *m*-cleavage pathway of catechol (HE and SPAIN, 1998). But not only the mineralization of aromatic compounds is the goal of microorganisms (or microbiologists!). The biologically mediated synthesis of hydroxylated aromatic compounds from nitroaromatics could be an interesting alternative to the expensive chemical synthesis. *Comamonas acidovorans* NBA-10 was isolated on 4-nitrobenzoate which it degraded via hydroxylaminobenzoate to 3,4-dihydroxybenzoate (for review, see MEULENBERG and DE BONT, 1995). A stoichio-metrical conversion was reached if ethanol was used for cofactor regeneration. Those catechols could be useful building blocks in the chemical industry (WEISSERMERL and ARPE, 1997).

Investigations on the metabolism of trinitrobenzene by a *Pseudomonas* consortium revealed deamination of this molecule to 5-nitrobenzene (BOOPATHY et al., 1994). Dinitroaniline, 1,5-dinitrobenzene, nitroaniline, 5-nitrobenzene, and ammonium were detected as intermediates and end products, respectively. Although the consortium uses trinitrobenzene as a source of nitrogen the carbon skeleton was not converted. These findings have been confirmed later in studies by using *Pseudomonas vesicularis* (DAVIS et al., 1997).

2.3.3 Aerobic Microbial Catabolism of Nitrophenols

The toxicity of nitrophenols is well documented. They can act as uncouplers in oxidative phosphorylation, and they are known to affect cell metabolism at concentrations lower than 10 μM (TERADA, 1981; KOZARAC et al., 1991). When entering biological environments they are able to alter the species balance in ecosystems so that the characteristics of natural relationships change drastically (ZIERIS et al., 1998). Nitrated phenols also received attention because of their atmospheric occurrence and their plant damaging potential (RIPPEN et al., 1987; HINKEL et al., 1989).

SIMPSON and EVANS (1953) provided first evidence of nitrophenol degradation. They demonstrated that a *Pseudomonas* strain converted 4-nitrophenol to hydroxyquinone with the concomitant liberation of nitrite. But the detailed pathway was elucidated almost 40 years later by SPAIN and GIBSON (1991).

Enzymes responsible for the oxygenolytic removal of nitrite have been identified in recent years. A 2-nitrophenol monooxygenase with broad substrate specificity has been purified from *Pseudomonas putida* which degrades *o*- as well as *m*-nitrophenol (SPAIN et al., 1979; ZEYER and KOCHER, 1988). The soluble monomeric NADH-dependent enzyme catalyzes the conversion of 2-nitrophenol via 1,2-benzoquinone to catechol and nitrite. Its activity was stimulated by cations like Mg^{2+}, Mn^{2+}, and Ca^{2+}. Activity is inducible with 4-methyl-2-nitrophenol, 4-chloro-2-nitrophenol, and 2-nitrophenol (FOLSOM et al., 1993; ZEYER and KEARNEY, 1984, ZEYER and KOCHER, 1988). The pathway of *m*-nitrophenol degradation has been elucidated by MEULENBERG et al. (1996). Evidence was presented that *Pseudomonas putida* B2 converts 3-nitrophenol to 1,2,4-benzenetriol and ammonium anaerobically including the participation of a hydroxylamino lyase activity. 1,2,4-Benzenetriol was not further metabolized under these conditions. Reduction of 3-nitrophenol was NADPH-dependent and the involved reductase could be characterized. The enzyme exhibits broad substrate specificity. 17 nitroaromatics tested were reduced, but the hydroxylamino compounds remain in cellfree extracts, suggesting that the hydroxylamino lyase has a very narrow substrate specificity.

A 4-nitrophenol oxygenase with a similar function was found in a *Moraxella* sp., but in contrast to the 2-nitrophenol oxygenase it is a membrane-bound enzyme (SPAIN et al., 1979; SPAIN and GIBSON, 1991). 2 moles of NADPH are consumed during oxidation of 1 mole 4-nitrophenol, and hydroquinone and nitrite were released. Experiments with $^{18}O_2$ provided strong evidence that the reaction is a monohydroxylation performed by a flavoprotein monooxygenase. This is analogous to pathways found in other flavoprotein monooxygenases. The hydroquinone produced serves as a carbon source and undergoes a ring-opening attack by a dioxygenase. The product was further oxidized and canalized into the β-ketoadipate pathway for mineralization (for review, see SPAIN 1995b).

A variation of 4-nitrophenol degradation is found in an *Arthrobacter* sp. which converts nitrophenols to nitrocatechols before removing the nitro substituent (JAIN et al., 1994). In an early report this has been proposed as the initial step in the degradation of nitrophenol by *Flavobacterium* sp. (RAYMOND and ALEXANDER, 1971). The authors observed a slow accumulation of 4-nitrocatechol from 4-nitrophenol. Later a 4-nitrophenol 2-hydroxylase able to perform such a reaction has been purified from a *Nocardia* sp. (MITRA and VAIDYANA-

THAN, 1984). But only the studies of JAIN et al. (1994) confirmed this early finding.

KADIYALA and SPAIN (1998) reported on a novel two-component monooxygenase from *Bacillus sphaericus* JS905. The components are a flavoprotein reductase and a monooxygenase that are performing two sequential monooxygenations which convert 4-nitrophenol via 4-nitrocatechol to 1,2,4-trihydroxybenzene with the concomitant release of nitrite, but nitrite was interestingly not released directly from 4-nitrophenol. This was the first report of a monooxygenase which performs a sequential hydroxylation of a nitroaromatic compound (KADIYALA and SPAIN, 1998).

Dioxygenase enzymes which incorporate oxygen atoms from molecular oxygen into the oxidized products, are also involved in the degradation of nitrophenols. Aromatic hydrocarbons are oxidized to dihydrodiols, and if the benzene ring is substituted the reaction may lead to the spontaneous elimination of these functional groups. Especially oxidative dehalogenations of haloaliphatic as well as haloaromatic compounds mediated by dioxygenases are very important in the biodegradation of such molecules (for review, see FETZNER, 1998). That bacterial dioxygenases could catalyze replacement of nitro groups substituted to aromatic ring systems was first shown in a *Pseudomonas* species during growth on 2,4-dinitrophenol (SPANGGORD et al., 1991). A phototrophic bacterium, *Rhodobacter capsulatus*, reduces 2,4-dinitrophenol to 2-amino-4-nitrophenol which is further metabolized by an unknown light-dependent pathway. Although no stable aromatic intermediates were detected the compound is used neither as carbon feedstock nor as a source of nitrogen indicating a cometabolic degradation pathway (WITTE et al., 1998).

Rhodococcus sp. strain RB1 mineralizes 2,4-dinitrophenol and the nitro groups were released as nitrite in a two-step process. The following pathway was hypothezised according to the experimental evidence: First, two successive hydride transfers take place forming a hydride–Meisenheimer complex, first shown by LENKE and KNACKMUSS (1992), intermediates followed by an *o*-ring fission, and concomitant release of the *o*-nitro group, producing 3-nitroadipate (BLASCO et al., 1999). The further metabolism of this intermediate is correlated with the release of the other nitro group. Whether the β-ketoadipate pathway is used for further breakdown or not, is not yet clear (BLASCO et al., 1999). An O_2-dependent degradation of 2,6-dinitrophenol by *Alcaligenes eutrophus* JMP134 includes the formation of 4-nitropyrogallol as an intermediate (ECKER et al., 1992). According to their results the authors concluded that first a nitro group is eliminated by a dioxygenase reaction (resulting in the formation of 4-nitropyrogallol), whereas the second nitro group was released after ring cleavage.

As discussed before the electron withdrawing effect of the nitro group changes the susceptibility to oxygenases, it is, therefore, obvious that polynitroaromatic compounds are more electrophilic and thus impede attack by oxygenases (RIEGER and KNACKMUSS, 1995). It would be expected that a reductive mechanism is more likely. Indeed, a novel reductive pathway for the utilization of 2,4,6-trinitrophenol (picric acid) by a *Rhodococcus erythropolis* strain has been demonstrated (LENKE and KNACKMUSS, 1992). Nucleophilic addition of a hydride ion forms an orange-red hydride–Meisenheimer complex which is rearomatized by releasing nitrite and forming 2,4-dinitrophenol (RIEGER and KNACKMUSS, 1995). An analogous pathway was also confirmed for 2,4,6-trinitrotoluene and will be discussed in Sect. 2.3.4 (VORBECK et al., 1994).

2.3.4 Aerobic Microbial Catabolism of Nitrotoluenes

As in the case of mononitrophenols and -benzoates, several bacteria relatively easily degraded mononitrotoluenes (SPAIN, 1995b).

The transformation and degradation of nitrotoluenes into more oxidized products are analogous to the well-characterized route for toluene catabolism. This means that enzymes encoded by the TOL plasmid are involved (DELGADO et al., 1992; RHYS-WILLIAMS et al., 1993; JAMES and WILLIAMS, 1998). These findings support the proposition that such aromatic degradation pathways evolved by the coacquisition of genetic modules or parts of it by

different strains (WILLIAMS and SAYERS, 1994; JAMES and WILLIAMS, 1998).

Intermediates appear in the same order as in the toluene pathway (hydrocarbon→alcohol→aldehyde→carboxylic acid). Two types of dehydrogenase activities have been detected which convert 4-nitrobenzyl alcohol to the corresponding aldehyde and subsequently to 4-nitrobenzoate (RHYS-WILLIAMS et al., 1993). But surprisingly the authors could not find significant DNA homology with probes of the TOL plasmid. A novel degradative pathway for 4-nitrobenzoate which is a possible intermediate in nitrotoluene degradation is found in *Comamonas acidovorans* in which the nitro group is reduced through 4-nitrosobenzoate to 4-hydroxylaminobenzoate by the action of NADP(H)-dependent reductases, and the hydroxylamino group is eliminated subsequently in a single-step mechanism (GROENEWEGEN et al., 1992; GROENEWEGEN and DE BONT, 1992).

2,4-dinitrotoluene is a very important compound in the production of diisocyanates, a component of polyurethanes. But dinitrotoluenes are also precursors as well as degradation products of the explosive 2,4,6-trinitrotoluene (WEISSERMERL and ARPE, 1997). So production of building blocks might be an alternative to mineralization (MEULENBERG and DE BONT, 1995). Several bacteria able to mineralize 2,4-dinitrotoluene have been isolated from a variety of contaminated soils, and the biodegradation of 2,4-dinitrotoluene has been studied in detail (SPAIN, 1995b).

A dioxygenase enzyme that catalyzes the initial attack of 2,4-dinitrotoluene is constitutive and has a broad substrate range, a property similar to that of naphthalene oxygenases (SPANGGORD et al., 1991; HAIGLER et al., 1994). These dioxygenases have been identified as new three-component dioxygenase systems: 2-nitrotoluene dioxygenase from *Pseudomonas* sp. strain JS42 and 2,4-dinitrotoluene dioxygenase from *Burkholderia* (formerly *Pseudomonas*) sp. strain DNT (AN et al., 1994; SUEN et al., 1996). Recently, their properties and molecular features that determine their composition and specificity have been thoroughly investigated (PARALES et al., 1998a, b).

As mentioned above dinitro-substituted aromatics could also be degraded by oxygenation as shown for 2,4-dinitrotoluene degradation by *Pseudomonas* sp. (SPANGGORD et al., 1991). SUEN and SPAIN (1993) were able to clone the genes responsible for the nitro group removing enzymes. A dinitrotoluene dioxygenase, a 4-methyl-5-nitrocatechol monooxygenase and a 2,4,5-trihydroxytoluene oxygenase have been detected and tentatively characterized. Trihydroxytoluenes could serve as substrates of dioxygenases. The 2,3-dioxygenase gene family shows a high DNA homology to the recently characterized gene of the 2,4,5-trihydroxytoluene dioxygenase (SUEN et al., 1996).

But today most of all the biodegradation of TNT was investigated, because this compound is a major nitroaromatic contaminant in the environment due to its extensive usage in explosives (WALKER and KAPLAN, 1992; GORONTZY et al., 1994; SPAIN, 1995a). The worldwide production of TNT is estimated of around $1 \cdot 10^6$ kg a year (HARTTER, 1985). TNT or TNT-derived chemicals are, therefore, abundant in industrial waste streams, soils, surface, and even ground waters. Consequently, TNT and its metabolites are of environmental concern, and a strong demand for remediation exists. The chemical basics of the bioavailability and the biodegradability have been discussed thoroughly by RIEGER and KNACKMUSS (1995). In summary, the molecule has a low water solubility (140 mg L^{-1} at 25 °C), it is resistant to an oxygenolytic attack and exhibits high toxicity against different organisms. According to this basic knowledge and findings, it is unlikely that this recalcitrant molecule will undergo an initial oxidative reaction mechanism. Therefore, an initial oxidative attack of a polynitro-substituted aromatic ring through bacteria has not been described so far. Although many reports of TNT degradation exist, in most cases a transformation or a partial reduction of nitro substituents with the concomitant accumulation of aminotoluenes is meant. Nevertheless, DUQUE et al. (1993) were able to isolate a *Pseudomonas* sp. strain C1S1 which is able to use besides TNT, 2,4- and 2,6-dinitrotoluene, 2-nitrotoluene, nitrate, nitrite, and ammonium as a source of nitrogen. Because aromatic metabolites lacking one or more nitro groups were detected the authors concluded that nitro elimination of the first two groups takes place according to the more

recently described MEISENHEIMER complex formation (LENKE and KNACKMUSS, 1992; DUQUE et al., 1993; VORBECK et al., 1994). In additional experiments a hybrid strain of a toluene degrader, *Pseudomonas putida*, and of the *Pseudomonas* sp. strain C1S1 was constructed. The conjugates obtained were able to grow on TNT as the sole nitrogen and carbon sources, but accumulation of nitrite in the growth medium was observed (DUQUE et al., 1993). Addition of fructose avoids further nitrite release. In contrast to these findings VORBECK et al. (1998) could demonstrate that the proposed mechanism of denitration of TNT, i.e., the transient formation of a hydride–Meisenheimer complex, is not realized in this *Pseudomonas* sp. clone A. Additionally they found out that unlike the degradation of picric acid in the corresponding TNT metabolism neither nitrite elimination nor re-aromatization into dinitrotoluenes occurred (VORBECK et al., 1998). Therefore, it was concluded that reductive denitration via hydride addition is not a key reaction in TNT enriched bacteria. VANDERBERG et al. (1995) claimed TNT ring cleavage reactions by *Mycobacterium vaccae*, but could not provide any evidence for the occurrence of non-aromatic intermediates or formation of $^{14}CO_2$ with radiotracer experiments employed. Further attempts to degrade TNT were made with *Serratia marcescens* (MONTPAS et al., 1997). Facilitated growth in the presence of Tween 80 and benzyl alcohol was reported, but only reduced intermediates such as 4-amino-2,6-dinitrotoluene and 2-amino-4,6-dinitrotoluene were detected. A limited degree of NO_2 removal from TNT was observed during cometabolic growth of a *Bacillus* sp. (KALAFUT et al., 1998). But it remains questionable whether the small amounts of nitrite measured stem from TNT denitration, because no labeling experiments were performed, and nitrate served as an additional nitrogen source. More promising results are obtained with *Pseudomonas savastanoi*. TNT denitration could be demonstrated and was enhanced by the addition of NO_2^-, resulting in the production of easier degradable dinitrotoluenes, because they are more susceptible to oxygenase attack (MARTIN et al., 1997). However, no growth on TNT alone could be detected. In contrast cell density decreased and sorption of

TNT to the cells was observed. *Enterobacter cloacae* PB2 able to utilize nitrate esters such as glycerol trinitrate and pentaerythritol tetranitrate (PETN) as the sole nitrogen source, was found to be capable of slow aerobic growth on TNT (FRENCH et al., 1998). Responsible for this ability is a NADPH-dependent PETN reductase (BINKS et al., 1996; FRENCH et al., 1996). The enzyme reduces the TNT molecule via the transient evolution of hydride–Meisenheimer complexes and dihydride–Meisenheimer complexes, thereby causing release of nitrite even in its purified state (FRENCH et al., 1998). Therefore, this enzyme is the basis for the ability of *Enterobacter cloacae* to utilize TNT as the sole nitrogen source. As in all other studies performed so far, the benzene ring is not metabolized.

In summary, aerobic bacteria evolved a variety of strategies to transform or remove nitro groups from organic molecules which includes

(1) mono- and dioxygenase elimination as nitrite (see Fig. 2 and 3)
(2) reduction to a hydroxylamine, with subsequent elimination as NH_4^+
(3) addition of a hydride ion, forming a Meisenheimer complex and release of nitrite (see Fig. 4).

To the best of our knowledge, a complete mineralization of polynitrated organics by bacteria has not been observed to date.

But what is learned from the presented data concerning the oxidative catabolism of TNT is that oxygenolytic denitration appears to be the crucial step in the aerobic mineralization of this molecule. Due to their chemical nature it is more likely that especially polynitroaromatics are nucleophilicly attacked in biological systems. A reductive transformation of the aromatic ring is reached by the formation of a hydride–Meisenheimer complex described above, but pathways leading directly to the formation of aminoaromatic molecules are more common.

2.4 Anaerobic Microbial Degradation of Nitroaromatic Compounds

2.4.1 Abiotic and Biotic Reduction of Nitroaromatic Compounds

In natural ecosystems oxic and anoxic conditions coexist. They could even be created in a time-dependent manner by oxygen consumption and production through microorganisms within very small particles or space of soil, sediments, and other life inhabiting material. Therefore, it is important to consider (biotic/abiotic) transformation reactions occurring under anoxic conditions.

The reduction of the nitro substituent by microbial enzymes is the first reported modification of several classes of nitro compounds. The capacity of biological systems catalyzing the reduction of nitro substituents is due to the chemical properties of the nitro group.

It is assumed that the reduction of the nitro group occurs by sequential addition of 2 electrons and is the privileged reaction in anoxic habitats. This type of conversion of nitro groups to hydroxylamines or amines is mediated by "nitroreductases" (nitroreductase-like reactions) with pyridine nucleotides as electron donors in most cases. The reduction of nitro groups is probably not the physiological role of the involved enzymes which are highly conserved in several prokaryotes (CERNIGLIA and SOMERVILLE, 1995). Exceptions are, e.g., nitroreductases isolated from *Comamonas* sp. and *Pseudomonas* sp., because they are induced by nitroaromatic substances (GROENEWEGEN et al., 1992; SOMERVILLE et al., 1995). Several microorganisms degrade nitroorganic molecules along this common pathway for nitrobenzoates, nitrotoluenes, and nitrophenols (for review, see MARVIN-SIKKEMA and DE BONT, 1994). But most bacteria are not capable of mineralizing nitroaromatics completely after reduction of the nitro group. Often the resulting amino compounds are dead-end metabolites or were partly used as source of nitrogen after a subsequent deamination (OREN et al., 1991; BOOPATHY and KULPA, 1993; GORONTZY et al., 1993; PREUSS et al., 1993).

It should be stated here that nearly every redox system with a low redox potential is able to reduce the nitro substituents. In addition, abiotic electron donors like iron species or reduced sulfur compounds are also able to reduce nitro groups (HEIJMAN et al., 1993). The abiotic reduction of nitroaromatic compounds has recently been investigated carefully (HADERLEIN and SCHWARZENBACH, 1995; KRUMHOLZ et al., 1997). It was also found that dissolved organic constituents such as exudates or porphyrines from bacteria as well as reduced quinones act as electron transfer mediators for the reduction of nitroaromatics (SCHWARZENBACH et al., 1990; GLAUS et al., 1992; HASAN et al., 1992).

Hence, additional evidence is also provided that besides hydrogenases and ferredoxin, which are able to establish one-electron reduction of nitroaromatics, other cell-free components mediate similar reactions (McCORMICK et al., 1976; ANGERMAIER and SIMON, 1983a, 1983b; GORONTZY et al., 1993; VAN BEELEN and BURRIS, 1995).

But besides these many reports about unspecific reductions, only few enzymes expressing nitroreductase-like activities could be identified until now. Transformation of nitrobenzene to phenylhydroxylamine is achieved by an oxygen-sensitive ferredoxin–NADP oxidoreductase from spinach leaves (SHAH and CAMPBELL, 1997). Interestingly, this enzyme is also capable of eliminating nitrite from the explosive 2,4,6-trinitrophenylmethyl nitramine (tetryl) (SHAH and SPAIN, 1996). A mechanism is proposed in which tetryl is reduced to N-methylpicramide with the nitro anion radical as an intermediate. The flavoenzyme NADPH–thioredoxin reductase from *Arabidopsis thaliana* catalyzed single-electron reductions of TNT, tetryl, the herbicide 3,5-dinitro-*o*-cresol and the evolution of nitro anion radicals (MISKINIENĖ et al., 1998). Tetryl conversion was accompanied by the formation of N-methylpicramide, similar to the observations of SHAH and SPAIN (1996). But a loss of the original thioredoxin reductase activity was detected in the presence of tetryl or 2,4-dinitro-chlorobenzene, indicating a mechanism of nitroaromatic toxicity towards plant cells.

Nitroreductases have also been detected in *Escherichia coli, Bacteroides fragilis, Entero-*

bacter cloacae, and *Clostridium* sp. (PETERSON et al., 1979; KINOUCHI and OHNISHI, 1983; BRYANT and DELUCA, 1991; RAFII and CERNIGLIA, 1995; RAFII and HANSEN, 1998). The major oxygen-sensitive nitroreductase of *Escherichia coli* and *Vibrio harveyi* belongs to the newly identified nitroreductase–flavine reductase superfamily (ZENNO et al., 1996). A close relationship between flavine reductases and nitroreductases was revealed. It was shown that a single amino acid substitution causes the transformation from a nitroreductase into a FMN reductase in *Escherichia coli*, where nitrofurazone and methyl-4-nitrobenzoate function as electron acceptors for the reductase activity (ZENNO et al., 1996, 1998).

2.4.2 Conversion and Degradation of Nitro Compounds by Anaerobic Bacteria

To our knowledge no reports of anaerobic biotransformations of non-aromatic nitro compounds are available in the present scientific literature. Therefore, present research is focused on investigations of the bioconversion of nitro-substituted aromatic molecules by anaerobic bacteria or consortia and will be discussed here in more detail. In anoxic environments where nucleophilic reactions are favored the presence of electron withdrawing groups, such as $-NO_2$ or $-Cl$, are welcome for an initial reductive attack. Catabolism beyond simple reduction pathways, where 3 moles of hydrogen are required, $R-NO_2 \rightarrow R-NO \rightarrow R-NHOH \rightarrow R-NH_2$, is seldom reported. But studies with species of sulfate-reducing bacteria and *Clostridium* sp. provide new insights that may lead to a new picture of the capabilities of anaerobes. But all reaction pathways observed with anaerobes so far depend exclusively on cometabolism in the degradation of xenobiotics.

2.4.3 Biodegradation of Nitrated Compounds by Sulfate-Reducing Bacteria

The versatility of sulfate-reducing bacteria concerning substrate utilization is well documented. One of the most frequently encountered genus is *Desulfovibrio* (HANSEN, 1994).

BOOPATHY and KULPA (1992, 1993) studied a *Desulfovibrio* species that uses different nitroaromatic compounds including TNT as the source of nitrogen.

During their metabolic studies pyruvate served as the carbon source, whereas TNT is the nitrogen supply and is completely reduced to 2,4,6-triaminotoluene (TAT). Because toluene was detected in the culture fluid and nitrate could also be assimilated, analogous to the findings of SCHNELL and SCHINK (1991) with *Desulfobacterium anilini* a reductive deamination mechanism for TAT was proposed. But in contrast to SCHNELL and SCHINK (1991) the authors could not provide any experimental evidence for such an enzymatic process. Deamination of aromatic molecules by anaerobic bacteria was first reported for 2-aminobenzoate (TSCHECH and SCHINK, 1988). In additional studies consortia of sulfate-reducing bacteria were described, capable of converting TNT to fatty acids (BOOPATHY and MANNING, 1996; COSTA et al., 1996).

Another *Desulfovibrio* strain was isolated and described by PREUSS et al. (1993). It is able to grow with pyruvate as its carbon source and sulfate as the electron acceptor, while TNT is the source of nitrogen. In growth experiments it was shown that TNT is chemically reduced to diamino nitrotoluenes (DANT) by the reductant sulfide present in the medium, but DANT were further reduced to TAT via nonspecific enzyme reactions in cell-free extracts and with hydrogen, carbon monoxide, and pyruvate as electron donors. Further investigations on the conversion mechanism lead to the assumption that ferredoxin-reducing enzymes may be involved. The authors gather strong evidence that hydrogenase, pyruvate–ferredoxin oxidoreductase, carbon monoxide dehydrogenase and even sulfite reductase are involved in this reaction sequence (PREUSS et al., 1993). Because in the presence of CO 2,4-diamino-6-

hydroxylaminotoluene (2,4-DAHAT) accumulates, the authors concluded that the observed rate limiting reduction of 2,4-DAHAT is due to the inhibitory effect of CO. Furthermore, direct application of 2,4-DAHAT yielded no further reduction in the presence of CO. This led to the assumption that sulfite reductase, known to be active on hydroxylamine by *Desulfovibrio vulgaris*, is responsible for 2,4-DAHAT reduction (PREUSS et al., 1993).

Recently reactions of *O*-demethylation and reduction of nitro to amino substituents at aromatic molecules were observed in physiological studies with *Desulfitobacterium frappieri* which expand possibilities for new reactions in bacterial metabolism (DENNIE et al., 1998).

2.4.4 Biodegradation of Nitrated Compounds by Fermenting Bacteria

As mentioned in Sect. 2.4.3, complete reduction of the nitro groups has been widely reported. In this respect it is not unexpected that because of their ubiquity, versatility, and relative easily manipulable properties fermenting bacteria have come into consideration more in the recent years. Their ability to reduce nitro compounds had been established 40 years ago. Cell-free preparations of the fermentative bacterium *Veillonella alcalescens* catalyzed the reduction of different nitroaromatics in the presence of hydrogen (CARTWRIGHT and CAIN, 1959). These first findings were confirmed and extended to TNT and several other nitroaromatics by MCCORMICK et al. (1976). They also demonstrated the formation of aminonitrotoluenes and azoxy compounds in an anoxic environment producing more recalcitrant molecules.

Recently, *Clostridium* sp. were studied in more detail because of their known ability to reduce nitroaromatics (EDERER et al., 1997; KHAN et al., 1997). RAFII et al. (1991) found that several *Clostridium* strains isolated from human feces were able to reduce different nitroaromatics. Although they could characterize some enzymes which might be involved in the electron transfer, the first biochemical work was done much earlier by ANGERMAIER and SIMON (1983a, b) who clearly demonstrated that ferredoxin and hydrogenases from *Clostridium kluyveri* were responsible for nitro group reduction. The importance of ferredoxin in this reduction mechanism has also been confirmed in studies with *Clostridium thermoaceticum* and *C. pasteurianum* (PREUSS et al., 1993).

From an anaerobic fermentative consortium pure cultures of *C. bifermentans* were obtained which could degrade the explosive TNT among other components (REGAN and CRAWFORD, 1994). Although the reductive pathway of TNT to TAT was not different from that reported before, the authors provided evidence that TAT undergoes further catabolism. Sequential accumulation of intermediates other than the usual aminotoluenes was observed. The occurrence of methylphloroglucinol (2,4,6- trihydroxytoluene) and *p*-cresol lead to the assumption that hydrolytic displacements of the amino groups forming polyphenols took place (FUNK et al., 1993; LEWIS et al., 1996; EDERER et al., 1997). Reductive dehydroxylation of these chemicals will yield simpler aromatics like *p*-cresol (4-hydroxytoluene). If confirmed, this could be the most promising way for future treatment of soil and water contaminated with nitro compounds, since there are several aerobic and anaerobic bacteria capable of mineralizing these molecules.

In summary, it must be stated that degradation processes of nitro-substituted molecules are non-specific by anaerobes as far as we know. This is in contrast to some aerobic bacteria which can use nitroaromatic compounds as growth substrates and energy source. Reports on biotransformation of nitro compounds by methanogenic archaea have no biotechnological value (GORONTZY et al., 1993; BOOPATHY, 1994). The crucial point in anaerobic metabolism of TNT is the fate of TAT. Evidence for a complete deamination of TAT by enzymatic processes is weak, so one may ask: Is there a (micro)biology behind TAT? If these preliminary results of TAT deamination to toluene could be confirmed this would be a promising point of application, because several pure cultures of anaerobic toluene-degrading bacteria were described in the past years (ANDERS et al., 1995; COLBERG and YOUNG, 1995; BELLER et al., 1996).

2.5 Biotransformation of Minor Occurring Nitro Compounds

Substances with a minor potential of environmental concern or which are less studied are besides a few others: nitrated polycyclic hydrocarbons (NO_2-PAHs), herbicides, and nitrodiphenylamines or naturally occurring nitro compounds. 4-nitroanisole, e.g., is a secondary metabolite of the fungus *Lepista diemii* (THALLER and TURNER, 1972). It is also of great importance in the chemical industry. Unspecific transformation to nitrophenol is a known fact, but complete mineralization and usage as the sole source of carbon and energy were first observed with two strains of the genus *Rhodococcus* (SCHÄFER et al., 1996). The authors proposed a degradation pathway according to their findings which involves *O*-demethylation of 4-nitroanisole to 4-nitrophenol, transformation of this intermediate to 4-nitrocatechol and 1,2,4-trihydroxybenzene prior to ring cleavage. Less is known about catabolism of heterocyclic nitrated compounds which are often applied as antibiotics (CARLIER et al., 1998; RAFII and HANSEN, 1998).

2.5.1 Metabolism of Polycyclic Nitroaromatic Hydrocarbons

Nitrated polycyclic aromatic hydrocarbons are potent mutagens (BELAND et al., 1985). They have been identified in airborne particulate matter, diesel engine exhaust, carbon black, and certain photocopier toners, and they were also generated by incomplete combustion of fossil fuels (CERNIGLIA and SOMERVILLE, 1995). This means in a few words, that nitro PAHs are ubiquitous. Among the more than 200 known congeners, 1-nitropyrene and 2-nitroflourene predominate (RITTER and MALEIA-GIGANTI, 1998). Nitro PAHs require metabolic activation to develop their mutagenic and carcinogenic potential. Reduction of the nitro group by certain bacterial nitroreductases to arylamines implies the formation of electrophilic species such as nitroso intermediates and hydroxylamino intermediates which could bind to cellular nucleophiles

(MANNING et al., 1988; RITTER and MALEJKA-GIGANTI, 1998). Because these substances bear a carcinogenic risk to humans and nitroreduction has a strong influence on this potential, many studies were performed with consortia derived from the intestinal microflora; less is known about their behavior in terrestrial and aquatic environments. The bioavailability is low due to the large molecular size, low water solubility, and their low concentrations in natural habitats. From an oil contaminated site, *Mycobacterium* sp. PYR-1 was isolated expressing a low 1-nitropyrene mineralization capability (HEITKAMP et al., 1991).

2.5.2 Catabolism of 2-*sec*-Butyl-4,6-Dinitrophenol (Dinoseb)

Maintaining anoxic conditions achieved the successful removal of 2-*sec*-butyl-4,6-dinitrophenol (Dinoseb) from contaminated soils (KAAKE et al., 1992). Dinoseb is a herbicide persisting in the environment for decades. Under reducing conditions ($E_h \leq -200$ mV) the complete degradation of Dinoseb by an anaerobic microbial consortium was observed (KAAKE et al., 1995). The degradation pathway involves the subsequential reduction of the nitro groups to the corresponding amino substituents followed by hydroxylation with the transient formation of quinone-like structures. HAMMILL and CRAWFORD (1996) achieved the first demonstration of Dinoseb degradation with a pure culture of *Clostridium bifermentans*. The authors succeeded to show a partial degradation of U-ring ^{14}C labeled Dinoseb during cometabolic growth of *C. bifermentans* KMR-1. A subsequent aerobic incubation of anaerobic metabolism products with soil bacteria led to further liberation of ^{14}CO$_2$, all together during the anaerobic and aerobic phase of nearly 39%.

2.5.3 Reductive Biodegradation of Nitrodiphenylamines by Anaerobic Bacteria

Diphenylamine (DPA) is the most commonly used stabilizer for nitrocellulose containing explosives and propellants. Stabilization involves reactions of diphenylamine and nitration products with chemicals such as NO, NO$_2$, and HNO$_2$ resulting from the decomposition of nitrocellulose under normal storage conditions. Further reaction results in the production of *N*-nitroso-diphenylamine, mononitro diphenylamine, di- or trinitro diphenylamines which then could be found at contaminated sites. Concomitantly the explosive hexanitro diphenylamine (HEXYL) often can be found at such sites (GORONTZY et al., 1994). DRZYZGA et al. (1995b) reported on the anaerobic metabolism of nitro diphenylamines in sediment–water enrichments. Their first findings of complete reduction of the nitro groups were confirmed by a subsequent study with sulfate-reducing bacteria (DRZYZGA et al., 1996). Cometabolic growth with lactate or benzoate in the presence of nitro diphenylamines leads to the formation of breakdown components such as aniline or methylaniline as dead-end products. Additionally, condensation reactions took place that led to the formation of phenazines and acridine derivatives. Disappearance of 2-nitro diphenylamine during cometabolic growth of two newly isolated *Clostridium* sp. was described recently, but no intermediates could be detected with the methods applied (POWELL et al., 1998). Although these results are preliminary it seems that nature evolved a broad spectrum of possibilities to combat such xenobiotic chemicals.

3 Biodegradation of Nitro-Substituted Compounds by Fungi

The interest in using fungi for bioremediation strategies arises from the ability of especially white-rot fungi to metabolize a diverse range of persistent and toxic environmental pollutants. Their capability to degrade the very complex and persistent plant polymer lignin is well studied and makes them the organisms of choice for developing new methods of bioremediation with fungi (BARR and AUST, 1994).

Several research groups have reported on the catabolism of nitroaromatics by *Phanerochaete chrysosporium, Phlebia radiata*, and other fungi (PARRISH, 1977; FERNANDO et al., 1990; SPIKER et al., 1992; STAHL and AUST, 1993a, b; VAN AKEN et al., 1997; SCHEIBNER et al., 1997).

Under ligninolytic conditions mineralization of 2,4-dinitrotoluene (2,4-DNT) by *P. chrysosporium* was detected (VALLI et al., 1992). As reported with several other microorganisms the initial step is a reduction of 2,4-DNT to 2-amino-4-nitrotoluene. This molecule is oxidized by manganese peroxidase to yield methanol and 4-nitro-1,2-benzoquinone which is further converted to 4-nitrocatechol. The manganese peroxidase then removes the nitro group oxidatively. The lignin peroxidase can also remove nitro functions from the benzene ring, yielding methylated and hydroxylated products like the key intermediate 1,2,4-trihydroxybenzene which subsequently undergoes ring cleavage. As observed with prokaryotes most fungi catalyze the reduction of nitro substituents, and under ligninolytic conditions

extensive mineralization of aminoaromatics occurs (STAHL and AUST, 1993a, b). The presently known pathways for the degradation of 2,4,6-TNT by *P. chrysosporium* are shown in Fig. 5. There is disagreement about the initial reduction steps. It is not clear whether these take place in or outside the cell and which enzymes are involved. Non-ligninolytic fungi degrade nitroaromatics to a lesser extent (VALLI et al., 1992; STAHL and AUST, 1993a, b; MICHELS and GOTTSCHALK, 1994). Because concentrations above 20 ppm of TNT are inhibitory, this might be a limiting factor for the application of fungi in bioremediation (SPIKER et al., 1992). Recently, it could be shown that 4-hydroxylamino-2,6-dinitrotoluene and 2-hydroxylamino-4,6-dinitrotoluene inhibited an essential process of the ligninolytic pathway, i.e., the oxidation of veratryl alcohol by lignin peroxidase yielding organic radicals necessary for the unspecific oxidation of chemicals (MICHELS and GOTTSCHALK, 1994; BUMPUS and TATARKO, 1994). Another reaction not observed in fungi so far, is the formylation of the amino group of 4-amino-2,6-dinitrotoluene by *P. chrysospori-*

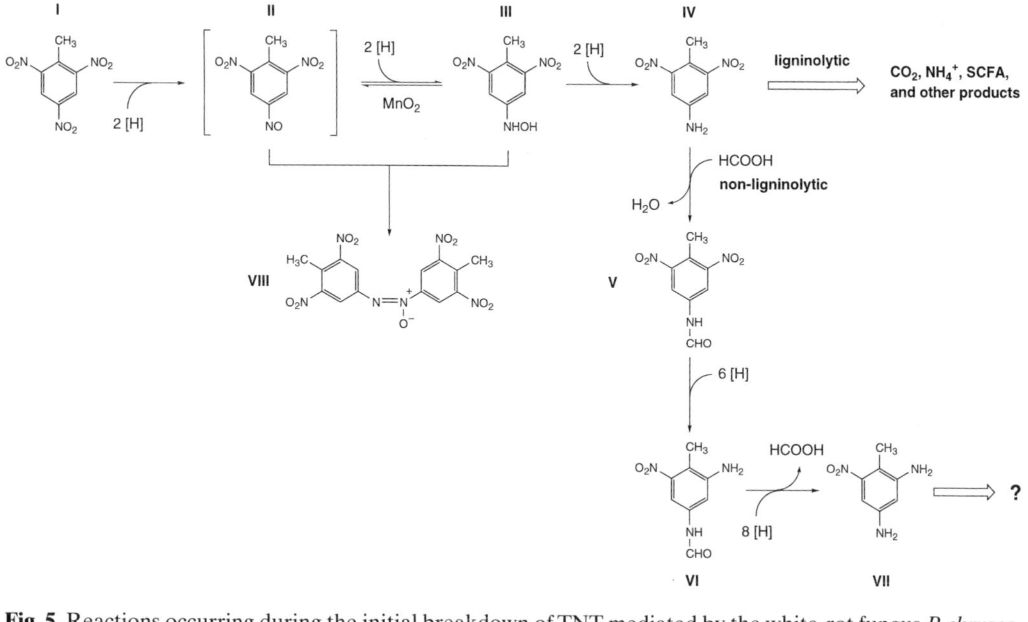

Fig. 5. Reactions occurring during the initial breakdown of TNT mediated by the white-rot fungus *P. chrysosporium*. I: 2,4,6-trinitrotoluene, II: 4-nitroso-2,6-dinitrotoluene, III: 4-hydroxylamino-2,6-dinitrotoluene, IV: 4-amino-2,6-dinitrotoluene, V: 4-formamido-2,6-dinitrotoluene, VI: 2-amino-4-formamido-6-nitrotoluene, VII: 2,4-diamino-6-nitrotoluene, VIII: 4,4′-azoxy-2,2′,6,6′-tetranitrotoluene.

um. Often acetylation and formylation reactions were assumed to be detoxifying or antibiotic resistance mechanisms. Here the formylated compound serves as a substrate for the lignin peroxidase. Mineralization of TNT under different growth conditions varies but in all cases it remains below 25%. More than 90 fungal strains were tested for their capability to mineralize explosives (SCHEIBNER et al., 1997). Micromycetes proved to be unable to degrade a compound like TNT, whereas wood and litter decaying basidiomycetes such as *Clitocybula dusenii* and *Stropharia rugosa-annulata* mineralized 42% and 36%, respectively, of the added labeled TNT within 64 d. Due to the research results with *Nematoloma frowardii* an extracellular enzymatic combustion system of aromatics and aliphatics with manganese peroxidase could be established (SCHEIBNER and HOFRICHTER, 1998; HOFRICHTER et al., 1998).

The authors concluded that such a manganese peroxidase dependent process has general significance in the biodegradation of recalcitrant pollutants. The widespread genotoxic nitrated polycyclic aromatic hydrocarbons can be converted to less toxic products by the fungus *Cunninghamella elegans* (POTHULURI et al., 1998a, b). The fungus formed sulfate conjugates from mixtures of 2-nitro fluoranthene and 3-nitro fluoranthene. In addition glycolysations, hydroxylations, and glucuronidations may occur. This is similar to that reported for other polycyclic hydrocarbons.

4 Biotechnological Applications of Remediation Techniques for Nitrated Pollutants – State of the Art

Because of the extraordinary relevance of the contamination of soils and ground water with explosives (i.e., nitroaromatics) published techniques mostly deal with the treatment of such compounds. At the beginning often mixed cultures were used to develop new strategies for biodegradation processes or to simulate more natural conditions in the laboratory. Generally microorganisms appear to utilize only those enzyme catalyzed reactions that produce the greatest possible energy return for the cell (SHELLEY et al., 1996). As discussed before nitroorganic compounds posess a xenobiotic character for several reasons, and although they are very energy rich molecules their degree of mineralization by microorganisms seems to be low. The reductive transformation of such compounds (e.g., see Fig. 6) is assumed to be "gratuitous". This implies that there is no specific pathway for metabolizing the compound, but conversion occurs cometabolically more or less by accident (RIEGER and KNACKMUSS, 1995). Consequently, most attempts for the clean-up of contaminated soils and water were performed with mixed cultures (mainly the authochtonous microflora) and the addition of a supplemental source of carbon and energy.

Over the last years the effectiveness of composting for the treatment of nitroaromatics impacted soils has been confirmed in several studies (KAPLAN and KAPLAN, 1982; ISBISTER et al., 1984; WILLIAMS et al., 1992; BREITUNG et al., 1995, 1996). The outcome of these studies is that a rapid decrease of the pollutants (nitrotoluenes) was detected with the concomitant formation of minor amounts of monodinitro and diamino nitrotoluenes. By using ring-labeled ^{14}C-TNT no $^{14}O_2$ was produced, indicating that no mineralization occurred. In an aerated compost system a loss of 92% of the initial TNT concentration within 28 d was found, but approximately 25% of the original concentration could be recovered by an acidic treatment (BREITUNG et al., 1996). If the compost was pretreated anaerobically a slow TNT decrease occurred and monoamino dinitrotoluenes were formed in detectable amounts which disappeared during the subsequent aerobic phase. This indicates a possible binding of the amino compounds to the soil constituents like the humus matrix (Fig. 7). In a follow-up of these studies the transformation of 90% of the TNT to mono- and dinitrotoluenes in an anaerobic/aerobic compost system could be observed (BRUNS-NAGEL et al., 1998). Additionally, the occurrence of formylated and acetylated intermediates, such as 4-acetylamino-2-

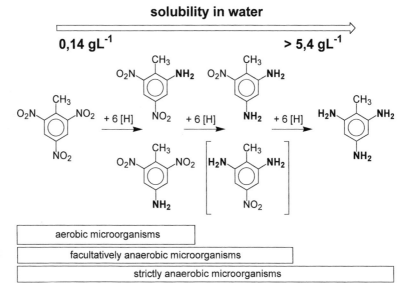

Fig. 6. Microbial transformation of trinitrotoluene to triaminotoluene via the sequential reduction of each nitro group correlating with increase of the solubility in water.

hydroxylamino-6-nitrotoluene and 4-formamido-2-amino-6-nitrotoluene, was detected which were completely degraded in the ongoing process. Similar observations have been made earlier during degradation studies of 2,4-dinitrotoluene by a *Pseudomonas aeruginosa* strain and TNT degradation by *Phanerochaete chrysosporium* (NOGUERA and FREEDMAN, 1996; MICHELS and GOTTSCHALK, 1994). Nearly 20% of the initial concentration of nitroaromatics (20 mM) were washed out during the treatment procedure, whereas the rest remained in the compost. Only 10% of the total amount of the nitroaromatics could be recovered from compost with harsh chemical treatment procedures. An oxygen-dependent, covalent binding of aromatic amines to the soil–humus matrix might, therefore, cause disappearance of TNT and its metabolites. Aromatics play a significant role in the genesis of humic substances. They are involved in the generation of reactive radicals which are very important for the entry of non-aromatic organics into the humification process (ZIECHMANN, 1994).

To evaluate the level of incorporation of TNT into the organic soil matrix of anaerobic and sequential anaerobic/aerobic treatments of soil–molasses mixtures, experiments with ^{14}C labeled TNT were conducted (DRZYZGA

et al., 1998). Nearly 84% of the initially applied radioactivity were immobilized into the soil during the subsequential anaerobic/aerobic treatment, whereas only 57% were bound to the soil matrix during an anaerobic treatment of a similar mixture for 5 weeks. This supports the hypothesis that the reduced intermediates of the nitro compounds (amino and hydroxylamino groups) were subjected to oxidative soil coupling processes (PENNINGTON et al., 1995; DAWEL et al., 1997). According to these results several researchers favored a bioremediation process enhancing humification of nitroaromatics instead of mineralization, but in such a case the question of the stability of the "bound residues" will arise.

From this point of view a study of PALMER et al. (1997) is very important which demonstrated that TNT residues in composts are bioavailable following introduction into lungs of rats. Furthermore, JARVIS et al. (1998) demonstrated that extracts from finished compost treatments indicated reduced toxicity to microorganisms, whereas the mutagenicity was markedly increased. This was attributed to the attenuation of the explosives during composting. The increased mutagenicity, however, seemed to be the result of the transient formation of reactive metabolites during the composting process. Other research groups prefer a bio-

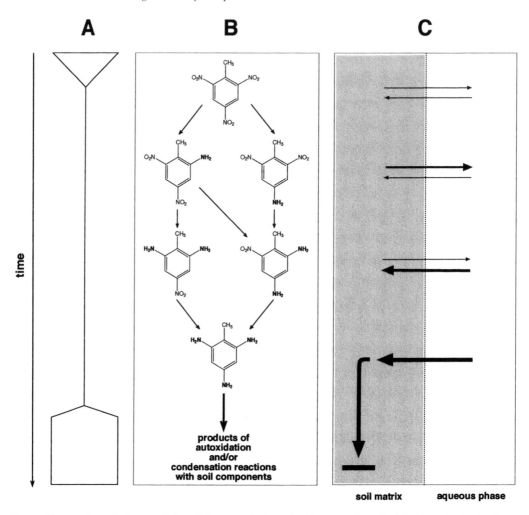

Fig. 7. Time scale and characteristics of the remediation of soils contaminated with the explosive TNT by immobilization. (**A**) Concentration of oxygen, (**B**) fate of TNT, (**C**) desorption and adsorption behavior of intermediates and products in soil.

slurry treatment process (FUNK et al., 1993; DAUN et al., 1994; SHEN et al., 1998). A well-established anaerobic bioslurry technology is the so-called J. R. Simplot Anaerobic Bioremediation (SABRE™). The technology depends on the enhancement of naturally selected anaerobic soil bacteria with organic rich supplements like potato starch water in the presence of xenobiotics. Water was added to obtain 1 L water to 1 kg of contaminated soil ratio. pH regulation is necessary to avoid acidification (EPA, 1995). Optimum operating conditions were found at

37°C, a pH below 8, and a redox potential ≤ -200 mV. The reduction efficiency of the initial TNT concentration was reported as 99.4%. Biotransformation of 2,4,6-trinitrotoluene with anaerobic sludge (10% vol/vol supplemented with 3.3 g L^{-1} molasses) revealed only 0.1% mineralization of TNT, although TNT disappeared in less than 15 h (HAWARI et al., 1998). Besides azo compounds triaminotoluene was also detected during the anaerobic treatment procedure. In contrast to the findings of LEWIS et al. (1996) the authors observed no denitration or deamination of TAT thus forming methyl phloroglucinol or *p*-cresol. This indicates that TAT acted as a dead-end metabolite. They called TAT a "killer metabolite" that misrouted TNT from mineralization under anoxic conditions.

In accordance with the results of LEWIS et al. (1996) GORONTZY (1997) also observed a further metabolization of TAT to more polar yet unknown substances by *Clostridium* sp. strain CWT3 in liquid culture, indicating that the degradation potential of anaerobic bacteria is not already exhausted. If these results could be confirmed in the future it would be a desirable biodegradation pathway, because *p*-cresol, toluidines, and toluene can be mineralized under oxic as well as anoxic conditions.

The removal of the nitro and amino groups from the TNT molecule and derived metabolites, as well as the cleavage of the benzene ring are the major rate controlling steps in biological conversion systems. Therefore, another possibility to treat soils contaminated with explosives was suggested and described as a "biologically enhanced washing of soil" (KAHL et al., 1996). Maintaining anoxic conditions while percolating water through soil columns resulted in the production of polar amino compounds which were eluted from the soil matrix after several washings on a laboratory scale. For a detailed information on the experimental setup see Fig. 8. The recovery of the initial TNT concentration (2,5 g TNT kg^{-1} dry weight of soil) was approximately 95% within 40 weeks. This results in a mobilization of aminoaromatic compounds instead of an immobilization, but it makes a subsequent treatment of the process water necessary, e.g., by photocatalytic degradation (HESS et al., 1998; KLAPPROTH et al., 1998; NAHEN et al., 1997) and/or by biological destruction via TAT as a reactive intermediate.

It has also been suggested to treat soils by phytoremediation including mycorrhiza phenomena. It has been known for a long time that plants successfully combat xenobiotics, the metabolism is divided into 3 phases. At first there is a phase of transformation, followed by conjugation, and a storage process (SANDERMANN, 1992). Phytoremediation technology, however, depends on the synergism between microbes and plants, plant enzyme-

mediated conversion of the contaminating substances, and subsequent bioaccumulation in plant tissues (see chapter 16, this volume; SCHNOOR et al., 1995; LARSON, 1997). But a major problem is how and where the products are incorporated in the plant matrix and their possible release under certain circumstances. Deposition of the xenobiotics or derived transformation products in root tissue or the slow transport of water insoluble substances will be other limiting factors in the phytoremediation technology. Here is a special need for future research concerning these questions.

5 Outlook

As a consequence of the overview given here biological treatment procedures of soils contaminated with nitroaromatic compounds could be definitively an alternative to conventional and often very destructive processes like the incineration of soil (EPA, 1995; WANNINGER, 1995). Especially polynitrated substances may be trapped in the soil matrix rather than degraded. But nevertheless, due to a long turnover of humic substances this might be an alternative to mineralization. But to establish such a method the following questions have to be addressed:

(1) What nature are the geological site characteristics? An important point for choosing a suitable method, because, e.g., composting is difficult to apply to contaminations of water or extremely stony material.

(2) What is the cheapest and best choice for a compost mixture (supplementation of a carbon source to enhance cometabolism) to establish covalent binding of a pollutant during metabolism?

(3) Is there a release of toxic materials to be expected in the future due to remobilization processes in the (organic) soil matrix?

(4) Could preceding anaerobic treatments achieve a better bioavailability of certain pollutants due to the formation of aminoaromatics which, e.g., reveal a higher solubility in water?

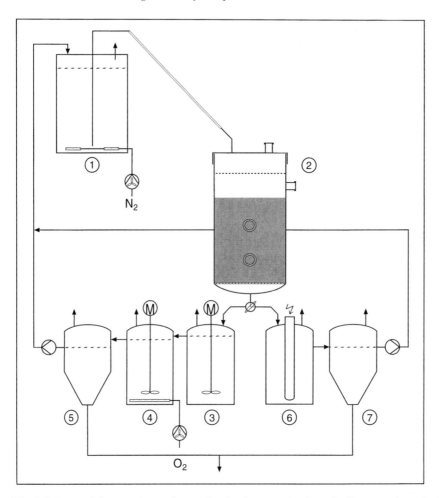

Fig. 8. Scheme of the experimental setup for the decontamination of soils contaminated with the explosive TNT by mobilization. (1) Reservoir of the washing solution, (2) soil column, (3) bioreactor (anoxic), (4) bioreactor (oxic), (5) device for the sedimentation of particulate matter, (6) photo reactor, (7) device for the sedimentation of particulate matter.

(5) How stable are cell-free applications with enzymes such as the manganese peroxidase treatment in natural and contaminated environments?

New molecular biological methods and the clarification of the biochemical pathways will answer the central question in the near future, whether to change the contaminated site to suit the organisms or to change the organisms to suit the site or even a combination of both. The race is on.

6 References

Accashian, J. V., Vinopal, R. T., Kim, B.-J., Smets, B. F. (1998), Aerobic growth on nitroglycerin as the sole carbon, nitrogen, and energy source by a mixed bacterial culture, *Appl. Environ. Microbiol.* **64**, 3300–3304.

Alexander, M. (1994), *Biodegradation and Bioremediation.* San Diego, CA: Academic Press.

An, D., Gibson, D. T., Spain, J. C. (1994), Oxidative release of nitrite from 2-nitrotoluene by a three-component enzyme system from *Pseudomonas* sp. strain JS42, *J. Bacteriol.* **176**, 7462–7467.

ANDERS, H. J., KAETZKE, A., KÄMPFER, P., LUDWIG, W., FUCHS, G. (1995), Taxonomic position of aromatic-degrading denitrifying *Pseudomonas* strains K-172 and kb-740 and their description as new members of the genera *Thauera*, as *Thauera aromatica* sp-nov, and *Azoarcus*, as *Azoarcus evansii* sp-nov, respectively, members of the *beta*-subclass of the proteobacteria, *Int. J. Syst. Bacteriol.* **45**, 327–333.

ANGERMAIER, L., SIMON, H. (1983a), On the reduction of aliphatic and aromatic nitro compounds by Clostridia, the role of ferredoxin and its stabilization, *Hoppe-Seyler's Z. Physiol. Chem.* **364**, 961–975.

ANGERMAIER, L., SIMON, H. (1983b), On nitroaryl reductase activities in several Clostridia, *Hoppe-Seyler's Z. Physiol. Chem.* **364**, 1653–1664.

BACHOFER, R., LINGENS, F., SCHÄFER, W. (1975), Conversion of aniline into pyrocatechol by a *Nocardia* sp.: Incorporation of oxygen-18, *FEBS Lett.* **50**, 288–290.

BARR, D. P., AUST, S. D. (1994), Mechanisms white rot fungi use to degrade pollutants, *Environ. Sci. Technol.* **28**, 78A–87A.

BELAND, F. A., HEFLICH, R. H., HOWARD, P. C., FU, P. P. (1985), The *in vitro* metabolic activation of nitro polycyclic aromatic hydrocarbons, in: Polycyclic hydrocarbons and carcinogenesis (HARVEY, R. G., Ed.), pp. 317–396. Washington, DC: American Chemical Society.

BELLER, H. R., SPORMANN, A. M., SHARMA, P. K., COLE, J. R., REINHARD, M. (1996), Isolation and characterization of a novel toluene-degrading, sulfate-reducing bacterium, *Appl. Environ. Microbiol.* **62**, 1188–1196.

BINKS, P. R., FRENCH, C. E., NICKLIN, S., BRUCE, N. C. (1996), Degradation of pentaerythritol tetranitrate by *Enterobacter cloacae, Appl. Environ. Microbiol.* **62**, 1214–1219.

BLASCO, R., MOORE, E., WRAY, V., PIEPER, D., TIMMIS, K. N., CASTILLO, F. (1999), 3-Nitroadipate, a metabolic intermediate for mineralization of 2,4-dinitrophenol by a new strain of a *Rhodococcus* species, *J. Bacteriol.* **181**, 149–152.

BOOPATHY, R. (1994), Transformation of nitroaromatic compounds by a methanogenic bacterium *Methanococcus* sp. (strain B), *Arch. Microbiol.* **162**, 131–137.

BOOPATHY, R., KULPA, C. F. (1992), Trinitrotoluene (TNT) as a sole source for a sulfate-reducing bacterium *Desulfovibrio* sp. (B strain) isolated from an anaerobic digester, *Curr. Microbiol.* **25**, 235–241.

BOOPATHY, R., KULPA, C. F. (1993), Nitroaromatic compounds serve as a nitrogen source for *Desulfovibrio* sp. (B strain), *Can. J. Microbiol.* **28**, 131–137.

BOOPATHY, R., MANNING, J. (1996), Characterization of partial anaerobic metabolic pathway for 2,4,6-trinitrotoluene degradation by a sulfate-reducing bacterial consortium, *Can. J. Microbiol.* **42**, 1203–1208.

BOOPATHY, R., MANNING, J., MONTEMAGNO, C., RIMKUS, K. (1994), Metabolism of trinitrobenzene by a *Pseudomonas* consortium, *Can. J. Microbiol.* **40**, 787–790.

BREITUNG, J., BRUNS-NAGEL, D., VON LÖW, E., STEINBACH, K., KAMINSKI, L. et al. (1995), Mikrobielle Sanierung von 2,4,6-Trinitrotoluol (TNT) kontaminierten Böden, *UWSF-Z. Umweltchem. Ökotox.* **7**, 195–200.

BREITUNG, J., BRUNS-NAGEL, D., STEINBACH, K., KAMINSKI, L., GEMSA, D., VON LÖW, E. (1996), Bioremediation of 2,4,6-trinitrotoluene-contaminated soils by two different aerated compost systems, *Appl. Microbiol. Biotechnol.* **44**, 795–800.

BRUNS-NAGEL, D., DRZYZGA. O., STEINBACH, K., SCHMIDT, C., VON LÖW, E. et al. (1998), Anaerobic/aerobic composting of 2,4,6-trinitrotoluene-contaminated soil in a reactor system, *Environ. Sci. Technol.* **32**, 1676–1679.

BRYANT, C., DELUCA, M. (1991), Purification and characterization of an oxygen-insensitive NAD(P)H nitroreductase from *Enterobacter cloacae, J. Biol. Chem.* **266**, 4119–4125.

BUMPUS, J. A., TATARKO, M. (1994), Biodegradation of 2,4,6-trinitrotoluene by *Phanerochaete chrysosporium:* Identification of initial degradation products and the discovery of a TNT metabolite that inhibits lignin peroxidases, *Curr. Microbiol.* **28**, 185–190.

BUTRUILLE, D., DOMINGUEZ, X. A. (1972), Un nouveau produit naturel: Dimethoxy-1,4-nitro-2-trichloro-3,5,6-benzene, *Tetrahedron Lett.* **3**, 211–212.

CARLIER, J. P., SELLIER, N., RAGER, M. N., REYSSET, G. (1998), Metabolism of a 5-nitroimidazole in susceptible and resistant isogenic strains of *Bacteroides fragilis, Antimicrob. Agents Chemother.* **41**, 1495–1499.

CARTWRIGHT, N. J., CAIN, R. B. (1959), Bacterial degradation of the nitrobenzoic acids, *Biochem. J.* **71**, 248–261.

CERNIGLIA, C. E., SOMERVILLE, C. C. (1995), Reductive metabolism of nitroaromatic and nitropolycyclic aromatic hydrocarbons, in: *Biodegradation of Nitroaromatic Compounds* (SPAIN, J. C., Ed.), pp. 99–115. New York: Plenum Press.

COLBERG, P. J., YOUNG, L. Y. (1995), Anaerobic degradation of nonhalogenated homocyclic aromatic compounds coupled with nitrate, iron, or sulfate reduction, in: *Microbial Transformation and Degradation of Toxic Organic Chemicals* (YOUNG, L.

Y., CERNIGLIA, C. E., Eds.), pp. 307–330. New York: John Wiley & Sons.

CORBETT, M. D., CORBETT, B. R. (1995), Bioorganic chemistry of the aryl hydroxyl amine and nitrosoarene functional groups, in: *Biodegradation of Nitroaromatic Compounds* (SPAIN, J. C., Ed.), pp. 151–182. New York: Plenum Press.

COSTA, V., BOOPATHY, R., MANNING, J. (1996), Isolation and characterization of a sulfate-reducing bacterium that removed TNT (2,4,6-trinitrotoluene) under sulfate- and nitrate-reducing conditions, *Biores. Technol.* **56**, 273–278.

DAUN, G., LENKE, H., KNACKMUSS, H.-J., REUß, M. (1994), Vergleich zweier Verfahren zur biologischen Sanierung von Rüstungsaltlasten aus der TNT-Produktion, *Chem.-Ing.-Tech.* **66**, 1246–1247.

DAVIS, E. P., BOOPATHY, R., MANNING, J. (1997), Use of trinitrobenzene as a nitrogen source by *Pseudomonas vesicularis* isolated from soil, *Curr. Microbiol.* **34**, 192–197.

DAWEL, G., KÄSTNER, M., MICHELS, J., POPPITZ, W., GÜNTHER, W., FRITSCHE, W. (1997), Structure of laccase-mediated product coupling of 2,4-diamino-6-nitrotoluene to guaiacol, a model for coupling of 2,4,6-trinitrotoluene metabolites to a humic organic soil matrix, *Appl. Environ. Microbiol.* **63**, 2560–2565.

DELGADO, A. M., WUBBOLTS, G., ABRIL, M.-A., RAMOS, J. L. (1992), Nitroaromatics are substrates for the TOL plasmid upper-pathway enzymes, *Appl. Environ. Microbiol.* **58**, 415–417

DENNIE, D., GLADU, I., LEPINE, F., VILLEMUR, R., BISAILLON, J. G., BEAUDET, R. (1998), Spectrum of the reductive dehalogenation activity of *Desulfitobacterium frappieri* PCP-1, *Appl. Environ. Microbiol.* **64**, 4603–4606.

DHAWALE, M. R., HORNEMANN, U. (1979), Nitroalkane oxidation by streptomyces, *J. Bacteriol.* **137**, 916–924.

DRZYZGA, O., GORONTZY, T., SCHMIDT, A., BLOTEVOGEL, K. H. (1995a), Toxicity of explosives and related compounds to the luminescent bacterium *Vibrio fischeri* NRRL-B-11177, *Arch. Environ. Contam. Toxicol.* **28**, 229–235.

DRZYZGA, O., SCHMIDT, A., BLOTEVOGEL, K. H. (1995b), Reduction of nitrated diphenylamine derivatives under anaerobic conditions, *Appl. Environ. Microbiol.* **61**, 3282–3287.

DRZYZGA, O., SCHMIDT, A., BLOTEVOGEL, K. H. (1996), Cometabolic transformation and cleavage of nitrodiphenyl amines by three newly isolated sulfate-reducing bacterial strains, *Appl. Environ. Microbiol.* **62**, 1710–1716.

DRZYZGA, O., BRUNS-NAGEL, D., GORONTZY, T., BLOTEVOGEL, K. H., GEMSA, D., VON LÖW, E. (1998), Incorporation of ^{14}C-labeled 2,4,6-trinitrotoluene metabolites into different soil fractions after anaerobic and anaerobic-aerobic treatment of soil/molasses mixtures, *Environ. Sci. Technol.* **32**, 3529–3535.

DUQUE, E., HAÏDOUR, A., GODOY, F., RAMOS, J. L. (1993), Construction of a *Pseudomonas* hybrid strain that mineralizes 2,4,6-trinitrotoluene, *J. Bacteriol.* **175**, 2278–2283.

ECKER, S., WIDMANN, T., LENKE, H., DICKEL, O., FISCHER, P. et al. (1992), Catabolism of 2,6-dinitrophenol by *Alcaligenes eutrophus* JMP 134 and JMP 222, *Arch. Microblol.* **158**, 149–154.

EDERER, M. M., LEWIS, T. A., CRAWFORD, R. L. (1997), 2,4,6-Trinitrotoluene (TNT) transformation by Clostridia isolated from a munitions-fed bioreactor: comparison with non-adapted bacteria, *J. Ind. Microbiol. Biotechnol.* **18**, 82–88.

EPA (1995), *J. R. Simplot ex-situ Anaerobic Bioremediation Technology: TNT, EPA Site Technology Capsule*. EPA 540/R-95/529a.

FERNANDO, T., BUMPUS, J. A., AUST, S. D. (1990), Biodegradation of TNT (2,4,6-trinitrotoluene) by *Phanerochaete chrysosporium*, *Appl. Environ. Microbiol.* **56**, 1667–1671.

FETZNER, S. (1998), Bacterial dehalogenation, *Appl. Microbiol. Biotechnol.* **50**, 633–657.

FIELD, J. A., STAMS, A. J. M., KATO, M., SCHRAA, G. (1995), Enhanced biodegradation of aromatic pollutants in cocultures of anaerobic and aerobic bacterial consortia, *Antonie van Leeuwenhoek* **67**, 47–77.

FOLSOM, B. R., STIERLI, R., SCHWARZENBACH, R. P., ZEYER, J. (1993), Comparison of substituted 2-nitrophenol degradation by enzyme extracts and intact cells, *Environ. Sci. Technol.* **28**, 306–311.

FRENCH, C. E., NICKLIN, S., BRUCE, N. C. (1996), Sequence and properties of pentaerythritol tetranitrate reductase from *Enterobacter cloacae* PB2, *J. Bacteriol.* **178**, 6623–6627.

FRENCH, C. E., NICKLIN, S., BRUCE, N. C. (1998), Aerobic degradation of 2,4,6-trinitrotoluene by *Enterobacter cloacae* PB2 and by pentaerythritol tetranitrate reductase, *Appl. Environ. Microbiol.* **64**, 2864–2868.

FUNK, S. B., ROBERTS, D. J., CRAWFORD, D. L., CRAWFORD, R. L. (1993), Initial-phase optimization for bioremediation of munition compound-contaminated soils, *Appl. Environ. Microbiol.* **59**, 2171–2177.

GADDA, G., EDMONDSON, R. D., RUSSELL, D. H., FITZPATRICK, P. F. (1997), Identification of the naturally occurring flavin of nitroalkane oxidase from *Fusarium oxysporum* as a 5-isobutyl-FAD and conversion of the enzyme to the active FAD-containing form, *J. Biol. Chem.* **272**, 5563–5570.

GLASBY, J. G. (1976), *Encyclopedia of Antibiotics*. Chichester: John Wiley & Sons.

GLAUS, M. A., HEIJMAN, C. G., SCHWARZENBACH, R. P., ZEYER, J. (1992), Reduction of nitroaromatic compounds mediated by *Streptomyces* sp. exudates, *Appl. Environ. Microbiol.* **58**, 1945–1951.

GORLATOVA, N., TCHORZEWSKI, M., KURIHARA, T., SODA, K., ESAKI, N. (1998), Purification, characterization, and mechanism of a flavin mononucleotide-dependent 2-nitropropane dioxygenase from *Neurospora crassa*, *Appl. Environ. Microbiol.* **64**, 1029–1033.

GORONTZY, T. (1997), Untersuchungen zur biologischen Sanierung von Rüstungsaltlasten, *Thesis*, University of Oldenburg, Germany.

GORONTZY, T., KÜVER, J., BLOTEVOGEL, K. H. (1993), Microbial transformation of nitroaromatic compounds under anaerobic conditions, *J. Gen. Microbiol.* **139**, 1331–1336.

GORONTZY, T., DRZYZGA, O., KAHL, M. W., BRUNS-NAGEL, D., BREITUNG, J. et al. (1994), Microbial degradation of explosives and related compounds, *Crit. Rev. Microbiol.* **20**, 265–284.

GROENEWEGEN, P. E. J., DE BONT, J. A. M. (1992), Degradation of 4-nitrobenzoate via 4-hydroxylaminobenzoate and 3,4-dihydroxybenzoate in *Comamonas acidovorans* NBA-10, *Arch. Microbiol.* **158**, 381–386.

GROENEWEGEN, P. E. J., BREEUWER, P., VAN HELVOORT, J. M. L. M., LANGENHOFF, A. A. M., DE VRIES, F. P., DE BONT, J. A. M. (1992), Novel degradative pathway of 4-nitrobenzoate in *Comamonas acidovorans* NBA-10, *J. Gen. Microbiol.* **138**, 1599–1605.

HAAG, W. R., SPANGGORD, R. J., MILL, T., PODOLL, R. T., CHOU, T. W. et al. (1991), Fate of diethylene glycol dinitrate in surface water, *Chemosphere* **23**, 215–230.

HADERLEIN, S. B., SCHWARZENBACH, R. P. (1993), Adsorption of substituted nitrobenzenes and nitrophenols to mineral surfaces, *Environ. Sci. Technol.* **27**, 316–326.

HADERLEIN, S. B., SCHWARZENBACH, R. P. (1995), Environmental processes influencing the rate of abiotic reduction of nitroaromatic compounds in the subsurface, in: *Biodegradation of Nitroaromatic Compounds* (SPAIN, J. C., Ed.), pp. 199–225. New York: Plenum Press.

HADERLEIN, S. B., WEISSMAHR, K. W., SCHWARZENBACH, R. P. (1996), Specific adsorption of nitroaromatic explosives and pesticides to clay minerals, *Environ. Sci. Technol.* **30**, 612–622.

HAIGLER, B. E., SPAIN, J. C. (1993), Biodegradation of 4-nitrotoluene by *Pseudomonas* sp. strain 4NT, *Appl. Environ. Microbiol.* **59**, 2239–2243.

HAIGLER, B. E., WALLACE, W. H., SPAIN, J. C. (1994), Biodegradation of 2-nitrotoluene by *Pseudomonas* sp. strain JS42, *Appl. Environ. Microbiol.* **60**, 3466–3469.

HALL, D. R., BEEVOR, P. S., CAMPION, D. G., CHAMBERLAIN, D. J., CORK, A. et al. (1992), Nitrate esters, novel sex pheromone components of the cotton leaf perforator, *Bucculatrix thurberiella* Busck (Lepidoptera, Lyonetildae), *Tetrahedron Lett.* **33**, 4811–4814.

HAMMILL, T. B., CRAWFORD, R. L. (1996), Degradation of 2-*sec*-butyl-4,6-dinitrophenol (Dinoseb) by *Clostridium bifermentans* KMR-1, *Appl. Environ. Microbiol.* **62**, 1842–1846.

HANSEN, T. A. (1994), Metabolism of sulfate-reducing prokaryotes, *Antonie van Leeuwenhoek* **66**, 165–185.

HARTTER, D. R. (1985), The importance of nitroaromatic chemicals in the chemical industry, in: *Toxicity of Nitroaromatic Chemicals: Chemical Industry Institute of Toxicology Series* (RICKERT, D. E., Ed.), pp. 1–14. New York: Hemisphere Publishing Corp.

HASAN, S., CHO, J. G., SUBLETTE, K. L., PAK, D. MAULE, A. (1992), Porphyrin-catalyzed degradation of chlorinated and nitro-substituted toluenes, *J. Biotechnol.* **24**, 195–201.

HASSPIELER, B. M., HAFFNER, G. D., ADELI, K. (1997), Roles of DT diaphorase in the genotoxicity of nitroaromatic compounds in human and fish cell lines, *J. Toxicol. Environ. Health* **52**, 137–148.

HAWARI, J., HALASZ, A., PAQUET, L., ZHOU, E., SPENCER, B. et al. (1998), Characterization of metabolites in the biotransformation of 2,4,6-trinitrotoluene with anaerobic sludge: role of triamino toluene, *Appl. Environ. Microbiol.* **64**, 2200–2206.

HE, Z., SPAIN, J. C. (1997), Studies of the catabolic pathway of degradation of nitrobenzene by *Pseudomonas pseudoalcaligenes* JS45: removal of the amino group from 2-aminomuconic semialdehyde, *Appl. Environ. Microbiol.* **63**, 4839–4843.

HE, Z., SPAIN, J. C. (1998), A novel 2-aminomuconate deaminase in the nitrobenzene degradation pathway of *Pseudomonas pseudoalcaligenes* JS45, *J. Bacteriol.* **180**, 2502–2506.

HEASLEY, C. J., FITZPATRICK, P. F. (1996), Kinetic mechanism and substrate specificity of nitroalkane oxidase, *Biochem. Biophys. Res. Comm.* **225**, 6–10.

HEIJMAN, C. G., HOLLIGER, C., GLAUS, M. A., SCHWARZENBACH, R. P., ZEYER, J. (1993), Abiotic reduction of 4-chloronitrobenzene to 4-chloroaniline in a dissimilatory iron-reducing enrichment culture, *Appl. Environ. Microbiol.* **59**, 4350–4353.

HEITKAMP, M. A., FREEMAN, J. P., MILLER, D. W., CERNIGLIA, C. E. (1991), Biodegradation of 1-nitropyrene, *Arch. Microbiol.* **156**, 223–230.

HESS, T. F., LEWIS, T. A., CRAWFORD, R. L., KATAMENI, S., WELLS, J. H., WATTS, R. J. (1998), Combined photocatalytic and fungal treatment for the destruction of 2,4,6-trinitrotoluene (TNT), *Water*

Res. **32**, 1481–1491.

HIGSON, F. K. (1992), Microbial degradation of nitro-aromatic compounds, *Adv. Appl. Microbiol.* **37**, 1–19.

HINKEL, M., REISCHL, A., SCHRAMM, K. W., TRAUTNER, F., REISSINGER, M., HUTZINGER, O. (1989), Concentration levels of nitrated phenols in conifer needles, *Chemosphere* **18**, 2433–2439.

HOFRICHTER, M., SCHEIBNER, K., SCHNEEGAß, I., FRITSCHE, W. (1998), Enzymatic combustion of aromatic and aliphatic compounds by manganese peroxidase from *Nematoloma frowardii*, *Appl. Environ. Microbiol.* **64**, 399–404.

HONEYCUTT, M. E., JARVIS, A. S., MACFARLAND, V. A. (1996), Cytotoxicity and mutagenicity of 2,4,6-trinitrotoluene and its metabolites, *Ecotoxicol. Environ. Safety* **35**, 282–287.

IPPEN, H. (1994a), *Nitromoschus.* Teil I. Bundesgesundheitsblatt 6/94, 255–260.

IPPEN, H. (1994b), *Nitromoschus.* Teil II. Bundesgesundheitsblatt 7/94, 291–294.

ISBISTER, J. D., ANSPACH, G. L., KITCHENS, J. F., DOYLE, R. C. (1984), Composting for decontamination of soils containing explosives, *Microbiologica* **7**, 47–73.

JAIN, R. K., DREISBACH, J. H., SPAIN, J. C. (1994), Biodegradation of *p*-nitrophenol via 1,2,4-benzenetriol by an *Arthrobacter* sp., *Appl. Environ. Microbiol.* **60**, 3030–3032.

JAMES, K. D., WILLIAMS, P. A. (1998), *ntn* genes determining the early steps in the divergent catabolism of 4-nitrotoluene and toluene in *Pseudomonas* sp. strain TW3, *J. Bacteriol.* **180**, 2043–2049.

JARVIS, A. S., MCFARLAND, V. A., HONEYCUTT, M. E. (1998), Assessment of the effectiveness of composting for the reduction of toxicity and mutagenicity of explosive-contaminated soil, *Ecotoxicol. Environ. Safety* **39**, 131–135.

KAAKE, R. H., ROBERTS, D. J., STEVENS, T. O., CRAWFORD, R. L., CRAWFORD, D. L. (1992), Bioremediation of soils contaminated with the herbicide 2-*sec*-butyl-4,6-dinitrophenol (Dinoseb), *Appl. Environ. Microbiol.* **58**, 1683–1689.

KAAKE, R. H., CRAWFORD, D. L., CRAWFORD, R. L. (1995), Biodegradation of the nitroaromatic herbicide Dinoseb (2-*sec*-butyl-4,6-dinitrophenol) under reducing conditions, *Biodegradation* **6**, 329–337.

KADIYALA, V., SPAIN, J. C. (1998), A two-component monooxygenase catalyzes both the hydroxylation of *p*-nitrophenol and the oxidative release of nitrite from 4-nitrocatechol in *Bacillus sphaericus* JS905, *Appl. Environ. Microbiol.* **64**, 2479–2484.

KAHL, M. W., GORONTZY, T., BLOTEVOGEL, K. H. (1996), Reinigung TNT-kontaminierter Böden durch anaerobe Prozeßführung, *gwf-Wasser/Abwasser* **137**, 140–146.

KALAFUT, T., WALES, M. E., RASTOGI, V. K., NAUMOVA, R. P., ZARIPOVA, S. K., WILD, J. R. (1998), Biotransformation patterns of 2,4,6-trinitrotoluene by aerobic bacteria, *Curr. Microbiol.* **36**, 45–54.

KAPLAN, D. L., KAPLAN, A. M. (1982), Thermophilic biotransformation of 2,4,6-trinitrotoluene under simulated composting conditions, *Appl. Environ. Microbiol.* **44**, 757–760.

KARRER, W. (1985), *Konstitution und Vorkommen der organischen Pflanzenstoffe*, Ergänzungsband 2, Teil 2. Basel: Birkhäuser Verlag.

KHAN, T. A., BHADRA, R., HUGHES, J. (1997), Anaerobic transformation of 2,4,6-TNT and related nitroaromatic compounds by *Clostridium acetobutylicum, J. Ind. Microbiol. Biotechnol.* **18**, 198–203.

KIDO, T., SODA, K., SUZUKI, T., ASADA, K. (1976), A new oxygenase, 2-nitropropane dioxygenase of *Hansenula mrakii, J. Biol. Chem.* **251**, 6994–7000.

KINOUCHI, T., OHNISHI, Y. (1983), Purification and characterization of 1-nitropyrene nitroreductases from *Bacteroides fragilis, Appl. Environ. Microbiol.* **46**, 596–604.

KLAPPROTH, A., LINNEMANN, S., BAHNEMANN, D., DILLERT, R., FELS, G. (1998), ^{14}C-trinitrotoluene: synthesis and photocatalytical degradation, *J. Labeled Comp. Radiopharm.* **41**, 337–343.

KOZARAC, Z., COSOVIC, B., GASPAROVIC, B., DHATHATHREYAN, A., MÖBIUS, D. (1991), Interaction of *p*-nitrophenol with lipids at hydrophobic interfaces, *Langmuir* **7**, 1076–1081.

KRUMHOLZ, L. R., LI, J., CLARKSON, W. W., WILBER, G. G., SUFLITA, J. M. (1997), Transformations of TNT and related aminotoluenes in ground water aquifer slurries under different electron-accepting conditions. *J. Ind. Microbiol. Biotechnol.* **18**, 161–169.

KUHN, E. P., COLBERG, P. J., SCHNOOR, J. L., WANNER, O., ZEHNDER, A. J. B., SCHWARZENBACH, R. P. (1985), Microbial transformations of substituted benzenes, *Environ. Sci. Technol.* **19**, 961–968.

LARSON, S. L. (1997), Fate of explosive contaminants in plants, *Ann. N. Y. Acad. Sci.* **829**, 195–201.

LENKE, H., KNACKMUSS, H.-J. (1992), Initial hydrogenation during catabolism of picric acid by *Rhodococcus erythropolis* HL24-2, *Appl. Environ. Microbiol.* **58**, 2933–2937.

LENKE, H., PIEPER, D. H., BRUHN, C., KNACKMUSS, H.-J. (1992), Degradation of 2,4-dinitrophenol by two *Rhodococcus erythropolis* strains, HL 24-1 and HL 24-2, *Appl. Environ. Microbiol.* **58**, 2928–2932.

LEWIS, T. A., GOSZCZYNSKI, S., CRAWFORD, R. L., KORUS, R. A., ADMASSU, W. (1996), Products of anaerobic 2,4,6-trinitrotoluene (TNT) transformation by *Clostridium bifermentans, Appl. Environ. Microbiol.* **62**, 4669–4674.

LITTLE, H. N. (1951), Oxidation of nitroethane by extracts from *Neurospora, J. Biol. Chem.* **193**, 347–358.

MANNING, B. W., CAMPELL, W. L., FRANKLIN, W., DELCLOS, K. B., CERNIGLIA, C. E. (1988), Metabolism of 6-nitrochrysene by intestinal microflora, *Appl. Environ. Microbiol.* **54**, 197–203.

MARTIN, J. L., COMFORT, S. D., SHEA, P. J., KOKJOHN, T. A., DRIJBER, R. A. (1997), Denitration of 2,4,6-trinitrotoluene by *Pseudomonas savastanoi, Can. J. Microbiol.* **43**, 447–455.

MARVIN-SIKKEMA, F. D., DE BONT, J. A. M. (1994), Degradation of nitroaromatic compounds by microorganisms, *Appl. Microbiol. Biotechnol.* **42**, 499–507.

McCORMICK, N. G., FEEHERRY, F. E., LEVINSON, H. S. (1976), Microbial transformation of 2,4,6-trinitrotoluene and other nitroaromatic compounds, *Appl. Environ. Microbiol.* **31**, 949–958.

MENG, M., SUN, W. Q., GEELHAAR, L. A., KUMAR, G., PATEL, A. R. et al. (1995), Denitration of glycerol trinitrate by resting cells and cell extracts ot *Bacillus thuringiensis/cereus* and *Enterobacter agglomerans, Appl. Environ. Microbiol.* **61**, 2548–2553.

MEULENBERG, R., DE BONT, J. A. M. (1995), Microbial production of catechols from nitroaromatic compounds, in: *Biodegradation of Nitroaromatic Compounds* (SPAIN, J. C., Ed.), pp. 37–52. New York: Plenum Press.

MEULENBERG, R., PEPI, M., DE BONT, J. A. M. (1996), Degradation of 3-nitrophenol by *Pseudomonas putida* B2 occurs via l,2,4-benzenetriol, *Biodegradation* **7**, 303–311.

MICHELS, J., GOTTSCHALK, G. (1994), Inhibition of lignin peroxidase of *Phanerochaete chrysosporium* by hydroxylamino dinitrotoluene, an early intermediate in the degradation of 2,4,6-trinitrotoluene, *Appl. Environ. Microbiol.* **60**, 187–194.

MISKINIÈNÈ, V., SARLAUSKAS, J., JACQUOT, J.-P., CENAS, N. (1998), Nitroreductase reactions of *Arabidopsis thaliana* thioredoxin reductase, *Biochim. Biophys. Acta* **1366**, 275–283.

MITRA, D., VAIDYANATHA, C. S. (1984), A new 4-nitrophenol 2-hydroxylase from a *Nocardia* sp., *Biochem. Int.* **8**, 609–615.

MIX, D. B., GUINAUDEAU, H., SHAMMA, M. (1982), The aristolochic acids and aristolactams, *J. Nat. Prod.* **45**, 657–666.

MONTPAS, S., SAMSON, J., LANGLOIS, E., LEI, J., PICHÉ, Y., CHÊNEVERT, R. (1997), Degradation of 2,4,6-trinitrotoluene by *Serratia marcescens, Biotechnol. Lett.* **19**, 291–294.

NAHEN, M., BAHNEMANN, D., DILLERT, R., FELS, G. (1997), Photocatalytic degradation of trinitrotoluene: reductive and oxidative pathways, *J. Photochem. Photobiol.* **110**, 191–199.

NISHINO, S. F., SPAIN, J. C. (1995), Oxidative pathway for the biodegradation of nitrobenzene by *Comamonas* sp. strain JS675, *Appl. Environ. Microbiol.* **61**, 2308–2313.

NOGUERA, D. R., FREEDMAN, D. L. (1996), Reduction and acetylation of 2,4-dinitrotoluene by a *Pseudomonas aeruginosa* strain, *Appl. Environ. Microbiol.* **62**, 2257–2263.

OREN, A., GUREVICH, P., HENIS, Y. (1991), Reduction of nitro substituted aromatic compounds by the halophilic anaerobic eubacteria *Haloanaerobium praevalens* and *Sporohalobacter marismortui, Appl. Environ. Microbiol.* **57**, 3367–3370.

PARALES, J. V., PARALES, R. E., RESNICK, S. M., GIBSON, D. T. (1998a), Enzyme specificity of 2-nitrotoluene 2,3-dioxygenase from *Pseudomonas* sp. strain JS42 is determined by C-terminal region of the α-subunit of the oxygenase component, *J. Bacteriol.* **180**, 1194–1199.

PARALES, R. E., EMIG, M. D., LYNCH, N. A., GIBSON, D. T. (1998b), Substrate specificity of hybrid naphthalene and 2,4-dinitrotoluene dioxygenase enzyme systems, *J. Bacteriol.* **180**, 2337–2344.

PALMER, W. G., BEAMAN, J. R., WALTERS, D. M., CREASIA, D. A. (1997), Bioavailability of TNT residues in composts of TNT-contaminated soil, *J. Toxicol. Environ. Health* **51**, 97–108.

PARRISH, F. W. (1977), Fungal transformation of 2,4-dinitrotoluene and 2,4,6-trinitrotoluene, *Appl. Environ. Microbiol.* **34**, 232–233.

PENNINGTON, J. C., HAYES, C. A., MYERS, K. F., OCHMAN, M., GUNNISON, D. et al. (1995), Fate of 2,4,6-trinitrotoluene in a simulated compost system, *Chemosphere* **30**, 429–438.

PETERSON, F. J., MASON, R. P., HOVESEPIAN, J., HOLTZMAN, J. L. (1979), Oxygen-sensitive and -insensitive nitroreduction by *Escherichia coli* and rat hepatic microsomes, *J. Biol. Chem.* **254**, 4009–4014.

POTHULURI, J. V., DOERGE, D. R., CHURCHWELL, M. I., FU, P. P., CERNIGLIA, C. E. (1998a), Fungal metabolism of nitrofluoranthenes, *J. Toxicol. Environ. Health* **53**, 153–174.

POTHULURI, J. V., SUTHERLAND, J. B., FREEMAN, J. P., CERNIGLIA, C. E. (1998b), Fungal biotransformation of 6-nitrochrysene, *Appl. Environ. Microbiol.* **64**, 3106–3109.

POWELL, S., FRANZMANN, P. D., CORD-RUWISCH, R., TOZE, S. (1998), Degradation of 2-nitrodiphenyl amine, a component of Otto fuel II, by *Clostridium* spp., *Anaerobe* **4**, 95–102.

PREUSS, A., FIMPEL, J., DIEKERT, G. (1993), Anaerobic transformation of 2,4,6-trinitrotoluene (TNT), *Arch. Microbiol.* **159**, 345–353.

RAFII, F., CERNIGLIA, C. E. (1995), Reduction of azo dyes and nitroaromatic compounds by bacterial enzymes from the human intestinal tract, *Environ. Health Perspec.* (Suppl.) **103**, 17–19.

RAFII, F., HANSEN, E. B. (1998), Isolation of nitrofurantoin-resistant mutants of nitroreductase-producing *Clostridium* sp. strains from the human intestinal tract, *Antimicrob. Agents Chemother.* **42**, 1121–1126.

RAFII, F., FRANKLIN, W., HEFLICH, R. H., CERNIGLIA, C. E. (1991), Reduction of nitroaromatic compounds by anaerobic bacteria isolated from the human gastrointestinal tract, *Appl. Environ. Microbiol.* **57**, 962–968.

RAYMOND, D. G. M., ALEXANDER, M. (1971), Microbial metabolism and cometabolism of nitrophenols, *Pest. Biochem. Physiol.* **1**, 23–30.

REGAN, K. M., CRAWFORD, R. L. (1994), Characterization of *Clostridium bifermentans* and its biotransformation of 2,4,6-trinitrotoluene (TNT) and 1,3,5-triaza-1,3,5-trinitro cyclohexane (RDX), *Biotechnol. Lett.* **16**, 1081–1086.

RHYS-WILLIAMS, W., TAYLOR, S. C., WILLIAMS, P. A. (1993), A novel pathway for the catabolism of 4-nitrotoluene by *Pseudomonas, J. Gen. Microbiol.* **139**, 1967–1972.

RICKERT, D. E., CHISM, J. P., KEDDERIS, G. L. (1986), Metabolism and carcinogenicity of nitrotoluenes, *Adv. Exp. Med. Biol.* **197**, 563–571.

RIEGER, P.-G., KNACKMUSS, H.-J. (1995), Basic knowledge and perspectives on biodegradation of 2,4,6-trinitrotoluene and related nitroaromatic compounds in contaminated soil, in: *Biodegradation of Nitroaromatic Compounds* (SPAIN, J. C., Ed.), pp. 1–19. New York: Plenum Press.

RIPPEN, G., ZIETZ, E., FRANK, R., KNACKER, T., KLOEPFFER, W. (1987), Do airborne nitrophenols contribute to forest decline, *Environ. Technol. Lett.* **8**, 475–482.

RITTER, C. L., MALEJKA-GIGANTI, D. (1998), Nitroreduction of nitrated and C-9 oxidized fluorenes *in vitro*, *Chem. Res. Toxicol.* **11**, 1361–1367.

SANDERMANN, H. (1992), Plant metabolism of xenobiotics, *Trends Biochem.* **17**, 82–84.

SCHÄFER, A., HARMS, H., ZEHNDER, A. J. B. (1996), Biodegradation of 4-nitroanisole by two *Rhodococcus* spp., *Biodegradation* **7**, 249–255.

SCHEIBNER, K., HOFRICHTER, M. (1998), Conversion of aminonitrotoluenes by fungal manganese peroxidase, *J. Basic Microbiol.* **38**, 51–59.

SCHEIBNER, K., HOFRICHTER, M., HERRE, A., MICHELS, J., FRITSCHE, W. (1997), Screening for fungi intensively mineralizing 2,4,6-trinitrotoluene, *Appl. Microbiol. Biotechnol.* **47**, 452–457.

SCHMEISSER, H. H., WIESSLER, M. (1995), Carcinogene Naturstoffe. *BIOforum* **18**, 306–311.

SCHNELL, S., SCHINK, B. (1991), Anaerobic aniline degradation via reductive deamination of 4-aminobenzoyl-CoA in *Desulfobacterium anilini,* *Arch. Microbiol.* **155**, 183–190.

SCHNOOR, J. L., LICHT, L. A., MCCUTCHEON, S. C.,

WOLFE, N. L., CARREIRA, L. H. (1995), Phytoremediation of organic and nutrient contaminants, *Environ. Sci. Technol.* **29**, 318A–323A.

SCHWARZENBACH, R. P., STIERLI, R., LANZ, K., ZEYER, J. (1990), Quinone and iron porphyrin mediated reduction of nitroaromatic compounds in homogeneous aqueous solution, *Environ. Sci. Technol.* **24**, 1566–1574.

SHAH, M. M., SPAIN, J. C. (1996), Elimination of nitrite from the explosive 2,4,6-trinitrophenyl methyl nitramine (Tetryl) catalyzed by ferredoxin NADP oxidoreductase from spinach, *Biochem. Biophys. Res. Comm.* **220**, 563–568.

SHAH, M. M., CAMPBELL, J. A. (1997), Transformation of nitrobenzene by ferredoxin NADP oxidoreductase from spinach leaves, *Biochem. Biophys. Res. Comm.* **241**, 794–796.

SHELLEY, M. D., AUTENRIETH, R. L., WILD, J. R., DALE, B. E. (1996), Thermodynamic analyses of trinitrotoluene biodegradation and mineralization pathways, *Biotechnol. Bioeng.* **50**, 198–205.

SHEN, C. F., GUIOT, S. R., THIBOUTOT, S., AMPLEMAN, G., HAWARI, J. (1998), Fate of explosives and their metabolites in bioslurry treatment processes, *Biodegradation* **8**, 339–347.

SHOJI, J. H., HINOO, H., TERUI, Y., KIKUCHI, J., HATTORI, T. et al. (1989), Isolation of azomycin from *Pseudomonas fluorescens, J. Antibiot.* **42**, 1513–1514.

SIMPSON, J. R., EVANS, W. C. (1953), The metabolism of nitrophenols by certain bacteria, *Biochem. J.* **55**, 24.

SMITH, M. R. (1990), The biodegradation of aromatic hydrocarbons by bacteria, *Biodegradation* **1**, 191–206.

SNAPE, J. R., WALKLEY, N. A., MORBY, A. P., NICKLIN, S., WHITE, G. F. (1997), Purification, properties, and sequence of glycerol trinitrate reductase from *Agrobacterium radiobacter, J. Bacteriol.* **179**, 7796–7802.

SOMERVILLE, C. C., NISHINO, S. F., SPAIN, J. C. (1995), Isolation and characterization of nitrobenzene nitroreductase from *Pseudomonas pseudoalcaligenes* JS45, *J. Bacteriol.* **177**, 3837–3842.

SPAIN, J. C. (1995a), *Biodegradation of Nitroaromatic Compounds,* Environmental Science Research Vol. 49. New York: Plenum Press.

SPAIN, J. C. (1995b), Bacterial degradation of nitroaromatic compounds under aerobic conditions, in: *Biodegradation of Nitroaromatic Compounds* (SPAIN, J. C., Ed.), pp. 19–35. New York: Plenum Press.

Spain, J. C. (1995c), Biodegradation of nitroaromatic compounds, *Ann. Rev. Microbiol.* **49**, 523–555.

SPAIN, J. C., GIBSON, D. T. (1991), Pathway for biodegradation of *p*-nitrophenol in a *Moraxella* sp., *Appl. Environ. Microbiol.* **57**, 812–819.

SPAIN, J. C., WYSS, O., GIBSON, D. T. (1979), Enzymatic oxidation of p-nitrophenol, Biochem. Biophys. Res. Comm. 88, 634–641.

SPANGGORD, R. J., SPAIN, J. C., NISHINO, S. F., MORTELMANS, K. E. (l991), Biodegradation of 2,4-dinitrotoluene by a Pseudomonas sp., Appl. Environ. Microbiol. 57, 3200–3205.

SPIKER, J. K., CRAWFORD, D. L., CRAWFORD, R. L. (1992), Influence of 2,4,6-trinitrotoluene (TNT) concentration on the degradation products of TNT in explosive-contaminated soils by the white rot fungus Phanerochaete chrysosporium, Appl. Environ. Microbiol. 58, 3199–3202.

STAHL, J. D., AUST, S. D. (1993a), Metabolism and detoxification of TNT by Phanerochaete chrysosporium, Biochem. Biophys. Res. Comm. 192, 477–482.

STAHL, J. D., AUST, S. D. (1993b), Plasma membrane dependent reduction of 2,4,6-trinitrotoluene by Phanerochaete chrysosporium, Biochem. Biophys. Res. Comm. 192, 471–476.

SUEN, W.-C., HAIGLER, B. E., SPAIN, J. C. (1996), 2,4-dinitrotoluene dioxygenase from Burkholderia sp. strain DNT: similarity to naphthalene dioxygenase, J. Bacteriol. 178, 4926–4934.

SUN, W. Q., MENG, M., KUMAR, G., GEELHAAR, L. A., PAYNE, G. F. et al. (1996), Biological denitration of propylene glycol dinitrate by Bacillus sp. ATCC 51912, Appl. Microbiol. Biotechnol. 45, 525–529.

TERADA, H. (1981), The interaction of highly active uncouplers with mitochondria, Biochem. Biophys. Acta 639, 225–242.

THALLER, V., TURNER, J. L. (1972), Natural Acetylenes. Part XXXV. Polyacetylenic acid and benzenoid metabolites from cultures of the fungus Lepista diemii Singer, J. Chem. Soc. Perkin Trans. 1, 2032–2034.

TSCHECH, A., SCHINK, B. (1988), Methanogenic degradation of anthranilate (2-aminobenzoate), Syst. Appl. Bacteriol. 11, 9–12.

VALLI, K., BROCK, B. J., JOSHI, D. K., GOLD, M. H. (1992), Degradation of 2,4-dinitrotoluene by the lignin-degrading fungus Phanerochaete chrysosporium, Appl. Environ. Microbiol. 58, 221–228.

VAN AKEN, B., SKUBISZ, K., NAVEAU, H., AGATHOS, S. N. (1997), Biodegradation of 2,4,6-trinitrotoluene (TNT) by the white-rot basidiomycete Phlebia radiata, Biotechnol. Lett. 19, 813–817.

VAN BEELEN, P., BURRIS, D. R. (1995), Reduction of the explosive 2,4,6-trinitrotoluene by enzymes from aquatic sediments. Environ. Toxicol. Chem. 14, 2115–2123.

VANDERBERG, L. A., PERRY, J. J., UNKEFER, P. J. (1995), Catabolism of 2,4,6-trinitrotoluene by Mycobacterium vaccae, Appl. Microbiol. Biotechnol. 43, 937–945.

VORBECK, C., LENKE, H., FISCHER, P., SPAIN, J. C.,

KNACKMUSS, H.-J. (1998), Initial reductive reactions in aerobic microbial metabolism of 2,4,6,-trinitrotoluene, Appl. Environ. Microbiol. 64, 246–252.

VORBECK, C., LENKE, H., FISCHER, P., KNACKMUSS, H.-J. (1994), Identification of a hydride–Meisenheimer complex as a metabolite of 2,4,6-trinitrotoluene by a Mycobacterium strain, J. Bacteriol. 176, 932–934.

WALKER, J. E., KAPLAN, D. L. (1992), Biological degradation of explosives and chemical agents, Biodegradation 3, 369–385.

WANNINGER, P. (1995), Konversion von Explosivstoffen, Chem. uns. Zeit 29, 135–140.

WEISSERMERL, K., ARPE, H.-J. (1997), Industrial Organic Chemistry. Weinheim: Wiley-VCH.

WHITE, G. F., SNAPE, J. R. (1993), Microbial cleavage of nitrate esters: defusing the environment, J. Gen. Microbiol. 139, 1947–1957.

WHITE, G. F., SNAPE, J. R., NICKLIN, S. (1996), Biodegradation of glycerol trinitrate and pentaerythritol tetranitrate by Agrobacterium radiobacter, Appl. Environ. Microbiol. 62, 637–642.

WILLIAMS, P. A., SAYERS, J. R. (1994), The evolution of pathways for aromatic hydrocarbon oxidation in Pseudomonas, Biodegradation 5, 195–217.

WILLIAMS, M. C., BARNEBY, R. C. (1977), The occurrence of nitro toxins in North American Astragalus (Fabaceae), Brittonia 29, 310–326.

WILLIAMS, R. T., ZIEGENFUSS, P. S., SISK, W. E. (1992), Composting of explosives and propellant contaminated soils under thermophilic and mesophilic conditions, J. Ind. Microbiol. 9, 137–144.

WITTE, C.-P., BLASCO, R., CASTILLO, F. (1998), Microbial photodegradation of aminoarenes and metabolism of 2-amino-4-nitrophenol by Rhodobacter capsulatus, Appl. Biochem. Biotechnol. 69, 191–202.

WYMAN, J. F., GUARD, H. E., COLEMAN, W. M. (1984), Environmental chemistry of 1,2-propanediol dinitrate: azeotrope formation, photolysis and biodegradability, Arch. Environ. Toxicol. 13, 647–652.

YINON, Y. (1990), Toxicity and Metabolism or Explosives. Boca Raton, FL: CRC Press.

ZENNO, S., KOLKE, H., TANOKURA, M., SAIGO, K. (1996), Conversion of NfsB, a minor Escherichia coli nitroreductase, to a flavin reductase similar in biochemical properties to Frase I, the major flavin reductase in Vibrio fischeri, by a single amino acid substitution, J. Bacteriol. 178, 4731–4733.

ZENNO, S., KOBORI, T., TANOKURA, M., SAIGO, K. (1998), Conversion of NfsA, the major Escherichia coli nitroreductase, to a flavin reductase with an activity similar to that of Frp, a flavin reductase in Vibrio harveyi, by single amino acid substitution, J. Bacterlol. 180, 422–425.

ZEYER, J., KEARNEY, P. C. (1984), Degradation of

o-nitrophenol and *m*-nitrophenol by a *Pseudomonas putida*, *J. Agric. Food. Chem.* **32**, 238–242.

ZEYER, J., KOCHER, H. P. (1988), Purification and characterization of a bacterial nitrophenol oxygenase which converts *ortho*-nitrophenol to catechol and nitrite, *J. Bacteriol.* **170**, 1789–1794.

ZIECHMANN, W. (1994), *Humic Substances*. Mannheim: BI Wissenschaftsverlag.

ZIERIS, F. J., FEIND, D., HUBER, W. (1998), Long-term effects of 4-nitrophenol in an outdoor synthetic aquatic ecosystem, *Arch. Environ. Contam. Toxicol.* **31**, 165–177.

III Processes for Soil Clean-Up

12 Thermal Processes, Scrubbing/Extraction, Bioremediation and Disposal

MICHAEL KONING
KARSTEN HUPE
RAINER STEGMANN

Hamburg, Germany

List of Abbreviations

BTEX   benzene, toluene, ethylbenzene, xylene
NAPL   non-aqueous phase liquids
ORC   oxygen release compounds
PAH   polycyclic aromatic hydrocarbons
PCB   polychlorinated biphenyls
PCDD   polychlorinated dibenzodioxin
PCDF   polychlorinated dibenzofuran
PCP   pentachlorophenol
SVE   soil vapor extraction
TAT   2,4,6-triaminotoluene
TNT   2,4,6-trinitrotoluene
TPH   total petroleum hydrocarbons

1 Introduction

For the treatment of contaminated sites securing as well as remediation methods are applied. While remediation achieves decontamination or reduction of the pollutant content, securing only includes technical barriers for danger defence. Thus, securing methods only represent provisional measures in general as the old hazardous site remains and a further remediation is normally inevitable.

Remediation methods are classified according to the location as well as to process aspects. Thus *ex situ* and *in situ* processes are available on the one hand and thermal, chemical, physical, and biological processes on the other hand. The *ex situ* processes require an excavation of the contaminated soil and a soil treatment either on the site (on-site remediation) or in a soil treatment center (off-site remediation) after an additional transport. In contrast to that *in situ* treatment takes place in the contaminated soil so that an excavation is not necessary.

Thermal processes are applied for the treatment of highly concentrated organic pollutants, but are only suitable to a small extent for the elimination of heavy metals. With soil scrubbing, the coarse grain fraction >63 μm is purified transferring the pollutants into the water phase and/or into the fine grain fraction. This fine fraction is highly loaded with pollutants so that it has to be treated and disposed afterwards. The biopile process is applied on a large scale as state of the art. It is an effective process for the treatment of biologically degradable pollutants such as mineral oil and its derivatives, aliphatic hydrocarbons, phenols, formaldehyde, and other soil contaminants.

In the practice of remediation there is a trend towards actions with a "minimum of requirements". Contaminated sites are remediated or secured depending on the intended after-use, e.g., housing, development of commercial or industrial utilization, or as recreational areas. Due to the kind of use specific target values have to be met after treatment. If the contaminated site is not used and so far no major dangerous contamination of the groundwater, surface water, etc. has occurred one relies more and more on natural attenuation processes in soil and groundwater. Economic active (e.g., selective biostimulation) and passive *in situ* measures (reactive walls, funnel & gate systems) are currently developed and have to be investigated with regard to their long-term effectiveness.

For the *ex situ* purification of pollutants and soils which are hardly amenable to treatment (chlorinated hydrocarbons, polycyclic aromatic hydrocarbons, tar oil, silty soils) combined processes with soil scrubbing as basic process are increasingly considered. For the subsequent treatment of problematic polluted residues (fine grain fraction, process waters, etc.) a variety of processes is available, e.g., suspension bioreactors, oxidation with ozone, H_2O_2, and thermal treatment.

The experiences dealing with hazardous old sites have fundamentally shown that balancing pre-investigations are of essential importance regarding the process evolution as well as the evaluation of remediation measures. Far-reaching investigations have been carried out in the field of microbiological (*ex situ*) treatment and standardizations have been made (DECHEMA, 1992; GDCh, 1996). In the future, emphasis should be placed on a process evaluation for the *in situ* treatment in order to predict the processes of natural attenuation by balancing as fas as possible the fate of pollutants.

2 Thermal Processes

Thermal soil purification is mainly based on the pollutant transfer from the soil matrix into the gas phase. The pollutants are released from the soil by vaporization and burned. The polluted gas is treated further.

The different processing concepts for the thermal purification of contaminated soils are characterized by variations of the process parameters (e.g., range of temperature, retention time for solids and exhaust gas in certain temperature zones, supply of oxygen, supply of reactive gases for gasification, supply of inert gas, kind of heat input and optimum heat utilization, etc.). The large variability led to a multitude of different *ex situ* and *in situ* process combinations. Some of them are presented in Sects. 2.1–2.2.

2.1 Thermal *ex situ* Processes

The basic principle of a thermal soil purification plant includes the following processing steps:

(1) soil conditioning,
(2) thermal treatment,
(3) exhaust gas purification.

In soil conditioning, the contaminated soil is broken and sieved and thus prepared at the particle size which is necessary for the following thermal treatment [< 20 (fluid bed) – 80 mm]. During thermal treatment the soil is heated so that volatile pollutants are stripped from the soil. In the gas phase above the soil the combustion of pollutants takes place, but in this phase no complete destruction of volatile pollutants can be achieved. For this reason the gases are burned in an after-burner chamber at high temperatures ($\sim 1,200\,°C$) and a given retention time. Under these conditions dioxins are also destroyed.

Two different technologies are mainly applied in thermal soil treatment:

(1) *processes with an exclusive thermal effect:* directly and indirectly heated processes (pyrolysis) with following gas treatment (after-burning, condensation) (practiced rarely),
(2) *processes with a thermal effect and additional measures* (common practice).

The *processes with an exclusive thermal effect* (directly and indirectly heated) are conventional processes available on a large scale, applying, e.g., rotary kiln plants, fluid bed plants, and sintering strand plants where long-term experiences in practice were gained. Most of the processes work in the low temperature range of 350–550 °C where the structure of the soil is not fundamentally changed and where humic components are only partly destroyed. In the early days of thermal soil treatment high temperature processes were practiced where the soil is heated to temperatures of 800 to 1,100 °C. In this case partial liquefying or sintering of the soil particles is possible. At the same time, the organic components and the clay minerals of the soil are destroyed to a large extent, hydroxides are changed into oxides and primary minerals are crushed by gritting. Furthermore, the quantity of NO_x increases rapidly at temperatures $>1,000$ °C so that special equipment could be necessary for the NO_x reduction in the exhaust gas.

The processes with an exclusive thermal effect show significant differences regarding the exhaust gas purification systems applied. These are decisively influenced by the regulations in force in each country concerning the reduction of emissions and the immission protection. Usually, aggregates are used which are normally applied for flue gas purification in large scale power plants and in waste incineration plants. The exhaust gas purification in these processes mainly contains three partial units:

(1) high-temperature after-burning,
(2) dedusting,
(3) flue gas purification.

The separation of fine soil particles from the gas stream (dedusting) is done by means of cyclones, by hot gas filters, or by combinations of different filter techniques – specifically adapted to the process. While the organic pollutants (hydrocarbons) are completely oxidized during the after-burning step (high temperature after-burning) at 900–1,300 °C, flue gas purification minimizes the inorganic pollutants such as hydrogen chloride, hydrogen fluoride, and sulfur dioxide, the dust and heavy metal emissions down to the target values. Fundamentally, flue gas purification processes can be divided into a dry sorption process, a wet cleaning process (wet scrubbers), and a combination of both, the semidry process. If reduction of nitrogen oxides is required, special measures are necessary. The efficiency of

the different processes increases from the dry sorption via the semidry up to the wet scrubber process – and so do the total costs. But at the same time, the degree of pollutant separation guaranteed by these processes increases so that – especially regarding the wet cleaning processes – high exhaust gas requirements can be fulfilled. To guarantee the target values of the exhaust gas, activated carbon filters are applied as a final step in almost all processes. An operation without excess water can be achieved by treatment of the washing water (heavy metal precipitation and stabilization) and a subsequent recirculation for semidry and wet cleansing processes (FORTMANN and JAHNS, 1996).

The processes with thermal effect and additional measures are different developments using a temperature range of 60–350 °C where the pollutants (especially low boiling and medium boiling hydrocarbons) are stripped out of the soil by the influence of heat and additional measures (e.g., steam stripping, vacuum). The efficiency of the process not only depends on the effect of the temperature, but also on the physical properties of the pollutants.

For these processes the exhaust gas treatment normally includes no high temperature after-burning of the exhaust gas. The purification is carried out via condensation and/or by means of absorption (FORTMANN and JAHN, 1996; ITVA, 1997).

2.2 Thermal *in situ* Processes

Thermal *in situ* processes are still in the stage of development. They mainly differ in the kind of energy input for heating the soil matrix in order to transfer the pollutants into the gas phase. In order to capture and treat the gas phase, a combination of soil vapor extraction (SVE) and subsequent gas treatment is necessary. Compared with pure soil vapor extraction, the required treatment period can be reduced.

In the steam injection process, a hot steam–air mixture is passed into the unsaturated soil zone by steam-air injection (60–100 °C). As a consequence the volatile as well as the semivolatile compounds (NAPL) pass into the gas phase. The soil gas phase is extracted by

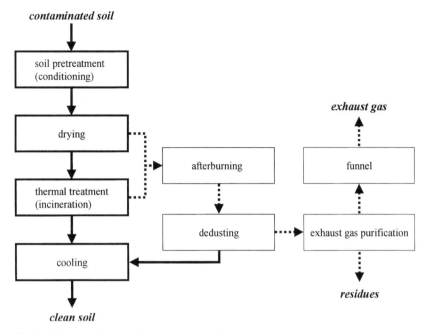

Fig. 1. Principal scheme of the thermal *ex situ* treatment process.

means of gas extraction systems and is treated afterwards. A transport of the mobilized pollutants in the direction of the groundwater is to be avoided by specific temperature control and by specific adjustment of the steam–air mixture. Limitations for this process can be stated for soils with a low permeability or for soils with very high inhomogeneities, as these soils require long periods of time for heating up (BETZ et al., 1998). As an alternative, the temperature within the unsaturated soil zone can be increased by installing high frequency electromagnetic fields using electrodes (JÜT-TERSCHENKE and PATZAK, 1996).

2.3 Application of Thermal Processes

In principle, all kinds of pollutants which can be stripped from the soil under the influence of thermal energy can be treated by means of thermal processes. The operation temperatures and the retention times depend on the type and concentration of the pollutants as well as on the intended use of the treated soil material.

Thermal *ex situ* processes are preferably applied when high initial concentrations of organic compounds are found and a high degree of purification is required. They can mainly remove petroleum hydrocarbons (TPH), polycyclic aromatic hydrocarbons (PAH), benzene, toluene, ethylbenzene, xylenes (BTEX), phenolic compounds, cyanides, and chlorinated compounds like polychlorinated biphenyls (PCB), pentachlorophenol (PCP), chlorinated hydrocarbons, chlorinated pesticides, polychlorinated dibenzodioxins (PCDD), and polychlorinated dibenzofurans (PCDF). Furthermore, thermal *ex situ* processes can be applied as a pretreatment step or after-treatment in conjunction with other *ex situ* processes.

Compared to thermal *ex situ* processes with additional measures, the processes with an exclusive thermal effect are mainly applied to single substance class contaminations with volatile pollutants. Due to the relatively low evolution of heat the *in situ* processes are only suitable for pollutants which can be stripped in the lower temperature range (e.g., BTEX).

Basically, soil materials of all particle size distributions can be purified in thermal *ex situ* processes. A limitation of the silt part (<30 to 50%) can be suitable for economic reasons. The application of thermal *in situ* processes can be restricted due to inhomogeneities or unsuitable water contents of the soil.

The efficiency of thermal treatment processes for the removal of organic pollutants from contaminated soils approaches almost 100% and is in most cases higher than the efficiency of biological or chemical/physical *ex situ* processes. However, to evaluate the application of different treatment processes further aspects, e.g., the necessary energy input, technical requirements, treatment costs, possibilities for the reuse of the treated soils, and other aspects have to be considered. Usually, the costs of thermal soil treatment are higher than the costs of biological or chemical/physical processes.

3 Chemical/Physical Processes

Chemical/physical soil treatment processes are mainly extraction and/or wet classification processes. The principle of *ex situ* soil scrubbing technologies is to concentrate the contaminants in a small residual fraction by separation. In general, water (with or without additives) is used as an extracting agent. For the transfer of contaminants from the soil to the extracting agent two mechanisms are of importance:

(1) strong shearing forces induced by: pumping, mixing, vibrations, high pressure water jets (break-up of agglomerates of polluted and non-polluted particles and dispersion of contaminants in the extracting phase),
(2) dissolution of contaminants by extracting agents.

In situ extraction basically consists of percolation of an aqueous extracting agent into the contaminated site. Percolation can be achieved by means of surface trenches, horizontal drains, or vertical deep wells. Soluble contaminants present in the soil will dissolve in the percolate which is pumped up and treated on-site.

3.1 Chemical/Physical *ex situ* Processes

During *soil scrubbing*, the pollutants are detached from the soil particles by means of mechanical energy and/or solubilizing effects often supported by surfactants. As a consequence a concentration of pollutants takes place in the liquid phase and in the solid fine fraction of the soil where the pollutants are sorbed at the surface. Water, optionally enhanced by additives, serves as a dissolving agent and as a medium for transportation. In general, soil scrubbing consists of the following steps:

(1) soil pretreatment,
(2) soil washing,
(3) separation by gravity (classification),
(4) separation of dispersed particles,
(5) separation of process water and rinsing of the purified soil fraction,
(6) process water recirculation,
(7) wastewater purification,
(8) exhaust air purification.

First of all, mechanical preparation of the contaminated soil takes place to separate coarse substances which might disturb subsequent process steps. The soil which includes rubble is granulated to <30–100 mm in diameter by crushing and sieving. This also serves as a homogenization step. In the next step the solids are dispersed in the liquid phase and heavily agitated, so that the detachment of the pollutants as well as the separation of the fine particles from the coarse particles takes place. In many cases chemical additives are added to the water (acids, bases, surface active substances/surfactants) to overcome the binding force between pollutants and soil particles. By means of grading machines (jigs, hydraulic classifiers, spiral separators) a separation of low and high density solids (grading) takes place.

During this procedure the pollutants are relocated, the surface charge of the pollutants increasing with a decreasing particle size. Depending on the process, the forces necessary for the detachment of the contaminants are restricted to a certain range of particle size due to economical reasons. These fine particles have to be removed and disposed off or have to be treated using other processes. The separation of the fine particles from the soil takes place in three subsequent steps:

(1) separation of the highly polluted fine particles from coarser components that are only polluted at a low level, employing hydrocyclones (separation down to 0,1 to 0,01 mm),

(2) separation of the fine particles from the process water by means of coagulation, flotation, and sedimentation,

(3) dewatering of the fine particles using screens and filter systems, e.g., filter presses.

After the separation of the polluted fine particles the purified coarse fraction has to be separated from the process water. To this aim, drain screens or vacuum band filters are applied according to the particle size. For the removal of the remaining process water rinsing with uncontaminated water is done followed by another dewatering step. To reduce the consumption of rinsing water the water is recirculated (process water recirculation). To keep the concentration of the pollutants in the process water at a certain level, parts of the water – normally up to 10% – are separated and treated in a wastewater purification plant. The treatment of this water is practised to an extent where the water can be reused as rinsing water instead of taking fresh water. Depending on the kind of pollution, different processes or process combinations of the wastewater treatment technology are applied (flotation, coagulation/ flocculation, oxidation, neutralization, emulsion breaking, heavy metal precipitation, sand filtration, active carbon adsorption). The

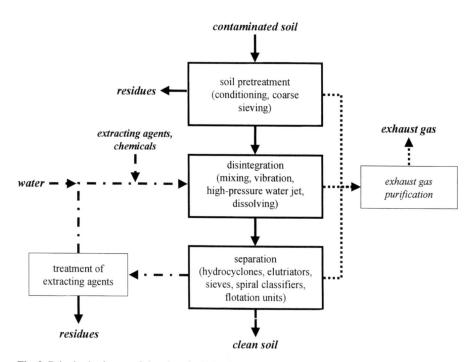

Fig. 2. Principal scheme of the chemical/physical *ex situ* process.

temperature during the treatment may be high enough that – supported by heavy mixing – the organic compounds with a low vapor pressure may volatilize. For this reason the relevant areas are encapsulated, the air is captured and purified subsequently, by means of separators for solids, drop separators, or activated carbon filters (HEIMHARD et al., 1996).

Principally, the chemical/physical soil treatment plants differ with regard to the kind of energy input for the detachment of pollutants from the soil particles and with regard to the separation of the purified soil particles from the liquid phase. The following aggregates are mainly applied for the input of kinetic energy: washing drum, attrition washing drum, or centrifugal accelerator (ITVA, 1994a; HEIMHARD et al., 1996).

The particle size separation cut – below which the soil particles of the contaminated fine grain fraction are separated – is at 25 to 63 μm. There are technical and economical limitations for the fine grain portion to be treated . The treatment of high silt portions is uneconomical since the portion of highly polluted residues increases drastically and consequently the disposal costs increase. Depending on the plant configuration, the treatable fine grain portion is between 25 and 40%. Referred to the plant input, the average amount of residual material is between 2 and 30% (LOTTER, Stand der Technik in der Praxis der Bodensanierung, VDI-Seminar, Sanierung kontaminierter Böden, November 1996, Berlin).

3.2 Chemical/Physical *in situ* Processes

During *pump & treat processes* water is supplied to the soil in order to leach out the contaminants. The contaminated water is pumped back to the surface and treated. Surfactants which increase the solubility of the pollutants may be added to the water. The treatment of the extracted washing water is carried out by standard wastewater treatment technologies.

The possible applications of chemical/physical *in situ* processes are especially limited by the permeability of soil. For hydraulic *in situ* measures, the soil permeability needs a permeability factor k_f of at least $5 \cdot 10^{-4}$ m s^{-1}. Side effects such as bio-clogging can be responsible for a further reduction of the natural permeability of the soil. The problem always remains that the entire amount of the injected water is not recaptured by pumping so that there is the risk of pollutant transport into the groundwater. In addition the added surfactants may be a source of secondary pollution.

Soil vapor extraction (SVE) is an effective and economic process for the reduction of highly volatile pollutants (e.g., BTEX) in the unsaturated zone of permeable soils (k_f $<10^{-3}$ m s^{-1}). Perforated pipes are placed in the contaminated soil area. The volatile pollutants are sucked out of the soil using low vacuum blowers. The extracted pollutants and condensates are treated on-site using activated carbon filters, compost filters, etc. The kind of treatment system used depends on the amount of air to be treated as well as the kind of pollutants. The time needed for the extraction of the pollutants to an acceptable degree lasts from months to years. In many cases complete decontamination of the soil is not achieved (ITVA, 1997).

The efficiency of the soil vapor extraction process is influenced by the characteristics of the soil (permeability, moisture content, temperature, homogeneity), and the kind of pollutants (vapor pressure). In many cases the pollutants are in the liquid phase in the soil so that volatilization takes place at the border of liquid plume, which prolongs the extraction process. In some cases the extraction can be enhanced by means of increasing temperature in the soil (e.g., by steam addition).

3.3 Application of Chemical/Physical Processes

There is no limitation to the kind of pollutants that can be treated in soil washing processes as long as they can be detached from soil particles and solubilized in the washing water. Therefore, all kinds of pollutant groups have been treated using the *ex situ* soil scrubbing process: BTEX, TPH, PAH, PCB, heavy metals, and even dioxins (LOTTER, Stand der Technik in der Praxis der Bodensanierung, VDI-Seminar, Sanierung kontaminierter Böden, November 1996, Berlin). A systematic approach for estimating the prospects of recycling contaminated soil by soil washing processes was shown by FEIL et al. (1997). The actual purification of the polluted liquid phase as well as the fine particle fraction may take place outside the soil washing plant in separate treatment facilities. But in most cases the polluted fine fraction is landfilled. In this case no actual treatment but only a separation process is achieved.

4 Biological Processes

During biological treatment soil microorganisms convert organic pollutants (e.g., hydrocarbons) mainly into CO_2, water, and biomass. Furthermore, a part of the pollutants can be immobilized as bound residues in the humic substance fraction. Degradation may take place under aerobic as well as under anaerobic conditions. In the practice of soil treatment, the aerobic process is predominantly applied. For an efficient biological treatment of contaminated soils the adjustment of optimal milieu conditions for the microorganisms (oxygen supply, water content, etc.) is of decisive significance. For the stimulation of the biological activity measures like addition of nutrients, mixing of the soil, addition of structure agents, or inoculation with microorganisms can be taken (COOKSON, 1995). In comparison to thermal or chemical/physical treatment processes, bioremediation processes usually require longer treatment periods.

4.1 Biological *ex situ* Processes

Landfarming was practiced first where the contaminated soil is excavated and mechanically pretreated (mechanical desagglomeration, sieving). After removal of disruptive materials the soil is placed at a layer thickness of maximally 0.4 m on the ground and is sealed by means of a membrane and/or concrete or clay layer. Enhanced oxygen supply as well as mixing are carried out by plowing, harrowing,

or by milling. By this means additives such as wood chips, compost, etc. can also be added in order to improve the soil structure. In addition water can be applied (COOKSON, 1995).

In the biopile process, biological treatment is carried out after mechanical preparation in which the soil material has been homogenized by sieving and crushing, and disruptive materials have been removed. By homogenization of the soil material, loosening of the soil structure is achieved as well as an improvement of the oxygen supply of the soil particles. To activate the biological degradation of the contaminants water, nutrients, and cosubstrates may be added. As an option, substances improving the soil structure or, in very special cases, microorganisms can also be added. Organic additives like compost, bark, or straw serve as cosubstrate or nutrient source for the microorganisms and as structure material.

In the static biopile process, biopiles are built up similar to the composting of organic waste substances. Trapezeoeder forms, oblong, or pyramidal forms are applied. The height of the biopiles varies between 0.8 and 2.0 m. Passive and active aeration systems can be installed with active aeration increasing the biopile height and reducing the surface. The principle of the dynamic biopile process is the decomposition of the soil by a repeated ploughing and turning of the biopiles. This increases the bioavailability of the pollutants and the contamination is brought into an intensive contact with microorganisms, nutrients, water, and air. The water contents should not exceed 50% of the maximum water holding capacity of the soil.

In the dynamic biopile process the soil of the biopiles is taken up in certain intervals, it is homogenized and built up in biopiles again. According to the soil type and to the kind of treatment, the height of the biopile can vary between 0.5 and 3.0 m. If necessary, water and nutrients are added once again during the turnover process (COOKSON, 1995).

In the bioreactor process soils are treated in the solid or liquid phase. The principle of solid phase reactors is the mechanical decomposition of the soil by attrition and by intensive mixture of the components in a closed container. This guarantees that contamination, microorganisms, nutrients, water, and air are brought into permanent contact. The solid phase reactor normally is operated at a water content of not more than 60% of the maximum water holding capacity. Mostly, additives to improve the soil structure are not necessary. The aeration can be carried out actively via an air blast system or via exhaustion. The exhaust air has to be treated in activated carbon filters or biofilters. By the adjustment of optimal milieu conditions, an enhancement of the degradation processes as well as an improvement of the degree of degradation especially for high concentrations and readily degradable pollutants can be achieved. In contrast to the biopile process only a few bioreactors are used, since the advantages gained do not justify the high technical input.

Soils – even clay and silt soils – can be treated as slurry in suspension bioreactors . In addition to the solid reactor processes, the application is also suitable for the residual fine particle fractions of the soil scrubbing process which are highly loaded with contaminants. After treatment the slurry has to be dewatered and the contaminated water fraction has to be treated. In most cases there is a high degree of water recirculation (ITVA, 1994b; COOKSON, 1995).

4.2 Biological *in situ* Processes

In biological *in situ* treatment, the contaminated soil stays in place and the milieu conditions for biological treatment are optimized as far as possible. In most cases oxygen has to be supplied, which can be done by artificial aeration or addition of electron acceptors like nitrate or oxygen release compounds (ORC). In some cases also H_2O_2 or O_3 dissolved in water are added. By these means the organic contaminants are degraded. If the oxygen is supplied via the water phase also nutrients and, in a few cases, bacteria are added. However, in most cases the authochtonous microflora is adapted to the present contaminants and the addition of cultivated microorganisms is not necessary.

The basis of intrinsic bioremediation as a passive remedial measure, is the utilization of natural degradation processes during long-term treatment periods. Recent investigations concern the question in how far natural degradation processes are able to contribute to the remediation of contaminated soils without significant technical actions. While in some cases natural degradation processes can be observed under suitable milieu conditions (e.g., NEWMAN and BARR, 1997) other applications show

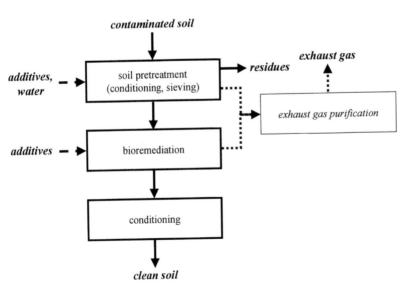

Fig. 3. Principal scheme of the biological *ex situ* process.

that under limiting conditions without active biostimulating measures the natural biological degradation processes come nearly to a standstill (PEASLEY and GUERIN, 1998).

The supply of oxygen into the unsaturated or saturated soil zone also is the basis for active aeration processes as there are bioventing (pressure aeration of the unsaturated soil zone), air sparging (pressure aeration of the saturated soil zone), and bioslurping (combined soil atmosphere and groundwater exhaustion) (GIDARAKOS and SCHACHTEBECK, 1996). Passive aeration processes as, e.g., the application of oxygen release compounds are at present in a testing stage [USEPA (1998), Cost and performance report on oxygen release compounds (ORC). EPA innovative site clean-up technologies: EPA 542-R-98-015, www.frtr.gov].

4.3 Application of Biological Processes

The biological turnover of organic pollutants mainly depends on the bioavailbility and biodegradability of the contaminants, as well as on the milieu conditions for the degrading microorganisms. Therefore, the degree of biological degradation achieved in a technical process is influenced by many factors (e.g., type, concentration and physical state of the contaminants, soil type, content of organic substances, adjustment of the milieu conditions) and can be limited due to biological, physicochemical or technical reasons (DECHEMA, 1991).

Bioremediation methods are applied to a number of different contaminants. The biological treatment of TPH contaminated soils can be considered as state-of-the-art technology. Biological degradation can also be expected for BTEX and phenols, although the release of the volatile substances has to be taken into consideration. PAH are biodegradable only to a certain extent (up to 4-ring PAH) and the degradation rate is relatively slow. Furthermore, PAH can be located in coal particles and in these cases are not bioavailable (WEIßEN-FELS et al., 1992). Regarding soils contaminated by TNT, an irreversible fixing of the conver-

sion product TAT (2,4,6-triaminotoluene) in the soil components can be achieved by means of anaerobic biological treatment with very low redox potentials (REIS and HELD, 1996). PCBs are relatively inert. Nevertheless, the biological degradation by anaerobic dechlorination and aerobic oxidation is possible (AB-RAMOVICZ, 1990). However, PCBs are in general not treated biologically due to their low biodegradation rates.

The literature statements about possible biodegradation endpoints vary for different types of contamination and treatment methods. Basically, residues from the contamination have to be expected.

5 Disposal

In the practice of remediation contaminated soils are often excavated and landfilled today. In addition sludges as residues from wet soil scrubbing are also often landfilled, although at many landfills this is only allowed for low contaminated soils and sludges. When soils are landfilled, the pollutants are not removed, but only translocated. To save landfill volume for other wastes and to utilize the treated soil again, preference should be given to purification methods.

6 Utilization of Decontaminated Soil

A major aspect in *ex situ* soil remediation is the reuse of decontaminated soil. During the different treatment processes the soil materials change their chemical and physical properties in different ways. Residual concentrations of contaminants and of organic materials originating from organic additives (e.g., compost, bark, wood chips) can restrict the reuse of biologically treated soil. This soil normally is not suitable to be used as filling material or for agricultural uses. Therefore, it is often applied for landscaping. Thermally treated soil can be used as filling material (i.e., refill where excavated), but is not suitable for vegetation due to its inert status. Soil from wet scrubbers may be used in a similar way. In many cases it is not easy to find adequate possibilities for the utilization of the treated soil.

A crucial factor for the reuse of decontaminated soils is the toxicological/ecotoxicological assessment. For this purpose bioassays are conducted to measure the possible impacts of treated soils. Bioassays could be an appropriate tool, if treated soils have to be tested with regard to their hazard potential. They integrate the effects of all relevant substances, including those not considered or recorded in chemical analysis (DECHEMA, 1995; KLEIN, 1999).

7 Conclusions

The selection of a suitable remediation process depends on the kind and concentration of pollutants, on the soil type, on the local availability of remediation processes and on economical aspects. Adequate treatment processes are available for all kinds of situations: biologically degradable pollutants should preferably be treated biologically. Soils contaminated with non-biodegradable organic pollutants can be treated using thermal processes. Heavy metals can be concentrated in the fine soil fraction by means of wet scrubbers (soil washing). Wet scrubbers can also be used for organically polluted soils as a concentration step where the fine fraction may be subsequently treated thermally or biologically (e.g., as a slurry) (MANN et al., 1995; KLEIJNTJENS, 1999; KONING et al., 1999).

In order to treat soils under controlled conditions, *ex situ* treatment, where the soil is excavated and treated in specialized plants, should be preferred. Of course those processes are costly, since in addition to the actual treatment costs, excavation, transport as well as pretreatment of the soil (sorting out of bulky materials, homogenization, etc.) must be considered. Today the costs are very much influenced by strong competition and are often not real costs.

In situ treatment avoids excavation and is, therefore, less costly, but often less effective and less controllable due to ubiquitous soil inhomogeneities. Additionally, it has to be assured that during *in situ* remediations no secondary pollution takes place and uncontrolled movements of the pollutants into uncontaminated areas are prevented. Therefore, extensive monitoring and securing measures could be necessary. This is especially true, if natural attenuation processes were taken into consideration (BARCELONA et al., 1999), [USEPA (1999), Use of monitored natural attenuation at superfund RCRA. Corrective action, and underground storage tank sites. OSWER directive 9200.4-170, www.epa.gov/oust/directive/d9200417.htm].

The excavation of significantly polluted soil and disposal on landfills should be abolished. For treated soils the possibilities for utilization should be improved. It is essential that as a first step of soil treatment preinvestigations should be done in order to predict as far as possible the treatment efficiency of the selected process. This is especially true for biological soil treatment (DECHEMA, 1992; HUPE et al., 1998).

8 References

ABRAMOVICZ, D. A. (1990), Aerobic and anaerobic biodegradation of PCBs: A review, *Crit. Rev. Biotechnol.* **10**, 241–251.

BARCELONA, M. J., FANG, J., VARLJEN, M. D. (1999), Mixed contaminated source magnitude and plume stability: Fuels/chlorinated solvents, *Proc. 5th Int. Symp. In situ and On-Site Bioremediation*, April 19–22, San Diego, CA, USA. Columbus, OH: Battelle Press.

BETZ, C., FÄRBER, A., GREEN, C. M., KOSCHITZKY, H.-P., SCHMIDT, R. (1998), Removing volatile and semi-volatile contaminants from the unsaturated zone by injection of steam/air-mixture, in: *Contaminated Soil '98*, pp. 575–584. London: Thomas Telford.

COOKSON, J. T., JR. (1995), *Bioremediation Engineering: Design and Application*. McGraw-Hill, Inc.

DECHEMA e.V. (1991), *Einsatzmöglichkeiten und Grenzen mikrobiologischer Verfahren zur Bodensanierung*. Frankfurt/Main: DECHEMA.

DECHEMA e.V. (1992), *Labormethoden zur Beurteilung der Biologischen Bodensanierung*. Frankfurt/Main: DECHEMA.

DECHEMA e.V. (1995), *Bioassays for Soils*. Frankfurt/Main: DECHEMA.

FEIL, A., NEEßE, T., HOBERG, H. (1997), Washability of contaminated soil, *Aufbereitungs-Technik* **38**, 399–409.

FORTMANN, J., JAHNS, P. (1996), Thermische Bodenreinigung, in: *Altlasten – Erkennen, Bewerten, Sanieren* 3rd Edn. (NEUMAIER, H., WEBER, H. H., Eds.), pp. 272–303. Berlin: Springer-Verlag.

GDCh (Gesellschaft Deutscher Chemiker) (1996), Leitfaden – Erfolgskontrolle bei der Bodenreinigung; Arbeitskreis "Bodenchemie und Bodenökologie" der Fachgruppe Umweltchemie und Ökotoxikologie, *GDCh Monographien* Vol. 4. Frankfurt/Main.

GIDARAKOS, E., SCHACHTEBECK, G. (1996), *In-situ*-Sanierung von Mineralölschäden durch Bioventing, Bioslurping und Air-Sparging, *TerraTech* **3**, 50–54.

HEIMHARD, H.-J., FELL, H. J., WEILANDT, E. (1996), Waschen, in: *Altlasten – Erkennen, Bewerten, Sanieren*, 3rd Edn. (NEUMAIER, H., WEBER, H. H., Eds.), pp. 303–339. Berlin: Springer-Verlag.

HUPE, K., KONING, M., LÜTH, J.-C., HEERENKLAGE, J., STEGMANN, R. (1998), Einsatz von Testsystemen zur bilanzierenden Untersuchung der biologischen Schadstoffumsetzung im Boden, *Altlastenspektrum* **7**, 360–366.

ITVA (1994a), Arbeitshilfe H 1–1, *Dekontamination durch Bodenwaschverfahren*. Berlin.

ITVA (1994b), Arbeitshilfe H 1–3, *Mikrobiologische Verfahren zur Bodendekontamination*. Berlin.

ITVA (1997), Arbeitshilfe H 1–6, *Thermische Verfahren zur Bodendekontamination*. Berlin.

JÜTTERSCHENKE, P., PATZAK, A. (1996), Thermische *In-situ*-Bodensanierung mit Hilfe hochfrequenter elektromagnetischer Felder, *TerraTech* **5**, 53–57.

KLEIJNTJENS, R. (1999), The slurry decontamination process: Bioprocessing of contaminated solid waste streams, *Proc. 9th Eur. Congr. Biotechnol. ECB9*, July 11–15, Brussels, Belgium.

KLEIN, J. (1999), Biological soil treatment – Status, development and perspectives. Bioremediation 1999: State of the art and future perspectives, *Proc. 9th Eur. Congr. Biotechnol. ECB9*, July 11–15, Brussels, Belgium.

KONING, M., LÜTH, J.-C., COHRS, I., JESSE, M., QUANDT, C., STEGMANN, R. (1999), Combined chemical and biological treatment of mixed contaminated soils in slurry reactors. Bioreactor and *ex situ* biological treatment technologies, *Proc. 5th Int. Symp. in situ and On-Site Bioremediation* Vol. 5 (ALLEMAN, B. C., LEESON, A., Eds.), pp. 25–30. Columbus, OH: Battelle Press.

MANN, V., KLEIN, J., PFEIFFER, F., SINDER, C., NITSCHKE, V., HEMPEL, D. C. (1995), Bioreaktorverfahren zur Reinigung feinkörniger, mit PAK kontaminierter Böden, *TerraTech* **3**, 69–72.

NEWMAN, A. W., BARR, K. D. (1997): *Assessment of Natural rates of Unsaturated zone Hydrocarbon Bioattenuation; in situ and on-site Bioremediation* Vol. 1 (ALLEMAN, B. C., LEESON, A., Eds.), pp. 1–6. Columbus, OH: Battelle Press.

PEASLEY, B. J., GUERIN, T. F. (1998), An *in situ* biostimulation strategy for intractable shoreline sediments contaminated with diesel fuel, in: *Contaminated Soil '98*, pp. 995–996. London: Thomas Telford.

REIS, K.-H., HELD, T. (1996), Mikrobiologische *In-situ*-Verfahren zur Dekontaminierung 2,4,6-Trinitrotoluol(TNT)-kontaminierter Böden mittels kontrollierter Humifizierung; *In-situ*-Sanierung von Böden; Resümee und Beiträge des 11. DECHEMA-Fachgespräches Umweltschutz, pp. 323–328. Frankfurt/Main: DECHEMA.

WEIßENFELS, W. D., KLEWER, H.-J., LANGHOFF, J. (1992), Adsorption of PAHs by soil particles: influence on biodegradability and biotoxicity, *Appl. Microbiol. Biotechnol.* **36**, 689–696.

13 Bioremediation with Heap Technique

VOLKER SCHULZ-BERENDT

Ganderkesee, Germany

List of Abbreviations

BTEX   benzene, toluene, ethylbenzene, xylene
HDPE   high-density polyethylene
PAH   polycyclic aromatic hydrocarbons
RDX   cyclotrimethylene trinitramine
TAT   triaminotoluene
TNT   2,4,6-trinitrotoluene
whc   water holding capacity

1 Introduction

Although the potential of microorganisms to degrade contaminants like petroleum hydrocarbons has been known for more than 100 years, the technical application of this knowledge has a history of only about 15 years. During this time biological soil remediation has made a strong development, marked by great efforts for research and development, manyfold conceptual and technical innovations as well as economical ups and downs.

Today the biological treatment of contaminated soil is the technology mostly used in large scale soil remediation (SCHMITZ and ANDEL, 1997), with global proliferation and an expanding international market (COOKSON, 1995). Especially the heap technique has a high potential for widespread use, because this technology is easy to handle and needs only a low technical and economical input.

A large number of investigations and case studies all over the world have shown the potentials and limits of soil remediation. The biological treatment of contaminated soil by heap technique is considered to be the most effective and competitive technology for pollution by petroleum hydrocarbons (SCHULZ-BERENDT, 1999). Nevertheless there is some need for further development, especially for technical solutions to enhance the height of soil heaps or to establish thermophilic conditions to use the high metabolic potential of extremophilic microorganisms (SORKOH et al., 1993; FEITKENHAUER, 1998).

This chapter describes the principles of heap technique and the different approaches and technological solutions. It will also show the advantages and limits of this technique and the actual research to overcome these problems. Additionally, it will discuss some economical and legal features.

2 Principles of Heap Technique

Heap technique for biological soil treatment is an *ex situ* technology, that means the contaminated soil is excavated and separated from the uncontaminated material. In contrast to the so called "landfarming", where the contaminated material is spread over a large area in a comparable thin layer and mixed with the existing soil cover, with the heap technique the contaminated soil is prepared by homogenizing and mixing with additives and then set up in heaps. Fig. 1 shows the different steps of treatment, demonstrated by the Terraferm system (HENKE, 1989).

Before starting the treatment samples of the excavated contaminated soil are tested for biological degradation of the pollutants using a standardized laboratory procedure (DECHEMA, 1992). After a positive result of these tests the first step of soil preparation is the separation of non-soil material like plastics metals, etc. and stones with a particle size of more than 40–60 mm. While non-soil material has to be disposed of or treated with other methods, the stones can be crushed and added to the soil again. In many cases the stones are not contaminated to the same degree as the fine particles because the pollutants did not penetrate into this material. Then the separated stones can be reused directly. Therefore, separation is not only important for the degradation process but also a possibility to reduce the volume of material that has to be treated.

The most important steps of pretreatment are homogenization of soil material and mixing with the additives. Homogenization means that the normally inhomogenous distribution of the pollution is transferred to an average concentration of contaminants in the total volume of soil. Statistical evaluation of 35 large scale remediation projects using the heap technique (KRASS et al., 1998) shows that standard deviation of the hydrocarbon concentration is reduced significantly after homogenization and that the average level of concentration decreases to about 50% of that calculated from the original material. This second effect is due to an overestimation by the analytic results from original material because samples are normally taken from higher concentrated parts of the soil.

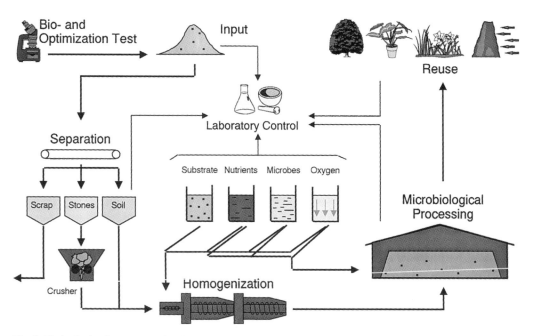

Fig. 1. Biological soil treatment by the Terraferm technology.

Determination and inoculation of additives of suitable quality and quantity is another important step for successful soil treatment with the heap technique. As shown in Fig. 1 additives can be divided into different categories. The so-called "substrate" stands for all additives that improve physical and chemical soil structure. Depending on soil quality like particle size, pH value, organic matter, etc., and the results of laboratory testing materials like compost, bark, lime, tensides, etc. are added to create optimal environmental conditions for the degrading microflora. "Substrate" can also be used to enhance the soil temperature by adding a high amount of easy degradable organics with the risk of a high level of competitive carbon sources to the contaminant. Otherwise, the added carbon source can be used as cosubstrate for energy supply or inducer for degrading enzymes which has been investigated for the degradation of chlorinated hydrocarbons like tetrachloroethylene or trichloroethylene (EWERS et al., 1990; KOZIOLLEK et al., 1999; MEYER et al., 1993).

The second class of additives are the nutrients for the degrading microflora. Because contaminated soil in most cases has its origin on industrial sites and is often excavated from layers of some meters depth it normally has no significant contents of nutrients like nitrogen, phosphorus, or potassium. Of course degrading microorganisms need these substances for growing and metabolizing. In most cases mineral fertilizers are used as liquids or in granules to supplement the soil with these compounds. Because nutrients normally will remain in the soil after treatment it is important not to overdose these additives. The level of fertilizer should not exceed conventional agricultural practice.

As a third element specialized microorganisms can be added to the soil during the mixing and homogenization procedure. The inoculation of bacteria, fungi or enzymes to enhance the degradation process has been argued in the scientific community. In contrast to many approaches in the United States investigations in Germany have shown no significant effects of added cultures of specialized microorganisms on petroleum hydrocarbon degradation (DOTT and BECKER, 1995) which is the prior application of the heap technique. The poten-

tial of the autochthonous microflora is normally sufficient for an effective degradation process.

Good results have been observed by adding complex mixtures of "substrates", nutrients and microorganisms like compost (HUPE et al., 1998) or activated sludge from wastewater treatment plants. The overall target of the mixing and homogenization process is to obtain optimum conditions for an aerobic metabolism of the contaminating substances.

For sufficient homogenization and mixing of contaminated soil special machines and aggregates have been developed which have to combine powerful homogenization and mixing units with controlled and sophisticated dosing of the different additives. Today specialized equipment is available (Fig. 2) with capacities of $50 \ t \ h^{-1}$ and more and adapted to different soil qualities from sand to clay. Partly crushing units are integrated directly in this machinery.

After the pretreatment the soil is transferred to the degradation area and set up to heaps. According to the environmental regulations for treating hazardous wastes in Germany the heaps must be located in a closed space. Depending on local climatic conditions locating the process in a closed system is not only necessary to meet the environmental regulations but is also an important tool for the control of the degradation process, especially water content and temperature.

The underground must be prepared to prevent contaminated seepage water from penetrating into the subsoil. This can be done by compacting the soil or installation of an area sealed with concrete or asphalt. In many cases plastic layers of 1.0–1.5 mm high density polyethylene (HDPE) are used to ensure a safe and sustainable enclosure of the contaminated material.

Fig. 2. Soil preparation unit type "mole".

To minimize emissions of volatile compounds to the air the heaps are set up in buildings like tents or halls or covered with plastic layers or membranes. With this measure it is also possible to protect the heaps from unsuitable weather conditions like rainfall or extreme temperatures. Design, construction and material of the cover depend on the kind of heap technique which is used. In the last 5 years in Germany soil treatment has changed from on-site to off-site installations. Therefore, the heap technique is used mainly in stationary treatment centers which are permanent installations and normally equipped with a treatment hall or a similar building in which the heaps are set up.

During the process of degradation the soil is monitored permanently by analyzing samples from different parts of the heap. Main control parameters are:

(1) concentration of the contaminants,
(2) water content,
(3) concentration of available nutrients,
(4) biological activity (soil respiration).

Depending on the monitoring results the degradation conditions are optimized by aeration, addition of water or nutrients, and further homogenization. The treatment ends after reaching the target values which are sufficient for a reuse of the cleaned soil. The time of treatment differs very much depending on the kind and concentration of the contaminants, the target values that have to be reached and the soil quality. The normal residence time is in the range of some months. KRASS et al. (1998), e.g., detected average halftimes of 85 d with a 95% confidence interval in a range from 75.4 to 94.6 d.

After treatment with the heap technique the soil has a quality to be used as topsoil for landscaping or as dump-site cover. Due to the treatment procedure the soil is free of larger stones, very homogeneous and enriched in nutrients and humic substances.

3 Different Heap Techniques

Besides the different technical solutions for underground preparation and heap cover different treatment systems have been developed to establish and maintain suitable conditions for microbiological degradation of toxic compounds. Fig. 3 shows the different approaches concerning humidity, agitation, aeration and temperature and their impact on the area needed for the treatment.

The first technologies used for large scale biological soil remediation were open air heap installations with the installation of a water cycle (ALTMANN et al., 1988). The humidity of the heaps is above the maximum water holding capacity (whc) and the seepage water is collected via a drainage system in a pond at one end of the heap (Fig. 4).

From the collecting pond the water is pumped to the surface of the heap and spread over the heap. Nutrients and other soluble additives are mixed with the water and the microorganisms are supplied with the added substances and oxygen via the water phase.

In contrast to these "wet" systems comparable "dry" technologies have been developed and are mainly used today. The humidity of the dry heaps is below maximum water holding capacity so that seepage water is avoided and the soil pores are filled with water and air.

The systems with low water content can be operated with (dynamic) and without agitation (static), while the wet systems are all static. By the agitation, i.e., turning and mixing the soil at time intervals of some days or weeks depending on the level of biological activity the heap is aerated and the water or nutrient content can be adjusted again. With the static and dry systems no additional supply with additives is possible so that all ingredients must be added during pretreatment. Therefore, special additives like slow release fertilizers must be used with these technologies. It is also necessary to install some kind of aeration system to supply the microorganisms with oxygen.

One advantage of the static and dry system is the height of heaps which can be up to 3–5 m, whereas in dynamic systems the heaps are not higher than about 2 m because the spe-

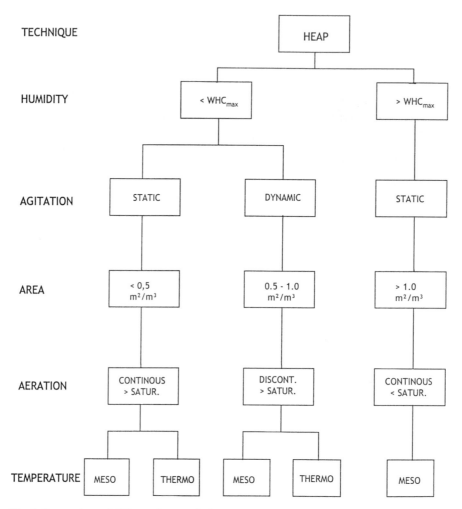

TECHNIQUE

HUMIDITY

AGITATION

AREA

AERATION

TEMPERATURE

HEAP

< WHC$_{max}$

> WHC$_{max}$

STATIC

DYNAMIC

STATIC

< 0,5 m²/m³

0.5 - 1.0 m²/m³

> 1.0 m²/m³

CONTINOUS > SATUR.

DISCONT. > SATUR.

CONTINOUS < SATUR.

MESO   THERMO   MESO   THERMO   MESO

Fig. 3. Comparison of different heap techniques.

Fig. 4. Open air heap.

Fig. 5. Turning machine "Wendy" on the surface of a dynamic heap.

cial turning machines (Fig. 5) have a limited turning depth.

The most area is needed with the "wet" systems because the limited capacity of water to transport oxygen leads to anaerobic layers if the height of the heaps is more than 0.5 m.

Supply with oxygen is the most critical factor of biological soil remediation using the heap technique. Because of the high oxygen consumption during aerobic degradation of hydrocarbons an aeration system must be installed in all static heaps. The "wet" technologies supply oxygen through the water phase. Therefore, the efficiency of oxygen transport to the different parts of the heap is limited by the concentration of dissolved oxygen and cannot exceed saturation. Investigations of the development of oxygen concentrations in different parts of the heaps show a rapid decrease during the first phase of degradation with the creation of anaerobic zones and methane production inside the heaps (KONING et al., 1999). Recent results of large scale experiments show that high pressure injection of air can solve this problem and may be an efficient alternative to dynamic treatment technologies.

If oxygen supply is sufficient the temperature in the heaps increases by biological activity to a level of 30°–35°C. Especially in closed systems this temperature level can be maintained independent of outside conditions so that optimum mesophilic conditions can be established by the heap technique. The research concerning extremophilic microorganisms and

their practical biotechnological application results in a high potential of thermophilic bacteria for hydrocarbon degradation (SORKOH et al., 1993). The establishment of thermophilic conditions in the heaps is one approach to increase degradation efficiency. Besides using the existing climatic conditions in some regions of the world like Arabia, Africa, or South America it is possible to enhance the temperature by adding well degradable organic matter. This leads to an enhanced oxygen consumption which must also be ensured by the technical design.

Today the dry and dynamic solution is the common technology for treating petroleum hydrocarbons with the heap technique. "Wet" solutions failed because of their long degradation times of 1–2 a and the large demand for space. If the dry and static approach could overcome its limited oxygen supply, it may be the technology of the future because of the small area needed for installation and the possibility of thermophilic process design which together can enhance the efficiency of the heap technique.

4 Efficiency and Economy

As mentioned in Sects. 1–3 the heap technique has been used in large scale bioremediation for more than 10 a and is a standard technology for the clean-up of soil contaminated with petroleum hydrocarbons. Depending on their origin petroleum hydrocarbons can differ very much. Successful bioremediation with the heap technology has been described for contamination with gasoline, light and heavy heating oil, and crude oil from different exploration sites. For the success of bioremediation not only the quality of the pollutant is important but also its concentration and the target value that has to be reached. With the heap technique it is possible to achieve degradation rates of about 80–90% in a reasonable time depending on the quality of the pollutant. That means starting concentrations should not exceed 10,000–20,000 mg kg^{-1} to reach end concentrations of 1,000–2,000 mg kg^{-1} which is sufficient for reuse on industrial sites. As mentioned in Sect. 2, the average concentration in the heap is significantly lower than calculated by the average value from single soil samples, therefore, concentrations of up to 100,000 mg kg^{-1} can be treated due to homogenization.

Besides petroleum hydrocarbons the following classes of pollutants can be treated by the heap technique:

(1) BTEX aromatics,
(2) phenols,
(3) polycyclic aromatic hydrocarbons (PAH) with up to 4 aromatic rings,
(4) explosives (TNT, RDX).

Each of the different compounds needs some modifications of the basic technique. Especially if volatile substances have to be treated, the emissions must be controlled and the waste air has to be treated. An example how to manage this problem is shown in Fig. 6 for the remediation of a soil contaminated with phenols and aromats.

For an efficient extraction of volatiles from the soil, extraction pipes are installed in the heap for soil air extraction. During the setup of the heap and the turning procedure of the dynamic system instead of the soil air extraction the air of the tent is sucked off and treated. The treatment system consists of 4 biofilter units and 2 filters with activated carbon. In addition to the normal tent material another plastic cover with a thin layer of aluminium is installed inside the tent as an effective barrier for the volatile compounds. This example demonstrates that the heap technique is very flexible and can be modified from a very simple installation to a highly performed technology.

Another modification has been applied for the bioconversion of explosives, especially TNT and its derivates. For biological detoxification TNT (trinitrotoluene) is converted to TAT (triaminotoluene) under anaerobic conditions (LENKE et al., 1997). Under aerobic conditions TAT is fixed to the soil matrix. To establish strictly anaerobic conditions in the heap organic material is added to the soil to a high degree. During degradation of the organic material all oxygen is consumed and the temperature reaches levels of about 60 °C. By turning the soil aerobic conditions appear for a short time so that TAT is bound to the soil matrix and cannot be detected after treatment. With this technology the treatment time can be reduced to less than 4 weeks to reach the target values (Fig. 7). Besides TNT the concentration of PAH is also reduced in the heaps. This example shows that the heap technology can be used for thermophilic degradation processes and that under thermophilic conditions degradation times can be reduced significantly.

Efficiency has to be discussed in connection with the economical effort. Because of the low investment costs for installation and the simple operation procedures the heap technique is of course a low cost technology compared

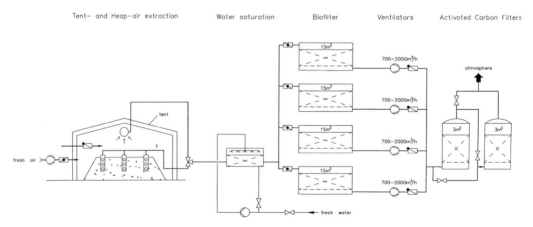

Fig. 6. Schematic representation of soil air extraction from heaps.

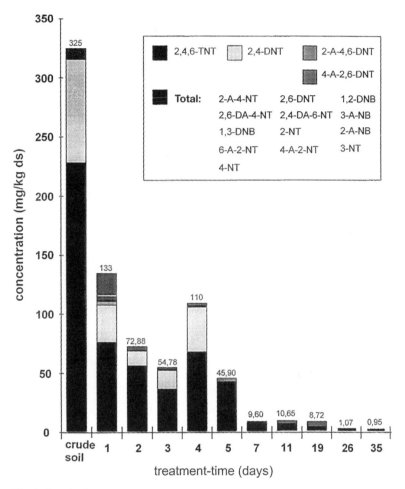

Fig. 7. Graph of TNT degradation.

with other biological technologies like bioreactors or soil washing and incineration. Therefore, about 80% of the soil treatment plants installed in Germany are biological treatment plants (SCHMITZ and ANDEL, 1997) most of them working with the heap technique. But because of the high number of treatment plants in Germany with an annual capacity of about 2,500,000 t the competition is extremely strong. During the last decade prices for soil remediation have decreased from nearly 100 US$ per ton to prices of 30–50 US$ per ton. Even for the simple heap technique this is the absolutely lowest limit for any reliable operation.

On the other hand the legislative regulations for soil handling and plant operation have become stricter so that nearly half of the costs for soil bioremediation with the heap technique are due to measures set by the authorities. In contrast to this high level of environmental safety and health care for treatment plants which destroy the contaminants and lead to a recycling of soil for different uses, it is still allowed to dump contaminated soil and conserve the problem for the next generation.

Nevertheless, soil treatment by heap technique is suitable for a lot of different situations concerning contaminants, soil quality or climatic conditions. It is adjustable by technical modifications to meet any requirement of the degradation process. Therefore, although it has been the first large scale remediation technique, it still needs further developement to become the biological treatment technology of the future.

5 References

ALTMANN, B. R. et al. (1988), *DGMK Forschungsbericht 396-02* – Erfahrungsbericht über die biologische *Ex-situ*-Sanierung ölverunreinigter Böden. Hamburg: DGMK.

COOKSON, J. T., Jr. (1995), *Bioremediation Engineering: Design and Application.* New York: McGraw-Hill.

DECHEMA (1992), *Labormethoden zur Beurteilung der biologischen Bodensanierung.* Frankfurt/Main: DECHEMA.

DOTT, W., BECKER, P. M. (1995), Functional analysis of communities of aerobic heterotrophic bacteria from hydrocarbon-contaminated soils, *Microb. Ecol.* **30**, 285–296.

EWERS, J., FREIER-SCHRÖDER, D., KNACKMUSS, H.-J. (1990), Selection of trichloroethene (TCE) degrading bacteria that resist inactivation by TCE, *Arch. Microbiol.* **154**, 410–413.

KOZIOLLEK, P., BRYNIOK, D., KNACKMUSS, H.-J. (1999), Ethene as an auxiliary substrate for co-oxidation of *cis*-1,2-dichloroethene and vinyl chloride, *Arch. Microbiol.* **172**, 240–246.

FEITKENHAUER, H. (1998), Biodegradation of Aliphatic and Aromatic Hydrocarbons at High Temperatures: Kinetics and Applications, Thesis, Technical University Hamburg-Harburg.

HENKE, G. A. (1989), Experience reports about on-site bioremediation of oil-polluted soils, in: *Recycling International* (THOMÉ-KOZMIENSKY, K. J., Ed.), pp. 2178–2183. Berlin: EF-Verlag.

HUPE, K., KONING, M., LEMKE, A., LÜTH, J.-C., STEGMANN, R. (1998), Steigerung der Reinigungsleistung bei MKW durch die Zugabe von Kompost, *TerraTech* **1**, 49–52.

KONING, M., BRAUCKMEIER, J., LÜTH, J.-C., RUIZ-SAUCEDO, U., STEGMANN, R. et al. (1999), Optimization of the biological treatment of TPH-contaminated soils in biopiles, *Proc. 5th Int. Symp. In Situ and On Site Bioremediation*, San Diego, USA.

KRASS, J. D., MATHES, K., SCHULZ-BERENDT, V. (1998), Scale up of biological remediation processes: evaluating the quality of laboratory derived prognoses for the degradation of petroleum hydrocarbons in clumps, in: *Proc. SECOTOX 99*, 5th Eur. Conf. Ecotoxicol. Environ. Safety (KETTRUP, A., SCHRAMM, K.-W., Eds.). March 15–17, München.

LENKE, H., WARRELMANN, J., DAUN, G., WALTER, U., SIEGLEN, U., KNACKMUSS, H.-J. (1997), Bioremediation of TNT contaminated soil by an anaerobic/aerobic process, in: *In situ and On-site Bioremediation* Vol. 2 (ALLEMAN, B. C., LEESON, A., Eds.), pp. 1–2. Columbus, OH: Battelle Press.

MEYER, O., REFAE, R. I., WARRELMANN, J., REIS, H., VON (1993), Development of techniques for the bioremediation of soil, air and groundwater polluted with chlorinated hydrocarbons: the demonstration project at the model site in Eppelheim, *Microb. Releases* **2**, 2–11.

SCHMITZ, H.-J., ANDEL, P. (1997), Die Jagd nach dem Boden wird härter, *TerraTech* **5**, 17–31.

SCHULZ-BERENDT, V. (1999), Biologische Bodensanierung – Praxis und Defizite, in: *Bödenökologie: interdisziplinäre Aspekte* (KÖHLER, H., MATHES, K., BRECKLING, B., Eds.), Berlin, Heidelberg, New York: Springer-Verlag.

SORKOH, N. A., IBRAHIM, A. S., GHANOUM, M. A., RADWAN, S. S. (1993), High-temperature hydrocarbon degradation by *Bacillus stearothermophilus* from oil polluted Kuwait desert, *Appl. Microbiol. Biotechnol.* **39**, 123–126.

14 Bioreactors

RENÉ H. KLEIJNTJENS

Gravenhage, The Netherlands

KAREL CH. A. M. LUYBEN

Delft, The Netherlands

List of Abbreviations

BTEX   benzene, toluene, ethylbenzene, and xylene
CSTR   continuous stirred tank reactor
DITS reactor   dual injected turbulent separation reactor
FORTEC   fast organic removal technology
ISB   interconnected suspension bioreactor
NPP   nitrogen, potassium, phosphate
PAH   polycyclic aromatic hydrocarbons
PCB   polychlorinated biphenyls
SDP   slurry decontamination process

1 Introduction

1.1 Contaminated Solid Waste Streams (Soils, Sediments, and Sludges)

Waste recycling plays a key role in the development of a sustainable economy (SUZUKI, 1992). The classical approach, remediation without the production of recycled materials, does not contribute to durable material flows. Moreover, the production of reusable materials is a necessity to make waste treatment an attractive economic solution. Recycling, however, can not be done regardless of the effort and costs needed. The overall environmental benefits should be positive and fit within the local economic and legal framework.

A practical way to qualify these benefits is found in three issues:

(1) the quality of the recycle products,
(2) the amount of energy required per ton,
(3) the costs per ton.

Solid waste streams (contaminated soils, sediments, and sludges) can be recycled. The solids have to be transformed into usable products while the contaminants are removed or destroyed. In case the contaminants are organic (such as mineral oil, PAH, solvents, BTEX, PCB) the use of bioreactors can result in environmental benefits (RISER-ROBERTS, 1998). In bioreactors populations of soil organisms degrade the contaminants to yield carbon dioxide, water, and harmless byproducts (SCHLEGEL, 1986).

A prerequisite for the implementation of new technologies such as bioprocessing is the definition of recycling and treatment targets. In The Netherlands legal targets for recycled materials were set in the Dutch building material decree (Building Material Act, 1995). Depending on the content and the leaching of components, two different ways for using recycled materials in plants are defined:

(1) category 1 products, needing no further isolation,
(2) category 2 products, needing further isolation and monitoring.

In the Dutch practice treatment processes are aimed at the production of category 1 recycle products. Although each country still has its own standards, and standardization is far away (NORTHCLIFF et al., 1998), it is clear that practical recycle standards accelerate the development needed to create sustainable technologies.

1.2 Characteristics of the Contaminated Solids

In soils the solid matrix is frequently dominated by sand, while the water content may be <25%; levels of debris can be found depending on the history of the site, but rarely exceed 10%. River, harbor and canal sediments contain a majority of water (frequently >60 to 70%) while the fine fraction (<63 μm) dominates the solids. The latter holds if dredging takes place in the upper sediment layers. Industrial and municipal sludges are mostly very humid (>95% water) and have a large content of organics (>60%) (INES, 1997). Industrial sludges mostly originate directly from corrosion and wear from installations or water treatment units on the site.

Disregarding the heterogeneous nature of the waste, the contaminant behavior is largely determined by the fines (WERTHER and WILICHOWSKI, 1990). This is due to the fact that submicron particles such as humic–clay structures and clay agglomerates have an extremely high adsorption capacity (BRADY, 1984). The solid waste, therefore, basically contains a contaminated fine fraction, a less contaminated sand–gravel fraction, cleaner debris, and a contaminated water phase.

In Fig. 1 the drawn line shows a typical particle size distribution for soils having a dominant fraction of fines (the step between 10 and 20 μm) and a dominant sand fraction (the step between 150 and 250 μm).

The bars show the measured mineral oil concentration as distributed over the fraction. For efficient processing of this waste only the fractions <150–200 μm should be treated.

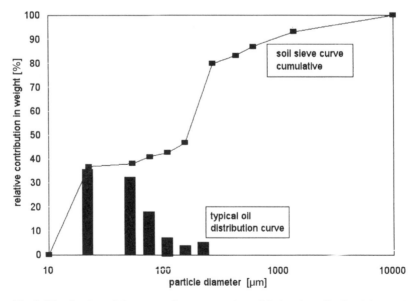

Fig. 1. Distribution of the contaminants over the solids fractions (in % of the total amount of oil present) and particle size distribution (cumulative).

In general it can be stated that in order to develop an appropriate recycling technology the particle features of the solids have to match the type of process operation. For bioprocessing this implies the integration of the separation technology and bioreactors in sequence of operation in which the clean fractions are removed from the feed before entering the reactor (KLEIJNTJENS et al., 1999).

2 Bioreactors

2.1 Reactor Configurations

The aerated bioreactor for solids processing is a 3-phase (solid–liquid–gas) multiphase system. The solids phase contains the adsorbed contaminants, the liquid phase (process water) provides the medium for microbial growth, aeration complicates the system. Nutrients and adapted bio-mass may be added to enhance breakdown. Furthermore, process conditions (temperature, pH, O_2 level, etc.) can be monitored and to some extent controlled.

Regarding the bioreactor configuration there are two major topics:

(1) physical state of the multiphase system:
  - (•) bioreactors with a restricted solids hold-up: slurry reactors (typical solids hold up <40 wt%),
  - (•) bioreactor with restricted humidity: solid state fermentation (solids content >50 wt%);

(2) operation mode:
  - (•) batch operation; no fresh material is introduced to the bioreactor during processing, the composition of the content changes continuously;
  - (•) continuous operation (plug flow); fresh material is introduced and treated material removed during processing, the composition in the reactor remains unchanged with time (LEVENSPIEL, 1972); in practice semi-continuous operation is often used (interval-wise feeding and removal giving small fluctuations in the reactor).

Three basic reactor configurations exist:

(1) slurry bioreactors
(2) solid state fixed bed bioreactors
(3) rotating drum dry solid bioreactors.

Characteristic for all types of slurry bioreactors (Fig. 2) is the need of energy input to sustain a 3-phase system in which the solid particles are suspended; the gravity forces acting on the solids have to be compensated by the drag forces executed by the liquid motion (HINZE, 1959). In a properly designed slurry system the energy input is used to establish three phenomena:

(1) suspension,
(2) aeration,
(3) mixing.

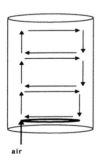

air

bubble column

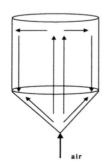

air

conical system

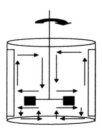

stirred reactor

Fig. 2. Common configurations of slurry bioreactors.

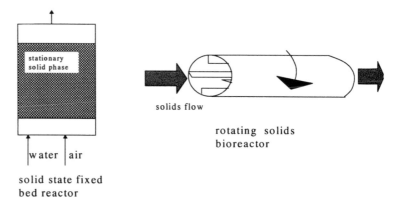

Fig. 3a, b. Bioreactors for solid state processing, (**a**) fixed bed reactor, (**b**) rotating drum reactor.

A slurry bioreactor can only work properly if these three measures are balanced. For each reactor configuration, the appropriate process conditions depend on parameters such as the reactor scale, particle size distribution, slurry density, slurry viscosity, oxygen demand of the biomass, and the solids hold-up (KLEIJNTJENS, 1991).

For solid state fermentations there is no need to maintain a solids–liquid suspension; a compact moist solid phase determines the system. Both the fixed bed reactor as well as the rotating drum bioreactor are suited for solid state fermentation (Fig. 3). In the fixed bed reactor the contaminated solids rest on a drained bottom as a stationary phase. Forced aeration and the supply of water are mostly applied as a continuous phase (RISER-ROBERTS, 1998). Fixed bed reactors are mostly batch operated. Although landfarming might be considered as a solid state batch treatment under fixed bed conditions (HARMSEN, 1991) this technique offers limited control options (in comparison to other solid state treatment) and, therefore, is not considered to be a bioreactor within the present context.

Continuous solid state processing is possible in the rotating system. Here the solid phase (as a compact moist material) is "screwed and pushed" through the reactor. In line with slurry processing energy is required to maintain the transport of the solids through the system.

2.2 Diffusion of the Contaminants out of the Solid Particles

Regardless of the bioreactor type the presence of sufficient water is crucial. Not only is a water activity around 100% a necessity for microbial activity (VAN BALEN, 1991), biodegradation fully depends on the availability of the components in the water phase. At microlevel diffusion of the adsorbed contaminants into the bulk phase is the rate determining step (Fig. 4). Mathematical models focus on capturing of the different physical processes into single diffusion parameters (WU and GESCHWEND, 1988). The overall diffusion process depends also on the flow conditions around the particles (CRANK, 1975).

An illustration of the importance of the flow around the particles was given by KONING et al. (1998). Comparing the microbial breakdown in a petroleum contaminated soil a conversion level of 80% was reached within about 10 d in the slurry reactor while under fixed bed conditions some 150 d were required to reach the same level. It was concluded that "the enhanced breakdown is based on an increasing bioavailability of the contaminant due to suspending of and the mechanical strain on the soil material in the slurry reactors". In another slurry experiment it was proven that manipulation of reactor parameters such as the energy input resulted in a faster breakdown (REYNAARTS et al., 1990).

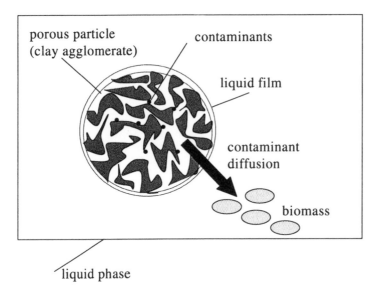

porous particle
(clay agglomerate)

contaminants

liquid film

contaminant
diffusion

biomass

liquid phase

Fig. 4. Diffusion of the contaminants.

The reactor type, scale, energy input, and aeration rate each have an impact on the diffusion. The better these parameters can be handled, the faster the breakdown proceeds.

Having established proper diffusion conditions, the presence of adapted biomass, capable of degrading the desorbed contaminants, is essential. In almost all cases the use of organisms which are already present in the waste is the most practical source to find adapted biomass.

Of the three different bioreactors, the slurry system offers the best features to substantially influence the diffusion rate. Intense multiphase mixing allows for the exchange of desorbed contaminants, nutrients, and biomass through the medium. In addition, particle deagglomeration has been measured (OOSTEN-BRINK et al., 1995) which results in smaller particles and a faster diffusion rate (smaller particles lead to shorter diffusion distances).

For solid state processes in fixed bed reactors there are fewer factors to influence the microenvironment of the pollution. Only air and liquid (nutrients) may be forced into the system while the solids are packed. At the microlevel the rotating solids bioreactor holds the middle between fixed bed and a slurry system, deagglomeration of the moist mass and phase exchange may take place.

3 Slurry Bioreactors

3.1 Slurry Processing

A slurry bioreactor only functions with a pretreated feedstock, therefore, the bioreactor is necessarily integrated with washing–separation operations and a dewatering operation at the end of the process (KLEIJNTJENS, 1991; ROBRA et al., 1998).

A typical set-up of an integrated (slurry) bioprocess is depicted in Fig. 5. First, the feedstock is screened using a wet vibrating screen to remove the debris (typical size >2–6 mm). Second, sand fractions are removed by one or more separation techniques such as sieves, hydrocyclones, Humphrey spirals, flotation cells, jigs, and upflow columns; a typical separation diameter (the so-called cutpoint) for the depicted hydrocyclone is 63 µm (CULLINANE et al., 1990). In the cyclone the slurry flow is split into a sand fraction (particle size >63 µm) at the bottom and a fine fraction at the top (<63 µm).

The top flow of the cyclone, containing the contaminated fines, is being fed to the bioreactor (depicted is a stirred tank, but any of the three types from Fig. 2 might be chosen). The final operation results in a dewatered product containing the fines and a flow of process water.

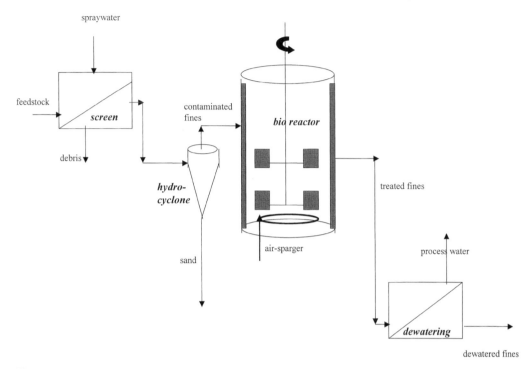

Fig. 5. Typical set-up of a (slurry) bioprocess using a batch operated aerated stirred tank reactor (typical solids hold-up is 20 wt%).

3.2 Batch Operation

A slurry bioreactor is often designed as a standard continuously stirred tank reactor (CSTR) having a mechanical stirrer, baffles, and a sparger at the bottom (Fig. 5). In the system a 3-phase suspension of contaminated solids, water and air is maintained. Batch degradation experiments in aerated stirred slurry bioreactors have been carried out frequently.

Fig. 6a shows the experimental result of a batch experiment at a 4 L scale. Using a conventional stirred bioreactor with baffles (stirrer speed 600 rpm, aeration rate 1.5–3 L min^{-1}, temperature kept at 30 °C), adapted biomass and nutrients were added (ammonium, phosphate, and potassium) while the pH was kept at 7–8. The soil was an oil contaminated sandy soil starting at an oil concentration of 3,500 mg kg^{-1}. Within 30 d the concentration dropped to about 500 mg kg^{-1}, while during the following period (up to day 56) no significant degradation was measured.

Fig. 6b shows the batch degradation curve for a heavily polluted harbor sediment as measured at the pilot scale (650 L) in an air agitated slurry reactor (the DITS reactor). The DITS-reactor originally is the first reactor part in the Slurry Decontamination Process (see Fig. 7). For this experiment the DITS reactor was deconnected and used as an individual batch reactor. For mineral oil the batch process showed a decrease from 10,000 mg kg^{-1} down to levels of 6,000–7,000 mg kg^{-1}.

Other batch processes have been carried out (see also Sect. 3.3). Resuming these batch results a "typical" batch curve might be constructed (Fig. 8): a significant breakdown in the first few days is followed by slow down of the process in the second stage. In this stage only a small percentage of the contaminant is degraded over a relatively long period.

The exact shape of the curve and the conversion level at which breakdown stops vary with the composition and age of the solids and type of contaminant.

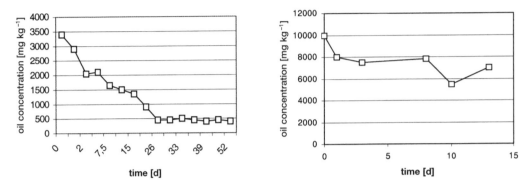

Fig. 6. (a) Batch curve (4 L); soil, **(b)** batch curve (650 L); sediment.

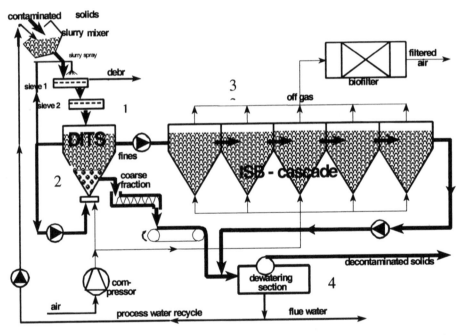

Fig. 7. Slurry decontamination process (SDP), dual injected turbulent separation reactor (DITS), inter-connected suspension bioreactor cascade (ISB-cascade).

To understand the typical batch curve three major phenomena have to be considered:

(1) In batch processing the biomass, growing on a complex substrate, has to adapt continuously to different components, since the easy parts are degraded first (TESCHNER and WEHNER, 1985). This pattern leads inevitably to increas-

ing difficulties for the microbial population present. The adaption rate of the population to feed on the available contaminants is insufficient.

(2) The contaminant desorption kinetics are of such an order that at lower concentrations there is a limited "driving force" for the adsorbed contaminants to be released from the solids (DI TORO

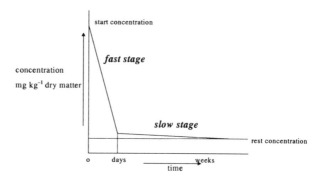

Fig. 8. Generalized batch curve.

and HORZEMPA, 1982). Since microbial breakdown only proceeds, if the contaminant is dissolved in the water phase, this type of kinetics does not allow for the easy breakdown at lower concentrations.

(3) During batch processing inhibiting side products of the microbial breakdown may increasingly be released into the medium; in addition, the physical conditions of the solids may change (e.g., attrition) which may lead to unfavorable microbial circumstances. An example might be the drop in pH due to humification processes which can join the contaminant breakdown.

3.3 Full Scale Batch Processes

3.3.1 The DMT–BIODYN Process

A full scale batch process has been realized with the DMT-BIODYN process (SINDER et al., 1999). The process consists of a fluidized bed slurry reactor in which the fined grained contaminated soils are treated. For aeration the slurry circulation loop is being treated in an external bubble column (Fig. 9). The reactor configuration allows a high solids loading up to 50 wt%. The DMT-BIODYN process has been designed by the Deutsche Montan Technologie GmbH to treat PAH polluted sites at former coal facilities in the Federal Republic of Germany (NITSCHKE, 1994).

At pilot scale (a 1.2 m³ reactor) various technological parameters were investigated and the hydraulic feed system at the bottom optimized. Experiments show that the PAH in the tests rapidly degraded without a noticeable lag phase; the measured oxygen consumption rate correlated to the PAH degradation rate. For this specific soil (starting concentrations about 250 mg kg^{-1}) after 6 d the target levels were reached. Comparison of the pilot results with results at laboratory scale and a respirometer showed that the degradation for this test soil in each of these systems was similar. For another PAH contamination (starting values at 1,100 mg kg^{-1}) the target levels could not be reached (treatment time was 30 d). It was concluded that for this specific soil "adsorption of the PAH to the organic matrix results in a reduced bioavailability". For this latter contaminated soil experiments combining bio-

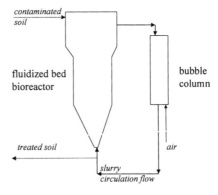

Fig. 9. DMT–BIODYN process configuration.

processing and soil washing resulted in final levels below target values.

In Sweden a full scale installation (55 m³ of the DMT-BIODYN process has been established to treat a contaminated gas plant area. The first trials at this scale have been successfully completed (SINDER et al., 1999).

3.3.2 The FORTEC Process

In The Netherlands the FORTEC process (fast organic removal technology) treating contaminated sediments combines several separation technologies with bioreactor treatment (TEN BRUMMELEN et al., 1997). First step in the process is the use of hydrocyclones to remove the sand fractions. The efficiency of the separation step is optimized by a multi-stage hydrocyclone configuration, separation was achieved down to a particle size of 20 μm. After the sand separation the contaminated fine fraction is fed to the bioreactor section. Based on initial experiments with aerated stirred tank bioreactors, a full scale bioreactor section with a total volume of 300 m³ (several reactors) was constructed (RIZA-Report 97.067).

A heavily contaminated harbor sediment (port of Amsterdam) was treated with the FORTEC process. After sand removal the fine fraction (<20 μm) was batchwise treated in the full scale bioreactors. At this scale 85% of the PAH and 78% of the mineral oil were removed after 15 d batch processing (TEN BRUM-MELEN et al., 1997).

3.3.3 The OMH Process

In the USA a large scale slurry bioreactor (750 m³) was used to treat creosote contaminated lagoon solids stabilized with fly ash (total PAH was 11 g kg⁻¹). An extensive pretreatment to classify the material was combined with areated and stirred (900 rpm) bioreactor with a 20% solids load. Conversion for the PAH was measured at 82–99% remediation for the 3- and 4-ring PAH and 34–78% for the higher PAH (JERGER et al., 1993).

3.3.4 The Huber Process

At a scale of 30 m³ a batch process treating the fine fraction (<200 μm) of a diesel contaminated soil has been carried out in an air lift bioreactor. The oil concentration dropped from 12,000 to 2,000 mg kg⁻¹ after 2 weeks residence time (BLANK-HUBER et al., 1992). Stripping effects of the diesel contaminant were determined at about 20% by laboratory experiments.

3.4 Sequential Batch Operation (Semi-Continuous)

To overcome some of the limitations of batch processing, continuous operation in a plug flow system has to be considered. As the contaminated solids travel through the system, specific conditions in the successive reactor develop. One practical way of achieving a plug flow system is a sequence of batch reactors (Fig. 10). In an experimental cascade of three stirred batch slurry reactors the contaminated solids (oil contamination) are periodically transferred to the next step (only part of the reactor slurry content is transferred, the so called "slug") (APITZ et al., 1994).

In Fig. 11 the breakdown pattern in the transferred slug during its residence in each individual step is depicted. It is explained "that each successive stage of the cascade maintains a microbial consortium that is optimized to consume organic compounds of increasing complexity. When compared to the batch process the biocascade was shown to be more effective, both in terms of the rate and degree of degradation" (APITZ et al., 1994).

3.5 Continuous Operation

In The Netherlands a continuous plug flow system has been achieved with the Slurry Decontamination Process (SDP) as depicted in Fig. 7. This process contains four major unit operations (KLEIJNTJENS, 1991).

(1) The contaminated solids are mixed with (process) water into a slurry and

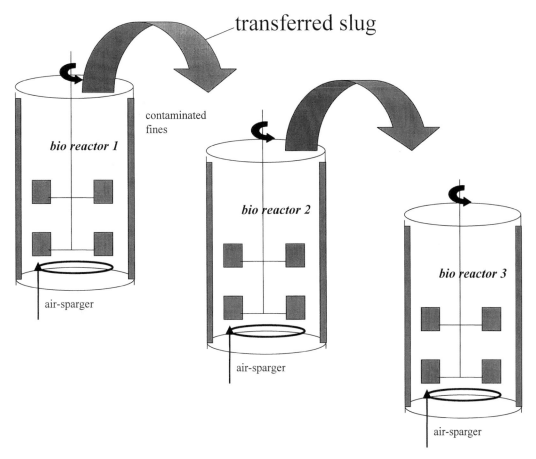

Fig. 10. Cascade of three sequencing batch reactors.

sized over a vibrating screen. In this wet sieving step, the debris is removed and a slurry prepared having the proper density (about 30 w/w%).

(2) In the first reactor/separator (a tapered air lifted bioreactor: the DITS reactor) the sand fractions are removed by means of a fluidized bed. Extensive organic material is removed by fine screening of light material. In addition, the agglomerates of the contaminated fines are demolished due to the power input and, therefore, opened to biological breakdown (also inoculation with the active biomass takes place).

(3) In a second reactor stage the contaminated fine fraction is subsequently treated. The second stage consists of a cascade of interconnected bioreactors (ISB cascade).

(4) A dewatering stage completes the process, the water released is partly recirculated as process water to mix the fresh solids into a slurry.

Fig. 12 shows the configuration of the DITS bioreactor (LUYBEN and KLEIJNTJENS, 1988). In the tapered bioreactor system energy is introduced at the bottom by the simultaneous injection of compressed air and slurry which was re-

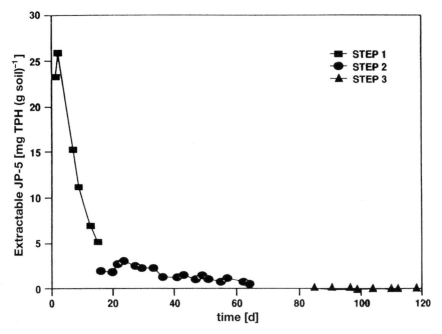

Fig. 11. Detail of one slug of material traveling through the sequencing batch process (3-step biocascade). Successive steps reflect the degradation of the more recalcitrant components (APISTZ et al., 1994; with permission).

cycled from the reactor content itself. Depicted are the dual injector, the settlers (used in the recycle flow), and the tapered vessel. This system has been built at a scale of 400 L, 800 L, and 4 m³. It was operated for 2.5 years to test various solids waste streams. The integral process was operated was semi-continuously at a pilot scale (3 m³ working reactor volume).

Fig. 13 shows the experimental results for a heavily polluted sediment (harbor) over a steady state period of 6 weeks. Nutrients (nitrogen, phosphorus, and potassium) were added and the temperature was kept at 30 °C. The steady state PAH concentration in the solids is depicted as a function of time. The upper symbols show the feed concentration in the slurry mill with an average of about 350 mg kg⁻¹ (the input data are scattered due to the heterogeneity of the feedstock). In the DITS reactor the steady state concentration dropped to values around 100 mg kg⁻¹ (first part of the microbial breakdown).

In the ISB cascade the average concentration dropped to values of 30–40 mg kg⁻¹. After dewatering the final concentration increased somewhat in the filter cake. The overall PAH degradation level was about 92% for an overall residence time of 16 d. No significant evaporation was measured in the off-gas. In this sediment the mineral oil degradation pattern showed a similar trend (Fig. 14), however, an overall degradation level of only 65% was established. Final concentrations after dewatering were around 3,000 mg kg⁻¹. In this sediment mineral oil seems to be less available for microbial breakdown than the PAH. Recycling standards for PAH could be reached, oil levels remained at a too high level.

A contaminated soil treated in the slurry process resulted in a steady state breakdown pattern as depicted in Fig. 15. Starting at moderate contamination levels of 1,200–1,300 mg kg⁻¹ mineral oil, final concentrations were reached at levels <50 mg kg⁻¹, well below the recycling standards.

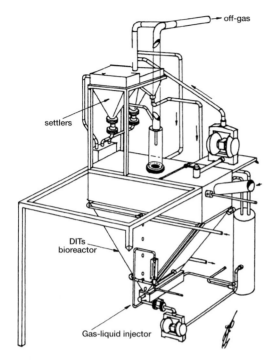

Fig. 12. Technical impression of the DITS bioreactor.

Comparing the steady state continuous results (Fig. 14) with the batch results (Fig. 6b) found for the same sediment and contaminant, it clearly is demonstrated that in the continuous mode the conversion significantly improved (final concentration continuous mode is ±3,000 mg kg^{-1} vs. 7,000 mg kg^{-1} in batch). This improvement due to continuity is in accordance with the APITZ observations on cascade breakdown (APITZ et al., 1994). Despite the improvement Fig. 14 showed, the Dutch recycling standard (mineral oil = 500 mg kg^{-1}) for this specific sediment was not reached.

For a less contaminated solid waste (Fig. 15) the recycling standard was reached without difficulty a mineral oil concentration of 900 mg kg^{-1} gives an output concentration <50 mg kg^{-1}. If this oil conversion is compared with the results from Fig. 14 (8,000 mg kg^{-1} input of mineral oil gives 3,000 mg kg^{-1} output) it is shown that at higher concentrations bioavailability seriously restricts the microbial breakdown. As a consequence it is often difficult to achieve complete conversion and reach low end concentrations if started with high input concentrations.

It should be noted that analytical difficulties (especially for oil), in the handling of wet samples, may easily lead to incorrect figures. It has been found that, besides a standard deviation

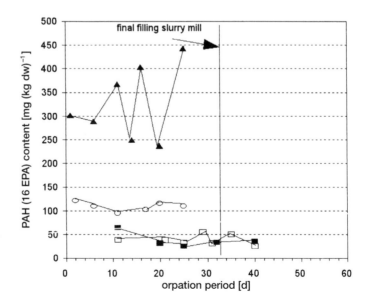

Fig. 13. Steady state PAH breakdown pattern in the SDP (sediment), ▲ slurry mill, ○ DITS reactor, ■ ISB reactor, □ press cake.

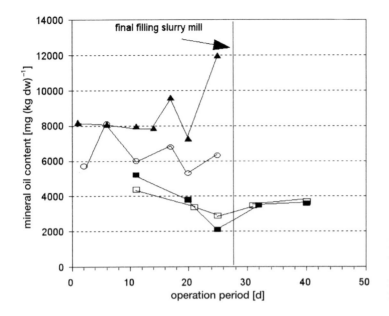

Fig. 14. Steady state mineral oil breakdown pattern in the SDP (sediment), ▲ slurry mill, ○ DITS reactor, ■ ISB reactor, □ press cake.

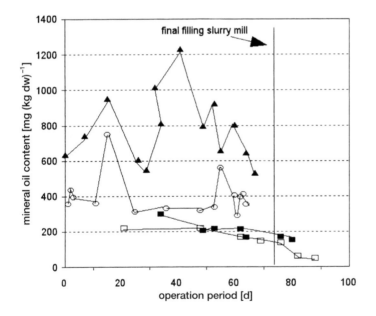

Fig. 15. Steady state mineral oil breakdown pattern in the SDP (soil), ▲ slurry mill, ○ DITS reactor, ■ cascade, □ press cake.

of 20% within one laboratory, ring investigations between various laboratories showed a variation of more than 75% (WARBOUT and OUBOTER, 1988). Bearing in mind these variations, analytical results should be regarded carefully. This holds the more when samples of processed solids are compared to recycle standards.

4 Solid State Bioreactors

4.1 Process Configuration

Systems operating at restricted humidity conditions may be referred to as solid state fermentations. A solid state fermentation includes the growth of microorganisms, both fungi and bacteria on a moist solid. In contrast to the slurry systems, solid state systems operate with:

(1) limited pretreatment,
(2) restricted energy input (no intense multiphase mixing),
(3) a limited amount of unit operations (the solids remain as a "solid phase").

In a solid state fermenter, process conditions are maintained by controlling temperature, humidity, and aeration. In contrast to slurry systems a compact moist solid phase dominates the reactor physics. Due to the fact that the handling of the moist mass is a tedious task, solid state fermentation typically takes place as a batch operation. Continuous operation depends on the possibility of a practical and smooth loading and unloading procedures.

4.2 Batch Operation: Composting

In line with its traditional role as a soil fertilizer, compost has been used as an additive in soil remediation. Composting mostly takes place in fixed bed reactors (Fig. 3a). Basically the compost addition is used to stimulate microbial breakdown. In experiments soil contaminated with hydrocarbons has been mixed with compost in various ratios (soil–compost ratios 2:1, 3:1, and 4:1). In 3 L test batch reactors the hydrocarbon degradation was >90% after a period of 44 d. Compared to the breakdown result without compost addition, the soil–compost systems showed a much faster degradation rate and a lower end concentration (LOTTER et al., 1990). In addition to composting, experiments focusing on the use of white-rot fungi have also been carried out (SCHAEFFER et al., 1995).

Another example of composting contaminated soil (in a column) was given by GOROSTIZA et al. (1998). In a soil column the degradation of pentachlorophenol with and without compost addition was tested. It was shown that compost addition significantly enhanced the degradation rate (residence time about 60 d).

At larger scale (in Finland) composting systems have been realized by mixing the contaminated soil with spruce bark chips as a bulking agent, lime, and nutrients beforehand. The bed is constructed as a biopile and may be turned by a tractor drawn screw mixer. Degradation temperature rose during composting as high as 44 °C. During 5 months the mineral oil concentration dropped from 2,400 to 700 mg kg^{-1} (PUUSTINEN et al., 1995).

In another experiment four biopiles of 10 m^3 each were created to treat chlorophenol contaminated soil. Chalk, commercial fertilizer (NPK), and bark chips (as bulk aeration agent) were added. After 2 months 80% of the contaminant were removed (LAINE and JORGENSEN, 1995).

4.3 (Semi-)Continuous Operation: the Rotating Drum Bioreactor

In a rotating drum bioreactor (or revolving tubular reactor) the reactor itself performs as a rotating conveyer in which the contaminated solids are being mixed (Fig. 3). In a (semi-)continuous operation the solids are transported (screwed) through the system. Loading at the head of the process is combined with unloading at the back. The solids residence time in the system is determined by the shape of the screw blades, the rotation speed, the scale and the amount of material present. Experiments have been carried out in a bioreactor with a length of 25 m and a diameter of 3.5 m. A warm air blower provided an appropriate temperature and oxygen, a sprinkler system was installed and nutrients added; the residence time was 14 d. For starting concentrations (mineral oil) ranging from 1,000–6,000 mg kg^{-1}, at a temperature of 22 °C the breakdown was measured down 50–350 mg kg^{-1}. Most activity took place in the first few days (MUNCKHOF and VEUL, 1990).

In a revolving tubular reactor, developed (by BioteCon in Germany), with a length of 18 m and a diameter of 2.5 m, between 0.5 and 1 t h^{-1} of soil can be treated (KIEHNE et al., 1995). Model drums were tested at diameters of 0.5 m and 1 m. The typical moisture content was 15%, the revolution speed was 1 h^{-1} and the aeration rate 0.125 VVM. In the trials the breakdown of motor fuel, kerosene and phenol was investigated. Under batch conditions kerosene was degraded from 8,300 mg kg^{-1} to 500 mg kg^{-1} in 300 h. Treating the same soil in a repeated batch (25% of the treated soil were returned and mixed with the new soil) within 180 h end concentrations of 180 mg kg^{-1} were reached. The increase of microbial activity due to the reuse of biomass was held responsible for the enhanced breakdown.

In an experiment a combination between heap leaching and reactor processing was tested. After treatment in the reactor for 120 h, the kerosene contaminated soil was deposited in a heap for 4 weeks. The treatment was finalized by a second reactor treatment for 100 h. With a total residence time in the reactor of 220 h, an overall conversion of 98.5% was achieved. It was concluded that for "contaminated soils in the order of 500–3,000 t the revolving tubular reactor especially demonstrates its advantage" (KIEHNE et al., 1995). The technology has been developed to operate on-site.

Another solid state process has been developed by Umweltschutz-Nord (Bremen, Germany). Using a steel tube a length of 45 m and an diameter of 3 m, the contaminated soil is screwed through the Terranox system. For mineral oil contaminated soils a treatment period of 6–8 weeks is taken, the system can also be used for the combination of anaerobic and aerobic treatment.

At a smaller scale (3, 6 and 90 L) a blade mixing bioreactor was operated as a solid state process (HUPE et al., 1995). It was concluded that besides the addition of compost moisture will enhance the process, the water content should, however, be <65% of the maximum moisture content to obtain solid state process conditions. The aeration rate is also a key parameter.

5 Comparison of the Bioreactors

Tab. 1 gives a technological and economical characterization of the three bioreactor systems.

It is clear that composting is the "low technological" type of bioreactor requiring only a limited technological infrastructure and investment. Slurry processing demands a technological infrastructure, while rotating drum systems take an intermediate position.

Experiments show that during batch processing the microbial breakdown slows down at lower concentrations and that full conversion is not reached. Continuous processing significantly improves these conditions.

The economic feasibility of bioprocessing largely depends on the nature of waste streams to be treated. Basically waste streams with a large water content, a considerable fine fraction (which is predominantly contaminated with organic components) are suitable for bioprocessing.

From the economic point of view continuous slurry processing, in an off-site installation, can only be beneficial at a larger scale. Capacities >40,000–60,000 t a^{-1} (depending on the local conditions) are needed to benefit from the "economy of scale" and thus to perform at acceptable cost levels. For batch operated composting processes smaller volumes, handled close to or on the site, might offer solutions if areas are available and time is not limited. Rotating drum systems in which solid state processing is achieved are considered to perform well for smaller volumes (up to 3,000 t) in a mobile plant used on-site.

6 Conclusion and Outlook

6.1 Conclusions

(1) In terms of product quality, energy consumption, and costs bioprocessing has the capability to function as a useful waste recycling technology.

Tab. 1. Comparison of Bioreactors

Technology	Economics
Biological slurry processing	
Batch and (semi-)continuous	substantial investment (equipment)
Water addition	complex plant or trained operators required
Energy input	large scale needed (economy of scale)
Addition of nutrients and/or biomass	use as recycle technology
Controlled conditions in closed systems	costs per ton determined by surface, machinery, energy and labor costs
Composting	
Batch operation	low investment
Limited use of technology and machinery	larger surface required
Low energy input	use as decontamination technology only
Addition of compost to the solids	might be used at smaller scale
Limited control options	costs per ton determined by surface and labor
Longer treatment times	costs
Rotating solids bioreactor	
(Semi-)continuous	substantial investment (equipment)
Limited control options	mobile plant on site
Moderate scale required	smaller volumes
Moderate energy input	costs per ton similar to slurry systems

(2) Comparing the three bioreactor types considered, it can be concluded that
- (•) slurry processing offers the best option for controlled and fast integral treatment,
- (•) composting can be used as a batch approach (biopiles) for smaller amounts,
- (•) rotating drum bioreactors are a solid state alternative for slurry processing.

(3) In the decontamination market bioreactor processing should focus on wet waste streams with a large content of contaminated fines, the contamination basically should have an organic character.

(4) Unambiguous analytical techniques, procedures and clear (international) recycle standards are needed to properly evaluate the various technologies, the end products, and calibration standards.

(5) Research has to identify the reasons for the hampered breakdown as measured in various field experiments. Especially for larger input concentrations the bioavailability (or even toxicity) of the contaminants seriously may interfere with sound microbial breakdown conditions. Ways have to be found to overcome these setbacks.

(6) To fully explore the possibilities of bioprocessing notice should be taken of the benefits of (semi-)continuous processing.

6.2 Outlook

(1) The trend which shows an increasing use of environmental biotechnology will result in the availability of organisms able to degrade almost any (organic) contaminant. To transform the potentials of these organisms into feasible bioprocesses the proper combination of bioreactors with separation techniques (to remove the clean fractions such as sand) is crucial.

(2) Various options focusing on the integration of bioreactors with landfarming, ripening, or phytoremediation will be explored. In addition, *in situ* treatment based on bioreactor research, such as biorestoration or bioscreens will be fully developed.

(3) Biotechnological processes will be explored and used at a larger scale. Owners of contaminated sites which might be treated by bioprocesses, have to benefit from legal and financial stimulation tools such as eco taxes on waste disposal.

7 References

APITZ, S. E., PICKWELL, G. V., MEYER-SCHULTE, K. J., KIRTAY, V., DOUGLASS, E. (1994), A slurry biocascade for the enhanced degradation of fuels in soils, *Fed. Environ Restoration and Waste Minimization Conf.* Vol. 2, New Orleans, pp. 1288–1299.

BALEN VAN, A. (1991), Influence of the water activity on *Zymomonas mobils, Thesis*, Technical University of Delft, The Netherlands.

BLANK-HUBER, M., HUBER, E., HUBER, S., HUTTER, J., HEISS, R. (1992), Development of a mobile plant, in: *Proc. Soil Decontamination Using Biological Processes*, Karlsruhe. Frankfurt/Main: DECHEMA.

BRADY, N. C. (1984), *Nature and Properties of Soils*. New York: Macmillan.

BRUMMELER TEN, E., OOSTRA, R., PRUIJN, M., WELLER, B. (1997), Bioremediation with the Fortec process: a save harbour for sediment, *Int. Conf. Contaminated Sediments*, preprints Vol. 1, p. 405, Rotterdam, The Netherlands.

Building Material Act (1995), *Bouwstoffenbesluit boden- en oppervlaktewaterbescherming*, TK 22683, Staatsdrukkerij, Den Haag, The Netherlands.

CRANK, J. (1975), *The Mathematics of Diffusion*, 2nd Edn. Oxford: Clarendon.

CULLINANE, M. J., AVERETT, D. E., SHAFER, R. A., MALE, J. M., TRUITT, C. L., BRADBURY, M. R. (1990), *Contaminated Dredged Material*, New Yersey: Noyes Data Corporation.

GOROSTIZA, I., SUSAETA, I., BIBAO, V., DIAZ, A. I., SAN VICENTE, A. I., SALAS, O. (1998), Biological removal of wood preservative (PCP) waste in soil, autochtonous microflora and effect of com-

post addition, in: *Proc. Contaminated Soil*, p. 1149. London: Thomas Telford.

HARMSEN, J, (1991), in: *Possibilities and Limitations of Landfarming, On-Site Bioreclamation* (HINCHEEVE, OLFENBUTTEL, Eds.). London: Butterworth Heinemann.

HINZE, J. O. (1959), *Turbulence*. New York: McCraw-Hill.

HUPE, K., HEERENKLAGE, J., LUTH, J., STEGMANN, R. (1995), Enhancement of the biological degrdation processes in contaminated soils, in: *Proc. Contaminated Soil* (VAN DE BRINK, W. J., BOSMAN, R., ARENDT, F., Eds.), p. 853. Dordrecht: Kluwer.

INES (1997), *Project bioslib*, Stichting Europort Botlek Belangen, The Netherlands.

JERGER, D., CADY, D. J., BENTJEN, S., EXNER, J. (1993), Full scale bioslurry reactor treatment of creosote-contaminated material at South Eastern wood preserving Superfund site, in: *Speaker Abstract* of the In-situ and On-site Bioreclamation, the 2nd Int. Symp. San Diego, CA.

KIEHNE, M., BERGHOF, K., MULLER-KUHRT, L., BUCHHOLZ, R. (1995), Mobile revolving tubular reactor for continuous microbial soil decontamination, in: *Proc. Contaminated Soil* (VAN DE BRINK, W. J., BOSMAN, R., ARENDT, V., Eds.), p. 873. Dordrecht: Kluwer.

KLEIJNTJENS, R. H. (1991), Biotechnological slurry process for the decontamination of excavated polluted soils, *Thesis*, Technical University of Delft, The Netherlands.

KLEIJNTJENS, R. H., KERKHOF, L., SCHUTTER, A. J., LUYBEN, K. C. A. M. (1999), The Slurry Decontamination Process, bioprocessing of contaminated solid waste streams, *Abstract* 2489 in the Abstract Book of the 9th European Congress of Biotechnology, July 1999, Brussels.

KONING, M., HUPE, K., LUTH, C., COHRS, C., STEGMANN, R. (1998), Comparative investigation into biological degradation in fixed bed and slurry reactors, in: *Proc. Contaminated Soil*, p. 531. London: Thomas Telford.

LAINE, M., JORGENSEN, S. (1995), Pilot scale composting of chlorophenol-contaminated saw mill soil, in: *Proc. Contaminated Soil* (VAN DE BRINK, W. J., BOSMAN, R., ARENDT, F., Eds.), p. 1273. Dordrecht: Kluwer.

LEVENSPIEL, O. (1972), *Chemical Reaction Engineering*. New York: John Wiley & Sons.

LOTTER, S., STREGMANN, R., HEERENKLAGE, J. (1990), Basic investigation on the optimization of biological treatment of oil contaminated soils, in: *Proc. Contaminated Soil* (VAN DE BRINK, W. J., BOSMAN, R., ARENDT, F., Eds.), p. 967. Dordrecht: Kluwer.

LUYBEN, K., KLEIJNTJENS, R. (1988), Werkwijze en inrichting voor het scheiden van vaste stoffen, *Dutch Patent* 8802728, Den Haag.

MUNCKHOF, P., VEUL, F. (1990), Production scale trials on the decontamination of oil polluted soil in a rotaing bioreactor at field capacity, in: *Proc. Contaminated Soil* (VAN DE BRINK, W. J., BOSMAN, R., ARENDT, F., Eds.). Dordrecht: Kluwer.

NITSCHKE, V. (1994), Entwicklung eines Verfahrens zur mikrobiologischen Reinigung feinkörniger mit PAH belasteter Boden, *Thesis*, Universität Gesamthochschule Paderborn, Germany.

NORTHCLIFF, S., BANNICK, C., PAETZ, A. (1998), International standardization for soil quality, in: *Proc. Contaminated Soil*, May 1998, Edinburgh, pp. 83–91. London: Thomas Telford.

OOSTENBRINK, I., KLEIJNTJENS, R., MIJNBEEK, G., KERKHOF, L., VETTER, P., LUBEN, K. (1995), Biotechnological decontamination using a 4 m³ pilot plant of the Slurry Decontamination Process, in: *Proc. Contaminated Soil* (VAN DE BRINK, W. J., BOSMAN, R., ARENDT, F., Eds.). Dordrecht: Kluwer.

PUUSTINEN, K., JORGENSEN, S., STRANDBERG, T., SUORTTI, M. (1995), Bioremediation of oil contaminated soil from service stations: evaluations of biological treatment, in: *Proc. Contaminated Soil* (VAN DE BRINK, W. J., BOSMAN, R., ARENDT, F., Eds.), p. 1325. Dordrecht: Kluwer.

REYNAARTS, H., BACHMANN, A., JUMELET, J., ZEHNDER, A. (1990), Effect of desorption and mass transfer on the aerobic mineralization of α-HCH in contaminated soils, *Environ. Sci. Technol.* 24, 1493.

RISER-ROBERTS, E. (1998), *Remediation of Petroleum Contaminated Soils*. Boca Raton, FL: Lewis Publishers (CRC-Press).

RIZA-Report 97-067 (1997), *Hoofdrapport pilot sanering Petroleumhaven Amsterdam: monitoring en evaluatie*. SDU-Den Haag, The Netherlands.

ROBRA, K., SOMITSCH, W., BECKER, J., JERNEJ, J., SCHNEIDER, M., BATTISTI, A. (1998), Off-site bioremediation of contaminated soil and direct re-utilization of all oil fractions, in: *Proc. Contaminated Soil*. London: Thomas Telford.

SCHAEFFER, G., HATTWIG, S., UNTERSTE, M., HUPE, K., HEERENKLAGE, J. et al. (1995), PAH-degradation in soilmicrobial activity or inocculation, in: *Proc. Contaminated Soil* (VAN DE BRINK, W. J., BOSMAN, R., ARENDT, F., Eds.), p. 415. Dordrecht: Kluwer.

SCHEGEL, H. G. (1986), *Allgemeine Mikrobiologie*. Stuttgart: Thieme.

SINDER, C., KLEIN, J., PFEIFER, F. (1999), The DMT-BIODYN process, a suspension reactor for biological treatment of fine grained soil contaminated with PAH, *Abstract* 2812 in the Abstract Book of the 9th European Congress of Biotechnology, July 1999, Brussels.

SUZUKI, M. (1992), Waste management according to Japanese experience, *4th World Congr. Chemical Engineering*, June 1991, DECHEMA, Frankfurt, Germany.

TESCHNER, M., WEHNER, H. (1985), Chromatographic investigations on biodegraded crude oils, *Chromatographia* 2, 407–416.

TORO, DI D. M., HORZEMPA, L. (1982), Reversible and resistent components of PCB adsorption-desorption isotherms, *Environ. Sci. Technol.* 16, 594–602.

WARBOUT, J., OUBOTER, P. S. H. (1988), Een vergelijking tussen verschillende methode om het gehalte aan minerale olie in grond te bepalen, H_2O 21, 1.

WERTHER, J., WILICHOWSKI, M. (1990), Investigations in the physical mechanisms involved in washing processes, in: *Proc. Contaminated Soil* (ARENDT, F., HINSENVELD, M., VAN DE BRINK, W. J., Eds.), pp. 907–920. Dordrecht: Kluwer.

WU, S., GESCHWEND, P. (1988), Numerical modelling of sorption kinetics of organic compounds to soil and sediment particles, *Water Res. Des.* 24, 1373–1383.

15 *In situ* Remediation

THOMAS HELD
HELMUT DÖRR

Darmstadt, Germany

List of Abbreviations

BOC   biochemical oxidation capacity
BTEX   benzene, toluene, ethylbenzene, xylenes (monoaromatic hydrocarbons)
DNAPL   dense non-aqueous phase liquid
DOC   dissolved organic carbon
HRC   hydrogen release compounds
LC-OCD   liquid chromatography with organic carbon detection
MHC   mineral oil hydrocarbons
NAPL   non-aqueous phase liquid
ORC   oxygen release compounds
PAH   polycyclic aromatic hydrocarbons

1 Introduction

The input of contaminants into soil and groundwater may lead to a persistant pollution. The methods for the remediation of the environmental compartments contaminated with organic contaminants comprise besides physical and chemical also biological technologies. The technologies are subdivided into *ex situ* and *in situ* methods. *Ex situ* means the excavation of the soil and subsequent treatment at the site (on-site) or elsewere (off-site). *In situ* means that the soil remains in its natural condition during treatment. Generally, *in situ* technologies comprise also *ex situ* components, e.g., water or vapor treatment plants.

The goal of *in situ* technologies is to mineralize the contaminants microbiologically. Many degradation mechanisms, especially mineral oil hydrocarbons require aerobic conditions. However, some hydrocarbon compounds are also mineralizable under denitrifying conditions. These processes are productive, i.e., the contaminants serve as source of carbon and energy. Additional nutrient elements like nitrogen and phosphate as well as electron acceptors are lacking in many cases. Besides this mineralization other biochemical processes that result in a reduction of the toxicity can also applied in *in situ* remediation processes. These are

(1) cometabolic aerobic or anaerobic transformation,
(2) humification,
(3) precipitation,
(4) solubilization,
(5) volatilization.

Applying these processes a wide variety of contaminants seem to be treatable *in situ*, including:

(1) mineral oil hydrocarbons,
(2) monoaromatic and polyaromatic compounds,
(3) chlorinated or nitrated aliphatics and aromatics,
(4) inorganic ions including simple and complex cyanides,
(5) heavy metals.

However, up to now only a limited number of possible techniques are developed for practical application. The main reason for the actual state of development is that the *in situ* degradability of contaminants is limited by numerous factors (Fig. 1). Most of these factors like low solubility, strong sorption on solids, sequestration with high molecular weight matrices, diffusion into macropores of soils and sediments, scavenging by non-soluble and lipophilic phases leads to a limited mass transfer. This lack of bioavailability may prevent sufficient degradation and causes the persistence of the contaminants.

During planning an *in situ* remediation it has to be considered that the technology covers processes working at different scales (from nm to km) (Fig. 2). Whereas import and export of contaminants by bacterial cells and the induction of the degradative enzymes takes place at the nm scale, surface processes are subject of the µm scale. In the range of µm to mm diffusion processes (soil micropores) as well as micro-inhomogeneities are located. The m scale represents small inhomogeneities, e.g., silt lenses in a sandy aquifer. Finally, the km scale comprises local aquifer systems. From scale to scale, different constraints are limiting the processes. In general, the constraints of the largest scale, i.e., hydraulic site

conditions are the limiting processes. Hydraulically favorable conditions are offered by course sediments with hydraulic conductivities of $>10^{-4}\,\mathrm{m\,s^{-1}}$.

One of the main features of *in situ* technologies is to transport supplements (nutrient salts, electron acceptors or donors) to the contaminants and remove reaction products as well as to induce proper environmental conditions for contaminant degradation. This may, e.g., be realized by the pumping of reaction product-enriched groundwater, water cleaning and reinfiltration of water, enriched with nutrients.

The *in situ* processes significantly depend on the distribution of the nutrients in adequate concentrations within the aquifer. Biodegradation processes that require aerobic conditions may be limited by the lack of oxygen. Oxygen can be supplied by dissolution of gaseous O_2 (air or technical oxygen) in the groundwater before infiltration. The maximum O_2 concentration that can be infiltrated is given by its solubility of about 50 mg L^{-1}. The oxygen supply can be increased by the application of hydrogenperoxide (H_2O_2) which is decomposed to O_2 and water enzymatically or by geogenic components like Fe^{2+} or Mn^{2+}. However, the concentrations of H_2O_2 in the infiltration water are limited by the toxicity to bacteria at concentrations above 1,000 mg L^{-1} and the

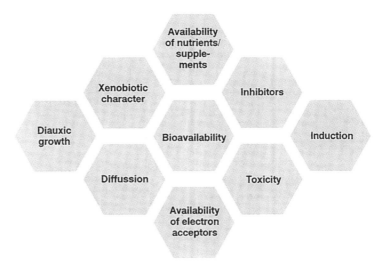

Fig. 1. Limiting factors influencing the degradability of contaminants.

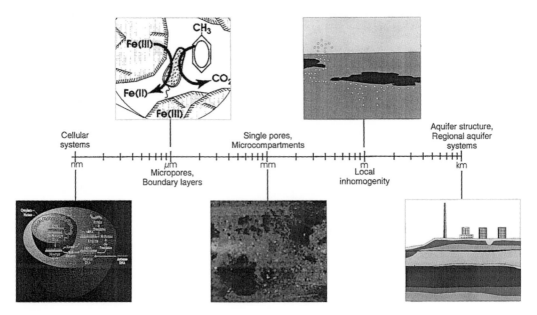

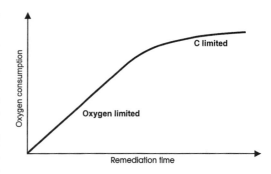

Fig. 2. Processes at different scales.

formation of O_2 bubbles leading to a decrease of the hydraulic conductivity and thus to a decreased nutrient and oxygen supply when too much H_2O_2 is added.

The oxygen demand is decreasing in the course of the remediation (Fig. 3). In the beginning, in most cases the oxygen consumption is limited by the supply rate. In a later phase, when all easily available compounds are degraded the oxygen consumption is limited by the rate of contaminant dissolution or diffusion out of micropores. Especially in this phase, it is important to use proper O_2 infiltration concentrations to avoid bubble formation.

The application of denitrifying conditions for contaminant degradation by the infiltration of nitrate may also lead to the formation of bubbles, because the end product of the denitrification is nitrogen (N_2) which has a comparably low solubility. In the case of overdosing of nitrate or too slow exfiltration rates this may result in a N_2 outgasing and bubble formation.

In *in situ* processes only a limited number of remediation-specific activities can be installed (Fig. 4). The aim is to control specific processes (condition parameters) within the aquifer with

the result that, e.g., biodegradation is optimized. The interactions are manyfold. The influence on one of the parameters may result in numerous effects. For example, to increase groundwater temperature may lead to a higher solubility of contaminants and to an enhanced biodegradation. On the other hand, higher contaminant concentrations may be toxic to bacteria. Furthermore, increased degradation may result in excess biomass formation which may block the aquifer and reduce the nutrient supply. Finally, the degradation capabilities

Fig. 3. Oxygen consumption.

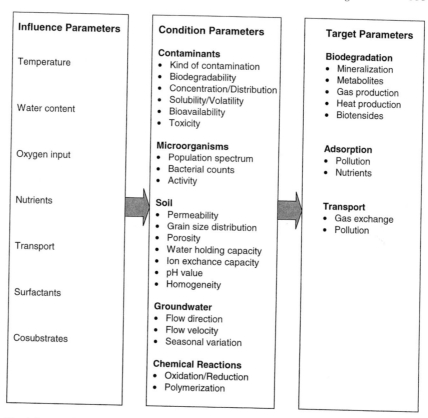

Fig. 4. Interactions during biological *in situ* remediations.

and kinetics as well as the impact of the chosen activities can only be determined in laboratory investigations to obtain a reliable basis for the planning of the technical scale remediation.

2 Investigations

Each site has to be investigated not only for its geological, hydrogeological, and contaminant situation, but also for the site-specific degradability. The following description solely refers to the laboratory methods for the investigation of the microbiological degradability of the contaminants (Fig. 5). These methods are subdivided into two phases. During the first phase the general degradability is determined, i.e., it is investigated whether the contamination is sufficiently degradable at the given site. A conclusive proof of the fate of the contaminants can only be obtained from tracer (e.g., ^{14}C) labeled substances. However, this is only possible in closed laboratory systems and not at a technical scale. Hence, it is necessary to use indirect parameters to demonstrate biodegradation.

Investigations of numerous sites have shown that the indigenous microflora in many cases comprises specific contaminant degraders. Therefore, the degradation potential is principally given. Usually the environmental conditions are unfavorable for a fast degradation. In this first phase the maximum degradation rate can be determined instead of a realistic rate.

During the second phase basic data for planning of the remediation such as nutrient demand, remediation duration, achieveable end

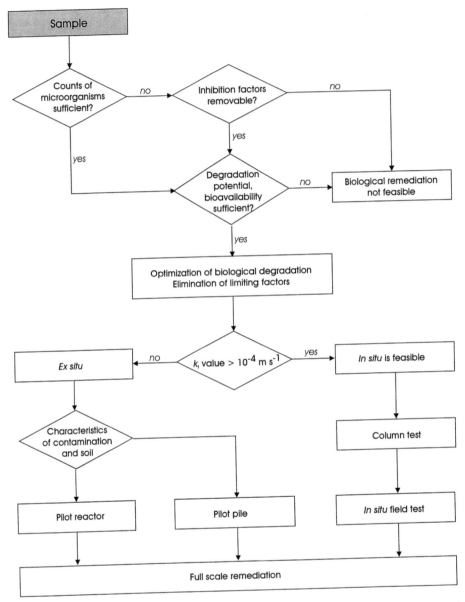

Fig. 5. Laboratory test methods for the investigation and optimization of the microbial degradability of contaminants (according to KLEIN and DOTT, 1992).

concentrations are determined by test methods which already simulate the remediation technology (bench–top scale).

The results of all site investigations and of the risk assessment are considered during the planing of the remediation. Today sufficient experience is availabe to scale up the results of the bench–top scale investigations to a technical scale. Pilot scale investigations are carried out, when new technologies without sufficient practical experience are applied, recalcitrant contaminants are treated or the site exhibits special complex conditions.

Prior to the start of the remediation the derivation of remediation target values is necessary. These values have to consider toxicological, ecological, site-specific, and site-use aspects. In case of remediation of the saturated soil zone, it is only reasonable to fix remediation target values for the groundwater not for the soil. This takes the highly inhomogeneous soil bound contaminant distribution into consideration. Concentrations of contaminants in groundwater represent an integration over a larger volume and are more representative to quantify possible risks arising from the contamination.

The data collected for a specific site are used for the description of the site (site model) (Fig. 6). From this site model the laboratory investigations are designed. The results thereof are used to plan the technical scale remediation. In case of *in situ* remediation it is reasonable to insert an additional planning instrument, the modeling. The used models are fed with parameters determined during site and laboratory investigations.

3 Remediation Technologies

3.1 General Considerations

Fig. 7 shows the typical characteristics of an old pollution. The contaminants enter the soil and migrate downwards. The amount of contaminants remaining in the unsaturated zone are controlled by sorption, diffusion into soil pores, and retention by capillary forces. Generally, the contaminants have low solubility, hence high amounts of contaminants may form – when they reach the groundwater table – a separate phase on top of the groundwater, called NAPL (non-aqueous phase liquid) or at the basis of the aquifer (groundwater zone), called DNAPL (dense non-aqueous phase liquid). Minor amounts of the contaminants are dissolved in the groundwater and are transported with the natural groundwater flow forming the contaminant plume. The spatial extent of the plume and the level of the contaminant concentrations depend on the age of the pollution, the sorption and transport characteristics as well as on the efficiency of the natural degradation.

Depending on the characteristics of the contamination and on the site conditions different technologies may be chosen. The technologies are subdivided into technologies for the treat-

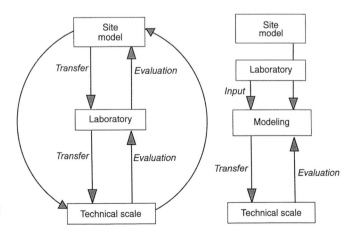

Fig. 6. Process of upscaling during remediation planning.

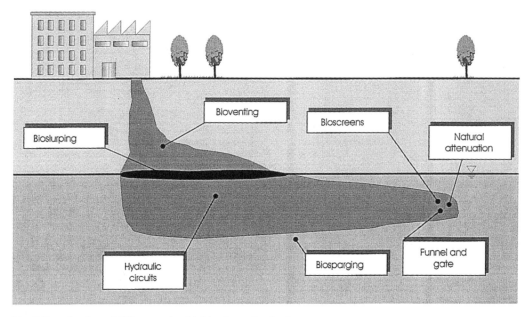

Fig. 7. Localization of different microbial *in situ* technologies.

ment of the unsaturated and saturated soil. A further subdivision comprises highly active as well as active and passive technologies. All these technologies may be combined with *bioaugmentation*, which is the addition (i.e., infiltration) of specific contaminant degrading bacteria previously isolated and propagated in the laboratory. However, bioaugmentation is still controversial because the majority of the infiltrated cells become attached to the soil within a few centimeters. Furthermore, establishing of the added microflora in an environmental compartment requires highly specific conditions (ANDREWS and HARRIS, 1987). Usually, the infiltration of nutrients during the remediation causes the added bacteria to be overgrown by the autochthonous microflora. If bioaugmentation actually works, it usually requires a continuous addition of the specific degraders.

Most of the technologies include not only the degradation of the contaminants *in situ* but also the physical removal of the contaminants (e.g., together with exhausted soil vapor or pumped groundwater), which requires an additional treatment.

3.2 Treatment of the Unsaturated Soil (Bioventing)

The only available technology to treat the unsaturated soil with biotechnological *in situ* methods is bioventing. The process scheme is shown in Fig. 8 (SUTHERSAN, 1996). The process includes vacuum-enhanced soil vapor extraction. The pressure difference in the soil causes an inflow of atmospheric air and, therefore, an oxygen supply, necessary for aerobic contaminant degradation. In some cases nutrients are not added, however, it seems necessary to add at least nitrogen salts. This may be achieved by sprinkling nutrient solutions on top of the soil or by installing horizontal infiltration drainages above the contaminant soil zone.

One of the most important tasks of process design is to ensure a sufficient air flow regime within the soil. To be considered in particular are the geometry of the exfiltration wells, the necessity of active or passive air injection wells as well as ground sealings. High contaminant concentrations may result in a clogging of the

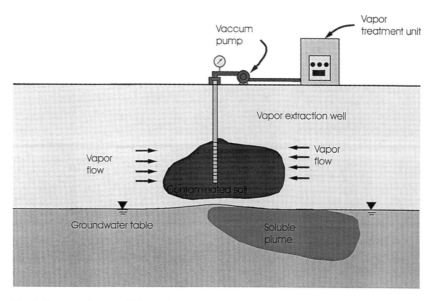

Fig. 8. Process scheme of bioventing.

soil pores followed by a reduced efficiency of the oxygen supply. In this case, a pulsed soil vapor extraction may be of advantage.

If the contaminants to be treated are volatile, the extracted soil vapor has to be treated, e.g., by sorption of the contaminants on activated carbon or by biodegradation within a biofilter. The bioventing is applicable to treat petroleum hydrocarbons, aromatic hydrocarbons and non-volatile hydraulic oils and other comparable contaminants.

The water budget of the soil may also be controlled with the nutrient additon. An optimum biodegradation requires a water content of 40–60% of the maximum water holding capacity. Lower water contents reduce the biodegradation rates, higher water contents lead to water-saturated zones where air flow is not possible. In such zones aerobic biodegradation is prevented. On the other hand, soil vapor extraction leads to a drying of the soil. By choosing appropriate nutrient infiltration rates and nutrient concentrations, drying may be compensated.

Bioventing is easy to monitor (Sect. 4), therefore, in this case achieveable degradation rates of about 0.2–20 mg kg^{-1} d^{-1} for medium permeable soils and petroleum products may be stated.

The achieveable degradation rates in the processes for the treatment of the saturated soil are influenced by many more parameters, hence these rates diverge to a much higher extent.

3.3 Treatment of the Saturated Soil

3.3.1 Hydraulic Circuits

Hydraulic circuits consist of groundwater pumping, cleaning, addition of nutrients, and reinfiltration. The contaminants are degraded in the underground (*in situ*) or are removed with the groundwater and eliminated in the groundwater treatment plant. This technology is the first one applied *in situ* and thus the most experienced. The technique was mainly used to treat mineral oil contaminations. Mineral oil hydrocarbons (MHC) are only extractable to a very low degree. During the complete remediation less than 1% of the MHC were removed with the exfiltrated groundwater; the rest was biodegraded *in situ*. BTEX, however, a co-contamination of mineral oil products may be washed out to a higher degree. The distribution of *in situ* degradation and exfiltration (as well as elimination within the water treatment plant) depends on the solubility of the contaminants, the kinetics of biodegradation and on the process technology. Because pumping and treatment of groundwater are expensive, the process may be designed to minimize the amount of groundwater to be pumped. The process as well as the complete *in situ* infrastructure including positioning and size of pumping and infiltration wells may be chosen on the basis of process modeling.

An example of a hydraulic circuit is shown in Fig. 9. In this case, a gravel filter (0.5 m) was installed at the level of the groundwater table (in the course of the excavation and on-site treatment of the unsaturated soil), containing drainage pipes which collect and transport the groundwater to the pumping wells. The pumped groundwater was treated in a water treatment plant, where Fe and Mn were removed to prevent clogging by Fe hydroxide as well as the contaminants. The cleaned water was divided into two cycles. Both were supplemented with nutrients (urea, phosphoric acid). The first one received hydrogen peroxide (H_2O_2) as electron acceptor for aerobic degradation. Nitrate (NO_3^-) for the degradation under denitrifying conditions was added to the

second cycle. The two different water circuits were infiltrated into two of four plots. After specific time intervals the infiltration was switched into the neighboring plots. The combination of aerobic degradation and degradation under denitrifying conditions resulted in substantial cost savings.

The supplemented waters were infiltrated at the bottom of the aquifer and pumped from the groundwater table. The resulting vertical groundwater flow direction increases the quality of the nutrient supplementation because otherwise flow channels are favored because the horizontal permeability is 10 times higher than the vertical permeability. The locations of the pumping wells have to be chosen in such a way that no contaminants may escape from the site with the natural groundwater flow. This may also be achieved by enclosing the site by a slurry wall reaching down to the aquiclude (groundwater impermeable layer).

3.3.2 Special Groundwater Wells

A variety of special groundwater wells were developed which have two common features.

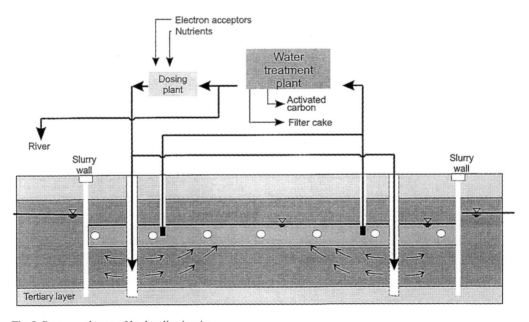

Fig. 9. Process scheme of hydraulic circuits.

They cause groundwater circulation and stripping within the well, resulting in an intensive throughput of groundwater and, therefore, an efficient supply of nutrients and electron acceptors (air oxygen or others, e.g., H_2O_2). Furthermore, volatile compounds are stripped within the well. The waste air is extracted and cleaned on-site.

There are some common features of the individual techniques, called *groundwater circulation wells, in-well-stripping* or *BioAirlift®*. The wells consist of a combined system of groundwater removal and infiltration within the wells. For that purpose the well is screened at the bottom and at the groundwater table. Both areas are separated by a cover pipe and a bentonite sealing. A lance is used to inject atmospheric air at the bottom of the well. This causes an upstream of the water according to the principle of a mammoth pump and simultaneously a stripping of volatile compounds. The elevation of the water table within the well leads to an infiltration of the oxygen-enriched groundwater at the top of the groundwater level. After circulating within the aquifer, the water enters the well again at the bottom.

Additional elements may be nutrient infiltration pipes within the well. An electric water pump may be installed instead of the mammoth pump. In this case no stripping and no oxygen enrichment of the groundwater occurs. The electric pump may also be installed in addition to the mammoth pump. Furthermore, a permeable bioreactor containing immobilized contaminant degrading bacteria may be installed between the points of water input and output. However, the water flow velocity usualy is too high to allow significant degradation of the contaminants within the residence time in the bioreactor. Therefore, the reactor material also contains activated carbon. The contaminants are then sorbed onto the carbon which is reactivated by the biodegradation of the contaminants within the reactor. The reactor may also contain other materials like ion exchanger (to remove heavy metals) or only activated carbon, if the contaminants are not sufficiently biodegradable (e.g., PAH).

In case of contamination of the unsaturated soil, the outlet of the groundwater may be installed above the groundwater table. With this type of operation the unsaturated zone can be flushed with nutrient enriched water. On the other hand, the well may be combined with a vapor extraction system. In this case, bioventing of the unsaturated soil may also be induced. These special wells have been applied to remediate sites polluted with petroleum hydrocarbons, aromatic hydrocarbons, or volatile chlorinated hydrocarbons.

The highly active technologies like the hydraulic circuits have the goal of a quick remediation closure. However, the processes determining the duration of the remediation are principially slow. Furthermore, these highly active technologies require not only high investment costs for the treatment plants in the beginning of the remediation but also high operating costs. The transport of large amounts of water (e.g., pumping and reinfiltration of groundwater) requires intensive monitoring and excessive use of process chemicals and high energy costs.

3.3.3 Biosparging and Bioslurping

Biosparging and bioslurping technologies were developed to reduce the energy consumption. Both technologies are not only restricted to the treatment of the saturated zone, but are also used for the treatment of the unsaturated zone.

Biosparging means the injection of atmospheric air into the aquifer (Fig. 10) (BROWN et at., 1994; REDDY et al., 1995). This results in the formation of branched small channels through which air moves to the unsaturated zone. Outside of these channels all processes are limited by diffusion. Therefore, highly branched channels are desired which may be achieved by pulsing the air injection. Biosparging enhances the *in situ* stripping of volatile contaminants, the desorption of the contaminants and their degradation by the enrichment of the groundwater with oxygen. Because the contaminants are transported to the unsaturated zone, the biosparging usually is combined with a soil vapor extraction.

Biosparging is applicable if the sparging point can be installed below the zone of contamination because air always flows upward forming a cone. The angle of the cone and also

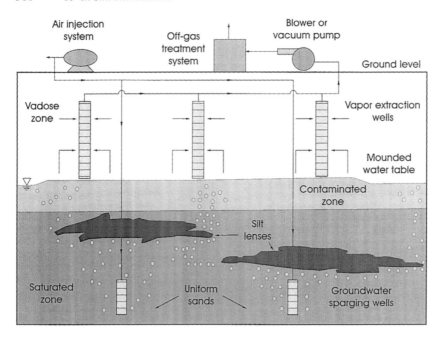

Fig. 10. Biosparging process scheme.

the degree of branching of the channels in a given soil depend mainly on the injection pressure, which should be only a little higher than the pressure of the water column. Usually the radius of influence of a biosparging well is determined by a pilot test at the site. At different distances from the sparging point monitoring wells are installed, in which the groundwater level and the oxygen saturation are measured. An increase of the groundwater table is only observed at the beginning of the treatment; after the air channels are established, the groundwater returns to its original level.

Biosparging is very sensitive to soil inhomogeneities. Zones of lower permeability may deflect the air channels. Zones of high permeability may act like open pipe channeling the air flow. In both cases the soil above these zones remains untreated.

Additional nutrients to enhance the microbiological degradation may be infiltrated with the same biosparging wells or with separate infiltration wells.

Bioslurping is the only technology which also treats free product phases floating on top of the groundwater (KITTEL et al., 1994). The bioslurping treatment wells are mainly screened from the lower to the upper limit of the phase. By supplying a vacuum, a mixture of free product, soil vapor and groundwater is extracted. The free product and the water are separated from each other. Free product, water, and soil air are cleaned separately. The treatment plant can be small at comparatively low cost because only a small amount of groundwater and soil air is pumped. The main advantage of bioslurping is the horizontal flow of the free product. Compared to conventional recovery systems, e.g., by hydraulic gradients toward a pumping well, bioslurping does not enhance the smearing of the free phase to greater depths of the aquifer.

The simultaneous extraction of soil vapor leads to the enrichment of the soil with oxygen comparable to bioventing. Hence the biodegradation in this zone is enhanced. Bioslurping may be combined with infiltration of nutrient salts into the vadose zone or into the saturated zone.

3.3.4 Passive Technologies

Experience shows that all active technologies require homogenous geological conditions (no inhomogeneities). If these conditions are not met passive technologies may be of advantage. Furthermore, the low solubility of hydrophobic organic contaminants and the slow back diffusion of the contaminants which have entered the soil micropores for decades may result in a low efficiency of remediation technologies based on induced groundwater flow.

The passive technologies are applied at or near the end of the contaminant plume. They consist of constructed zones (reactors) in which the contaminants are degraded. If the zones cover the complete cross section of the plume the technologies are called *activated zone, bioscreen, reactive wall*, or *reactive trench*. The main difference between these single techniques is the need for soil excavation. Whereas activated zones or bioscreens are arranged without any soil management, the lat-

ter techniques require the construction of an underground bioreactor.

Activated zones may be arranged by a line of narrow wells, installed perpendicular to the groundwater flow direction (Fig. 11). Incomplete passive technologies comprise the alternated pumping and reinfiltration of groundwater in closed, directly linked loops combined with an in-line nutrient amendment system consisting of a water-driven proportional feeder and a reservoir (SPUIJ et al., 1997). With this system the autochthonous microbial population is stimulated to adapt to a new and suitable redox situation and to develop the appropriate contaminant-degrading activity (RIJNAARTS et al., 1997).

In such systems the hydraulic conductivity of the activated zone is the same as in the surrounding aquifer. Any alteration, e.g., by Fe hydroxide precipitation leads to a reduction of the hydraulic conductivity with a changed groundwater flow regime, which may result in a deficiency of contaminant treatment.

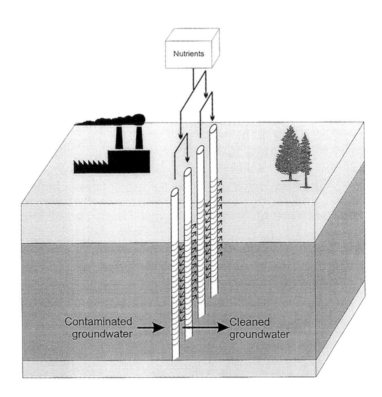

Fig. 11. Process scheme of activated zones.

Such loops still consume significant amounts of energy. Therefore, complete passive systems were developed. The advantage of these systems is the highest, when the energy demand is the lowest. This is the case in complete passive system, where the remediation time is long and the contaminant freight to be treated is usually low. In these complete passive systems the wells are, e.g., charged with solid cylinders consisting of highly permeable structure material (sand/cement) and so-called oxygen release compounds (ORC®), which represent a proprietary MgO_2 formulation. This compound may release oxygen over a period of about 300 d. If the ORC® are exhausted the cylinders may easily be replaced by new ones. Typical designs are PVC wells with a diameter of 15 cm, spaced at 1.5 m. The ORC® loaded wells are constructed in a fashion that the ORC® cylinder has a much higher permeability than the surrounding aquifer, resulting in a convergence of groundwater flow upstream and a divergence downstream in the well (Fig. 12). The wells may be installed in a staggered way in two adjacent rows perpendicular to the direction of the groundwater flow (SMYTH et al., 1995). ORC® have been shown to be effective in degradation of BTEX (BORDEN et al., 1997).

Because the degradation of numerous contaminants requires anaerobic conditions, it was necessary to supply electron donors (e.g., hydrogen) in completely passive systems. Hence, *hydrogen release compounds* (HRC®) were developed. This product is used especially to enhance the *in situ* transformation of volatile highly chlorinated hydrocarbons.

Current investigations are concerned with the electrochemical generation of hydrogen as electron donor:

$$2\,H_2O \rightarrow O_2 + 2\,H_2 \tag{1}$$

The cathode, where H_2 is generated may be located within an anaerobic treatment zone, whereas the anode, where O_2 is generated is located within an aerobic treatment zone. Laboratory column experiments demonstrated complete conversion of tetrachloromethane in a residence time of 1 d. This technique might be applicable in activated zones as well as in bioscreens.

Bioscreens or comparable systems include the construction of a "local zone in a natural porous medium with high contaminant retention capacity and increased bioactivity". Systems are desirable with high longevity and without significant maintenance or the necessity of nutrient replenishment. Bioscreens may be composed of a mixture of organic waste (compost, wood chips, sewage sludge, etc.) and of, e.g., limestone for pH correction. The organic waste serves as nutrient source, structure material to establish high permeability, and as a source of bacteria. Carrier (e.g., activated carbon) coated with specific contaminant degrading microorganisms may also be used. The bioscreens may be constructed for the complete duration of the treatment. In this case the necessary amount of nutrients is calculated on the basis of mass balances. However, it is difficult to estimate the fraction of nutrient mass, that will be available for the contaminant removal. On the other hand, bioscreens may be constructed in such a way that the screen material is exchangeable or restorable. The materials have to be homogenized prior to the installation to avoid channeling within the bioscreen.

Usually the bioscreens are designed to have a hydraulic conductivity 10-fold higher than the surrounding aquifer. Nevertheless, online monitoring of the permeability is necessary to avoid changes in the predicted groundwater flow regime. The thickness of the bioscreen depends on groundwater flow velocity within the bioscreen, contaminant concentra-

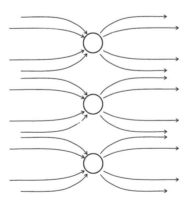

Fig. 12. Process scheme of bioscreens.

tion, degradation rates and the required concentrations at the downstream side of the bioscreen. Long-time changes of these parameters have to be considered. In some cases the groundwater flow through the bioscreen is adjusted by a minimum of pumping. The application of a numeric groundwater flow model for the wall design is helpful.

Currently a large number of different types of bioscreens have been developed, however, the experience on a technical scale is limited. Bioscreens are carried out as denitrification zones (ROBERTSON and CHERRY, 1997) or metal barriers with bioprecipitation (BENNER et al., 1997) (see Sect. 3.3.6). The most adverse effect during metal precipitation are high concentrations of Fe(II) and Mn(II), because they consume most of the precipitation capacity. Futhermore, because the metals stay within the bioscreen as metal sulfides, the permeability of the screen will decrease with time. If environmental conditions will not change and the long-time stability of the insoluble metal sulfides is given, they may remain in the subsoil, otherwise the screen material has to be removed.

In general, bioscreens may involve all biochemical processes that eliminate the pollutants. However, in any case the consideration of site-specific conditions shows which technology is feasible and also economical. For example, plumes with widths of about 100 m or more seem not to be treatable in an economical way with a design that requires 1.5 m spaced wells. Furthermore, processes creating solids (bioprecipitation) are questionable for use in bioscreens.

Funnel-and-Gate™ is a system that channels the contaminated groundwater (at the edge of the plume) by impermeable walls (the so-called funnel) towards gates within the wall, where a reactor is located (Fig. 13). To design such systems the application of a groundwater model which also considers inhomogeneities of the underground is essential. Within the gates the same bioprocesses may be installed as in bioscreens. However, the smaller width of the bioreactors is favorable. Constructive measures to exchange the reactor material are easier to realize. On the other hand, the groundwater flow velocity is increased within the gates and the groundwater table may rise.

Full scale experience with passive technologies up to now is rare. Especially, the longevity can only be calculated from projection of short-term monitored processes.

The advantage of passive systems is in low operation costs over long periods where only monitoring at the edge of the contamination plume is necessary. Thus, contaminated sites can be reused by avoiding high remediation costs of the site itself. In any case the elimination of the contaminant source (e.g., leaking tanks or free product phases) is favorable for the remediation duration of passive systems.

3.3.5 Natural Attenuation

Contaminant transport in the groundwater results in a dilution of the contaminants. In a certain distance from the contaminant source concentration ranges can be reached in which nutrients and electron acceptors supplied, e.g., with rain, are sufficient to allow a complete degradation of the residual contaminants. First of all, this results in a steady state of contaminant spreading and degradation. At a decreasing source intensity (e.g., by removal of the primary contamination) the plume extention also decreases with time. The application of this technology is only possible if the extention of the plume in time and space can be forecast reliably. Futhermore, it is necessary that sensitive receptors (targets) are not affected.

An essential part of this technology is the knowledge of the hydrogeology, e.g., by a groundwater model. The geological and hydrogeological parameters of the model (e.g., sorption coefficients, hydraulic conductivity of the aquifer, groundwater flow velocity) have to be determined by site-specific investigations.

The biological parameters, i.e., natural biodegradation rate must also be investigated in the laboratory. However, low degradation rates often require the application of radioactively labeled contaminants. On the other hand, indirect methods also allow the estimation of the degradation. During the degradation of the contaminants the biochemical oxidation capacity (BOC) based on the concentrations of electron acceptors is reduced. With the determination of the BOC in the ground-

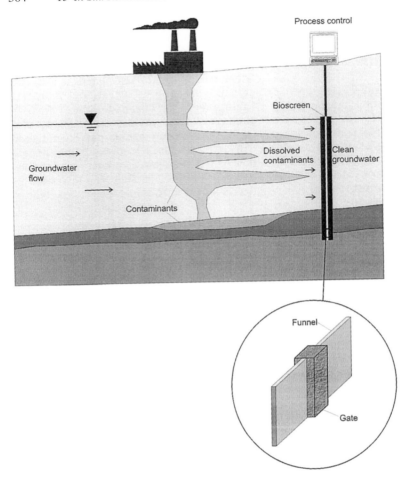

Fig. 13. Process scheme of funnel and gate.

water at the site the degradation rate can be estimated.

These parameters as well as the determined distribution of the contaminants are used to feed the model. With the model the future extention of the plume may be estimated allowing the decision whether intrinsic bioremediation is possible or not.

The experience showed that natural attenuation is applicable only in a limited number of cases. However, it might be suitable as a post-treatment phase after active remediation. Natural attenuation already has been applied to remove chlorinated ethenes, monoaromatic hydrocarbons (BTEX) and petroleum hydrocarbons (WIEDEMEIER et al., 1995).

3.3.6 Evolving Technologies

Most of the evolving technologies are only investigated on a laboratory scale. Some are already tested on the pilot or even on the technical scale. However, detailed experience and a wide commercial use are not available.

Up to now especially heavy metals are treated with physico-chemical technologies. Although these elements are not "degradable", they are not biochemically inert. Several microbiological transformations are known that mainly alter the physico-chemical behavior of the metals (SINGH and WILSON, 1997), including:

(1) solubilization and sorption (bioleaching, biosorption, phytoremediation),
(2) precipitation (bioprecipitation),
(3) volatilization by alkylation.

The solubilization was originally developed to support the mining of metals (Me) by leaching (*bioleaching*) (RAWLINGS and SILVER, 1995). Various metabolites play an important role in bioleaching, such as surfactants, chelators, and organic acids. A multitude of bacteria, e.g., *Thiobacillus* sp. and *Leptospirillum ferrooxidans* are reported to be able to leach metals. The metals which mainly occur in solid ore deposits (MS_2) are solubilized by two indirect mechanisms via thiosulfate or via polysulfides and sulfur. Some metal sulfides are chemically attacked by Fe(III) hexahydrate ions resulting in the formation of thiosulfate, which is oxidized to sulfuric acid. Other metal sulfides are attacked by Fe(III) and by protons. This leads to the formation of polysulfides as intermediates and finally elemental sulfur which is biooxidized to sulfuric acid. The two mechanisms may be simplified by the following equations (SCHIPPERS and SAND, 1999):

- Thiosulfate mechanism (FeS_2, MoS_2, WS_2):

$$FeS_2 + 6\,Fe^{3+} + 3\,H_2O \rightarrow \\ S_2O_3^{2-} + 7\,Fe^{2+} + 6\,H^+ \qquad (2)$$

$$S_2O_3^{2-} + 8\,Fe^{3+} + 5\,H_2O \rightarrow \\ 2\,SO_4^{2-} + 8\,Fe^{2+} + 10\,H^+ \qquad (3)$$

- Polysulfide mechanism (e.g., ZnS, $CuFeS_2$, PbS):

$$MeS + Fe^{3+} + H^+ \rightarrow \\ M^{2+} + 0.5\,H_2S_n + Fe^{2+}\,(n \geq 2) \qquad (4)$$

$$0.5\,H_2S_n + Fe^{3+} \rightarrow 0.125\,S_8 + Fe^{2+} + H^+ \qquad (5)$$

$$0.125\,S_8 + 1.5\,O_2 + H_2O \rightarrow SO_4^{2-} + 2\,H^+ \qquad (6)$$

The bioleaching method requires pumping of the groundwater and removal of the solubilized metals in a groundwater treatment plant.

For the treatment biosorption may be used. In summary, although bioleaching is a pump-and-treat process, the higher solubility of the "contaminants", i.e., of the solubilized metals

may result in an enhanced removal of the contaminants compared to the classical pump-and-treat.

Another technology, where dissolved contaminants are removed from the soil and groundwater is *phytoremediation* where the contaminants are taken up through roots of plants and trees. The selection of the plant depends on the characteristics of the contaminants and the soil as well as on the three-dimensional distribution of the contaminants, because the functionality of this technology is restricted to the depth of root growth (Chapter 16, this volume). Some contaminants, e.g., nitroaromatics, are taken up, but are not transported within the plant. Transformation of the contaminants within the plant is in most cases not effective enough for final elimination and, therefore, of minor importance. For example, heavy metals may be accumulated without significant transformation by some plants to a very high extent. Hence, for final removal of the contaminants, the plants have to be harvested and eliminated. Due to the high water demand of some trees phytoremediation may also be used for a hydraulic limitation of a pollution plume.

A main problem of the mining industry is the acidification and contamination of groundwater with dissolved metal ions together with sulfate. With *bioprecipitation* the problems can be solved all at once. The addition of an organic substrate (represented by "CH_2O") leads to the formation of sulfide and an increase of the pH. The metals (Me^{2+}) are precipitated as a non-toxic metal sulfide in an abiotic reaction:

$$SO_4^{2-} + CH_2O \rightarrow H_2S + HCO_3^- \qquad (7)$$

$$Me^{2+} + H_2S \rightarrow MeS\downarrow + H^+ \qquad (8)$$

The reactions are disturbed by the occurrence of oxygen. However, O_2 is readily consumed, when the organic substrate is added. The formed metal sulfides are very stable. A remobilization occurs only when the pH drops below pH 3. Due to the excess formation of H_2S, in some cases a second, aerobic treatment phase is required, in which surplus H_2S is re-oxidized to sulfate. With bioprecipitation at least the following metal contaminations can be treated: Pb, Zn, Cu, Cd, Ni.

Pollution with chromium is very frequent at mining industry sites, galvanizations, deposits, and tanneries. It is reported that the mobile very toxic hexavalent chromium ion (Cr^{6+}) may serve as electron acceptor alternative to dissimilatory Fe(III) reducing bacteria which use the reaction for the production of energy. Cr^{6+} is thereby reduced to less mobile non-toxic trivalent chromium (Cr^{3+}) which is stable in aerobic environments. Comparable reactions are reported for uranium (U^{6+}). Fe(II) and S^{2-} may also reduce Cr(VI) in non-enzymatic reactions. Hence, the role of the bacteria is questionable. Furthermore, microorganisms prefer nitrate as electron acceptors. Recent investigations try to find conditions for preferred chromium reduction.

Besides these redox reactions the microbiological formation and degradation of organometals are known. Alkylations and dealkylations are carried out by a wide range of microorganisms. By these reactions a number of important parameters such as toxicity, volatility, and water solubility are altered (WHITE and GADD, 1988).

Arsenic contaminations in the unsaturated soil zone can also be treated by this method. For the soil and soil microorganisms such a process is a detoxification, because the volatile transformation products evaporate into the atmosphere. However, these volatile products are rather toxic. Therefore, when this bioprocess is used as a remediation technology, the volatile products have to be extracted, e.g., by soil vapor extraction. The exhaust air can be treated chemically, the compound is dealkylated and the product is removed in a gas scrubber.

In detail, the following microbial reduction pathways of inorganic arsenic towards volatile organic compounds may occur. Arsenate (As^{5+}) as well as arsenite (As^{3+}) are transformed to arsine (AsH_3) by bacteria, preferably under anaerobic conditions. Fungi and some bacteria are known to transform arsenate and arsenite to di- and trimethylarsine in an aerobic environment. The process can only take place when an appropriate C source is available.

$$As^{5+}, As^{3+} + C_{org} \rightarrow (CH_3)_2As + (CH_3)_3As \quad (9)$$

In addition, several other biotransformations forming a variety of other products may occur. For example, As^{5+} may be consumed as electron acceptor during the oxidation of organic matter resulting in a reduction to the more soluble and more toxic As^{3+} (ANDERSON and LOVELY, 1997). Up to now, the necessary environmental conditions for a controlled microbial arsenic reduction have not been investigated. Thus, if a technique could be found to stimulate the desired bioprocess, this might be a useful strategy for an *in situ* treatment of metal contaminated sites.

Although such reduction processes leading to gaseous As^{3-} compounds are well documented from laboratory experiments with significant transformation rates, substantial natural arsenic losses in the field have not yet been observed (LÈONARD, 1991).

The bioprocess that adds methyl groups to metal ions forming volatile organometals which volatilize from the environmental compartment is a general detoxification process used by bacteria. It is well documented, e.g., for mercury (Hg) (MATILAINEN, 1995).

The bacterium *Deinococcus radiodurans* transforms toluene and chlorobenzene partly also under high doses of γ radiation. Although in this case only the organic part of the pollution is degraded, the radionuclide remains unaffected. Because the bacterium survives in the radioactive environment, probably other bacteria may be isolated which carry out some of the discussed other biochemical reactions for the elimination of contaminants including bioleaching of radionuclides.

Future progress may be achieved by the construction of genetically engineered microorganisms (GEM) which contain, e.g., a completed degradation sequence for final mineralization of the contaminants, specific genes to resist unfavorable environments or other improved biochemical features like an adhesion deficiency to be used for bioaugmentation.

4 Monitoring

The conditions for the success of a remediation measure are the controllability of the bioprocesses and the hydraulics. The reactions occurring within an aquifer during microbiological *in situ* remediations are manifold including abiotic and biotic processes. The *in situ* bioremediation can only be successful, if the transport of nutrients and electron acceptors and donors to the contaminants and the removal of metabolic end products (e.g., CO_2, N_2) is sufficient. Otherwise the formation of gas bubbles may occur resulting in a change of the transport process. For this reason hydrogeological parameters have to be included in the monitoring. Although there are numerous successfully completed *in situ* measurements reported, up to now the *in situ* methods cannot be regarded as *state-of-the-art* (KREYSA and WIESNER, 1996). The success can only be achieved, if the remediation is accompanied by intensive monitoring. The tasks of monitoring are, therefore, to obtain

(1) information for controlled addition of electron acceptors and donors and nutrient salts,
(2) information on the geobiochemical site conditions,
(3) a proof of the functionality of the remediation measures,
(4) a proof that the remediation target is reached.

In a continuous remediation process the monitoring data are used to optimize the bioprocess and to correct malfunctions.

In the case of treatment of the unsaturated soil, the most important monitoring instrument is the *in situ respiration test*. During this test, the consumption of oxygen and formation of carbon dioxide in the soil vapor are determined without operation of the soil vapor extraction. Although CO_2 formation may be influenced by abiotic CO_2 fixation within the soil (e.g., as carbonates), the O_2 consumption allows a determination of the actual degradation rate. In the case of mineral oils, a representative molecule (e.g., C_n-alkane) is assumed for calculation of the degradation according to the following equation:

$$C_nH_m + \left(n + \frac{m}{4}\right) O_2 \rightarrow nCO_2 + \frac{m}{2} H_2O \quad (10)$$

When the net increase of biomass is zero, the respiratory ratio $[O_2(CO_2)^{-1}]$ equals $[(n+m/4)(n)^{-1}]$. When all hydrocarbons are transformed to biomass the respiratory ratio is 0. The relation between these two points is linear. Therefore, from the actual measured respiratory ratio the biomass production [as mol C biomass (mol C hydrocarbon)$^{-1}$] can be estimated. The estimation is only valid when the formed CO_2 is not precipitated.

The change of the degradation rate over time allows an assessment of the quality of the installed remediation measures. A decrease of the degradation rate may indicate the lack of water, nutrients, or degradable contaminants.

The monitoring of remediation measures of the saturated soil is often more complex. Usually, groundwater samples are taken and the following parameters are determined:

(1) contaminants,
(2) metabolites (as dissolved organic carbon; DOC),
(3) degradation end products (CO_2, CH_4),
(4) nutrient salts (e.g., NH_4^+, PO_4^{3-}),
(5) electron acceptors (O_2, NO_3^-, NO_2^-),
(6) electron donors,
(7) geogenic substances (Fe^{2+}, Fe^{3+}, Mn),
(8) field parameters (pH, redox potential, electrical conductivity, temperature),
(9) bacterial counts (total counts, contaminant degraders, denitrifiers).

Due to the variety of possible metabolites, it is not possible to determine specific compounds. Because the metabolites are more polar and, therefore, better soluble than the parental compounds they are rather summarized as dissolved organic carbon. In some cases (e.g., when water treatment plants are involved in the remediation process) more detailed information on the character of DOC is required. This information may be supplied by LC-OCD analytics (liquid chromatography with organic carbon detection) where the DOC is subdivided into fractions of humic substances, building blocks, low molecular weight acids, amphiphilic substances, and polysaccharides.

The monitoring of degradation end products is only necessary if a balancing of the remediation is required. However, up to now no suitable methods for balancing technical scale *in situ* processes are available. In many cases degradation of end products is only monitored in research projects. In commercial projects a more pragmatical approach is chosen, i.e., only a very limited number of parameters are determined.

These parameters include monitoring of the nutrients and electron acceptors or electron donors in the case of anaerobic processes to ensure a sufficient supply of these supplements. The field parameters are monitored to ensure that the remediation measures have led to environmental conditions optimal for contaminant degradation. In this case an increase of the bacterial counts of specific contaminant degraders is expected. However, investigations have shown that the remediation measures strongly interfere with the indigenous microflora in the sense of a complex alteration of the biodiversity. The effects are not yet well understood.

Because many *in situ* technologies are based upon controlled hydraulic conditions, it is evident that the maintenance of the designed infiltration rate or groundwater flow rate is of utmost importance. Most pollutants exhibit low solubility and a high trend to bind to the soil matrix resulting in a strong retardation. The elimination of the contaminants is enhanced by biodegradation at the place where the contaminants are located. Hence, a transport of nutrients and electron acceptors and donors to the contaminants is necessary. Once the pollutants are solubilized they are degraded prior to resorption. Thus biodegradation increases the concentration gradient between sorbed/solid and dissolved substances which accelerate the desorption, dissolution and in addition the diffusion of pollutants out of micropores where they are not available for biodegradation.

An alteration of the hydraulic conductivity (expressed as k_f value) of the aquifer has significant influence on the functionality of the remediation. Hence, monitoring of the hydraulic conductivity is necessary. This can be done, if the *in situ* technology includes infiltration (of supplements or recharge of groundwater) or pumping of groundwater as a technology component. With the monitoring of the infiltration rate Q and the water level within the infiltration well N with the help of pressure

probes located at the bottom of the infiltration well the hydraulic conductivity can be calculated according to the law of DARCY (I: length of the hydraulic gradient):

$$Q = k_f \cdot \frac{N}{I} \qquad (11)$$

A decrease of the hydraulic conductivity may be caused by the formation of gas bubbles (in the case of excess infiltration of H_2O_2 or denitrification), of Fe oxide precipitations (in the case of altering the redox condition), by clogging with produced biomass or by shifting of fine soil particles caused by fast infiltration. By correlating an observed decrease of the hydraulic conductivity with actual remediation measures, the influence of the remediation on the hydraulic conductivity can be minimized.

Generally, there is a partitioning of soluble gases between groundwater and soil vapor. Soil vapor samples can be taken and investigated for:

(1) volatile contaminants,
(2) degradation end products (CO_2, CH_4),
(3) electron acceptor (O_2),
(4) natural tracer radon (^{222}Rn).

Degradation end products are monitored to balance the *in situ* processes. With the monitoring of O_2 an oversaturation of this electron acceptor can be avoided. With the monitoring of the natural tracer radon transport processes from groundwater to soil vapor and from soil vapor to the atmosphere can be determined.

Simplified, the required rate of nutrient supply is limited by the rate of contaminant mobilization. Since the mobilization and the biodegradation rate cannot be determined directly *in situ*, the supply of nutrients is done on the basis of the nutrient consumption determined during monitoring. Usually, at the point of infiltration the nutrient concentration is quite high and at the point of exfiltration or at the edge of the treatment zone almost zero. The concentrations to be added are limited. Compounds like technical oxygen have only a low solubility. Ammonium and hydrogen peroxide are toxic at higher concentrations. Furthermore, overdosing of hydrogen peroxide leads to the formation of oxygen bubbles.

5 Outlook

Up to now microbiological *in situ* remediation technologies cannot be regarded as *state-of-the-art*. Many of the processes occurring in the underground cannot be predicted during remediation planning. Nevertheless, the reported success demonstrates that *in situ* remediation can be a good technology if sufficient experience is available. The uncertainty in the process course can be minimized by proper investigations and suitable monitoring. In this case *in situ* technologies may be the most economic remediation technologies. However, there are still potential cost savings, if the processes are optimized. One of the optimization procedures is the consideration of natural limitations within the meaning of slower processes like the passive technologies. However, if they are not applicable active technologies may be designed.

There is still a large potential for development within the *in situ* technologies. Although today more efforts are made in environment protection management, the current environmental situation shows that *end-of-the-pipe* technologies will still be needed for decades.

6 References

ANDERSON, R. T., LOVLEY, D. R. (1997), Ecology and biogeochemistry of *in situ* groundwater bioremediation, in: *Advances in Microbial Ecology* Vol. 15 (JONES, Ed.), pp. 289–350. New York: Plenum Press.

ANDREWS, J. H., HARRIS, R. F. (1987), r- and K-selection and microbial ecology, in: *Advances in Microbial Ecology* Vol. 9 (MARSHALL, K. C., Ed.), pp. 99–147. New York: Plenum Press.

BENNER, S. G., BLOWES, D. W., PTACEK, C. J. (1997), A full-scale porous reactive wall for prevention of acid mine drainage, *Ground Water Monit. Rem.* **17**, 99–107.

BORDEN, R. C., GOIN, R. T., KAO, C.-M. (1997), Control of BTEX migration using a biological enhanced barrier, *Ground Water Monit. Rem.* **17**, 70–80.

BROWN, R. A., HICKS, R. J., HICKS, P. M. (1994), Use of air sparging for *in situ* bioremediation, in: *Air Sparging for Site Remediation* (HINCHEE, R. E., Ed.), pp. 38–55. Boca Raton, FL: CRC Press.

KITTEL, J. A., HINCHEE, R. E., HOEPPEL, R., MILLER, R. (1994), Bioslurping – vacuum-enhanced free-product recovery coupled with bioventing: a case study, *Proc. Petroleum Hydrocarbons and Organic Chemicals in Groundwater: Prevention, Detection, Remediation Conf. Groundwater Management*, Dublin, OH, pp. 255–270.

KLEIN, J., DOTT, W. (1992), *Laboratory Methods for the Evaluation of Biological Soil Cleanup Processes*. Frankfurt: DECHEMA.

KREYSA, G., WIESNER, J. (Eds.) (1996), *In-situ-Sanierung von Böden*. Frankfurt: DECHEMA.

LÉONARD, A. (1991), Arsenic, in: *Metals and Their Compounds in the Environment. Occurrence, Analysis and Biological Relevance* (MERIAN, E., Ed.), pp. 751–773. Weinheim: VCH.

MATILAINEN, T. (1995), Involvement of bacteria in methyl mercury formation in anaerobic lake waters, *Water Air Soil Pollut.* **80**, 757–764.

RAWLINGS, D. E., SILVER, S. (1995), Mining with microbes. *Biotechnology* **13**, 773–778.

REDDY, K. R., KOSGI, S., ZHOU, J. (1995), A review of *in situ* air sparging for the remediation of VOC-contaminated saturated soils and groundwater, *Hazardous Waste – Hazardous Materials* **12**, 97–117.

RIJNAARTS, H. H. M., BRUNIA, A., VAN AALST, M. (1997), *In situ* bioscreens, in: *Pap. 4th Int. In Situ On-Site Biorem. Symp.* 4, pp. 203–208. Columbus, OH: Battelle Press.

ROBERTSON, W. D., CHERRY, J. A. (1997), Long-term performance of the Waterloo denitrification barrier, *Land Contam. Reclam.* **5**, 183–188.

SCHIPPERS, A., SAND, W. (1999), Bacterial leaching of metal sulfides proceeds by two indirect mechanisms via thiosulfate or via polysulfides and sulfur, *Appl. Environ. Microbiol.* **65**, 319–321.

SINGH, B., WILSON, M. J. (1997), Geochemistry of acid mine waters and the role of microorganisms in such environments: a review, *Adv. Geoecol.* **30**, 159–192.

SMYTH, D. J. A., BYERLEY, B. T., CHAPMAN, S. W., WILSON, R. D., MACKAY, D. M. (1995), Oxygen-enhanced *in situ* biodegradation of petroleum hydrocarbons in groundwater using a passive interception system, *5th Annu. Symp. Groundwater Soil Remediation*, pp. 23–34. Toronto, Canada.

SPUIJ, F., ALPHENAAR, A., DE WIT, H., LUBBERS, R., V/D BRINK, K. et al. (1997), Full-scale application of *in situ* bioremediation of PCE-contaminated soil, in: *Pap. 4th Int. In Situ On-Site Biorem. Symp.* 4, pp. 431–437. Columbus, OH: Battelle Press.

SUTHERSAN, S. S. (1996), *Remediation Engineering: Design Concepts*. Boca Raton, FL: CRC Lewis Publishers.

WHITE, C., GADD, G. M. (1998), Reduction of metal cations and oxyanions by anaerobic and metal-resistant microorganisms: chemistry, physiology, and potential for the control and bioremediation of toxic metal pollution, in: *Extremophiles: Microbial Life in Extreme Environments* (HORIKOSHI, K., GRANT, W. D., Eds.), pp. 233–254. New York: Wiley-Liss.

WIEDEMEIER, T. H., WILSON, J. T., KAMPBELL, D. H., MILLER, R. N., HANSEN, J. E. (1995), Technical protocol for implementing intrinsic remediation with long-term monitoring for natural attenuation of fuel contamination dissolved in groundwater, *AFC Environ. Excellence Books.* AFB: San Antonio, TX.

16 Degradation by Plants – Phytoremediation

JERALD L. SCHNOOR

Iowa City, IA 52242, USA

List of Abbreviations

BEHP	bis-ethyl hexyl phthalate
BTEX	benzene, toluene, ethyl benzene, and xylenes
DNT	dinitrotoluene
GSH	glutathione *S*-transferase
HSRC	Hazardous Substances Research Center
NIEHS	National Institute of Environmental Health Sciences
PAH	polynuclear aromatic hydrocarbons
PCB	polychlorinated biphenyls
RCF	root concentration factor
RDX	hexahydro-1,3,5-trinitro-1,3,5-triazine
TCE	trichloroethylene
TNT	2,4,6-trinitrotoluene
TPH	total petroleum hydrocarbon
TSCF	transpiration stream concentration factor
VOC	volatile organic chemical

1 Introduction

Phytoremediation is the use of vegetation for *in situ* treatment of contaminated soils, sediments, and water. It has been utilized at a number of pilot- and full-scale field tests and is popular because of its cost effectiveness, esthetic advantages, and long-term applicability. Vegetation is well-suited for use

- at large field sites where other methods of remediation are not cost-effective or practicable,
- at low-level contaminated sites where only "polishing treatment" is required over long periods of time,
- in conjunction with other technologies where plants are used as a final cap and closure at the site.

There are limitations to the technology that need to be considered carefully before it is selected for site remediation. These include limited regulatory acceptance, lengthy time periods that are sometimes required for clean-up to below action levels, potential contamination of the vegetation and food chain, and difficulty establishing and maintaining vegetation at some toxic waste sites.

Different species of plants have been used in various phytoremediation applications for treatment of organic contaminants including:

- phreatophytes, which establish roots deep into the ground water table (*Populus* spp., hybrid poplar, cottonwoods, and willow),
- grasses for their fibrous root systems that hold soils in place (*Festuca* spp., *Cynodon dactylon*, and *Phalaris arundinacea*),
- legumes for nitrogen enrichment of poor soils (*Trifolium* spp., *Medicago sativa*, clover, alfalfa, and cowpeas),
- aquatic plants in created wetlands for degradation of contaminants in sediments and water (*Myriophyllum spicatum*, *Saggitaria latifolia*, *Potamogeton nodosus*, and *Ceratophyllum*).

Phytoremediation has been applied in riparian zone buffer strips for

- pesticide removal from agricultural run-off,
- trichloroethylene (TCE) removal from ground water,
- total petroleum hydrocarbon (TPH) and BTEX (benzene, toluene, ethyl benzene, and xylenes) transformations in rhizosphere soil,
- pentachlorophenol and polynuclear aromatic hydrocarbons (PAH) degradation from wood preservative and manufactured gas plant sites,
- 2,4,6-trinitrotoluene (TNT) and hexahydro-1,3,5-trinitro-1,3,5 triazine (RDX) removal in a created wetlands at ammunition waste sites (SCHNOOR et al., 1995; SCHNOOR, 1997).

In addition, many successful applications have occurred at small sites, such as agricultural cooperatives with pesticide and ammonia spills, where state agencies have jurisdiction. At these small sites, few funds are available for long-term compliance monitoring. Therefore, long-term monitoring and evaluation of phytoremediation technology are needed still to demonstrate efficacy, to further define suitable plants and applications, and to gain acceptance from regulatory agencies. It is important to demonstrate that chemicals and metabolites do not accumulate in plant tissues or enter the food chain.

Plants are able to take up contaminants directly from soil water and either store them in special plant tissues as conjugates, metabolize them, or volatilize them to the atmosphere with transpiration water. In addition, the roots of plants help to foster large numbers and a wide diversity of microorganisms that biodegrade organic contaminants. Release of plant exudates helps to degrade organic pollutants by supplying organic carbon as a primary substrate for microorganisms and by cometabolism in the rhizosphere (Fig. 1). The rhizosphere is defined here as the soil environment in the immediate vicinity of roots (within 1 mm).

Actively growing plants increase soil organic carbon, bacterial activity, and mycorrhizae fungi, all factors that encourage degradation of organic chemicals in soil. Rhizosphere bioremediation is also known as phytostimulation or plant-assisted bioremediation. Leakage of sugars, alcohols, and acids from plants and root turnover can amount to 10–20% of plant photosynthesis on an annual basis (SCHNOOR et al., 1995). Researchers have characterized the molecular weight distribution of organic exudates from root systems of hybrid poplar trees (SCHNOOR et al., 1995). Exudates include short chain organic acids, phenolics, and small concentrations of high molecular weight compounds (enzymes and proteins). Some exudates may act as cosubstrates which induce enzyme systems in rhizosphere bacteria to degrade compounds that would otherwise be recalcitrant.

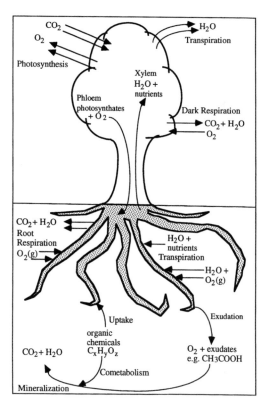

Fig. 1. Schematic of the transport of chemicals, oxygen, and water in a phytoremediation application.

Plants help with microbial transformations in many ways:

(1) Mycorrhizal fungi associated with plant roots metabolize the organic pollutants.
(2) Plant exudates stimulate bacterial transformations (enzyme induction).
(3) Build-up of organic carbon increases microbial mineralization rates (substrate enhancement).
(4) Plants provide habitat for diverse microbial populations and activity.
(5) Oxygen is pumped to roots ensuring aerobic transformations.

Fungi, growing in symbiotic association with the plant, have unique enzymatic pathways that help to degrade organics that could not be transformed solely by bacteria. Mycorrhizal fungi and bacteria are shown in the vicinity of fine roots from spruce trees in Fig. 2. The rapid decay of fine root biomass can become an important addition of organic carbon to soils which serves to retard organic chemical transport. Microbial mineralization is directly related to the fraction of organic carbon in the soil, and microbial assemblages are abundant in the rhizosphere with typical communities comprising $5 \cdot 10^6$ bacteria, $9 \cdot 10^5$ actinomycetes, and $2 \cdot 10^3$ fungi per gram of air dried soil. Bacteria live in colonies that cover as much as 4–10% of the root surface area (SCHNOOR et al., 1995).

Plants have shown the capacity to withstand relatively high concentrations of organic chemicals without toxic effects, and they can take up and convert chemicals to less toxic metabolites in many cases. Vegetation stimulates microbial degradation of organic chemicals in the rhizosphere by the release of root exudates, enzymes, and the build-up of organic carbon in the soil. In this chapter, processes and pathways leading to degradation of organics in phytoremediation systems are presented, and an estimation of the time required for transformation is provided.

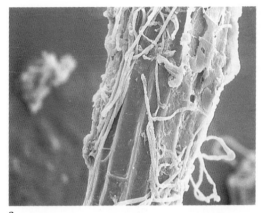

a

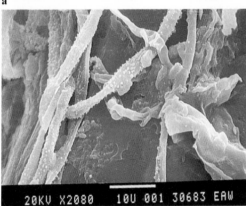

b

Fig. 2. Scanning electron micrographs (SEM) of mycorrhizal fungi and bacteria in association with fine roots in the rhizosphere, **a** magnified 500 times, **b** magnified 2,080 times, a 10 μm bar is shown for scale.

2 Degradation Processes

2.1 Degradation by Plants

Degradation of organic contaminants by plants for the purpose of phytoremediation is a relatively new research area. Most literature relates to the uptake and mode of action of herbicides in crops and weed species. Uptake of organic chemicals by plants from hydroponic solution has been studied by BRIGGS et al. (1982) and others (BURKEN and SCHNOOR, 1998). Direct uptake by plants is a surprisingly efficient removal mechanism of chemicals from shallow ground water and from soils contaminated with moderately hydrophobic organics (octanol–water partition coefficients, log $K_{ow} = 1.0–3.5$). This includes most BTEX chemicals, chlorinated solvents, and short-chain aliphatic chemicals. Hydrophobic chemicals (log $K_{ow} > 3.5$) are bound so strongly to the surface of roots and soils that they cannot be easily translocated within the plant; and most chemicals which are water soluble (log $K_{ow} < 1.0$) are not sufficiently sorbed to roots nor actively transported through plant membranes (BRIGGS et al., 1982). Very hydrophobic chemicals (log $K_{ow} > 3.5$) bind so strongly to roots and soils that they are not bioavailable for plant uptake. The ratio of the contaminant concentration sorbed onto roots to that in aqueous solution is known as the root concentration factor (RCF, mL g^{-1}), and it is directly related to the log K_{ow} (BRIGGS et al., 1982).

The direct uptake of chemicals into plants through roots depends on the plant uptake efficiency, transpiration rate, and the concentration of chemicals in soil water (BURKEN and SCHNOOR, 1998). Uptake efficiency is known as the transpiration stream concentration factor (TSCF), the ratio of chemical concentration in the plant stem flow to the concentration in soil solution (BRIGGS et al., 1982). It depends on physical-chemical properties, chemical speciation, and plant membrane selectivity. Transpiration is a key variable that determines the rate of chemical uptake – it depends on the plant type, leaf area, nutrients, soil moisture, wind conditions, and relative humidity.

Plants are known to take up many soluble organic chemicals and store the parent compounds or metabolites in lignin, cell wall materials, or plant vacuoles. Generally, the first step in degradation by plants requires root uptake and translocation to the shoots or leaves for transformation (Fig. 1). Volatile organic chemicals (VOC) can be taken up by plants and transpired to the atmosphere without transformation (phytovolatilization).

BURKEN and SCHNOOR (1996, 1998) reported on the uptake, translocation, and metabolism of atrazine by hybrid poplar trees. Atrazine, an herbicide, was formulated to be rapidly taken up by rooted plants. Its herbicidal properties are exerted via inhibition of photosynthesis following transport in xylem to apical meristems and leaves. Most crops and tolerant plants detoxify atrazine by metabolizing it to hydroxylated, dealkylated products. Metabolites that were observed in hybrid poplar cuttings, suitable for phytoremediation, are shown in Fig. 3 (BURKEN and SCHNOOR, 1997).

Once an organic chemical is translocated, the plant may store the chemical and its fragments into new plant structures via conjugation or lignification (covalent bonding of chemical or its fragments into lignin of the plant), or it can volatilize, metabolize, or mineralize the chemical all the way to carbon dioxide and water. Major metabolic products for trichloroethylene are oxidation products: trichloroethanol, trichloroacetic acid, and dichloroacetic acid reported by NEWMANN et al. (1997). A small fraction of TCE has been reported to be mineralized to CO_2 and HCl, and some may be volatilized through leaves (NEWMAN et al., 1997). One strategy in phytotransformation is *phytovolatilization*, whereby volatile chemicals or their metabolic products are released to the atmosphere through the plant. Many compounds are recalcitrant in soil or ground water but may be rapidly broken down in the atmosphere by oxidation with hydroxyl radicals.

A proposed pathway for TCE transformation in hybrid poplars is shown by Fig. 4. Cytochrome P-450 enzymes (monooxygenases) could be responsible for formation of the epoxide metabolite shown in the center, top of Fig. 4. This pathway is similar to that which occurs in mammalian livers. The epoxide is a short-lived toxic species which does not accumulate due to rapid transformation to the products shown. These products can be recovered from plant tissues by extraction and derivatization (NEWMAN et al., 1997). Glycolysis and formation of various conjugates are also possible. Plants have considerable glutathione S-transferase (GSH in Fig. 4) which is also capable of forming adducts with trichloroethylene. Such reactions eventually lead to binding with amino acids and the formation of "bound residues", defined here as the incorporation of the chemical into plant tissues from which the parent compound can no longer be extracted by chemical means (SCHNABEL et al., 1997). A covalent bond is formed which essentially

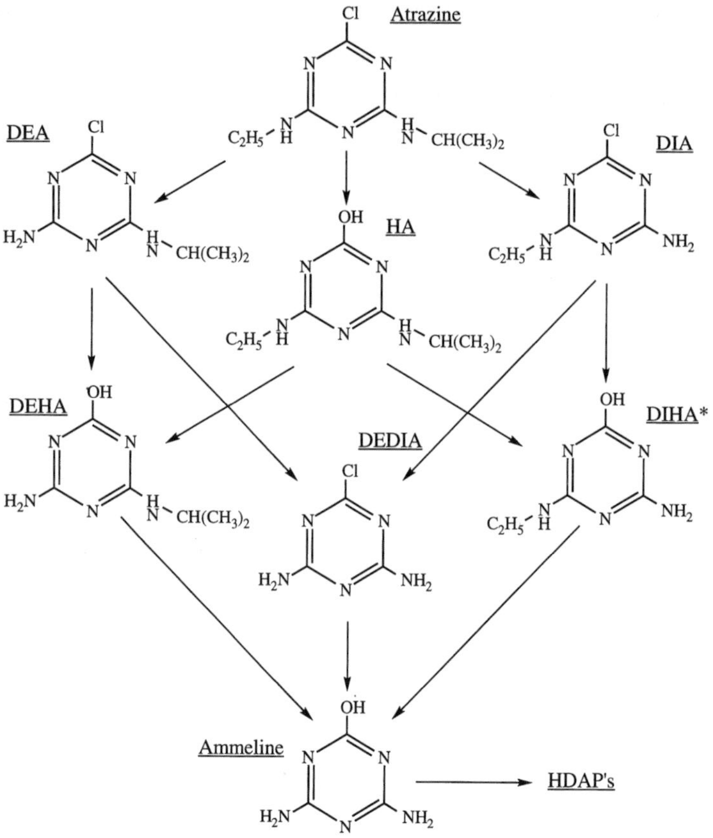

Fig. 3. Atrazine transformations and degradation in hybrid poplar leaves; star: not detected, DEA: deethyl atrazine, DIA: deisopropyl atrazine, HA: hydroxy atrazine, DEHA: deethyl hydroxyatrazine, DEDIA: deethyl deisopropyl atrazine, HDAP: hydroxylated, dealkylated products (adapted from BURKEN and SCHNOOR, 1997).

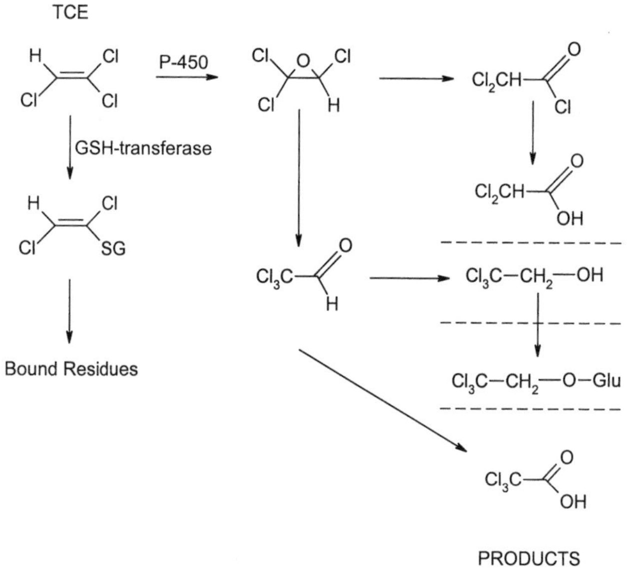

Fig. 4. Trichloroethylene transformations in plant tissues (adapted from NEWMAN et al., 1997, and SCHNABEL et al., 1997).

transforms the organic chemical into part of the plant structure.

Nitroaromatic compounds are particularly reactive with plant enzymes at the root–water and leaf–water interfaces (THOMPSON et al., 1998; HUGHES et al., 1997; RIEGER and KNACK-MUSS, 1995). Cellular and extracellular enzymes are efficient at transforming TNT to various products (Fig. 5), and partially reduced metabolites of TNT are formed at every step in the pathway. Irreversible binding to cellular components has been demonstrated, but it is not known to what extent these metabolites are bioavailable. These "bound residues" are no longer toxic to hybrid poplars used in phytoremediation, but they may be toxic to macroinvertebrates in the soil, to birds, or to herbivores as they graze on plant tissues. Azoxy compounds which are condensates of 2-hydroxyamino-4,6-dinitrotoluene and 4-hydroxyamino-2,6-dinitrotoluene (2-HA 4,6-

DNT and 4-HA 2,6-DNT) are known to be toxic to some organisms should they enter the food chain. Thus, there is a concern about food chain toxicity in the metabolism of certain xenobiotic chemicals by plants.

2.2 Degradation by Rhizosphere Bacteria

Biostimulation of rhizosphere bacteria to degrade organic chemicals is widely recognized (ANDERSON et al., 1993). Transformation by microorganisms (bacteria, fungi, and protozoa) in the rhizosphere is the key mechanism by which hydrophobic organic chemicals can be degraded in phytoremediation systems. Waste sites containing petroleum hydrocarbons, PAH, PCB, and dioxin can be transformed only via microbial processes because

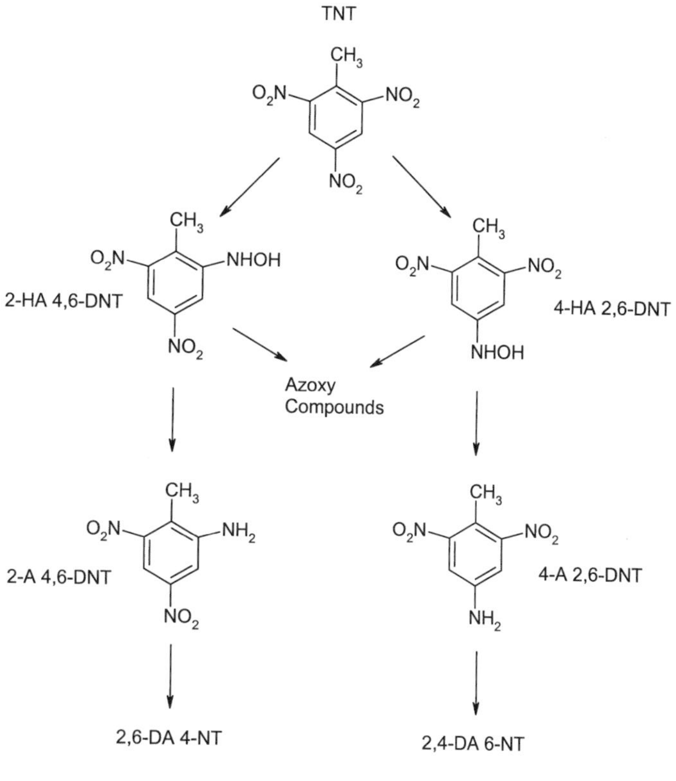

Fig. 5. TNT transformation in hybrid poplar tissues (adapted from THOMPSON et al., 1998).

the amount of plant uptake is negligible for these hydrophobic compounds. Other chemicals such as BTEX aromatics are rapidly degraded in the rhizosphere under aerobic conditions. Plants help to maintain aerobic conditions in the subsurface by transferring oxygen to the root zone and by physically enhancing the porosity and permeability of the soil.

As depicted in Fig. 1, the release of exudates from roots in the rhizosphere serves as a food source for heterotrophs and as an inducing agent for enzyme systems in cometabolism. Some enzyme systems are constitutive in plants, but are not expressed except in the presence of a model organic compound. HEDGE and FLETCHER (1996) have shown that phenolic compounds are released during fine root turnover and induce PAH and PCB degrading enzyme systems in bacteria in the rhizosphere. Flavonoids and coumarin are released by root turnover from trees like mulberry, osage orange, and apple (HEDGE and FLETCHER, 1996). Organic herbicides such as alachlor and atrazine have been degraded in rhizosphere soils at agricultural cooperatives and in riparian zone buffer strips where plantations of hybrid poplars have been established. Mineralization to carbon dioxide is greatly enhanced when atrazine is present in the rhizosphere compared to unplanted soils (BURKEN and SCHNOOR, 1996; ANDERSON et al., 1993).

A general equation for the role of plant exudates is presented below (Eq. 1). A fraction of photosynthesis is shunted towards exudates (represented here as acetic acid, CH_3COOH). The fraction of simple sugars (such as glucose, $C_6H_{12}O_6$) that is used to produce exudates is represented as p. 10–20% of the primary production of a plant may be routed towards fine root production, turnover, and leakage of exudates.

Photosynthesis and exudate production is described by Eq. (1):

$$6\,CO_2 + 6\,H_2O \rightarrow (1\text{-}p)\,C_6H_{12}O_6 \atop + 3p\,CH_3COOH + 6\,O_2 \tag{1}$$

Aerobic microbial processes in the rhizosphere oxidize these exudates, carbohydrates, and other xenobiotic organic compounds (represented here as $C_xH_yO_z$) and convert them to carbon dioxide and water.

Aerobic cometabolism in the rhizosphere follows Eq. (2):

$$3p\,CH_3COOH + n\,C_xH_yO_z + (6p + nx \atop + ny/2 + ny/4 - nz/2)\,O_2 \rightarrow \atop (6p + nx)\,CO_2 \atop + (6p + ny/2)\,H_2O \tag{2}$$

Thus, cometabolism of exudates provides an important substrate for bacteria in the rhizosphere and the potential for enzyme transformation reactions that would not occur in an unplanted system.

A summary of phytoremediation mechanisms is presented in Tab. 1 for some common organic contaminants at waste sites. Moderately hydrophobic chemicals (such as atrazine, TCE, and BTEX) are readily taken up and translocated by plants. Volatile organic chemicals (VOC), like TCE and benzene, may be transpired from the plant and/or metabolized

Tab. 1. Summary of Phytoremediation Mechanisms for Selected Organic Contaminants

Chemical	Rhizosphere Biodegradation	Root Binding and Transformation	Translocation to the Leaves	Transformation in Plant	Transpiration through Leaves
Atrazine	x		x	x	
TCE	?		x	x	x
BTEX	x		x		x
TNT	x	x			
RDX			x	x	
PAH	x				
PCB	?				
BEHP	x				

in the leaf tissues. TNT is not translocated much, but it binds strongly to roots and is transformed there. Eventually bound residues are formed which should be assessed for possible food chain effects. For highly hydrophobic chemicals (e.g., PAH, PCB, and bis-ethyl hexyl phthalate BEHP) (Hsu et al., 1993), biostimulation and degradation in the rhizosphere are the primary mechanisms for phytoremediation. These are enhanced by the release of plant exudates. Still, aerobic mineralization of PAH and PCB compounds is very slow.

3 Practical Considerations

Design of phytoremediation systems varies according to the contaminant(s), the conditions at the site, the level of clean-up required, and the plant(s) used. Some practical considerations are following that are necessary to establish a successful phytoremediation system for the degradation of organic chemicals:

(1) plant selection,
(2) treatability,
(3) planting density and pattern,
(4) irrigation, agronomic inputs, and maintenance,
(5) ground water capture zone and transpiration rate,
(6) contaminant uptake rate and clean-up time required.

3.1 Plant Selection

Plants are selected according to the needs of the application and the contaminants of concern. For phytotransformation of organics, the plants must be fast growing and hardy, easy to plant and maintain, capable of utilizing a large quantity of water by evapotranspiration (if ground water capture is the goal), and capable of transforming the contaminants to non-toxic or less toxic products. In temperate climates, phreatophytes (e.g., hybrid poplar, willow, cottonwood, aspen) often are selected because of fast growth, a deep rooting ability to the ground water table, large transpiration rates, and a wide range throughout temperate regions of the world.

Grasses are often planted in tandem with trees at sites with petroleum hydrocarbons, or as the primary remediation method when contaminants are located at the surface of the soil. They provide a tremendous density of fine roots in the surface soil which is effective at binding and transforming hydrophobic contaminants such as TPH and PAH. Grasses also aid in soil stabilization and protection against wind-blown dust that can move contaminants off-site. Legumes (such as alfalfa, alsike clover, and cowpeas) can be used to restore nitrogen to poor soils. Fescue, rye, and reed canary grass have been used successfully at several sites, especially petrochemical wastes. The grasses are harvested periodically and disposed to compost or burned. Hydrophobic contaminants do not translocate, so the top-portion of grasses does not become contaminated. The system achieves phytoremediation via rhizosphere processes and sorption to roots.

Aquatic plants are used in created wetlands applications. They fall into two categories: emergent and submerged species. Emergent vegetation transpires water, and it is easier to harvest the vegetation if desired. Submerged species do not transpire water, but they provide more biomass within the aquatic portion of the system for uptake and sorption of contaminants. Aquatic species in created wetlands have included bullrush, cattail, coontail, duckweed, arrowroot, pondweed, parrot feather, Eurasian water milfoil, stonewort, and *Potamogeton* spp.

3.2 Treatability

Treatability studies should be conducted prior to full-scale phytoremediation. Toxicity and transformation data are obtained in these studies. There is a large variation in toxicity and transformation rates that can be expected from one plant species to another, even from one variety or cultivar to another. Boron, zinc, ammonium, some metals, and salts are especially toxic to plants. Some organic contaminants are toxic as well. Thus, it is critical to obtain treatability information in the laboratory or greenhouse, if prior knowledge has not been reported for the waste with that plant. The sequence of design information that is required typically ranges from hydroponic studies, to small plot studies with soils from the site in a greenhouse, to plot studies at the site (up to 15 m · 15 m). Different strengths of contaminant can be analyzed for toxicity, and plant tissues can be harvested for metabolite or parent compound analyses. Regulators may require total mass balance information which necessitates use of radiolabeled compounds in laboratory studies.

Laboratory studies are utilized to assess the fate of the contaminant(s) in the plant system. Volatile organic chemicals (VOC) are sometimes transpired to the atmosphere by plants, in which case, air toxics calculations would be needed to estimate the atmospheric emissions and whether these chemicals would be considered objectionable. Similarly, moderately hydrophobic organics (log K_{ow} = 1–3.5) are often translocated to the leaves of the plant and metabolized. Measurements of leaf concentrations of parent compound and metabolites would be needed to determine if actionable levels are reached.

3.3 Plant Density

Plant density depends on the application. For hybrid poplar plantations, 1,000–1,500 trees per acre are typically planted with a conventional tree planter at 12–18 inches depth or in trenched rows 1–6 ft deep. Rooted or unrooted cuttings can be planted. Poplars have the ability to root along the entire buried depth. If a row conformation is used, the trees may be spaced with 3–4 ft between trees and 10 ft between rows. Unrooted poplars are planted simply as "whips", long cuttings that will root and grow rapidly in the first season. Several phreatophytes in the *Salix* family, such as willow and cottonwood, can be planted in a similar manner. Hardwood trees and evergreens may require a lower planting density initially. A high initial planting density assures a significant amount of evapotranspiration in the first year which is normally desirable, but the trees will naturally thin themselves by competition to 600–800 trees per acre over the

first six years. If desirable, hybrid poplars can be harvested on a 6-year rotation and sold for fuelwood or pulp and paper, and the trees will grow back from the cut stump (coppicing trait). The dense, deep root system stays in place to sustain growth for the next year. The lifetime of hybrid poplars such as *Populus deltoides* x *nigra* DN-34 (Imperial Carolina) is on the order of 30 a which is usually sufficient as the design life of the project.

Grasses are usually drilled or broadcast for planting at waste sites. Biomass densities (above ground) of 200–600 g m^{-2} are achieved by the second crop, with 1–3 crops per year depending on climate and water availability.

The initial planting density of aquatic species in a created or natural wetland is normally 3 plants to a pod, located on 3 ft centers. Replanting and maintenance should be estimated in the cost of the project. One should consider that at least 30% of the plants may need to be replanted in the second or third year, as a contingency.

3.4 Irrigation, Agronomic Inputs

For terrestrial phytoremediation applications, it is desirable to include irrigation costs in the design, about 10–20 inches of water delivery capacity per year. Irrigation of the plants ensures a vigorous start to the system even in a drought. On the other hand, hydrologic modeling may be required to estimate the rate of percolation to ground water under irrigation conditions. Over time, irrigation should be withdrawn from the site, provided the area receives sufficient rainfall to sustain the plants. Maintenance costs should be considered in the design of phytoremediation systems. Mowing, replanting, pruning, harvesting, monitoring vegetation for contaminants, and fertilizer costs should all be included in the initial estimated costs if they are needed (O & M: operation and maintenance).

Agronomic inputs include the nutrients necessary for vigorous growth of vegetation and rhizosphere bacteria. These include N, P, and K from commercial fertilizer mixes, and carbon addition and soil conditioners such as aged manure, sewage sludge, compost, straw, or mulch. Typical application rates of fertilizer include 50 lbs P per acre and 50 lbs N per acre per application, 1–3 applications per year, especially for production of grasses and fine roots at petrochemical sites. It is critical that the site soils have sufficient water holding capacity to sustain vegetation. Amendments are necessary to improve soil conditions and to allow water to be absorbed in poor soils. Sometimes it is desirable to neutralize pH by lime addition – a standard agronomic analysis of site soils will allow assessment.

Biomass production can be estimated at 7 t dry matter per acre and year for fast growing trees. The amount of nitrogen stored in woody tissue is typically 0.5–1.0%, so nitrogen uptake can be calculated. Stoichiometries of woody tissue and leaf tissue are available in the literature to estimate major nutrient uptake requirements. Nutrients should be replaced in the soil environment each year by fertilizer inputs in a properly functioning phytoremediation system.

3.5 Groundwater Capture and Transpiration

It must be understood where the water is moving at a site in order to estimate contaminant fate and transport. For applications involving ground water remediation, a simple capture zone calculation (DOMENICO and SCHWARTZ, 1990) can be used to estimate whether the phytoremediation "pump" can be effective at containing the plume of contaminants. Trees can be grouped for consideration as average withdrawal points or each tree assumed as a "point ground water well" withdrawal. The goal of such a phytoremediation effort is to create a water table depression where contaminants will flow into the vegetation for uptake and treatment. It is important to realize that organic contaminants are not taken up at the same concentration as they are present in the soil or ground water, rather there is a transpiration stream concentration factor (a fractional efficiency of uptake) that accounts for the partial uptake of contaminant (due to membrane barriers at the root surface).

$$U = (TSCF)\,(T)\,(C) \tag{3}$$

where U [mg d^{-1}] is the uptake rate of contaminant, $TSCF$ is the transpiration stream concentration factor (dimensionless), T [L d^{-1}] is the transpiration rate of vegetation, and C [mg L^{-1}] is the aqueous phase concentration in soil water or ground water.

If the contaminant plume is not taken up by the vegetation, the plume that emerges will be evapoconcentrated, i.e., the mass of contaminant in the plume will be less (due to uptake by vegetation), but the concentration remaining will actually be greater. This is a potential concern for phytoremediation of ground water plumes or created wetlands, where a relatively hydrophilic contaminant can be concentrated on the downstream side of the phytoremediation system.

A method for estimating the transpiration stream concentration factor in Eq. (3) is given by Tab. 2. The root concentration factor is also defined as the ratio of the contaminant sorbed onto roots to the concentration dissolved in soil water (μg kg^{-1} root per μg L^{-1}). It is important in estimating the mass of contaminant sorbed to roots in phytoremediation systems.

Mature phreatophyte trees (poplar, willow, cottonwood, aspen, ash, alder, eucalyptus, mesquite, bald cypress, birch, and river cedar) typically can transpire 3–5 acres per ft of water per year (36–60 inches of water per year). This is equivalent to about 600–1,000 gallons of water per tree per year for a mature species planted at 1,500 trees per acre. Transpiration rates in the first two years would be somewhat less, about 200–500 gallons per tree per year, and hardwood trees would transpire about half the water of a phreatophyte. 2 m of water per year are a practical maximum for transpiration in a system with complete canopy coverage (a theoretical maximum would be 4 m a^{-1} based on the solar energy supplied at 40°N on a clear day that is required to evaporate water). If evapotranspiration of the system exceeds precipitation, it is possible to take up all water that is moving vertically through soil. Areas that receive precipitation in the winter time (dormant season for deciduous trees) must be modeled to determine if the soil will be sufficiently dry to hold water for the next spring's growth period. Mathematical models have been used to estimate vertical water movement and percolation to ground water.

Tab. 2. Estimating the Transpiration Stream Concentration Factor (TSCF) and Root Concentration Factor (RCF) for Selected Organic Contaminants (according to BURKEN and SCHNOOR, 1998)

Chemical	Log K_{ow}	Solubility– log C_w^{sat} @ 25°C	Henry's constant $k_{H'}$, @ 25°C	Vapor Pressure– log P^0 @ 25°C	TSCF[a]	RCF[b]
		[mol L^{-1}]	[dimensionless]	[Pa]		[L kg^{-1}]
Benzene	2.13	1.64	0.2250	$9.12 \cdot 10^4$	0.71	3.6
Toluene	2.69	2.25	0.2760	$1.44 \cdot 10^5$	0.74	4.5
Ethylbenzene	3.15	2.80	0.3240	$1.92 \cdot 10^5$	0.63	6.0
m-Xylene	3.20	2.77	0.2520	$2.01 \cdot 10^5$	0.61	6.2
TCE	2.33	2.04	0.4370	$1.02 \cdot 10^5$	0.74	3.9
Aniline	0.90	0.41	$2.2 \cdot 10^{-5}$	$2.93 \cdot 10^5$	0.26	3.1
Nitrobenzene	1.83	1.77	0.0025	$3.72 \cdot 10^5$	0.62	3.4
Phenol	1.45	0.20	$>1.0 \cdot 10^{-5}$	$3.64 \cdot 10^5$	0.47	3.2
Pentachlorophenol	5.04	4.27	$1.5 \cdot 10^{-4}$	$6.84 \cdot 10^5$	0.07	54
Atrazine	2.69	3.81	$1 \cdot 10^{-7}$	$9.52 \cdot 10^5$	0.74	4.5
1,2,4-Trichlorobenzene	4.25	3.65	0.1130	$3.25 \cdot 10^5$	0.21	19
RDX	0.87	4.57	–	–	0.25	3.1

[a] TSCF: $0.75 \exp\{-[(\log K_{ow} - 2.50)^2/2.4]\}$, BURKEN and SCHNOOR (1998).
[b] RCF: $3.0 + \exp(1.497 \log K_{ow} - 3.615)$, BURKEN and SCHNOOR (1998).

3.6 Contaminant Uptake and Clean-Up Time

From Eq. (3), it is possible to estimate the uptake rate of the contaminant(s). First order kinetics can be assumed as an approximation for clean-up time. The uptake rate should be divided by the mass of contaminant remaining in the soil:

$$k_{phyto} = U M_0^{-1} \qquad (4)$$

with k_{phyto} [a^{-1}] first-order rate constant for uptake, U [kg a^{-1}] contaminant uptake rate, and M_0 [kg] mass of contaminant initially.

Then, an estimate for mass remaining at any time is expressed by Eq. (5).

$$M = M_0 e^{-(k_{phyto})t} \qquad (5)$$

where M [kg] is the mass remaining, and t [a] is the time.

Solving for the time required to achieve clean-up of a known action level gives Eq. (6):

$$t = -(\ln M M_0^{-1})(k_{phyto})^{-1} \qquad (6)$$

with t [a^{-1}] time required for clean-up to action level, M [kg] mass allowed at action level, and M_0 [kg] initial mass of contaminant.

Eqs. (1–4) can be applied to most sites where soil clean-up regulations are known for organic contaminants.

3.6.1 Organics – Example

TCE residuals have been discovered in an unsaturated soil profile at a depth of 3 m. From lysimeter samples, the soil water concentration is approximately 100 mg L^{-1}. Long cuttings of hybrid poplar trees will be planted through the waste at a density of 1,500 trees per acre for uptake and phytotransformation of the TCE waste. By the second or third year, the trees are expected to transpire 3 acre ft a^{-1} of water (36 inches a^{-1}) or about 600 gal per tree per year. The time required for clean-up is estimated as follows, if the mass of TCE per acre is estimated to be 1,000 kg per acre, and the clean-up standard has been set at 100 kg per acre (90% clean-up).

$$U = (TSCF)\,(T)\,(C) \qquad (3)$$

where

- $TSCF = 0.74$ from Tab. 2
- $T = (600$ gal per tree) (1,500 tree per acre) $(3.89$ L gal$^{-1}) = 3.5 \cdot 10^6$ L per acre and year
- $C = 100$ mg L^{-1} (given)
- $U = 2.59 \cdot 10^8$ mg per acre and year $= 259$ kg per acre and year
- $k = U\,(M_0)^{-1}$ (4)
- $k = (259$ kg a$^{-1})\,(1,000$ kg$)^{-1}$
- $k = 0.259$ a^{-1}
- $t = -(\ln M\,M_0^{-1})\,k^{-1}$ (5)
- $t = -[\ln 100\,(1,000)^{-1}]\,k^{-1}$
- $t = 8.9$ a

Some of the TCE that is taken up is expected to volatilize to the atmosphere. A portion will be metabolized by the leaves and woody tissue of the trees (NEWMAN et al., 1997).

4 Final Remarks

Hydrophobicity of chemical contaminants (as measured by the octanol–water partition coefficient) largely controls the sorption of organic contaminants to roots, uptake by plants, and translocation to leaves. However, biochemistry and enzyme interactions determine whether the contaminant can be metabolized in the plant or at the surface of plant tissues. Volatile organic chemicals may be metabolized within the plant or transpired to the atmosphere. Uptake of moderately hydrophobic chemicals is controlled by the transpiration rate of the plant, so rapidly growing phreatophytes are particularly desirable in many phytoremediation applications. Leakage of exudates to the rhizosphere greatly enhances the potential for microbial transformations and mineralization.

Phytoremediation is an emerging technology for contaminated sites that is attractive due to its low cost and versatility. It is not a pana-

cea for hazardous waste problems, but it shows tremendous potential in several applications for treatment of organics at shallow contaminated sites. We need to understand better the role of enzymes, metabolites, and the selection of plant systems for various wastes. Plants have the ability to withstand relatively high concentrations of pollutants; they can sometimes take up the chemicals and convert them to less toxic products, and they are known to stimulate degradation of organics in the rhizosphere.

Acknowledgments
This chapter would not have been possible without the efforts of graduate students and colleagues who first documented the important role of plants in transforming organic chemicals, including: JOEL BURKEN, PHIL THOMPSON, CRAIG JUST, ANNETTE DIETZ, and PEDRO ALVAREZ. The EPA Hazardous Substances Research Center (HSRC), located at Kansas State University, the Environmental Health Sciences Research Center (NIEHS), and the Center for Biocatalysis and Bioprocessing at the University of Iowa were essential in facilitating this research.

5 References

ANDERSON, T. A., GUTHRIE, E. A., WALTON, B. T. (1993), Bioremediation of the rhizosphere, *Environ. Sci. Technol.* **27**, 2630–2636.

BRIGGS, G. G., BROMILOW, R. H., EVANS, A. A. (1982), Relationships between lipophilicity and root uptake and translocation of non-ionized chemicals by barley, *Pestic. Sci.* **13**, 495–504.

BURKEN, J. G., SCHNOOR, J. L. (1996), Phytoremediation – Plant uptake of atrazine and role of root exudates, *J. Environ. Eng. ASCE* **122**, 958–963.

BURKEN, J. G., SCHNOOR, J. L. (1997), Uptake and metabolism of atrazine by poplar trees, *Environ. Sci. Technol.* **31**, 1399–1406.

BURKEN, J. G., SCHNOOR, J. L. (1998), Predictive relationships for uptake of organic contaminants by hybrid poplar trees, *Environ. Sci. Technol.* **32**, 3379–3385.

DOMENICO, P. A., SCHWARTZ, F. W. (1990), *Physical and Chemical Hydrogeology*. New York: John Wiley & Sons.

HEDGE, R. S., FLETCHER, J. S. (1996), Influence of plant-growth stage and season on the release of root phenolics by mulberry as related to development of phytoremediation technology, *Chemosphere* **32**, 2471–2479.

HSU, S. M., SCHNOOR, J. L., LICHT, S. A., ST. CLAIR, M. A., FANNIN, S. A. (1993), Fate and transport of organic compounds in municipal solid waste compost, *Compost Sci. Util.* **1**, 36–48

HUGHES, J. B., SHANKS, J., VANDERFORD, M., LAURITZEN, J., BHADRA, R. (1997), Transformation of TNT by aquatic plants, *Environ. Sci. Technol.* **21**, 266–271.

NEWMAN, L. A., STRAND, S. E., CHOE, N., DUFFY, J., EKUAN, G. et al. (1997), Uptake and biotransformation of trichloroethylene by hybrid poplars, *Environ. Sci. Technol.* **31(4)**, 1062–1067.

RIEGER, P. G., KNACKMUSS, H. J. (1995), Biodegradation of 2,4,6-trinitrotoluene and related nitroaromatic compounds, in: *Biodegradation of Nitro-aromatic Compounds* (SPAIN, J. C., Ed.), New York: Plenum Press.

SCHNABEL, W. E., DIETZ, A. C., BURKEN, J. G., SCHNOOR, J. L., ALVAREZ, P. J. (1997), Uptake and transformation of trichloroethylene by edible garden plants, *Water Res.* **31**, 816–824.

SCHNOOR, J. L. (1997), *Phytoremediation. Technology Evaluation Report TE-98-01*. Pittsburgh, PA: Ground-Water Remediation Technologies Analysis Center. (http://www.gwrtac.org).

SCHNOOR, J. L., LICHT, L. A., MCCUTCHEON, S. C., WOLFE, N. L., CARRIERA, L. H. (1995), Phytoremediation of organic and nutrient contaminants, *Environ. Sci. Technol.* **29**, 318–323A.

THOMPSON, P. L., RAMER, L. A., SCHNOOR, J. L. (1998), Uptake and transformation of TNT by hybrid poplar trees, *Environ. Sci. Technol.* **32**, 975–980.

17 Phytoremediation of Metals

DAVID E. SALT
Flagstaff, AZ 86011, USA

ALAN J. M. BAKER
Sheffield, UK

List of Abbreviations

CEC   Commission of the European Communities
EDAX   energy dispersive X-ray microanalysis
EDTA   ethylene diamine tetraacetic acid
EGTA   ethylene glycol-0,0′-bis(2-aminoethyl)-N,N,N′,N′-tetraacetic acid
EPA   Environmental Protection Agency
XAS   X-ray absorbance spectroscopy

1 Introduction

Phytoremediation, the use of green plants for environmental remediation started to attract interest in the 1980s and early 1990s. However, as so many discoveries in human history have, this intriguing idea developed out of the novel synthesis of research which already existed in the scientific literature. Early fieldwork by BAUMANN (1885), BYERS (1935, 1936), and MINGUZZI and VERGNANO (1948) identified particular plant species with the extraordinary capacity to hyperaccumulate Zn, Se, and Ni in their shoots. These discoveries provided the primary backdrop for the development of phytoremediation. However, it was the discovery by PRAT (1934) and later BRADSHAW (1952) of certain plant species with the capacity to tolerate elevated soil concentrations of certain metals such as Zn, Pb, Cu, Ni, and Cd, that lead to the creation of research groups focused on understanding the mechanisms involved in this tolerance (THURMAN,

1981 and references therein). It was the focus these groups provided that leads to the suggestion, first by CHANEY (1983) and then later by BAKER et al. (1988), that plants could be used to remove pollutant metals from contaminated soils and waters (SALT et al., 1998 and references therein).

Phytoremediation has been divided into at least six different functional areas (SALT et al., 1998), however, in this review only two of the most well developed areas will be addressed: phytoextraction and rhizofiltration. Phytoextraction has generally been defined as the use of pollutant-accumulating plants to remove metals and organics from soil by concentrating them in the harvestable portions of the plant. Rhizofiltration is generally defined as the use of plant roots to absorb/adsorb pollutants from water and aqueous waste streams (see Chapter 16, this volume). Although these technologies can be applied to the removal of organic compounds this chapter will focus on removal of pollutant metals.

2 Phytoextraction

For phytoextraction to be a viable alternative to existing soil remediation strategies it will require the existence of high biomass, rapidly growing metal-accumulating plants. It is also of critical importance that the concentration of metal in the harvestable plant tissue be higher than in the soil. This will ensure that the volume of contaminated plant material generated by the phytoextraction processes is less than the original volume of the contaminated soil. Unfortunately, plants do not exist at present which have these desirable characteristics. To generate this type of plant requires detailed information on the rate-limiting steps in the phytoextraction process.

2.1 Requirements for Efficient Phytoextraction

For over 400 million years plants have evolved to accumulate micronutrient metals, such as Fe, Zn, and Mn from the soil solution. However, plants are unable to acquire elements directly from the insoluble mineral phase of the soil. Therefore, for phytoextraction to be successful target pollutant metals must be present in significant concentrations in the soil solution. Uptake of these soluble target metals into the plant requires passage of the metal across the plasma membrane – a selectively permeable barrier surrounding cells. Passage across this membrane is mediated by ion transport proteins embedded in the membrane. To allow uptake of pollutant metals such as Cd, Pb, and U plants will, therefore, require transport proteins with the capacity to select for the target pollutant metal. Another key component of the phytoextraction process is that the extracted metal(s) are accumulated in the harvestable portion of the plant, in general the above-ground shoot biomass. To achieve this, it is important that the metal be efficiently transported from the root to the shoot. Finally, to simultaneously achieve high concentrations of metal(s) in shoots and a high growth rate, plants will have to be capable of tolerating the toxic effects of the accumulated

metal. Two very different strategies are being taken to achieve these objectives; chelate assisted phytoextraction termed *induced phytoextraction*, and long-term *continuous phytoextraction*.

2.2 Induced Phytoextraction

Induced phytoextraction addresses both the requirements of high soil metal solubility and efficient metal transport from the soil solution to the shoots, by using metal chelates to solubilize and transport the target metal from the soil into the plant. Application of chelates "induces" the uptake of the target metal into previously established plant biomass, thereby removing the requirements of a metal-tolerant plant, and making it feasible to use rapidly growing, high biomass crop plants such as corn and Indian mustard for the phytoextraction process.

Studies by JØRGENSEN (1993) initially indicated that application of synthetic metal chelates such as ethylene diamine tetraacetic acid (EDTA) to soils enhances Pb accumulation by plants. HUANG and CUNNINGHAM (1996), HUANG et al. (1997) and BLAYLOCK et al. (1997), building on these findings, were able to achieve rapid accumulation of Pb from soil into shoots of corn and Indian mustard to concentrations greater than 1% of shoot dry biomass. BLAYLOCK et al. (1997) also demonstrated the simultaneous accumulation of Pb, Cd, Cu, Ni, and Zn in Indian mustard plants after application of EDTA to soil, showing that induced phytoextraction is applicable to other metals in addition to Pb. The efficiency of metal accumulation in these experiments was directly related to the affinity of the applied chelate for the metal, suggesting that for efficient phytoextraction, synthetic chelates having a high affinity for the target metal should be used. For example, EDTA for Pb, EGTA for Cd (BLAYLOCK et al., 1997), citrate for uranium (HUANG et al., 1998), and ammonium thiocyanate for gold (ANDERSON et al., 1998).

Based on this information, a hypothetical protocol for the treatment of a contaminated site by induced phytoextraction can be outlined:

(1) the site is evaluated and the appropriate chelate and crop combination are determined based upon target metal, climate, and soil type,

(2) the site is prepared, planted, and the crop is cultivated,

(3) once the crop has reached optimal biomass the appropriate chelate is applied, and

(4) after a short metal accumulation phase of several days or weeks, the crop is harvested.

Depending on the crop and the season, the site could be replanted for further phytoextraction. Following harvest, the weight and volume of contaminated material can be reduced by air drying, composting or ashing. Metal-enriched plant residue can then be disposed of as hazardous waste or, if economically feasible, used for metal recovery.

In order to optimize this process it is important to understand the chemical and biological mechanisms involved. Induced phytoextraction can be divided into two basic processes: release of bound metals from the solid soil matrix into the soil solution and transport of these solubilized metals from the soil solution into the plant. The role of chelates in increasing the solubility of metals in the soil can be explained using well established chemical equilibrium principles. However, the biological mechanisms involved in "induction" of plant metal uptake by chelates are not clear though recent experiments are starting to shed some light on the process.

Following EDTA application, Pb accumulation in shoots is directly correlated with accumulation of EDTA (Vassil et al., 1998; Epstein et al., 1999), suggesting that Pb is transported within the plant as a Pb–EDTA complex. This was recently confirmed after Pb, complexed to EDTA, was directly observed in the xylem sap of Pb–EDTA exposed Indian mustard (Vassil et al., 1998). Therefore, solubilization of Pb as a Pb–EDTA complex appears to allow Pb to be freely transported from roots to shoots, where it accumulates as Pb–EDTA.

2.2.1 Field Applications of Chelate-Assisted Phytoextraction

At present, induced phytoextraction is being developed commercially by Phytotech Inc., Monmouth Junction, NJ, USA, for the remediation of Pb and U contaminated sites. Phytotech has performed a number of induced phytoextraction field trials during the last 4 years. Recent trials at brownfield sites at Trenton and Bayonne, NJ, have both been reported as being successful. The majority of the treated areas of these sites had Pb levels reduced to below industrial standards after one summer of treatment (Watanabe, 1997). In the summer of 1997, the US EPA (Environmental Protection Agency) coordinated the Phytotech trial at Trenton, NJ, and they reported that after the application of EDTA the plants in the field had in excess of 800 mg kg^{-1} Pb in shoots on a dry weight basis (Anonymous, 1997). This trial was continued in 1998 and both Phytotech and the EPA are awaiting the results to establish if phytoextraction at this site can bring the Pb concentrations to below 600 mg kg^{-1} (NJ EPA limits for soil Pb concentrations in industrial zones). Of major concern to both regulators and potential users of this technology is the possibility for chelate-induced leaching of metals from the soil and into the groundwater. In order to establish the likelihood of this, Phytotech have been performing extensive leaching studies during the field trial outlined above. The data from these studies are very important in establishing the validity of this approach and the research community awaits the results.

2.3 Continuous Phytoextraction

Continuous phytoextraction involves the application of specialized plants which have the physiological capacity to accumulate metals during their normal growth cycle.

The optimal plants for this type of process should be capable of solubilizing metals in the soil, efficiently transporting them to the shoot where the toxic effects of the metals are tolerated, and where they accumulate to high concentrations. At present there are limited numbers of plants, collectively termed hyperaccumulators, which grow on soils enriched in various metals including Zn, Ni, and Se. These plants have the capacity to naturally accumulate metals to between 0.1 and 3% of their shoot dry biomass (BAKER and BROOKS, 1989; BAKER et al., 1999; BROWN and SHRIFT, 1982). This unique ability of hyperaccumulators to accumulate high foliar metal concentrations makes these plants suitable as a starting point for the development of phytoremediation crops for continuous phytoextraction (CHANEY, 1983; BAKER et al., 1988). For continuous phytoextraction to be a practical technology plants will need to be developed which are able to grow on metal-polluted soils to a high biomass, and accumulate and resist high concentrations of metal in their shoots. Unfortunately, at present there are no known strong hyperaccumulators of pollutant metals of interest such as Pb, Cd, As, and U. Also, most hyperaccumulator plants have a relatively low biomass and growth rates, though there are some exceptions to this (ROBINSON et al., 1997). Therefore, in order to develop a versatile range of plants for use in phytoextraction applications it will be necessary to understand in detail the molecular mechanisms involved in metal hyperaccumulation.

The question of metal bioavailability in the soil is of key importance to these developments. Will it be possible to develop plants with the ability to "solubilize" pollutant metals such as Pb from the soil? If this is not possible, then continuous phytoextraction will be limited to the removal of pollutant metals which are relatively soluble, such as Cd and Zn. One possible solution to this problem would be the development of a hybrid system in which certain amendments would be applied continuously to the soil in order to mobilize the metals of interest, (e.g., phosphate to mobilize arsenate). Specialized plants would then accumulate the "solubilized" metal continuously over their full growth period. However, to achieve this level of biological sophistication in a "designer plant" it is clear that we will have to develop the capacity to manipulate the genes involved in metal uptake, translocation, and resistance in plants. At present, we are a long way from this goal, though important strides are now being taken in the investigation of these processes (for review, see SALT and KRÄMER, 1999).

A prerequisite for metal hyperaccumulation must be the ability to efficiently resist high concentrations of metals within plant tissues and cells (KRÄMER et al., 1997a). Recently, a number of interesting discoveries have started to shed some light on the mechanism(s) involved in metal resistance in various hyperaccumulator species.

The tissue, cellular, and subcellular localization of metals in hyperaccumulator species is critical to understanding the mechanism(s) of metal resistance in these plants. Scanning proton microscopy and energy-dispersive X-ray microanalysis have recently been used to localize the metal in the Ni hyperaccumulators *Senecio coronatus* (MESJASZ-PRZYBYLOWICZ et al., 1994), *Alyssum lesbiacum* (KRÄMER et al., 1997b), *Thalspi montanum* var. *siskiyouense* (HEATH et al., 1997), and the Zn hyperaccumulator *T. caerulescens* (VÁZQUEZ et al., 1992, 1994). The highest Ni concentrations were found in the unicellular stellate trichomes covering the leaf surface in *A. lesbiacum*, and in the subsidiary cells that surround guard cells in *T. montanum* var. *siskiyouense*.

Using energy dispersive X-ray microanalysis (EDAX), VÁZQUEZ et al. (1992, 1994) reported high Zn concentrations in the vacuoles of epidermal and subepidermal leaf cells of *T. caerulescens* and some Zn in the leaf apoplast. By directly isolating intact protoplasts and vacuoles, KRÄMER and SALT (unpublished data) have recently found that 75% of the intracellular Ni in leaves of the Ni hyperaccumulator *T. goesingense* is localized within the vacuole. In the same studies it was observed that *T. goesingense* was able to compartmentalize Ni into the vacuole 2-fold more efficiently than the non-accumulator *T. arvense*. The efficient compartmentalization of metals in the shoots of hyperaccumulators appears to be contrasted by reduced vacuolar compartmentalization in the roots. Based on efflux analysis it has recently been suggested that roots of the non-accumulator *T. arvense* more efficiently accumulate Zn in the vacuole than the hyperaccumulator *T. caerulescens* (LASAT et al., 1998). Instead of transporting Zn into vacuoles in the root cells, *T. caerulescens* appears to more efficiently export the Zn out of the root to the shoot. This reflects the two different strategies of the hyperaccumulator and non-accumulator plants: the hyperaccumulator exports Zn to the shoots; whereas the non-accumulator sequesters Zn in the root (BAKER, 1981).

Chelation of metal ions by specific high affinity ligands reduces the solution concentration of the free metal ions thereby reducing their phytotoxicity. Researchers have, therefore, been engaged in an effort to identify any high-affinity, low-molecular weight metal chelators which may confer metal tolerance (REEVES, 1992), and possibly metal specificity, to hyperaccumulator plants (STILL and WILLIAMS, 1980; PANCARO et al., 1978). Concentrations of free histidine in the xylem sap of several Ni hyperaccumulators in the genus *Alyssum* have recently been shown to increase 30-fold when the plants are exposed to Ni. Based on this and other evidence it has been proposed that free histidine acts as a metal chelator involved in Ni tolerance and translocation in these Ni hyperaccumulator species (KRÄMER et al., 1996). However, free histidine concentrations in the xylem sap of the Zn hyperaccumulator *T. caerulescens* did not increase on exposure to Zn (LASAT et al., 1998). Histidine concentrations also do not appear to change in the xylem sap, root or shoot tissues of the Ni hyperaccumulator *T. goesingense* (KRÄMER and SALT, unpublished results). Nickel also does not appear to regulate the expression of ATP phophoribosyl transferase (GenBank accession AF003347), imidazol glycerol phosphate dehydratase or histidinol dehydrogenase (PERSANS et al., 1998), key genes involved in histidine biosynthesis in *T. goesingense* (PERSANS and SALT, unpublished results).

In an attempt to verify some of the *in vitro* observations on metal speciation in plants, X-ray absorbance spectroscopy (XAS) has recently been used (SALT et al., 1995, 1997, 1999; KRÄMER et al., 1996; DE SOUZA et al., 1998; LYTLE et al., 1998; PILON-SMITS et al., 1999). Prior to analysis at approximately 10 K, tissues are simply frozen and ground to a fine powder in liquid nitrogen, avoiding many of the problems associated with the extraction and purification of metal complexes. Though not necessary for analysis, grinding of the tissue before

analysis is preferable to obtain consistent signals. SALT et al. (1999a) have used this technique to determine the ligand environment of Zn in different tissues of the Zn hyperaccumulator *T. caerulescens* (J. & C. Presl.). Approximately 70% of intracellular Zn in roots of *T. caerulescens* was found to be coordinated with histidine. In the xylem sap, 79% of the Zn was found to be transported as the free hydrated Zn^{2+} cation with the remaining proportion coordinated with citrate. In the shoots, Zn coordination occurred mainly via citrate, with a smaller proportion present as the hydrated cation, coordinated to oxalate, histidine, and the cell wall. This supports the idea that histidine plays an important role in Zn homeostasis in the roots, whereas organic acids are involved in xylem transport and Zn storage in shoots. This technique has also been used to demonstrate that a proportion of the Ni in roots, xylem sap and shoots of the Ni hyperaccumulator *A. lesbiacum* is also coordinated with histidine (KRÄMER et al., 1996). The use of XAS has also conclusively established that phytochelatins (RAUSER, 1995) are not involved in Ni binding in *A. lesbiacum* (KRÄMER et al., 1996), and *T. goesingense* (KRÄMER and SALT, unpublished data) or Zn binding in *T. caerulescens* (SALT et al., 1999a).

2.3.1 Field Applications of Continuous Phytoextraction

There are a number of examples of continuous phytoextraction field trials, and the results are promising and informative. BROWN et al. (1995) performed a field study during the 1991 and 1992 growing seasons on the use of hyperaccumulator and crop plant species for the continuous phytoextraction of Zn and Cd from sewage sludge amended plots at the Central Maryland Research and Education Facility in Beltsville, MD, USA (BROWN et al., 1995). The Zn hyperaccumulator *T. caerulescens* grown at this field site accumulated approximately 4,000 mg kg^{-1} Zn in its shoots, generally 10-fold higher than in the non-accumulator plants tested. Based on their data the authors tentatively concluded that it would take 18 growing seasons to reduce the soil Zn concentration from 400 mg kg^{-1} to background levels of 40 mg kg^{-1}.

In 1991 a similar field trial was performed on sewage sludge amended plots at Rothamsted, UK. This also revealed that *T. caerulescens* effectively accumulates Zn under field conditions (BAKER et al., 1994). From this study the authors predicted that 13 croppings of the Zn hyperaccumulator *T. caerulescens* would be required to bring the Zn loading in the soil to below the limit of the Commission of the European Communities (CEC), a reduction from 444 mg kg^{-1} to below 300 mg kg^{-1}. This was compared to 800 cropping of oilseed rape. The ability of *T. caerulescens* to remove Zn and Cd from contaminated soils in Switzerland was also tested during the 1993–1996 growing seasons (FELIX, 1997). Here the author concluded that it would take approximately 54 crops of *T. caerulescens* to decrease the Zn concentration from 810 mg kg^{-1} to the Swiss target of 200 mg kg^{-1}.

It is clear from these field trials that plants identified as hyperaccumulators in their natural habitat maintain that ability during remediation-type field trials. However, the time required for effective remediation of Zn polluted soils was predicted to range from 13–54 years, making phytoextraction with hyperaccumulator plants impractical in most cases. This limitation appears to be directly related to the low biomass yield of the hyperaccumulator plants used in these studies. For example, *T. caerulescens* was found to have a biomass yield of 8–16 t ha^{-1} over a single growing season (BROWN et al., 1995; FELIX, 1997), compared to 21–24 t ha^{-1} for *Brassica juncea* (Indian mustard), a non-accumulating crop plant grown at the same site. However, recent biomass production studies have shown that yields of certain hyperaccumulators may be substantially higher. Trials conducted on experimental field plots amended with crushed serpentine rock at Massey University, NZ, showed that yields of 22 t ha^{-1} are possible with the recently discovered serpentine endemic Ni hyperaccumulator *Berkheya coddii* (MORREY et al., 1989; ROBINSON et al., 1997). Based on these results it was predicted that four successive croppings of *B. coddii* would be sufficient to reduce a soil Ni concentration of 250 mg kg^{-1} to below the European Union guidelines of 75 mg kg^{-1}. This

suggests that at least for Ni continuous phytoextraction with certain wild hyperaccumulator plants may be a feasible remediation strategy.

To establish the potential for using plants to phytoextract B and Se from soil, an extensive field trial was carried out over two growing seasons during 1990 and 1991. In this trial, four rapidly growing, high biomass plants including *B. juncea*, *Festuca arundinacea* (tall fescue), *Lotus corniculatus* (birdsfoot trefoil) and *Hibiscus cannibinus* (kenaf) were tested for their ability to accumulate B and Se from soil in central California naturally enriched in these elements (BAÑUELOS et al., 1993). All plant species were found to cause significant reductions in both B and Se concentrations in the soil. Mass balance calculations revealed that between 57 and 92% of the removed B was accumulated in the plants. However, only 10% of the Se removed from the soil was found in the plants. The authors suggested that the loss of Se could be accounted for by Se volatilization by the plant or by soil microorganisms associated with the plants. Of the four species tested, *B. juncea* contained the highest concentrations of Se. However, *H. cannibinus* accumulated the highest total amount of Se due to its larger total biomass (BAÑUELOS et al., 1993). These trials proved that, in principle, continuous phytoextraction of Se and B is feasible. Work is now in progress to optimize this process (BAÑUELOS et al., 1997).

None of these field trials to date have provided a clean-up. However, they all contributed to a better understanding of the potential and limitations of the continuous phytoextraction process. Since these early field trials, several research groups have been investigating ways to overcome some of the limitation they highlighted. Application of modern techniques in plant physiology, biochemistry, and molecular biology should yield large advances over the next 5–10 years, with the expectation that continuous phytoextraction of polluted soils will become a reality in the not-too-distant future.

3 Rhizofiltration

Adsorption or precipitation are the processes of choice for the removal of pollutant metals from water. A number of different methods have been applied to achieve this, including the use of ion exchange resins, and various living and non-living biological systems including plant cell cultures (JACKSON et al., 1990), bacteria and algae (SUMMERS, 1992), fungi (BLACKWELL et al., 1995), and various aquatic plants including water hyacinth (*Eichhornia crassipes*) (TURNQUIST et al., 1990), pennywort (*Hydrocotyle umbellata*) (DIERBERG et al., 1987), duckweed (*Lemna minor*) (MO et al., l989), and water velvet (*Azolla pinnata*) (JAIN et al., 1989). Recently, the roots of hydroponically cultivated terrestrial plants were also found to be effective at removing metals from water (DUSHENKOV et al., 1995), and this approach was termed rhizofiltration. It was suggested that using this approach should be more effective than other plant-based systems because of the high growth rates and large surface area of the root system developed by terrestrial plants (RASKIN et al., 1997). An ideal plant for rhizofiltration should, therefore, be able to rapidly produce large amounts of fine root biomass, which have a high capacity to accumulate pollutant metals from solution. For example, sunflower roots can accumulate Pb up to 140 mg g^{-1} dry weight of roots (DUSHENKOV et al., 1995), and hydroponically grown sunflowers are capable of producing as much as 1.5 kg root dry biomass m^{-2} month^{-1} (RASKIN et al., 1997). It has also been proposed that rhizofiltration is potentially more effective than conventional ion exchange systems because of the role of high affinity, biologically active transport systems in the accumulation of metals by plant roots. However, the cost of cultivating plants and the large footprint of a hypothetical rhizofiltration system are significant drawbacks which may make it difficult for rhizofiltration to compete with more conventional ion exchange technologies.

In an effort to overcome some of these drawbacks it was recently discovered that young seedlings of certain species grown in aerated water (aquacultured) are effective at removing metals from water (SALT et al., 1997). Seeds germinated and grown in this way rapidly generate a large seedling biomass which is capable of accumulating various metals including Cd, Pb, Sr, Ni, and Cr. Because these seedlings require no nutrients or light for their growth and development there is no requirement for greenhouse facilities. Seedlings, therefore, potentially provide a more economic and efficient alternative to live plant roots for water treatment applications.

In order to assess the competitiveness of each of these biological and non-biological matrices in water treatment it is important to compare their effectiveness under controlled conditions. To quantify this performance we have recently compared the initial rates of metal removal from solution and the affinity of various matrices for 7 different metals (Tabs. 1, 2). In these studies the ability of *Brassica juncea* (Indian mustard) seedlings, sunflower roots, peat moss and ion exchange resins to remove Ni, Cd, Pb, Cr(VI), U, Sr and Cs from water were compared.

From this comparison it is clear that all the biological matrices out-perform the ion exchange resins when the initial rates of metal ion removal from solution are compared. The only exception to this is for the anionic metal ion Cr(VI), present as dichromate. Biological material has only a limited capacity to remove this anion. We hypothesize that the higher initial rates of metal removal in the biologically-derived material are due to their larger, more porous active surface for metal ion adsorption. Another measure of effectiveness is the affinity of the matrix for a particular metal. This is defined as the concentration of metal on the matrix divided by the concentration of metal remaining in solution at equilibrium. In simple terms, this factor is an indicator of the "strength" with which a particular matrix can attract an ion from solution. Using this measure, ion exchange resins out-perform all the biological matrices for removal of Ni, Cd, Pb, and Cr(VI). The exception to this are sunflower roots which appear to be superior to the ion exchange resin for U accumulation. Sr and Cs were removed by all matrices equally.

Based upon the above criteria it would appear that in general biological materials would be expected to out-perform ion exchange resins in treatment systems with a high throughput of

Tab. 1. Initial Rates of Metal Ion Removal from an Artificial Groundwater Using Various Matrices

Matrix	Ni	Cd	Pb	Cr(VI)	U	Sr	Cs
	Initial rate of metal removal [μg g^{-1} dry biomass h^{-1}]						
Live seedlings	73	46	127	2	194	20	11
Dried seedlings	2.5	101	75	6	177	20	13
Live sunflower roots	16.5	186	214	16	179	47	27
Dried sunflower roots	71	126	170	2	198	5	11
Peat moss	172	75	157	13	131	20	11
Ion exchange resin[a]	30	37	22	31	77	18	12

[a] The ion exchange resin IRC-718 was used for all experiments except for Cr(VI) where IRA-400 was used.

Brassica juncea seeds were germinated in 50 mL of standard artificial groundwater containing 0.7 mM Ca(NO$_3$)$_2$, 0.5 mM MgSO$_4$, 1.5 mM KNO$_3$, and 1 μM Na$_2$HPO$_4$. Seeds were allowed to germinate and grow in the dark for 7 d. Solutions were aerated by shaking on an orbital shaker. Sunflower roots were harvested from 3 weeks old sunflowers hydroponically grown in the greenhouse. Dead *B. juncea* seedlings and sunflower roots were dried at 52 °C overnight. Equal quantities (40 mg final dry weight) of the various matrices under investigation were added to a flask containing 50 mL of standard artificial groundwater, spiked with 200 μg L^{-1} of the particular metal being investigated. The flasks were shaken on an orbital shaker, and water samples taken at time 0 and 1, 2, 4, 8, 24 and 48 h. Three replicates for each treatment were performed. Water samples were analyzed for their metal content by ICP, graphite furnace, γ-scintillation counting or β-scintillation counting depending on the metal or radioactive isotope being measured.

Tab. 2. Affinity of Various Matrices for Metal Ion Accumulation from an Artificial Groundwater

Matrix	Ni	Cd	Pb	Cr(VI)	U	Sr	Cs
	Affinity [metal on matrix/metal in solution][a]						
Live seedlings	2,393	1,016	8,831	220	6,513	1,205	
Dried seedlings	117	5,525	8,058	140	37,852	910	10
Live sunflower roots	2,313	1,874	12,059	542	27,450	1,232	16
Dried sunflower roots	855	2,288	17,287	357	5,124	588	13
Peat moss	6,425	465	16,392	413	4,933	1,078	20
Ion exchange resin	139,263	89,840	42,533	2,405	10,411	1,188	5

[a] The affinity of each matrix was calculated by dividing the concentration of metal on the matrix (estimated by acid digestion of the matrix, and calculation of the metal concentration assuming a density of 1) by the concentration of metal in solution at equilibrium (at 48 h). All experimental conditions were as described in Tab. 1.

water which contains low concentrations of metal contaminants. However, because of the higher affinities ion exchange resins would be expected to be able to remove metals to a much lower concentration than the biological materials tested. Depending on the target metal and the discharge standards this property may have a significant impact on the decision to use either ion exchange resins or biological material in a treatment system. Interestingly, for U removal live sunflower roots and dried seedlings out-performed ion exchange resin both in initial rates of U removal and in their affinity for U accumulation (Fig. 1). Based upon these findings Phytotech Inc. have recently completed a successful rhizofiltration field trial at Ashtabula, OH. In this trial Phytotech Inc. were able to demonstrate the large-scale removal of U from groundwater using a pump-and-treat rhizofiltration system (DU-SHENKOV et al., 1997). These types of experiments should provide valuable data allowing the accurate assessment of both the efficiency and economic feasibility of large-scale rhizofiltration systems.

In order to improve the performance of biological matrices for rhizofiltration applications it is clear that their affinity for metals will need to be improved. One approach to this is to use modern molecular biological techniques to engineer plants with the capacity to express high-affinity, metal-binding peptides on their roots (RASKIN, 1996). These genetically engineered roots would then have the capacity to bind, with high affinity, selected pollutant metals of interest. A number of potential strategies exist for the identification of such peptides. For example, high-affinity, metal-binding peptides could be identified by screening recombinatorial phage display libraries.

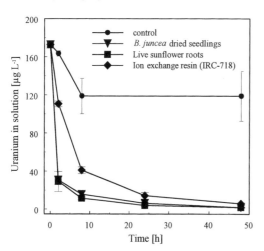

Fig. 1. Comparison of the uranium removing abilities of various matrices. The ability to remove uranium from an artificial ground by either dried *B. juncea* seedlings, intact live sunflower roots or an ion exchange resin (IRC-718) were compared. Equal quantities of each matrix (40 mg final dry weight) were added to water containing 0.7 mM $Ca(NO_3)_2$, 0.5 mM $MgSO_4$, 1.5 mM KNO_3, 1 μM Na_2HPO_4 which had been spiked with uranium. Each point represents the mean of 3 duplicate experiments (±SE).

Recruiting microorganims to assist in the process of metal accumulation by plant roots may also be a possible avenue in which rhizofiltration could by improved. In nature plant roots are coated with microorganisms adapted to live in the microenvironment of the root surface. Recently, it was demonstrated that these natural microbial populations can be manipulated to improve the metal-accumulating abilities of plant roots (SALT et al., 1999b). Screening of 300 different rhizobacterial isolates identified several which, when inoculated onto roots of Indian mustard, significantly enhanced the total amount of Cd removed from solution. Further investigations revealed that this enhancement was due to an overall increase in root biomass in the rhizobacterial treated plants. Rhizobacteria were found to accumulate Cd from solution, and ultrastructural observations suggested that rhizobacteria promote the precipitation of Cd on the root surface. By precipitating Cd at the root surface, rhizobacteria reduce the amount of Cd taken up into roots, thereby protecting the plants, and in particular the roots, from the toxic effects of Cd. This reduced Cd toxicity allows for the increased proliferation of roots observed when plants are inoculated with certain rhizobacteria. This proof of principle suggests that it should also be possible to inoculate plant roots with rhizobacteria which have been genetically engineered to have special properties, e.g., the ability to immobilize certain metals by ligand binding, precipitation, oxidation or reduction, etc. This approach may hold great promise for the future.

The major advantage biological systems have over chemically synthesized resins is that biological systems are able to take advantage of the incredible diversity inherent in the genetic code, allowing the isolation and utilization of complex peptide structures for the selective binding of target metals. Identification and synthesis of similar molecules by conventional chemical means would be very difficult, if not impossible. Because of this advantage, it should be possible in the future to design a biological matrix with the capacity to selectively bind target metals with very high rates and affinities. These developments could make rhizo-filtration a very effective technology for water treatment in the future.

Acknowledgements

We would like to extend our thanks to DE-LORESSA GLEBA and ILYA RASKIN for kindly providing the data presented in Tabs. 1 and 2, and Fig. 1.

4 References

ANDERSON, C. W. N., BROOKS, R. R., STEWART, R. B., SIMCOCK, R. (1998), Harvesting a crop of gold in plants, *Nature* **395**, 553–554.

Anonymous (1997), Phyto-cleanup of lead progressing at NJ site, *Environ. Sci. Technol.* **31**, 496A–497A.

BAKER, A. J. M. (1981), Accumulators and excluders-strategies in the response of plants to heavy metals, *J. Plant Nutr.* **3**, 643–654.

BAKER, A. J. M., BROOKS, R. R. (1989), Terrestrial higher plants which hyperaccumulate metallic elements – a review of their distribution, ecology and phytochemistry, *Biorecovery* **1**, 81–126.

BAKER, A. J. M., BROOKS, R., REEVES, R. (1988), Growing for gold ... and copper ... and zinc, *New Scientist* **10**, 44–48.

BAKER, A. J. M., MCGRATH, S. P., SIDOLI, C. M. D., REEVES, R. D. (1994), The possibility of *in situ* heavy metal decontamination of polluted soils using crops of metal-accumulating plants, *Res. Conserv. Recycl.* **11**, 41–49.

BAKER, A. J. M., MCGRATH, S. P., REEVES, R. D., SMITH, J. A. C. (1999), Metal hyperaccumulator plants: a review of the biological resource for possible exploitation in the phytoremediation of metal-polluted soils, in: *Phytoremediation of Contaminated Soil and Water* (TERRY, N., BAÑUELOS, G. S., Eds.), Chapter 5. Boca Raton, FL: CRC Press.

BAÑUELOS, G. S., CARDON, G., MACKEY, B., BEN-ASHER. J., WU, L. et al. (1993), Plant and environment interactions – Boron and selenium removal in boron-laden soils by four sprinkler irrigated plant species, *J. Environ. Qual.* **22**, 786–792.

BAÑUELOS, G. S., AJWA, H. A., TERRY, N., ZAYED, A. (1997), Phytoremediation of selenium laden soils: A new technology, *J. Soil Water Cons.* **52**, 426–430.

BAUMANN, A. (1885), Das Verhalten von Zinksalzen gegen Pflanzen und im Boden, *Landwirtschaftl. Versuchsstation* **31**, 1–53.

BLACKWELL, K. J., SINGLETON, I., TOBIN, J. M. (1995), Metal cation uptake by yeast: a review, *Appl. Microbiol. Biotechnol.* **43**, 579–584.

BLAYLOCK, M. J., SALT, D. E., DUSHENKOV, S., ZAK-HAROVA, O., GUSSMAN, C. et al. (1997), Enhanced accumulation of Pb in Indian mustard by soil-applied chelating agents, *Environ. Sci. Technol.* **31**, 860–865.

BRADSHAW, A. D. (1952), Populations of *Agrostis tenuis* resistant to lead and zinc poisoning, *Nature* **169**, 1089.

BROWN, T. A., SHRIFT, A. (1982), Selenium: Toxicity and tolerance in higher plants, *Biol. Rev.* **57**, 59–84.

BROWN, S. L., CHANEY, R. L., ANGLE, J. S., BAKER, A. J. M. (1995), Zinc and cadmium uptake by hyperaccumulator *Thlaspi caerulescens* and metal tolerant *Silene vulgaris* grown on sludge-amended soils, *Environ. Sci. Technol.* **29**, 1581–1585.

BYERS, H. G. (1935), Selenium occurrence in certain soils in the United States, with a discussion of related topics, *US Dept. Agr. Tech. Bull.* **482**, 1–47.

BYERS, H. G. (1936), Selenium occurrence in certain soils in the United States, with a discussion of related topics. Second report, *US Dept. Agr. Tech. Bull.* **530**, 1–78.

CHANEY, R. L. (1983), Plant uptake of inorganic waste, in: *Land Treatment of Hazardous Wastes* (PARR, J. E., MARSH, P. B., KLA, J. M., Eds.), pp. 50–76. Park Ridge: Noyes Data Corp.

DE SOUZA, M. P., PILON-SMITS, E. A. H., LYTLE, C. M., HWANG, S., TAI, J. et al. (1998), Rate-limiting steps in selenium assimilation and volatilization by Indian Mustard, *Plant Physiol.* **117**, 1487–1494.

DIERBURG, F. E., DEBUSK, T. A., GOULET, N. A. JR. (1987), Removal of copper and lead using a thin-film technique, in: *Aquatic Plants for Water Treatment and Resource Recovery* (REDDY, K. R., SMITH, W. H., Eds.), pp. 497–504. Orlando, FL: Magnolia Publishing Inc.

DUSHENKOV, V., KUMAR, P. B. A. N., MOTTO, H., RASKIN, I. (1995), Rhizofiltration – the use of plants to remove heavy metals from aqueous streams, *Environ. Sci. Technol.* **29**, 1239–1245.

DUSHENKOV, S., VASUDEV, D., KAPULNIL, Y., GLEBA, D., FLEISHER, D. et al. (1997), Removal of uranium from water using terrestrial plants, *Environ. Sci. Technol.* **31**, 3468–3474.

EPSTEIN, A. L., GUSSMAN, C. D., BLAYLOCK, M. J., YERMIYAHU, U., HUANG, J. et al. (1999), EDTA and Pb-EDTA accumulation in *Brassica juncea* grown in Pb contaminated soils, *Plant Soil.* **208**, 87–94.

FELIX, H. (1997), Field trials for *in situ* decontamination of heavy metal polluted soils using crops of metal-accumulating plants, *Z. Pflanzenernaehr. Bodenk.* **160**, 525–529.

HEATH, S. M., SOUTHWORTH, D., D'ALLURA, J. A. (1997), Localization of nickel in epidermal subsidiary cells of leaves of *Thlaspi montanum* var. *siskiyouense* (Brassicaceae) using energy-dispersive X-ray microanalysis, *Int. J. Plant Sci.* **158**, 184–188.

HUANG, J. W., CUNNINGHAM, S. D. (1996), Lead phytoextraction: species variation in lead uptake and translocation, *New Phytol.* **134**, 75–84.

HUANG, J. W., CHEN, J., BERTI, W. B., CUNNINGHAM, S. D. (1997), Phytoremediation of lead-contaminated soils: Role of synthetic chelates in lead phytoextraction, *Environ. Sci. Technol.* **31**, 800–805.

HUANG, J. W., BLAYLOCK, M. J., KAPULNIK, Y. (1998), Phytoremediation of uranium-contaminated soils: role of organic acids in triggering uranium hyperaccumulation by plants, *Environ. Sci. Technol.* **32**, 2004–2008.

JACKSON, P. J., TORRES, A. P., DELHAIZE, E., PACK, E., BOLENDER, S. L. (1990), The removal of barium ion from solution using *Datura innoxia* Mill. Suspension culture cells, *J. Environ. Qual.* **19**, 644–648.

JAIN, S. K., VASUDEVAN, P., JHA, N. K. (1989), *Azolla pinnata* R. Br. and *Lemna minor* L. for removal of lead and zinc from polluted water, *Biol. Wastes* **28**, 115–126.

JØRGENSEN, S. E. (1993), Removal of heavy metals from compost and soil by ecotechnological methods, *Ecol. Eng.* **2**, 89–100.

KRÄMER, U., COTTER-HOWELLS, J. D., CHARNOCK, J. M., BAKER, A. J. M., SMITH, J. A. C. (1996), Free histidine as a metal chelator in plants that accumulate nickel, *Nature* **379**, 635–638.

KRÄMER, U., SMITH, R. D., WENZEL, W., RASKIN, I., SALT, D. E. (1997a), The role of nickel transport and tolerance in nickel hyperaccumulation by *Thlaspi goesingense* Halácsy, *Plant Physiol.* **115**, 1641–1650.

KRÄMER, U., GRIME, G. W., SMITH, J. A. C., HAWES, C. R., BAKER, A. J. M. (1997b), Micro-PIXE as a technique for studying nickel localization in leaves of the hyperaccumulator plant *Alyssum lesbiacum*, *Nucl. Instr. Methods Physiol. Res.* **B130**, 346–350.

LASAT, M. M., BAKER, A. J. M., KOCHIAN, L.V. (1998), Altered Zn compartmentation in the root symplast and stimulated Zn absorption into the leaf as mechanisms involved in Zn hyperaccumulation in *Thlaspi caerulescens*, *Plant Physiol.* **118**, 875–883.

LYTLE, C. M., LYTLE, F. W., YANG, N., QIAN, J.-H., HANSEN, D. et al. (1998), Reduction of Cr(VI) to Cr(III) by wetland plants: Potential for *in situ* heavy metal detoxification, *Environ. Sci. Technol.* **32**, 3087–3093.

MESJASZ-PRZYBYLOWICZ, J., BALKWILL, K., PRZYBYLOWICZ, W. J., ANNEGARN, H. J. (1994), Proton microprobe and X-ray fluorescence investigations of nickel distribution in serpentine flora from South Africa, *Nucl. Instr. Methods Physiol. Res.* **B89**, 208–212.

MINGUZZI, C., VERGNANO, O. (1948), Il contenuto di nichel nelle ceneri di *Alyssum bertolonii* Desv., *Atti Soc. Toscana Sci. Naturali Mem. Ser.* **A55**, 49–77.

MO, S. C., CHOI, D. S., ROBINSON, J. W. (1989), Uptake of mercury from aqueous solution by duckweed: the effect of pH, copper and humic acid, *J. Environ. Sci. Health* **A24**, 135–146.

MORREY, D. R., BALKWILL, K., BALKWILL, M.-J. (1989), Studies on serpentine flora: Preliminary analysis of soils and vegetation associated with serpentinite rock formations in the south-eastern Transvaal, *S. Afr. J. Bot.* **55**, 171–177.

PANCARO, L., PELOSI, P., VERGNANO GAMBI, O., GALOPPINI, C. (1978), Ulteriori indagini sul rapporto tra nichel e acidi malico e malonico in *Alyssum*, *Giorn. Bot. Ital.* **112**, 141–146.

PERSANS, M., YAN, X., SMITH, R., SALT, D. E. (1998), Cloning of two cDNAs from the Ni hyperaccumulator *Thlaspi goesingense*: histidinol dehydrogenase (Accession No. AF023141) and imidazolglycerol-phosphate dehydratase (Accession No. AF023140), two enzymes in the histidine biosynthetic pathway, *Plant Physiol. Plant Gene Reg.* **117**, 332.

PILON-SMITS, E. A. H., HWANG, S., LYTLE, C. M., ZHU, Y., TAI, J. C. et al. (1999), Overexpression of ATP sulfurylase in Indian mustard leads to increased selenate uptake, reduction, and tolerance, *Plant Physiol.* **119**, 123–132.

PRAT, S. (1934), Die Erblichkeit der Resistenz gegen Kupfer, *Ber. Dtsch. Bot. Ges.* **102**, 65.

RASKIN, I. (1996), Plant genetic engineering may help with environmental cleanup, *Proc. Natl. Acad. Sci. USA* **93**, 3164–3166.

RASKIN, I., SMITH, R. D., SALT, D. E. (1997), Phytoremediation of metals: using plants to remove pollutants from the environment, *Curr. Opin. Biotechnol.* **8**, 221–226.

RAUSER, W. E. (1995), Phytochelatins and related peptides. Structure, biosynthesis, and function, *Plant Physiol.* **109**, 1141–1149.

REEVES, R. D. (1992), The hyperaccumulation of nickel by serpentine plants, in: *The Vegetation of Ultramafic (Serpentine) Soils* (BAKER, A. J. M., PROCTOR, J., REEVES, R. D., Eds.), pp. 253–277, Andover, Hampshire, UK: Intercept.

ROBINSON, B. H., BROOKS, R. R., HOWES, A. W., KIRKMAN, J. H., GREGG, P. E. H. (1997), The potential of the high-biomass nickel hyperaccumulator *Berkheya coddii* for phytoremediation and phytomining, *J. Geochem. Explor.* **60**, 115–126.

SALT, D. E., KRÄMER, U. (1999), Mechanisms of metal hyperaccumulation in plants, in: *Phytoremediation of Toxic Metals: Using Plants to Clean-Up the Environment* (ENSLEY, B. D., RASKIN, I., Eds.),

New York: John Wiley & Sons.

SALT, D. E., PRINCE, R. C., PICKERING, I. J., RASKIN, I. (1995), Mechanisms of cadmium mobility and accumulation in Indian mustard, *Plant Physiol.* **109**, 1427–1433.

SALT, D. E., PICKERING, I. J., PRINCE, R. C., GLEBA, D., SMITH, R. D., RASKIN, I. (1997), Metal accumulation by aquacultured seedlings of Indian mustard, *Environ. Sci. Technol.* **31**, 1636–1644.

SALT, D. E., SMITH, R., RASKIN, I. (1998), Phytoremediation, *Ann. Rev. Plant Physiol. Plant Mol. Biol.* **49**, 643–668.

SALT, D. E., PRINCE, R. C., BAKER, A. J. M., RASKIN, I., PICKERING, I. J. (1999a), Zinc ligands in the metal hyperaccumulator *Thlaspi caerulescens* as determined using X-ray absorption spectroscopy, *Environ. Sci. Technol.* **33**, 713–717.

SALT, D. E., BENHAMOU, N., LESZCZYNIECKA, M., RASKIN, I., CHET, I. (1999b), A possible role for rhizobacteria in water treatment by plant roots, *Int. J. Phytoremediation* **1**, 67–79.

STILL, E. R., WILLIAMS, R. J. P. (1980), Potential methods for selective accumulation of nickel(II) ions by plants, *J. Inorg. Biochem.* **13**, 35–40.

SUMMERS, A. O. (1992), The hard stuff: metals in bioremediation, *Curr. Opin. Biotechnol.* **3**, 271–276.

THURMAN, D. A. (1981), Mechanism of metal tolerance in higher plants, in: *Effect of Heavy Metal Pollution On Plants* (LEPP, N. W., Ed.), pp. 239–249. Englewood, NJ: Applied Science Publishers.

TURNQUIST, T. D., URIG, B. M., HARDY, J. K. J. (1990), Nickel uptake by the water hyacinth, *J. Environ. Sci. Health* [A] **A25**, 897–912.

VASSIL, A. D., KAPULNIK, Y., RASKIN, I., SALT, D. E. (1998), The role of EDTA in lead transport and accumulation by Indian mustard, *Plant Physiol.* **117**, 447–453.

VÁZQUEZ, M. D., BARCELÓ, J., POSCHENRIEDER, C. et al. (1992), Localization of zinc and cadmium in *Thlaspi caerulescens* (Brassicaceae), a metallophyte that can hyperaccumulate both metals, *J. Plant Physiol.* **140**, 350–355.

VÁZQUEZ, M. D., POSCHENRIEDER, C., BARCELÓ, J., BAKER, A. J. M., HATTON, P., COPE, G. H. (1994), Compartmention of zinc in roots and leaves of the zinc hyperaccumulator *Thlaspi caerulescens*, J. & C. Presl., *Bot. Acta* **107**, 243–250.

WATANABE, M. E. (1997), Phytoremediation on the brink of commercialization, *Environ. Sci. Technol.* **31**, 182A–186A.

18 Advanced *in situ* Bioremediation – A Hierarchy of Technology Choices

RONALD UNTERMAN
MARY F. DEFLAUN
ROBERT J. STEFFAN
Lawrenceville, NJ 08648, USA

List of Abbreviations

BOD	biological oxygen demand
COD	chemical oxygen demand
CPT	cone penetrometer
DCE	dichloroethylene
DO	dissolved oxygen
FOC	fraction organic carbon
ITRD	innovative treatment remediation demonstration
ORP	oxidation–reduction potential
PCE	tetrachloroethylene
PHB	poly-β-hydroxy butyric acid
PLFA	phospholipid fatty acid
SRNL	Savanna River National Laboratory
T2MO	toluene 2-monooxygenase
TCE	trichloroethylene
TMO	toluene 4-monooxygenase
TOC	total organic carbon
TSCA	Toxic Substance Control Act
USEPA	US Environmental Protection Agency
VC	vinyl chloride
VOC	volatile organic compounds

1 Introduction

The widespread use of xenobiotic compounds and a long history of poor disposal practices or accidental spillage have led to pollution of the natural environment and a need for remedial technologies. In many cases biological approaches for remediating polluted environments provide significant advantages over alternative technologies. Most notably, biological treatment technologies can facilitate the complete destruction of hazardous chemicals without generating toxic emissions or by-products. *In situ* treatment systems provide the added advantage of destroying the contaminant in place, thereby further reducing environmental and human exposures. Ultimately, the cost of remediation is also reduced because contaminated media do not have to be excavated, relocated, or otherwise disposed, thereby resulting in lower capital and operating expenditures. Within the field of bioremediation, a hierarchy of *in situ* treatment approaches has emerged to address the varying complexities of contaminant types and distribution, site-specific concerns (e.g., cost and time constraints), and regulatory factors which ultimately dictate how remediation is performed.

The hierarchical approach to bioremediation may best be illustrated by considering the biological treatment technologies that have emerged to remediate volatile organic compounds, in particular chlorinated compounds (VOC), in aquifers. These VOC include a variety of short-chain aliphatic halogenated hydrocarbons including the solvents tetrachloroethylene (PCE), trichloroethylene (TCE), and dichloroethylene (DCE) which are among the most common groundwater pollutants in the industrialized world. These compounds are chemically stable and persistent in the environment, but numerous bacteria have been found to degrade them via either oxidative (typically cometabolic degradation) or reductive mechanisms. Because the organisms that anaerobically degrade VOC (via reductive dehalogenation) are indigenous to most aquifers, they are the focus of the first choice in the treatment hierarchy, natural attenuation. Organisms that cometabolically oxidize the VOC are typically the focus of the second and third levels of the hierarchy: biostimulation and bioaugmentation, respectively. Anaerobic biostimulation and bioaugmentation technologies, however, are also available and suitable in some cases. Each of these mechanisms and examples of their application are discussed in this review.

2 Natural Attenuation

The least invasive bioremediation approach is the identification of intrinsic bioremediation whereby it can be demonstrated that an indigenous microbial population exists under suitable environmental conditions and that degradation has already occurred and is continuing. This is a logical first choice for remediation because it requires no intervention besides monitoring of the natural progress of biodegradation. The potential for this technology to reduce the economic burden of environmental cleanup is so substantial that the US Environmental Protection Agency's (USEPA) Office of Research and Development and Office of Solid Waste and Emergency Response drafted a policy directive specifically for monitored natural attenuation (USEPA, 1999). In addition, the USEPA has also issued a Technical Protocol for Evaluating Natural Attenuation of Chlorinated Solvents in Groundwater (WIEDEMEIER et al., 1998). The USEPA policy directive defines monitored natural attenuation as "the reliance of natural attenuation processes (within the context of a carefully controlled and monitored site cleanup approach) to achieve site-specific remediation objectives within a time frame that is reasonable compared to that offered by more active methods" (USEPA, 1999).

Natural attenuation processes are often the dominant factors controlling the fate and transport of contaminants. They result from the integration of several subsurface attenuation mechanisms that are classified as either destructive or non-destructive (USEPA, 1999). The non-destructive mechanisms include sorption, dispersion, dilution, and volatilization which are directly influenced by the hydrogeological characteristics of the aquifer and the chemical properties of the contaminants. A fundamental understanding of these processes is required in order to assess natural attenuation. Biodegradation is the most important destructive mechanism involved in natural attenuation, and may represent a permanent solution to soil and groundwater contamination, thus constituting "remediation" for a particular site. Demonstration and quantification of *in situ* biodegradation depend upon the interaction between the environment, the contaminants of interest and indigenous microorganisms. The typical biodegradation pathways for common VOC are well understood (BOUWER, 1994; VOGEL, 1994; LEE et al., 1998), and the occurrence of biodegradation intermediates or end products are indicators that the process is occurring at a given site. For example, the common dry cleaning agent, PCE, is reductively dehalogenated to TCE, DCE (typically *cis*-1,2-DCE), vinyl chloride (VC), and ultimately ethene and ethane (MAJOR et al., 1991; DE BRUIN et al., 1992; KOENE-COTAAR and SCHRAA, 1998). In particular, the predominance of *cis*-DCE, which is a much lower percent component of the manufactured solvent than *trans*-DCE, and the occurrence of the gases, VC, ethene, and ethane which cannot be introduced to the subsurface, are demonstrative indicators of natural attenuation.

The natural attenuation approach is conducted in stages including:

(1) preliminary evaluation
(2) site investigation
(3) receptor analysis
(4) modeling and
(5) documentation

Experience has shown that it is imperative to involve all interested parties (e.g., responsible parties or site owners, regulators, oversight consultants, etc.) as early as possible in the process of evaluating natural attenuation at a particular site. Such an approach imparts ownership in the process and technology and prevents concerns about deception or intent to avoid remediation costs at a risk to potential receptors.

2.1 Preliminary Evaluation

Preliminary site evaluations involve either reviewing existing data for a site, or collecting new data as needed. The goals of the investigation include:

(1) assess the extent and distribution of contamination
(2) identify indications that natural attenuation is occurring at a site
(3) collect hydrogeological data to help assess the risk of contaminant movement
(4) identify factors that may limit natural attenuation and
(5) evaluate other remediation alternatives

For sites that have not already been investigated and do not have existing monitoring wells, plume delineation and site hydrology are most economically investigated with direct push technologies (GeoProbe or Cone Penetrometer) wherever practical. Direct push technologies can be used to delineate both the vertical and horizontal extent of the plume, and provide information on lithology, contaminant concentrations, and groundwater chemistry. The small spatial resolution of these techniques allows for a much more accurate three-dimensional conceptual model of the plume and its constituents than can be obtained by typical monitoring wells. The results of a Geo-Probe or Cone Penetrometer (CPT) investigation allow the accurate placement of monitoring wells in source areas and along the flow-path of the plume.

In most cases, existing site data can be used in a preliminary assessment of the eligibility of a site for a monitored natural attenuation remediation. Examples of the evidence that is used to justify further investigation of a natural attenuation remedy include:

- decreasing concentration of contaminants with time
- stable or shrinking plume
- occurrence of daughter products of degradation and
- absence of potential receptors

The data required for evaluating a natural attenuation remedy are also very valuable for evaluating engineered remedies for the site, including other *in situ* biological remedies such as biostimulation and bioaugmentation. Furthermore, monitored natural attenuation may be applicable to an entire site or only a part of a site and removal of the contaminant source or free product contamination may be required. Therefore, the preliminary evaluation is important to delineate the extent of the source area, and to determine if source area remedies are compatible with the monitored natural attenuation remedy applied to the rest of the plume. For example, if the monitored natural attenuation remedy relies on maintaining anaerobic conditions in the aquifer, it would be unwise to implement a source remedy that could oxygenate the aquifer.

2.2 Investigation of Natural Attenuation

In general, the site-specific case for natural attenuation should be built along three lines of evidence:

2.2.1 Stable or Decreasing Concentrations and/or Plume Size

Temporal and spatial trends in groundwater contaminant concentrations can be used to demonstrate natural attenuation. Historical trends in the concentrations of target compounds found at a site are useful for calculating degradation rates and for approximating mass removal. Contaminant concentrations measured over time can be statistically analyzed and determined to be increasing, decreasing, or stable with time.

A stable or shrinking plume is further evidence of destruction of contaminants. Changes in the size and shape of the contaminant plume with time can also be used to demonstrate degradation. A comparison of the distance of the plume from the presumed source compared to the distance that would be expected based upon groundwater flow rates (advection) and other physical parameters (dispersion, dilution, and sorption), can indicate that degradation is occurring. Both decreasing or stable groundwater concentrations and stable or decreasing plume size are an indication that natural attenuation is occurring.

2.2.2 Biogeochemical Conditions

Biological and geochemical evidence can be used to develop a strong case for degradative activity. Biogeochemical characterization includes the analysis of parameters that document subsurface conditions, and the presence of degradation products. Geochemical factors, e.g., pH, oxidation–reduction potential (ORP), dissolved oxygen (DO) concentration, and temperature define subsurface environmental conditions that affect the extent of *in situ* biodegradation processes, and thus correlate with the observed biodegradative activity. Collectively, these parameters indicate degradation potential, and offer signs of past and current biological or chemical transformations. Biochemical parameters include the concentration of compounds required for microbial activity, and the concentration of metabolic intermediates (daughter products). Specifically, the concentration of potential substrates necessary to sustain bacterial activity (electron donors) and electron acceptors (e.g., O_2, NO_3^-, Fe^{+2}, SO_4^{-2}) are important when evaluating biodegradation potential. The presence and concentration of daughter products are indicative of the extent of degradation that has occurred in the past. For example, biodegradation can be suggested in historical data by observing decreasing concentrations of the compound in the absence of a continuing source, and by the presence of metabolic breakdown products.

2.2.3 Presence of Microorganisms

Microbiological laboratory assays and laboratory microcosm studies can be used to detect the presence of a viable microbial community capable of degrading the target compounds under natural conditions. A laboratory microcosm study involves collecting groundwater and sediment from the contaminated portion of the aquifer, preserving the site *in situ* geochemical conditions, in particular dissolved oxygen and redox potential, and incubating these in bottles so that the concentrations of the contaminants of concern and daughter products can be monitored over time. Microcosm studies can provide direct evidence that microbial degradation is occurring at the site and provide rates of degradation. However, microcosm studies are expensive and time-consuming (typically a minimum of 24 weeks is required) and can provide equivocal results since it can be difficult to mimic natural field conditions in the laboratory. If degradation is observed, the degradation rates are in most cases higher than what is occurring at the site (RIFAI et al., 1995). Degradation rates obtained from field data are more predictive than rates obtained in microcosm studies and are, therefore, the preferred method for obtaining degradation rates.

Technologies now exist that permit a direct measurement of relevant microbial populations in soil and groundwater by phospholipid fatty acid (PLFA) analysis (SMITH et al., 1986). PLFA analysis is a simple, yet useful bacterial enumeration and differentiation technique. This analysis provides an estimate of total bacterial concentrations and also identifies relevant microbial groups (e.g., sulfate reducers) for both aerobic and anaerobic biotransformations. Therefore, the types of organisms that can potentially degrade the contaminants can be identified by a straightforward test directly on site samples, satisfying the requirements of the third line of evidence.

2.3 Sampling Protocol for Monitored Natural Attenuation

Sampling for monitored natural attenuation should be performed at a variety of locations at the site including:

- upgradient of the source area
- in the source area
- at several locations downgradient within the plume and
- downgradient outside of the plume

It is imperative that sampling is performed in a manner that preserves the *in situ* geochemical conditions in the aquifer. In general this means sampling by a method that prevents aeration of the groundwater. This involves:

- low flow sampling to prevent drawdown in the well,
- pumps that prevent contact of the groundwater with air,
- enclosed sampling device containing probes (flow cell) for the collection of field parameters, and
- accurate calibration of field instruments

The USEPA has protocols for both low flow purging and sampling procedures (USEPA Region I: http://www.epa.gov/region01/).

Parameters typically used to evaluate natural attenuation are listed in Tab. 1. Not all of these parameters are necessary for all sites. The parameters evaluated at any particular site are selected based on the contaminants present and the availability of aquifer sediment.

Tab. 1. Parameters Used to Assess Natural Attenuation

Groundwater Parameters

Field Parameters	Inorganics	Organics	Gases	Microbiological Parameters
Dissolved oxygen	ammonia/TKN	VOC (*cis* and *trans*)	methane	PLFA (phospholipid fatty acid analysis)
Redox potential	major cations (filtered)	semi VOC	ethane	total heterotrophic plate counts
Conductivity	major anions (filtered)	TOC	ethene	contaminant-specific plate counts
Temperature pH Turbidity Salinity	sulfate, sulfide nitrate, nitrite *o*-phosphate manganese (total and dissolved, field) iron (total and dissolved, field) alkalinity (carbonate and bicarbonate)	COD/BOD hydrocarbons	CO_2	

Sediment Parameters	Hydrogeological Parameters
TOC/FOC	subsurface and surficial geology including lithology, stratigraphy and structure
Humic/fulvic acids	hydraulic conductivity and gradient (to calculate groundwater velocity and direction)
Grain size analysis Porosity PLFA (phospholipid fatty acid analysis) Total heterotrophic and contaminant-specific plate counts	

2.4 Receptor Analysis

Receptor analysis is a very important component of evaluating any site for natural attenuation because it determines whether there is a risk of impact on human health or the environment. This analysis defines any potential receptors and the distance from the plume to the receptor. An understanding of the hydrogeology of the contaminated aquifer is essential to this evaluation. Aquifers with fracture flow or complex hydrogeology are particularly difficult to evaluate based upon the uncertainty associated with the rate and direction of plumes in these systems.

Modeling is used to determine whether or not the contaminants can potentially impact the receptor and estimate the time it will take for this impact to occur. Risk to human health or the environment can preclude the use of a monitored natural attenuation remediation for a site. However, removing potential receptors, such as drinking water wells, is one strategy to eliminate this risk.

2.5 Modeling Natural Attenuation

Two classes of mathematical models (screening and advanced) can be used to demonstrate that natural attenuation is a viable remedial option. Simple analytical screening models are primarily designed to determine the feasibility of using natural attenuation as part of a remedial strategy. At smaller sites with apparently limited impacts, it is often appropriate to use a screening model as the primary groundwater model to simulate natural attenuation, and predict the extent and duration of contaminant migration. One such model is BIOSCREEN which has been developed and endorsed by the US Air Force.

At more complicated sites, with complex hydrogeology or multiple contaminant source areas, the use of a more advanced numerical groundwater model to simulate natural attenuation and predict the extent and duration of contaminant migration may be warranted. Examples of advanced numerical models include: RT3D, BIOMOD 3-D, BioF&T 3-D, and BioRedox. RT3D, BIOMOD 3-D, BioF&T 3-D, and BioRedox are the most advanced three-dimensional numerical models and are used in conjunction with the USGS finite difference groundwater flow model, MODFLOW 3-D. These models are capable of simulating groundwater flow and contaminant transport in the saturated and unsaturated zones in complex porous media or fractured media. RT3D, BIOMOD 3-D, BioF&T 3-D, and BioRedox can be used at sites with the most complex hydrogeology (e.g., interbedded sands and clay, fractured bedrock, and multiple aquifers) and complex contaminant distribution (e.g., multiple source areas and non-aqueous phase contamination), and are applicable to most contaminants (e.g., petroleum hydrocarbons, chlorinated solvents, explosives, and heavy metals). Each of these models can simulate complex microbial processes based on oxygen limited, anaerobic, first order, or Monod type biodegradation, as well as first order sequential degradation involving multiple daughter species. BioRedox can also simulate spatially varying oxidation–reduction zones that are indicative of many natural attenuation sites.

2.6 Documentation of Natural Attenuation

The presentation of a monitored natural attenuation strategy should include an explanation of the physical, geochemical, and biological factors that contribute to protection of human health and the environment both at the present time and into the future until the site is remediated. The three lines of evidence outlined above, including the mechanisms of biodegradation occurring at the site, combined with a receptor analysis and modeling to provide the timeframe for the remediation, include all the site-specific information required to demonstrate the efficacy of a monitored natural attenuation. Of particular value are illustrations of plume size and contaminant concentrations with time, as well as projections of plume size and concentrations in the future.

The concerns of the regulators and questions that a natural attenuation remedy should answer include:

- Will natural attenuation remediate the contaminants at this site in a manner that is protective of human health and the environment?

- What are the contingency plans in the case that natural attenuation does not work? What would be the impact of remedy failure? A comprehensive natural attenuation remediation plan will include one or more monitoring results that would serve as active remediation triggers.
- What is the future of the groundwater plume in the context of the site boundaries and potential receptors?
- What is the time frame for the natural attenuation remedy compared to an active system?
- What is the cost of the natural attenuation remedy compared to an active remediation system?
- Is the long-term monitoring plan sufficient to be protective of human health and the environment?

Of all the *in situ* bioremediation technologies, monitored natural attenuation requires the most complete understanding of the biogeochemistry of the environment for successful implementation. There is no direct way of proving that natural attenuation is remediating a site, therefore, several lines of evidence are used to demonstrate that this process is occurring in an efficient and protective manner. If this remedy is not sufficiently protective, however, more highly engineered solutions may be required. In addition, depending on the required documentation and the length of time needed to monitor this process, natural attenuation may or may not be the most cost-effective remedy. Evaluating the relative costs of the various *in situ* strategies should be one of the first steps in the remediation process.

3 Biostimulation

The second choice in the bioremediation hierarchy is the stimulation of the degradative activity of indigenous populations to remediate the target chemicals. Biostimulation, as this process is called, is applied in cases where a degradative population exists within the contaminated zone, but where nutrient or other conditions are insufficient for microbial activity. Oxygen is often the limiting substrate, and introducing oxygen either as hydrogen peroxide, magnesium peroxide, pure oxygen, or air, can in many cases induce natural microbial populations to degrade the target chemicals. The application of bioventing and biosparging approaches for introducing atmospheric oxygen to the subsurface is generally a cost-effective approach for remediation. Biostimulation often can be accelerated by nutrient addition or physical processes such as:

(1) adding vapor phase carbon (cosubstrate), nitrogen, or phosphorus
(2) heating the subsurface to volatilize free product at high temperatures, and/or where lower ambient temperatures are limiting, to help stimulate bacterial activity
(3) utilizing hydro- or pneumatic fracturing to increase permeability and zones of influence and/or
(4) applying soil vapor extraction, air sparging, or multiphase extraction to remove free product

The addition of appropriate nutrients or cometabolites to the contaminated subsurface can stimulate degradation by indigenous microorganisms, and the list of chemicals for which this technology is suitable is increasing. For example, biostimulation has been used extensively for many years as an *in situ* treatment for hydrocarbon contamination (WILSON et al., 1986), and more recently, biostimulation protocols have been developed for more recalcitrant contaminants such as chlorinated solvents and gasoline oxygenates (Envirogen, Inc., unpublished data). Furthermore, biostimulation can be applied under aerobic conditions (HOPKINS et al., 1993; ROBERTS et al., 1990; SEMPRINI and MCCARTY, 1991), or it can be performed under anaerobic conditions (BEEMAN et al., 1994; DEBRUIN et al., 1992; GIBSON and SEWELL 1992; SEWELL et al., 1998), depending on the target contaminants and existing conditions.

In early examples of biostimulation for remediating VOC contaminated aquifers, methane and air were injected into the subsurface to stimulate the activity of methanotrophic bacteria. In a well-publicized field demonstration of the technology, researchers at the Savanna River National Laboratory (SRNL) installed horizontal wells below and above a TCE plume (LOMBARD et al., 1994). Methane and air were injected into the lower well and a vacuum was applied to the upper well to recover fugitive methane and TCE stripped from the groundwater. The greatest rates of TCE degradation were achieved when methane was pulsed into the aquifer (HAZEN et al., 1994), presumably because pulsing allowed methane depletion, thereby minimizing competitive inhibition of TCE degradation. Total numbers of methanotrophic microorganisms remained relatively low throughout the study, but increased significantly after the addition of gaseous nitrogen and phosphorus (BROCKMAN et al., 1995). Estimates of TCE degradation at the site indicated that approximately 700 kg of chlorinated solvents were degraded during the study, and cost estimates have predicted that the technology can be applied at a cost of approximately $ 16 per pound of TCE removed (TERRY HAZEN, SRNL, personal communication).

In another series of studies, continuous injection of methane or propane in the presence of oxygen was shown to lead to excessive microbial growth at the injection site, increasing the potential for clogging and reducing the formation of a uniform biomass distribution. SEMPRINI and MCCARTY (1991) showed that this limitation could be overcome by pulsed addition of methane and oxygen into the saturated zone. Since methane utilization requires the simultaneous presence of methane and oxygen, bacterial growth can only occur at the interface between the injected electron acceptor and donor where they are brought together through mixing. The longer the pulse cycle, the further methane and oxygen could travel before mixing and consumption, resulting in a more uniform biomass distribution (ROBERTS et al., 1990). Indigenous microorganisms stimulated by methane or propane (SEMPRINI and MCCARTY; 1992, SEMPRINI et al., 1994) had the capacity to cometabolically transform VC, *trans*-DCE, *cis*-DCE, and TCE. In both cases, oxidation of the chlorinated sol-

vents was closely coincident with the concentration oscillations of the injected cosubstrate and followed a mathematical model simulation using competitive inhibition kinetics and rate-limited sorption. During laboratory studies by DOLAN and MCCARTY (1995), the substrate pulse duration also affected VC degradation rates. If short pulses (6 h) were used, VC degradation was slow, presumably due to competitive inhibition by residual methane. If 12 h pulses were used, high rates of VC degradation were achieved.

In a more recent field study of biostimulation, MCCARTY et al. (1998) injected oxygen and toluene into a TCE contaminated aquifer at Edwards Air Force Base to stimulate the growth and activity of toluene oxidizing bacteria that can cometabolically degrade VOC. The site contained two TCE plumes separated by a layer of clay. A pumping system was installed to circulate water and stimulants between and through the two formations. The system was optimized by installing injection and recovery wells nearly perpendicular to regional groundwater flow, thereby allowing capture and treatment of an 80 m wide portion of the aquifer, and created an *in situ* mixing and biological treatment zone. The system operated efficiently for more than 1 year, resulting in a TCE concentration reduction from 1,000 to <50 µg L^{-1}. A similar system design is currently being utilized by Envirogen, Inc. to inject propane for stimulating the growth and activity of TCE degrading propane oxidizing bacteria (Lakehurst Naval Air Station, NJ, unpublished data). In this case, however, there is only a single plume, and water is being pumped above ground between injection and recovery wells to create the *in situ* biologically active treatment zone.

Anaerobic biostimulation, particularly for chlorinated compounds, is also gaining widespread acceptance as an effective remedial technology. A pilot scale demonstration of electron donor injection to stimulate *in situ* bioremediation of chlorinated solvents was performed at a US Department of Energy Site for the Innovative Treatment Remediation Demonstration (ITRD) program. The addition of benzoate, lactate, and methanol to an aquifer, where there was some evidence that natural attenuation of the VOC was occurring, was found to greatly enhance the rate of degradative activity. All of the contaminants of concern at the site (VC, *cis*-DCE, methylene chloride, TCE) were significantly reduced in concentration during the 5 months test (SEWELL et al., 1998). A similar anaerobic biostimulation pilot test for remediating chlorinated solvents is currently being performed by Envirogen, Inc. at a landfill site. The addition of benzoate at this site is effectively stimulating the degradation of *cis*-DCE and vinyl chloride to ethene and ethane (LIZOTTE et al., 1999).

Clearly, other biostimulation remediation approaches have been applied, and they range from relatively simple systems where oxygen is the only required additive, to the more complex cometabolic systems described above that require balancing the types and amounts of multiple additives applied to stimulate specific and specialized microbial populations. In either case, the robustness of natural microbial consortia can result in a stable and efficient treatment system. The relatively low cost of biostimulation systems and their proven track record for treating petroleum hydrocarbons and other contaminants make biostimulation the treatment of choice at many sites where some type of active treatment is required.

4 Bioaugmentation

The third choice in the treatment hierarchy is bioaugmentation. Bioaugmentation involves adding exogenous organisms to contaminated environments to promote the degradation of target contaminants. The technique can be suitable where natural attenuation or biostimulation are inappropriate or do not work. For example, although naturally occurring organisms can degrade most contaminants, biostimulation can be ineffective where

(1) the contaminant is a cometabolic target (i.e., not a suitable microbial growth substrate) and the necessary nutrients or inducers are not present and cannot be added to the environment or

(2) competent degradative organisms are not present among the indigenous population.

In these cases, bioaugmentation may be an effective treatment strategy. There are also cases where although the natural microbial population may be sufficient to achieve a biostimulation remediation, it may be desirable to choose a bioaugmentation approach to increase the rate of degradation and thereby shorten the time frame for full scale remediation. Unfortunately, bioaugmentation is a technology that has historically been oversold by vendors who claim to have a proprietary organism or mixture of organisms that can degrade any chemical known to mankind. Such misuse of the technology has created barriers to the rational application of bioaugmentation, despite its suitability for particular sites or contaminants.

Two distinct approaches have been developed in the area of bioaugmentation for pollution remediation. In the first approach, degradative organisms are added to complement the existing native microbial population. In effect, the added organisms act as a seed culture that will utilize the contaminant of concern as a growth substrate and maintain or grow its population while cleaning up the contaminated site. The added microorganisms can be selected for long-term survival and the ability to occupy a selective niche within the contaminated environment, and stimulants or selective cosubstrates can be added to aid survival. Thus, the goal of this approach is to achieve prolonged survival and growth of the added organisms and the concomitant prolonged degradation of the target pollutants. This approach can only be effective where the contaminant is a suitable growth substrate, or where a selective growth substrate can be added to support the survival of the added organisms. The technique, however, is plagued by millions of years of evolution that dictates that native organisms generally will be better suited for the *in situ* environments and outcompete the added organisms.

In the second basic bioaugmentation approach, large numbers of degradative bacteria are added to a contaminated environment as catalysts that will degrade a significant amount of the target contaminant before becoming inactive or perishing. Attempts can be made to increase the expression of the degradative enzymes, or to maximize catalytic efficiency or stability of the enzyme or organism, but the long-term survival, growth, and establishment of the biocatalyst are not the primary goals of the treatment approach. Additional microbial inocula can be added as needed to further the degradation process. This approach is driven by the cost of culturing a sufficient mass of degradative organisms, and the ability to effectively distribute the organisms throughout the area of contamination (DEFLAUN and STEFFAN, 1999). As with biostimulation, bioaugmentation can be combined with other technologies to improve its effectiveness.

In the case of contaminants in the subsurface, bacteria can be added directly to the vadose zone or aquifer in an attempt to target the contaminant plume in place, or they can be immobilized on solid substrates and then introduced into the subsurface to form a permeable barrier or "biocurtain" that intercepts the migrating contaminant plume. Similarly, degradative bacteria can be applied in a cylindrical reactor and placed in a recirculating well to treat groundwater *in situ* in an adaptation of recirculating well technology designed to strip volatiles out of groundwater. Although these barrier remedial approaches can be very effective containment strategies, they are in essence a subsurface pump and treat bioremediation. Therefore, they are constrained by the same problem observed for above ground pump and treat remediation. Specifically, they require long operational times to effect desorption, transport, and destruction of the target chemicals.

Among the VOC, TCE has become the model target compound for bioaugmentation because of its ubiquity at contaminated sites. Although many organisms that degrade TCE have been identified (ENSLEY, 1991), no known organism is capable of using TCE as a sole source of carbon and energy. Biological attack is cometabolic, meaning degradation is facilitated by an enzyme that has evolved to react with a different substrate, but fortuitously degrades the TCE. Thus, biological destruction of TCE requires the presence of a cosubstrate that acts as both a food source and an inducer to maintain enzymatic activity. Unfortunately, the cosubstrate/inducer is generally also a toxic or hazardous compound (e.g., toluene or phenol), and, therefore, several

factors need to be considered when selecting or developing organisms for use as biocatalysts to destroy VOC. For example, most known VOC degrading oxygenases are inducible by a primary substrate (e.g., methane, toluene, or phenol). Thus, organisms with constitutive activity (no inducer requirement) may be advantageous because they limit the need to add additional substrates. Likewise, because most natural microorganisms are highly adhesive, their transport *in situ* is limited. By selecting adhesion-deficient organisms, distribution of the biocatalyst and contact with the target contaminants can be improved. Several organisms that can potentially be used as biocatalysts for *in situ* destruction of VOC are shown in Tab. 2. The use and application of these organisms is discussed below.

4.1 Constitutive Variants of Degradative Microorganisms

In situ bioremediation of TCE (and related VOC) has been limited by the inability of naturally occurring TCE degrading microorganisms to degrade TCE in the absence of inducing cosubstrates. All known naturally occurring TCE degrading microorganisms require the presence of inducing chemicals to stimulate TCE biodegradation (ENSLEY, 1991). However, bacteria capable of constitutive expression of TCE degrading genes/enzymes have been developed. For example, constitutive toluene 4-monooxygenase (TMO) variants of *Pseudomonas mendocina* KR1 were developed by first treating the cells with mutagenizing agents, then selecting variants that constitutively produced TMO. The variants were extremely stable during >125 generations of growth (STEFFAN and TUGUSHEVA, 1993). SHIELDS and REAGIN (1992) first developed constitutive toluene 2-monooxygenase (T2MO) mutants of *Burkholderia cepacia* G4, but those mutants were modified by transposon mutagenesis, and they may be classified as "genetically engineered". Such a classification may require their approval in the USA under Toxic Substance Control Act (TSCA) regulations before environmental release. These researchers have since developed a constitutive

Tab. 2. Bacterial Strains for *in situ* Remediation of TCE Contaminated Aquifers

TCE Degrading Strain	Key Characteristics	Reference
B. cepacia G4	wild-type; induced by toluene or phenol; TCE degradation for approximately 30 h in the absence of induction; rapid growth to $OD_{550} > 65$; PHB production to >60% dwt; >90% adhesion	NELSON et al., 1986
B. cepacia PR1$_{301c}$	G4 derivative; constitutive TCE degradation; non-genetically engineered; growth and PHB production as for G4	MUNAKATA-MARR et al., 1996
B. cepacia ENVBF1	wild-type; low level constitutive TCE degradation; 3–5 times greater fully-induced TCE degradation activity than G4; very rapid high density growth and PHB production	McCLAY et al., 1996
B. cepacia ENV425	adhesion-deficient derivative of PR1$_{301c}$, <35% adheson	Envirogen Inc., STEFFAN et al., 1999
P. mendocina KR1	wild-type; TCE degradation activity induced by toluene; TCE degradation at low cell density	RICHARDSON and GIBSON 1984
P. mendocina ENV 1018–1027	genetically engineered constitutive TCE degrading variants of KR1	Envirogen Inc., unpublished data
Pseudomonas sp. ENVPC5	wild-type, TCE degradation induced by TCE	McCLAY et al., 1995
M. trichosporium OB3b	methanotroph, broad substrate specificity soluble methane monooxygenase	OLDENHUIS et al., 1989
M. vaccea JOB5	broad specificity propane monooxygenase	WACKETT et al., 1989

T2MO variant of *B. cepacia* G4 without the use of recombinant DNA techniques (MUNAKATA-MARR et al., 1996). When properly grown, a *B. cepacia* constitutive strain (PR1$_{301c}$) will continue to degrade TCE 5 times longer than wild-type *B. cepacia* G4 (STEFFAN et al., unpublished results). *B. cepacia* G4 degrades TCE for only about 30 h after which time enzyme activity ceases. Because the constitutive variants continue to express T2MO in the absence of inducing substrates, the cells continue to degrade TCE as long as the cells have sufficient energy reserves to catalyze the reaction.

4.2 Adhesion-Deficient Strains

One of the major difficulties when using bioaugmentation for remediation of subsur-

face contamination is delivering degradative microorganisms to the zone of contamination. Bacteria often adhere to solid surfaces and plug porous materials rather than pass through them (COSTERTON, 1992). Thus, *in situ* bioaugmentation technologies are currently limited by poor migration of bacteria through aquifer solids. In an *in situ* field study involving *B. cepacia* G4 injection into a TCE plume (NELSON et al., 1990), the data strongly suggested that movement of the microorganisms was severely retarded by the aquifer material. The hydraulic flow of the aquifer was approximately 15 m d^{-1}, however, none of the injected microorganisms were observed in a monitoring well 3 m away from the injection well until 6 d after injection. Non-retarded organisms would have appeared in a recovery well at about the same time after injection as the tracer, that is, within 8–10 h. Within 8 h of injecting the strain into the aquifer TCE concentrations were reduced, and they remained low for the following 10 d. Over the course of the study TCE concentrations were reduced from a high of 3,000 µg L^{-1} to a mean of 78 µg L^{-1} during a 20 d period. This field pilot demonstrated that *B. cepacia* could sustain TCE degrading activity under *in situ* conditions, and reduce a TCE contaminant plume in a very short time. Clearly, more aquifer material could be treated if the organisms or the aquifer conditions were altered to reduce adsorption and retardation of bacterial movement.

Previous studies have demonstrated the feasibility of generating adhesion-deficient variants of a number of different bacterial strains by transposon mutagenesis, chemical mutagenesis, or by selection of naturally occurring adhesion-deficient variants from a population (BAYER et al., 1983; ANDERSON, et al., 1988; GONG and FORSBERG, 1989; CREWS et al., 1990). 91% of a highly adhesive strain of *Pseudomonas fluorescens* (Pf0-1) were consistently retained by a 3 cm tall, 12 g sand column (DEFLAUN et al., 1990). With this degree of attenuation *in situ* dispersion would be limited to approximately 0.2 m in an unconsolidated sand aquifer. However, an adhesion-deficient variant (Pf0-5) with 40% retention in the same sand column would be capable of dispersing approximately 1 m in a sandy aquifer.

By using this simple adhesion assay, both transposon generated mutants and natural variants of a number of TCE degrading strains have been selected which are adhesion-deficient in comparison to the wild-type strains, and include: *B. cepacia* G4, *Methylosinus trichosporium* OB3b, *B. cepacia* PR1$_{301c}$, *P. mendocina* KR1, and *B. cepacia* ENVBF1. Some of the variants adhere to the sand columns at <10%, whereas the wild-type strains adhere at >90% (DEFLAUN et al., 1994). Model aquifer studies have demonstrated that some of these strains migrate through a variety of different aquifer sediments at a rate comparable to that of a chloride tracer (DEFLAUN et al., 1999).

When an adhesion-deficient variant of *B. cepacia* PR1$_{301}$, strain ENV435, was added to a silty sand aquifer, the organism was transported more than 12 m *in situ* at a rate comparable to that of a bromide tracer. The ratio of brom-

ide velocity to ENV435 velocity was 1.3, and the strain moved at a maximum rate of 0.54 m d^{-1} (STEFFAN et al., 1999). By injecting the same organism at monitoring wells distributed throughout the test plot, the organism was rapidly distributed across the site and facilitated the rapid degradation of VOC. During a related full scale remediation project, the strain was injected into several monitoring wells in glacial till soils at a TCE contaminated site and it was recovered more than 30 m downgradient at approximately the same time as a bromide tracer (Envirogen, unpublished data). Again the strain degraded a significant amount of TCE, thereby aiding in obtaining regulatory closure of the site.

4.3 Enhancing Energy Reserves in Degradative Organisms

In addition to problems of gene/enzyme induction and adhesion, microbial biocatalysts for cometabolic targets usually require the addition of nutrients to stimulate and prolong their activity. However, adding a cosubstrate to an aquifer can stimulate the growth of other non-TCE degrading organisms and create problems with biofouling and oxygen depletion. Adding methane to an aquifer can be difficult and supports the growth of non-degrading organisms, and the addition of toluene or phenol may reduce water quality. Furthermore, the use of cosubstrates adds to the cost of the treatment process and may be operationally inefficient if the microbes are transported or migrate away from the cosubstrate.

An alternative approach for enhancing and maintaining biological activity *in situ* is to utilize biocatalysts that are enriched in energy reserves. The production of energy storage polymers, most commonly poly-β-hydroxybutyric acid (PHB), by bacteria is a long studied phenomenon (LEMOIGNE, 1926). PHB is produced by many bacteria, usually under conditions of a nutrient limitation in the presence of excess carbon, and may account for up to 80% of the bacterial cell dry weight. Utilization of PHB as a source of reducing substrate for increasing the TCE transformation rate and capacity has been demonstrated (CHU and ALVAREZ-CO-

HEN, 1998, HENRY and GRBIĆ-GALIĆ, 1991; HENRYSSON and McCARTY, 1993).

Fermentation protocols have been developed for several TCE degrading organisms (*B. cepacia* G4, PR1$_{301c}$, and ENV BF1) that allow growth in 5 d to cell densities of greater than 20 g L^{-1} dry weight (O$_{550}$ = 65), with cells containing as much as 60% dry weight of PHB (e.g., STEFFAN et al., 1999). TCE degradation studies with these energy enriched constitutive variants have demonstrated that under certain conditions the cells can continuously degrade TCE for > 168 h. Non-energy enriched cells degraded TCE for only about 48 h (STEFFAN et al., unpublished results). The exact utility of the storage polymers during bioaugmentation in the field, however, is difficult to measure and is still uncertain (STEFFAN et al., 1999).

5 Summary

The selection of an appropriate remedial technology for a given site is controlled by a number of factors including:

(1) cost
(2) local and national regulations
(3) characteristics and distribution of contaminants
(4) site hydrogeology
(5) risk to sensitive receptors and
(6) the sensitivity of the surrounding community to given technologies

Clearly, no single remediation technology is suitable for all sites given the complexity of the factors involved. Consequently, a number of potential remedial technologies have emerged, and a treatment hierarchy has evolved. We have discussed only 3 levels of this hierarchy, but at actual sites the hierarchy expands rapidly as numerous *in situ* and *ex situ* chemical, physical, and biological approaches are considered within the umbrella of the above-listed controlling factors. In some cases a single remedial approach is sufficient, but more often multiple treatment technologies are needed to meet the needs of the particular site and to provide the most efficient and cost-effective

site restoration. More field experience with the biotreatment technologies described herein will help to better define their limitations as well as to broaden their range of applications.

6 References

ANDERSON, A. J., HABIBZADEGAH-TARI, P., TEPPER, C. S. (1988), Molecular studies on the role of a root surface agglutinin in adherence and colonization by *Pseudomonas putida*, *Appl. Environ. Microbiol.* **54**, 375–380.

BAYER, E. A., KENIG, R., LAMED, R. (1983), Adherence of *Clostridium thermocellum* to cellulose, *J. Bacteriol.* **156**, 818–827.

BEEMAN, R. E., HOWELL, J. E., SHOEMAKER, S. H., SALAZAR, E. A., BUTTRAM, J. R. (1994), A field evaluation of *in situ* microbial reductive dehalogenation by the biotransformation of chlorinated ethenes, in: *Bioremediation of Chlorinated and Polycyclic Aromatic Hydrocarbon Compounds* (HINCHEE, R. E., LEESON, A., SEMPRINI, L., KEE ONG, S., Eds.). Boca Raton, FL: Lewis Publishers.

BOUWER, E. J. (1994), Bioremediation of chlorinated solvents using alternate electron acceptors, in: *Handbook of Bioremediation* (NORRIS, R. D., HINCHEE, R. E., BROWN, R., McCARTY, P. L., SEMPRINI, L., WILSON, J. T., Eds.), pp. 149–175. Boca Raton, FL: Lewis Publishers.

BROCKMAN, F. J., PAYNE, W., WORKMAN, D. J., SOONG, A., MANLEY, S., HAZEN, T. C. (1995), Effect of gaseous nitrogen and phosphorous injection on *in situ* bioremediation of a trichloroethylene-contaminated site, *J. Haz. Material* **41**, 287–298.

CHU, K.-H., ALVAREZ-COHEN, L. (1998), Effects of nitrogen source on growth and trichloroethylene degradation by methane-oxidizing bacteria, *Appl. Environ. Microbiol.* **64**, 3451–3457.

COSTERTON, J. W. (1992), Ultramicrobacteria: Improving access to underground pollutants, *Abstracts*, SIM Annual Meeting, San Diego, CA.

CREWS, J. L., COLBY, R. S., MATTHYSSE, A. G. (1990), *Agrobacterium rhizogenes* mutants that fail to bind to plant cells, *J. Bacteriol.* **172**, 6182–6188.

DE BRUIN, W. P., KOTTERMAN, M. J. J., POSTHUMUS, M. A., SCHRAA, G., ZEHNDER, A. J. B. (1992), Complete biological reductive transformation of tetrachloroethene to ethane, *Appl. Environ. Microbiol.* **58**, 1996–2000.

DEFLAUN, M. F., STEFFAN, R. J. (1999), Strategies for the successful application of bioaugmentation, *5th Int. Symp. In Situ and On-Site Bioremediation*, San Diego, CA, April 19–22.

DeFlaun, M. F., Tanzer, A. F., McAteer, A. L., Marshall, B., Levy, S. B. (1990), Development of an adhesion assay and characterization of an adhesion-deficient mutant of *Pseudomonas fluorescens, Appl. Environ. Microbiol.* **56**, 112–119.

DeFlaun, M. F., Condee, C. W., Ensley, B. D. (1994), Enhanced transport of degradative bacteria for *in situ* bioremediation, in: *In situ Remediation: Scientific Basis for Current and Future Technologies*, Part I, (Gee, G. W., Wing, N. R., Eds.), pp. 249–258. Columbus, OH: Battelle Press.

DeFlaun, M. F., Oppenheimer, S. R., Streger S., Condee, C. W., Fletcher, M. (1999), Alterations in adhesion, transport, and membrane characteristics in an adhesion-deficient pseudomonad, *Appl. Environ. Microbiol.* **65**, 759–765.

Dolan, M. E., McCarty, P. L. (1995), Methanotrophic chloroethene transformation capacities and 1,1-dichloroethene transformation product toxicity, *Environ. Sci. Technol.* **29**, 2741–2747.

Ensley, B. D. (1991), Biochemical diversity of trichloroethylene metabolism, *Ann. Rev. Microbiol.* **56**, 283–299.

Gibson, S. A., Sewell, G. W. (1992), Stimulation of reductive dechlorination of tetrachloroethene in anaerobic aquifer microcosms by addition of short-chain organic acids or alcohols, *Appl. Environ. Microbiol.* **58**, 1392–1393.

Gong, J., Forsberg, C. W. (1989), Factors affecting adhesion of *Fibrobacter succinogenes* ssp. *succinogenes* S-85 and adherence-defective mutants to cellulose, *Appl. Environ. Microbiol.* **55**, 3039–3044.

Hazen, T. C., Lombard, K. H., Looney, B. B., Enzien, M. V., Dougherty, J. M. et al. (1994), Summary of *in situ* bioremediation demonstration (methane biostimulation) via horizontal wells at the Savannah River Site Inegrated Demonstration Project, in: *In situ Remediation: Scientific Basis for Current and Future Technologies* (Gee, G. W., Wing, N. W., Eds.), pp. 137–150. Columbus, OH: Battelle Press.

Henry, S. M., Grbić-Galić, D. (1991), Influence of endogenous and exogenous electron donors and trichloroethylene oxidation toxicity on trichloroethylene oxidation by methanotrophic cultures from a ground water aquifer, *Appl. Environ. Microbiol.* **57**, 236–244.

Henrysson, T., McCarty, P. L. (1993), Influence of the endogenous storage lipid poly-beta-hydroxybutyrate on the reducing power availability during cometabolism of trichloroethylene and naphthalene by resting methanotrophic mixed cultures, *Appl. Environ. Microbiol.* **59**, 1602–1606.

Hopkins, G. D., Semprini, L., McCarty, P. L. (1993), Microcosm *in situ* field studies of enhanced biotransformation of trichloroethylene by phenol-utilizing microorganisms, *Appl. Environ. Microbiol.* **59**, 2277–2285.

Koene-Cottaar, F. H. M., Schraa, G. (1998), Anaerobic reduction of ethene to ethane in an enrichment culture, *FEMS Microbiol. Ecol.* **25**, 251–256.

Lee, M. D., Odom, J. M., Buchanan, Jr., R. J. (1998), New perspectives on microbial dehalogenation of chlorinated solvents: Insights from the field, *Annu. Rev. Microbiol.* **52**, 423–452.

Lemoigne, M. (1926), Produit de déshydration et de polymerization de l'acide béta-oxybutyrique, *Bull. Soc. Chim. Biol.* **8**, 770–782.

Lizotte, C. C., Marley, M. C., Crawford, S. C., Lee, A. M., Steffan, R. J. (1999), Demonstration of a sequential anaerobic/aerobic *in situ* treatment system at a Superfund site, *5th Int. Symp. In Situ and On-Site Bioremediation*, San Diego, CA, April 19–22.

Lombard, K. H., Borthen, J. W., Hazen, T. C. (1994), The design and management of system components for *in situ* methanotrophic bioremediation of chlorinated hydrocarbons at the Savannah River Site, in: *Air Sparging for Site Remediation* (Hinchee, R. E., Ed.), pp. 81–96. Boca Raton, FL: Lewis Publishers.

Major, D. W., Hodgins, E. W., Butler, B. J. (1991), Field and laboratory evidence of *in situ* biotransformation of tetrachloroethene to ethene and ethane at a chemical transfer facility in North Toronto, in: *On-Site Bioreclamation Processes for Xenobiotic and Hydrocarbon Treatment* (Hinchee, R. E., Olfenbuttel, R. F., Eds.), pp. 147–171. Boston, MA: Butterworth-Heinemann.

McCarty, P. L., Goltz, M. N., Hopkins, G. D., Dolan, M. E., Allan, J. P. et al. (1998), Full-scale evaluation of *in situ* cometabolic degradation of trichloroethylene in groundwater through toluene injection, *Environ. Sci. Technol.* **32**, 88–100.

McClay, K., Streger, S. H., Steffan, R. J. (1995), Induction of toluene oxidation activity in *Pseudomonas mendocina* KR1 and *Pseudomonas* ENVPC5 by chlorinated solvents and alkanes, *Appl. Environ. Microbiol.* **61**, 3479–3481.

McClay, K., Fox, B. G., Steffan, R. J. (1996), Chloroform mineralization by toluene-oxidizing bacteria, *Appl. Environ. Microbiol.* **62**, 2716–2722.

Munakata-Marr, J., McCarty, P. L., Shields, M. S., Reagin, M., Francesconi, S. (1996), Enhancement of trichloroethylene degradation in aquifer microcosms bioaugmented with wild-type and genetically-altered *Burkholderia* (*Pseudomonas*) *cepacia* G4 and PR1, *Environ. Sci. Technol.* **30**, 2045–2052.

Nelson, M. J. K., Montgomery, S. O., O'Neill, E. J., Pritchard, P. H. (1986), Aerobic metabolism of trichloroethylene by a bacterial isolate, *Appl. Environ. Microbiol.* **52**, 383–384.

NELSON, M. J., KINSELLA, J. V., MONTOYA, T. (1990), *In situ* biodegradation of TCE contaminated ground water, *Environ. Progress* **9**, 190–196.

OLDENHUIS, R., VINK, R. L. J. M., JANSEN, D. B., WITHOLT, B. (1989), Degradation of chlorinated alophatic hydrocarbons by *Methylosinus trichosporium* OB3b expressing soluble methane monooxygenase, *Appl. Environ. Microbiol.* **55**, 2819–2826.

RICHARDSON, K. L., GIBSON, D. T. (1984), A novel pathway for toluene oxidation in *Pseudomonas mendocina*, *Abstr. Annu. Meet. Am. Soc. Microbiol.* **84**, K54.

RIFAI, H. S., BORDEN, R. C., WILSON, J. T., WARD, H. (1995), Intrinsic bioattenuation for subsurface restoration, in: *Intrinsic Bioremediaton* (HINCHEE, R. E., WILSON, J. T., DOWNEY, D. C., Eds.), pp. 1–29. Columbus, OH: Batelle Press.

ROBERTS, P. V., HOPKINS, G. D., MACKAY, D. M., SEMPRINI, L. (1990), A field evaluation of *in situ* biodegradation of chlorinated ethenes: Part I, Methodology and field site characterization, *Ground Water* **28**, 591–604.

SEMPRINI, L., MCCARTY, P. L. (1991), Comparison between model simulations and field results from *in situ* biorestoration of chlorinated aliphatics: Part 1, Biostimulation of methanotropic bacteria, *Ground Water* **29**, 365–374.

SEMPRINI, L., MCCARTY, P. L. (1992), Comparison between model simulations and field results for *in situ* biorestoration of chlorinated aliphatics: Part 2, Cometabolic transformations, *Ground Water* **30**, 37–44.

SEMPRINI, L., HOPKINS, G. D., MCCARTY, P. L. (1994), A field and modeling comparision of *in situ* transformation of trichloroethylene by methane utilizers and phenol utilizers, in: *Bioremediation of Chlorinated and Polycyclic Aromatic Hydrocarbon Compounds* (HINCHEE, R. E., LEESON, A., SEMPRINI, L., KEE ONG, S., Eds.), pp. 248–254. Boca Raton, FL: Lewis Publishers.

SEWELL, G. W., DEFLAUN, M. F., BAEK, N. H., KUTZ, E., WEESNER, B., MAHAFFEY, B. (1998), Performance evaluation of an *in situ* anaerobic biotreatment system for chlorinated solvents, in: *Designing and Applying Treatment Technologies Remediation of Chlorinated and Recalcitrant Compounds* (WICKRAMANAYAK, G. B., HINCHEE, R. E., Eds.), pp. 15–20. Columbus, OH: Battelle Press.

SHIELDS, M. S., REAGIN, M. J. (1992), Selection of a *Pseudomonas cepacia* strain constitutive for the degradation of trichloroethylene, *Appl. Environ. Microbiol.* **58**, 3977–3983.

SMITH, G. A., NICKELS, J. S., KERGER, B. D., DAVIS, J. D., COLLINS, S. P. et al. (1986), Quantitative characterization of microbial biomass and community structure in subsurface material: a prokaryotic consortium responsive to organic contamination, *Can. J. Microbiol.* **32**, 105–111.

STEFFAN, R. J., TUGESHEVA, M. (1993), Construction and selection of constitutive toluene monooxygenase (TMO) mutants of *Pseudomonas mendocina* KR1, *4th Int. Symp. Pseudomonas: Biotechnology and Molecular Biology*. Vancouver, BC, Canada, August 8–12.

STEFFAN, R. J., SPERRY, K. L., WALSH, M. T., CONDEE, C. W. (1999), Field-scale evaluation of *in situ* bioaugmentation for remediation of chlorinated solvents in groundwater, *Environ. Sci. Technol.* **33**, 2771–2781.

U.S. Environmental Protection Agency (1999), Use of monitored natural attenuation at Superfund, RCRA corrective action and underground storage tank sites, *OSWER Directive 9200.4-17P*.

VOGEL, T. M. (1994), Natural bioremediation of chlorinated solvents, in: *Handbook of Bioremediation* (NORRIS, R. D., HINCHEE, R. E., BROWN, R., MCCARTY, P. L., SEMPRINI, L., WILSON, J. T., Eds.), pp. 201–225. Boca Raton, FL: Lewis Publishers.

WACKETT, L. P., BRUSSEAU, G. A., HOUSEHOLDER, S. R., HANSON, R. S. (1989), Survey of microbial oxygenases: trichloroethylene degradation by propane-oxidizing bacteria, *Appl. Environ. Microbiol.* **55**, 2960–2964.

WIEDEMEIER, T. H., SWANSON, M. A., MOUTOUX, D. E., GORDON, E. K., WILSON, J. T. et al. (1998), Technical protocol for evaluating natural attenuation of chlorinated solvents in ground water, *EPA/600/R-98/128*.

WILSON, J. T., LEACH, L. E., HENSON, M., JONES, J. N. (1986), *In situ* biorestoration as a ground water remediation technique, *Ground Water Monit. Remed.* **6**, 56–64.

19 Application of Immobilized Microorganisms in Soil Decontamination

HANS-JÜRGEN REHM
Münster, Germany

1 Introduction

Soil contamination with recalcitrants and xenobiotics is a serious problem not only in industrial countries, but also in developing countries. Many of these substances can be degraded by the autochthonic microflora. These are microorganisms, which are normally present in contaminated soil.

This autochthonic microflora can be increased by the recalcitrants or by the xenobiotics. An adaptation of different microorganisms to the substances, which should be degraded can take place.

For a review of the effects of xenobiotic chemicals on microorganisms in soil, see HICKS et al. (1990). Much worthwhile information about the degradation of xenobiotics, especially pesticides in soil, can be obtained from DOMSCH (1992). For basic information, see FRITSCHE (1990) and other books on microbiology and applied microbiology.

Furthermore, an increased development of the degrading microflora by addition of suitable fertilizers for the microorganisms and of inducers for a degradation of the recalcitrant substances or of cosubstrates for a degradation may be successful for soil decontamination (see McLOUGHLIN, 1994).

Many doubts exist for a successful addition of living microorganisms into the soil without any stabilization factors. Only a few successful scientific results are available for this method (KORDA et al., 1997).

FORD et al. (1999) have described a containment of a genetically engineered strain of *Pseudomonas fluorescens* during a field bioremediation application. This strain contained a fusion between the *lux* genes of *Vibrio fischeri* and the promotor for the lower pathway of naphthalene degradation, enabling the strain to become bioluminescent when degrading naphthalene. In six lysimeters filled with soil, which was contaminated with a mixture of naphthalene, phenanthrene, and anthracene, (10^{14}) bacteria were sprayed during the loading process; but these bacteria could only sporadically be detected after some time.

These experiments demonstrate, that the direct introduction of microorganisms into soil is only a minor success regarding the degradation. Many attemps of direct addition of special microorganisms to decontaminate soil have been published (see DECHEMA, 1992), but mostly without control of the microorganisms, which survived in the soil.

The reason of this failure is often a very quick overgrowth of the inoculated microorganisms by the autochthonic microflora (see SLATER et al., 1983), especially the fact, that the inoculated microorganisms have been partly degenerated by cultivation in the laboratories. These inoculated microorganisms have been changed in their metabolism to special degrading behavior. They are mutants, or perhaps they can be strains, which are constructed by genetic engineering. They are not adapted to the conditions in the soil and need a longtime process for adaptation to the soil conditions.

Such new, special microorganisms for soil decontamination are needed, because many substances cannot be degraded by the autochthonic microflora, but only by newly developed special microorganisms.

The problem will be to introduce these microorganisms from their very protected environment in the laboratories into the environment of the soil, which is a dangerous environment for them. One of the methods which can be applied here is immobilization.

Many applications of immobilized microorganisms, plant and animal cells have been realized in practice, e.g., in enzyme techniques, in medicine, in industrial production, in treatment of wastewater (see HEMPEL and LINDERT, 1990; WINTER, 1999), in waste gas treatment (KIRCHNER et al., 1991), and in other fields (for some examples, see BONT et al., 1990). About the altered physiology of cells by immobilization, see BONT et al. (1990) and REHM and OMAR (1993). Direct practical applications of immobilized cells in the soil are lacking, but numerous results of laboratory scale experiments are available, and some of these results will be presented here.

2 Methods of Immobilization

Many methods for immobilization exist (FU-KUI and TANAKA, 1984; MATTIASSON, 1983a, b; HARTMEIER, 1988; REHM and OMAR, 1993; for further literature, see also McLOUGHLING, 1994).

The following list shows the most important methods:

(1) covalent coupling, including cross-linking,
(2) adsorption,
(3) affinity (specific) immobilization,
(4) entrapment in a three-dimensional polymer network,
(5) confinement in a liquid–liquid emulsion,
(6) capture behind semipermeable membranes.

Among others these six different methods for immobilization of microorganisms are important. For application of microorganisms in soil decontamination two methods are of special interest:

(1) immobilization of microorganisms by adsorption and
(2) immobilization of microorganisms by entrapment.

2.1 Immobilization of Microorganisms by Adsorption

A broad spectrum of different supports have been used until now. Some of them are important for soil decontamination, e.g.,

(1) wood chips,
(2) modified cellulose,
(3) granular activated carbon,
(4) anthracite,
(5) granular clay,
(6) lava,
(7) synthetic polymers.

These and other natural substances have been used successfully for the immobilization of microorganisms and also partly for an application in wastewater treatment and sometimes in soil decontamination.

The application of synthetic polymers for adsorption and then for localization in soil should be discussed very carefully, before an application can be considered.

2.2 Immobilization of Microorganisms by Entrapment

Many synthetic polymers can be used for microbial entrapment, but often it may not be useful to bring these polymers into the soil. Otherwise, natural polymers which are not too expensive, should be discussed for an application to decontaminate soil. Among others, the following natural polymers are of interest:

(1) collagen as an entrapment material,
(2) carbohydrate-based polymers,
(3) agar and agarose,
(4) cellulose,
(5) κ-carrageenan,
(6) chitosan,
(7) alginate,
(8) polymers of polyhydroxy fatty acids.

Of these substances alginate is of main interest, because it is not too expensive in comparison with most of the other natural polymers mentioned.

A problem is the gel stability in soil. The major factors influencing this stability are (McLOUGHLING, 1994):

(1) gel properties:
  • polymer properties, gelation and hardening techniques employed,
  • characteristics and heterogeneity of the gel formed, e.g., shape (surface/volume ratio), strength, pore size, etc.,
  • influence of drying/lyophilization, rehydration, etc.;
(2) mechanism and rate of polymer breakdown, e.g., chelation or counter ion exchange, enzymatic breakdown, disruption due to expanding biomass;

(3) nature of cell leakage or cell outgrowth from the gel matrix, e.g., single cells/mycelia, influence of biomass loading, growth rate, etc.

Comparing entrapment methods with adsorption methods, immobilization by adsorption may be preferred for application in soil decontamination because of the much lower price and the much easier techniques.

Besides many bacteria the following molds immobilized in polymer beads have been recommended as biocontrol agents (for literature, see McLOUGHLING, 1994): *Trichoderma* ssp., *Gliocladium* spp., *Talaromyces flavus, Laetisaria arvalis, Penicillium oxalicum, Beauveria bassiana.*

Recently, LEIDIG et al. (1999) have discussed PVAL hydrogel beads of 1–2 mm diameter for encapsulation of mycelia, e.g., of mycelia of *Trametes versicolor.* The authors also recommended these immobilized mycelia for use in bioremediation.

3 Results with Immobilized Microorganisms in Soil Decontamination

It is obvious, that immobilized microorganisms have been applied in the practice of soil decontamination, especially when these microorganisms – mostly bacteria – are adsorbed on wooden materials. No exact results about the degradation of special substances with special microorganisms are available. Therefore, it is only possible to describe results which have been obtained in laboratories or in pilot plant dimensions.

Many results are available concerning the degradation of recalcitrants and xenobiotics (e.g., EHRHARD and REHM, 1985, 1989; BETTMANN and REHM, 1985; MÖRSEN and REHM, 1990, and many others). It is also possible to degrade recalcitrants with immobilized microorganisms with continuous cultures in waste-

water (WESTMEIER and REHM, 1986) or in waste gas treatment (KIRCHNER et al., 1991). In the following examples some results on the degradation of chlorinated or aromatic substances in soil by special immobilized microorganisms will be described.

In one example a degradation of 4-chlorophenol by an *Alcaligenes* sp. A7-2 which was immobilized on granular clay (BALFANZ and REHM, 1991) was shown. The experiments were made with a thermostated, packed-bed percolator in which the immobilized microorganisms were packed in little boxes in the interior of the packed sand. During fed batch fermentations 4-chlorophenol concentrations up to 160 mg L^{-1} d^{-1} were degraded (BALFANZ and REHM, 1991). The results are shown in Fig. 1.

These investigations were continued with a mixed culture of *Alcaligenes* sp. A7-2 which was able to degrade phenol, monochlorophenols, and dichlorophenols (2,3-, 2,4-, 2,5-, and 3,4-dichlorophenol) and a *Pseudomonas putida* P8 strain which was able to degrade phenols and monochlorophenols. This bacterium formed the 5-Cl-2-hydroxy muconic acid semialdehyde from 2- and 3-chlorophenol and chlorocatechols and from 4-chlorophenol. It could not degrade dichlorophenols.

The mixed culture was immobilized by adsorption onto granular clay and brought into the soil columns. In the course of long-term experiments (4 months) the titer of both bacteria was determined. The results are shown in Tab. 1.

If phenol was the only substrate, *Alcaligenes* was overgrown. 4-Chlorophenol was a better substrate for *Alcaligenes* than for *Pseudomonas*, which was overgrown. If the two substances were applied *Pseudomonas* grew better than *Alcaligenes*, but no strain was totally suppressed, if the cultures were incubated with a non-optimal growth substrate. This demonstrates that both bacteria could be established by immobilization in the soil model system (BALFANZ and REHM, 1992).

In other experiments immobilized *Alcaligenes* could be established in sandy soil even when the soil was inoculated with activated sludge.

Xanthobacter autotrophicus which had a good short-chain dehalogenase activity was

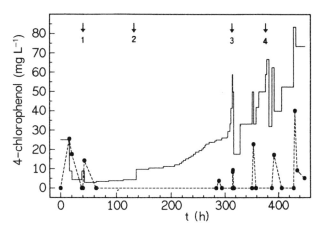

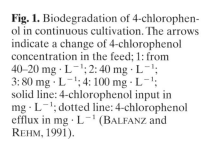

Fig. 1. Biodegradation of 4-chlorophenol in continuous cultivation. The arrows indicate a change of 4-chlorophenol concentration in the feed; 1: from 40–20 mg · L^{-1}; 2: 40 mg · L^{-1}; 3: 80 mg · L^{-1}; 4: 100 mg · L^{-1}; solid line: 4-chlorophenol input in mg · L^{-1}; dotted line: 4-chlorophenol efflux in mg · L^{-1} (BALFANZ and REHM, 1991).

Tab. 1. Ratio of *Pseudomonas*:*Alcaligenes* in the Fermentation Medium at the End of Each Batch

Batch No.	Phenol (5.0 M)	4-CP (0.3 mM)	Phenol/4-CP
1	10:1	1:4	3:1
2	35:1	1:8	3:1
3	4:1	1:2	16:1
4 (2,4-DCP)[a]	9:1	1:14	14:1
5	3:1	1:14	10:1
6	8:1	1:19	19:1
7	17:1	1:29	24:1
8	15:1	1:29	24:1
9	19:1	1:25	24:1
10	17:1	1:24	23:1
11[b]	16:1	1:8	12:1

[a] Batch   4:0.05 mM 2,4-DCP
[b] Batch 11:phenol and 4-CP

immobilized on granulated lecaton particles. The bioreactor with sand and immobilized microorganisms was the same as described above. Fig. 2 shows the degradation of dichloroacetic acid by the adsorbed *Xanthobacter autotrophicus* under non-sterile conditions.

Under these conditions the bacteria were stabilized sufficiently so that the repeated addition of activated sludge samples did not interfere with their ability to degrade recalcitrant substances. Bacteria which had not been immobilized were washed out in a very short time, and hence could not degrade recalcitrant substances.

In other experiments the degradation rate of hydrocarbons in oily sludge obtained from a flotation unit by free cells and cells which were immobilized on granular clay was investigated (OMAR et al., 1990). Free cells of *Candida parapsilosis* needed 7–8 weeks to use 30% of a 3.3% hydrocarbon containing sludge, whereas with immobilized cells the same result was ob-

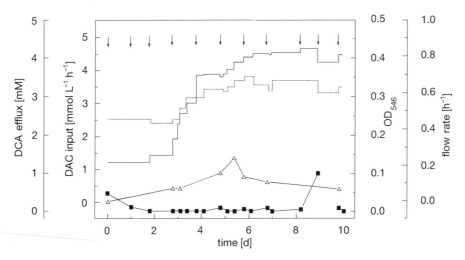

Fig. 2. Degradation of dichloroacetic acid by adsorbed *Xanthobacter autotrophicus* GJ10 in continuous culture under non-sterile conditions. Arrows indicate addition of activated sludges samples; (—) DCA-input; (···) dilution rate; ■ DCA-efflux; △ OD_{546} (optical density at 546 nm).

tained after only 3–4 weeks. Free cells degraded 50% of the 5% hydrocarbon containing oily sludge in 7 weeks, whereas the immobilized cells gave the same result after only 4 weeks. Fig. 3 shows the degradation of 8% hydrocarbon containing oily sludge by free and immobilized cells.

Biodegradation with an immobilized mixed culture was studied in long-term outdoor experiments at a scale of 50 L (working volume) and an artificial contamination of sandy soils with 1% diesel fuel. The mixed culture contained *Pseudomonas putida* P8, *Candida guilliermondii*, *C. oleophila*, and *C. lipolytica*. The microorganisms were adsorbed on lava granulate. The fermentor was a PVC 70 L container with a working volume of 50 L. Fig. 4 shows an example of the results for the diesel fuel degradation by free and immobilized cells (GRUNDMANN and REHM, 1991).

The results show that an addition of immobilized microorganisms to the soil, which contained the natural microflora, had a positive effect on the degradation, especially during the first period of soil treatment.

Granular clay may be a better adsorption substance for yeasts than aquifer sand, as it could be demonstrated for the degradation of *n*-alkanes (OMAR and REHM, 1988) (see Tab. 2).

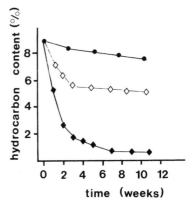

Fig. 3. Degradation of 8% hydrocarbon containing oily sludge by free and immobilized cells of *Candida parapsilosis:* ● control; ◇ free cells; ◆ immobilized cells (OMAR et al., 1990).

CASSIDY et al. (1997) could demonstrate, that it was possible to degrade pentachlorophenole by a κ-carrageenan-encapsulated *Pseudomonas* sp. Almost double the PCB concentration could be metabolized by the encapsulated cells in comparison with free cells.

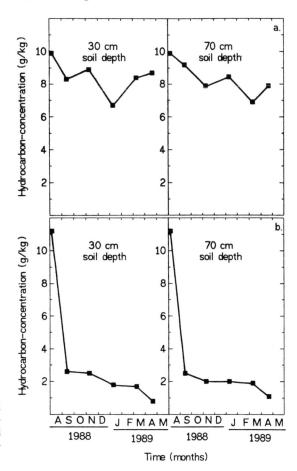

Fig. 4a, b. Degradation of 1% diesel fuel in soil in 30 cm and in 70 cm soil depth. (**a**) control experiment; (**b**) treated test variant (immobilized cells + NPK fertilizer C:N = 60:1 + active aeration) (GRUNDMANN and REHM, 1991).

Tab. 2. Percentage of Residual n-Alkane after Action by Cells or Spores Immobilized on Different Carriers. The Cells or Spores Were Allowed to Grow on 1% n-Alkane for 8 d in a Trickle Flow Column

	Carrier			
	Granular Clay		Granular Clay and Aquifer Sand	Aquifer Sand
Chain Length	C. para-psilosis	P. frequen-tans	C. para-psilosis	C. para-psilosis
C_{12}	0.0	0.0	0.0	N.D.[a]
C_{13}	16.1	20.0	58.5	85.9
C_{14}	21.6	30.0	63.1	98.9
C_{15}	21.4	32.3	67.5	93.2
C_{16}	20.9	31.0	67.6	89.7
C_{17}	18.8	27.1	67.2	89.7
C_{18}	14.5	13.4	52.3	98.4

[a] N.D.: not detected.

These results and other studies of HEKMAN et al. (1994), and VAN ELSAS et al. (1992) also show better degradation of target compounds with immobilized cells than with free cells.

4 Conclusions and Final Discussion

All these results demonstrate that the immobilization of microorganisms will be a suitable method to establish special microorganisms in the soil. These microorganisms are for a long time protected against the autochthonic microflora of the soil and are able to degrade the recalcitrants or xenobiotics against which they have been constructed. This is in contrast to the simple inoculation of microorganisms into the soil, which are only rarely able to establish themselves and to degrade the recalcitrant substances.

In a review KORDA et al. (1997) have listed the *in situ* treatments in petroleum hydrocarbon bioremediation with inocula of microorganisms and the available microbial strains used in bioremediation. The analytical techniques have also been described. They concluded that each of these strains may yield spectacular results *in vitro* for specific target compounds. However, the overall success of such strains in treating a wide range of contaminants *in situ* remains very limited. They recommend supplements of oxygen and fertilizers to get a more effective bioremediation.

Petroleum hydrocarbons belong to compounds, which can be degraded more readily and faster by microorganisms than other compouds like dioxins or polychlorinated compounds. Therefore, the microorganisms must be located in the soil for a longer time to degrade these substances, than for petroleum remediation. This establishes a special need for the immobilization of these degrading microorganisms.

It is obvious that engineers should develop a suitable method with which special immobilized microorganisms can be introduced into the contaminated soil. At present these immobilized microorganisms can be applied especially in soil composting methods where the immobilized microorganisms can be mixed

easily. The application of immobilized microorganisms may be easy and successful also in bioreactors for soil decontamination.

BASHAN and GONZALES (1999) described long-term survival of the plant growth promoting bacteria *Azospirillum brasiliense* and *Pseudomonas fluorescens* in drug alginate inoculant and recommended inoculation of these immobilized bacteria in soil. BASHAN (1998) repeated once more the main advantage of alginate inoculants because of the non-toxic nature of alginate and its degradation in nature. He showed the slow release of the entrapped microorganisms into the soil.

Genetically engineered *Pseudomonas fluorescens*, encapsulated and later released into soil microcosm, showed significantly longer survival rates than those of non-encapsulated cells after 3 months (VAN ELSAS et al. 1992).

It has been suggested that plasmids which carry the information on degradation of the inoculated microorganisms may be transferred to other microorganisms, which are normally present in the soil and which were not able to degrade the recalcitrants before capturing the new plasmids. This is only a theory and not proved enough. Until now we must try to place the special microorganisms into the soil and give them the chance to live and to grow. It may also be possible, that these microorganisms will transfer their plasmids to other microorganisms in the soil according to the theory described above.

Much work has to be done in this field. But it must be done, if we want to apply our special microorganisms successfully in soil decontamination and the investigations of many groups, who construct new xenobiotic degrading strains, will get practical application.

5 References

BALFANZ, J., REHM, H.-J. (1991), Biodegradation of 4-chlorophenol by adsorptive immobilized *Alcaligenes* sp. A7-2 in soil, *Appl. Microbiol. Biotechnol.* **35**, 662–668.
BALFANZ, J., REHM, H.-J. (1992), Biodegradation of phenol and chlorinated phenols by an immobilized mixed culture in a sandy soil, *DECHEMA Biotechnology Conferences*, Vol. 5, Part B, pp. 1005–1008.

BASHAN, Y. (1998), Inoculants for the plant growth-promoting bacteria in agriculture, *Biotechnol. Adv.* **16**, 729–770.

BASHAN, Y., GONZALES, L. E. (1999), Long-term survival of the plant-growth-promoting bacteria *Azospirillum brasiliense* and *Pseudomonas fluorescens* in drug alginate inoculant, *Appl. Microbiol. Biotechnol.* **51**, 262–266.

BETTMANN, H., REHM, H.-J. (1985), Continuous degradation of phenol(s) by *Pseudomonas putida* P8 entrapped in polyacrylamide-hydrazide, *Appl. Microbiol. Biotechnol.* **22**, 389–393.

CASSIDY, M. B., SHAW, K. W., LEE, H., TREVORS, J. T. (1997), Enhanced mineralization of pentachlorophenol by κ-carrageenan-encapsulated *Pseudomonas* sp. UG 30, *Appl. Microbiol. Biotechnol.* **47**, 108–113.

DE BONT, J. A. M., VISSER, J., MATTIASSON, B., TRAMPER, J. (1990), Physiology of immobilized cells. Amsterdam, Oxford, New York, Tokyo: Elsevier.

DECHEMA (1992), Soil decontamination using biological processes, preprints, *Int. Symp. Karlsruhe*, 6–9 Dec, 1992.

DOMSCH, K. H. (1992), *Pestizide im Boden, Mikrobieller Abbau und Nebenwirkungen auf Mikroorganismen*. Weinheim: VCH.

EHRHARDT, H. M., REHM, H.-J. (1985), Phenol degradation by microorganisms adsorbed on activated carbon, *Appl. Microbiol. Biotechnol.* **21**, 32–36.

EHRHARDT, H. M., REHM, H.-J. (1989), Semicontinuous and continuous degradation of phenol by *Pseudomonas putida* P8 adsorbed on activated carbon, *Appl. Microbiol. Biotechnol.* **30**, 312–317.

FORD, C. Z., SAYLER, G. S., BÄRLAGE, R. S. (1999), Containment of genetically engineered microorganisms during a field bioremediation application, *Appl. Microbiol. Biotechnol.* **51**, 397–400.

FRITSCHE, W. (1990), *Mikrobiologie*. Jena: Gustav Fischer Verlag.

FUKUI, S., TANAKA, A. (1984), Application of biocatalysts immobilized by prepolymer methods, *Adv. Biochem. Eng. Biotechnol.* **29**, 1–33.

GRUNDMANN, R., REHM, H.-J. (1991), Biodegradation of diesel fuel, *Erdöl und Kohle-Erdgas-Petrochmie* **44**, 149–150.

HARTMEIER, W. (1988), *Immobilized Biocatalysts*. Berlin, Heidelberg: Springer-Verlag.

HEKMAN, W. B., HEIJNEN, C. E., TREVORS, J. T., VAN ELSAS, J. D. (1994), Waterflow-induced transport of *Pseudomonas fluorescens* cells through soil columns as affected by inoculum treatment, *FEMS Microbiol. Ecol.* **13**, 313–324.

HEMPEL, D. C., LINDERT, M. (1990), Behandlung von Abwässern mit schwer abbaubaren Inhaltsstoffen: Reaktionskinetiken zum Einsatz immobilisierter Spezialkulturen, *Wasser, Abwasser GWF* **131**, 52.

HICKS, R. J., STOTZKY, G., VAN VORIS, P. (1990), Review and evaluation of the effects of xenobiotic chemicals on microorganisms in soil, *Adv. Appl. Microbiol.* **35**, 195–253.

KIRCHNER, K., GOSSEN, C. A., REHM, H.-J. (1991), Purification of exhaust air containing organic pollutants in a trickle-bed bioreactor, *Appl. Microbiol. Biotechnol.* **35**, 396–400.

KORDA, A., SANTAS, P., TENENTE, A., SANTAS, R. (1997), Petroleum hydrocarbon bioremediation: Sampling and analytical techniques, *in situ* treatments and commercial microorganisms currently used, *Appl. Microbiol. Biotechnol.* **48**, 677–686.

LEIDIG, E., PRÜSSE, U., VORLOP, K. D., WINTER, J. (1999), Biotransformation of Poly R-478 by continuous cultures of PVAL-encapsulated *Trametes versicolor* under non-sterile conditions, *Bioprocess Eng.* **21**, 5–12.

MATTIASSON, B. (1983a), *Immobilized Cells and Organelles*, Vol. I. Boca Raton, FL: CRC Press.

MATTIASSON, B. (1983b), *Immobilized Cells and Organelles*, Vol. II. Boca Raton, FL: CRC Press.

McLOUGHLING, A. J. (1994), Controlled release of immobilized cells as a strategy to regulate ecological competence of inocula, *Adv. Biochem. Eng./Biotechnol.* **51**, 1–46.

MÖRSEN, A., REHM, H.-J. (1990), Degradation of phenol by a defined mixed culture immobilized by adsorption on activated carbon and sintered glass, *Appl. Microbiol. Biotechnol.* **33**, 206–212.

OMAR, S. H., REHM, H.-J. (1988), Degradation of n-alkanes by *Candida parapsilosis* and *Penicillium frequentans* immobilized on granular clay and aquifer sand, *Appl. Microbiol. Biotechnol.* **28**, 103–108.

OMAR, S. H., BÜDECKER, U. W., REHM, H.-J. (1990), Degradation of oily sludge from a flotation unit by free and immobilized microorganisms, *Appl. Microbiol. Biotechnol.* **34**, 259–263.

REHM, H.-J., OMAR, S. H. (1993), Special morphology and metabolic behavior of immobilized microorganisms, in: *Biotechnology* 2nd Edn., Vol. 1 (REHM, H.-J., REED, G., PÜHLER, A., STADLER, P., Eds.), pp. 223–248. Weinheim: VCH.

SLATER, J. H., WHITTENBURG, R., WIMPENNY, J. W. T. (Eds.) (1983), *Microbes in Their Natural Environments*. Cambridge: Cambridge University Press.

VAN ELSAS, J. D., TREVORS, J. T., JAIN, D., WOLTERS, A. C., HEIJNEN, C. E., VAN OVERBECK, L. S. (1992), Survival of and colonization by alginate-encapsulated *Pseudomonas fluorescens* cells following introduction into soil, *Biol. Fertil. Soils* **14**, 14–22.

WESTMEIER, F., REHM, H.-J. (1986), Einsatz von Immobilisationsverfahren zum biologischen Abbau chlorierter Phenole, *Chem. Ind.* **38**, 158–160.

WINTER, J. (1999), in: *Biotechnology* 2nd Edn., Vol. 11a (REHM, H.-J., REED, G., PÜHLER, A., STADLER, P., Eds.). Weinheim: WILEY-VCH.

20 Bacterial Activity Enhancement and Soil Decontamination

FU-MIN MENN
JAMES P. EASTER
GARY S. SAYLER
Knoxville, TN 37996-1605, USA

List of Abbreviations

BTEX   benzene, toluene, ethylbenzene, and xylene
2,4-D   2,4-dichlorophenoxyacetic acid
GEM   genetically engineered microorganisms
PAH   phenanthrene, pyrene, and benzo[a]pyrene
PCB   polychlorinated biphenyls
PCP   pentachlorophenol
TCE   trichloroethylene
TNT   2,4,6-trinitrotoluene

1 Introduction

Bioremediation can be cost-effective using natural capabilities of microorganisms, including fungi and bacteria, to degrade and detoxify certain hazardous wastes existing either in the soil or groundwater. A recent survey showed that the cost of soil bioremediation on TCE is significantly cheaper (up to 60%) than conventional clean-up methods (KATO and DAVIS, 1996).

The ultimate goal of bioremediation is to transform all organic wastes into small molecules, such as CO_2, CH_4, or intermediates that can be re-incorporated into microbial anabolic pathways. Microorganisms used in bioremediation can be either indigenous bacteria or other known biodegradative organisms. Bioremediation technology is divided into two categories based on whether the degradation activity is accomplished *in situ* or is carried out *ex situ* of the contaminated sites. *In situ* bio-

remediation involves treatment of the contaminated materials in place via either biostimulation or bioaugmentation. Biostimulation involves enhancement of microbial metabolic activities by optimizing environmental factors affecting microbial growth. Bioaugmentation is accomplished by addition of nonindigenous bacteria (degraders) to the pollutant site. The *ex situ* treatment requires physical removal of the contaminated materials to another place (maybe within the site), followed by decontamination processes under defined conditions. However, it may not be economically feasible to excavate soil or pump contaminated groundwater for a large scale decontamination treatment. Consequently, *in situ* bioremediation technology is the choice for decontamination in situations where conventional physical-chemical technology is ineffective. However, growth conditions in the natural environment are not as favorable as in the laboratory, which often cause very low survival rates for inocula, and low cell density of indigenous bacteria. In order to obtain the maximum biodegradation rate, it is necessary to optimize bacterial growth and activity by altering conditions in the natural environment.

Both aerobic and anaerobic reactions play important roles in the biodegradation of pollutants. Anaerobic degradation commonly involves electron donor–acceptor reactions and usually takes place in subsurface soils and groundwater. Compounds transformed by these reactions range from BTEX, chlorinated solvents, PCB and PAH, etc. Numerous studies have been reported on these topics. However, aerobic reduction of contaminants in the soil is the focus of this chapter.

In order to facilitate pollutant reduction, biostimulation and bioaugmentation are the most common practices to enhance biodegradation in recent years. Biostimulation enhances biodegradation by increasing indigenous bacterial populations. Addition of mineral nutrients, electron acceptors, organic nutrients, or adjusting environmental properties can achieve successful *in situ* remediation. Some examples of biostimulation are listed in Tab. 1.

The application of bioaugmentation has been used in soil remediation. Bioaugmentation remains uncertain in economics and effectiveness to enhance *in situ* bioremediation. It is carried out by injecting exogenous bacteria or consortia into contaminanted sites for degrading specific pollutant(s). In contaminated subsurface soil environments the major difficulty arises in the delivery of the bioaugmentative populations to the contaminants to be degraded. A number of microorganisms applied for different target pollutants in bioaugmentation are summarized in Tab. 2.

Successful *in situ* bioremediation depends not only on a combination of biostimulation and bioaugmentation, but also on the bioavailability of the target contaminant.

Petroleum and fuel hydrocarbon bioremediation has been the focus of many processes at both the research and the commercial scale. A field demonstration of *in situ* bioremediation was carried out in the Exxon Valdez oil spill in Prince William Sound, Alaska. Adding fertilizer directly to the surface of oil contaminated beaches accelerated the degradation of oil by indigenous microorganisms (PRITCHARD and COSTA, 1991; PRITCHARD et al., 1992). This ultimately stimulated renewed interest in natural bioremediation processes. The cost of remediation in the Kuwait desert is expected to be enormous due to the large volume of oil satu-

Tab. 1. Examples of Biostimulation Activities in Bioremediation

Type	Amendment	Target Contaminants	Organism	Reference
Mineral nutrients	C–N ratio	diesel oil	indigenous	MARGESIN and SCHINNER, 1997b
	nitrogen	toluene	indigenous	ALLEN-KING et al., 1994
Electron acceptor	nitrate, sulfate	chlorophenols	indigenous	HAGGBLOM et al., 1993
	nitrate, sulfate	BTEX	indigenous	BOROLE et al., 1997
	2,5,3′4′-tetra-chlorobiphenyl	PCB	indigenous	BEDARD et al., 1996
Organic nutrients	fertilizer	petroleum oil	indigenous	PRITCHARD et al., 1992
Physical property	temperature	pentachlorophenol	*Pseudomonas* sp. UG30	SHAW et al., 1997
	temperature	petroleum oil	*B. stearothermophilus*	SORKOH et al., 1993
	pH	oil sludge	indigenous	DIBBLE and BARTHA, 1979

Tab. 2. Examples of Bioaugmentation Application in Bioremediation

Organisms	Target Contaminants	Reference
P. stutzeri KC	carbon tetrachloride	DYBAS et al., 1998
A. eutrophus TCP	2,4,6-trichlorophenol	ANDREONI et al., 1998
R. chlorophenolicus	pentachlorophenol	MCBAIN et al., 1997
Flavobacterium sp. ATCC 39723	pentachlorophenol	CRAWFORD and MOHN, 1985; SEECH et al., 1991
P. cepacia DBO1	2,4-D	JACOBSEN and PEDERSEN, 1992
P. aeruginosa JB2	PCB	HICKEY et al., 1993
A. xylosoxidans	PCB	HALUŠKA et al., 1995
Anthrobacter sp. B1B	PCB	GILBERT and CROWLEY, 1998
P. fluorescens HK44	naphthalene, phenanthrene, anthracene	SAYLER et al., 1999

rated soil. From feasibility and economic aspects, *in situ* bioremediation technology was suggested for clean-up of Kuwait contaminated soil and ultimate cost reduction. A laboratory scale demonstration of *ex situ* bioremediation of Kuwait oil contaminated desert soil was conducted. Addition of nutrients and coconut charcoal to the contaminated soil enhanced the biodegradation rate most effectively (CHO et al., 1997).

There are a number of review articles published that discuss the bioremediation on various pollutants (LEAHY and COLWELL, 1990; MORGAN and WATKINSON, 1989a, b; THOMAS and WARD, 1989; THOMAS and LESTER, 1993; RISER-ROBERTS, 1998), and on limiting factors of recalcitrants (PROVIDENTI et al., 1993). The focus of this chapter is on soil decontamination by bacteria, including improvement of bacterial activity in contaminated soil and enhancement of biodegradation via bioaugmentation.

The prospect of success for *in situ* bioremediation of contaminants is limited if the conditions in the soils are not suitable for the growth and metabolism of microorganisms. Carbon and energy are major growth limiting factors in nature for heterotrophic bacteria that play an important role in the degradation of xenobiotic compounds. These complications can be overcome by manipulating soil conditions or by adding organic and inorganic supplements.

The following methods have been applied to enhance microbial degradation activities on various contaminated sites:

(1) physical property amendment to provide more homogeneous laboratory-like growth conditio,
(2) nutrient and electron donor and acceptor supplementation to increase the population and activity of the degradative microbial community,
(3) chemical inducers of a cometabolic transformation process of organic compounds,
(4) bioaugmentation, adding known degraders to a contaminated site for degradation of particular contaminants.

The modification of environmental factors and nutrient compositions is the most common and efficient methods to achieve an optimal growth environment for bacteria in nature. Bioaugmentation is a direct and effective way to decontaminate hazardous wastes that exist in the environment. These treatment methods vary in different contaminated sites because each site has its own unique characteristics. The influences on bioremediation by these methods will be discussed in Sects. 2, 3, and 5. The consequences of application of surfactants (to increase bioavailability and solubility of contaminants) and of inducers (specific carbon sources to stimulate a specific group of microbes) to increase biodegradative activity will not be reviewed here.

In soil bioremediation, soil type can affect the efficacy of biodegradation because of its influence on microbial movement and contaminant sorption. However, this issue will not be discussed here.

2 Physical Property Amendment

Environmental factors affect microbial activities, growth, and metabolism, which substantially influence the results of soil bioremediation. Some factors can be manipulated to provide optimal conditions for biotreatment processes, whereas others are difficult to modify in the contaminated site. The effects of environmental factors are highly site-specific and should be considered individually. However, it is important to note that many effects are interactive, e.g., both pH value and temperature can affect contaminant solubility. Soil temperature, moisture, oxygen concentration and pH value affect microbial activities substantially and are discussed in Sects. 2.1–2.4.

2.1 Temperature

This is one of the most important environmental parameters that influence microbial metabolism and effect the rate of biodegradation. Temperature also can affect solubility, sorption, viscosity, and volatilization of contaminants (MORGAN and WATKINSON, 1989b). The maximum rate of hydrocarbon metabolism by microorganisms in oil sludge treated soil was reached at 20 °C (DIBBLE and BARTHA, 1979). The optimal temperature for dechlorination and degradation of pentachlorophenol by *Pseudomonas* sp. UG30 was between 25 °C and 30 °C. Strain UG30 was able to degrade pentachlorophenol at 10 °C and degradation rates increased linearly with increasing temperature to 30 °C (SHAW et al., 1997). The optimal temperature for biodegradation varies by climate and consortia of bac-

teria. Most *in situ* bioremediation applications have been performed under mesophilic conditions (20–40 °C). Only microorganisms that can adapt their metabolism to extreme environments are expected to work effectively under low temperatures (MARGESIN and SCHINNER, 1994). There are reports on decontamination by bacteria on subarctic soils (SPARROW and SPARROW, 1988). Microbial degradation of diesel oil was detected at 10 °C in alpine subsoils when supplemented with an inorganic fertilizer (MARGESIN and SCHINNER, 1997a). Another study showed similar biodegradation rates were obtained at 10 °C and 25 °C, indicating the adaptation of the indigenous soil microorganisms to the low temperatures of alpine soils (MARGESIN and SCHINNER, 1997b). WHYTE et al. (1999) have reported biodegradation activities at 5 °C in arctic soils on hexadecane by cold adapted microorganisms and by a *Rhodococcus* sp., Q15 (WHYTE et al., 1998). *Rhodococcus* sp. Q15 also exhibited ability to mineralize alkanes and aromatics (WHYTE et al., 1996). Both alkane and naphthalene catabolic pathways were found to coexist in the *Pseudomonas* spp. isolated from petroleum contaminated arctic soils (WHYTE et al., 1997).

Temperature was the limiting factor for bioremediation of highly petroleum contaminated cryic soil (WALWORTH et al., 1999). Maximum microbial activity was observed between 21° and 31 °C. Addition of nitrogen enhanced biodegradation rates when soil temperature increased. A separate study also showed that additions of ammonium nitrate (25 mg N kg^{-1} soil) and phosphate (2.5 mg P kg^{-1} soil) to contaminated arctic gravel pads between 5 °C and 10 °C can increase oxygen uptake rate substantially (CHAOBAL et al., 1999).

At the other extreme, *Bacillus stearothermophilus*, isolated from the Kuwaiti desert, had an optimal growth temperature of 60 °C and was able to use petroleum oil as sole energy and carbon source (SORKOH et al., 1993). Temperature manipulation may not be technically or economically feasible in contaminated sites, and the fluctuation in temperature can be an obstacle to *in situ* bioremediation. However, application of radiowave frequency to heating soil has been reported (HÜTTMANN et al., 1999). Microbial biomass and viable cell counts were reduced dramatically after long-term radiowave exposure at temperatures < 40 °C.

2.2 Oxygen

Aerobic degradation is the most favored route of microorganisms to degrade most organic compounds. The catabolic rate is determined by oxygen availability. Oxygen not only serves as the terminal electron acceptor in aerobic reactions by bacteria, but also as a substrate in the oxygen catalyzed reactions. The initial reactions of catabolism of environmental pollutants, such as aromatic hydrocarbons (GIBSON, 1984) and polychlorinated biphenyls (FURUKAWA et al., 1990), by microorganisms are catalyzed by either a dioxygenase or monooxygenase in which molecular oxygen is required for oxidation. Therefore, adequate aeration is necessary for aerobic degradation by microorganisms. Tilling is the most common and effective way to improve the oxygen concentration in surface contaminated soils (ATLAS, 1991). Composting is another cost-effective method for decontamination of surface soils. This method has been demonstrated successfully in clean-up of different contaminants, such as chlorophenol (VALO and SALKINOJA-SALONEN, 1986), hexachlorocyclohexane (DOELMAN et al., 1988), and petroleum-based oil wastes (KIRCHMANN and EWNETU 1998). In general, the efficiency of composting is proportional to the temperature in the natural environment. General guidelines for land treatment on petroleum contaminated soil were reviewed by HUESEMANN (1994).

Hydrogen peroxide injection has been broadly used in municipal and industrial wastewater treatment and in contaminated subsoils to increase oxygen concentration. An in-depth review on hydrogen peroxide from biochemistry, chemistry, and toxicity aspects was done by PARDIECK et al. (1992). Being relatively inexpensive and not persisting in the environment are attractive characteristics for choosing hydrogen peroxide as source of oxygen in in situ bioremediation. Inoculation of Pseudomonas putida (ATCC 12633) into a gasoline contaminated soil showed the greatest total degradation of toxic BTEX constitu-

ents, and highest bacterial growth at 0.1 mM of hydrogen peroxide (DA CUNHA and GOMES FERREIRA LEITE, 1997). Inhibition of bacterial growth (PARDIECK et al., 1992) and decomposition reactions of hydrogen peroxide solutions resulting in soil pore blockage and reduction in permeability (SPAIN et al., 1989) are hindrances to the use of this compound. Concentrations of hydrogen peroxide > 100 mg L^{-1} significantly reduced the permeability of sandy loam soil which decreased oxygen transfer in the soil (MORGAN and WATKINSON, 1992).

2.3 Soil Moisture

Water is a fundamental requirement for bacterial growth and metabolism. The water activity (a_w) is the availability of water in the medium (environment), and it ranges from 0.0–0.99 in the soils. The minimum a_w values for bacterial growth vary widely. Inadequate moisture can lower microbial metabolism and bacterial movement through soil, and reduces the transport of nutrient and contaminant. Excess water in soil limits oxygen conveyance. An optimal biodegradation rate was obtained between 30% and 90% of water content in an oil sludge study (DIBBLE and BARTHA, 1979). The influence of soil moisture on the movement, survival, and degradative activity of a 2,4-D degrading bacterium, genetically engineered bioluminescent strain Burkholderia cepacia BRI6001L, was examined in unsaturated soil columns. No degradation activity was detected when the moisture level drops below 6%. An increase of 4 orders of magnitude in bacterial density, and a 10-fold increase in the 2,4-D degradation rate were monitored when the moisture was > 40% (CATTANEO et al., 1997).

Other studies suggested the optimal moisture content for aerobic biodegradation is commonly between 50% and 70% of the soil water holding capacity (LAJOIE and STROM, 1994; PRAMER and BARTHA, 1972).

2.4 pH

The variation of pH value affects not only microbial activity, but also the solubility and sorption of contaminants. In general, heterotrophic bacteria have growth optima at or near neutral pH values (pH 6–8), and fungi prefer acidic conditions (pH 4–5). Extremes of pH are often not well tolerated by microorganisms (HANSTVEIT et al., 1988). The optimum pH for petroleum hydrocarbon degradation is pH 7.4–7.8 (DIBBLE and BARTHA, 1979). However, the pH values for most soils are acidic. Increasing soil pH by amendment can increase biodegradation rate significantly. VERSTRAETE et al. (1976) reported a doubling of the degradation rate of gasoline by increasing soil pH from 4.5–7.4. Addition of ground limestone or calcium and magnesium containing compounds to the soils is a common method to neutralize soil acidity (BAKER, 1994). Addition of ferrous sulfate can reduce the alkalinity in the soil. Recently, PAH metabolism under highly acidic conditions has been reported (STAPLETON et al., 1998).

It is important to note that a change of soil pH may affect bioavailability and solubility not only of organic pollutants but also of nutrients in the soil. The solubilities of nitrogen, phosphorus, ammonia, calcium, iron, and magnesium in the soil are inversely proportional to pH value increase, i.e., solubility increases under acidic conditions. This is in contrast to nitrates and chlorides where the lower soil pH results in lower solubility (SHARPLEY, 1991).

3 Nutrient Supplements

All bacteria require an exogenous energy source for growth, and the source of energy is quite different among species. Autotrophic bacteria acquire their energy from light or by oxidation of inorganic element(s) or compound(s). Heterotrophic bacteria derive energy from oxidation of hydrocarbons. In addition, trace elements, such as Ca^{2+}, Na^+, Fe^{2+}, PO_4^{3-}, and SO_4^{2-}, are also essential for bacterial growth. These elements usually are not limiting factors because sufficient amounts are present in the natural environment. The requirement of nutrient supplementation is site-dependent and is used in both *in situ* and *ex situ* bioremediation. Typically, concentrations of nitrogen and phosphorus are limited in the soil (U.S. EPA, 1985). In general, the goal of nutrient amendment is to increase the populations of total indigenous heterotrophic microorganisms and to maximize degradation rates of organic pollutants. The rate of toluene degradation in a sandy soil was accelerated after addition of nitrogen as either ammonium or nitrate (ALLEN-KING et al., 1994). The adaptation lag periods for microbial transformation of *p*-cresol were decreased significantly after samples were amended with nitrogen or phosphorus (LEWIS et al., 1986). However, the disadvantage of this method is that the biodegradation rate may not always be proportional to the increase of bacterial population density. This is due to non-selective enrichment of microorganisms.

It is well documented that addition of nitrogen and phosphorus can stimulate the biodegradation rate of soil and groundwater contaminants. The quantity to be added often is estimated from C–N ratios. The ratio is used to quantify the nitrogen requirement in terms of the amount of carbon substrate available. The recommended C–N ratios for hydrocarbon degradation in soils vary widely. The U.S. Environmental Protection Agency suggested a range of 10:1–100:1 C–N ratio for bioremediation of petroleum hydrocarbons from underground storage tank leakage (U.S. EPA, 1995). A higher indigenous biodegradation rate was obtained at C–N ratios of 10:1 at diesel oil contaminated soil (MARGESIN and SCHINNER, 1997b). The optimal ratios reported for biodegradation of petroleum compounds fall within this range, such as a 60:1 ratio in degradation of oil sludge hydrocarbons (DIBBLE and BARTHA, 1979) and a 50:1 ratio of gasoline biodegradation in a clay–loam soil (ZHOU and CRAWFORD, 1995). A separate study showed an optimum C–N ratio of 9:1 for the refinery sludge and of 124:1 for petrochemical sludge (BROWN et al., 1983). ALEXANDER (1994) suggested 33:1 as an appropriate C–N ratio. MORGAN and WATKINSON (1989b) reported a wide range of optimal C–N ratios between 9:1 and 200:1 for waste oil and sludge. These results suggest that there is no general rule for the C–N ratio and the optimal ratio has to be determined individually for each contaminated site.

In general, the addition of nutrients can increase the indigenous biodegradation activity significantly. However, there are studies that have shown either no effect or mixed impacts on addition of nutrients to the soils. Inhibition of mineralization of phenanthrene and glucose was reported in an oil contaminated soil when inorganic nitrogen was added (MORGAN and WATKINSON, 1992). Excessive nitrogen (C–N ratio 1.8:1) impairs gasoline biodegradation, the ammonia toxicity may contribute to this result (ZHOU and CRAWFORD, 1995). An almost complete inhibition of biodegradation activities was observed at a C–N ratio of 10:1 in diesel contaminated soil (MARGESIN and SCHINNER, 1997b). Inhibition of microbial activity and a reduction of degradation was observed in hydrocarbon contaminated soils at nitrogen levels of 200–300 mg kg^{-1} of soil (BRADDOCK et al., 1997). Degradation of oil sludge by landfarming was inhibited due to an excessive dosage (4,000 mg kg^{-1} of soil) of nitrogen (GENOUW et al., 1994). Nutrient supplements only increased total populations of heterotrophic microorganisms and showed little effect on the mineralization of ^{14}C-phenanthrene and ^{14}C-pyrene in five tested soils (CARMICHAEL and PFAENDER, 1997).

A different approach to determine the amount of nitrogen required in contaminated soils was proposed by WALWORTH et al. (1997). They suggested that the maximum nitrogen amount required for biodegradation is related to soil moisture (mg N kg^{-1} soil water). An optimum nitrogen level of 2,000 mg N kg^{-1} H$_2$O for the degradation of petroleum contaminated soil was recommended.

Slow release oleophilic fertilizers that have affinity to hydrocarbons are widely used in oil decontamination. The Inipol EPA® 22, an oil-in-water microemulsion, has been studied most intensively and is composed of oleic acid, tri(laureth-4) phosphate (surfactant and source of phosphorus), 2-butoxyl-1-ethanol, urea, and water. This fertilizer was used in large quantities to successfully treat oil contaminated beaches and to enhance biodegradation rates (LINDSTROM et al., 1991; PRITCHARD and COSTA, 1991; PRITCHARD et al., 1992; ATLAS, 1995). ROSENBERG et al. (1992) developed a controlled release, hydrophobic fertilizer (F1). F1 is a modified urea–formaldehyde polymer containing 18% of nitrogen and 10% phosphorus, and is insoluble in water. It binds to the oil–water interface with micro-

organisms and provides nutrients for bacterial growth and degradation activities.

Soil temperature can influence the result of nutrient amendment (WALWORTH and REYNOLDS, 1995). At 10 °C, bioremediation rate was not changed by addition of nitrogen or phosphorus to petroleum contaminated cryic soils. However, degradation rate was increased only by phosphorus addition at 20 °C. Addition of nitrogen did not affect the degradation rate since the untreated soil contained enough nitrogen for microbial activity.

It is worthy to note that the benefit of nutrient amendment can be limited immensely due to chemical interaction with minerals in the soil. Ammonium ions bind to clay minerals by cation exchange and phosphate anions react with calcium, iron and aluminum to form precipitates of phosphate salts (MORGAN and WATKINSON, 1992). Furthermore, an overdose of nitrogen fertilizers (ammonia, nitrate) could result in high concentrations of nitrate in the soil that might leach into groundwater.

4 Cometabolism

This is a transformation process of organic compounds by microorganisms, but these compounds and their products do not serve as a source of carbon or energy for growth. Stimulation of aerobic cometabolism has been used widely and successfully in soil and groundwater to degrade different environmental contaminants (reviewed by ALEXANDER, 1994).

Lately, a cometabolism process has been demonstrated successfully to degrade trichloroethylene (TCE) under aerobic conditions (ENSLEY, 1991). Methane (FOGEL et al., 1986; LITTLE, et al., 1988; BRIGMON et al., 1999), phenol (NELSON et al., 1987, 1988; HARKER and KIM, 1990), formate (McFARLAND et al., 1992), and toluene (WACKETT and GIBSON, 1988) were reported to serve as primary (growth) substrates for TCE degradation. Methane monooxygenase, phenol hydroxylase, and toluene dioxygenase were involved in aerobic cometabolism of TCE. McCARTY et al. (1998) reported that more than 80% of TCE was removed through *in situ* cometabolic degradation from groundwater by periodic toluene injection. A significant reduction of TCE (30–60%) was also observed in biofiltration columns with inoculation of *P. putida* F1 (COX, et al., 1998). Other chlorinated solvents, such as chloroethenes, also can be degraded aerobically by cometabolism (SEGAR and SPEITEL, 1995; SIPKEMA et al., 1997).

Cometabolism can take place anaerobically and has been studied intensively on environmental pollutants. Reductive dehalogenation, which involves removal of halogen atoms from the contaminant and replacement with hydrogen, is the major route for the transformation of halogenated pollutants under anaerobic conditions. The halogenated compound serves as electron acceptor, not as growth substrate, in reductive dehalogenation reactions. BOUWER and MCCARTY (1983) first reported reductive dehalogenation halogenated aliphatic compounds (TCE and carbon tetrachloride) which were degraded under methanogenic conditions when acetate served as growth substrate.

Anaerobic cometabolism can also be applied to a variety of aromatic environmental pollutants in sulfate reducing, denitrifying, and methanogenic environments. A sulfate reducing bacterium, *Desulfobacula toluolica* strain Tol2, cometabolically degrades *p*-toluidine to *p*-amino phenylacetate and phenylacetate as dead end products while using toluene as growth substrate (RABER et al., 1998). Nitrodiphenylamines were anaerobically transformed to the corresponding amino diphenylamines by sulfate reducing bacteria using either lactate or benzoate as growth substrates (DRZYZGA et al., 1996). The *o*-xylene can be cometabolically transformed by denitrifying bacteria using toluene as growth substrate. Nitrate was the electron acceptor for the cometabolic process (ARCANGELI and ARVIN, 1995). Some cometabolic degradation examples of other pollutants, such as TNT (DAUN et al., 1998), and chloroform (WEATHERS et al., 1997), were reported lately. Glucose and cathodic hydrogen, produced by anaerobic corrosion of elemental iron, were used as electron donor in cometabolism of TNT and chloroform, respectively.

ADRIAENS and VOGEL (1995) reported a comprehensive review on the bioremediation of chlorinated organics, including biodegradation, cometabolic processes, dehalogenation, and chlorinated solvents and chlorinated aromatics.

5 Bioaugmentation

It is a general assumption that one of the limitations of *in situ* bioremediation is that indigenous microorganisms are insufficient, either in capability or population to degrade specific contaminant(s). The introduction of natural isolates (bacteria or fungi) or genetically engineered microorganisms (GEM) to contaminated sites for enhancing the decomposition of contaminants is called bioaugmentation or inoculation. This method has been used for decades in municipal wastewaster treatment plants. Recently, this technology has been applied broadly to degradation of various hazardous wastes by using either pure or mixed cultures of microorganisms. A collection of papers from the *Third International in situ and On-Site Bioreclamation Symposium* were published specifically on the topic of bioaugmentation (HINCHEE et al., 1995).

Successful bioaugmentation is dependent on the following criteria:

(1) the microorganisms must be able to survive and maintain enough population, i.e., compete with indigenous organisms, adapt to the natural environment, and resist toxin(s) in hazardous waste sites,

(2) the microorganisms must be able to preserve their degradative ability, i.e., express appropriate enzymes capable of degrading target compounds,

(3) the microorganisms must have contact with the contaminant for degradation reactions to take place.

In addition, suitable growth conditions and appropriate nutrient supplements are essentials for added microorganisms to adapt to the environment. Some successful cases of bioaugmentation in activated sludge were reviewed lately by VAN LIMBERGEN et al. (1998). Positive results were also reported on a carbon tetrachloride contaminated aquifer by addition of a denitrifying bacterium, *Pseudomonas stutzeri* KC (DYBAS et al., 1998).

In 1968, degradation of oil in contaminated soil by addition of a natural isolate, *Cellumonas* sp., probably was among the first reported bioaugmentation tests (SCHWENDINGER, 1968). Other examples of bioaugmentation are to enhance degradation of phenanthrene in an oil–tar contaminated soil by *Alcaligenes* sp. (MOLLER and INGVIRSEN, 1993), and of PAH in a coal–tar contaminated soil by *Arthrobacter* sp. and *Pseudomonas* sp. (AAMAND et al., 1995). However, bioaugmentation with commercial products did not benefit the clean-up of Alaska North Slope crude oil (VENOSA et al., 1992). Another negative result was reported on degradation of petroleum hydrocarbons after 15 months of inoculation (MAXWELL and BAGAI, 1995). Laboratory cultured microorganisms (*Bacillus, Pseudomonas, Serratia*, and *Aztobacter*) were injected (once) into contaminated soil. Nutrient limitation may have been responsible for the failure of the task.

Bioaugmentation has been demonstrated successfully in clean-up of soils contaminated with chlorinated organics in laboratory and field studies. The degradation of chlorophenol contaminated soil was enhanced by the addition of *Rhodococcus chlorophenolicus* under laboratory conditions (VALO and SALKINOJA-SALONEN, 1986). Inoculation of an indigenous bacterium, *Alcaligenes eutrophus* TCP, to soil microcosms accomplished degradation of 2,4,6-trichlorophenol up to 500 mg L^{-1} (ANDREONI et al., 1998). Microbial degradation of pentachlorophenol (PCP) has been studied extensively and has been reviewed by MCALLISTER et al. (1996). There are several PCP degrading bacteria that have been used in soil remediation studies including *Arthrobacter* sp. ATCC 33790 (EDGEHILL and FINN, 1983), *Flavobacterium* sp. ATCC 39723 (CRAWFORD and MOHN, 1985; SEECH et al., 1991), *Pseudomonas resinovorans* (SEECH et al., 1994), *Pseudomonas* sp. strain SR3 (RESNICK and CHAPMAN, 1994), and *Rhodococcus chlorophenolicus* PCP-1 (BRIGLIA et al., 1994b; MIDDELDORP et al., 1990). The genus of some of the PCP degrading strains was reclassified as *Mycobacterium* (BRIGLIA et al., 1994a) and *Sphingomonas* (EDERER et al., 1997) due to phylogenetic evidence. SHAW et al. (1997) demonstrated that the degradation rate of PCPL linearly increased with increasing initial cell density up

to 1.1×10^9 CFU cm^{-3}. 7 d after inoculation *R. chlorophenolicus* showed higher adaptive ability (with a significant higher survival rate) in PCP contaminated soil than *Flavobacterium* sp. (MCBAIN et al., 1997). In a bench scale study, augmentation of PCP contaminated soil with activated soil, containing PCP acclimated consortium, resulted in a 99% (from 400 mg kg^{-1} s to 5 mg kg^{-1} oil) reduction of PCP concentration in 130 d (BARBEAU et al., 1997).

Degradation of TCE was also demonstrated by inoculating *B. cepacia* G4 and constitutive mutant strains PR1$_{301}$ and PEL–1010PC, under laboratory conditions (MUNAKATA-MARR et al., 1996; SWADLEY et al., 1999). The results suggested wild-type strain G4 has higher maximum enzyme activity, but mutant *PR1*$_{301}$ maintains activity for a longer period of time. A mathematical model on bioaugmentation for TCE remediation also suggested the biodegradation rate can be enhanced (WANG and CORAPCIOGLU, 1999).

Degradation of the herbicide 2,4-dichlorophenoxyacetic acid (2,4-D) was achieved by the inoculation of *P. cepacia* DBO1, *A. eutrophus* AEO106, and *A. eutrophus* JMP134 into the contaminated soil (JACOBSEN and PEDERSEN, 1992). More than 95% of 2,4-D were mineralized by augmentation of enriched indigenous microorganisms to contaminated soil (BAUD-GRASSET and VOGEL, 1995).

Degradation of polychlorinated biphenyls in various contaminated soils was enhanced by the inoculation with a PCB cometabolizer, *Acinetobacter* sp. P6 (BRUNNER et al., 1985), with chlorobenzoate degrading bacteria, *P. aeruginosa* JB2, and *P. putida* P111 (HICKEY et al., 1993), and a biphenyl degrading strain, *Alcaligenes xylosoxidans* (HALUŠKA et al., 1995). Bioremediation of PCB contaminated soil was also enhanced by repeated inoculation of carvone-induced bacterium, *Anthrobacter* sp. B1B (GILBERT and CROWLEY, 1998). It is interesting to note that carvone is a principal component of spearmint oil with moderate water solubility. However, repeated inoculation of a PCB degrader, *A. eutrophus* H850, failed to improve the biodegradation activity in PCB river sediments (HARKNESS et al., 1993).

In 1996, the first EPA approved field release of a genetically engineered microorganism (GEM) for bioremediation of PAH was car-

ried out by the Center for Environmental Biotechnology of the University of Tennessee in
collaboration with the Environmental Science
Division of Oak Ridge National Laboratory
(Sayler et al., 1999). The GEM, *Pseudomonas
fluorescens* HK44, was used in this investigation and designed as a bioluminescent reporter while it degrades naphthalene (KING et al.,
1990). This pilot study was performed in six
large soil lysimeters (4 m deep by 2.5 m in diameter) that were filled with contaminated
(containing naphthalene, phenanthrene, and
anthracene), and uncontaminated soil. After
two years, strain HK44 was capable of surviv-
ing in the lysimeter ecosystem and reporting
on the bioavailability of introduced PAH.

6 Conclusion

Biostimulation and bioaugmentation are accepted practices in enhancing bioremediation.
Improved knowledge on specific effects of enhancement practices and the toxicological profile of metabolites produced by indigenous organisms or bioaugmentative bioremediation
will further increase the applications and acceptance of this form of bioremediation. There
are indications that the chemical industry itself
may promote application of this technology
for difficult to degrade contaminants if they
can reduce ultimate chemical exposure at a
reasonable cost.

It is expected that stimulation and control of
anaerobic processes will likely continue to
grow. This is a result of new scientific advances
in documenting the mechanism of anaerobic
metabolism and cometabolism by defined and
mixed microbial processes; and through the
ability to manipulate electron donors and acceptors under field conditions.

7 References

AAMAND, J., BRUNTSE, G., JEPSEN, M., JØRGENSEN, C.,
JENSEN, B. K. (1995), Degradation of PAHs in soil
by indigenous and inoculated bacteria, in: *Bioaugmentation for Site Remediation* (HINCHEE, R.
E., FREDRICKSON, J., ALLEMAN, B. C., Eds.), pp.
121–127. Columbus, OH: Battelle Press.

ADRIAENS, P., VOGEL, T. M. (1995), Biological treatment of chlorinated organics, in: *Microbial Transformation and Degradation of Toxic Organic Chemicals* (YOUNG, L. Y., CERNIGLIA, C. E., Eds.), pp.
435–486. New York: Wiley-Liss.

ALEXANDER, M. (1994), *Biodegradation and Bioremediation*. San Diego, CA: Academic Press.

ALLEN-KING, R. M., BARKER, J. F., GILLHAM, R. W.,
JENSEN, B. K. (1994), Substrate- and nutrient-
limited toluene biotransformation in sandy soil,
Environ. Tox. Chem. **13**, 693–705.

ANDREONI, V., BAGGI, G., COLOMBO, M., CAVALCA,
L., ZANGROSSI, M., BERNASCONI, S. (1998), Degradation of 2,4,6-trichlorophenol by a specialized
organism and by indigenous soil microflora: bioaugmentation and self-remediability for soil restoration, *Lett. Appl. Microbiol.* **27**, 86–92.

ARCANGELI, J. P., ARVIN, E. (1995), Cometabolic
transformation of *o*-xylene in a biofilm system
under nitrate reducing conditions, *Biodegradation* **6**, 19–27.

ATLAS, R. M. (1991), Microbial hydrocarbon degradation, bioremediation of oil spills, *J. Chem. Technol.* **52**, 149–156.

ATLAS, R. M. (1995), Bioremediation of petroleum
pollutants, *Int. Biodeter. Biodegrad.* **35**, 317–327.

BAKER, K. H. (1994), *Bioremediation of surface and
subsurface soils*, in: *Bioremediation* (BAKER, K.
H., HERSON, D. S., Eds.), pp. 203–259. New York:
McGraw-Hill.

BARBEAU, C., DESCHÊNES, L., KARAMANEV, D., CO-
MEAU, Y., SAMSON, R. (1997), Bioremediation of
pentachlorophenol-contaminated soil by bioaugmentation using activated soil, *Appl Microbiol.
Biotechnol.* **48**, 745–752.

BAUD-GRASSET, F., VOGEL, T. M. (1995), Bioaugmen-
tation: biotreatment of contaminated soil by add-
ing adapted bacteria, in: *Bioaugmentation for Site
Remediation* (HINCHEE, R. E., FREDRICKSON, J.,
ALLEMAN, B. C., Eds.), pp. 39–48. Columbus, OH:
Battelle Press.

BEDARD, D. L., BUNNELL, S. C., SMULLEN, L. A.
(1996), Stimulation of microbial *para*-dechlorina-
tion of polychlorinated biphenyls that have persi-
sted in Housatonic river sediment for decades,
Environ. Sci. Technol. **30**, 687–694.

BOROLE, A. P., SUBLETTE, K. L., RATERMAN, K. T., JA-
VANMARDIAN, M., FISHER, J. B. (1997), The potential for intrinsic bioremediation of BTEX hydrocarbons in soil/ground water contaminated with
gas condensate, *Appl. Biochem. Biotechnol.* **63-65**,
719–730.

BOUWER, E. J., McCARTY, P. L. (1983), Transformation of 1- and 2-carbon halogenated aliphatic organic compounds under methanogenic conditions, *Appl. Environ. Microbiol.* **45**, 1286–1294.

BRADDOCK, J. F., RUTH, M. L., CATTERALL, P. H.,
WALWORTH, J. L., McCARTHY, K. A. (1997), Enhancement and inhibition of microbial activity in
hydrocarbon-contaminated arctic soils: implica-

tions for nutrient-amended bioremediation, *Environ. Sci. Technol.* **31**, 2078–2084.

BRIGLIA, M., EGGEN, R.I., VAN ELSAS, J. D., DE VOS, W. M. (1994a), Phylogenetic evidence for transfer of pentachlorophenol-mineralizing *Rhodococcus chlorophenolicus* strain PCP-1 to the genus *Mycobacterium*, *Int. J. Syst. Bacteriol.* **44**, 494–498.

BRIGLIA, M., MIDDELDORP, P. J. M., SALKINOJA-SALONEN, M. S. (1994b), Mineralization performance of *Rhodococcus chlorophenolicus* strain PCP-1 in contaminated soil simulating on site conditions, *Soil Biol. Biochem.* **26**, 377–385.

BRIGMON, R. L., ALTMAN, D. J., FRANCK, M. M., FLIERMANS, C. B., HAZEN, T. C., BOURQUIN, A. W. (1999), Evaluation of methanotrophic bacteria during injection of gaseous nutrients for *in situ* trichloroethylene bioremediation in a sanitary landfill, in: *Engineered Approaches for in situ Bioremediation of Chlorinated Solvent Contamination* (LEESON, L., ALLEMAN, B. C., Eds.), pp. 107–112. Columbus, OH: Battelle Press.

BROWN, K. W., DONNELLY, K. C., DEUEL, L. E. JR. (1983), Effects of mineral nutrients, sludge application rate, and application frequency on biodegradation of two oily sludges, *Microb. Ecol.* **9**, 363–373.

BRUNNER, W., SUTHERLAND, F. H., FOCHT, D. D. (1985), Enhanced biodegradation in soil by analog enrichment and bacterial inoculation, *J. Environ. Qual.* **14**, 324–328.

CARMICHAEL, L. M., PFAENDER, F. K. (1997), The effect of inorganic and organic supplements on the microbial degradation of phenanthrene and pyrene in soils, *Biodegradation* **8**, 1–13.

CATTANEO, M. V., MASSON, C., GREER, C. W. (1997), The influence of moisture on microbial transport, survival and 2,4-D biodegradation with a genetically marked *Burkholderia cepacia* in unsaturated soil columns, *Biodegradation* **8**, 87–96.

CHAOBAL, V. N., WOOLARD, C. R., WHITE, D. M. (1999), Effect of nutrients on microbial activity in contaminated arctic gravel pads, in: *In situ Bioremediation of Petroleum Hydrocarbon and Other Organic Compounds* (ALLEMAN, B. C., LEESON, A., Eds.), pp. 499–503. Columbus, OH: Battelle Press.

CHO, B.-Y., CHINO, H., TSUJI, H., KUNITO, T., NAGAOKA, K. et al. (1997), Laboratory-scale bioremediation of oil-contaminated soil of Kuwait with soil amendment materials, *Chemosphere* **35**, 1599–1611.

COX, C. D., WOO, H.-J., ROBINSON, K. G. (1998), Cometabolic biodegradation of trichloroethylene (TCE) in the gas phase, *Water Sci. Technol.* **37**, 97–104.

CRAWFORD, R. L., MOHN, W. W. (1985), Microbiological removal of pentachlorophenol from soil using a *Flavobacterium*, *Enzyme Microb. Technol.* **7**,

617–620.

DA CUNHA, C. D., GOMES FERREIRA LEITE, S. (1997), Optimization of some environmental conditions to enhance gasoline biodegradation in soil microcosms bioaugmented with *Pseudomonas putida*, *Revista De Microbiol.* **28**, 129–134.

DAUN, G., LENKE, H., REUSS, M., KNACKMUSS, H.-J. (1998), Biological treatment of TNT-contaminated soil. 1. Anaerobic cometabolic reduction and interaction of TNT and metabolites with soil components, *Environ. Sci. Technol.* **32**, 1956–1963.

DIBBLE, J. T., BARTHA, R. (1979), Effect of environmental parameters on the biodegradation of oil sludge, *Appl. Environ. Microbiol.* **37**, 729–739.

DOELMAN, P., HAANSTRA, L., VOS, A. (1988), Microbial sanitation of soil with alpha and beta HCH under aerobic glasshouse conditions, *Chemosphere* **17**, 489–492.

DRZYZGA, O., SCHMIDT, A., BLOTEVOGEL, K.-H. (1996), Cometabolic transformation and cleavage of nitrodiphenylamines by three newly isolated sulfate-reducing bacterial strains, *Appl. Environ. Microbiol.* **62**, 1710–1716.

DYBAS, M. J., BARCELONA, M., BEZBORODNIKOV, S., DAVIES, S., FORNEY, L. et al. (1998), Pilot-scale evaluation of bioaugmentation for *in situ* remediation of a carbon tetrachloride-contaminated aquifer, *Environ. Sci. Technol.* **32**, 3598–3611.

EDERER, M. M., CRAWFORD, R. L., HENWIG, R. P., ORSEN, C. S. (1997), PCP degradation is mediated by closely related strains of the genus *Sphingomonas*, *Mol. Ecol.* **6**, 39–49.

EDGEHILL, R. U., FINN, R. K. (1983), Microbial treatment of soil to remove pentachlorophenol, *Appl. Environ. Microbiol.* **45**, 1122–1125.

ENSLEY, B. D. (1991), Biochemical diversity of trichloroethylene metabolism, *Annu. Rev. Microbiol.* **45**, 283–299.

FOGEL, M. M., TADDEO, A. R., FOGEL, S. (1986), Biodegradation of chlorinated ethenes by a methane-utilizing mixed culture, *Appl. Environ. Microbiol.* **51**, 720–724.

FURUKAWA, K., HAYASE, N., TAIRA, K. (1990), Biphenyl/polychlorinated biphenyl catabolic gene (*bph* operon): organization, function, and molecular relationship in various pseudomonads, in: *Pseudomonas: Biotransformations, Pathogenesis, and Evolving Biotechnology* (SILVER, S., CHAKRABARTY, A. M., IGLEWSKI, B., KAPLAN, S., Eds.), pp. 111–120. Washington, DC: American Society for Microbiology.

GENOUW, G., DE NAEYER, F., VAN MEENEN, P., VAN DE WERF, H., DE NIJS, W., VERSTRAETE, W. (1994), Degradation of oil sludge by landfarming – a case study at the Ghent harbor, *Biodegradation* **5**, 37–46.

GIBSON, D. T. (Ed.) (1984), *Microbial Degradation of Organic Compounds*. New York: Marcel Dekker.

GILBERT, E. S., CROWLEY, D. E. (1998), Repeated application of carvone-induced bacteria to enhance biodegradation of polychlorinated biphenyls in soil, *Appl. Microbiol. Biotechnol.* **50**, 489–494.

HAGGBLOM, M. M., RIVERA, M. D., YOUNG, L. Y. (1993), Influence of alternative electron acceptors on the anaerobic biodegradability of chlorinated phenols and benzoic acids, *Appl. Environ. Microbiol.* **59**, 1162–1167.

HALUŠKA, L., BARANČÍKOVA, G., BALÁZ, Š., DERCOVA, K., VRANA, B. et al. (1995), Degradation of PCB in different soils by inoculated *Alcaligenes xylosoxidans, Sci. Tot. Environ.* **175**, 275–285.

HANSTVEIT, A. O., VAN GEMERT, W. J. T., JANSSEN, D. B., RULKENS, W. H., VAN VEEN, H. J. (1988), Literature study on the feasibility of microbiological decontamination of polluted soils, in: *Biotreatment Systems* Vol. 1 (WISE, D. L., Ed.), pp. 63–155. Boca Raton, FL: CRC Press.

HARKER, A. R., KIM, Y. (1990), Trichloroethylene degradation by two independent aromatic-degrading pathways in *Alcaligenes eutrophus* JMP134, *Appl. Environ. Microbiol.* **56**, 1179–1181.

HARKNESS, M. R., MCDERMOTT, J. B., ABRAMOWICZ, D. A., SALVO, J. J., FLANAGAN, W. P. et al. (1993), *In situ* stimulation of aerobic PCB biodegradation in Hudson River sediments, *Science* **259**, 503–507.

HICKEY, W. J., SEARLES, D. B., FOCHT, D. D. (1993), Enhanced mineralization of polychlorinated biphenyls in soil inoculated with chlorobenzoate-degrading bacteria, *Appl. Environ. Microbiol.* **59**, 1194-1200.

HINCHEE, R. E., FREDRICKSON, J., ALLEMAN, B. C. (Eds.) (1995), Bioaugmentation for site remediation, Columbus, OH: Battelle Press.

HUESEMANN, M. H. (1994), Guidelines for land-treating petroleum hydrocarbon-contaminated soils, *J. Soil Contam.* **3**, 299–318.

HÜTTMANN, S., ANGELMI, B., PETERS, H., JÜTTER-SCHENKE, P., BEYER, L. (1999), Enhancement on *in situ* and on-site bioremediation by radiowave frequency soil heating, in: *In situ Bioremediation of Petroleum Hydrocarbon and Other Organic Compounds* (ALLEMAN, B. C., LEESON, A., Eds.), pp. 487–491. Columbus, OH: Battelle Press.

JACOBSEN, C. S., PEDERSEN, J. C. (1992), Mineralization of 2,4-dichlorophenoxyacetic acid (2,4-D) in soil inoculated with *Pseudomonas cepacia* DBO1 (pRO101), *Alcaligenes eutrophus* AEO106 (pRO101), and *A. eutrophus* JMP134(pJP4): effects of inoculation level and substrate concentration, *Biodegradation* **2**, 253–263.

KATO, K., DAVIS, K. L. (1996), Current use of bioremediation for TCE cleanup: results of a survey. Remediation/Fall.

KING, J. M. H., DIGRAZIA, P. M., APPLEGATE, B., BURLAGE, R., SANSEVERINO, J. et al. (1990), Rapid, sensitive bioluminescent reporter technology for naphthalene exposure and biodegradation, *Science* **249**, 778–781.

KIRCHMANN, H., EWNETU, W. (1998), Biodegradation of petroleum-based oil wastes through composting, *Biodegradation* **9**, 151–156.

LAJOIE, C. A., STROM, P. F. (1994), Biodegradation of polynuclear aromatic hydrocarbons in coal tar oil contaminated soil, in: *Remediation of Hazardous Waste Contaminated Soil* (WISE, D. L., TRANTOLO, D. J., Eds.), pp. 149–162. New York: Marcel Dekker.

LEAHY, J. G., COLWELL., R. R. (1990), Microbial degradation of hydrocarbon in the environment, *Microbiol. Rev.* **54**, 305–315.

LEWIS, D. L., KOLLIG, H. P., HODSON, R. E. (1986), Nutrient limitation and adaptation of microbial populations to chemical transformations, *Appl. Environ. Microbiol.* **51**, 598–603.

LINDSTROM, J. E., PRINCE, R. C., CLARK, J. C., GROSSMAN, M. J., YEAGER, T. R. et al. (1991), Microbial populations and hydrocarbon biodegradation potentials in fertilized shoreline sediments affected by the T/V Exxon Valdez oil spill, *Appl. Environ. Microbiol.* **57**, 2514–2522.

LITTLE, C. D., PALUMBO, A. V., HERBES, S. E., LIDSTROM, M. E., TYNDALL, R. L., GILMER, P. J. (1988), Trichloroethylene biodegradation by a methane-oxidizing bacterium, *Appl. Environ. Microbiol.* **54**, 951–956.

MARGESIN, R., SCHINNER, F. (1994), Properties of cold-adapted microorganisms and their potential role in the biotechnology, *J. Biotechnol.* **33**, 1–14.

MARGESIN, R., SCHINNER, F. (1997a), Bioremediation of diesel-oil-contaminated alpine soils at low temperature, *Appl. Microbiol. Biotechnol.* **47**, 462–468.

MARGESIN, R., SCHINNER, F. (1997b), Laboratory bioremediation experiments with soil from a diesel-oil contaminated sitel significant role of cold-adapted microorganisms and fertilizers, *J. Chem. Tech. Biotechnol.* **70**, 92–98.

MAXWELL, C. R., BAQAI, H. A. (1995), Remediation of petroleum hydrocarbons by inoculation with laboratory-cultured microorganisms, in: *Bioaugmentation for Site Remediation* (HINCHEE, R. E., FREDRICKSON, J., ALLEMAN, B. C., Eds.), pp. 129–137. Columbus, OH: Battelle Press.

MCALLISTER, K. A., LEE, H., TREVORS, J. T. (1996), Microbial degradation of pentachlorophenol, *Biodegradation* **7**, 1–40.

MCBAIN, A. D. K., SALKINOJA-SALONEN, M. S., SENIOR, E., DU PLESSIS, C. A., PATERSON, A., WATSON-CRAIK, I. A. (1997), Pentachlorophenol-contaminated soil bioremediation: survival and efficacy of monoculture inoculants and enrichment of indigenous catabolic populations, *S. Afr. J. Plant Soil*, **14**, 81-88.

MCCARTY, P. L., GOLTZ, M. N., HOPKINS, G. D., DO-

LAN, M. E., ALLAN, J. P. et al. (1998), Full-scale evaluation of *in situ* cometabolic degradation of trichloroethylene in groundwater through toluene injection, *Environ. Sci. Technol.* **32**, 88–100.

MCFARLAND, M. J., VOGEL, C. M., SPAIN, J. C. (1992), Methanotrophic cometabolism of trichloroethylene (TCE) in a two stage bioreactor system, *Water Res.* **26**, 259–265.

MIDDELDORP, P. J. M., BRIGLIA, M., SALKINOJA-SALONEN, M. S. (1990), Biodegradation of pentachlorophenol in natural soil by inoculated *Rhodococcus chlorophenolicus, Microb. Ecol.* **20**, 123–139.

MOLLER, J., INGVIRSEN, H. (1993), Biodegradation of phenanthrene in soil microcosms stimulated by an introduced *Alcaligenes* sp., *FEMS Microbiol. Ecol.* **102**, 271–278.

MORGAN, P., WATKINSON, R. J. (1989a), Microbiological methods for the clean-up of soil and groundwater contaminated with halogenated organic compounds, *FEMS Microbiol. Rev.* **63**, 277–300.

MORGAN, P., WATKINSON, R. J. (1989b), Hydrocarbon degradation in soils and methods for soil biotreatment, *CRC Crit. Rev. Biotechnol.* **8**, 305–333.

MORGAN, P., WATKINSON, R. J. (1992), Factors limiting the supply and efficiency of nutrient and oxygen supplements for the *in situ* biotreatment of contaminated soil and groundwater, *Water Res.* **26**, 73–78.

MUNAKATA-MARR, J., MCCARTY, P. L., SHIELD, M. S., REAGIN, M., FRANCESCONI, S. C. (1996), Enhancement of trichloroethylene degradation in aquifer microcosms bioaugmented with wild-type and genetically altered *Burkholderia* (*Pseudomonas*) *cepacia* G4 and PR1, *Environ. Sci. Technol.* **30**, 2045–2052.

NELSON, M. J. K., MONTGOMERY, S. O., MAHAFFEY, W. R., PRITCHARD, P. H. (1987), Biodegradation of trichloroethylene: an involvement of an aromatic biodegradative pathway, *Appl. Environ. Microbiol.* **53**, 949–954.

NELSON, M. J. K., MONTGOMERY, S. O., PRITCHARD, P. H. (1988), Trichloroethylene metabolism by microorganisms that degrade aromatic compounds, *Appl. Environ. Microbiol.* **54**, 604–606.

PARDIECK, D. L., BOUWER, E. J., STONE, A. T. (1992), Hydrogen peroxide use to increase oxidant capacity for *in situ* bioremediation of contaminated soils and aquifers: a review, *J. Contam. Hydrol.* **9**, 221–242.

PRAMER, D., BARTHA, R. (1972), Preparation and processing of soil samples for biodegradation studies, *Environ. Lett.* **2**, 217–224.

PRITCHARD, P. H., COSTA, C. F. (1991), EPA's Alaska oil spill bioremediation project, *Environ. Sci. Technol.* **25**, 372–379.

PRITCHARD, P. H., MULLER, J. G., ROGERS, J. C., KREMER, F. V., GLASER, J. A. (1992), Oil spill bioremediation: experiences, lessons and results from the Exxon Valdez oil spill in Alaska, *Biodegradation* **3**, 315–335.

PROVIDENTI, M. A., LEE, H., TREVORS, J. T. (1993), Selected factors limiting the microbial degradation of recalcitrant compounds, *J. Ind. Microbiol.* **12**, 379–395.

RABER, T., GORONTZY, T., KLEINSCHMIDT, M., STEINBACH, K., BLOTEVOGEL, K.-H. (1998), Anaerobic degradation and transformation of *p*-toluidine by the sulfate-reducing bacterium, *Curr. Microbiol.* **37**, 172–176.

RESNICK, S. M., CHAPMAN, P. J. (1994), Physiological properties and substrate specificity of a pentachlorophenol degrading *Pseudomonas* species, *Biodegradation* **5**, 47–54.

RISER-ROBERTS, E. (1998), Remediation of petroleum contaminated soils: biological, physical, and chemical processes. New York: Lewis Publishers.

ROSENBERG, E., LEGMANN, R., KUSHMARO, A., TAUBE, R., ALDER, E., RON, E. Z. (1992), Petroleum biodegradation – a multiphase problem, *Biodegradation* **3**, 337–350.

SAYLER, G. S., COX, C. D., BURLAGE, R., RIPP, S., NIVENS, D. et al. (1999), Field application of a genetically engineered microorganism for polycyclic aromatic hydrocarbon bioremediation process monitoring and control, in: *Novel Approaches for Bioremediation of Organic Pollution* (FASS, R., FLASHNER, Y., REUVENY, S., Eds.), pp. 241–254. New York: Kluwer Academic/Plenum Publishers.

SCHWENDINGER, R. B. (1968), Reclamation of soil contaminated with oil, *J. Inst. Petrol.* (London) **54**, 182–197.

SEECH, A. G., TREVORS, J. T., BULMAN, T. L. (1991), Biodegradation of pentachlorophenol in soil: the response of physical, chemical, and biological treatment, *Can. J. Microbiol.* **37**, 440–444.

SEECH, A. G., MARVAN, I. J., TREVORS, J. T. (1994), On site/*ex situ* bioremediation of industrial soils containing chlorinated phenols and polycyclic aromatic hydrocarbons, in: *Bioremediation of Chlorinated and Polycyclic Aromatic Hydrocarbon Compounds* (HINCHEE, R. E., LEESON, A., SEMPRINI, L., ONG, S. K., Eds.), pp. 451–455. Boca Raton, FL: Lewis Publishers.

SEGAR, R. L. JR., SPEITEL, G. E. JR. (1995), Cometabolism of chloroethene mixtures by biofilm grown on phenol, in: *Bioremediation of Chlorinated Solvents* (HINCHEE, R., LEESON, A., SEMPRINI, L., Eds.), pp. 245–253. Columbus, OH: Battelle Press.

SHARPLEY, A. N. (1991), Effect of soil pH on cation and anion solubility, *Commun. Soil Sci. Plant Anal.* **22**, 1391–1399.

SHAW, K. W., LEE, H., TREVORS, J. T. (1997), Effect of initial cell density, substrate concentration and temperature on pentachlorophenol degradation by *Pseudomonas* sp. UG30, *J. Chem. Tech. Biotechnol.* **69**, 107–113.

SIPKEMA, E. M., MOCORA, J., DE KONING, W., VAN HYLCKAMA VLIEG, J. E. T., GANZEVELD, K. J. et al. (1997), Aerobic cometabolic degradation of chlorinated ethenes in a two step system, in: *Proc. 4th Symp. Int. in situ On-Site Biorem.*, pp. 73–78. Columbus, OH: Battelle Press.

SORKOH, N. A., IBRAHIM, A. S., GHANNOUM, M. A., RADWAN, S. S. (1993), High temperature hydrocarbon degradation by a *Bacillus stearothermophilus* from oil polluted Kuwaiti desert, *Appl. Microbiol. Biotechnol.* **39**, 123–126.

SPAIN, J. C., MILLIGAN, J. D., DOWNEY, D. C., SLAUGHTER, J. K. (1989), Excessive bacterial decomposition of H_2O_2 during enhanced biodegradation, *Ground Water* **27**, 163–167.

SPARROW, S. D., SPARROW, E. B. (1988), Microbial biomass and activity in a subarctic soil ten year after crude spills, *J. Environ. Qual.* **17**, 304–309.

STAPLETON, R. D., SAVAGE, D. C., SAYLER, G. S., STACEY, G. (1998), Biodegradation of aromatic hydrocarbons in an extremely acidic environment, *Appl. Environ. Microbiol.* **64**, 4180–4184.

SWADLEY, L. A., MUNAKATA-MARR, J., MAHAFFEY, R. W. (1999), Bioaugmentation for trichloroethylene remediation: wild-type and constitutive mutant strains, in: *Engineered Approaches for in situ Bioremediation of Chlorinated Solvent Contamination* (LEESON, A., ALLEMAN, B. C., Eds.), pp. 293–297. Columbus, OH: Battelle Press.

THOMAS, A. O., LESTER, J. N. (1993), The microbial remediation of former gasworks sites: a review, *Environ. Tech.* **14**, 1–24.

THOMAS, J. M., WARD, C. H. (1989), *In situ* biorestoration of organic contaminants in the subsurface, *Environ. Sci. Technol.* **23**, 760–766.

U.S. EPA (1985), Handbook for remedial action at waste disposal sites (revised). *EPA 625-6-85-006.*

U.S. EPA (1995), How to evaluate alternative cleanup technologies for underground storage tank sites: a guide for corrective action plan Reviewers. *EPA 510-B-95-007.*

VALO, R., SALKINOJA-SALONEN, M. (1986), Bioreclamation of chlorophenol-contaminated soil by composting, *Appl. Microbiol. Biotechnol.* **25**, 68–75.

VAN LIMBERGEN, H., TOP, E. M., VERSTRAETE, W. (1998), Bioaugmentation in activated sludge: current features and future perspectives, *Appl. Microbiol. Biotechnol.* **50**, 16–23.

VENOSA, A. D., HAINES, J. R., ALLEN, D. M. (1992), Efficacy of commercial inocula in enhancing biodegradation of weathered crude oil contaminating a Prince William Sound beach, *J. Ind. Microbiol.* **10**, 1–11.

VERSTRAETE, W. R., VANLOOCKE, R., DE BORGER, R., VERLINDE, A. (1976), Modeling of the breakdown

and the mobilization of hydrocarbons in unsaturated soil layers, in: *Proc. 3rd Int. Biodegradation Symp.* (SHARPLEY, J. M., KAPLAN, A. M., Eds.), pp. 98–112. London: Applied Science Publishers.

WACKETT, L. P., GIBSON, D. T. (1988), Degradation of trichloroethylene by toluene dioxygenase in whole-cell studies with *Pseudomonas putida* F1, *Appl. Environ. Microbiol.* **54**, 1703–1708.

WALWORTH, J. L., REYNOLDS, C. M. (1995), Bioremediation of a petroleum-contaminated cryic soil: effects of phosphorus, nitrogen, and temperature, *J. Soil Contam.* **4**, 299–310.

WALWORTH, J. L., WOOLARD, C. R., BRADDOCK, F., REYNOLDS, C. M. (1997), Enhancement and inhibition of soil petroleum biodegradation through the use of fertilizer nitrogen: an approach to determining optimum levels, *J. Soil Contam.* **6**, 465–480.

WALWORTH, J., WOOLARD, C., ACOMB, L., CHAOBAL, V., WALLACE, M. (1999), Nutrient and temperature interaction in bioremediation of petroleum-contaminated cryic soil, in: *In situ Bioremediation of Petroleum Hydrocarbon and Other Organic Compounds* (ALLEMAN, B. C., LEESON, A., Eds.), pp. 505–510. Columbus, OH: Battelle Press.

WANG, S., CORAPCIOGLU, M. Y. (1999), An analysis of exogenous bacteria injection for improved biodegradation, in: *Engineered Approaches for in situ Bioremediation of Chlorinated Solvent Contamination* (LEESON, A., ALLEMAN, B. C., Eds.), pp. 269–274. Columbus, OH: Battelle Press.

WEATHERS, L. J., PARKIN, G. F., ALVAREZ, P. J. (1997), Utilization of cathodic hydrogen as electron donor for chloroform cometabolism by a mixed, methanogenic culture, *Environ. Sci. Technol.* **31**, 880–885.

WHYTE, L. G., GREER, C. W., INNISS, W. E. (1996), Assessment of the biodegradation potential of psychrotrophic microorganisms, *Can. J. Microbiol.* **42**, 99–106.

WHYTE, L. G., BOURBONNIERE, L., GREER, C. W. (1997), Biodegradation of petroleum hydrocarbons by psychrotrophic *Pseudomonas* strains possessing both alkane (alk) and naphthalene (nah) catabolic pathways, *Appl. Environ. Microbiol.* **63**, 3719–3723.

WHYTE, L. G., HAWARI, J., ZHOU, E., BOURBONNIERE, L., INNISS, W. E., GREER, C. W. (1998), Biodegradation of variable-chain-length alkanes at low temperatures by a psychrotrophic *Rhodococcus* sp., *Appl. Environ. Microbiol.* **64**, 2578–2584.

WHYTE, L. G., BOURBONNIERE, L., BELLROSE, C., GREER, C. W. (1999), Bioremediation assessment of hydrocarbon-contaminated soils from the high arctic, *Can. Biorem. J.* **3**, 69–79.

ZHOU, E., CRAWFORD, R. L. (1995), Effects of oxygen, nitrogen, and temperature on gasline biodegradation in soil, *Biodegradation* **6**, 127–140.

21 Genetically Engineered Microorganisms and Bioremediation

FU-MIN MENN
JAMES P. EASTER
GARY S. SAYLER

Knoxville, TN 37996-1605, USA

List of Abbreviations

BBIC	bioluminescent bioreporter integrated circuit
BTEX	benzene, toluene, ethylbenzene, and xylene
2,4-D	2,4-dichlorophenoxyacetic acid
DBT	dibenzothiophene
FAV	field application vector
GEM	genetically engineered microorganisms
IC	integrated circuit
IGP	Igepal CO-720®
IPB	isopropylbenzene
IPTG	isopropyl thio-β-galactoside
PAH	phenanthrene, pyrene and benzo[a]pyrene
PCB	polychlorinated biphenyls
PCP	pentachlorophenol
SDS-PAGE	sodium dodecyl sulfate polyacrylamide gel electrophoresis
sMMO	soluble methane monooxygenase
2,4,5-T	2,4,5-trichloro phenoxy acetic acid
TCE	trichloroethylene
TMO	toluene monooxygenase
TNT	2,4,6-trinitrotoluene
TOL	toluene plasmid pWWO

1 Introduction

During the past 20 years, recombinant DNA techniques have been studied intensively to improve the degradation of hazardous wastes under laboratory conditions. Only one field test has been successfully implemented (SAYLER et al., 1999). Recombinant bacteria can be obtained by genetic engineering techniques or by natural genetic exchange between bacteria. Applications for genetically engineered microorganisms (GEM) in bioremediation have received a great deal of attention, but have largely been confined to the laboratory environment. This has been due to regulatory risk assessment concerns, and to a large extent the uncertainty of their practical impact and delivery under field conditions. There are at least four principal approaches to GEM development for bioremediation application. These include:

(1) modification of enzyme specificity and affinity,
(2) pathway construction and regulation,
(3) bioprocess development, monitoring, and control, and
(4) bioaffinity bioreporter sensor applications for chemical sensing, toxicity reduction, and end point analysis.

There are many reports on the degradation of environmental pollutants by different bacteria. However, only potential bioremediation related to GEM will be reviewed, and the construction and application of GEM will be the focus of this review. Some examples of the relevant use of genetic engineering technology to improve bioremediation are listed in Tab. 1. These genetically engineered microorganisms have higher degradative capacity and have been demonstrated successfully for the degradation of various pollutants under defined conditions. However, ecological and environmental concerns and regulatory constraints are major obstacles for testing GEM in the field. These problems must be overcome before GEM can provide an effective clean-up process at lower cost.

The use of genetically engineered microorganisms has been applied to bioremediation

Tab. 1. Genetic Engineering for Biodegradation of Contaminants

Microorganism	Modification	Contaminants	Reference
Pseudomonas sp. B13	pathway	mono/dichlorobenzoates	REINEKE and KNACKMUSS, 1979, 1980
P. putida	pathway	4-ethylbenzoate	RAMOS et al., 1987
P. putida KT2442	pathway	toluene/benzoate	PANKE et al., 1998
Pseudomonas sp. FR1	pathway	chloro-, methylbenzoates	ROJO et al., 1987
C. testosteroni VP44	substrate specificity	*o-, p*-monochlorobiphenyls	HRYWNA et al., 1999
Pseudomonas sp. LB400	substrate specificity	PCB	ERICKSON and MONDELLO, 1993
E. coli JM109(pSHF1003)	substrate specificity	PCB, benzene, toluene	KUMAMMRU et al., 1998
P. pseudoalcaligenes KF707-D2	substrate specificity	TCE, toluene, benzene	SUYAMA et al., 1996
E. coli FM5/pKY287	regulation	TCE, toluene	WINTER et al., 1989

process monitoring, strain monitoring, stress response, end point analysis, and toxicity assessment. Examples of these applications are listed in Tab. 2. The range of tested contaminants includes chlorinated compounds, aromatic hydrocarbons, heavy metals, and nonpolar toxicants, etc.

Development of effective and cost-efficient bioremediation processes is the goal for environmental biotechnology. The combination of microbiological and ecological knowledge, biochemical mechanisms, and field engineering designs are essential elements for successful *in situ* bioremediation using GEM.

2 Degradative (Catabolic) Genes

Extensive review information is available on the biochemical pathway analysis, operon structure, and molecular biology of biodegradative pathways important in bioremediation. Much of this information, critical to the development of appropriate GEM, is confined to aerobic catabolic and cometabolic pathways (SAYLER et al., 1998).

Tab. 2. Genetic Engineering for Biodegradation Process Efficacy

Microorganism	Application	Contaminants	Reference
A. eutrophus H850Lr	process monitoring	PCB	VAN DYKE et al., 1996
P. putida TVA8	process monitoring	TCE, BTEX	APPLEGATE et al., 1998
P. fluorescens HK44	process monitoring	naphthalene, anthracene, phenanthrene	SAYLER et al., 1999
B. cepacia BRI6001L	strain monitoring	2,4-D	MASSON et al., 1993
P. fluorescens 10586s/pUCD607	stress response	BTEX	SOUSA et al., 1998
P. fluorescens 10586s/pUCD607	toxicity assessment	chlorobenzenes, chlorophenols, BTEX	BOYD et al., 1998; SINCLAIR et al., 1999; GLOVER et al., 1999
Pseudomonas strain Shk1	toxicity assessment	Cd, 2,4-dinitrophenol, hydroquinone	KELLY et al., 1999
A. eutrophus 2050	end point analysis	nonpolar narcotics	LAYTON et al., 1999

2.1 Branched Aromatic Hydrocarbons

2.1.1 *Pseudomonas* spp.

The TOL plasmid pathway of *Pseudomonas putida* has been manipulated extensively to expand its catabolic capability on different branched compounds. RAMOS et al. (1987) modified the TOL metabolic pathway to use 4-ethylbenzoate as a substrate. These restructured strains were mutants isolated by either altering a pathway regulator XylS (RAMOS et al., 1986) or by modifying substrate specificity of enzymes (catechol-2,3-dioxygenase).

The complete upper TOL operon of plasmid pWW0 with its regulator gene, *xylR*, was reconstructed as a single gene cassette and yielded a hybrid mini-Tn5 [*upp* TOL] transposon. The transposon was further inserted into the chromosome of *P. puida* KT2442. The hybrid strain, KT2442::mini-Tn5 [*upp* TOL]Δ[*Km*r/*xylE*], was stable and grew on toluene as the sole carbon source. Toluene was degraded to benzoate by the TOL upper pathway and benzoate was further metabolized through an *o*-cleavage pathway (PANKE et al., 1988).

2.2 Chlorinated Compounds

2.2.1 Chlorobenzoates

2.2.1.1 *Pseudomonas* sp. B13 – Hybrid Pathways

The plasmid pWR1 encoding 3-chlorobenzoate and 4-chlorophenol degradation genes in strain B 13 have been modified extensively to generate hybrid pathways that extend vertical degradation of haloaromatic compounds. REINEKE and KNACKMUSS (1979, 1980) first reported the introduction of TOL plasmid pWW0, of *P. putida* mt-2 into strain B13 via conjugation. The transconjugants demonstrated the capability to utilize mono- and dichlorobenzoates as growth substrates. The inactivation of the TOL plasmid encoded *m*-pathway (by loss of catechol-2,3-dioxygenase activity)

was an obligatory requirement for the hybrid strain, *Pseudomonas* sp. WR211, to utilize 4-chlorobenzoate as the sole carbon source (REINEKE et al., 1982).

SCHWIEN and SCHMIDT (1982) reported that a benzoate and phenol degrader, *Alcaligenes* sp. A7, acquired chlorocatechol degrading capacity from strain B13 through conjugation. The recombinant strain, *Alcaligenes* sp. A7-2, exhibited the ability to utilize all three isomeric chlorophenols as the sole source of carbon and energy.

The *xylXYZ* (toluate-1,2-dioxygenase) and *xylL* (carboxylate dehydrogenase) genes from TOL plasmid (pWW0-161) of *P. putida* and *nahG* (salicylate hydroxylase) from NAH7 plasmid were cloned into strain B13 to extend its substrate range (LEHRBACH et al., 1984). A hybrid plasmid, including genes (*xylXYZL*), positive regulator (*xylS*), and their native promoter (Pm), was introduced into strain B13 by conjugation. The resulting *Pseudomonas* sp. B13(TOL) was capable of utilizing 3-chloro-, 4-chloro-, and 3,5-dichlorobenzoate. The transconjugants containing the *nahG* gene were able to degrade salicylate, 3-, 4-, and 5-chlorosalicylate.

ROJO et al. (1987) reported a modified *o*-cleavage pathway, in strain B13, for simultaneous degradation of chloro- and methylaromatic compounds. This hybrid pathway included:

(1) toluate 1,2-dioxygenase and carboxylate dehydrogenase from the TOL pathway to degrade 4-chlorobenzoate and transform methylbenzoates into methyl-2-enelactones (strain FR1),
(2) 4-methyl-2-enelactone isomerase from *Ralstonia eutropha* (formerly *Alcaligenes eutrophus*) JMP134 to transform 4-methyl-2-enelactones to 3-methyl-2-enelactones to complete the *o*-pathway for 4-methylbenzoate [strain FR1(pRFC20P)],
(3) phenol hydroxylase from B13 to mineralize chloro- and methylphenols (cresols) to corresponding catechols [strain FR1(pFRC20P)-1].

A hybrid plasmid, pFRC4P (Tn5::*xylXYZLS*), containing Pm promoter was introduced into strain B13 through triparental mating using *E.*

coli HB101(pFRC4P) as a donor, and *E. coli* HB101(pRK2013) as a helper. The hybrid Tn5 transposon that carried TOL genes was transposed into the chromosome of B13 and yielded strain FR1. Gene encoded 4-methyl-2-enelactone isomerase from *R. eutropha* JMP134 was cloned and transferred by conjugation into FR1. The resulting strain containing a hybrid plasmid, pFRC20P, was designated as B13FR1(pFRC20P) and was capable of growing on 4-methylbenzoate. Spontaneous mutants of FR1(pFRC20P), such as FR1-(pFRC20P)-1 and -2, were able to use 4-methylphenol as a sole source of carbon and energy. The engineered bacterium *Pseudomonas* sp. B13FR1(pFRC20P)-1 was capable of growing on and mineralizing mixtures of 3-chloro-, 4-chloro-, and 4-methylbenzoate and 4-chloro- and 4-methylphenol via the modified *o*-pathway. Studies on the survival rate of the engineered strain, the stability and expression of the recombinant pathways, and gene transfer in the environment were conducted in an activated sludge microcosm (NÜßLEIN et al., 1992). Strain FR1(pFRC20P) was also examined in two different aquatic sediments and the results showed enhancement of the degradation rate of contaminants (PIPKE et al., 1993).

2.2.1.2 *Pseudomonas aeruginosa* AC869(pAC31)

Strain *P. putida*, containing plasmid pAC25 that encodes genes for the degradation of 3-chlorobenzoate, has been shown to be structurally homologous to pWR1 (CHATTERJEE and CHAKRABARTY, 1983). Genetic rearrangements occurred between plasmids pAC25 and TOL under chemostat maintenance and yielded pAC31. The new plasmid contained *xylDGEF* genes located on the chromosome. The resulting strain AC869(pAC31) showed the capability to degrade 3,5-dichlorobenzoate and 3- and 4-chlorobenzoate (CHATTERJEE and CHAKRABARTY, 1982).

2.2.1.3 *Pseudomonas* sp. US1 ex.

A monochlorobenzoate and 2,4-D degrader, strain US1 ex. containing pJP4 was obtained through conjugation. The *E. coli* JMP397, harboring plasmid pJP4 (no expression), was used as a donor and the 4-chlorobenzoate degrader, US1, was the recipient. The new strain released stoichiometric amounts of chloride when grown on respective chloroaromatics as carbon source (SAHASRABUDHE and MODI, 1991).

There are reports on the horizontal transfer of 3-chlorobenzoate catabolic plasmid (pBRC60) in a freshwater ecosystem (FULTHTHROPE and WYNDHAM, 1991) and in an activated sludge unit (pD10; MCCLURE et al., 1989).

2.2.2 Polychlorinated Biphenyls (PCB) and Chlorobiphenyls

2.2.2.1 *Ralstonia eutropha* (formerly *Alcaligenes eutrophus*) AE707/AE1216

Chromosomally located PCB catabolic genes of *R. eutropha* A5 (SHIELDS et al., 1985), *Achromobacter* sp. LBS1C1 (PETTIGREW et al., 1990), and *Alcaligenes dentrificans* sp. JB1 (PARSONS et al., 1988) were transferred into a heavy metal resistant strain *R. eutropha* CH34 through natural conjugation. All donor strains degraded biphenyl and monochlorobiphenyls to corresponding benzoate and chlorobenzoates. Benzoate was further metabolized via the *o*-cleavage pathway in strains A5 and LBS1C1, and via the *m*-cleavage pathway in strain JB1. Strain A5 harbored a catabolic transposon, Tn4317, which carried biphenyl and 4-chlorobiphenyl degradation genes (SPRINGEAL et al., 1993a). The dehalogenase activity was mediated by plasmid pSS50 in strain A5 (LAYTON et al., 1992).

Transfer of PCB degradation genes from A5 to the heavy metal resistant strain CH34 was carried out by conjugation (SPRINGAEL et al., 1993b). The constructed strain, AE707, exhibited phenotype of Bph$^+$Cbp$^+$, and degraded 4-chlorobiphenyl to 4-chlorobenzoate in the presence of heavy metals. In resting cell assays (grown on biphenyl), strain AE707 cometabolized di- and trichlorinated congeners of Aroclor 1242® in the presence of heavy metals. The

PCB catabolic chromosomal genes of strain JB1 were transferred into CH34 through RP4::Mu3A mediated R-prime plasmid formation (SPRINGAEL et al., 1994). A transconjugant, strain AE1216, utilized 2-, 3- and 4-chlorobiphenyl and exhibited properties of metal resistance.

2.2.2.2 *Pseudomonas* sp. – Hybrid Strains

Pseudomonas putida BN10 grew on biphenyl and accumulated metabolites of 2-, 4-chlorobenzoate and 3-chlorocatechol from corresponding monochlorobiphenyls. 3-Chlorobiphenyl degraders were obtained from conjugation between *Pseudomonas* sp. B13 and strain BN10. The resulting *Pseudomonas* strain BN210 (gained chlorocatechol degradation genes) and B131 (acquired biphenyl degradation genes) were able to grow on 3-chlorobiphenyl. Both strains exhibited the capability to degrade monochlorobiphenyls and 2 of the dichlorobiphenyls found in Aroclor 1221® (MOKROSS et al., 1990).

Another hybrid *Pseudomonas* sp. strain UCR2 was isolated from multi-chemostat mating between a chlorobenzoate degrader, *P. aeruginosa* JB2, and a 2-chlorobiphenyl utilizer, *Arthrobacter* sp. strain B1Barc (HICKEY et al., 1992). Strain UCR2 exhibited ability to mineralize 2-chloro- and 2,5-dichlorobiphenyl. The UCR2 showed higher phenotypic similarity and higher genomic DNA homology to strain JB2. No hybridization was observed when the parental strains were probed against each other.

Recombinant *Pseudomonas* sp. strain CB15 was obtained by multi-chemostat mating by mixing *Pseudomonas* sp. strain HF1 (3-chlorobenzoate utilizer) and *Acinetobacter* sp. strain P6 (biphenyl utilizer) on ceramic beads (ADAMS et al., 1992). 3-chlorobiphenyl, 3-chlorobenzoate, and biphenyl could be utilized as growth substrates by strain CB15. Results of DNA hybridization suggested strain CB15 was closely related to parent strain HF1.

An *in vivo* recombinant plasmid, pDD530, containing the *bphABCD* genes from *Burkholderia* sp. strain LB400 was isolated from *P. putida* KT2442. The *bph* genes on pDD530

were further cloned into a pUT transposon vector to yield plasmid pDDPCB. Plasmid pDDPCB was mobilized into *Pseudomonas* sp. strain B13, and its genetically engineered derivative B13FR1 via conjugation. A transconjugant, *Pseudomonas* sp. strain B13FR1::*bph*, with a chromosomally integrated *bph* gene removed approximately 90% of added 4-chlorobiphenyl after 5 d in lake sediment microcosms (DOWLING et al., 1993). The plasmid, pDDPCB, was later transferred into a rhizosphere pseudomonad, *Pseudomonas fluorescens* F113, by conjugation to generate a genetically modified strain F113pcb (BRAZIL et al., 1995). The *bph* operon was chromosomally located and was stable in non-sterile soil microcosms for 25 d after inoculated onto sugar beet seeds. Strain F113pcb gained the ability to utilize biphenyl as a sole carbon source.

2.2.2.3 *Pseudomonas cepacia* JHR22

HAVEL and REINEKE (1991) reported a hybrid bacterium, *Pseudomonas cepacia* JHR2, was constructed by filter mating with a biphenyl-grown donor and a chlorobenzoate-grown recipient (*P. cepacia* JH230). Strain 230 is a hybrid strain that originated from the transfer of chlorocatechol degradative genes from strain B13 into *Pseudomonas* sp. WR401 (HARTMANN et al., 1989). Strain JHR2 was able to grow on 3- and 4-chlorobiphenyl. A 2-chlorobiphenyl degrader, JHR22, was obtained by growing JHR2 with 4-chlorobenzoate (1 mM) in the presence of 2-chlorobiphenyl (3 mM). The new hybrid strain, *P. cepacia* JHR22, showed capability to utilize 2-chloro-, 3-chloro-, 4-chloro-, 2,4-dichloro-, and 3,5-dichlorobiphenyl as sole source of carbon and energy. The strain JHR22 also exhibited ability to degrade all monochlorobiphenyls in Aroclor 1221® when tested with soils (HAVEL and REINEKE, 1993).

2.2.2.4 *Pseudomonas acidovorans* M3GY

Recombinant bacterium, strain M3GY, was produced within a multi-chemostat culture by mixing *P. acidovorans* CC1 (a chloroacetate and biphenyl degrader) and *Pseudomonas* sp. strain CB15 (a hybrid strain) on ceramic beads that were coated with 3,3′-dichlorobiphenyl (MCCULLAR et al., 1994). Strain M3GY expressed catabolic ability to utilize 3,4′-dichlorobiphenyl as a sole carbon and energy source. The recipient strain was determined by phenotypic similarity and genetic homology between strains M3GY and CC1.

In a recent study, 2,3′-dichloro- and 2,4′-dichlorobiphenyl were mineralized by a two-member consortium, *Burkholderia* (formerly *Pseudomonas*) sp. strain LB400 and *P. putida* mt-2a (POTRAWFKE et al., 1998). The strain mt-2a was obtained by intergeneric mating with the chlorocatechol genes transferred from LB400 to mt-2 (TOL). Strain mt-2a exhibited the ability to grow on 3-chloro- and 4-chlorobenzoate.

2.2.2.5 *Pseudomonas putida* (pDA261)

Different biphenyl degradative genes (*bphABCD*) from *Comamonas testosteroni* B-356 were subcloned into *P. putida* and *E. coli* separately, and their degradative capabilities were examined (AHMAD et al., 1992). The *hphC* and *bphD* genes were expressed well in both cells, however, *bphA* and *bphB* in *E. coli* were poorly expressed even though located downstream of the *tac* promoter. A review of the strain B-356 was published earlier (SYLVESTRE, 1995).

2.2.2.6 *Comamonas testosteroni* VP44(pE43)/VP44(pPC3)

Introduction of dehalogenase genes into the biphenyl degrading strain *C. testosteroni* VP44 resulted in complete mineralization of *o*- and *p*-substituted monochlorobiphenyls (HRYWNA et al., 1999). Plasmid pE43, containing the *ohbAB* gene, encodes iron–sulfur protein (ISPOHB) of the *o*-halobenzoate-1,2-dioxygenase in *P. aerugrinosa* 142. This plasmid, pE43, was transformed into strain VP44. The resulting recombinant strain VP44(pE43) mineralized 2-chlorobiphenyl and 2-chlorobenzoate. The other recombinant strain VP44-(pPC3) contained the *fcbABC* genes (from *Arthrobacter globiformis* KZT1) that catalyze hydrolytic *p*-dechlorination of 4-chlorobenzoate. This strain demonstrated the capability to grow on 4-chlorobiphenyl and 4-chlorobenzoate.

2.2.2.7 *Escherichia coli* JM109 (pSHF1003)/(pSHF1007) – Hybrid Biphenyl Dioxygenase

The *bphA1* genes of biphenyl dioxygenase in *Pseudomonas pseudoalcaligenes* KF707 and *Burkholderia cepacia* LB400 were recombined randomly by DNA shuffling (KUMAMARU et al., 1998). The shuffled *bphA1* genes were cloned into pJHF18 (MulI), upstream of *bphA2A3A4BC*, to replace the disrupted *bphA1* genes in the pJHF18 (HIROSE et al., 1994). The resulting plasmids containing *bphA1* (shuffled) and *bphA2A3A4BC* (KF707) were transformed into *E. coli* JM109 by electroporation. Some chimeric biphenyl dioxygenases in *E. coli* cells exhibited enhanced degradation of different biphenyl compounds. Biphenyl dioxygenase (pSHF 1003) showed 3 amino acid substitutions in KF707 *bphA1* (H255Q, V258I, and D303E), derived from strain LB400, and acquired degradation capabilities not only for PCB, but also for benzene and toluene. Plasmid pSHF1007 showed 4 amino acid substitutions (H255Q, V258I, G268A, and T376N) in KF707 *bphA1* gene and increased substrate affinities for some PCB congeners. The above results suggested the substitutions of His with Gln at 255 and of Val with Ile at 258 leading to the differences in substrate specificity and mode of oxygenation between the two enzymes.

Site-directed mutagenesis on the LB400 *bphA* gene was reported previously (ERICKSON and MONDELLO, 1993). 4 amino acids in LB400 were converted to the corresponding KF707 sequence. The modified biphenyl dioxygenase exhibited broader substrate specificity with PCB.

The *bphB* gene, encoding the *cis*-biphenyl dihydrodiol dehydrogenase from *P. putida* OU83 was cloned into the pQE31 vector and expressed in IPTG-induced *E. coli* recombinant cells (KHAN et al., 1997).

2.2.2.8 *Pseudomonas putida* IPL5 (Field Application Vectors)

Field application vectors (FAV) are a combination of a selective substrate that can be used easily by host (not indigenous) microorganisms, and a cloning vector to provide a temporary niche for the host bacterium in harsh environments (LAJOIE et al., 1992). FAV can stabilize and enhance the expression of foreign genes in contaminated sites. The chromosomally encoded PCB catabolic genes (*bphABC*) from *Pseudomonas* sp. strain ENV307 were cloned into broad host range plasmid pRK293. The resulting plasmid was transferred to the host *Sphingomonas paucimobilis* 1IGP that utilizes non-ionic surfactant Igepal CO-720® (IGP) as selective substrate. The recombinant strain 1IGP4(pCL3) exhibited ability to degrade individual PCB congeners in Aroclor 1242® without biphenyl as an inducer (LAJOIE et al., 1993). The transposon encoded PCB degradative genes (*bphABC*) were more stable than plasmid encoded after insertion into the surfactant utilizing strain, *Pseudomonas putida* IPL5 (LAJOIE et al., 1994).

2.2.3 Trichloroethylene (TCE)

2.2.3.1 *Escherichia coli* HB101/pMY402 and FM5/pKY287

A XhoI fragment (4.7 kb), containing toluene monooxygenase (TMO) genes from *Pseudomonas mendocina* KR-1, was subcloned into a broad host range vector, pMMB66EH, which contains an *E. coli tac* promoter to yield plasmid pMY402. The same fragment was also inserted into another *E. coli* expression vector, pCFM1146, containing the temperature inducible *E. coli* phage λ P_L promoter, to yield plasmid pKY287. Recombinant *E. coli* strains, *E. coli* HB101/pMY402 and FM5/pKY287, were able to oxidize both toluene and TCE (WINTER et al., 1989).

Recently, TMO genes in pMMB503EH, a broad host range vector, were introduced into *P. putida* KT2440 by electroporation (WARD et al., 1998).

2.2.3.2 *Pseudomonas pseudoalcaligenes* KF707-D2 and *Pseudomonas putida* KF715-D5

HIROSE et al. (1994) reported a hybrid dioxygenase gene cluster between the *tod* and the *bph* operons in *E. coli* JM109. Plasmid pJHF101, containing *todC1::bphA2orf3A3A4*, was constructed by deleting a 1.3 kb *PpuMI* fragment from *bphB* and *bphC* genes from pJHF10 (FURUKAWA et al., 1993). The vector pUC118 was used for the construction of pJHF10. This recombinant strain, *E. coli* (pJHF101), degraded TCE at an initial rate of 1.8 µg mL^{-1} h^{-1} which was much faster than *E. coli* cells carrying the toluene dioxygenase genes (*todC1C2BA*) or the biphenyl dioxygenase genes (*bphA1A2A3A4*, FURUKAWA et al., 1994). The hybrid gene cluster, *todC1: :phA2orf3A3A4*, was further inserted into the chromosomal *bph* operons by SUYAMA et al. (1996). The plasmid pJHF101 was subcloned into suicide vector, pSUPB30, to yield plasmid pASF101. The pASF101 was inserted into *E. coli* S17-1 (chromosomally integrated RP4-2-Tc::Mu-Km::Tn7) cells via transformation, and the resulting strain was used as a donor in the mating with the recipient biphenyl utilizing *P. pseudoalcaligenes* KF707 and *P. putida* KF715. The resulting double crossover strains, KF707-D2 and KF715-D5, maintained *todC1* in LB broth under no selective pressure. Both single and double crossover strains carrying *todC1* on chromosomes of KF707 and KF715 degraded TCE efficiently, and grew on toluene and benzene.

2.2.3.3 *Pseudomonas* sp. strain JR1A::*ipb* – Hybrid Strains

A multi-component isopropylbenzene (IPB) dioxygenase from strain *Pseudomonas* sp. JR1, growing on IPB exhibited the capability to co-oxidize TCE (PFLUGMACHER et al.,

1996). Organization of *ipb* genes encodes IPB dioxygenase (*ipbA1A2A3A4*), 2,3-dihydro-2,3-dihyroxy-IPB dehydrogenase (*ipbB*), and 3-IPB-2,3-dioxygenase (*ipbC*). Recently, a recombinant, *Pseudomonas* sp. JR1A, exhibited constitutive TCE oxidation activity was reported (BERENDES et al., 1998). A transposon vector pC8 (pUT/miniTn5Km::*ipbABC*) was constructed, then transformed into *E. coli* S17.1(*pir*). The pC8 vector in *E. coli* was conjugatively transferred into different recipients. The *ipb* genes were detected in following transconjugants: strain JR1A (spontaneous IPB negative mutants of strain JR1), *P. putida* 548, and *Pseudomonas* sp. strain CBS-3. Two of these hybrid strains, JR1A::*ipb* and CBS-3::*ipb*, were stable for more than 120 generations in antibiotic free medium and degraded TCE and IPB constitutively. Further studies showed that strain JRlA::*ipb* can oxidize TCE without inducer.

2.2.3.4 *Escherichia coli* JM109(pDTG601)

Toluene dioxygenase from *P. putida* F1 has been reported as an enzyme responsible for degradation of TCE (WACKETT and GIBSON, 1988). The structural genes (*todC1C2BA*) of toluene dioxygenase were cloned from plasmid pKK223-3 then transformed into strain JM109. The resulting strain *E. coli* JM109 (pDTG601) degraded TCE at a slower initial rate when compared to *P. putida* F39/D, a mutant strain of strain F1 that does not contain *cis*-toluene dihydrodiol dehydrogenase (ZYLSTRA et al., 1989).

2.2.3.5 *Pseudomonas putida* G786(pHG-2)

A recombinant *P. putida* G786(pHG-2), containing two multicomponent enzyme systems, cytochrome *P450*$_{cam}$, and toluene dioxygenase, capable of degrading pentachloroethane and TCE was reported by WACKETT et al. (1994, 1995). The toluene dioxygenase genes (*todC1C2BA*) and *tac* promoter (*Ptac*) were cloned into plasmid pKT230 to form plasmid pHG-1. The *lac*I^Q gene cassette from

plasmid pMMB24 was cloned into plasmid pHG-1 to yield pHG-2, then transformed into *E. coli* DH5. Plasmid pRK2073 was used as a helper to transconjugate plasmid pHG-2 into *P. putida* G786 containing cytochrome *P450*$_{cam}$ genes on the CAM plasmid. The new hybrid strain, G786(pHG-2), transformed pentachloroethane to TCE by the cytochrome *P450*$_{cam}$, then TCE was further mineralized by the toluene dioxygenase.

2.2.3.6 *Pseudomonas putida* F1/pSMMO20

Slow growth rate, copper repression of the *smmo* locus, and strong competition between TCE and methane for soluble methane monooxygenase (sMMO) are restrictions to the use of strain OB3b to degrade TCE. To overcome these impediments, the complete *smmo* cluster of *Methylosinus trichosporium* OB3b was cloned into a wide host range vector, pMMB277, to form pSMMO20 (JAHNG and WOOD, 1994). pSMMO20 is a 14.7 kb plasmid and contains the IPTG-inducible *tac* promoter upstream of the *smmo* cluster. Plasmid pSMMO20 was transformed into *P. putida* F1, P. putida KT2440, *P. mendocina* KR1, *B. cepacia* G4, and *B. cepacia* G4 PR1 through electroporation. *P. putida* F1/pSMMO20 was the only bacterium that was able to degrade TCE. However, inconsistent sMMO activity was a major drawback in the recombinant strain. Furthermore, the sMMO protein bands were detected on SDS-PAGE gels only when TPTG was present to induce the *tac* promoter. The constructed strain, *P. putida* F1/pSMMO20, showed a lower TCE degradation rate and a much higher growth rate than strain OB3b. This recombinant strain also demonstrated the capability to mineralize chloroform.

2.2.3.7 *Burkholderia cepacia* G4 5223-PR1

An aerobic bacterium, *Burkholderia cepacia* G4, cometabolically degrades TCE to CO_2 and non-volatile products by toluene *o*-monooxygenase (SHIELDS et al., 1989). Tn5 mutants of strain G4 were constructed via triparental mating among G4, *E. coli* C600(pRZ102), and *E. coli* HB101(pRK2013). The resulting Tn5 mutants could not express toluene *o*-monooxygenase activity, and were unable to degrade TCE, toluene, and phenol. A mutant spontaneously reverted to express toluene *o*-monooxygenase constitutively and was designated as G4 5223-PR1. This strain metabolized TCE and *m*-trifluoromethyl phenol without induction (SHIELDS and REAGIN, 1992).

2.2.3.8 *Ralstonia eutropha* AEK301/pYK3021

Strain AEK301, a Tn5 mutant that lost phenol hydroxylase activity, was derived from strain JMP134. Plasmid pYK3021 encoding phenol hydroxylase was subcloned into pMMB67EH (vector from pJP4) and exhibited TCE degradation capability without phenol induction. Triparental mating was used to transfer plasmid pYK3021 from *E. coli* to *R. eutropha* AEK301 with the helper plasmid pRK2013. The recombinant strain AEK301/ pYK3021 expressed phenol hydroxylase activity constitutively and degraded TCE efficiently (KIM et al., 1996). The removal rate of TCE by the strain could be influenced by growth substrates (AYOUBI and HARKER, 1998).

2.2.4 2,4-Dichlorophenoxyacetic Acid (2,4-D)

2.2.4.1 *Pseudomonas putida* PPO300(pRO101) and PPO301(pRO103)

Plasmid pJP4 encodes the degradation genes of 2,4-D and 3-chlorobenzoate in *R. eutropha* JMP134 (DON and PEMBERTON, 1985). GEM containing catabolic genes for the degradation of 2,4-D to 2-chloromaleylacetate were reported by HARKER et al. (1989). Plasmid pRO101 (pJP4::Tn1721) was constructed by insertion of transposon Tn1721 into pJP4, then further transferred by conjugation to different *Pseudomonas* strains. One of the resulting strains, *P. putida* PPO300(pRO101),

with chromosomally encoded phenol hydroxylase, also degrades phenoxyacetate in the presence of an inducer (2,4-D or 3-chlorobenzoate) of the 2,4-D pathway. A mutant plasmid, pRO103, derived from pRO101 by the spontaneous deletion of negative regulatory gene (*tfdR*) was isolated from strain PPO300-(pRO101). The deletion of *tfdR* resulted in the constitutive expression of the *tfdA* gene (encodes 2,4-D monooxygenase), and enabled mutant strain PPO300(pRO103) to grow on phenoxyacetate as the sole carbon source. Strain PPO301, derived from PPO300 (ATCC 17514) and resistant to nalidixic acid, was used to harbor plasmid pRO103 to yield a constitutive 2,4-D degrader, PPO301(pRO103), that was used in agricultural soil studies (SHORT et al., 1990). A dual substrate (2,4-D and succinate) chemostat study with *P. cepacia* DBO1-(pRO101) indicated that succinate can act as repressor of the 2,4-D catabolic pathway. But, this repression can be relieved with appropriate adjustments, such as lower 2,4-dichlorophenol accumulation or reduced succinate concentration (<2,4-D) in the media (DAUGHERTY and KAREL, 1994).

2.2.4.2 *Pseudomonas cepacia* RHJ1

Burkholderia cepacia (formerly *P. cepacia*) RH1 is a recombinant strain created by performing conjugation between *R. eutropha* JMP134 (2,4-D degrader) and *B. cepacia* AC1100 (2,4,5-trichlorophenoxyacetate degrader). The self-transmissible 2,4-D degradative plasmid, pJP4, in *A. eutrophus* JMP134 was transferred into a 2,4,5-trichlorophenoxyacetic acid (2,4,5-T) degrader, *B. cepacia* AC1100. This new strain, designated RHJ1, was capable of degrading mixtures of 2,4-D and 2,4,5-T simultaneously (HAUGLAND et al., 1990).

2.3 Nitroaromatic Compounds

2.3.1 *Pseudomonas fluorescens* 410PR(pWWOΔpm) – Hybrid Pathway

A hybrid pathway for the metabolism of 4-nitrotoluene was assembled in *P. fluorescens* 410PR(pWWOΔpm) via conjugation. The new pathway consists of a 4-nitrobenzoate pathway from strain *P. fluorescens* 410P and a TOL pWWOΔ pm plasmid, from *P. putida* 2440, which converts 4-nitrotoluene to 4-nitrobenzoate through its upper pathway. Strain 410PR(pWWOΔpm) was stable in a non-selective medium and able to metabolize 4-nitrotoluene to 3,4-dihydroxybenzoate and NH_4^+ (MICHÁN et al., 1997).

2.3.2 *Pseudomonas* sp. clone A (pWWO-Km) – Hybrid Pathway

RAMOS et al. (1995) constructed a hybrid pathway for degradation of 2,4,6-trinitrotoluene (TNT). The TOL plasmid (pWWO-Km) from *P. putida* was introduced into *Pseudomonas* sp. clone A by conjugation, which transforms TNT to toluene. Transconjugants were able to grow on TNT as the sole carbon and nitrogen source. The hybrid strain, *Pseudomonas* sp. clone A (pWWO-Km), metabolized TNT to toluene by release of NO_2^-, then further degraded toluene via the TOL pathway. A similar strain construction was reported earlier by DUQUE et al. (1993).

2.4 Sulfur Compounds

2.4.1 *Pseudomonas aeruginosa* EGSOX and *Pseudomonas putida* EGSOX

A gram-positive bacterium, *Rhodococcus erythropolis* IGTS8 (formerly *R. rhodochrous* IGT8, ATCC 53968), has been used extensively in biodesulfurization of fossil fuels due to its non-destruction of the carbon ring (GRAY et al., 1996). The catabolic genes, *dszABC*, encod-

ing the enzymes for desulfurization of dibenzothiophene (DBT) have been cloned and sequenced (DENOME et al., 1994; PIDDINGTON et al., 1995). DBT is converted into 2-hydroxybiphenyl and sulfite via a sulfur-specific pathway present in IGT8 (GRAY et al., 1996).

The *dsz* gene cluster from strain IGT8 was inserted into chromosomes of different *Pseudomonas* strains (GALLARDO et al., 1997). Plasmid pESOX4, a derivative from pBSL118 (ALEXEYEV et al., 1995), containing the *Ptac::dsz* fusion and a kanamycin resistance gene within mini-Tn5 transposon, was constructed in *E. coli*. Triparental filter mating was performed using *E. coli* CC118(pir(pESOX4) as donor, *E. coli* HB101(pRK600) as helper and *P. putida* KT2442 as recipient. The transconjugant, *P. putida* KT2442::miniTn5Km (*Ptac::dsz*), was able to use DBT as the sole sulfur source, and designated as *P. putida* EGSOX. The same plasmid, pESOX4, was also integrated into the chromosome of *P. aeruginosa* PG201, a rhamnolipid biosurfactant producer. The resulting strain was designated as *P. aeruginosa* EGSOX.

2.4.2 *Pseudomonas* sp. HL7bR(pC1) and *Pseudomonas alcaligenes* DM201(pC1)

The plasmid (pC1) encoding DBT degradative genes in *P. alcaligenes* DBT2 was cloned into *E. coli* DH1 (FINNERTY and ROBINSON, 1986). The pC1 was further transferred into two *Pseudomonas* species from the *E. coli* DH1(pC1) (FOGHT and WESTLAKE, 1990). *E. coli* HB101(pRK2013) was used as a helper in the triparental conjugation. *Pseudomonas* sp. HL7b(Dbt⁻) and *P. alcaligenes* DM201(Dbt⁻) were recipients. The transconjugant strains HL7bR(pC1) and DM201(pC1) exhibited the phenotype of Dbt⁺tet⁺, and retained limited degradative capability on DBT.

3 Reporter Genes

In conjunction with knowledge of the regulation of biodegradative operons and operons associated with metal tolerance, molecular knowledge of bioremediation has been used to advantage in development of bioreporter systems for contaminant detection, bioremediation process monitoring, and environmental stress responses.

The use of different molecular methods, such as marker genes or suicide vectors, has been reported to monitor or control GEM in the environment (ATLAS, 1992). Several genes have been utilized as reporter genes to monitor transcriptional activity in the bacteria to reveal environmental activity, and indirectly as an indicator for the present pollutants. Bioluminescent and chemiluminescent genes are frequently adapted for this purpose. BRONSTEIN et al. (1994) have reviewed the detection procedures of these reporter genes. β-Galactosidase (*lacZ*) is a widely used chemiluminescent reporter gene in molecular biology. Recently, the marine bacterial luciferase (*luxAB*), a bioluminescence gene, has been used extensively to monitor environmental responses. It is also convenient to use bioluminescence for monitoring cell activities in liquid culture (RATTRAY et al., 1990), soil (SILCOCK et al., 1992) and for tracking GEM in nature (SHAW et al., 1992). The bioluminescent technology and its applications have been reviewed previously (GRIFFITHS and HALL, 1993; VAN DER LELIE et al., 1994; STEINBERG et al., 1995).

Bacterial luciferase (EC 1.14.14.3) is a heterodimeric enzyme that catalyzes the oxidation of reduced riboflavin phosphate (FMNH$_2$) and a long-chain fatty acid aldehyde with the emission of blue-green light (490 nm). A comprehensive review on the molecular level of the *lux* system from luminescent bacteria has been given by MEIGHEN (1991).

3.1 Chlorinated Compounds

3.1.1 *Ralstonia eutropha* ENV307(pUTK60)

A biphenyl/PCB bioreporter strain was constructed by introduction of the *orf0-bphA1* genes from strain ENV307 into the front of promoterless *luxCDABE* genes from *Vibrio fischeri* (formerly *Photobacterium phosphoreum*) contained in vector pUCD615 to form pUTK60. Plasmid pUTK60 was transformed into strain ENV307 by triparental mating with helper pRK2073 (LAYTON et al., 1998). Under non-ionic surfactant treatment, the bioluminescence responses of ENV307(pUTK60) to biphenyl, monochlorinated biphenyl, and Aroclor 1242® were increased proportionately to contaminant concentration.

3.1.2 *B. cepacia* BRI6001L

A *lacZY-luxAB* reporter gene system was constructed and integrated into the chromosome of a 2,4-D degrader, *B. cepacia* BRI6001, via Tn7 transposition (MASSON et al., 1993). The resulting strain BRI6001L was isogenic to its parent strain and could be used as a reliable model to monitor the fate of strain BRI6001 in contaminated soil. Survival and fate of this reporter strain in different soils were monitored successfully through its decanal-dependent bioluminescence response. In addition to temperature and competition by indigenous bacterial effects, its population was also substrate (2,4-D) and contaminant-dependent (GREER et al., 1994). Another study reported soil moisture content can influence 2,4-D biodegradation in unsaturated soils (CATTANEO et al., 1997).

3.1.3 *Alcaligenes eutrophus* H850Lr

Reporter strain H850Lr was constructed by conjugal transfer of the *luxAB* genes from *V. harveyi* into a rifampicin resistant PCB degrading strain H850r (VAN DYKE et al., 1996). This *lux* reporter has successfully survived in PCB contaminated soil and sediment samples after 56 d of inoculation. The addition of rhamnolipid, a biosurfactant produced by *P. aeruginosa* UG2Lr, led to rapid increases in the H850Lr population.

3.1.4 *Pseudomonas fluorescens* 10586s/FAC510 and 10586/pUCD607

The luciferase genes (*lux*) from *V. fischeri* were integrated into the chromosome and plasmid of soil bacterium strain 10586s. The strain 10586/FAC510 contained *luxABE* genes and a kanamycin resistance gene on the chromosome (AMIN-HANJANI et al., 1993). Addition of *n*-decyl aldehyde was required for bioluminescence. The chromosomal encoded luminescence marker was stable in both liquid culture and in soil. The *luxCDABE* (pUCD607) genes were inserted in a multi-copy plasmid to give strain 10586s/pUCD607 (PATON et al., 1995a). Both chromosomal and plasmid encoded *lux* biosensors showed decline of bioluminescence as potentially toxic elements (heavy metals) concentration increased. The plasmid encoded biosensor exhibited much higher sensitivity to environmental stress due to the high copy number of plasmid.

The plasmid encoded biosensor, *P. fluorescens* 10586s/pUCD607, was applied to toxicity assessment of a variety of environmental contaminants: chlorobenzenes (BOYD et al., 1998), chlorophenols (SINCLAIR et al., 1999), organotins (BUNDY et al., 1997), benzene and its metabolites, catechol and phenol (BOYD et al., 1997a), semi-volatile organic compounds in groundwater (BOYD et al., 1997b), and metals (CHAUDRI et al., 1999; McGRATH et al., 1999). Determination of bioavailability of a contaminant (copper) in whisky distillery effluent was also reported (PATON et al., 1995b). However, PAH (phenanthrene, pyrene, and benzo[a]pyrene) showed no acute toxicity to the tested *lux* biosensor (REID et al., 1998).

Lux-based biosensors have also been used in pollutant component (Cd, Cu, and PCP) diagnosis (PALMER et al., 1998), and in identification of bioremediation limitations of BTEX contaminated sites (SOUSA et al., 1998) and in toxicity assessment of BTEX (Glover et al., 1999).

3.2 Aromatic Hydrocarbons

3.2.1 *Pseudomonas putida* TVA8

A single copy of *tod-luxCDABE* fusion was introduced into the chromosome of *P. putida* F1 via a mini-Tn5 delivery vector, and yielded the bioreporter TVA8 (APPLEGATE et al., 1998). The transposon insertion was stable and antibiotic selection was not needed. The TVA8 is not only responsive to BTEX compounds but can also be used for on-line monitoring of trichloroethylene cometabolism since both *lux* and *tod* operons are under the same regulation. Strain TVA8 also showed an increasing bioluminescence response to increasing TCE concentration (up to 80 µmol) and toluene (up to 140 µmol) concentrations (SHINGLETON et al., 1998). A plasmid based *tod–lux* bioreporter, *P. putida* B2, was reported previously (APPLEGATE et al., 1997).

Strain TVA8 was coupled directly onto an integrated circuit (IC) specifically for detecting toluene. This is called the bioluminescent bioreporter integrated circuit (BBIC) which uses a whole cell biosensor, and is currently under development at the Oak Ridge National Laboratory, and the Center for Environmental Biotechnology (SIMPSON et al., 1988).

3.2.2 *Pseudomonas fluorescens* HK44

A catabolic bioluminescent reporter plasmid (pUTK21) was constructed by transposon (Tn4431) insertion of the lux genes (*luxCDABE*) from *V. fischeri* into a naphthalene catabolic plasmid (pKA1) in *P. fluorescens* 5R (KING et al., 1990). The *lux* transposon (Tn4431) inserted into the *nahG* gene, salicylate hydroxylase, of the plasmid resulted in inactivation of the lower pathway for naphthalene degradation (MENN et al., 1993). The catabolic reporter plasmid (pUTK21) was transferred by conjugation to a Nah⁻Sal⁺ environmental strain, *P. fluorescens* HK9, to yield strain HK44 (Nah⁺Sal⁺). This strain showed a linear relationship between the naphthalene salicylate concentration and overall bioluminescent response (HEITZER et al., 1992). An optical whole cell biosensor has been developed by immobilization of strain HK44 in alginate for on line monitoring of naphthalene, salicylate, and JP-4 jet fuel in waste streams (HEITZER et al., 1994). The reporter strain, HK44, was successfully released to large scale, semicontained soil lysimeters for field testing over a 2-year period. Results indicated that HK44 was able to survive in the soil ecosystem, and report on the bioavailability of introduced contaminants, and on the bioremediation process activity through bioluminescent signaling (SAYLER et al., 1999). The bacterial population dynamic was significantly affected by nutrient, hydrocarbon, and O_2 augmentation.

Another *nah–lux* reporter strain, *Pseudomonas* sp. RB1351, was developed by subcloning the promoter from HAH7 upper pathway, and fusion of that with *luxCDABE* genes (BURLAGE et al., 1990).

3.2.3 *Escherichia coli* HMS174(pOS25)

An *ipb–lux* reporter was constructed to investigate regulation of the isopropylbenzene (cumene) catabolism operon (*ipb*) of *P. putida* RE204 (SELIFONOVA and EATON, 1996). Plasmid pOS25 was assembled by fusing the regulatory protein (*ipbR*) and operator–promoter (*ibpo/p*) to the upstream of *luxCDABE* genes from *V. fischeri*. Light was produced by strain HMS174 (pOS25) in the presence of inducers of the *ipb* operon. The level of luciferase expression was related to the concentration of the inducer. In addition, hydrocarbon mixtures, such as gasoline, jet fuels (JP-4 and JP-5), diesel fuel, and creosote also served as inducers for the *ipb–lux* bioreporter.

3.2.4 *Pseudomonas* sp. UG14Lr

A mini-Tn5 transposon containing *luxAB* genes from *V. harveii* was inserted into chromosome of a rifampicin resistant phenanthrene mineralizing strain UG14r via conjugation (WEIR et al., 1995). The survival rate of resulting bioreporter, UG14Lr, in the creosote contaminated soil slurries was increased by alginate encapsulation. However, there was no statistical difference between populations of encapsulated UG14Lr cells and free cells in sterile environment. The *lux* insert was stable and did not affect the rate of phenanthrene mineralization by UG14Lr in soil slurries. Nutrient amendment enhanced both survival and mineralization by free and encapsulated UG14Lr cells.

Six different *lux*-based microbial biosensors were tested successfully to assess toxicity and to monitor the process of crude oil bioremediation in soil microcosms (BUNDY et al., 1999).

3.3 Metals

3.3.1 *Escherichia coli* HMS174(pOS14/pOS15)

Biosensors for the detection of Hg(II) have been constructed and use strain HMS174 as the host to examine bioluminescent responses (SELIFONOVA et al., 1993). Construction of *mer–lux* fusion plasmids was accomplished by cloning the mercury resistance operon (*Tn21 mer*) upstream from a promoterless *lux* operon (*luxCDABE*) from *V. fischeri*. The plasmid pOS14 (*Ro/pTPC*), containing the regulatory gene *merR*, the operator–promoter (o/p), and the *merTPC*, which mediates active transport of Hg(II), exhibited light emission at 0.5 nM Hg(II). The sensitivity was reduced in pOS15 (*Ro/pTPCAD*) by introduction of *merA* that encodes mercuric reductase.

3.3.2 *Staphylococcus aureus* RN4220 (pC200)/(pC300)

Plasmid PI258 from *S. aureus* contains the *cad* and *ars* operons that confer resistance to cadmium and zinc, and to arsenate, and arsenite, respectively. A shuttle vector, pC101, among *E. coli*, *S. aureus*, and *Pseudomonas* containing *luxAB* genes from *V. harveyi* was constructed. Two transcriptional fusions, *arsB–luxAB* and *cadA–luxAB*, were constructed using the shuttle vector for the study of the regulation of *ars* and *cad* operons (CORBISIER et al., 1993). A fragment containing *ars* promoter, regulatory gene (*arsR*), and first 676 nucleotides of *arsB* (a membrane protein) was transformed into the upstream region of *luxAB* genes in pC101 to yield pC200. Plasmid pC200 was electroporated into strain RN4220. This bioluminescent reporter (*arsB–luxAB*) responded to arsenite induction only. The *cadA–luxAB* fusion (pC300) containing the promoter, the *cadC* (encodes a soluble protein), and the first 51 nucleotides of *cadA* (encodes a membrane ATPase protein) was also electroporated into recipient cells. Cadmium (Cd^{+2}) was the most efficient inducer to the reporter (*cadA–luxAB*), but Cd^{+2} also acted as an inhibitor at higher concentrations. The reporter was induced at low level when high concentrations of bismuth (Bi^{+3}) and lead (Pb^{+2}) were introduced.

The *ars* promoter (*arsR*) from strain RN4220 was inserted in front of firefly luciferase gene (*lucFF*) to generate plasmid pTOO21. Plasmid pTOO21 was transformed by electroporation into strain RN4220. This firefly luciferase biosensor, *S. aureus* RN4220(pTOO21), detects arsenite, antimonite, and cadmium at low concentration (nM). The expression of firefly luciferase is controlled by the *ars* promoter (TAURIAINEN et al., 1997).

The bioavailability and assessment of toxicity of heavy metals (Cu and Zn) in the soils were examined by *lux*-base biosensors, *E. coli* HB101(pUCD607) and *P. fluorescens* 10586 (pUCD607), under laboratory and field conditions. The increasing toxicity at lower pH values is due to an increase in free metal ion concentration (PRESTON et al., 1999).

3.4 Others

3.4.1 *Alcaligenesis eutrophus* 2050 and *Stenotrophomonas* 3664

Plasmid pUTK2 contains bioluminescent genes (*lux*) on transposon Tn4331 (SHAW et al., 1988). Genetically engineered bioreporters were constructed by transferring *lux* genes to environmental bacteria through filter mating with *E. coli* DH5(pUTK2). Two *lux*-based bioreporters, *A. eutrophus* 2050 and *Stenotrophomonas* 3664, were approximately 400 times more resistant to the nonionic surfactant poly-

oxyethylene-10-lauryl ether than *V. fischeri*. These surfactant resistant reporters would facilitate toxicological evaluation of xenobiotic compounds and their metabolites when surfactants were used for enhancement of solubilization of hydrophobic compounds. A structure–toxicity model on predication of nonpolar narcotics has been established using three bioluminescent bioreporters. Furthermore, the sensitivity among tested bioreporters does not vary (LAYTON et al., 1999).

3.4.2 *Pseudomonas aeruginosa* FRD1(pUTK50)

The *algD* gene (GDP-mannose dehydrogenase) and promoter were inserted into the promoterless *lux* reporter plasmid pUCD615. The resulting *algD–lux* fusion, pUTK50, was further transformed into *P. aeruginosa* FRD1 from *E. coli* cells with helper plasmid pRK2013 (WALLACE et al., 1994). The *alg-D* reporter, strain FRD1 (pUTK50) responded positively to NaCl and ethanol (0.5%) exposure, and showed no significant effect on initial response when either NH_4 or NO_3 were used as nitrogen source. This *algD–lux* bioluminescent reporter plasmid (pUTK50) was transconjugated into different environmental strains for study of corrosive biofilms (RICE et al., 1995).

3.4.3 *Pseudomonas* strain Shk1

E. coli DB5(pUTK2), containing *lux* genes on the transposon Tn4331, was mated with activated sludge obtained from an industrial wastewater treatment plant. An isolate emitted visible light and was identified as a *Pseudomonas* strain by 16S rDNA sequence comparison and designated as strain Shk1. The bioreporter, Shk1, was used for continuous, online toxicity monitoring in the wastewater treatment process. The intensity of bioluminescence decreased as the concentration of toxicants (cadmium, 2,4-dinitrophenol and hydroquinone) increased (KELLY, et al., 1999).

3.4.4 *Escherichia coli* DPD2511

An oxidative stress biosensor was constructed by fusion of the *E. coli katG* (catalase) promoter to a *V. fischeri lux* operon (*luxCDABE*). The luminescence of bioreporter DPD2511, containing *katG'::luxCDABE*, elevated significantly in the presence of hydrogen peroxide (up to 50 mg L^{-1}), organic peroxides, redox-cycling agents (menadione and methyl viologen) and cigarette smoke. The response of the strain was under positive control of the *oxyR*, and the induction of *katG* led to light emission (BELKIN et al., 1996).

3.4.5 *Escherichia coli* WM1202(pRY002)/TV1061 (pGrpElux.3 or pGrpElux.5)

Two heat shock promoters, *dnaK* and *grpE*, from *E. coli* were fused to the *luxCDABE* genes of *V. fischeri*. The resulting plasmids, pRY002 containing a *dnaKp::luxCDABE* fusion, and pGrpELux containing a *grpE'::luxCDABE* fusion, were transferred into *E. coli* cells by CaCl$_2$ transformation (VAN DYK et al., 1994). These biosensors responded positively by induction of light production when exposed to environmental stresses, such as copper, ethanol, 2,4-D, and phenol.

Another heat shock bioluminescent biosensor *E. coli* DPD1006 containing *lon'::lux* fusion plasmid (pLonLux2) was constructed by the same group (VAN DYK et al., 1995). The bioluminescent response was much more than additive induction for all heat shock promoter–*lux* fusions when combinations of cadmium chloride, copper sulfate, ethanol, formamide, 4-nitrophenol, and pentachlorophenol were added to the test medium.

4 Conclusion

A variety of bacterial species, and catabolic and regulatory genes, have been the target of genetic engineering for applications in bioremediation. These applications have been broadly targeted to improve the performance of biodegradation, control degradation processes, monitor process performance and organisms, as well as detection of chemical pollutants and their bioavailability.

Practical application of *in situ* bioremediation has not been realized. This is due to a host of factors ranging from competitive costs, clean-up efficacy, regulatory limits both on biotechnology as well as clean-up goals, public perception, and practitioner acceptance.

A variety of niches remain where genetically engineered microorganisms may still contribute significantly to overall bioremediation strategies and contaminant risk reduction. These applications may include *in situ* use for highly recalcitrant compounds, treatment reactor technology for solids, liquids, and gases, and end point assessment and risk reduction using engineered microbials as surrogates for other exposed biological populations. Clearly engineered organisms in bioremediation must be cost competitive, processes must meet clean-up objectives and regulatory constraints, or they must offer unusual speed, safety, or uniqueness in application.

5 References

ADAMS, R. H., HUANG, C.-M., HIGSON, F. K., BRENNER, V., FOCHT, D. D. (1992), Construction of a 3-chlorobiphenyl-utilizing recombinant from an intergeneric mating, *Appl. Environ. Microbiol.* **58**, 647–654.

AHMAD, D., SYLVESTRE, M., SONDOSSI, M. (1992), Subcloning of *bph* genes from *Pseudomonas testosteroni* B-356 in *Pseudomonas putida* and *Escherichia coli*: Evidence for dehalogenation during initial attack on chlorobiphenyls, *Appl. Environ. Microbiol.* **57**, 2880–2887.

ALEXEYEV, M. F., SHOKOLENKO, I. N., CROUGHAN, T. P. (1995), New mini-Tn5 derivatives for insertion mutagenesis and genetic engineering in gram-negative bacteria, *Can. J. Microbiol.* **41**, 1053–1055.

AMIN-HANJANI, S., MEIKLE, A., GLOVER, L. A., PROSSER, J. I., KILLHAM, K. (1993) Plasmid and chromosomally encoded luminescence marker systems for detection of *Pseudomonas fluorescense* in soil, *Mol. Ecol.* **2**, 47–54.

APPLEGATE, B. M., KELLY, C., LACKEY, L., MCPHERSON, J., KEHRMEYER, S. R. et al. (1997), *Pseudo-*

monas putida B2: a *tod–lux* bioluminescent reporter for toluene and trichloroethylene co-metabolism, *J. Ind. Microbiol.* **18**, 4–9.

APPLEGATE, B. M., KEHRMEYER, S. R., SAYLER, G. S. (1998), A chromosomally based *tod–luxCDABE* whole-cell reporter for benzene, toluene, ethylbenzene, and xylene (BTEX) sensing, *Appl. Environ. Microbiol.* **64**, 2730–2735.

ATLAS, R. M. (1992), Molecular methods for environmental monitoring and containment of genetically engineered microorganisms, *Biodegradation* **3**, 137–146.

AYOUBI, P. J., HARKER, A. R. (1998), Whole-cell kinetics of trichloroethylene degradation by phenol hydroxylase in a *Ralstonia eutropha* JMP134 derivative, *Appl. Environ. Microbiol.* **64**, 4353–4356.

BELKIN, S., SMULSKI, D. R., VOLLMER, A. C., VAN DYK, T. K., LAROSSA, R. A. (1996), Oxidative stress detection with *Escherichia coli* harboring a *katG'::lux* fusion, *Appl. Environ. Microbiol.* **62**, 2252–2256.

BERENDES, F., SABARTH, N., AVERHOFF, B., GOTTSCHALK, G. (1998), Construction and use of an ipb DNA module to generate *Pseudomonas* strains with constitutive trichloroethene and isopropylbenzene oxidation activity, *Appl. Environ. Microbiol.* **64**, 2454-2462.

BOYD, E. M., MEHARG, A. A., WRIGHT, J., KILLHAM, K. (1997a), Assessment of toxicological interactions of benzene and its primary degradation products (catechol and phenol) using a *lux*-modified bacterial bioassay, *Environ. Toxicol. Chem.* **16**, 849–856.

BOYD, E. M., KILLHAM, K., WRIGHT, J., RUMFORD, S., HETHERIDGE, M. et al. (1997b), Toxicity assessment of xenobiotic contaminated groundwater using *lux* modified *Pseudomonas fluorescens*, *Chemosphere* **35**, 1967–1985.

BOYD, E. M., MEHARG, A. A., WRIGHT, J., KILLHAM, K. (1998), Toxicity of chlorobenzenes to a *lux*-marked terrestrial bacterium, *Pseudomonas fluorescens*, *Environ. Toxicol. Chem.* **17**, 2134–2140.

BRAZIL, G. M., KENEFICK, L., CALLANAN, M., HARO, A., DE LORENZO, V. et al. (1995), Construction of a rhizosphere pseudomonad with potential to degrade polychlorinated biphenyls and detection of *bph* gene expression in the rhizosphere, *Appl. Environ. Microbiol.* **61**, 1946–1952.

BRONSTEIN, I., FORTIN, J., STANLEY, P. E., STEWART, G. S. A. B., KRICKA, L. J. (1994), Chemiluminescent and bioluminescent reporter gene assays, *Anal. Biochem.* **219**, 169–181.

BUNDY, J. G., WARDELL, J. L., CAMPBELL, C. D., KILLHAM, K., PATON, G. I. (1997), Application of bioluminescence-based microbial biosensors to the ecotoxicity assessment of organotins, *Lett. Appl. Microbiol.* **25**, 353–358.

BUNDY, J. G., CAMPBELL, C. D., PATON, G. I. (1999), Assessing crude oil bioremediation using catabolic and metabolic microbial biosensors, in: *In situ Bioremediation of Petroleum Hydrocarbon and Other Organic Compounds* (ALLEMAN, B. C., LEESON, A., Eds.), pp. 351–356. Columbus, OH: Battelle Press.

BURLAGE, R. S., SAYLER, G. S., LARIMER, F. (1990), Monitoring of naphthalene catabolism by bioluminescence with *nah–lux* transcriptional fusions. *J. Bacteriol.* **172**, 4749–4757.

CATTANEO, M. V., MASSON, C., GREER, C. W. (1997), The influence of moisture on microbial transport, survival and 2,4-D biodegradation with a genetically marked *Burkholderia cepacia* in unsaturated soil columns, *Biodegradation* **8**, 87–96.

CHATTERJEE, D. K., CHAKRABARTY, A. M. (1982), Genetic rearrangements in plasmids specifying total degradation of chlorinated benzoic acids, *Mol. Gen. Genet.* **188**, 279–285.

CHATTERJEE, D. K., CHAKRABARTY, A. M. (1983), Genetic homology between independently isolated chlorobenzoate-degradative plasmids, *J. Bacteriol.* **153**, 532–534.

CHAUDRI, A. M., KNOGHT, B. P., BARBOSA-JEFFERSON, V. L., PRESTON, S., PATON, G. I. et al. (1999), Determination of acute Zn toxicity in pore water from soils previously treated with sewage sludge using bioluminescence assays, *Environ. Sci. Technol.* **33**, 1880–1885.

CORBISIER, P., JI, G., NUYTS, G., MERGEAY, M., SILVER, S. (1993), *luxAB* gene fusions with the arsenic and cadmium resistance operons of *Staphylococcus aureus* plasmid pI258, *FEMS Microbiol. Lett.* **110**, 231–238.

DAUGHERTY, D. D., KAREL, S. F. (1994), Degradation of 2,4-dichlorophenoxy acetic acid by *Pseudomonas cepacia* DBO1(pRO101) in a dual-substrate chemostat, *Appl. Environ. Microbiol.* **60**, 3261–3267.

DENOME, S. A., OLDFIELD, C., NASH, L. J., YOUNG, K. D. (1994), Characterization of the desulfurization genes from *Rhodococcus* sp. strain IGT8, *J. Bacteriol.* **176**, 6707–6716.

DON, R. H., PEMBERTON, J. M. (1985), Genetic and physical map of the 2,4-dichlorophenoxy acetic acid-degradative plasmid pJP4, *J. Bacteriol.* **161**, 466–468.

DOWLING, D. N., PIPKE, R., DWYER, D. F. (1993), A DNA module encoding *bph* genes for the degradation of polychlorinated biphenyls (PCBs), *FEMS Microbiol. Lett.* **113**, 149–154.

DUQUE, E., HAIDOUR, A., GODOY, F., RAMOS, J. L. (1993), Construction of a *Pseudomonas* hybrid strain that mineralizes 2,4,6-trinitrotoluene, *J. Bacteriol.* **175**, 2278–2283.

ERICKSON, B. D., MONDELLO, F. J. (1993), Enhanced biodegradation of polychlorinated biphenyls after site-directed mutagenesis of a biphenyl dioxygenase gene, *Appl. Environ. Microbiol.* **59**, 3858–3862.

FINNERTY, W. R., ROBINSON, M. (1986), Microbial desulfurization of fossil fuels: a review, *Biotechnol. Bioeng. Symp.* **16**, 205–221.

FOGHT, J. M., WESTLAKE, D. W. S. (1990), Expression of dibenzothiophene-degradative genes in two *Pseudomonas* species, *Can. J. Microbiol.* **36**, 718–724.

FULTHTHROPE, R. R., WYNDHAM, R. C. (1991), Transfer and expression of the catabolic plasmid pBRC60 in wild bacterial recipients in a freshwater ecosystem, *Appl. Environ. Microbiol.* **57**, 1546–1553.

FURUKAWA, K., HIROSE, J., SUYAMA, A., ZAIKI, T., HAYASHIDA, S. (1993), Gene components responsible for discrete substrate specificity in the metabolism of biphenyl (*bph* operon) and toluene (*tod* operon), *J. Bacteriol.* **175**, 5224–5232.

FURUKAWA, K., HIROSE, J., HAYASHISA, S., NAKAMURA, K. (1994), Efficient degradation of trichloroethylene by a hybrid aromatic ring dioxygenase, *J. Bacteriol.* **176**, 2121–2123.

GALLARDO, M. E., FERRÁNDEZ, A., DE LORENZO, V., GARCÍA, J. L., DÍAZ, E. (1997), Designing recombinant *Pseudomonas* strains to enhance biodesulfurization, *J. Bacteriol.* **179**, 7156–7160.

GLOVER, L. A., WEITZ, H., KUAN, H. L., SOUSA, S., KILLHAM, K. et al. (1999), Project Rheingold: rapid biosensor analysis of soil and groundwater toxicity, in: *In situ Bioremediation of Petroleum Hydrocarbon and Other Organic Compounds* (ALLEMAN, B. C., LEESON, A., Eds.), pp. 369–374. Columbus, OH: Battelle Press.

GRAY, K. A., POGREBINSKY, O. S., MRACHKO, G. T., XI, L., MONTICELLO, D. J., SQUIRES, C. S. (1996), Molecular mechanisms of biocatalytic desulfurization of fossil fuels, *Nature Biotechnol.* **14**, 1705–1709.

GREER, C. W., MASSON, L., COMEAU, Y., BROUSSEAU, R., SAMSON, R. (1994), Monitoring the fate of bacteria released into the environment using chromosomally integrated reporter genes, in: *Applied Biotechnology for Site Remediation* (HINCHEE, R. E., ANDERSON, D. B., METTING, JR., F. B., SAYLES, G. D., Eds.), pp. 405–409. Boca Raton, FL: Lewis Publishers.

GRIFFITHS, D., HALL, G. (1993), Biosensors – what real progress is being made? *TIBTECH.* **11**, 122–130.

HARKER, A. R., OLSEN, R. L., SEIDLER, R. J. (1989), Phenoxyacetic acid degradation by the 2,4-dichlorophenoxyacetic acid (TFD) pathway of plasmid pJP4: mapping and characterization of the TFD regulatory gene, *tfdR, J. Bacteriol.* **171**, 314–320.

HARTMANN, J., ENGELBERTS, K., NORDHAUS, B., SCHMIDT, E., REINEKE, W. (1989), Degradation of 2-chlorobenzoate by *in vivo* constructed hybrid pseudomonads, *FEMS Microbiol. Lett.* **61**, 17–22.

HAUGLAND, R. A., SCHLEMN, D. J., LYONS III, R. P., SFERRA, P. R., CHAKRABARTY, A. M. (1990), Degradation of the chlorinated phenoxyacetate herbicides 2,4-dichlorophenoxyacetic acid and 2,4,5-trichlorophenoxyacetic acid by pure and mixed bacterial cultures, *Appl. Environ. Microbiol.* **56**, 1357–1362.

HAVEL, J., REINEKE, W. (1991), Total degradation of various chlorobiphenyls by co-cultures and *in vivo* constructed hybrid pseudomonads, *FEMS Microbiol. Lett.* **78**, 163–170.

HAVEL, J., REINEKE, W. (1993), Degradation of Aroclor 1221 in soil by a hybrid pseudomonad, *FEMS Microbiol. Lett.* **108**, 211–218.

HEITZER, A., WEBB, O. F., THONNARD, J. E., SAYLER, G. S. (1992), Specific and quantitative assessment of naphthalene and salicylate bioavailability by using a bioluminescent catabolic reporter bacterium, *Appl. Environ. Microbiol.* **58**, 1839–1846.

HEITZER, A., MALACHOWSKY, K., THONNARD, J. E., BIENKOWSKI, P. R., WHITE, D. C., SAYLER, G. S. (1994), Optical biosensor for environmental online monitoring of naphthalene and salicylate bioavailability with an immobilized bioluminescent catabolic reporter bacterium, *Appl. Environ. Microbiol.* **60**, 1487–1494.

HICKEY, W. J., BRENNER, V., FOCHT, D. D. (1992), Mineralization of 2-chloro- and 2,5-dichlorobiphenyl by *Pseudomonas* sp. strain UCR2, *FEMS Microbiol. Lett.* **98**, 175–180.

HIROSE, J., SUYAMA, A., ZAIKI, T., HAYASHIDA, S., FURUKAWA, K. (1994), Construction of hybrid biphenyl (*bph*) and toluene (*tod*) genes for functional analysis of aromatic ring dioxygenases, *Gene* **138**, 27–33.

HRYWNA, Y., TSOI, T. V., MALTSEVA, O.V., QUENSEN III, J. F., TIEDJE, J. M. (1999), Construction and characterization of two recombinant bacteria that grow on *ortho-* and *para-*substituted chlorobiphenyls, *Appl. Environ. Microbiol.* **65**, 2163–2169.

JAHNG, D., WOOD, T. K. (1994), Trichloroethylene and chloroform degradation by a recombinant pseudomonad expressing soluble methane monooxygenase from *Methylosinus trichosporium* OB3b, *Appl. Environ. Microbiol.* **60**, 2473–2482.

KELLY, C. J., LAJOIE, C. A., LAYTON, A. C., SAYLER, G. S. (1999), Bioluminescence reporter bacterium for toxicity monitoring in biological wastewater treatment systems, *Water Environ. Res.* **71**, 31–35.

KHAN, A. A., WANG, R.-F., NAWAZ, M. S., CERNIGLIA, C. E. (1997), Nucleotide sequence of the gene encoding *cis*-biphenyl dihydrodiol dehydrogenase (*bphB*) and the expression of an active recombinant His-tagged *bphB* gene product from a PCB degrading bacterium, *Pseudomonas putida* OU83, *FEMS Microbiol. Lett.* **154**, 317–324.

KIM, Y., AYOUBI, P., HARKER, A. R. (1996), Constitutive expression of the cloned phenol hydroxylase gene(s) from *Alcaligenes eutrophus* JMP134 and concomitant trichloroethylene oxidation, *Appl. Environ. Microbiol.* **62**, 3227–3233.

KING, J. M. H., DIGRAZIA, P. M., APPLEGATE, B., BURLAGE, R., SANSEVERINO, J. et al. (1990), Rapid, sensitive bioluminescence reporter technology for naphthalene exposure and biodegradation, *Science* **249**, 778–781.

KUMAMARU, T., SUENAGA, H., MITSUOKA, M., WATANABE, T., FURUKAWA, K. (1998), Enhanced degradation of polychlorinated biphenyls by directed evolution of biphenyl dioxygenase, *Nature Biotechnol.* **16**, 663–666.

LAJOIE, C. A., CHEN, S.-Y., OH, K.-C., STROM, P. F. (1992), Development and use of field application vectors to express non-adaptive foreign genes in competitive environments, *Appl. Environ. Microbiol.* **58**, 655–663.

LAJOIE, C. A., ZYLSTRA, G. J., DEFLAUN, M. F., STROM, P. F. (1993), Development of field application vectors for bioremediation of soils contaminated with polychlorinated biphenyl, *Appl. Environ. Microbiol.* **59**, 1735–1741.

LAJOIE, C. A., LAYTON, A. C., SAYLER, G. S. (1994), Cometabolic oxidation of polychlorinated biphenyls in soil with a surfactant-based field application vector, *Appl. Environ. Microbiol.* **60**, 2826–2833.

LAYTON, A. C., SANSEVERINO, J., WALLACE, W., CORCORAN, C., SAYLER, G. S. (1992), Evidence for 4-chlorobenzoic acid dehalogenation mediated by plasmids related to pSS50, *Appl. Environ. Microbiol.* **58**, 399–402.

LAYTON, A. C., MUCCINI, M., GHOSH, M. M., SAYLER, G. S. (1998), Construction of a bioluminescent reporter strain to detect polychlorinated biphenyls, *Appl. Environ. Microbiol.* **64**, 5023–5026.

LAYTON, A. C., GREGORY, B., SCHULTZ, T. W., SAYLER, G. S. (1999), Validation of genetically engineered bioluminescent surfactant resistant bacteria as toxicity assessment tools, *Ecotoxicol. Environ. Safety, Environ. Res.* Sect. **B 43**, 222–228.

LEHRBACH, P. R., ZEYER, J., REINEKE, W., KNACKMUSS, H.-J., TIMMIS, K. N. (1984), Enzyme recruitment *in vitro:* use of cloned genes to extend the range of haloaromatics degraded by *Pseudomonas* sp. strain B13, *J. Bacteriol.* **158**, 1025–1032.

MASSON, L., COMEAU, Y., BROUSSEAU, R., SAMSON, R., GREER, C. (1993), Construction and application of chromosomally integrated *lac–lux* gene markers to monitor the fate of a 2,4-dichlorophenoxy acetic acid-degrading bacterium in contaminated soils, *Microb. Releases* **1**, 209–216.

MCCLURE, N. C., WEIGHTMAN, A. J., FRY, J. C. (1989), Survival of *Pseudomonas putida* UWC1 containing cloned catabolic genes in a model activated-

sludge unit, *Appl. Environ. Microbiol.* **55**, 2627–2634.

MCCULLAR, M. V., BRENNER, V., ADAMS, R. H., FOCHT, D. D. (1994), Construction of a novel polychlorinated biphenyl-degrading bacterium: utilization of 3,4′-dichlorobiphenyl by *Pseudomonas acidovorans* M3GY, *Appl. Environ. Microbiol.* **60**, 3833–3839.

MCGRATH, S. P., KNIGHT, B., KILLHAM, K., PRESTON, S., PATON, G. I. (1999), Assessment of the toxicity of metals in soils amended with sewage sludge using a chemical speciation and a *lux*-based biosensor, *Environ. Toxicol. Chem.* **18**, 659–663.

MEIGHEN, E. A. (1991), Molecular biology of bacterial bioluminescence, *Microbiol. Rev.* **55**, 123–142.

MENN, F.-M., APPLEGATE, B. M., SAYLER, G. S. (1993), NAH plasmid-mediated catabolism of anthracene and phenanthrene to naphthoic acids, *Appl. Environ. Microbiol.* **59**, 1938–1942.

MICHÁN, C., DELGADO, A., HAÏDOUR, A., LUCCHESI, G., RAMOS, J. L. (1997), *In vivo* construction of a hybrid pathway for metabolism of 4-nitrotoluene in *Pseudomonas fluorescens, J. Bacteriol.* **179**, 3036– 3038.

MOKROSS, H., SCHMIDT, E., REINEKE, W. (1990), Degradation of 3-chlorobiphenyl by *in vivo* constructed hybrid pseudomonads, *FEMS Microbiol. Lett.* **71**, 179–186.

NÜßLEIN, K., MARIS, D., TIMMIS, K., DWYER, D. F. (1992), Expression and transfer of engineered catabolic pathways harbored by *Pseudomonas* spp. introduced into activated sludge microcosms, *Appl. Environ. Microbiol.* **58**, 3380–3386.

PALMER, G., MVFADZEAN, R., KILLHAM, K., SINCLAIR, A., PATON, G. I. (1998), Use of *lux*-based biosensors for rapid diagnosis of pollutants in arable soils, *Chemosphere* **36**, 2683–2697.

PANKE, S., SÁNCHEZ-ROMERO, J. M., DE LORENZO, V. (1998), Engineering of quasi-natural *Pseudomonas putida* strains for toluene metabolism through an *ortho*-cleavage degradation pathway, *Appl. Environ. Microbiol.* **64**, 748–751.

PARSONS, J. R., SIJM, D. T. H. M., VAN LAAR, A., HUTZINGER, O. (1988), Biodegradation of chlorinated biphenyls and benzoic acids by a *Pseudomonas* strain, *Appl. Microbiol. Biotechnol.* **29**, 81–84.

PATON, G. I., CAMPBELL, C. D., GLOVER, L. A., KILLHAM, K. (1995a), Assessment of bioavailability of heavy metals using *lux* modified constructs of *Pseudomonas fluorescens, Lett. Appl. Microbiol.* **20**, 52–56.

PATON, G. I., PALMER, G., KINDNESS, A., CAMPBELL, C., GLOVER, L. A., KILLHAM, K. (1995b), Use of luminescence-marked bacteria to assess copper bioavailability in malt whisky distillery effluent, *Chemosphere* **31**, 3217–3224.

PETTIGREW, C. A., BREEN, A., CORCORAN, C., SAYLER, G. S. (1990), Chlorinated biphenyl mineralization by individual populations and consortia of fresh water bacteria, *Appl. Environ. Microbiol.* **56**, 2136–2145.

PFLUGMACHER, U., AVERHOFF, B., GOTTSCHALK, G. (1996), Cloning, sequencing, and expression of isopropyl benzene degradation genes from *Pseudomonas* sp. strain JR1: Identification of isopropylbenzene dioxygenase that mediates trichloroethene oxidation, *Appl. Environ. Microbiol.* **62**, 3967–3977.

PIDDINGTON, C. S., KOVACEVICH, B. R., RAMBOSEK, J. (1995), Sequence and molecular characterization of a DNA region encoding the dibenzothiophene desulfurization operon of *Rhodococcus* sp. strain IGT8, *Appl. Environ. Microbiol.* **61**, 468–475.

PIPKE, R., WAGNER-DÖBLER, I., TIMMIS, K. N., DWYER, D. F. (1993), Survival and function of a genetically engineered pseudomonad in aquatic sediment microcosms, *Appl. Environ. Microbiol.* **58**, 1259–1265.

POTRAWFKE, T., LÖHNERT, T.-H., TIMMIS, K. N., WITTICH, R.-M. (1998), Mineralization of low-chlorinated biphenyls by *Burkholderia* sp. strain LB400 and by a two-membered consortium upon directed interspecies transfer of chlorocatechol pathway genes, *Appl. Microbiol. Biotechnol.* **50**, 440–446.

PRESTON, S., HALL, J. M., RATTRAY, E. A. S., CHAUDRI, A. M., MCGRATH, S. P. et al. (1999), Assessing metal bioavailability in soils using luminescence-based microbial biosensors, in: *Bioremediation of Metals and Inorganic Compounds* (LESSON, A., ALLEMAN, B. C., Eds.), pp. 25–30. Columbus, OH: Battelle Press.

RAMOS, J. L., STOLZ, A., REINEKE, W., TIMMIS, K.N. (1986), Altered effector specificities in regulators of gene expression: TOL plasmid *xylS* mutants and their use to engineer expansion of the range of aromatics degraded by bacteria, *Proc. Natl. Acad. Sci. USA* **83**, 8467–8471.

RAMOS, J. L., WASSERFALLEN, A., ROSE, K., TIMMIS, K.N. (1987), Redesigning metabolic routes: manipulation of TOL plasmid pathway for catabolism of alkylbenzoates, *Science* **235**, 593–596.

RAMOS, J. L., HAÏDOUR, A., DELGADO, A., DUQUE, E., FANDILA, M.-D. et al. (1995), Potential of toluene-degrading systems for the construction of hybrid pathways for nitrotoluene metabolism, in: *Biodegradation of Nitroaromatic Compounds* (Spain, J. C., Ed.), pp. 53–68. New York: Plenum Press.

RATTRAY, E. A. S., PROSSER, J. I., KILLHAM, K., GLOVER, A. (1990), Luminescence-based non-extractive technique for *in situ* detection of *Escherichia coli* in soil, *Appl. Environ. Microbiol.* **56**, 3368–3374.

REID, B. J., SEMPLE, K.T., MACLEOD, C. J., WEITZ, H. J., PATON, G. I. (1998), Feasibility of using pro-karyote biosensors to assess toxicity of polycyclic aromatic hydrocarbons, *FEMS Microbiol. Lett.* **169**, 227–233.

REINEKE, W., KNACKMUSS, H.-J., (1979), Construction of haloaromatic utilzing bacteria, *Nature* **277**, 385–386.

REINEKE, W., KNACKMUSS, H.-J. (1980), Hybrid pathway for chlorobenzoate metabolism in *Pseudomonas* sp. B13 derivatives, *J. Bacteriol.* **142**, 467–473.

REINEKE, W., JEENES, D. J., WILLIAMS, P. A., KNACKMUSS, H.-J. (1982), TOL plasmid pWW0 in constructed halobenzoate-degrading *Pseudomonas* strains: prevention of *meta* pathway, *J. Bacteriol.* **150**, 195–201.

RICE, J. F., FOWLER, R. F., ARRAGE, A. A., WHITE, D. C., SAYLER, G. S. (1995), Effects of external stimuli on environmental bacterial strains harboring an *algD–lux* bioluminescent reporter plasmid for the study of corrosive biofilms, *J. Ind. Microbiol.* **15**, 218–328.

ROJO, F., PIEPER, D. H., ENGESSER, K.-H., KNACKMUSS, H.-J., TIMMIS, K. N. (1987), Assemblage of *ortho* cleavage route for simultaneous degradation of chloro- and methylaromatics, *Science* **238**, 1395–1398.

SAHASRABUDHE, A. V., MODI, V. V. (1991), Degradation of isomeric monochlorobenzoates and 2,4-dichlorophenoxy acetic acid by a constructed *Pseudomonas* sp., *Appl. Environ. Biotechnol.* **34**, 556–557.

SAYLER, G., MATRUBUTHAM, U., MENN, F.-M., JOHNSTON, W. H., STAPLETON JR., R. D. (1998), Molecular probes and biosensors in bioremediation and site assessment, in: *Bioremediation: Principles and Practice* (SIKDAR, S. K., IRVINE, R. L., Eds.), pp. 385–434. Lancaster, PA: Technomic Publishing.

SAYLER, G. S., COX, C. D., BURLAGE, R., RIPP, S., NIVENS, D. E. et al. (1999), Field application of a genetically engineered microorganism for polycyclic aromatic hydrocarbon bioremediation process monitoring and control, in: *Novel Approaches for Bioremediation of Organic Pollution* (FASS, R., FLASHNER, Y., REUVEN, S., Eds.), pp. 241–254. New York: Kluwer Academic/Plenum Publishers.

SCHWIEN, U., SCHMIDT, E. (1982), Improved degradation of monochlorophenols by a constructed strain, *Appl. Environ. Microbiol.* **44**, 33–39.

SELIFONOVA, O. V., EATON, R. (1996), Use of an *ipb–lux* fusion to study regulation of the isopropylbenzene catabolism operon of *Pseudomonas putida* RE204 and to detect hydrophobic pollutants in the environment, *Appl. Environ. Microbiol.* **62**, 778–783.

SELIFONOVA, O. V., BURLAGE, R., BARKAY, T. (1993), Bioluminescent sensors for detection of bioavailable Hg(II) in the environment, *Appl. Environ.*

Microbiol. **59**, 3083–3090.

SHAW, J. J., SETTLES, L. G., KADO, C. I. (1988), Transposon Tn4431 mutagenesis of *Xanthomonas campestris* pv. *campestris:* Characterization of a nonpathogenic mutant and cloning of a locus for pathogenicity, *Mol. Plant Microb. Interact.* **1**, 39–45.

SHAW, J. J., DANE, F., GEIGER, D., KLOEPPER, J. W. (1992), Use of bioluminescence for detection of genetically engineered microorganisms released into the environment, *Appl. Environ. Microbiol.* **58**, 267–273.

SHINGLETON, J. T., APPLEGATE, B. M., NAGEL, A. C., BIENKOWSKI, P. R., SAYLER, G. S. (1998), Induction of the *tod* operon by trichloroethylene in *Pseudomonas putida* TVA8, *Appl. Environ. Microbiol.* **64**, 5049–5052.

SHIELDS, M. S., REAGIN, M. J. (1992), Selection of a *Pseudomonas cepacia* strain constitutive for the degradation of trichloroethylene, *Appl. Environ. Microbiol.* **58**, 3977–3983.

SHIELDS, M. S., HOOPER, S. W., SAYLER, G. S. (1985), Plasmid mediated mineralization of 4-chlorobiphenyl, *J. Bacteriol.* **163**, 882–889.

SHIELDS, M. S., MONTGOMERY, S. O., CHAPMAN, P. J., CUSKEY, S. M., PRITCHARD, P. H. (1989), Novel pathway of toluene catabolism in the trichloroethylene-degrading bacterium G4, *Appl. Environ. Microbiol.* **55**, 1624–1629.

SHORT, K. A., SEIDLER, R. J., OLSEN, R. H. (1990), Survival and degradative capacity of *Pseudomonas putida* induced or constitutively expressing plasmid-mediated degradation of 2,4-dichlorophenoxyacetate (TFD) in soil, *Can. J. Microbiol.* **36**, 821–826.

SILCOCK, D. J., WATERHOUSE, R. N., GLOVER, L. A., PROSSER, J. I., KILLHAM, K. (1992), Detection of a single genetically modified bacterial cell in soil by using charge couple device-enhanced microscopy, *Appl. Environ. Microbiol.* **58**, 2444–2448.

SIMPSON, M. L., SAYLER, G. S., APPLEGATE, B. M., RIPP, S., NIVENS, D. E. et al. (1998), Bioluminescent–bioreporter integrated circuits for novel whole-cell biosensors, *TIBTECH* **16**, 332–338.

SINCLAIR, G. M., PATON, G. I., MEHARG, A. A., KILLHAM, K. (1999), Lux-biosensor assessment of pH effects on microbial sorption and toxicity of chlorophenols, *FEMS Microbiol. Lett.* **174**, 273–278.

SOUSA, S., DUFFY, C., WEITZ, H., GLOVE, L. A., BÄR, E. et al. (1998), Use of *lux*-modified bacterial biosensor to identify constraints to bioremediation of BTEX-contaminated sites, *Environ. Toxicol. Chem.* **17**, 1039–1045.

SPRINGAEL, D., KREPS, S., MERGEAY, M. (1993a), Identification of a catabolic transposon, Tn*4317*, carrying biphenyl and 4-chlorobiphenyl degradation genes in *Alcaligenes eutrophus* A5, *J. Bacteriol.* **175**, 1674–1681.

SPRINGAEL, D., DIELS, L., HOOYBERGHS, L., KREPS, S., MERGEAY, M. (1993b), Construction and characterization of heavy metal-resistant haloaromatic-degrading *Alcaligenes eutrophus* strains, *Appl. Environ. Microbiol.* **59**, 334–339.

SPRINGAEL, D., DIELS, L., MERGEAY, M. (1994), Transfer and expression of PCB-degradative genes into heavy metal resistant *Alcaligenes eutrophus* strains, *Biodegradation* **5**, 343–357.

STEINBERG, S. M., POZIOMEK, E. J., ENGELMANN, W. H., ROGERS, K. R. (1995), A review of environmental applications of bioluminescence measurements, *Chemosphere* **30**, 2155–2197.

SUYAMA, A., IWAKIRI, R., KIMURA, N., NISHI, A., NAKAMURA, K., FURUKAWA, K. (1996), Engineering hybrid pseudomonads capable of utilizing a wide range of aromatic hydrocarbons and of efficient degradation of trichloroethylene, *J. Bacteriol.* **178**, 4039–4046.

SYLVESTRE, M. (1995), Biphenyl/chlorobiphenyls catabolic pathway of *Comamonas testosteroni* B-356: prospect for use in bioremediation, *Int. Biodeter. Biodegrad.* **35**, 189–211.

TAURIAINEN, S., KARP, M., CHANG, W., VIRTA, M. (1997), Recombinant luminescent bacteria for measuring bioavailable arsenite and antimonite, *Appl. Environ. Microbiol.* **63**, 4456–4461.

VAN DER LELIE, D., CORBISIER, P., BAEYENS, W., WUERTZ, S., DIELS, L., MERGEAY, M. (1994), The use of biosensors for environmental monitoring, *Res. Microbiol.* **145**, 67–74.

VAN DYK, T. K., MAJARIAN, W. R., KONSTANTINOV, K. B., YOUNG, R. M., DHURJATI, P. S., LAROSSA, R. A. (1994), Rapid and sensitive pollutant detection by induction of heat shock gene–bioluminescence gene fusion, *Appl. Environ. Microbiol.* **60**, 1414–1420.

VAN DYK, T. K., REED, T. R., VOLLMER, A. C., LAROSSA, R. A. (1995), Synergistic induction of the heat shock response in *Escherichia coli* by simultaneous treatment with chemical inducers, *J. Bacteriol.* **177**, 6001–6004.

VAN DYKE, M. I., LEE, H., TREVORS, J. T. (1996), Survival of *luxAB*-marked *Alcaligenes eutrophus* H850 in PCB-contaminated soil and sediment, *J. Chem. Tech. Biotechnol.* **65**, 115–122.

WACKETT, L. P., GIBSON, D. T. (1988), Degradation of trichloroethylene by toluene dioxygenase in whole-cell studies with *Pseudomonas putida* F1, *Appl. Environ. Microbiol.* **54**, 1703–1708.

WACKETT, L. P., SADOWSKY, M. J., NEWMAN, L. M., HUR, H.-G., LI, S. (1994), Metabolism of polyhalogenated compounds by a genetically engineered bacterium, *Nature* **368**, 627–629

WACKETT, L. P., LANGE, C. C., ORNSTEIN, R. L. (1995), Biodegradation: natural and engineered systems, in: *Bioaugmentation for Site Remediation* (HINCHEE, R. E., FREDRICKSON, J., ALLEMAN, B. C.,

Eds.), pp. 105–114. Columbus, OH: Battelle Press.

WALLACE, W. H., FLEMING, J. T., WHITE, D. C., SAYLER, G. S. (1994), An *algD–lux* bioluminescent reporter plasmid to monitor alginate production in biofilms, *Microb. Ecol.* **27**, 225–239.

WARD, T. E., BULMER, D., WALTON, M. R., APEL, W. A. (1998), Development of genetically engineered bacteria for trichloroethylene degradation, *J. Environ. Sci. Health* **A33**, 179–193.

WEIR, S. C., DUPUIS, S. P., PROVIDENTI, M. A., LEE, H., TREVORS, J. T. (1995), Nutrient-enhanced survival of and phenanthrene mineralization by alginate-encapsulated and free *Pseudomonas* sp. UG14Lr cells in creosote-contaminated soil slurries, *Appl. Microbiol. Biotechnol.* **43**, 946–951.

WINTER, R. B., YEN, K.-M., ENSLEY, B. D. (1989), Efficient degradation of trichloroethylene by a recombinant *Escherichia coli, Bio/Technology* **7**, 282–285.

ZYLSTRA, G. Z., WACKETT, L. P., GIBSON, D. T. (1989), Trichloroethylene degradation by *Escherichia coli* containing the cloned *Pseudomonas putida* F1 toluene dioxygenase genes, *Appl. Environ. Microbiol.* **55**, 3162–3166.

22 Possibilities, Limits and Future Developments of Soil Bioremediation

JÜRGEN KLEIN

Essen, Germany

List of Abbreviations

BTEX	benzene, toluene, ethylbenzene, xylene
HC	hydrocarbon
HM	heavy metal
PAH	polycyclic aromatic hydrocarbons
PCB	polychlorinated biphenyls
PCDD	polychlorinated dibenzodioxins
PCDF	polychlorinated dibenzofurans
TNT	trinitrotoluene
TPHC	total petrol hydrocarbons
VOCH	volatile organic hydrocarbons

1 Introduction

Environmental protection over the past few decades has meant primarily the protection of air and water. With the increasing use of land in industrialized societies and the highlighting of possible hazards from contaminated soil the public became aware of soil protection in the early 1980s. This has also prompted industry to take up this market segment. Engineers and scientists have thus been spurred on to look for technically optimized, ecologically sound, and economically appropriate solutions.

For more than a century natural biochemical processes (nature's self-cleaning abilities) have been utilized to treat effluent, and reactors and plant systems had been adapted with increasing effect to cope with the difficult conditions. Only in the 1980s work was begun on testing biological methods for cleaning up soil. The experience accumulated in biological soil clean-up in the first few years was character-ized by both success and failure, including those of unprofessional suppliers. Initially, therefore, the acceptance of biological soil clean-up was only limited. But now, because of intensive and interdisciplinary work, impressive success is evident. Consequently biological soil clean-up methods now enjoy a high technical level and a broad acceptance.

In 1998 approximately $2.2 \cdot 10^6$ t of soil were remediated in Germany in 109 stationary soil treatment facilities, of which $1.2 \cdot 10^6$ t or 60% were treated biologically (HENKE, 1998). In Germany the available capacity for on-site biological treatment is about $4 \cdot 10^5$ t a^{-1} (Tab. 1).

The aim is to maintain and apply this level, even if the prospects for soil clean-up in Germany are seen from a more modest point of view. In addition to the preference for securing techniques and to the competition from less expensive suppliers in neighboring countries, in recent times low-price dumping has represented a challenge to operators of soil decontamination facilities in Germany.

Tab. 1. Soil Treatment Plants in Germany 1997 (SCHMITZ and ANDEL, 1997)

Treatment Centers	Biological	Scrubbing	Thermal	Σ
Capacity				
$[10^6 \, t \, a^{-1}]$	1.9	1.38	0.17	3.45
[%]	55	40	5	100
Number	81	24	4	109
Throughput				
$[10^6 \, t \, a^{-1}]$	1.24	0.85	0.11	2.20
[%]	56.4	38.7	4.9	100
Mobile plants:				
Number	8	7	5	20
Capacity $[10^6 \, t \, a^{-1}]$				~1.0
Contaminants to be treated	TPHC	TPHC	TPHC	
	BTEX	BTEX	BTEX	
	phenol	phenol	phenol	
	PAH	PAH	PAH	
	VOCH	VOCH	VOCH	
		pesticides	pesticides	
		PCB	PCB	
		PCDD/F	PCDD/F	
		cyanide	cyanide	
		phosphorus		
		HM	HM	

2 Fundamental Phenomena of Bioremediation

The use of biological processes for treatment of liquid wastes from human activities is an established technology dating back at least 4,000 a. The knowledge of the biodegradation mechanisms of organic pollutants and particularly of synthetic compounds, however, is more recent and has only been developed in the second part of this century.

In biological processes, use is made of the capacity of microorganisms to consume organic matter as nutrients (substrate) and to convert them to harmless natural materials such as CO_2, water, and biomass. For degradation of noxious matter in contaminated soils, bacteria and fungi are of foremost importance.

The attack of organic matter in soils by microorganisms results either in full degradation (mineralization) or in a partial degradation process producing metabolites which may be used by other members of the biocenosis or which remain in the soil. Furthermore, the original substance and the metabolites can be cycled to the carbon depot of the soil. This is called humification. The decisive point in that case is that the noxious matter or the resulting metabolite are incorporated into the soil matrix reducing sharply their availability for biochemical reactions. At present this phenomenon is used in comprehensive development work aimed at a conversion of pollutants by microbiological attack. So the pollutants are converted to natural matter, such as humic compounds, which imply no further environmental risk. Detoxification can also be obtained by cometabolism using non-growth substances as cosubstrate. With increasing complexity and xenobiotic character of synthetic chemicals complete pathways are unlikely to exist within the indigenous soil population. To avoid that cometabolic reactions may prevail giving rise to the accumulation of dead-end metabolites, particular attention has been paid to the role of microbial consortia in synergistic cometabolism and mineralization of substrates poorly or incompletely degraded by single organisms. Furthermore, the use of non-indigenous organisms, whether native or engineered, is required if the complementary biocatalytic activity is lacking in the indigenous microflora. Although the application of engineered microorganisms opens another dimension of bioremediation, experience has been made that the natural soil possesses microorganisms which are capable of degrading a broad variety of xenobiotics, but which cannot be cultivated in the laboratory. This raises hope of sufficient potential of nature's self-cleaning abilities.

3 Necessary Preliminary Investigations

The decision in favor of a biological decontamination process depends on the following prerequisites:

(1) degradability of contaminants,
(2) bioavailability of contaminants in the soil matrix,

(3) adjustability of the biological, physical, and chemical conditions required for biological degradation in the soil.

3.1 Degradability of Contaminants

Laboratory methods for the evaluation of biological soil clean-up processes have been developed (DECHEMA, 1992) allowing to find out thoroughly the microbial degradability of contaminants. However, it has to be considered that the assessment of degradability of soil contaminants requires investigations on the original soil, either in suspension or by means of naturally moist samples.

Tab. 2 contains a list of the substance groups which count as relevant in terms of contaminated sites, together with an assessment of their microbial degradability (DECHEMA, 1991). It is not possible to define in general terms a concentration range within which microbial degradation is possible. This concentration range differs according to the substance class. Similarly, it is not possible to fix a generally applicable value for an achieveable residual concentration, since this depends on the

Tab. 2. Biodegradability of Soil Contaminants (DECHEMA, 1991)

Class of Contaminants	Basically Well Degradable	Basically Hardly Degradable
Aliphatic hydrocarbons (HC), petrol HC and derivates	+	
Monocyclic aromatic HC (e.g., BTEX) and heterocyclic HC (e.g., pyridine, chinolin)	+	
Polycyclic aromatic HC (PAH)	+[a]	+[b]
Volatile halogenated, especially chlorinated HC (VOCH)	+	
Alicyclic VOCH and derivates (e.g., HCH)	+	
Polychlorinated biphenyls (PCB)		+[c]
Polychlorinated dibenzodioxins and dibenzofurans (PCDD or PCDF)		+[c]
Pesticides and derivates		+[c]
Compounds with nitro functions		+
Heavy metals	not degradable[d]	

[a] Up to 4-ring PAH.
[b] 5- and 6-ring PAH.
[c] Some low chlorinated congeneres are principally degradable/dehalogenable, degradation of highly chlorinated congeneres cannot be demonstrated at present.
[d] In part removable by biosorption or phytoremediation, genesis of methylated metals possible.

type and starting concentration of the pollutants and the adjustable ambient conditions.

To date the most frequently cleaned soils have been those contaminated with oil. The degradability of *aliphatic hydrocarbons*, such as petrol, diesel, and other oil derivatives, can be described well. The persistence increases with the chain length and the degree of branching, and for compounds with more than 25 carbon atoms the degradation rate slows down, or comes to a complete standstill.

It is more difficult to clean soils from the coal byproduct domain (gas works, coking plants), where the contaminants are mainly *BTEX aromates* (benzene, toluene, ethyl benzene, xylenes) and *polycyclic aromatic hydrocarbons* (PAH).

Monocyclic aromatic hydrocarbons, e.g., BTEX aromatics can be described as highly degradable and for heterocyclic hydrocarbons, e.g., pyridine, different degraders have been isolated, while the degradation of polycyclic hydrocarbons shows a very different degradability depending on the number of aromatic rings they consist of. 4-Ring PAH can now be described as being well degradable, but it is not always possible to degrade PAH with 5 and more aromatic rings. If degradation is encountered here, this is due to cometabolism, which means that the microorganisms do not use these aromates as the sole source of carbon and energy, but can only use them jointly in the presence of other hydrocarbons (cosubstrates), especially in the presence of benzene. During the last years humification, i.e., the incorporation of PAH into the soil matrix, is investigated and in a few cases tried to be applied (see Chapter 4, this volume).

In many cases success is experienced with the biological treatment of soils contaminated by *highly volatile, halogenated hydrocarbons* (VOCH) and in some cases also with soils contaminated with *alicyclic VOCH*, e.g., HCH (see Chapter 10, this volume).

In contrast the degradation of *polychlorinated biphenyls* (PCB) and *polychlorinated dibenzodioxins and dibenzofurans* (PCDD/ PCDF) should in principle be regarded as very difficult. To date it has not been possible to provide a scientific verification of the complete dehalogenation of polychlorinated dioxins.

Military and armament sites are often contaminated with explosives, i.e., *nitro aromatics*. The main contaminant is TNT (trinitrotoluene). To eliminate TNT and its thereby generated metabolites microbially only a combination of aerobic and anaerobic treatment together with humification seems to be an efficient strategy (see Chapter 11, this volume).

In principle heavy metals are not degradable by microbial means. Under certain conditions they can be adsorbed on biomass by biosorption or taken up by plants from the soil. The latter is called phytoremediation (see Chapter 17, this volume). Biochemical reactions with heavy metals can result in water soluble compounds, e.g., methylated lead compounds.

3.2 Bioavailability

In many cases not the actual microbial degradability is decisive for degradation and degradation rate, but physico-chemical parameters, such as adsorption–desorption, diffusion, and solution properties of the contaminants found in the soil, frequently in solid phase. The availability of contaminants in the soil to microorganisms – the bioavailability – is decisively influenced by the configuration of the soil matrix, i.e., its material composition and particle size distribution, and also by the history of the contamination (see Chapter 3, this volume). For these cases, suitable bioavailability assessment methods are on hand (DECHEMA, 1992; ITVA, 1994).

Fig. 1 shows a sequence pattern for close-to-practice preliminary investigations. When following this scheme we arrive – with relatively low expenditure – first at the important decision whether biological decontamination is possible at all. If there is no bioavailability of contaminants in the soil or if a contamination which is toxic to microorganisms cannot be eliminated, this means that the soil concerned is out of question for microbiological treatment.

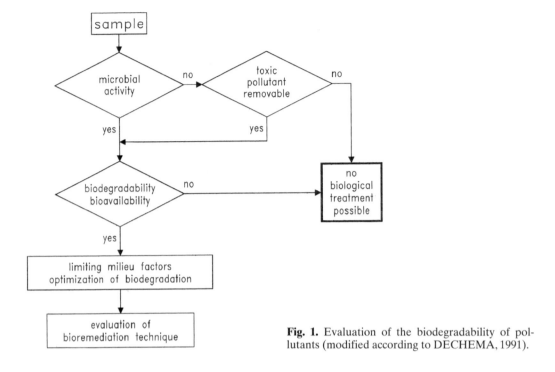

Fig. 1. Evaluation of the biodegradability of pollutants (modified according to DECHEMA, 1991).

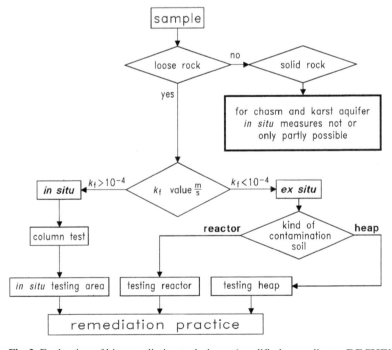

Fig. 2. Evaluation of bioremediation techniques (modified according to DECHEMA, 1991).

3.3 Adjustability of the Biological and Physico-Chemical Conditions for Biological Degradation in Soil

The physico-chemical and, of course, also the geological properties of the soil are decisive for the choice of methods for bioremediation.

Fig. 2 shows the sequence of investigations in view of the selection of methods. The decision in favor of an *in situ* or an *ex situ* method generally depends on the hydrogeological configuration of the soil, the permeability coefficient k_f, the homogeneity of the soil, and its silt and fines content (see Chapter 2, this volume). The k_f value is regarded as orientation parameter. Experience has shown that with k_f values $<10^{-5}$ m s^{-1}, *in situ* treatment is out of question. Only in relatively few cases favorable conditions for a microbiological *in situ* decontamination prevail.

4 Bioremediation Techniques

Fig. 3 shows the generally practicable process options. After the required preliminary investigations and most intense balancing of microbial degradation processes a decision in favor of an *ex situ* or *in situ* method is made.

4.1 *Ex situ* Processes

These processes require excavation of the contaminated soil and its treatment either in bio-heaps or in reactors, either on-site or – this requires transport capacity – off-site in a soil decontamination plant.

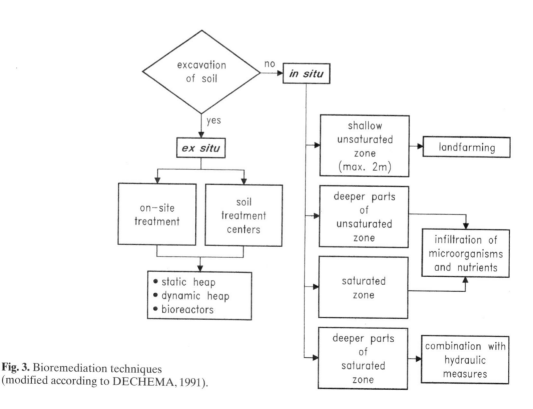

Fig. 3. Bioremediation techniques (modified according to DECHEMA, 1991).

4.1.1 Bio-Heap Treatment

As to heap treatment, a difference is made between *static* and *dynamic* methods (see Chapter 13, this volume). In static bio-heap methods, irrigation and aeration networks are installed for ensuring the necessary environmental conditions.

In case of dynamic heap methods the deposited soil masses are, in specified intervals, turned over, homogenized, and respread. During these turning over cycles, further water and nutrients are added, if necessary.

4.1.2 Reactor Processes

Reactor techniques are made use of for shortening the time of treatment of soils with high fines portions and high rates of contamination as well as for better process control with respect to exhaust air and wastewater treatment. According to the water content of the soil, a difference can be made between *solid state* and *slurry state* methods (see Chapter 14, this volume).

4.1.2.1 Solid State Processes

In solid state processes, the soil is treated at a humidity rate corresponding approximately to 50–70% of its maximum water capacity. The treatment is run in rotary reactors or in static reactors equipped with internal mixing systems. The principle of these methods is largely identical to that of the dynamic heap processes. In both cases the biological degradation of noxious matter contained in the soil is achieved by agitation and aeration of the naturally moist soil. Also the treatment of soil–compost blends is to be categorized as solid state process. These dry processes are advantageous in that the treated soil does not need to be dewatered.

4.1.2.2 Slurry State Processes

Extremely fine-particle or cohesive soils or sludges, e.g., products from pretreatment steps such as soil washing, cannot be purified by heap or solid state reactor techniques. Accordingly, slurry state techniques are run for the treatment of these soils. The soil portion per weight of slurry ranges between 30 and 50% by weight. Processes run in impeller type mixing vessels, airlift or fluidized bed reactors which provide sufficient and homogeneous nutrient and oxygen supply are particularly suitable (KLEIJNTJENS, 1999; SINDER et al., 1999).

The key criteria for these processes are:

(1) The particle sizes are <1,000 µm.
(2) All soil particles must be in full contact with the aqueous phase.
(3) In particular for PAH, long residence times are be catered for.
(4) Due to the long residence times the energy requirement must be low and the solids portion of a reactor fill must be high.

The advantages of the application of bioreactors summarized can be read as:

(1) Decontamination takes place under well controlled conditions easy to optimize.
(2) All kinds of soils, even with high ratios of fines, can be treated.
(3) All kinds of contaminants, also hazardous substances, can be handled.
(4) Aerobic and anaerobic as well as alternating aerobic–anaerobic treatment is possible.
(5) Bioavailability increases, especially in slurry reactors.
(6) Emissions can be controlled at optimum.

4.2 *In situ* Methods

The decisive viewpoint for an *in situ* treatment is that no soil masses need to be removed and that the saturated or unsaturated zones of a contaminated site itself are used as integral reactor for the microbial degradation of contaminants. Depending on the features of the contamination and the site different technologies may be chosen (see Chapter 18, this volume). The technologies are subdivided into technologies for the treatment of the unsaturated and the saturated zone. A further subdivision comprises highly active technologies (bioventing, bioslurping, biosparging, and hydraulic circuits) as well as active and passive technologies (funnel-and-gate, bioscreen, and natural attenuation).

For shallow unsaturated zones, landfarming is possible which allows the addition of nutrients and the adjustment of the environmental conditions. In deeper unsaturated zones and in parts of the saturated zone, infiltration of microorganisms (bioaugmentation) and nutrients by a flushing circuit, sometimes combined with hydraulic measures, is practiced (DECHEMA, 1996). The availability of molecular oxygen – dissolved in water and in this way cycled to the process – is decisive for rapid aerobic degradation. In such a process configuration, the addition of an oxygen carrier (such as hydrogen peroxide or industrial oxygen) may be advantageous.

For decontamination under anoxic or anaerobic conditions (e.g., in the unsaturated zone) another electron acceptor (such as nitrate) is necessary for the microbial metabolism.

4.3 Main Features of Soil Bioremediation

The main features of biological soil remediation differentiated between *in situ* and *ex situ* methods are shown in Tab. 3.

4.4 Reuse of Treated Soil

A major aspect of biological soil remediation is the reuse of treated soil. Structural use includes, e.g., backfilling noise baffle embankments, roadside fortification, or ditches and landfill measures. The vegetation related use ranges from roof greening and parks to agricultural use. The latter requires prior testing on large areas with grass as test seed. A number of positive results have already been obtained where it is possible to recultivate corn and potatoes after such a test (see Chapter 14, this volume).

4.5 Bioassays for Soils

A crucial factor for the reuse of soils is a toxicological/ecotoxicological assessment. For this purpose use is made of bioassays to measure the impact of contaminated or treated soils. Biological tests have proven particularly

Tab. 3. Main Features of Biological Soil Bioremediation

Features	Treatment	
	ex situ	*in situ*
Removal of bioavailable contaminants by mineralization to CO_2 and H_2O	+	+
Use of indigenous microorganisms	+	+
Pre-investigations needed for evaluation of biodegradability and remediation techniques	+	+
Operation time depends on kind and concentration of contaminants	+	+
Prediction of remediation success is difficult		+
Treatment of large amounts of soil possible		
– if suitable areas are available	+	
– if there are no remediation time limits		+
Flexible adaptation to site conditions, e.g., remediation under buildings or on built-up areas		+
Low energy demand	+	+
Control of degradation conditions required during the total remediation period	+	+
Suited for *ad hoc* measures, e.g., accidents	+	

beneficial, if chemical or environmental samples of complex composition have to be tested with respect to their hazard potential. They integrate the effects of all acting substances (see Chapter 5, this volume), including those not considered or recorded in chemical analysis (see Chapter 23, this volume). The state of the art of scientific development and experience has been reported, and an assessment strategy which meets practical requirements has been recommended (DECHEMA, 1995):

(1) preliminary assessment of contaminated sites; recognition of ecotoxicological effects caused by contaminants which might not have been detected by chemical analysis;
(2) ecotoxicological process control during biological soil clean-up;
(3) evaluation of a possibly residual toxic/ecotoxic potential of treated soils before their further use.

5 Perspectives

The enthusiasm that has accompanied soil decontamination in the last few years has now given way to a realistic attitude of more appropriate proportions. The discussion on the equivalence of securing and decontamination techniques must not lead to a clean-up on a low level, but rather to ecologically and economically appropriate solutions.

In most cases the original objective of multifunctional use through the restoration of a "natural" soil is not feasible for technical and financial reasons. In view of this biological clean-up techniques still have to be improved and optimized in order to provide cost-effective, technically simple and near-natural processes (REHM, 1999).

5.1 Trends

Solutions attempted for *ex situ* processes, e.g., are the combination of fixed bed and liquid phase bioreactors, which provide a high degree of flexibility in adapting the required ambient conditions to the relevant contaminant and also the special features of the soil matrix, or for *in situ* measures the dosing of tensides and heating of the soil by radio waves, e.g., by which it is possible to enhance the biological availability and hence the degree of degradation (STOTTMEISTER, 1998).

In Germany, the treatment principle of *humification* is currently subject to considerable attention by the clean-up of contaminated armament sites, primarily the contaminant TNT. To eliminate TNT the only reasonable approach seems to be cometabolic transformation with the subsequent binding of the products arising in the soil matrix. Both, certain bacteria and fungi, are capable of cometabolic transformation of TNT to reduced metabolic products, such as amino nitrotoluenes. These are then partly degraded, mineralized, or bound to the soil matrix in a process similar to natural humification. Both *ex situ* and *in situ* technologies are studied. For example, anaerobic–aerobic process combinations (modified composting) are tested as an *ex situ* technique (HENKE, 1998). In the development of suitable *in situ* methods, an investigation is conducted into whether an accumulation of TNT in the rhizosphere of plants and undergrowth can be exploited economically (WARRELMANN and WALTER, 1997). This method, known as *phytoremediation*, is mainly researched in the USA, Great Britain, and Germany with a view to the further development of near-natural, *in situ* clean-up processes for the accumulation of metals and organic compounds (see Chapter 16, this volume, concerning the phytoremediation of organics).

Successful research has also been conducted into the *activation of natural biocenoses* in the clean-up of sediments and acidic waters from uranium mining as demonstrated by the 3 years lasting operation of a pilot plant for heavy metal leaching (STOTTMEISTER, 1998).

Since all biological clean-up processes are based on nature's own self-purification abilities, they are subject of increasing attention and methods are propagated under the names of "*intrinsic bioremediation*" or "*natural attenuation*" (see Chapter 18, this volume).

The USEPA policy directive defines monitored natural attenuation as "the reliance of natural attenuation processes (within the context of a carefully controlled and monitored site clean-up approach) to achieve site-specific remediation objectives within a time frame that is reasonable compared to that offered by more active methods" (U.S. Environmental Protection Agency, 1999).

In neither case the terms are used to indicate that nothing is done and everything is left up to nature, which would not be acceptable anyway in the case of an actual hazard. Rather they involve more research work than is necessary for an engineering solution. The solution to the following questions is implied:

(1) How can suitable microorganisms most effectively be incorporated and distributed in the soil (*bioaugmentation*) and stimulated to optimum activity (*biostimulation*)?

(2) How must the groundwater be guided in order to render an *in situ* treatment possible and effective?

With regard to the latter, in addition to the common techniques for groundwater treatment (pump-and-treat) techniques propagated in the USA under the names *funnel-and-gate systems* and *reactive barriers* are increasingly tested (see Chapters 15 and 18, this volume).

The most important restriction of such near-natural processes is the time span available to clean up the site.

5.2 Future Outlook

The field of biodegradation and thereby bioremediation has experienced a dynamic evolution and remarkable developments over the past few years. It seems to have entered its most interesting and intense phase. Isolation and characterization of new microorganisms with novel catabolic activities continue unabated, and the use of plants and plant–microbe associations in bioremediation is strongly expanding. The continuously growing knowledge on catabolic pathways and critical enzymes provides the basis for the rational genetic design of new and improved enzymes and pathways for the development of more performant processes.

Suitable and sustainable solutions require a broader *interdisciplinary approach*; in other words: the involvement of microbiologists, geneticists, chemists, hydrogeologists, soil scientists, and process engineers (KNACKMUSS et al., 1998).

On the other hand *avoidance* of environmental contamination is the future challenge for which suitable and sustainable strategies also can only be achieved by an interdisciplinary collaboration between all protagonists in research and industry.

The wide-ranging experience accumulated with respect to the contamination of soils and groundwater must provide special impetus for testing the environmental impact of new chemical products before they are introduced, thus preventing subsequent contamination. A *benign-by-design chemistry* would, therefore, have to concentrate research on identifying forms of bonding which facilitate the development of biodegradable and environmentally sound substances in the circulation of chemical products.

6 References

DECHEMA (1991), *Einsatzmöglichkeiten und Grenzen der mikrobiologischen Verfahren zur Bodensanierung*. Frankfurt: DECHEMA.

DECHEMA (1992), *Laboratory Methods for the Evaluation of Biological Soil Clean-Up Processes*. Frankfurt: DECHEMA.

DECHEMA (1995), *Bioassays for Soils*. Frankfurt: DECHEMA.

DECHEMA (1996), *In-situ*-Sanierung von Böden, *11. DECHEMA-Fachgespräch Umweltschutz*. Frankfurt: DECHEMA.

HENKE, G. A. (1998), Biologische Bodensanierung, in: *Altlastensanierung* (HENKE, G. A., Ed.). Essen: Vulkan-Verlag.

ITVA (1994), Mikrobiologische Verfahren zur Bodendekontamination, *ITVA-Arbeitshilfe*. Berlin: Ingenieurtechnischer Verband Altlasten e.V. ITVA.

KLEIJNTJENS, R. (1999), The slurry decontamination process: Bioprocessing of contaminated solid waste streams, *Proc. ECB 9*, Brussels, July 1999.

KNACKMUSS, H.-J. (1998), "*The Take Home Message*", *Innovative Potential of Advanced Biological Systems for Remediation*. Workshop March

2–4, 1998. Hamburg-Harburg: Technical University.

REHM, H.-J. (1999), The development of biotechnology in environmental processes, *Acta Biotechnol.* **3**, 205–210.

SCHMITZ, H. J., ANDEL, P. (1997), Die Jagd nach dem Boden wird härter, *TerraTech* **5**, 17–31.

SINDER, C., KLEIN, J., PFEIFER, F. (1999), The DMT-BIODYN-Process: a suspension reactor for biological treatment of fine grained soil, *Proc. ECB 9*, Brussels, July 1999.

STOTTMEISTER, U. (1998), Trends in der Entwicklung der biologischen Methoden in der Sanierungsforschung, *Vom Wasser* **91**, 343–350.

U.S. Environmental Protection Agency (USEPA) (1999), Use of monitored natural attenuation at Superfund, RCRA corrective action and underground storage tank sites, *OSWER Directive* 9200.4-17P.

WARRELMANN, J., WALTER, U. (1997), Erprobung und Anwendung mikrobiologischer Verfahren bei der Sanierung von Rüstungsaltlasten, in: *Handbuch der Altlastensanierung* (FRANZIUS, V., WOLF, K., Eds.), pp. 1–18. Heidelberg: C. F. Müller Verlag.

23 Sampling and Investigation of Solid Matter

MICHAEL ROEMER

Aachen, Germany

List of Abbreviations

AAS	atomic absorption spectrometry
AED	atomic emission detector
AES	atomic emission spectrometry
AOX	adsorsbable organic chlorinated compounds
BOC	biochemical oxidation capacity
BTEX	benzene, toluene, xylenes, ethylbenzene
BTX	benzene, toluene, xylenes
BUS	Bundesamt für Umweltschutz der Schweiz
COD	chemical oxygen demand
DAD	diode array detector
DMSO	dimethyl sulfoxide
EC	electrochemical detector
ECD	electron capture detector
ELCD	electrolytic conductivity detector
EOX	extractable organic chlorinated compounds
FID	flame ionization detector
FL	fluorescence detector
FPD	flame-photometric detector
GC	gas chromatography
HPLC	high performance liquid chromatography
ICP-AES	inductively coupled plasma atom emission spectrometry
IR	infrared
IRD	infrared spectrometric detector
MS	mass spectrometric detector
MSD	mass selective detector
NPD (PND)	nitrogen–phosphorus detector
PAH	polycyclic aromatic hydrocarbons
PCB	polychlorinated biphenyls
PID	photoionization detector
PTV	programmable temperature vaporizer
RI	refractive index detector
SFE	extraction with supercritical fluids
TCD	thermal conductivity detector
TLC	thin layer chromatography
UV	ultraviolet
UV/VIS	UV/visible detectors with fixed/variable wavelength, diode array detector
VIS	visual
XAD	styrol-vinylbenzol resin

1 Introduction

Soil is the uppermost weathered layer of the earth and consists of broken down and chemically altered rock, air, water, and residues from living organisms. Because of its regulating properties, soil has an equalizing effect on the water, heat and substance distribution of natural systems and those influenced by human beings. An important property of soil is its ability to act as a filter and a buffer, and to effect transformations with regard to substances added to it. In areas of contaminated land, the deposition of potentially problematic substances can lead to exhaustion of the regulating function of the soil, with undesirable effects on the natural environment and on human beings.

As there is no mixing and no equalizing of concentrations, as in the case of groundwater or soil air, a soil sample only gives information on one particular point. The frequently occurring heterogeneities (with regard to space and time) of substances and substance mixtures in waste deposits and soils at disused sites are an important reason why a uniform procedure for sampling is so difficult, if not impossible. Stringent requirements are, therefore, made on the planning and carrying out of soil sampling, which are more demanding than in the investigation of other areas of the environment.

2 Investigation Strategy

2.1 Targets and Requirements of an Investigation Strategy

The aim of assessing areas which are suspected of contamination is to estimate the risks and hazards which can be caused by pollutants. The investigation strategy for sampling, sample processing and analysis forms the critical basis of this.

In practice it has frequently been shown that the determination and subsequent assessment of contamination using a small number of samples is only possible with considerable limitations.

The assessment strategy for areas suspected of contamination can be based on two different procedures

(1) A targeted investigation of particular substances at particular locations is possible using prior information.
(2) If there is insufficient prior information a screening investigation of the whole suspected area is necessary.

The tasks of the investigation strategy derived from the above targets can be divided up as follows:

(1) establishing the extent of the investigation,
(2) an adequate qualitative and quantitative determination of the inventory of pollutants,
(3) determination of the relevant dispersion routes of the substances,
(4) recognition and characterization of impurities from dispersing media according to the nature, quantity and spatial distribution,
(5) obtaining information to assess the most recent contamination and its availability,
(6) obtaining information to estimate the possible long-term effects on protective materials during the envisaged use.

From the aims of the investigation strategy cited and the tasks derived from them for an optimal investigation, in particular during the assessment, the program to be drawn up must satisfy the following general requirements:

(1) flexible structuring of the investigation grid,
(2) sensible and problem-orientated sampling,
(3) the investigation should take place in stages, and thus
(4) pooled samples should be obtained where possible,
(5) key parameters should be selected and screening analyses carried out.

Depending on the course of the work, the investigation strategy (Fig. 1) is divided into a

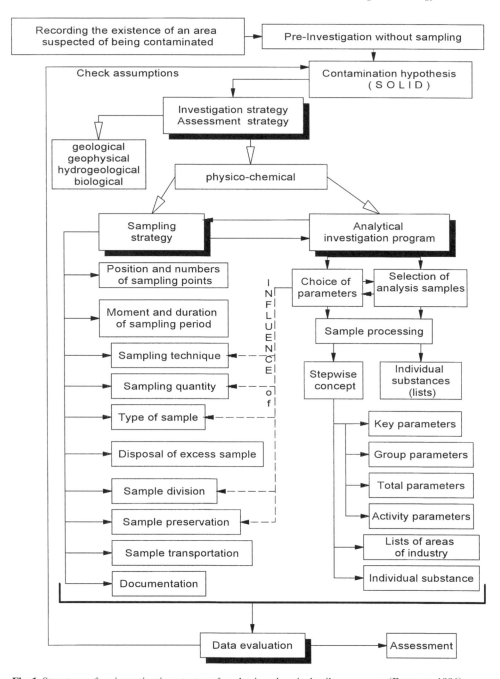

Fig. 1. Structure of an investigation strategy for physico-chemical soil assessment (ROEMER, 1996).

sampling strategy and an analytical investigation program. The subsequent data evaluation and assessment are affected by both components.

2.2 Course of the Chemical Investigation

The investigation of solid samples is divided into many smaller steps. Some will subsequently be described in more detail. These steps include:

(1) definition of the investigation task,
(2) establishment of the sampling strategy (location, duration, time, type of sampling),
(3) selection and preparation of equipment,
(4) sampling,
(5) dividing samples in the field,
(6) sample processing in the field (filtration, etc.),
(7) transport and storage,
(8) delivery of samples to the laboratory (identification of divided samples),
(9) sample pretreatment in the laboratory (homogenizing, taking aliquots, digestion, extraction, enrichment, grinding, drying, sieving, production of eluates, etc.),
(10) instrumental analysis (calibration, blank value control, measurements on the analysis samples, diluted if necessary),
(11) evaluations (e.g., of the measured value from the extinction or peak area),
(12) producing the results (rounding values, units),
(13) plausibility control,
(14) assessment.

3 Sampling Strategy

The sampling strategy is a systematic procedure which should ensure that the samples taken are reliable and representative of the contaminated or suspect area and that optimal sampling points are chosen (Anonymous, 1992). It should include plans of the location, number and quantity of samples to be taken, details of the sampling cycle and sampling technique, as well as transport, preservation and documentation.

The quality of the investigation is significantly determined by the quality of the sampling.

3.1 Importance of Sampling for the Investigation Results

For the investigation of solid samples a prior sampling strategy is usually indispensable. Pollutants can occur in soils in very varying concentrations and bound forms within a very small area.

The aim of sampling must be to achieve a representative determination of the sample material, as the analysis results obtained only refer to the sample analyzed. Transferability of the results to the original substance is only possible, if a representative sample is taken. From mathematical probability theory it is known that the variance [mean of the square of the difference between each value and the arithmetic mean (ROEMER, 1996)] of an investigation result where the variables are mutually independent is given by the sum of the individual squares (variance propagation law). The accuracy of the investigation results is thus determined by the large number of small operations. Possible sources of error can be assigned to sampling, sample processing and analysis (Eq. 1) (KNOOP, 1990).

$$S_{total}^2 = \sum S_{sampling}^2 \\ + \sum S_{sample\ processing}^2 \quad (1) \\ + \sum S_{analysis}^2$$

In the investigation of solids the greatest possibility for error lies in sampling and sample processing and less in the analysis. The following should, therefore, be taken into account: with solids in particular sampling is an extremely important step and should, therefore, only be carried out after prior planning and only by knowledgeable personnel.

Errors occurring during sampling cannot be compensated for significantly during the subsequent steps. The quality of an investigation is thus determined to a great extent by the quality of the sampling, particularly when inhomogeneous solid material is present.

3.2 Requirements of Soil Samples

With regard to possible contamination the soil sample should

(1) be representative of the area investigated or of the sampling point,
(2) give information on the extent of the contaminated area in all three dimensions,
(3) allow an estimation of the quantity of contaminated soil,
(4) allow possible hazards to ground water to be recognized,
(5) give information on the mobility at that particular time and the potential possibility of further spreading out of the pollutants (under the current or under different boundary conditions),
(6) allow geological conditions to be recognized,
(7) be obtained rapidly and economically.

A standardized and uniform procedure for soil sampling, particularly for areas suspected of being contaminated, is not available. However, the standardization of sampling procedures and sampling strategies is being worked on nationally and internationally.

PAETZ (1993) has drawn up a list of national (DIN, VDI, DVWK-leaflets, VDLUFA methods), European (CEN) and international (ISO) standards and guidelines, which are used in the investigation and assessment of soils.

Indications and instructions for the proper sampling of contaminated solid material for physico-chemical analysis can be found in Anonymous (1990, 1993, 1994), TOUSSAINT (1994), LfU BW (1993).

3.3 Components of the Sampling Strategy

Because of the importance of sampling for the results of the investigation, a systematic procedure for sampling is absolutely necessary. The sampling strategy involves establishing the following points:

(1) position of the sampling location (in area and depth),
(2) number of sampling points,
(3) quantity of sample,
(4) type of sample (individual sample, pooled sample),
(5) sampling technique,
(6) division of sample,
(7) sample preservation,
(8) documentation,
(9) quality assurance.

In a strategy for obtaining solid samples there are two basically different procedures for establishing sampling points, as shown in Fig. 2.

The extent and quality of prior information decides the type of sampling strategy to be used, either

(1) judgement-based or
(2) a statistical procedure.

Up to now international and national standards and guidelines for the investigation of contaminated soil have always been non-statistical (BOSMAN, 1993). With good prior information on the contamination and thus better knowledge of the expected spatial distribution of contaminants, the judgement-based procedure can be expedient. Fig. 3 shows examples of a possible spatial distribution of contaminants in the case of contaminated land.

3.3.1 Sampling Based on Judgement

When the possible causes of soil contamination or information on contaminated areas are known, definite clues concerning the pollutant, the point of deposition and the distribu-

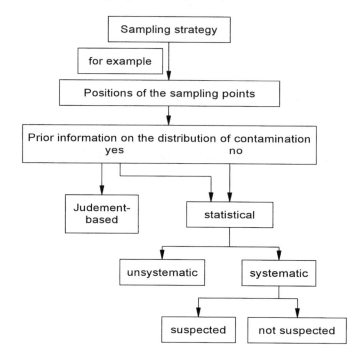

Fig. 2. Possible procedure to obtain samples.

tion can be derived. A target-orientated and judgement-based assessment is possible and the tasks involved in sampling become simpler. In the case of disused industrial sites possible contamination areas can be located through knowledge of the operation of a particular production process, transport routes and storage areas. Further contamination due to leakages, handling losses, accidents, and the effects of war can usually be determined only to the same small extent by judgement-based sampling as the former usual disposal of production residues around the industrial site.

3.3.2 Statistical Sampling

If insufficient prior information is available, statistically based sampling strategies need to be used. Here the aim is the highest possible probability of success so that the lowest possible number of sampling points is required.

With random sampling the sampling points are distributed over the whole area. This can result in a very irregular distribution and very

large unsampled areas. The transfer of the sampling points on to the site is also difficult.

Systematic or grid-based arrangements of sampling points are frequently used because of their simple planning and execution.

3.3.3 Sampling Points in Relation to Use

SIMMLEIT and ERNST (1994) suggested to establish the number of sampling points and grid distances, depending on the use and the total size of the area under suspicion. Three categories of use are first assigned with different potential exposures. The details are shown in Fig. 4.

In this approach the number of sampling points for a comparable use of the area increases from a minimum number with the logarithmic increase in the area (Fig. 4, above right). The type of distribution of contaminants and the size of the target area are not taken into account here.

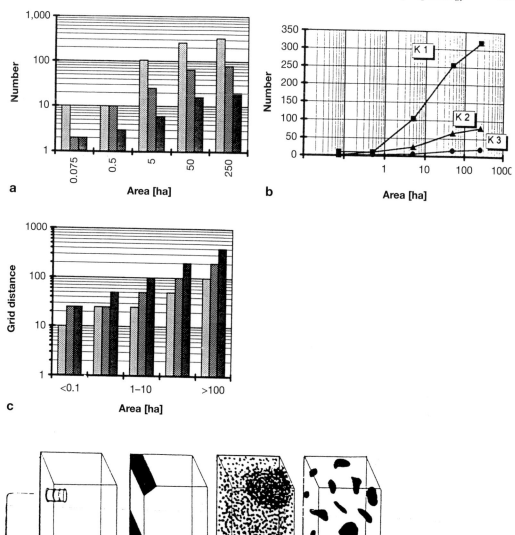

Fig. 3a–d. Examples of spatial distribution of contaminants in the case of contaminated land.
▦ K1: residential areas, sportsgrounds and playgrounds.
▨ K2: slag heaps, industrial and commercial areas in operation.
■ K3: former industrial and commercial sites, fallow areas, woods, agricultural areas, green areas.

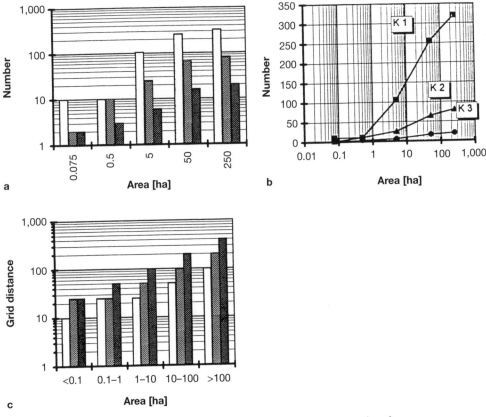

Fig. 4a–c. Number of sampling points and grid distances for different categories of use.
☐ K1: residential areas, sportsgrounds and playgrounds.
▨ K2: slag heaps, industrial and commercial areas in operation.
■ K3: former industrial and commercial sites, fallow areas, woods, agricultural areas, green areas.

3.3.4 Sampling Points for Determining the Location of the Center of Contamination

In areas where waste has been deposited localized epicenters of contamination with narrow boundaries can be responsible for a large proportion of the hazard. The number of sampling points for determining the position of such an epicenter depends on:

(1) the size of the contaminated area,
(2) the probability of success required,
(3) the shape of the contaminated area,
(4) the arrangement of the sampling points.

Besides the number of samples, the quantity of material in the individual sample also affects the distribution of the analytical results. If an area of 10,000 m^2 is examined, the weight of the upper 20 cm of soil alone amounts to about 3,000 t (density 1.5 t m^{-3}). A quantity of sample of, e.g., 1 kg gives the ratio of $1:3\cdot10^6$.

If a contaminant is present in the soil in particle form, the particle size affects the probability and the quantity with which a contaminated particle is determined in a soil sample. Besides the particle size

(1) the concentration of contaminants in the particle,
(2) the particle size distribution, and

(3) the number of particles containing contaminants

also have an effect on the total content of contaminants in the sample. BUNGE and BUNGE (1999) described tests for estimating the minimum quantity of sample necessary, taking these effects into account.

3.4 Sample Distortion and Preservation

When samples are taken, transported and stored distortion of the sample can occur, which leads to irreversible changes in the sample. Possible causes are:

(1) addition of foreign material to the sample,
(2) removal of material from the sample,
(3) changes in the sample material as a result of chemical and biological reactions caused by
    – abrasion of sampling equipment and tools,
    – adsorption, desorption, and diffusion effects,
    – spreading of contamination through equipment which has not been properly cleaned,
    – evaporation of contaminants during open storage,
    – impurities present in exhaust gases,
    – changes brought about by heat, light, and oxygen during transportation.

For volatile substances the production of pooled soil samples is associated with evaporation losses. The least affected area of a soil sample is the center of a soil sample which is as large as possible and which has been removed intact. Information on storage, transportation and preservation of sample material can be found in ISO 5667-3 (EN ISO, 1996).

The simplest and for many determinations also the most effective method of preservation involves cooling or freezing of the sample. Chemical preservation by addition of acid to pH <2 or of base to pH >11 is also possible. Some methods for sample preservation are listed in ISO 5667-3 (EN ISO, 1996). Some examples are given in Tab. 1.

4 Methods for Collecting Solid Samples

Through the use of soundings, bore holes and excavations, the possibility exists of gaining a direct view below the earth's surface. However, this gives information about one particular place or a very small area. The following criteria need to be taken into account when choosing a suitable boring procedure:

(1) the planned depth of the bore hole,
(2) the quantity of sample necessary,
(3) the suitability of the soil for boring techniques,
(4) avoidance of the spreading of contamination,

Tab. 1. Examples of Parameters for Different Types of Preservation

Preservation by	Suitable for
Acidification to pH <2	alkali and alkaline earth metals, Al, NH_4^+, COD, total hardness, mineral oils, nitrate, total P, silicates, heavy metals, As, AOX
Basification to pH >11	iodides, selenides
Cooling to 2–5 °C	acidity, alkalinity, NH_4^+, COD, iodides, conductivity, Cr(VI), nitrite, plant protection agents, orthophosphates, sulfates, phenols, dry residue, biological investigations
Deep freezing (-20 °C)	BOD_5, COD, biological investigations

(5) the parameters to be investigated,
(6) the consistency of the sample,
(7) the number of sampling points,
(8) a sufficiently large bore hole and reinforcement diameter for the measures planned (sampling, pumping experiments, bore hole measurements),
(9) the conditions imposed by buildings,
(10) the cost of the boring procedure,
(11) the work safety requirements.

The boring procedure can be classified depending on the nature of the boring movement (turning, pushing, pushing and turning) and the means of conveyance of the material bored (continuous, discontinuos). ISO 10381 (ISO/DIS 10381-1, 1996; ISO/DIS 10381-2, 1996; ISO/DIS 10381-3, 1996; ISO/DIS 10381-4, 1996; ISO/CD 10381-6, 1992; ISO/WD 10381-5, in preparation) contains information on obtaining and removing soil samples. ROTHE (1990) and the German Office of Environmental Protection (LfU BW, 1993), e.g., gave information on the special conditions in the case of contaminated land.

4.1 Procedures for Obtaining Samples

For the examination of areas under suspicion where there are loose rocks the method of ram core boring or the dry boring procedure using screws, augers, and clam grabs are particularly suitable.

Slit soundings serve to limit and localize the source of contamination together with qualitative and semiquantitative chemical methods of examination within the screening framework.

Bore holes with a possibility of the removal of the core of the bored material are more expensive, but, with a large enough bore diameter and sufficient bore depth, groundwater samples can also be obtained.

Excavation is always used if the location of the source of contamination can be determined beforehand with some certainty or where unambiguous assignment of the chemically determined parameters is not possible because of the inhomogeneity of the soil. However, with excavation only layers of soil near the surface can be reached. In addition

evaporation losses from open areas must be reckoned with in the course of time.

Compared with excavation, boring has, in principle, certain disadvantages. Among these are:

(1) Only limited information regarding geological parameters (e.g., fissures, areas of vibration, firmness) can be obtained from bored samples.
(2) Changes in the soil sample (water content, particle size distribution, distribution of contaminants) can occur during the boring procedure.
(3) Bore holes examine only a small section which is not necessarily representative.

Besides the advantages of a rapid and cost-effective sampling method, TOUSSAINT (1990) listed some sources of loss in ram core boring (given below), in particular in the investigation of volatile substances.

(1) Spreading contamination over the cross section is possible.
(2) Spreading contamination from one sampling point to another is possible if thorough cleaning is not carried out.
(3) Losses through evaporation of volatile substances are unfavorably high because of the small ratio of surface area to sample quantity.
(4) Through the application of energy during the ramming procedure (essentially causing shaking and warming) changes in the distribution of contaminants in the sample must be reckoned with.
(5) Desorption of weakly adsorbed pollutants is possible.
(6) In a gravelly soil and where building waste has been deposited, heat of abrasion at the tip of the ram must also be reckoned with.

4.2 Sampling Protocol

For every solid sample a sampling protocol must be completed as part of the documentation. An example of a sampling protocol is shown in Fig 5 (ECKHOFF, 1994).

Dienststelle: _____

Ortsbezeichnung: _____

Anlagenbezeichnung: _____ Betreiber: _____

Probenehmende Stelle:_____

Probennummer: Probenahmedatum: I__I__I__I__I__I__I

Uhrzeit Beginn: I__I__I : I__I__I   Uhrzeit Ende: I__I__I : I__I__I

Koordinaten der Probenahmestelle: Rechts: _____ Hoch: _____-Werte

Einzelprobe: I__I   Entnahmetiefe: I__I__I - I__I__I [m] / [cm]

Durchschnittsprobe: I_I aus I_I Einzelproben Entnahmetiefen: I__I__I - I__I__I [m] / [cm]

Art der Probengewinnung: I__I__I   Probenahmegeräte: I__I__I   Material: I__I__I
    01 = Handbohrung   01 = Kernsonde   01 = geh. Stahl
    02 = Rammkernsondierung   02 = Schappe   02 = Aluminium
    03 = Trockenbohrung   03 = Stechzylinder   03 = Kunststoff
    04 = Spülbohrung   04 = Probenstecher   04 = Edelstahl
    05 = Schürfe   05 = Schaufel

Bohrdurchmesser: I__I__I [mm] Probenmenge: I__I__I [l]   Anzahl der Gefäße: I_I_I

Art der Probengefäße: I__I__I   Material: I__I__I . Verschluß: I__I__I   Material: I__I__I
    01 = Flasche   01 = Glas, braun   01 = Schliffstopfen   01 = Glas
    02 = Beutel   02 = Glas, hell   02 = Schraubversch.   02 = Aluminium
    03 = Dose   03 = Edelstahl   03 = Klemmverschl.   03 = Kunststoff
    04 = Eimer   04 = Aluminium   04 = sonstige   04 = Edelstahl
                     05 = Kunststoff
                     06 = Papier alukasch.

<u>Probenansprache</u>

Abfallart: _____ Abfall-Reststoffschl.-Nr.: I__I__I__I__I__I__I__I

Geruch: I__I__I   Farbe:_____
00 = ohne   03 = faulig (H_2S)
02 = aromatisch   12 = Mineralöl

Feuchtezustand: I__I Zusammensetzung der Probe: I__I   Korngrößensortierung: I__I
    1 = trocken   1 = homogen   1 = gut
    2 = erdfeucht   2 = inhomogen   2 = mittel
    3 = naß   3 = schlecht
max. Korngröße ca.: I__I__I__I [mm]

Niederschlag: _____Lufttemperatur: ± I__I__I [°C]

Bemerkungen:_____

_____

Probenübergabe an: _____ am: _____ um: _____Uhr

Unterschrift des Probenehmers: Unterschrift des Empfängers:

Untersuchungsparameter:_____

Fig. 5. Sampling protocol.

5 Eluate Investigation

5.1 Aim of an Eluate Investigation

Elution is defined as the washing out of a substance from a solid adsorbent by the elution agent (eluant). Material is transferred from the solid to the liquid, usually aqueous phase, thus producing the liquid eluate.

An important target of soil investigations is on the one hand a reliable knowledge of the hazard potential in order to recognize effects which are actually dangerous and to avoid ecological contamination, and on the other hand to be able to select suitable reclamation measures. The potential hazard of soil contamination is not primarily determined by the total amount of contaminants, but rather by an assessment of the different mobilities of individual substances, depending on how they are bound to the solid material.

To assess the short- and long-term behavior of inorganic and organic substances in solid matter, a targeted application of the most widely differing elution procedures is attempted in order to determine the leaching potential under fixed conditions and thus estimate the mobility and predict the behavior. The long-term elution behavior is usually of particular interest.

The investigation of an eluate must involve:

(1) the determination of the nature and quantity of leachable substances (potential emissions),
(2) testing the efficiency of measures for lowering the availability (fixing, immobilizing),
(3) the estimation of the *in situ* concentration in the unsaturated zone,
(4) the assessment of the hazard to groundwater,
(5) the production of aqueous solutions for further processing steps:
 – obtaining eluates for biological investigations,
 – algae test, *Daphnia* test, luminous bacteria tests (luminescence, growth), determination of gene modification potential (Ames, umu, SOS tests) (KLEIN and DOTT, 1995),
 – operation of treatment equipment in the laboratory.

The behavior of substances in the saturated soil zones through which ground water flows and in the unsaturated zone through which the percolation water flows can be investigated in elution tests. Biological tests are predominantly not carried out directly with the soil material, but with the eluates. Thus the choice of the elution procedure and the treatment of the eluate can be important for the final result of the investigation.

Eluate tests can be used to investigate various transmission processes:

(1) from solid to groundwater (saturated soil zone),
(2) from solid to percolation water (unsaturated soil zone),
(3) from solid to plant (determination of the absorbable portion),
(4) from solid to humans (availability in the stomach and intestinal tract).

To estimate the effects of leaching of material into water, the following questions need to be answered:

(1) which substances,
(2) in what quantities and concentrations,
(3) in which bound forms,
(4) over what period of time?

The hazard can be estimated from these data.

An elution test can be carried out from two different starting points. One allows for a transfer which is as near as possible to the actual situation in the way the elution is carried out; in the other the most unfavorable possible conditions are used as a basis for simulation in the elution tests. Nevertheless, in this case the actual conditions in the field should always be taken into account.

5.2 Factors Affecting Elution Behavior

Leaching is not a defined parameter which is dependent on a single variable. Rather, leaching is a type of total parameter which is dependent on a large number of processes. A complex interaction of physical, chemical and biological mechanisms controls the mobility of contaminants in the soil. To achieve a sufficiently accurate picture of the processes, it is necessary to know the most important parameter affecting the elution behavior and to be able to estimate its effect with regard to the whole result. Tab. 2 gives an overview of the parameters affecting the mobility of pollutants during elution processes in the laboratory.

5.3 Elution Liquid

The nature and composition of the elution liquid can have a significant effect on the elution rate. It consists of the elution agent and possible additives. In order to correspond to the conditions present in the soil, water or aqueous solutions are the most frequently used elution agents. The use of particular additives can help to simulate the field conditions if these are known.

The following are used as elution agents:

(1) water: twice distilled water,
(2) synthetic rainwater: deionized water acidified with nitric acid to pH 4,
(3) synthetic percolation water: pH 4.6 and base capacity 15 mmol L^{-1} (household waste dump),
(4) synthetic percolation water: pH 4.6 and base capacity 150 mmol L^{-1} (waste building materials dump),
(5) solutions: NH_4NO_3, $CaCl_2$, $NaNO_3$, EDTA, NTA, NaOH, HCl and acetic acid (DOMINIK and PAETZ, 1995).

Special elution agents are used to determine the parts which are available to human beings via the digestion tract:

(1) stomach juices: pH 2.0 hydrochloric acid, test temperature 37 °C, enzyme for protein digestion, foodstuff (milk powder),
(2) intestinal medium: pH 7.0 Na_2CO_3 digestive enzyme mixture and gall, duration of test: 6 h, test temperature: 37 °C.

Tab. 2. Factors Affecting the Mobility of Pollutants in Laboratory Elution Processes, Amended and Extended; → effect on (after FRIEGE and LEUCHS, 1990)

Elution Liquid	Solid	Experimental Method
Throughput quantity	particle size distribution	sample processing
Composition	(→ adsorption of contaminants)	continuity and duration of leaching
– pH (→ precipitation/solution and adsorption/ desorption of heavy metal compounds)	pore structure	physical boundary conditions:
	density on storage	– change of eluents
	wettability (→ reaction rate)	– liquid–solid ratio
	permeability (water quantity in time)	– means of mixing (solid and/or eluent)
– Ionic strength (→ salts, heavy metal compounds)	swelling capacity (→ permeability)	– type of flow
– Redox potential (→ multivalent metals)	buffering capacity (→ variation of pH with time)	– temperature
– Buffer content (→ variation of pH with time)	ion exchange capacity (→ ion exchange of metals)	chemical boundary conditions
– Complexing agents (→ solubility of metals)	presence of microorganisms	– free establishment of equilibrium
– oil content, solubilizers (→ springly soluble organic substances)	composition	– defined conditions of medium
	– readily soluble salts	biological boundary conditions
	– heavy metal compounds	– colonization – growth
	– polar organic substances	– change in medium through biological activity
	– nonpolar organic substances	

Besides the use of a single elution agent, the latter can be changed during sequential extraction in the course of tests in order to differentiate between fractions with different types of chemical binding. Different types of environmental conditions can be simulated by changing the elution agent. In particular in the case of heavy metal compounds it should be noted that the mobility and the resulting hazard potential are not determined by the total content, but essentially by the nature of the chemical binding between the metal and the matrix. According to SCHROERS and FÖRSTNER (1987) the main types of binding of heavy metals in predominantly inorganic sludges (e.g., filter sludges, dredger sludges) are:

(1) absorptive binding to fine particle components,
(2) precipitation as single compounds,
(3) coprecipitation with iron and manganese oxides as well as with carbonate, sulfide and phosphate minerals,
(4) binding to organic solid phases,
(5) fixing in chemically inert bound forms.

5.4 pH Value

The pH value is an important parameter for the mobility in the elution test. Particularly in the case of metals the pH established during or before the elution is known to have a significant effect on the elution rate. The way the test is carried out can affect the pH as follows:

(1) self-adjustment of the pH of the elution liquid,
(2) defined pH adjustment of the elution liquid at the beginning of the test,
(3) stepwise adjustment during the test,
(4) continuous adjustment during the test.

Elution tests should generally be differentiated depending on whether inorganic or organic components are to be investigated. Organic compounds can be subdivided into two classes:

(1) polar, relatively water soluble compounds
(2) nonpolar, hydrophobic compounds.

5.5 Elution Procedure Depending on the Methods and Techniques Used in the Test

Extensive overviews of elution procedures have been written by REICHERT and ROEMER (1996) and by the Baden-Württemberg Environmental Protection Office (LfU BW, 1994a, b). The various elution tests can be classified according to the method used in the test. In static elution the elution liquid is not changed during the whole period of the test, while in dynamic tests the eluate is removed after a defined period of time (continuously or stepwise) and fresh elution liquid is added.

From the type of test technique used, a further differentiation can be made between:

(1) bottle tests
   – static (shaking tests, standing tests, saturation tests), e.g., DIN 38414-S4, pH stat, bath or
   – dynamic (serial batch test), e.g., cascade test, BUS test,
(2) column tests
   – static (circulating elution) or
   – dynamic (flowing around, flowing through), e.g., percolation columns, triaxial cells, lysimeters,
(3) other test procedures (Soxhlet).

Bottle tests are always used as a screening procedure. A wide necked bottle is often used as the elution container.

In column tests (Fig. 6) the sample material to be leached is placed in the column and the elution liquid is passed through it and/or around it. The flow can be regulated. In static tests the liquid is circulated, but in dynamic tests the elution liquid only passes through the sample material once so that the sample is continually treated with new elution liquid. Depending on the nature of the binding of the contaminant to the soil material, the result of the test is affected by the flow rate, the duration of the test and the distance through which the liquid has to flow. In order to guarantee comparability of test results, it is necessary to define the test parameters.

Under field conditions the liquid–solid ratio increases with increasing exposure and in the

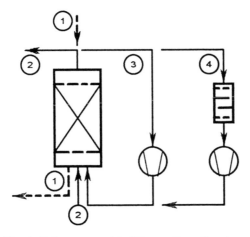

Fig. 6. Column test with different flow directions, frequency of passage and influence on the eluate, 1 sprinkling: flow direction from top to bottom, 2–4 flowing through: flow direction from bottom to top, 2 single passage, 3 circulation without influence, 4 circulation with influence.

tests can be used instead of a time axis. Thus by choosing the liquid–solid ratio in the elution test, the time required to gather information can be reduced taking into account the exposure of the material being tested.

The liquid–solid ratio represents a relative time axis, in particular in column tests, which allows an estimation of the passage of real time, taking into account the proportion of percolating rainwater and the strength of the layer through which it has to percolate. Only the proportion of rainwater penetrating below the topsoil (percolation water) is important for spreading out contamination.

All shaking tests and procedures involving continuous elution do not take into account the natural variations in the water content of the unsaturated soil zone.

With the determination of the highest concentration (in column tests this is the first percolate in practice) information on the effect on biological systems can be deduced as these usually react against a high, acute contamination with greater sensitivity than to a chronic, but less severe one.

Because of the time required, multiple elutions, column tests and lysimeters are not suitable for routine investigations. However, with

these methods important information with regard to long-term hazards can be derived. To assess long-term behavior the buffering capacity against action of acid or base and possible changes in the redox potential are important. If no suitable procedures are available for determining the proportion of potentially mobile contaminants, the total content of contaminants can be determined and the mobilizable proportion can be estimated on the basis of factors at the location which control the mobility. For metal, e.g., these factors include:

(1) pH value,
(2) the redox potential,
(3) the acid production capacity and the quantity of acids entering the soil,
(4) the buffering capacity,
(5) the content of organic substances,
(6) the content of fine particles,
(7) the permeability and length of the filtration layer, and
(8) the possible quantity of percolation water.

A standardized procedure for the determination of contaminant elution potentials should consist of several different modules. The information from one module should supplement that from another and together they should permit an overall assessment. The individual advantages of particular test methods should be made use of. As an example the combination of the following test procedures is recommended:

(1) shaking test (two-stage, rapid information, screening test),
(2) rapid column test (approximation of subterranean conditions and indication of time required),
(3) elution under the conditions of the most unfavorable case likely, to estimate the highest possible contamination,
(4) diffusion controlled test to determine the proportion diffusing.

In order to be able to compare the different elution tests with one another and to present the process more clearly, the course of an elution test is shown schematically in Fig. 7 which

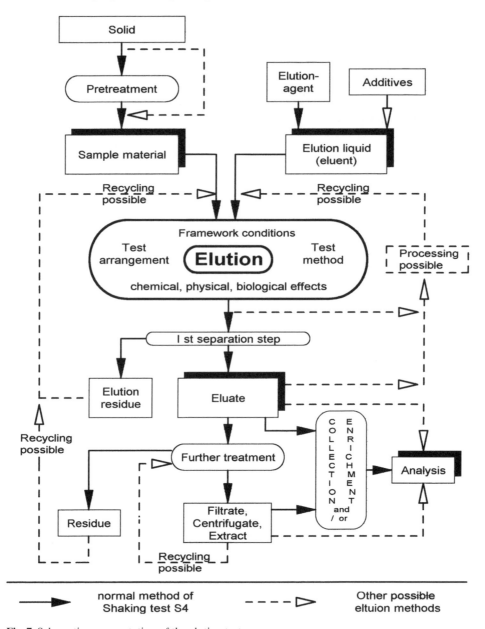

Fig. 7. Schematic representation of the elution tests.

shows the usual course of a shaking test according to DIN 38414-S4, as well as the possible course of an alternative elution procedure. The removal and pretreatment of the sample and preparation of the elution liquid are carried out before the actual elution. The elution itself is carried out under the conditions given in the test instructions. For an analytical investigation a solid–liquid phase separation (first separation step) is necessary after the elution. It should be recognized that the eluate is often not analyzed directly, but only after further steps when a combined, treated or enriched fraction is obtained. Direct testing of the eluate is only possible in a few cases.

With volatile substances (volatile halogenated hydrocarbons, BTX) evaporation during filtration and, for lipophilic substances, absorption effects during centrifugation and filtration must be reckoned with during the elution and in the separation of the eluate from the solid phase (LfU BW, 1994a).

6 Analytical Investigation Program

Work in the analytical laboratory is summarized in the analytical investigation program. In drawing up this program decisions must be made concerning:

(1) the analysis parameters required,
(2) the sample processing procedures planned,
(3) the criteria for selection of partial samples for the analysis (analytical samples) and
(4) the choice of the analysis procedure.

6.1 Choice of Analysis Parameters

Because of the large number of possible contaminants and emissions, it is sensible to limit the extent of the investigation to one that is feasible. Relevant substances and groups of substances are thus those which can pose a hazard even at low concentrations and which are or were often in use. Furthermore, intrinsic properties of a particular substance affect the choice of a group of substances. Among the substance-specific properties are mobility, persistence, accumulation, and degradability. In association with local conditions, such as soil properties and nature of the exploitation of the land, these account for much of the potential hazard in an individual case.

The analysis parameters which are useful in a standardized investigation should fulfill the following criteria:

(1) The parameter must be specific to the contamination.
(2) The parameter should allow comparison with available data.
(3) It should be possible to check the results of the investigation by a plausibility control.
(4) The results should be able to be determined with a total parameter within the framework of a screening test.
(5) The ratio of information to the costs of determination should be favorable.

The investigation parameters for particular samples can be established in advance using available lists or using a stepwise procedure in the course of the investigation. For the analysis of individual substances within a defined parameter range, lists of substances and groups of substances are available which have been known to be present and to represent the risks associated with contaminated land in previous cases. There is a disadvantage in that these lists are sometimes incomplete and do not take account of individual cases. Thus substances which are present but not contained in the lists are not recognized. Furthermore, parameters may be investigated which have no relevance for a particular situation.

Examples of such lists are:

(1) US EPA priority pollutants list (113 parameters) (US EPA, 1976),
(2) LAGA list of relevant substances and groups of substances in contaminated land (76 parameters) (LAGA, 1991).

Commercial lists give an indication as to what substances might be present based on a chemical analysis of the previous use. Examples of lists of parameters for substances suspected of causing contamination of disused sites are:

(1) determination of possible soil contamination in disused sites – assistance in obtaining and evaluating information on soil contamination on disused industrial and commercial sites resulting from production processes (KVR, 1989),
(2) compilation of an inventory of soil contamination according to the type of industry (KINNER and KÖTTER, 1986),

(3) compilation of an inventory of soil contamination on areas previously used for
service industries (NICLAUß et al.,
1989),

(4) contaminated land analysis – list of parameters for the choice of analysis parameters for disused sites specific to a
particular industry (GIHR and DANIEL,
1993).

6.2 Assessment Parameters

Reference parameters are necessary for
assessing test results. As a hazard potential is
also always determined by the characteristics
of use, the reference values to be used (orientation values, reclamation values) should defi-
nitely be dependent on the substance, the location, and the usage. In the case of protective
materials dangers to human health should be
differentiated between absorption by the
stomach and intestinal tract (oral), by the
lungs (inhalation), and by the skin (dermal).

Orientation, test and guide values for soil
contaminants have been collected by EWERS et
al. and presented as tables in which they are
classified according to area of use. Background
values for inorganic and organic parameters
are also given.

6.3 Concept of Stepwise Procedures

Stepwise investigation programs are currently being used. The first stage in such a program is a screening investigation giving total
and group parameters. After this, parameter
lists specific to a particular type of application
within limited areas are then used. For this second stage meaningful screening parameters
are still required:

(1) total parameters: determination of organic substances using a common structure component (e.g., C) or the common property of oxidizability with O_2
consumption,

(2) group parameters: determination of
substances which are similar in their

constitution or activity without differ-
entiating between individual substances,

(3) key parameters: determination of classes of substance or mixtures of isomers
or of an individual substance as the
representative of a particular class.

6.3.1 Activity Parameters

Biological tests have so far hardly been used
in the investigation of solid material, as uncontaminated comparison soil is difficult to create
and also variations in the composition of the
soil itself can lead to deviations in the results.
On the other hand, for investigation of the mobile phase, i.e., the aqueous eluate which can
be separated from the solid material, standardized biotests can indeed be used (DIN 38412).
Using different investigation parameters (respiration, luminescence, lethal doses, reproduction, growth) damage resulting from material added or its degradation products can be
determined. The advantage lies in the overall
description of the activity, which cannot be derived directly from a chemical investigation.
Examples of biological test procedures for the
estimation of pollutants in aqueous media are
given in Tab. 3. Information on the use of biological test procedures in the investigation of
contaminated soil is given in GUNKEL (1993),
LfU BW (1994c), NN (1993).

6.4 Selection of Analysis Samples for Solid Matter

For analytical investigations analysis samples need to be selected from the total quantity
of sample material. Information concerning
the whole area is derived from these samples.
This selection can be targeted using prior information or sensory observations or using a
rigid grid (e.g., a sample for every meter
bored). In the case of obvious visible or
odorous contamination the area where the
concentration of contaminants is suspected to
be highest and the boundaries of the contaminated area should be determined.

For all analyses the conditions used for digestion, processing and the method of meas-

Tab. 3. Examples of Biological Test Procedures for the Quantitative Estimation of Pollutants in Water and Wastewater Samples

Test Organism	DIN 38412	Measurement Criterion	Test Duration
Fish	L 15	mortality	48 h
Water fleas	L 11	inability to swim	24 or 48 h
Luminescent bacteria	L 34	decrease in luminescence	1 h
Algae	L 9	inhibition of growth	72 h

urement should be given with the analysis results in order to be able to assess the results themselves and to be able to compare the results with those taken at different times or determined by other institutions, and to evaluate them.

6.5 Sample Processing

6.5.1 Aims of Sample Processing

To convert the original sample into one that is suitable for technical analysis (e.g., GC) various sample processing steps are necessary. The aims of sample processing are (WEBER and WEBER, 1992):

(1) the preparation of a representative laboratory sample from one or more large samples,
(2) the removal of most of the matrix components in order to obtain a sample which is compatible with measuring instruments,
(3) the removal of matrix components which interfere with quantitative determination,
(4) the concentration or dilution of the substances to be determined to match the sensitivity range of the method of measurement.

In the case of solids the sample material is mixed, divided into four, half is discarded and the rest ground up. This takes place in several steps. Particle size reduction is necessary for the subsequent division of samples.

6.5.2 Particle Size Reduction

Machines for particle size reduction can be divided into (PITSCH, 1990):

(1) crushers (jaw crushers),
(2) mills (cutting mills) and
(3) pulverizing mills (centrifugal ball mill, rapid planetary mill, vibration grinding mill, mortar mill, ultracentrifuge mill).

Direct determinations on the solid phase are usually not possible, so the sample substances first have to be extracted and thus brought into solution before further processing steps can be carried out.

The quantity of sample necessary for subsequent analysis is determined by the particle size distribution and the proportions of the different components in the starting material (degree of heterogeneity). Specifications of minimum sample quantities are frequently based on prior experience together with considerations of economy. It is generally true that with increasing heterogeneity of a material

(1) the size of the sample selected and
(2) the number of samples must increase and
(3) the pooled samples must be homogenized.

While for material which is actually completely homogeneous the sample quantity has no effect on the results of the investigation, with heterogeneous distribution of pollutants the quantity of sample is significant.

There is a discrepancy between the effort required for the determination of the actual content by justifiable methods and the justifiable economic expenditure for the work and this must first be established.

People working in areas suspected of being contaminated and in sample collection must use protective measures (respiratory protection, protection from explosions, skin and body protection).

6.6 Digestion Procedures

The aim of these procedures is the determination of substances adsorbed on and contained in solid material (total determination):

(1) acidic digestion: strongly oxidizing acids and heat
    – open systems,
    – closed systems (digestion under pressure);
(2) UV digestion: energy rich radiation with addition of oxidizing agents, e.g., H_2O_2, HNO_3;
(3) microwave digestion;
(4) ultrasound digestion.

It is usually not possible to carry out direct determinations on the solid phase so the sample first has to be extracted and substances it contains brought into solution. The solution is then processed further.

6.7 Purification and Enrichment Methods

The aims of this step are removal of accompanying substances and the selective enrichment of the substances being investigated. The nature and quantity of accompanying substances are determined by the matrix and the extraction method. The accompanying substances contaminate the measuring instruments (injector, chromatography columns), can interfere with the separation and produce interference signals.

6.7.1 Inorganic Substances

(1) Solvent extraction: liquid–liquid extraction, shaking with a solvent which is immiscible with water, phase separation,
(2) ion exchange column: selective binding and desorption,
(3) precipitation of matrix components,
(4) evaporation, concentration: normal pressure, rotary evaporator,
(5) chromatography columns: solid–liquid extraction: single pass columns for selective removal and enrichment.

Extraction and elution procedures for heavy metals are listed in Tab. 4.

6.7.2 Organic Substances

Organic substances can, e.g., be extracted by (see also Tab. 5).

(1) adsorption on to activated charcoal or other adsorbents with subsequent elution or thermodesorption,
(2) solvent extraction,
(3) degassing methods: purge and trap procedure: degassing with carrier gas and subsequent collection,
(4) headspace methods: determination in the vapor phase,
(5) chromatography columns: solid–liquid extraction: single pass columns for selective removal and enrichment,

Tab. 4. Extraction/Elution Procedures for Heavy Metals in Soil. Used for: Photometric Analysis, X-Ray Fluorescence, ICP-AES, AAS

Extraction Solution	Information
KCl, NaCl, CaCl$_2$, NH$_4$NO$_3$	part available for plants and mobilizable heavy metals
Phthalate buffers	effect of the acidic reagent, complex forming metals in the solid
Hydrochloric/nitric acid (heating/digestion)	total content of heavy metals (acid soluble metals)

Tab. 5. Examples of the Possibilities for Extracting Organic Compounds

Organic Compounds	Extraction Method
Polycyclic aromatic hydrocarbons (PAH)	extraction with cyclohexane, toluene (shaking, soxhlet) ultrasound extraction with butanol, hexane extraction with supercritical fluids (SFE), e.g., CO_2 + modifier – GC/MS – HPLC/UV and fluorescence detection – chemiluminescence – immunoassay
Volatile halogenated hydrocarbons	extraction with pentane–water or methyl glycol headspace enrichment on XAD – GC/MS – GC/ECD
Benzene, toluene, xylenes, ethylbenzene (BTEX)	extraction with pentane–water or methyl glycol headspace enrichment on XAD – GC/MS – GC/ECD
Polychlorinated biphenyls (PCB)	extraction with hexane extraction with DMSO – GC/ECD
Hydrocarbons	extraction with 1,1,2-trichlorotrifluoroethane (substitute necessary) extraction with supercritical fluids (SFE), e.g., CO_2 – IR detection

(6) gel chromatography: separation according to molecular weight or size.

Ultrasound extraction has the following advantages:

(1) time saving,
(2) low solvent consumption,
(3) reduced loss of volatile and thermolabile compounds and
(4) higher extraction yields.

In order not to purify unnecessarily large quantities of soil material and yet to obtain a representative sample, the extraction step should be carried out on a large quantity of sample and then only an aliquot should be taken for further purification. This has the advantage that enough sample material with identical pretreatment is available for subsequent measurements.

6.8 Choice of Method

The choice of suitable methods of investigation is made on the basis of a sufficient or highest possible gain in information at the lowest possible expenditure. The expenditure includes the necessary effort in terms of time, personnel, instrumentation, materials consumed, as well as pre- and post-processing.

An improvement in the detection limit can be achieved by:

(1) increasing the sample size,
(2) optimizing the parameters used in the method,
(3) removal of the matrix, and
(4) preconcentration.

The analysis procedures are classified as follows in order of increasing precision, but also increasing expenditure:

(1) orientation test, screening analysis,
(2) field methods,
(3) laboratory comparison procedures and
(4) reference procedures, precision analysis.

SCHWEDT (1992) gave an overview of the analysis methods currently available. The analytical methods can be divided into:

(1) sensory tests: odor, color, turbidity;
(2) (wet) chemical methods: gravimetric, titrimetric,
(3) physical instrumental methods: electrochemical, chromatographic and spectrometric methods.

6.9 Physical or Instrumental Methods

The physical instrumental methods can be divided into four main groups (Fig. 8). These methods are mainly used in the determination of organic and/or inorganic trace elements and require special measuring techniques and instruments and often the use of computers. In order to improve the selectivity and sensitivity to detection for individual substances in multi-substance mixtures, a combination of two different procedures is often used.

Electrochemical procedures use reactions in which ions, electrons and the phase boundaries between the electrodes and conducting media are involved. In potentiometry substance con-

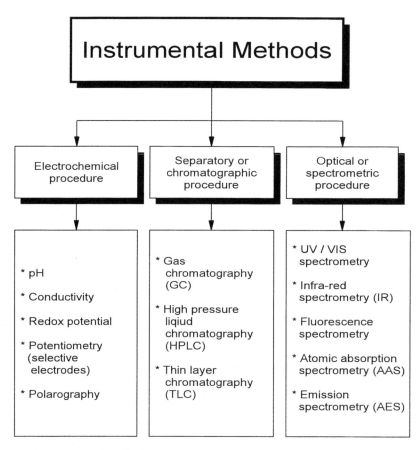

Fig. 8. Instrumental methods.

centrations are determined by measuring the voltage differences between the measuring electrode and a reference electrode in an electrochemical or galvanic cell without a current.

6.10 Spectrometric Procedures

Spectrometric procedures are based on the interaction between atoms, molecules and, electromagnetic radiation. The electromagnetic radiation is divided up into various spectral regions:

(1) ultraviolet (UV) spectra: 200–400 nm,
(2) visual (VIS) spectra; 400–800 nm,
(3) infrared (IR) spectra 0.8–500 μm.

In these wavelength regions both absorption and emission spectra can be taken. Fig. 9 shows the analysis parameters which can be investigated spectrometrically (HEIN and KUNZE, 1995).

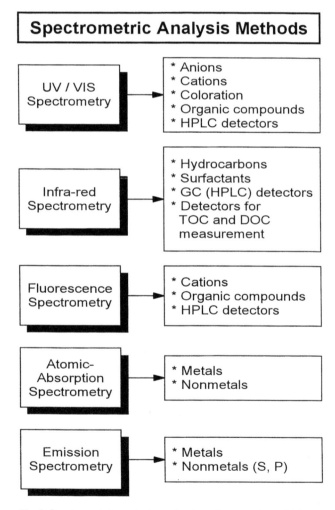

Fig. 9. Spectrometric analysis methods and parameters which can be determined (SCHWEDT, 1993).

6.11 Chromatographic Procedures

Chromatography involves a mobile and a stationary phase and can be classified, depending on the nature of the mobile phase, as liquid chromatography (TLC or HPLC) and gas chromatography (GC). Overviews of the procedures have been written by SCHWEDT (1993) and HEIN and KUNZE (1995).

6.11.1 Gas Chromatography (GC)

Analysis by gas chromatography (GC) is possible in principle with all substances which can be evaporated without decomposition. Limits are imposed by the thermal stability of the column material and stationary phase and by the maximum temperature which can be reached by the column oven.

Most nonpolar organic substances with no or few functional groups and molecular weights <400 are readily analyzed by GC (WEBER and WEBER, 1992). Tab. 6 gives an overview of the most important sample injection techniques.

6.11.2 High Performance Liquid Chromatography (HPLC)

The following substances can be analyzed by HPLC:

(1) involatile,
(2) thermally labile or
(3) ionogenic substances.

The phase system in HPLC consists of a mobile and a stationary phase which are in equilibrium with each other. After applying the sample to the column the dissolved components are transported by the mobile phase through the column whereby the sample molecules and the eluent interact with the stationary phase with the result that over time the individual components of the sample are separated. Tab. 7 shows possible separating mechanisms in HPLC (HEIN and KUNZE, 1995). UNGER (1995), ENGELHARDT (1986) and MEYER (1986) have written introductions to the use of HPLC.

6.12 Detectors

In order to detect the sample components obtained after chromatographic separation a detector is required. The principle of measurement by GC detectors is based on the change in a chemical or physical parameter of the carrier gas, whereby the measured parameter is converted into an electrical signal in the detector and is emitted. Various detectors have been developed for GC, each of which is particularly suitable for a certain type of analysis. Tab. 8 lists detectors divided into two basic groups (WEBER and WEBER, 1992) and Tab. 9 lists detectors used for GC and HPLC.

Detectors can also be classified according to whether they are universal, selective or specific:

(1) universal detectors: show all or at least most of the possible substances present in the sample with the same or similar sensitivity,
(2) selective detectors: show certain substances or substance classes with much higher sensitivity,

Tab. 6. Classification of Sample Injection Techniques (WEBER and WEBER, 1992)

Evaporation before the Column			Evaporation on the Column	
Evaporation on injection		cold injection		
Direct injection	split injection	cold injection system	cool injection	on-column
	splitless injection	programmed evaporation (PTV)		
Headspace				
Purge and trap				

Tab. 7. Separation Processes in HPLC (HEIN and KUNZE, 1995, amended)

Type of Chromatography	Carrier or Stationary Phase	Eluent or Mobile Phase
Adsorption chromatography	adsorbents such as silica gel, aluminium oxide, dextranal	organic solvents
Adsorption chromatography (normal and reversed phase)	silanized silica gel (reversed phase material)	solvent mixtures of organic solvents (e.g., methanol, acetonitrile, dioxan) and water or buffer
Partition chromatography (liquid–liquid)	carrier liquid stationary phase (hydrophilic or lipophilic)	organic solvent mixtures
Ion exchange chromatography	ion exchanger (cationic or anionic)	buffer solution
Affinity chromatography	affinity carrier with specific ligands	buffer solution
Size exclusion chromatography	porous particles with a defined pore size	solvent mixtures of organic solvents and water

Tab. 8. Types of Detectors (according to WEBER and WEBER, 1992)

Type of Detector	Examples	Characteristics
Concentration detectors	TCD, PID, IRD	– the signal is dependent on the concentration in the test substance, – the test substance is neither destroyed nor altered – a second detector can be connected in series – the gas flow must be held constant
Mass flow detectors	FID, NPD, FPD, MSD, AED	– the signal is dependent on the sample quantity flow per unit time – in a first approximation it is independent of the gas flow – includes all ionizing and destructive detectors

(3) specific detectors: exclusively show substances with particular structures or containing particular chemical elements.

6.13 Field Measurements

If the period of time between sampling and obtaining the results of the investigation is too long, field measurements need to be carried out. Comparatively rapid availability of information on contamination using many if less precise analysis results is often more important in the case of assessment and reclamation of old sites than a more exact determination of pollutants at selected points with higher analysis precision from a few samples (REICHERT and KANITZ, 1992).

SCHWEDT (1995) gave an overview of rapid test procedures and field measuring techniques.

MATZ (1993) considered that the advantage of field analysis is that the rapidly available results allow building site managers to react immediately to them by:

(1) changing the boring or sampling grid to narrow down the heaviest areas of contamination while the area under suspicion is assessed,

(2) controlling building work,

Tab. 9. Detectors for GC and HPLC (according to WEBER and WEBER, 1992 and to HEIN and KUNZE, 1995)

Detector		GC	HPLC
Thermal conductivity detector	TCD	x	
Flame ionization detector	FID	x	
Electron capture detector	ECD	x	
Nitrogen–phosphorus detector	NPD (PND)	x	
Flame-photometric detector	FPD	x	
Photoionization detector	PID	x	
Electrolytic conductivity detector	ELCD	x	
Mass selective detector	MSD	x	
Infrared spectrometric detector	IRD	x	
Atomic emission detector	AED	x	
UV/visible detectors with fixed/variable	UV/VIS		x
wavelength, diode array detector	DAD		x
Fluorescence detector	FL		x
Refractive index detector	RI		x
Conductivity detector			x
Electrochemical detector	EC		x
Polarimetric detector			x
Radioactivity detector			x
Mass spectrometric detector	MS		x

(3) sorting contaminated soil while it is being dug out and assigning suitable treatment methods,

(4) immediate recognition of pollutants for protection of the workforce from hazardous substances.

By using field measurements rapid, but also reliable, screening tests for relevant pollutant concentrations can be achieved. For obtaining an analysis result rapidly a short analysis time is necessary and sample processing should be omitted or be as simple as possible. According to REICHERT and KANITZ (1992) 1 h should be sufficient for sample processing, analysis and evaluation. The suitability of a field measuring instrument should be tested before it is used.

The following testing criteria should be considered:

(1) nature and expected concentration of substances to be measured (detection limit),

(2) parameters for accompanying and interfering substances,

(3) effects of temperature, moisture, dust, and vibrations on the instruments,

(4) requirements and extent of sampling and sample processing,

(5) sample throughput and time requirement,

(6) safety requirements (explosion protection),

(7) simplicity of use,

(8) investment and operating costs.

Field measurements have so far only been carried out on gas and water phases. The examination of solid samples requires the preparation of an eluate and represents the rate determining step of field analysis. Rapid eluates are prepared using heat, mechanical energy, ultrasound or microwaves with water in a period of up to 20 min. In the examination of organic substances solubilizers can be added (carboxylic acids, alcohols, surfactants). Besides determining substances dissolved in the aqueous phase, volatile substances can be determined after stripping from aqueous solution with simultaneous analysis of the extraction gas using a suitable gas test tube (Drägerwerk, 1992).

In the field analysis of water, soil, and soil air, parameters should be differentiated according to whether they can be determined by portable systems or whether they need to be determined more extensively in field laboratories (laboratory vans or containers). Examples of the latter are X-ray fluorescence analysis (20–50 elements, metals) and GC/MS (for organic, chromatographable substances). There are no simple methods for field analysis of organic substance groups (AOX, EOX, PAH, and PCB). Depending on the procedure involved, the different methods of field examination of liquid samples can be differentiated as follows:

(1) test rods (visual, reflectometric evaluation),
(2) colorimetric test procedures (color charts, comparators),
(3) gas test tubes (after stripping),
(4) titrimetric procedures,
(5) rapid photometric tests (cuvette test, reagent test, microtiter plates, etc.),
(6) electrometric and electrochemical procedures (conductometry, potentiometry, voltametry; ion-selective and -sensitive electrodes),
(7) physical, instrumental methods (GC, GC/MS, ion chromatography, IR, X-ray fluorescence),
(8) biotests and biosensors.

6.14 Perspectives

The tendency to evaluate the pollution of contaminated areas *in situ* where possible requires a determined optimization of the meaningfulness of the results found. Important steps to make the results more representative for areas of contaminated land involve working out a sensible sampling strategy, the subsequent sampling procedure, appropriate sample processing depending on the problem and a sufficiently sensitive analysis, which can also be increasingly carried out in the field. Only in this way it will be possible to determine potential hazards and distribution of pollutants in contaminated areas even at different depths taking economic forces into account and to be able to plan and monitor safeguarding and decontamination procedures better.

The miniaturization of analysis techniques, which is continually advancing with semiconductor laser diodes, microelectrodes in nanometer dimensions coupled with chip technology to give powerful measuring systems, and also the development of fiberoptic sensors will make it possible, e.g., to continuously monitor the pollution load in percolation water streams. With the continuing further developments in immunoassay techniques, organic pollutants, such as plant protection agents, BTEX, PCB, PCP and PAH, can be determined in the field. Gene probes and biochemical methods of pollutant detection, partially coupled with chromatographic techniques, are being developed.

7 References

Anonymous (1993), Hinweise zur Ermittlung und Sanierung von Altlasten 2nd Edn., 2nd and 3rd Installment (Ministerium für Umwelt, Raumordnung und Landwirtschaft Nordrhein-Westfalen, Ed.). Düsseldorf, Germany.

Anonymous (1990), Analytik von Feststoffen – Abfall – Altlasten (Probenahme – Probenvorbereitung – Bewertung) (14. Aachener Seminar in Roetgen), *Gewässerschutz Wasser Abwasser* **118**.

Anonymous (1992), *Altlasten-ABC*, erweiterte Fassung vom Juni 1992 (Ministerium für Umwelt, Raumordnung und Landwirtschaft des Landes Nordrhein-Westfalen, Ed.).

Anonymous (1993), Erfahrungen mit biologischen Wirkungstesten bei der Untersuchung von Wasser- und Bodenverunreinigungen (Freie und Hansestadt Hamburg, Umweltbehörde, Ed.), *Hamburger Umweltberichte 43/93*.

Anonymous (1994), *Merkblätter zur Qualitätssicherung* Nr. 6, Entnahme von Bohrproben bei Bohr- und Sondierarbeiten für die chemische analytik (Freie und Hansestadt Hamburg, Umweltbehörde, Ed.).

BOSMAN R. (1993), Probenahmestrategien und die Rolle der Geostatistik bei der Untersuchung von Bodenkontaminationen, in: *Altlastensanierung '93* 4. Int. KfK/TNO-Kongreß über Altlastensanierung, 3.–7. Mai 1993, Berlin (ARENDT, F., Ed.), pp. 581–595. Dordrecht: Kluwer Academic Publishers.

BUNGE, R., BUNGE K. (1999), Probename auf Altlasten: minimal notwendige Probenmasse, *Z. Altlastenspektrum* **3**, 174ff.

DIN 38412, *Testverfahren mit Wasserorganismen*. Berlin: Beuth Verlag.

DIN 38414-S4 (1984), *Bestimmung der Eluierbarkeit mit Wasser* (S4); Deutsche Einheitsverfahren zur Wasser-, Abwasser- und Schlammuntersuchung. Berlin: Beuth Verlag.

DOMINIK, P., PAETZ, A. (1995), *Methodenhandbuch Bodenschutz I* UBA-Texte 10/95. Berlin: Umweltsbundesamt.

Drägerwerk (1992), *Schadstoffmessung in flüssigen Proben, Verfahren zur Schnellanalyse von Wasser-, Mehrphasen- und aufgeschlämmten Bodenproben mit Dräger-Röhrchen*. Lübeck: Drägerwerk AG.

ECKHOFF, U. (1994), *Abfallprobe, quo vadis?* 18. Aachener Werkstattgespräch 28./29. 09. 1994. Düsseldorf: Landesumweltamt Nordrhein-Westfalen.

EN ISO 5667-3 (1996), Probenahme, Teil 3: Anleitung zur Konservierung und Handhabung von Proben. Berlin: Beuth Verlag.

ENGELHARDT, H. (1986), *Practice of High Performance Liquid Chromatography*. Heidelberg: Springer-Verlag.

EWERS, U., VIERECK, L., HERGET, J. (1994), *Bestandsaufnahme der vorliegenden Richtwerte zur Beurteilung von Bodenverunreinigungen und synoptische Darstellung der diesen Werten zugrundeliegenden Ableitungskriterien und -modelle* UBA-Texte 35/94. Berlin: Umweltbundesamt.

FRIEGE, H. LEUCHS, W. (1990), Zur Aussgekraft von Elutionsversuchen (14. Seminar in Roetgen 1990, Institut für Siedlungswasserwirtschaft, RWTH Aachen), *Gewässerschutz Wasser Abwasser* **118**, 34–52.

GIHR, R., DANIEL, B. (1993), Altlasten-Analytik, Parameterlisten zur branchenspezifischen Auswahl von Analysenparametern für Altstandorte. Landsberg/Lech: ecomed Verlag.

GUNKEL, J. (1993), Optimierung einer Biotest-Kombination zur ökotoxikologischen Beurteilung organisch kontaminierter Böden, *Thesis*, TU Hamburg-Harburg, Germany.

HEIN, H., KUNZE W. (1995), Umweltanalytik mit Spektrometrie und Chromatographie, 2nd Edn. Weinheim: VCH.

ISO/DIS 10381-1: Bodenbeschaffenheit; Probenahme: Anleitung zur Aufstellung von Probenahmeprogrammen (2/96). Berlin: Beuth Verlag.

ISO/DIS 10381-2, Bodenbeschaffenheit, Probenahme: Anleitung für Probenahmeverfahren (2/96). Berlin: Beuth Verlag.

ISO/DIS 10381-3, Bodenbeschaffenheit, Probenahme: Anleitung für Probenahmeverfahren (2/96). Berlin: Beuth Verlag.

ISO/DIS 10381-4, Bodenbeschaffenheit, Probenahme: Anleitung für das Vorgehen bei der Untersuchung von natürlichen, naturnahen und Kulturstandorten (2/96). Berlin: Beuth Verlag.

ISO/DIS 10381-6, Bodenbeschaffenheit, Probenahme: Anleitung zur Entnahme, Behandlung

und Lagerung von Böden für die Bestimmung aerober Prozesse unter Laboratoriumsbedingungen (1992). Berlin: Beuth Verlag.

KINNER, U. H., KÖTTER, L., NICLAUß, M. (1986), *Branchentypische Inventarisierung von Bodenkontaminationen – Ein erster Schritt zur Gefährdungsabschätzung für ehemalige Betriebsgelände*, UBA-Texte 31/86. Berlin: Umweltbundesamt.

KLEIN, J., DOTT, W. (1995), *Biologische Testmethoden für Böden*, DECHEMA Fachgespräche Umweltschutz. Frankfurt/M: DECHEMA.

KNOOP, G. (1990), Optimierung der Probenahmeverfahren bei der Gewinnung von Bodenproben, *Gewässerschutz Wasser Abwasser* **118**, 116–130.

KVR (1989), Erfassung möglicher Bodenverunreinigungen auf Altstandorten, *Arbeitshefte Ruhrgebiet, A 039* (Kommunalverband Ruhrgebiet Essen, Ed.).

LAGA (1991), LAGA-Informationsschrift Altablagerungen und Altlasten, in: *Abfallwirtschaft in Forschung und Praxis* Vol. 37 (zugleich LAGA-Mitteilung 15; insb. Anhang 3-1). Bonn: Erich Schmidt Verlag.

LfU BW (1993), Verfahrensempfehlungen für die Probenahme bei Altlasten (Boden, Abfall, Grund-, Sickerwasser, Bodenluft), *Texte und Berichte zur Altlastenbearbeitung*, 6/93. Karsruhe: Landesamt für Umweltschutz Baden-Württemberg.

LfU BW (1994a), Literaturstudie: Elutionsverfahren für schwer lösliche organische Schadstoffe in Boden- und Abfallproben, *Texte und Berichte zur Altlastenbearbeitung*, 12/94. Karlsruhe: Landesanstalt für Umweltschutz Baden-Württemberg.

LfU BW (1994b), Derzeitige anwendung und Entwicklung von Elutionsverfahren, *Texte und Berichte zur Altlastenbearbeitung*, 11/94. Karlsruhe: Landesanstalt für Umweltschutz Baden-Württemberg.

LfU BW (1994c), 13 Altlastenerkundung mit biologischen Methoden, in: *Handbuch Altlasten und Grundwasserschadensfälle, Materialien zur Altlastenbearbeitung* Vol. 13. Karlsruhe: Landesanstalt für Umweltschutz Baden-Württemberg.

MATZ, G. (1993), *Erfahrungen, Grenzen und Ausblicke der schnellen Vorortanalytik von organischen Schadstoffen*, Bodenreinigung (STEGMANN, R., Ed.), Hamburger Berichte Vol. 6, Abfallwirtschaft TU Hamburg-Harburg, pp. 27–38. Bonn: Economica Verlag.

MEYER, V. (1986), *Praxis der Hochleistungs-Flüssigkeits-Chromatographie*. Frankfurt/M.: Verlag Diesterweg Salle, Verlag Sauerländer.

NICLAUß, M. (1989), *Inventarisierung von Bodenkontaminationen auf Geländen mit ehemaliger Nutzung aus dem Dienstleistungsbereich*, UBA-Texte 16/89. Berlin: Umweltbundesamt.

PAETZ, A. (1993), Vereinheitlichung von Bodenun-

tersuchungsverfahren; Statusseminar des Haupt-
ausschuß I der Fachgruppe Wasserchemie „Nor-
mung in der Wasseranalytik – Hilfe und Hürde",
Frankfurt a.M. 15.–16. Dez. 1993, *GdCh-Seminar
323/93*.

PITSCH, H. (1990), Mechanische Aufbereitung von
Feststoffen, *Gewässerschutz Wasser Abwasser*
118, 229–246.

REICHERT, J. K., KANITZ, J. (1992), Vor-Ort-Analytik
bei kontaminierten Böden und Altlasten, (25. Es-
sener Tagung), *Gewässerschutz Wasser Abwasser*
135, 647–660.

REICHERT, J., ROEMER, M. (1996), Eluatuntersu-
chung, in: *Chemie und Biologie der Altlasten*.
Weinheim: VCH.

ROEMER, M. (1996), Grundlagen zur Aufstellung einer optimierten Untesrsuchungsstrategie für Alt-
lastverdachtsflächen; *Thesis*, RWTH Aachen [pu-
blished in: Abfall – Recycling – Altlasten, Vol. 10
(DOHMANN, Ed.) ISA der RWTH Aachen].

ROTHE, U. (1990), Erkundung von Verdachts-
flächen – Methoden und Geräte (14. Aachener
Seminar), *Gewässerschutz Wasser Abwasser* **118**,
95–115.

SCHROER, J., FÖRSTNER, U. (1987), Abschätzung der
Langzeitbelastung von Grundwasser durch die
Ablagerung metallhaltigser Feststoffe, in: *Vom
Wasser* Vol. 69, pp. 23–32. Weinheim: VCH.

SCHWEDT, G. (1992), *Taschenatlas der Analytik*. Stuttgart, New York: Thieme Verlag.

SCHWEDT, G. (1993), *Chromatographische Trennme-
thoden*, 3rd Edn. Stuttgart: Thieme Verlag.

SCHWEDT, G. (1995), *Mobile Umweltanalytik –
Schnelltestverfahren und Vor-Ort-Meßtechniken*.
Weinheim: Vogel Buchverlag.

SIMMLEIT, N., ERNST, A. (1994), *Handbuch: Kommu-
nales Altlastenmanagement – Ein praktischer
Leitfaden*. Umweltbundesamt Berlin, Berichte
3/94.

TOUSSAINT, B. (1990), Kritische Anmerkungen zur
Plausibilität der Gehalte leichtflüchtiger Halo-
genkohlenwasserstoffe in beprobten Umweltme-
dien, in: *Angewandte Geologie Karlsruhe* Vol. 9,
Schriftenreihe des Lehrstuhls für angewandte
Geologie TH Karlsruhe, Germany.

TOUSSAINT, B. (1994), Umweltproblematik und Hydrogeologie der Erkundung von Boden- und
Grundwasser-Kontaminationen durch leicht-
flüchtige halogenierte Kohlenwasserstoffe, *Um-
weltplanung, Arbeits- und Umweltschutz* Heft Nr.
168. Wiesbaden: Hessische Landesanstalt für
Umwelt.

UNGER, K. K. (1995), *Handbuch der HPLC*, Part 1 –
Leitfaden für Anfänger und Praktiken, 2nd Edn.,
Part 2 – Präparative Säulenflüssig-Chromatogra-
phie. Darmstadt: GIT-Verlag.

US EPAS (1976), *Quality Criteria for Water*, US Environmental Protection Agency, EPA-440/9-76-
023, Washingtson, DC 20460.

WEBER, E., WEBER, R. (1992), *Buch der Umweltanalytik* Vol. 4, Methodik und Applikationen in
der Kapillargaschromatographie. Darmstadt:
GIT-Verlag.

Index

X

Xanthobacter autotrophicus, degradation of dichloroethane 253
xenobiotics, bacterial degradation of, dioxygenase
   reaction 149
– – monooxygenase reactions 149
– bioavailability of 61ff
– biodegradation by fungi 156ff, 160ff
– complete metabolization and mineralization
   221
– formation of bound residues 110f
– – binding modes 111ff
– humification processes 90, 101f
– interactions, with soil 91
– – with black carbon 92f
– microbial degradation of 101f, 221
– – threshold concentrations 96
– oxidative coupling reactions 113

– release of their transformation products 114
– solubility in water 64ff
– sorption of hydrophobic – 218
xylenes, anaerobic degradation of 184
– extraction of 499
– microbial degradation of 214

Y

yeasts, use of aromatic compounds as growth substrates 159

Z

zinc, bioprecipitation 365
– biosensors for depletion of 456
– hyperaccumulator 389ff
– removal by phytoextraction 389ff